T4-AQM-100

TK 2000 .D45 1985

Del Toro, Vincent.

Electric machines and power systems

DATE DUE

SEP 19 1995		
MAR 0 0 1996		
MAR 0 5 1996		
APR 2 1 2006		

DEMCO 38-297

ELECTRIC MACHINES AND POWER SYSTEMS

ELECTRIC MACHINES AND POWER SYSTEMS

Vincent Del Toro

Department of Electrical Engineering
The City College
The City University of New York

Prentice-Hall, Inc., Englewood Cliffs, N.J. 07632

Library of Congress Cataloging in Publication Data

Del Toro, Vincent.
  Electric machines and power systems.

  Bibliography: p. 699
  Includes index.
  1. Electric machinery. 2. Electric power systems.
I. Title.
TK2000.D45 1985   621.31′042 84–26550
ISBN 0–13–248709–8

Editorial/production supervision: Fred Dahl
Cover design: Diane Saxe
Manufacturing buyer: T. Caruso

Printed in the United States of America

10 9 8 7 6 5 4 3 2 1

ISBN 0-13-248709-8

Prentice-Hall International, Inc., *London*
Prentice-Hall of Australia Pty. Limited, *Sydney*
Editora Prentice-Hall do Brasil, Ltda., *Rio de Janeiro*
Prentice-Hall Canada Inc., *Toronto*
Prentice-Hall Hispanoamericana, S.A., *Mexico*
Prentice-Hall of India Private Limited, *New Delhi*
Prentice-Hall of Japan, Inc., *Tokyo*
Prentice-Hall of Southeast Asia Pte. Ltd., *Singapore*
Whitehall Books Limited, *Wellington, New Zealand*

To
My Students . . .
All Ten Thousand

Contents

PART ONE

CONVENTIONAL ELECTRIC MACHINES, 49

Transformers, 51

Fundamentals of
Electromechanical Energy Conversion, 113

Three-Phase Induction Motors, 164

Three-Phase Synchronous Machines, 238

Direct-Current Machines, 324

PART TWO

SPECIAL ELECTRIC MACHINES, 385

Two-Phase Servomotors, 385

Single-Phase Induction Motors, 409

Stepper Motors, 433

Synchros, 450

PART THREE

DYNAMICS OF ELECTRIC MACHINES, 477

Dynamics, 479

PART FOUR

INTRODUCTION TO
ELECTRIC POWER SYSTEMS, 537

Transmission Line Parameters, 539

Parameters of Electric Power Systems:
A Description in Terms of Symmetrical Components, 588

14

Equivalent Circuits for Fault Calculations, 629

Appendices, 681

Index, 701

Preface

There are three principal reasons for writing this book. Two relate to subject matter and the third deals with meeting various course objectives in treating this topic. The first and most important purpose of the book, of course, is to provide a coherent, highly motivated, and pedagogically sound exposition of the theory and characteristics of electric machines when operating as individual units in the steady state and in the dynamical state. This subject matter is covered in Chapters 1 through 11. With one exception these chapters are the same as those that appear in *Electro-mechanical Devices for Energy Conversion and Control System* but with rather extensive revisions.

Substantial additions are included in Chapters 2 through 6 in order to achieve several objectives.

1. In the interest of removing some rough spots in those parts of the original book that are difficult conceptually (as in Chapter 3), a generous number of illustrative examples are now included. These should serve to remove obstacles to learning and to enhance comprehension generally. Chapter 3, for instance, includes five such examples as against none in the original version. A total of fifteen additional examples are contained in these first few chapters.

2. The subject matter is extended to provide appropriate updating as well as to cover topics that furnish the background to gain an easy transition to the study of electric power systems. For this reason Chapter 2 now covers three-phase transformer connections in their various configurations while Chapter 5, on synchronous machines, has been expanded to include material on the parallel operation of synchronous generators, the derivation and equivalent circuits of transient and subtransient reactances, as well as the detailed analysis of the dynamical behavior of the

prime mover–synchronous generator coupled system. The dynamical response of the synchronous motor and its attached load is also treated. Furthermore, the treatment of electric machinery is brought up to date by the coverage of a number of germane topics. These include the solid-state electronic control of a-c induction motors as well as d-c motors for speed adjustments. The electronic control of d-c motors is treated both for the case where the source is a-c, thereby requiring thyrister drives, and where the source is d-c, in which case chopper drives are used as the means of achieving speed adjustments over a wide range. The role played by electronics is further extended through the treatment of brushless excitation for synchronous generators using solid-state devices. Moreover, in recognition of the enormous importance of stepper motors in the application of computer peripheral devices such as tape drives, printers, robots, and the like, a new chapter (Chapter 9) is included to deal with the constructional and operational features of these omnipresent motors. In a similar vein the material on d-c machines now also includes permanent magnet motors, which are especially ubiquitous in automobiles and special-purpose industrial applications.

3. The number of problems in Chapters 2 through 6 is increased by approximately 27 percent.

A second important reason for writing this book is to direct the study of electric machines in a way that prepares the student to understand the behavior of electric power systems when these systems are placed under stress by line-to-line and line-to-ground short circuits. The aim here is to present an accurate description of the behavior of the principal power system components (i.e., generators, motors, transformers, and transmission lines) when operating under these stressful conditions. Because such conditions involve unbalanced modes of operation, they are best described by symmetrical components as they apply to three-phase systems. When the entire electric power system operates in an unbalanced fashion following the occurrence of a fault, it is especially important to know how to represent these power system components in terms of their positive-, negative-, and zero-sequence symmetrical components. The limited knowledge of machine parameters that apply for balanced operation, and that is studied in the first eleven chapters of this book, is not sufficient to permit the study of transients in electric power systems. To achieve this goal, it is necessary to embark on a more detailed investigation of power system components. This is accomplished by a study of Chapters 12 through 14. In Chapter 12 the transmission line is characterized in terms of suitable parameters that are applicable for unbalanced operation. Chapter 13 then fulfills the important mission of identifying suitable representations of synchronous generators, synchronous motors, and three-phase transformer configurations when they, too, operate under severe conditions of unbalance occasioned by various faults that can occur anywhere along the electric power system. Finally, Chapter 14 is included to describe the manner in which the various positive-, negative-, and zero-sequence components of the power system equipment come into play in establishing the short-circuit currents that flow throughout the power system for various fault types. Chapter 14 concludes with a description of the protective relaying that can be used to prevent serious damage to the electrical equipment and to help avoid extensive power outages. It is significant to note in this treatment that the emphasis is on power system component descriptions and their applications in fault analysis rather than on a complete and thorough power systems study. Accordingly, topics such as power flow analysis, optimum operating strategies, and emergency control of power systems are left for a subsequent course for those students who have the interest.

The third essential reason for writing this book is to provide a degree of flexibility in meeting the needs of various engineering curricula as regards the conduct of the associated "power" course or courses. Listed below are suggestions that can serve as guidelines in the use of the text material.

1. *One-semester course on electric machines:*
   Chapter 1 (if needed)
   Chapter 2 (Secs. 2-1 to 2-8)
   Chapter 3 (Secs. 3-1 to 3-4)
   Chapter 4 (all)
   Chapter 5 (Secs. 5-1 to 5-7, 5-11 to 5-15, 5-17)
   Chapter 6 (all)
   Chapter 7 [Secs. 7-1, 7-2 (needed for Chapter 8)]
   Chapter 8 (all)
   Chapter 9 (all)
   Chapter 11 (Secs. 11-1 to 11-5)

2. *Two-course sequence on electric machines and power systems:*
   (a) First semester
       Same as course 1 above
   (b) Second semester
       Chapter 2 (Sec. 2-9)
       Chapter 5 (Secs. 5-8 to 5-10, 5-16, 5-18)
       Chapter 12 (all)
       Chapter 13 (all)
       Chapter 14 (all)

3. *One-semester course on electric machines and power systems:*
   Chapter 1 (if needed)
   Chapter 2 (Secs. 2-1 to 2-8)
   Chapter 3 (Secs. 3-1 to 3-3)
   Chapter 4 (Secs. 4-1 to 4-4, 4-6, 4-9, 4-11)
   Chapter 5 (Secs. 5-1 to 5-3, 5-6, 5-7, 5-11, 5-12, 5-14, 5-15)
   Chapter 6 (Secs. 6-1, 6-2, 6-5, 6-7 to 6-11)
   Chapter 8 (Secs. 8-1, 8-6, 8-7)
   Chapter 9 (Secs. 9-1, 9-2, 9-5)
   Chapter 11 (Secs. 11-1, 11-2)
   Chapter 12 (Secs. 12-1, 12-2, 12-4, 12-6)
   Chapter 13 (Secs. 13-1 to 13-3: discuss results of Secs. 13-4 to 13-6)
   Chapter 14 (Secs. 14-1, 14-3, 14-7)

Option 1 is appropriate for those curricula where a single three-credit course is available for this subject matter and where the goal is to stress a deeper understanding of the steady-state and dynamical performance of electric machines per se. Option 2 is intended for those schools where the curricula is so constructed that it allows students with an interest in power to elect two or more courses in this subject area. In such cases it is suggested that the first-semester course be the same as that of option 1. The second course then picks up the specialty topics of Chapters 2 and 5 as a preparation for a complete study of Chapters 12 through 14. Option 3 is intended for the majority of the electrical engineering curricula, where it is more likely than not that a single course is devoted to this subject area. The outline here is designed to expose the student to more than a familiarity with just conventional machines. The author is of the opinion that, although this is a worthy objective in itself, nevertheless it is also important that baccalaureates in electrical engineering possess some knowledge of electric power systems and specialty machines as well. Time for these topics is made available by forgoing some of the in-depth study of conventional machines.

VINCENT DEL TORO

ELECTRIC MACHINES AND POWER SYSTEMS

INTRODUCTION

1

Magnetic Theory and Circuits

An understanding of electromagnetism is essential to the study of electrical engineering because it is the key to the operation of a great part of the electrical apparatus found in industry as well as the home. All electric motors and generators, ranging in size from the fractional horsepower units found in home appliances to the 25,000-hp giants used in some industries, depend upon the electromagnetic field as the coupling device permitting interchange of energy between an electrical system and a mechanical system and vice versa. Similarly, static transformers provide the means for converting energy from one electrical system to another through the medium of a magnetic field. Transformers are to be found in such varied applications as radio and television receivers and electrical power distribution circuits. Other important devices—for example, circuit breakers, automatic switches, relays, and magnetic amplifiers—require the presence of a confined magnetic field for their proper operation. This chapter provides the reader with background so that he or she can identify a magnetic field and its salient characteristics and more readily understand the function of the magnetic field in electrical equipment.

The science of electrical engineering is founded on a few fundamental laws derived from basic experiments. In the area of electromagnetism it is Ampere's law that concerns us and, in fact, serves as the starting point of our treatment.† On the basis of the results obtained by Ampere in 1820, in his experiments on the forces existing between two current-carrying conductors, such quantities as magnetic flux

---

†It should not be here inferred that this is the only starting point in developing a quantitative theory of the magnetic circuit. Faraday's law of induction is equally valid as a starting point and is preferred when the goal is the development of an electromagnetic wave theory rather than a theory leading to the treatment of electromechanical energy conversion.

density, magnetic field intensity, permeability, and magnetic flux are readily defined. Once this base is established, attention is then directed to a discussion of the magnetic properties of certain useful engineering materials as well as to the idea of a "magnetic circuit" to help simplify the computations involved in analyzing magnetic devices.

1-1 AMPERE'S LAW—DEFINITION OF MAGNETIC QUANTITIES

Appearing in Fig. 1-1 is a simplified modification of Ampere's experiment. The configuration consists of a very long conductor 1 carrying a constant-magnitude current I_1 and an elemental conductor of length l carrying a constant-magnitude current I_2 in a direction opposite to I_1. When taken together the elemental conductor and the current I_2 constitute a *current element* I_2l. The elemental conductor 2 is actually part of a closed circuit in which I_2 flows, but for simplicity and convenience the details of the circuit are omitted except for the length l. Moreover, it is assumed that conductors 1 and 2 lie in the same horizontal plane and are parallel to each other. In accordance with Ampere's law it is found that with this configuration there exists a force on the elemental conductor directed to the right. Furthermore, the magnitude of the force is found to be directly proportional to I_1, I_2, l, and the medium surrounding the conductors as well as inversely proportional to the distance between them. In mks units† the magnitude of this force can be shown to be given by

$$F = \frac{\mu I_1}{2\pi r} I_2 l \qquad \text{newtons (N)} \qquad (1\text{-}1)$$

where I_1 and I_2 are expressed in amperes, l and r in meters, and μ is a property of the medium. A further interesting revelation about this experiment is that, if the elemental conductor 2 is used as an exploring device to find those points in space where the force is of constant magnitude and outwardly directed, the locus is found to be a circle of radius r and centered along the axis of conductor 1. In other words it is possible to identify a field having constant *lines of force*. In this connection it is useful at this

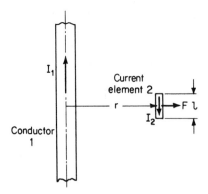

Figure 1-1 Illustrating the force existing between a current element I_2l and a very long conductor carrying current I_1 as described by Ampere's law.

†Meter-kilogram-second system of units.

point to rewrite Eq. (1-1) as follows:

$$F = I_2 l B \qquad (1\text{-}2)$$

where

$$B \equiv \frac{\mu I_1}{2\pi r} \qquad (1\text{-}3)$$

Thus, B is defined† as the magnetic field or, still better, the *magnetic flux density* existing in the region where the elemental conductor 2 lies and has the units of *webers/meter²* or *teslas* (T). As Eq. (1-2) reveals, the magnetic flux density is defined as a measured force F per known current element $I_2 l$. Equation (1-3) is merely a means of identifying the manner in which the current I_1 influences the force field about the current element. It is significant to note that, as long as I_2 is not zero, the force field and the magnetic field have the same characteristics—both have a circular locus and both are vector quantities possessing magnitude and direction. However, because of the way B is defined, the magnetic field exists as long as I_1 is not zero irrespective of the value of I_2. Magnetic flux density thus represents the effects caused by the moving charges that make up I_1.

In our study of electrical machinery, which comes later, conductor 1 will be referred to as the *field winding* because it sets up the working magnetic field, whereas the total circuit of which the elemental conductor is a part is called the *armature winding*.

The direction of the magnetic field is readily determined by the *right-hand rule* which states that, if the field winding conductor (1 in this case) is grasped in the right hand with the thumb pointing in the direction of current flow, the lines of flux (or flux density) will be in the direction in which the fingers wrap around the conductor.

Permeability

A glance at Eq. (1-1) shows that all the factors are known in this equation with the exception of the proportionality factor μ, which is a characteristic of the surrounding medium. Upon repeating the experiment of Fig. 1-1 in iron rather than air, it is found that the force is many times greater for the same values of I_1, I_2, l, and r. Therefore, it follows that μ may be defined from Eq. (1-1) for various media because it is the only unknown quantity. Moreover, because of the way magnetic flux density was defined, Eq. (1-3) indicates that the effect of the surrounding medium may be described in terms of the degree to which it increases or decreases the magnetic flux density for a specified current I_1. Thus when iron rather than free space is the medium it can be said that the iron provides a greater penetration of the magnetic field in a given region, i.e., there is a greater flux density. This property of the surrounding medium in which the conductors are embedded is called *permeability*.

When the conductors of Fig. 1-1 are assumed placed in a vacuum (free space) and the force is measured for specified values of I_1, I_2, l, and r, the solution for the

---

†This definition of B is correct for the configuration of Fig. 1-1. For other configurations the forms will be similar, but the constants will differ generally.

permeability of free space obtained from Eq. (1-1) and expressed in mks units comes out to be

$$\mu_0 = 4\pi \times 10^{-7}$$

(1-4)

The unit of permeability also follows from Eq. (1-1). Thus

$$\mu_0 \propto \frac{\text{newtons}}{\text{ampere}^2}$$

where $\propto$ denotes "is proportional to." But

$$\text{newton-meter} = \text{joule} = \text{volt} \times \text{ampere} \times \text{second}$$

or

$$\mu_0 \propto \frac{\text{volt} \times \text{ampere} \times \text{second}}{\text{ampere}^2 \times \text{meter}} = \frac{\text{volt} \times \text{second}}{\text{ampere} \times \text{meter}}$$

However, volt $\times$ seconds/ampere is the unit of inductance expressed in henrys. Accordingly, permeability is expressed in units of henrys/meter (H/m).

In those cases where the surrounding medium is other than free space, the absolute permeability is again readily found from Eq. (1-1). A comparison with the result obtained for free space then leads to a quantity called *relative permeability*, μ_r. Expressed mathematically we have

$$\mu_r = \frac{\mu}{\mu_0}$$

(1-5)

Equation (1-5) clearly indicates that relative permeability is simply a numeric which expresses the degree to which the magnetic flux density is increased or decreased over that of free space. For some materials, such as Deltamax, the value of μ_r can exceed one hundred thousand. Most ferro-magnetic materials, however, have values of μ_r in the hundreds or thousands.

Magnetic Flux Φ

Before considering the manner in which this quantity is defined, let us first investigate more closely the dimensions of the magnetic flux density B as described by its defining expression Eq. (1-2). Thus

$$B \propto \frac{\text{newtons}}{\text{amperes} \times \text{meter}}$$

$$B \propto \frac{\text{volts} \times \text{seconds} \times \text{amperes}}{\text{amperes} \times \text{meter}^2} = \frac{\text{volts} \times \text{seconds}}{\text{meter}^2}$$

(1-6)

A study of this expression reveals that B is in fact a density because it involves meters squared in the denominator. Accordingly, it is reasonable to expect that, since B is defined as a magnetic flux density, then multiplication by the effective area that B penetrates should yield the total *magnetic flux*. To illustrate this point refer to Fig. 1-2, which shows a coil of area ab lying in the same horizontal plane containing conductor

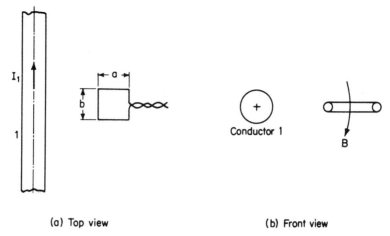

(a) Top view   (b) Front view

Figure 1-2 Associating an area with a magnetic field B to identify a magnetic flux.

1. We already know that when a current I_1 flows through this conductor a magnetic field is created in space and specifically is described by Eq. (1-3). To find the total flux penetrating the coil it is necessary merely to perform an integration of B over the surface area involved. Of course if B were a constant over the area of concern, the flux would be simply the product of B and the area ab. Next consider that the plane of the coil is tilted with respect to the plane of conductor 1 by 60°, as depicted in Fig. 1-3. Clearly now the total flux penetrating the coil is reduced by a factor of $\frac{1}{2}$. If the coil is oriented to a position of 90° with respect to the horizontal plane, no flux threads the coil.

Figure 1-3 Same as Fig. 1-2(b) except that the coil is tilted 60° relative to the horizontal plane.

On the basis of these observations, then, the magnetic flux through any surface is more rigorously defined as the surface integral of the normal component of the vector magnetic field B. Expressing this mathematically we have

$$\Phi = \int_s B_n \, dA \qquad (1\text{-}7)$$

where s stands for surface integral, A represents the area of the coil, and B_n is the normal component of B to the coil area. From expression (1-6) we know that magnetic flux must have the dimensions of *volt-seconds*. However, this is more commonly called *webers* and denoted by Wb. The volt-seconds unit of flux is better understood in terms of Faraday's law of induction.

Magnetic Field Intensity H

Often in magnetic circuit computations it is helpful to work with a quantity representing the magnetic field which is independent of the medium in which the magnetic flux exists. This is especially true in situations such as are found in electrical machinery where a common flux penetrates several different materials, including air. A glance at Eq. (1-3) reveals that division of B by μ identifies such a quantity. Accordingly, magnetic field intensity is defined as

$$H \equiv \frac{B}{\mu}$$
(1-8)

and has the units of

$$\frac{\dfrac{\text{newtons}}{\text{amperes} \times \text{meter}}}{\dfrac{\text{newtons}}{\text{ampere}^2}} = \frac{\text{amperes}}{\text{meter}} = \frac{\text{A}}{\text{m}}$$

Thus H is dependent upon the current that produces it and also on the geometry of the configuration but not the medium. For the system of Fig. 1-1 the value of the magnetic field intensity immediately follows from Eq. (1-3) and is given by

$$H = \frac{B}{\mu} = \frac{I_1}{2\pi r}$$

Because H is independent of the medium it is frequently looked upon as the intensity that is responsible for driving the flux density through the medium. Actually, though, H is a derived result.

More generally the units for H are ampere-turns/meter rather than amperes/meter. A little thought should make this apparent whenever the field winding is made up of more than just a single conductor.

Ampere's Circuital Law

Now that the magnetic field intensity has been defined and shown to have dimensions of ampere-turns per meter, we can develop a very useful relationship. Recall that H is a vector having the same direction as the magnetic field B. For the configuration of Fig. 1-1 H has the same circular locus as B. A line integration of H along any given closed circular path proves interesting. Of course the line integral is considered because H involves a per unit length dimension. Thus

$$\oint H \, dl = \int_0^{2\pi r} \frac{I_1}{2\pi r} \, dl = I_1 \qquad \text{amperes}$$
(1-9)

(Again keep in mind that the units here would be ampere-turns if more than one conductor were involved in Fig. 1-1.) Equation (1-9) states that the closed line integral of the magnetic field intensity is equal to the enclosed current (or ampere-turns) that produces the magnetic field lines. This relationship is called *Ampere's circuital law* and is more generally written as

$$\oint H \, dl = \mathfrak{F}$$

(1-10)

where $\mathfrak{F}$ denotes the ampere-turns enclosed by the assumed closed flux line path. The quantity $\mathfrak{F}$ is also known as the *magnetomotive force* and frequently abbreviated *mmf*. This relationship is useful in the study of electromagnetic devices and is referred to in subsequent chapters.

Derived Relationships

In the preceding pages the fundamental magnetic quantities—flux density, flux, field intensity, and permeability—are defined starting with Ampere's basic experiment involving two current-carrying conductors. By the appropriate manipulation of these quantities additional useful results can be obtained.

Equation (1-8) is a vector equation describing the magnetic field intensity for a given geometry and current. If the total path length of a flux line is assumed to be l, then the total magnetomotive force (mmf) associated with the specified flux line is

$$\mathfrak{F} = Hl = \frac{B}{\mu} l$$

(1-11)

Now in those situations where B is a constant and penetrates a fixed, known area A, the corresponding magnetic flux may be written from Eq. (1-7) as

$$\Phi = BA$$

(1-12)

Inserting Eq. (1-12) into Eq. (1-11) yields

$$\mathfrak{F} = Hl = \Phi\left(\frac{l}{\mu A}\right)$$

(1-13)

The quantity in parentheses in this last expression is interesting because it bears a very strong resemblance to the definition of resistance† in an electric circuit. Recall that the resistance in an electric circuit represents an impediment to the flow of current under the influence of a driving voltage. An examination of Eq. (1-13) provides a similar interpretation for the magnetic circuit. We are already aware that $\mathfrak{F}$ is the driving mmf which creates the flux Φ penetrating the specified cross-sectional area A. However, this flux is limited in value by what is called the *reluctance* of the magnetic circuit, which is defined as

$$\mathfrak{R} = \frac{l}{\mu A}$$

(1-14)

†The resistance of a copper wire of length l and cross-sectional area A is given by $R = \rho l/A$, where ρ is the resistivity of the copper.

No specific name is given to the dimension of reluctance except to refer to it as so many units of reluctance.

Equation (1-14) reveals that the impediment to the flow of flux which a magnetic circuit presents is directly proportional to the length and inversely proportional to the permeability and the cross-sectional area—results which are entirely consistent with physical reasoning.

Inserting Eq. (1-14) into Eq. (1-13) yields

$$\mathcal{F} = \Phi \mathcal{R} \qquad (1\text{-}15)$$

which is often referred to as Ohm's law of the magnetic circuit. It is important to keep in mind, however, that these manipulations are permissible as long as B and A are fixed quantities.

In most practical ferromagnetic materials the functional relationship between $\mathcal{F}$ and Φ (or the corresponding quantities $H = \mathcal{F}/l$ and $B = \Phi/A$) is nonlinear, which indicates that the reluctance $\mathcal{R}$ varies with the applied mmf. A glance at Fig. 1-10 makes this apparent. In some applications it is the change in reluctance about an operating point that is important rather than the total value of the reluctance. This leads naturally to the notion of *differential reluctance*, which is defined as

$$\mathcal{R}_d = \frac{d\mathcal{F}}{d\Phi} = \frac{d(Hl)}{d(BA)} = \frac{l}{A} \frac{dH}{dB} = \frac{l}{\mu_d A} \qquad (1\text{-}16)$$

Here μ_d is the differential permeability and is represented graphically as the slope of a tangent drawn to the B versus H curve.

Ampere's Law for Various Orientations of the Current Element

In Fig. 1-1 the assumption was made that the current element was located parallel to conductor 1 and lying in the same plane. Because this orientation was sufficient for the purpose at hand—to define the fundamental magnetic quantities—it was pursued as a matter of convenience. However, in the interest of furnishing a more complete picture of the experiment we shall now consider the effect on the force of placing the current element $I_2 l$ in two additional different orientations. Consider first that the current element is no longer placed parallel to conductor 1 but continues to be located in the same horizontal plane. Refer to Fig. 1-4. The dots in this figure indicate that

Figure 1-4 Showing the direction of the force when the current element is no longer located parallel to conductor 1 but remains in the same place.

MAGNETIC THEORY AND CIRCUITS CHAP. 1

the magnetic field is directed outward on the left side of conductor 1 and inward (with respect to the plane of the paper) on the right side of the conductor as revealed by the right-hand rule. The results of this experiment show that the magnitude of the force is the same as that found by using the configuration of Fig. 1-1.† This conclusion is not surprising because the value of the magnetic flux density as well as I_2 and l remain unchanged so that Eq. (1-2) is still valid in describing the force. The only change is the direction of the force. However, as Fig. 1-4 indicates, the force continues to be normal to the current element. It is worthwhile to keep this point in mind. Presently a general rule for establishing the force direction for all configurations is described.

Next let us consider that orientation of the current element which places it parallel to conductor 1 but inclined at an angle $\theta = 30°$ with respect to the vertical. A side-view projection of the configuration is depicted in Fig. 1-5. Note that the magnetic field is directed downward along the vertical for this view. Actually, of course, the locus of B is circular, but in Fig. 1-5 we are looking at just that small portion of the B field about the plane containing conductor 1. With this configuration the force on the current element is found to have the same direction but one-half the magnitude of that obtained with the orientation of Fig. 1-1. It follows, then, that the angle between the current vector $I_2 l$ and the flux density B affects the magnitude of the force. As a matter of fact, further experimentation reveals that the general expression for the force is

$$\boxed{F = I_2 l B \sin \theta} \qquad \text{N} \qquad (1\text{-}17)$$

Figure 1-5 Side-view projection of Fig. 1-1 but with the current element tilted relative to the horizontal plane. The force is directed out of the paper.

Equation (1-17) conveys information solely about the magnitude of the force and not its direction. It is possible, however, by employing the notation of vector analysis, to rewrite Eq. (1-17) so that information about magnitude as well as direction is present. This result is readily accomplished by the use of the *cross-product* notation between two vectors, yielding a third vector having magnitude and direction. Thus Eq. (1-17) is more completely expressed as

$$\boxed{\bar{F} = I_2 \bar{l} \times \bar{B}} \qquad \text{N} \qquad (1\text{-}18)$$

†Because of the infinitesimal nature of the current element, the value of B is assumed to be constant about $I_2 l$ irrespective of its orientation.

The cross notation must always be understood to involve the *sine* of the angle between the two vectors $\bar{B}$ and $\bar{l}$ (or the direction of I_2, which is determined by the orientation of l). Moreover, wherever the cross product is involved, the direction of the resultant vector is always normal to the plane containing the vectors $\bar{B}$ and $\bar{l}$ in the sense determined by the direction of advance of a right-hand screw as $\bar{l}$ is turned into $\bar{B}$ through the smaller of the two angles made by the vectors. Accordingly, in the configuration of Fig. 1-5 the direction of the force is found by turning $I_2\bar{l}$ into $\bar{B}$ and then noting that this would cause a right-hand screw to advance out of the plane of the paper. Hence the force is directed outward, and this corresponds with the experimentally established result.

It is also possible to determine the direction of the force by means of another right-hand rule, which requires that the forefinger be put in the direction of the current and the middle finger in the direction of $\bar{B}$, with the two assumed to be lying in the same plane. The thumb of the right hand then points in the direction of the force when placed perpendicular to the other two fingers.

1-2 THEORY OF MAGNETISM

In order to understand the magnetic behavior of materials, it is necessary to take a microscopic view of matter. A suitable starting point is the composition of the atom, which Bohr described as consisting of a heavy nucleus and a number of electrons moving around the nucleus in specific orbits. Closer investigation reveals that the atom of any substance experiences a torque when placed in a magnetic field; this is called a *magnetic moment*. The resultant magnetic moment of an atom depends upon three factors—the positive charge of the nucleus spinning on its axis, the negative charge of the electron spinning on its axis, and the effect of the electrons moving in their orbits. The magnetic moment of the spin and orbital motions of the electron far exceeds that of the spinning proton. However, this magnetic moment can be affected by the presence of an adjacent atom. Accordingly, if two hydrogen atoms are combined to form a hydrogen *molecule*, it is found that the electron spins, the proton spins, and the orbital motions of the electrons of each atom oppose each other so that a resultant magnetic moment of zero should be expected. Although this is almost the case, experiment reveals that the relative permeability of hydrogen is not equal to 1 but rather is very slightly less than unity. In other words, the molecular reaction is such that when hydrogen is the medium there is a slight decrease in the magnetic field compared with free space. This behavior occurs because there is a precessional motion of all rotating charges about the field direction, and the effect of this precession is to set up a field opposed to the applied field regardless of the direction of spin or orbital motion. Materials in which this behavior manifests itself are called *diamagnetic* for obvious reasons. Besides hydrogen, other materials possessing this characteristic are silver and copper.

Continuing further with the hydrogen molecule, let us assume next that it is made to lose an electron, thus yielding the hydrogen ion. Clearly, complete neutralization of the spin and orbital electron motions no longer takes place. In fact, when a magnetic field is applied, the ion is so oriented that its net magnetic moment aligns itself with the field, thereby causing a slight increase in flux density. This behavior is described as *paramagnetism* and is characteristic of such materials as aluminum and platinum. Paramagnetic materials have a relative permeability slightly in excess of unity.

So far we have considered those elements whose magnetic properties differ only very slightly from those of free space. As a matter of fact the vast majority of materials fall within this category. However, there is one class of materials—principally iron and its alloys with nickel, cobalt, and aluminum—for which the relative permeability is very many times greater than that of free space. These materials are called *ferromagnetic†* and are of great importance in electrical engineering. We may ask at this point why iron (and its alloys) is so very much more magnetic than other elements. Essentially, the answer is provided by the *domain* theory of magnetism. Like all metals, iron is crystalline in structure with the atoms arranged in a space lattice. However, domains are subcrystalline particles of varying sizes and shapes containing about 10^{15} atoms in a volume of approximately 10^{-9} cubic centimeters. *The distinguishing feature of the domain is that the magnetic moments of its constituent atoms are all aligned in the same direction.* Thus in a ferromagnetic material, not only must there exist a magnetic moment due to a nonneutralized spin of an electron in an inner orbit, but also the resultant spin of all neighboring atoms in the domain must be parallel.

It would seem by the explanation so far that, if iron is composed of completely magnetized domains, then the iron should be in a state of complete magnetization throughout the body of material even without the application of a magnetizing force. Actually, this is not the case, because the domains act independently of each other, and for a specimen of unmagnetized iron these domains are aligned haphazardly in all directions so that the net magnetic moment is zero over the specimen. Figure 1-6 illustrates the situation diagrammatically in a simplified fashion. Because of the crystal lattice structure of iron the "easy" direction of domain alignment can take place in any one of six directions—left, right, up, down, out, or in—depending upon the direc-

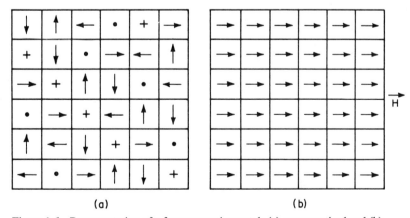

(a)   (b)

Figure 1-6 Representation of a ferromagnetic crystal: (a) unmagnetized and (b) fully magnetized by the field *H*.

†Derived from the Latin word for iron—*ferrum*.

tion of the applied magnetizing force. Figure 1-6(a) shows the unmagnetized configuration. Figure 1-6(b) depicts the result of applying a force from left to right of such magnitude as to effect alignment of all the domains. When this state is reached the iron is said to be *saturated*—there is no further increase in flux density over that of free space for further increases in magnetizing force.

Large increases in the temperature of a magnetized piece of iron bring about a decrease in its magnetizing capability. The temperature increase enforces the agitation existing between atoms until at a temperature of 750°C the agitation is so severe that it destroys the parallelism existing between the magnetic moments of the neighboring atoms of the domain and thereby causes it to lose its magnetic property. The temperature at which this occurs is called the *curie point*.

1-3 MAGNETIZATION CURVES OF FERROMAGNETIC MATERIALS

If the experiment of Fig. 1-1 is repeated with iron or steel as the medium for increasing values of the field winding current I_1 and the corresponding values of μ computed, it is found that the relative permeability varies considerably with the magnetizing force that establishes the operating flux density. A typical variation of μ_r for cast steel appears in Fig. 1-7. Here μ_r is plotted versus flux density rather than the magnetizing force because it is the onset of the realignment of more and more domains that brings about the change in permeability. Unfortunately, the state of development of the theory of magnetism is not so far advanced that it allows the prediction of the magnetic properties of a material on a purely theoretical basis even though the exact composition of the material is known. For example, with the present theory it is not possible to say exactly what the flux density will be in a given specimen of iron for a specified value of the magnetizing force. Rather, it is customary to obtain this information by consulting technical and descriptive bulletins where the measured magnetic properties of a representative sample of the specimen are published. These bulletins are made available to users by the manufacturers of magnetic steels and they include information on such varied shapes and forms as sheets, wires, bars, and even castings weighing up to hundreds of tons.

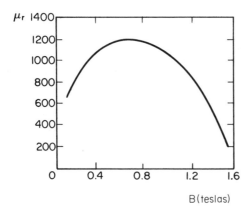

Figure 1-7 Graph of relative permeability versus flux density.

Cross section

l_m

Figure 1-8 Obtaining the magnetization curve of a sample steel specimen.

Sample steel specimen

Usually the published magnetic characteristics of the various iron and steel samples are presented as plots of flux density B as a function of the magnetic field intensity H. In the interest of presenting a complete graphical picture of the functional relationship existing between these two quantities as well as to define additional terms used in this connection, refer to Fig. 1-8. Assume that the steel specimen is initially unmagnetized and is in the form of a toroidal ring with a coil of N turns wrapped around it. Assume too that the coil can be energized from a variable voltage source capable of furnishing current flow in either direction in the coil. As the current I is increased from zero in the positive direction (current flowing into the top terminal of the coil), an increasing magnetic flux Φ can be measured taking place within the body of the toroid in the clockwise direction. For any fixed value of I there is a specific value of flux. Then by Eq. (1-12) the corresponding flux density is determined since the toroidal cross-sectional area is known. Moreover, the magnetomotive force NI can be replaced by Hl_m in accordance with Ampere's circuital law [Eq. (1-10)], where l_m is the mean length of path of the toroid (as shown by the broken-line circle in Fig. 1-8). The two fundamental quantities involved in this arrangement then are B in webers/meter^2 and H expressed in ampere-turns per meter. H is the quantity we want to deal with rather than magnetomotive force because, for the same flux density, doubling the mean magnetic length will not change H but will require doubling the magnetomotive force. The conclusion to be drawn is that a plot of B versus H is a universal plot for the given material because it can be extended to any geometry of cross-sectional area and length. In contrast, a plot of Φ versus NI is limited to a single geometrical configuration. Therefore, a plot of the magnetic characteristics of a material always involves plotting B versus H. For the virgin sample of Fig. 1-8 the graph of B versus H follows the curve Oa of Fig. 1-9 for field intensities up to H_a. Take note of the nonlinear relationship existing between these two quantities.

Another interesting characteristic of ferromagnetic materials is revealed when the field intensity, having been increased to some value, say H_a, is subsequently decreased. It is found that the material opposes demagnetization and, accordingly, does not retrace along the magnetizing curve Oa but rather along a curve located above Oa. See curve ab in Fig. 1-9. Furthermore, it is seen that, when the field intensity is returned to zero, the flux density is no longer zero as was the case with the virgin sample. This happens because some of the domains remain oriented in the direction of the originally applied field. The value of B that remains after the field intensity H

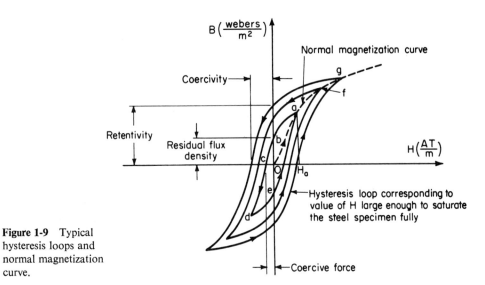

$B\left(\dfrac{webers}{m^2}\right)$

Normal magnetization curve

Coercivity

Retentivity

Residual flux density

$H\left(\dfrac{AT}{m}\right)$

H_a

Hysteresis loop corresponding to value of H large enough to saturate the steel specimen fully

Coercive force

Figure 1-9 Typical hysteresis loops and normal magnetization curve.

is removed is called *residual flux density*. Moreover, its value varies with the extent to which the material is magnetized. The maximum possible value of the residual flux density is called *retentivity* and results whenever values of *H* are used that cause complete saturation.

Frequently, in engineering applications of ferromagnetic materials, the steel is subjected to cyclically varying values of *H* having the same positive and negative limits. As *H* varies through many identical cycles, the graph of *B* versus *H* gradually approaches a fixed closed curve as depicted in Fig. 1-9. The loop is always traversed in the direction indicated by the arrows. Since time is the implicit variable for these loops, note that *B* is always lagging behind *H*. Thus, when *H* is zero, *B* is finite and positive, as at point *b*, and when *B* is zero, as at *c*, *H* is finite and negative, etc. This tendency of the flux density to *lag behind* the field intensity when the ferromagnetic material is in a symmetrically cyclically magnetized condition is called *hysteresis*† and the closed curve *abcdea* is called a *hysteresis* loop. Moreover, when the material is in this cyclic condition the amount of magnetic field intensity required to reduce the residual flux density to zero is called the *coercive force*. Usually, the larger the residual flux density, the larger must be the coercive force. The maximum value of the coercive force is called the *coercivity*.

A glance at the hysteresis loops of Fig. 1-9 makes it quite evident that the flux density corresponding to a particular field intensity is not single-valued. Its value lies between certain limits depending upon the previous history of the ferromagnetic material. However, since in many situations involving magnetic devices this previous history is unknown, a compromise procedure is used in making magnetic calculations by working with a single-valued curve called the *normal magnetization curve*. This curve is found by drawing a curve through the tips of a group of hysteresis loops generated while in a cyclic condition. Such a curve is *Oafg* in Fig. 1-9. Typical normal magnetization curves of commonly used ferromagnetic materials appear in Fig. 1-10.

†Derives from the Greek *hysterein* meaning to be behind or to lag.

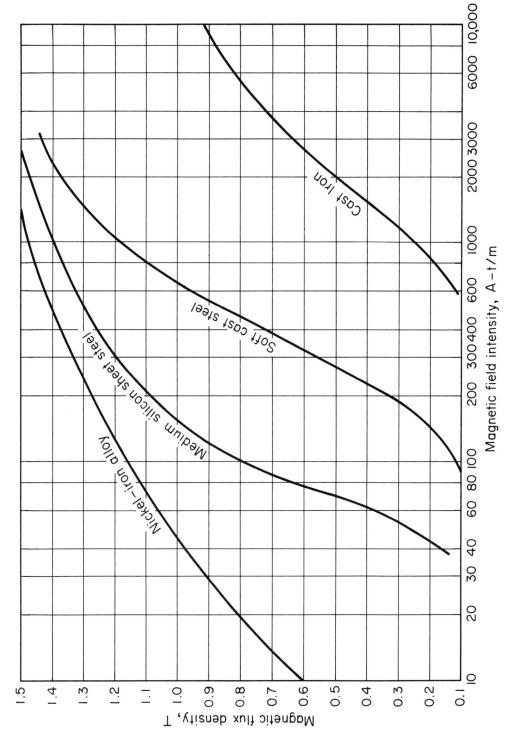

Figure 1-10 Magnetization curves of typical ferromagnetic materials.

A final observation is in order at this point. By Eq. (1-8) the permeability of a material may be expressed as a ratio of B to H. Coupling this with the nonlinear variation existing between B and H (see Fig. 1-9) bears out the variation of permeability with flux density as already cited in connection with Fig. 1-7. As a matter of fact, for a material in a cyclic condition the permeability is nothing more than the ratio of B to H for the various points along the hysteresis loop.

1-4 THE MAGNETIC CIRCUIT: CONCEPT AND ANALOGIES

In general, problems involving magnetic devices are basically field problems because they are concerned with quantities such as Φ and B which occupy three-dimensional space. Fortunately, however, in most instances the bulk of the space of interest to the engineer is occupied by ferromagnetic materials except for small air gaps which are present either by intention or by necessity. For example, in electromechanical energy-conversion devices the magnetic flux must permeate a stationary as well as a rotating mass of ferromagnetic material, thus making an air gap indispensable. On the other hand, in other devices an air gap may be intentionally inserted in order to mask the nonlinear relationship existing between B and H. But in spite of the presence of air gaps it happens that the space occupied by the magnetic field and the space occupied by the ferromagnetic material are practically the same. Usually this is because air gaps are made as small as mechanical clearance between rotating and stationary members will allow and also because the iron by virtue of its high permeability confines the flux to itself as copper wire confines electric current or a pipe restricts water. On this basis the three-dimensional field problem becomes a one-dimensional circuit problem and in accordance with Eq. (1-15) leads to the idea of a *magnetic circuit*. Thus we can look upon the magnetic circuit as consisting predominantly of iron paths of specified geometry which serves to confine the flux; air gaps may be included. Figure 1-11 shows a typical magnetic circuit consisting chiefly of iron. Note that the mmf of the coil produces a flux which is confined to the iron and to that part of the air having effectively the same cross-sectional area as the iron. Furthermore, a little thought reveals that this magnetic circuit may be replaced by a single-line equivalent circuit as depicted in Fig. 1-12. As suggested by Eqs. (1-14) and (1-15) the equivalent circuit consists of the magnetomotive force driving flux through two series-connected reluctances—$\mathcal{R}_i$, the reluctance of the iron, and $\mathcal{R}_a$, the reluctance of the air.

Figure 1-11 Typical magnetic circuit involving iron and air.

Figure 1-12 Single-line equivalent circuit of Fig. 1-11.

This analogy of the magnetic circuit with the electric circuit carries through in many other respects. For the sake of completeness these details are presented below for the case of a toroidal copper ring and a toroidal iron ring having the same mean radius r and cross-sectional area A.

ELECTRIC CASE	MAGNETIC CASE
The toroidal copper ring is assumed open by an infinitesimal amount with the ends connected to a battery; a current of I amperes flows through the ring.	The toroidal iron ring is assumed to be wound with N turns of wire so that with a current i flowing through it the mmf creates the flux Φ.

Driving Force

applied battery voltage $= E$	applied ampere-turns $= \mathfrak{F}$

Response

$\text{current} = \dfrac{\text{driving force}}{\text{electric resistance}}$	$\text{flux} = \dfrac{\text{driving force}}{\text{magnetic reluctance}}$

or

$$I = \frac{E}{R}$$

$$\Phi = \frac{\mathfrak{F}}{\mathfrak{R}} \qquad (1\text{-}19)$$

Impedance

Impedance is a general term used to indicate the impediment to a driving force in establishing a response.

$$\text{resistance} = R = \rho \frac{l}{A}$$

$$\text{reluctance} = \mathfrak{R} = \frac{l}{\mu A} \qquad (1\text{-}20)$$

where $l = 2\pi r$, the mean length of turn of the toroid, and A is the toroidal cross-sectional area.

where $l = 2\pi r$, the mean length of turn of the toroid, and A is the toroidal cross-sectional area.

Equivalent Circuit

$$E = IR$$

$$\mathfrak{F} = \Phi\mathfrak{R} \qquad (1\text{-}21)$$

Electric Field Intensity

With the application of the voltage E to the homogeneous copper toroid, there is produced within the material an electric potential gradient given by

$$\mathcal{E} \equiv \frac{E}{l} = \frac{E}{2\pi r} \quad \text{V/m}$$

This electric field must occur in a closed path if it is to be maintained. It then follows that the closed line integral of $\mathcal{E}$ is equal to the battery voltage E. Thus

$$\oint \mathcal{E} \, dl = E$$

Electric Potential Difference

If it is desired to find the voltage drop occurring between two points a and b of the copper toroid, we may write:

$$V_{ab} = \int_a^b \mathcal{E} \, dl = \frac{E}{l} \int_a^b dl = \frac{IR}{l} l_{ab}$$

$$= \frac{I}{l} \rho \frac{l}{A} l_{ab} = I\rho \frac{l_{ab}}{A} = IR_{ab}$$

i.e.,

$$V_{ab} = IR_{ab}$$

where R_{ab} is the resistance of the copper toroid between points a and b.

Current Density

By definition, current density is the number of amperes per unit area. Thus

$$J \equiv \frac{I}{A} = \frac{E}{AR} = \frac{\mathcal{E}l}{A\rho(l/A)} = \frac{\mathcal{E}}{\rho}$$

or

$$\mathcal{E} = \rho J$$

This last expression is often referred to as the *microscopic* form of Ohm's law.

Magnetic Field Intensity

When a magnetomotive force is applied to the homogeneous iron toroid, there is produced within the material a magnetic potential gradient given by

$$H \equiv \frac{\mathcal{F}}{l} = \frac{\mathcal{F}}{2\pi r} \quad \text{A-t/m} \quad (1\text{-}22)$$

As already pointed out in connection with Ampere's circuital law, the closed line integral of H equals the enclosed mmf. Thus

$$\oint H \, dl = \mathcal{F} \quad (1\text{-}23)$$

Magnetic Potential Difference

This is the mmf drop appearing between two points when flux flows. Thus the portion of the total applied mmf appearing between points a and b is found similarly:

$$U_{ab} = \int_a^b H \, dl = \frac{\mathcal{F}}{l} l_{ab} = \frac{\Phi \mathcal{R}}{l} l_{ab}$$

$$= \frac{\Phi}{l} \frac{l}{\mu A} l_{ab} = \Phi \frac{l_{ab}}{\mu A} = \Phi \mathcal{R}_{ab}$$

$$U_{ab} = \Phi \mathcal{R}_{ab} \quad (1\text{-}24)$$

where $\mathcal{R}_{ab}$ is the reluctance of the iron toroid between points a and b.

Flux Density

Flux density is expressed as webers per unit area. Thus

$$B = \frac{\Phi}{A} = \frac{\mathcal{F}}{A\mathcal{R}} = \frac{Hl}{A(l/\mu A)} = \mu H$$

or

$$H = \frac{B}{\mu} \quad (1\text{-}25)$$

MAGNETIC THEORY AND CIRCUITS   CHAP. 1

It should not be inferred from the foregoing that electric and magnetic circuits are analogous in *all* respects. For example, there are no magnetic insulators analogous to those known to exist for electric circuits. Also, when a direct current is established and maintained in an electric circuit, energy must be continuously supplied. An analogous situation does not prevail in the magnetic case, where a flux is established and maintained constant.

1-5 UNITS FOR MAGNETIC CIRCUIT CALCULATIONS

Magnetic circuit calculations can be carried out by use of any one of several different systems of units. These various systems arose initially because it was thought that the phenomena of electricity and magnetism were unrelated—which lead to the development of a separate system of units for each—and, second, because there was a need to deal with practical values of the units once the relationship was discovered. Up to now attention has been given exclusively to the mks (meter-kilogram-second) system of units as developed by Giorgi about the turn of the twentieth century. This policy is prompted by the acceptance in 1960 of the mks system of units as the standard for scientific work and now referred to as SI units (Système International Unités). However, a good part of the past literature is written in the units of the cgs (centimeter-gram-second) system. Furthermore, many of the present-day computations are carried on in the *mixed* system employing such units as ampere-turns/inch, maxwells/inch², and ampere-turns because of the convenience they offer in dealing with dimensions that are expressed in inches. For these reasons the units of all three systems are shown in Table 1-1. (See page 22.)

The weber, which is the unit of flux in the mks system, is equal to 10^8 *maxwells* (or lines), where the maxwell is the unit of flux in the cgs system. The *gilbert* is the cgs unit for mmf and is equal to 0.4π times the number of ampere-turns. The cgs unit for magnetic field intensity H is the *oersted* (or gilbert/cm) and the cgs unit for flux density B is the *gauss* (or lines/cm²). The relationships existing for the same quantity between the various systems of units are given in the last column of the table.

1-6 MAGNETIC CIRCUIT COMPUTATIONS

Basically magnetic circuit calculations involving ferromagnetic materials fall into two classes. In the first the value of the flux is known and it is required to find the mmf to produce it. This is the situation typical of the design of a-c and d-c electromechanical energy converters. On the basis of the desired voltage rating of an electric generator or the torque rating of an electric motor information about the required magnetic flux is readily obtained. Then with this knowledge and the configuration of the magnetic circuit the total mmf needed to establish the flux is determined straightforwardly. In the second case it is the flux for which we must solve, knowing the geometry of the magnetic circuit and the applied mmf. An engineering application in which this situation prevails is the magnetic amplifier, where it is often necessary to find the resultant magnetic flux caused by one or more control windings. Because the reluctance

TABLE 1–1 MAGNETIC UNITS

Quantity	Symbol	CGS Unit	CGS Relation	SI Unit	SI Symbol	SI Relation	Mixed English Unit	Mixed English Relation	Conversion factors
mmf	$\mathcal{F}$	gilberts	$\mathcal{F} = 0.4\pi NI$	ampere-turn	A-t	$\mathcal{F} = NI$	A-t	$\mathcal{F} = NI$	pragilbert = 10 gilberts; A-t = 0.4π gilberts = 1.257 gilberts
Permeability Free space, Abs. norm. perm.	μ_0 μ		$\mu_0 = 1$ $\mu = \mu_0\mu_r = \mu_r$			$\mu_0 = 4\pi 10^{-7}$ $= 4\pi 10^{-7}\mu_r$		$\mu_0 = 3.19$ $\mu = 3.19\mu_r$	
Length	l	cm		meter	m		inch		1 centimeter = 0.01 meter; 1 meter = 39.4 inches
Area	A	cm²		meter²	m²		inch²		1 meter² = 1550 inches²
Reluctance	$\mathcal{R}$		$\mathcal{R} = \dfrac{l}{\mu_r A}$			$\mathcal{R} = \dfrac{l}{4\pi 10^{-7}\mu_r A}$		$\mathcal{R} = \dfrac{l}{3.19\mu_r A}$	
Flux	Φ	maxwells = lines	$\Phi = \dfrac{0.4\pi NI}{\dfrac{l}{\mu_r A}}$	weber	Wb	$\Phi = \dfrac{NI}{\dfrac{l}{\mu_r 4\pi 10^{-7} A}}$	maxwells = lines	$\Phi = \dfrac{NI}{\dfrac{l}{3.19\mu_r A}}$	1 weber = 10^8 lines
Magnetic field intensity	H	gilb/cm = oersteds	$H = \dfrac{0.4\pi NI}{l}$	A-t / meter	A-t / m	$H = \dfrac{N\Phi}{l}$	A-t / inch	$H = \dfrac{NI}{l}$	1 oersted = 79.6 A-t/meter; 1 praoersted = 1000 oersteds; $1\dfrac{\text{A-t}}{\text{inch}} = \dfrac{0.4\pi}{2.54} = 0.495$ oersted; $1\dfrac{\text{A-t}}{\text{inch}} = \dfrac{2.02}{1000}$ praoersted
Flux density	B	lines/cm² = gausses	$B = \mu_r H$	webers / meter²	T	$B = \mu_r\mu_0 H$	lines/inch²	$B = 3.19\mu_r H$	1 gauss = 6.45 lines/inch²; $\dfrac{1 \text{ weber}}{\text{meter}^2} = \dfrac{64{,}500 \text{ lines/inch}^2}{} = 1$ tesla; $\dfrac{1 \text{ weber}}{\text{meter}^2} = \dfrac{10{,}000 \text{ gausses}}{} = 1$ tesla

(or permeability) of the ferromagnetic material is not constant, the solution of this problem is considerably more involved than that of the first type as illustrated by the examples below.

Example 1-1

A toroid is composed of three ferromagnetic materials and is equipped with a coil having 100 turns as depicted in Fig. 1-13. Material a is a nickel-iron alloy having a mean arc length l_a of 0.3 m. Material b is medium silicon steel and has a mean arc

Figure 1-13 Toroid composed of three different materials.

length l_b of 0.2 m. Material c is of cast steel having a mean arc length equal to 0.1 m. Each material has a cross-sectional area of 0.001 m^2.

(a) Find the magnetomotive force needed to establish a magnetic flux of $\Phi = 6 \times 10^{-4}$ Wb = 60,000 lines.

(b) What current must be made to flow through the coil?

(c) Compute the relative permeability and reluctance of each ferromagnetic material.

Solution (a) To obtain the total mmf of the coil all we need to do is to apply Ampere's circuital law. Thus

$$\mathcal{F} = U_a + U_b + U_c = H_a l_a + H_b l_b + H_c l_c$$

The unknown quantities here are H_a, H_b, and H_c. These can readily be found from a knowledge of the flux density, which here is the same for each section because the flux is common and the cross-sectional areas are the same. Hence

$$B_a = B_b = B_c = \frac{\Phi}{A} = \frac{0.0006}{0.001} = 0.6 \text{ T}$$

Now H_a is found by entering the B-H curve of the nickel-iron alloy of Fig. 1-10 corresponding to $B_a = 0.6$. This yields

$$H_a = 10 \text{ A-t/m}$$

Similarly

$$H_b = 77 \text{ A-t/m}$$

$$H_c = 270 \text{ A-t/m}$$

Accordingly, the total required mmf is

$$\mathcal{F} = H_a l_a + H_b l_b + H_c l_c$$
$$= 10(0.3) + 77(0.2) + 270(0.1)$$
$$= 3 + 15.4 + 27.0 = 45.4 \text{ A-t}$$

Note that, although the path length of cast steel is the smallest, it nonetheless requires the greatest portion of the mmf to force the specified flux through. This happens because of its much lower permeability as shown in part (c).

(b) In the SI system the mmf is equal to the number of ampere-turns. Hence

$$I = \frac{\mathcal{F}}{N} = \frac{45.4}{100} = 0.454 \text{ A}$$

(c) From Eq. (1-8)

$$\mu_a = \frac{B_a}{H_a} = \frac{0.6}{10} = 0.06 \text{ H/m}$$

Also

$$\mu_{ra}\mu_0 = \mu_a$$

$$\therefore \quad \mu_{ra} = \frac{\mu_a}{\mu_0} = \frac{0.06}{4\pi \times 10^{-7}} = 47,746$$

Furthermore, from Eq. (1-13) the reluctance is found to be

$$\mathcal{R}_a = \frac{\mathcal{F}_a}{\Phi} = \frac{3}{6 \times 10^{-4}} = 5000 \text{ rationalized mks units of reluctance}$$

Proceeding in a similar fashion for materials b and c leads to the following results:

$$\mu_{rb} = 6207 \qquad \mathcal{R}_b = 25{,}667$$
$$\mu_{rc} = 1768 \qquad \mathcal{R}_c = 45{,}000$$

Next we consider the more difficult problem: that of finding the flux in a given magnetic circuit corresponding to a specified mmf. The solution cannot be arrived at directly because, as a result of the nonlinear relationship between B and H, there are too many unknowns. The easiest way of finding the solution is to employ a cut-and-try procedure guided by a knowledge of the permeability characteristics of the materials such as appears in their magnetization curves. The following example illustrates the technique involved.

Example 1-2

For the toroid of Example 1-1, shown in Fig. 1-13, find the magnetic flux produced by an applied magnetomotive force of $\mathcal{F} = 35$ A-t.

The solution cannot be determined directly because to do so we must know the reluctance of each part of the magnetic circuit, which can be known only if the flux density is known—which means that Φ must be known right at the start. This is clearly impossible.

To obtain the solution by the cut-and-try procedure, we begin by first assuming that all of the applied mmf appears across the material having the highest reluctance. This yields an approximate value of Φ which can subsequently be refined. A glance at Fig. 1-10 shows that the poorest magnetic "conductor" is cast steel. Hence by assuming the entire mmf to appear across material c we can find H, from which B follows, which in turn yields Φ. Thus

$$H_c = \frac{U_c}{l_c} = \frac{\mathcal{F}}{l_c} = \frac{35}{0.1} = 350 \text{ A-t/m}$$

From Fig. 1-10

$$B_c = 0.65 \text{ T}$$
$$\therefore \quad \Phi_1 = B_c A_c = 0.65(0.001) = 0.00065 \text{ Wb}$$

This value represents the first approximation for the flux as indicated by the subscript.

Also, since the cross-sectional area is the same for each material, it follows that

$$B_a = B_b = B_c = 0.65 \text{ T}$$

Reference to the nickel-iron magnetization curve reveals that the value of H_a corresponding to B_a is negligibly small compared with H_c. Hence for all practical purposes its effect can be neglected. However, note that for medium silicon steel the value of H_b is almost 81 A-t/m. This, coupled with the fact that $l_b = 2l_c$, indicates that material b takes about half as much mmf as material c in maintaining the flow of flux. In other words, at this point in our analysis we can make a refinement on our original assumption of assigning the entire mmf to material c. Now we see that about 50% of that assigned to c should be assigned to b. Thus

$$U_c + U_b = \mathcal{F} \qquad \text{(assumed)}$$

but

$$U_b = 0.5U_c \quad \text{(assumed)}$$

Hence

$$1.5U_c = \mathfrak{F} = 35$$

$$\therefore \quad U_c = 23.3 \text{ A-t}$$

Accordingly a second approximation for the solution can be obtained. Therefore

$$H_c = \frac{23.3}{0.1} = 233 \text{ A-t/m}$$

which in turn yields

$$B_c = 0.4 \text{ T}$$

so that the value of the flux now becomes

$$\Phi_2 = B_c A_c = 0.0004 \text{ Wb}$$

To determine whether or not this is the correct answer we must at this point compute the mmf drops for each material and add to see whether they yield a value equal to the applied mmf. If not, the foregoing procedure must be repeated until Ampere's circuital law is satisfied. Making this check for the second approximation we have

$$H_b = 62 \text{ A-t/m} \qquad \text{corresponding to } B_b = 0.4 \text{ T}$$

and

$$H_a = 5.7 \text{ A-t/m} \qquad \text{for } B_a = 0.4 \text{ T}$$

Accordingly

$$\text{mmf} = H_a l_a + H_b l_b + H_c l_c$$

$$= 5.7(0.3) + 62(0.2) + 233(0.1)$$

$$= 1.7 + 12.4 + 23.3 = 37.4 \text{ A-t}$$

Obviously this is too high by about 7%. Hence, as a third try, reduce the biggest contributor to the mmf by a factor of 5%. That is, assume that now

$$U_c = 22 \text{ A-t}$$

Then

$$H_c = 220 \text{ A-t/m} \quad \text{and} \quad B_c = 0.375 \text{ T}$$

$$\therefore \quad \Phi_2 = 0.000375 \text{ Wb}$$

Corresponding to this flux we find

$$H_b = 59 \quad \text{and} \quad H_a = 5$$

Hence

$$\text{mmf} = 5(0.3) + 59(0.2) + 22 = 35.3 \text{ A-t}$$

Since this summation of mmf's agrees with the applied mmf of 35 A-t, the correct solution for the flux is

$$\Phi = 0.000375 \text{ Wb} = 37,500 \text{ lines}$$

When making magnetic circuit computations of the kind just illustrated, it is common practice to accept as valid any solution that comes within $\pm 5\%$ of the exact solution. The reason is that we are dealing with normal magnetization curves which neglect hysteresis and which are after all only *typical* of the material actually being used in the specified circuit. Deviations can and often do exist.

As a final example to illustrate magnetic circuit computations, we shall solve a problem involving parallel magnetic paths as well as the presence of an air gap. Moreover, we shall consider only the first type where the mmf needed to establish a specified flux is to be found. This is justified not only because the solution is straightforward but also because it is far more representative of the kind of magnetic circuit problem the engineer is likely to be concerned with.

Example 1-3

A magnetic circuit having the configuration and dimensions shown in Fig. 1-14 is made of cast steel having a thickness of 0.05 m and an air gap of 0.002 m length appearing between points *g* and *h*. The problem is to find the mmf to be produced by the coil in order to establish an air-gap flux of 4×10^{-4} Wb (or 40,000 lines).

Figure 1-14 Magnetic circuit for Example 1-3. All dimensions are in meters.

Figure 1-15 Equivalent circuit of Fig. 1-14.

The method of solution can be readily ascertained by referring to the equivalent circuit of this magnetic circuit as shown in Fig. 1-15. Knowledge of Φ_g enables us to find the mmf drop appearing across *b* and *c*. From this information the flux in leg *bc* can be determined and, upon adding it to Φ_g, we find the flux in leg *cab*. In turn, the mmf needed to maintain the total flux in leg *cab* can be computed, and when we add it to the mmf drop across *bc* we obtain the resultant mmf.

The computations involved for the various parts of the magnetic circuit are as follows:

Part gh. This is the air gap for which the flux is specified as 4×10^{-4} Wb. The cross-sectional area of the gap is $0.05(0.05) = 0.0025$ m². Normally, however, this area is slightly higher because of the tendency of the flux to bulge outward along the edges of the air gap—which is often referred to as *fringing*. For convenience this effect is neglected. Thus, the air-gap flux density is found to be

$$B_g = \frac{\Phi_g}{A_g} = \frac{4 \times 10^{-4}}{0.0025} = 0.16 \text{ T}$$

Since the permeability of air is practically the same as that of free space, we have

$$H_g = \frac{B_g}{\mu_0} = \frac{0.16}{4\pi \times 10^{-7}} = 127,300 \text{ A-t/m}$$

and
$$U_g = H_g l_g = 127{,}300(0.002) = 255 \text{ A-t}$$

Part bg and hc. The length of ferromagnetic material involved here is

$$l_{bg} + l_{hc} = 2\left(l_{bd} + l_{de} + l_{ef} + \frac{l_{fk}}{2}\right) - l_g$$
$$= 2(0.025 + 0.1 + 0.025 + 0.1) - 0.002$$
$$= 0.498 \text{ m}$$

Moreover, corresponding to a flux density in the cast steel of 0.16 T, the field intensity H is found to be 125 A-t/m. Hence

$$U_{bg+hc} = 125(0.498) = 62.2 \text{ A-t}$$

Part bc. Because path *bfkc* is in parallel with path *bc*, the total mmf across path *bfkc* also appears across path *bc*. Hence

$$U_{bc} = 255 + 62.2 = 317.2 \text{ A-t}$$

Also

$$l_{bc} = 0.1 + 0.075 = 0.175 \text{ m}$$
$$\therefore \quad H_{bc} = \frac{317.2}{0.175} = 1810 \text{ A-t/m}$$

And from Fig. 1-10 for cast steel the corresponding flux density is found to be

$$B_{bc} = 1.38 \text{ T}$$

Hence

$$\Phi_{bc} = 1.38(0.0025) = 0.00345 \text{ Wb}$$

Part cab. Accordingly, the total flux existing in leg *cab* is

$$\Phi_{cab} = \Phi_{bc} + \Phi_g = 0.00345 + 0.0004 = 0.00385 \text{ Wb}$$

Knowledge of this flux then leads to determination of the mmf needed in leg *cab* to sustain it. Thus

$$B_{cab} = \frac{0.00385}{0.00375} = 1.026 \text{ T}$$

From which

$$H_{cab} = 690 \text{ A-t/m}$$

Hence

$$U_{cab} = H_{cab} l_{cab} = 690(0.5) = 345 \text{ A-t}$$

Therefore the total mmf required to produce the desired air-gap flux is

$$\mathfrak{F} = U_{cab} + U_{bc} = 345 + 317.2 = 662.2 \text{ A-t}$$

1-7 MAGNETIC LEAKAGE FLUX

In the practical magnetic circuits used in electric machines not all the flux produced by the field winding is effective in the energy conversion process. This is readily illustrated by referring to the magnetic circuit depicted in Fig. 1-16, which is typical of most electric machines. The ring-type structure is usually the stator and carries the

Figure 1-16 Typical magnetic circuit found in energy conversion devices, showing leakage flux.

coil that provides the mmf to create the operating flux in the air gap. The cylindrical part is the rotating member of the machine, and it carries the armature winding in which the working emf is induced. The air-gap flux is the useful flux of the machine. Any flux produced by the field coil that does not link the armature coil is identified as *field leakage flux*. Three typical paths for the leakage flux are shown in Fig. 1-16. The existence of such flux paths is ensured by Ampere's circuital law. So long as a flux path can be drawn linking a net mmf different from zero, such a flux path does in fact exist. Of course the magnitude of this leakage flux is not very large because the reluctance of its path is chiefly that of air, as is apparent from Fig. 1-16.

To understand how one may compute the value of the leakage flux, consider the following data as they apply to the configuration of Fig. 1-16:

$$l_{abd} = 1 \text{ m}$$
$$l_{ef} = 0.25 \text{ m}$$
$$l_g = 0.0025 \text{ m}$$
$$r = 0.08 \text{ m} = \text{rotor radius}$$
$$A_s = 3 \times 10^{-3} \text{ m}^2 = \text{cross-sectional area of stator}$$
$$B_g = 0.62 \text{ T} = \text{flux density in air gap}$$
$$\Phi_g = 3 \times 10^{-3} \text{ Wb} = \text{air-gap flux}$$

The ferromagnetic material is medium silicon steel and of uniform cross section throughout.

First let us find the coil mmf needed to produce the specified gap flux, assuming no leakage. Thus

For the air gap:

$$U_g = H_g(2l_g) = \frac{B_g}{\mu_0}(2l_g) = \frac{0.62}{4\pi \times 10^{-7}}2(0.0025) = 2550 \text{ A-t}$$

For the steel:

$$B_s = \frac{\Phi_s}{A_s} = \frac{\Phi_g}{A_s} = \frac{3 \times 10^{-3}}{3 \times 10^{-3}} = 1 \text{ T}$$

From Fig. 1-10 we get

$$H_s = 154 \text{ A-t/m}$$

Hence

$$U_s = H_s(l_{abd} + l_{ef}) = 154(1 + 0.25) = 192.5 \text{ A-t}$$

Therefore the total required mmf is

$$\mathfrak{F} = U_g + U_s = 2550 + 192.5 = 2742.5 \text{ A-t}$$

Next assume that measurements on this circuit indicate that, to produce the specified air gap flux a coil having 560 turns draws a current of 5 A. Accordingly, the actual coil mmf is found to be 2800 A-t. The difference between this mmf and the computed value is due to the leakage flux.

To obtain an exact accounting of the leakage flux in computing the coil mmf is a difficult task because the coil mmf produces partial-linkage paths in addition to the full-linkage paths already shown in Fig. 1-16 for the leakage flux. However, a simplified approximate method may be used, which is satisfactory in most cases. It is based on the assumption that the total leakage flux may be treated in terms of an equivalent leakage flux that takes a path symmetrically placed with respect to the total magnetic circuit. Thus for the configuration of Fig. 1-16 the points c and b are so selected that the distance between them is the same whether the path taken is clockwise or counterclockwise. An estimate of the amount of leakage flux existing in this case can then be found as follows:

Compute the magnetic potential difference between points c and b. Thus

$$U_{cb} = U_g + U_{ef+cd+ab} = 2550 + l_{ef+ab+cd}$$

$$= 2550 + 154\left[0.25 + \frac{1 + 0.16 + 0.005}{2} - (0.16 + 0.005)\right]$$

$$= 2653.3 \text{ A-t}$$

The portion of steel that lies to the left of c and b is assumed to carry the leakage flux as well as the air-gap flux. Hence the magnetic potential difference between these points caused by this total flux must be equal to $2800 - 2653.3 = 146.7$ A-t. In short, this is the net mmf drop in the iron to the left of points c and b. The field intensity is correspondingly found to be

$$H_{cb} = \frac{146.7}{l_{cb}} = \frac{146.7}{0.5825} = 251.8 \text{ A-t/m}$$

Then from Fig. 1-10 it follows that

$$B_{cb} = 1.15 \text{ T}$$

and

$$\Phi_{cb} = \Phi_{\text{total}} = 1.15(3 \times 10^{-3}) = 3.45 \times 10^{-3} \text{ Wb}$$

Therefore the leakage flux is

$$\Phi_l = \Phi_{cb} - \Phi_g = (3.45 - 3)10^{-3} = 0.45 \times 10^{-3} \text{ Wb}$$

In this example the leakage flux is about 15% of the air-gap flux.

1-8 THE INDUCTANCE PARAMETER

Inductance is a characteristic of magnetic fields, and it was first discovered by Faraday in his renowned experiments of 1831. In a general way *inductance* can be characterized as that property of a circuit element by which energy is capable of being stored in a magnetic flux field. A significant and distinguishing feature of inductance, however, is that it makes itself felt in a circuit only when there is a *changing current* or flux. Thus, although a circuit element may have inductance by virtue of its geometrical and magnetic properties, its presence in the circuit is not exhibited unless there is a time rate of change of current. This aspect of inductance is particularly stressed when we consider it from the circuit viewpoint. However, for the sake of completeness, inductance is also treated from an energy and a geometrical view.

Circuit Viewpoint

The current-voltage relationship involving the inductance parameter is expressed by

$$v_L = L\frac{di}{dt} \qquad (1\text{-}26)$$

In general both v_L and i are functions of time. Figure 1-17 depicts the potential difference v_L appearing across the terminals of the inductance parameter when a changing current flows into terminal c. Note that the arrowhead on the v_L quantity is shown at terminal c, indicating that this terminal is instantaneously positive with respect to terminal d. In turn, this means that current is increasing in the positive sense, i.e., the slope di/dt in Eq. (1-26) is positive. Any circuit element that exhibits the property of inductance is called an *inductor* and is denoted by the symbolism shown in Fig. 1-17. In the ideal sense the inductor is considered to be resistanceless, although practically it must contain the wire resistance out of which the inductor coil is formed.

It follows from Eq. (1-26) that an appropriate defining equation for inductance is

$$\boxed{L = \frac{v_L}{di/dt}} \qquad \text{volt-sec/ampere, or henrys (H)} \qquad (1\text{-}27)$$

Figure 1-17 Illustrating the inductance parameter.

Thus by recording the potential difference at a given time instant across the terminals of an inductor and dividing by the corresponding derivative of the current time function we determine the inductance parameter. Note that the units of inductance are volt-seconds per ampere. For simplicity this is commonly called the *henry*.

A *linear inductor* is one for which the inductance parameter is independent of current. As current flows through an inductor it creates a space flux. When this flux permeates air, strict proportionality between current and flux prevails so that the inductance parameter stays constant for all values of current. A plot of the potential difference across the coil as a function of the derivative of the current then appears as shown in Fig. 1-18, which is a plot of Eq. (1-27). When the flux is made to penetrate iron, however, it is possible for large currents to upset the proportional relationship between the current and the flux it produces. In such a case the inductor is called *nonlinear* and a plot of Eq. (1-27) will then no longer be a straight line.

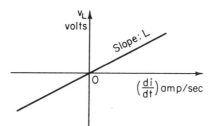

Figure 1-18 Graphical representation of the inductance parameter L, from the circuit viewpoint.

For the resistance parameter Ohm's law can be written to express either voltage in terms of current or current in terms of voltage. The same procedure may be followed for the inductance parameter. Equation (1-26) already expresses the voltage as a function of the current. However, to express the current in terms of the potential difference across the inductor, Eq. (1-26) must be transposed to read as follows:

$$di = \frac{1}{L} v_L \, dt \qquad (1\text{-}28)$$

In integral form this becomes

$$\int_{i(0)}^{i(t)} di = \frac{1}{L} \int_0^t v_L \, dt \qquad (1\text{-}29)$$

or

$$i(t) = \frac{1}{L} \int_0^t v_L \, dt + i(0) \qquad (1\text{-}30)$$

Equation (1-30) thus reveals that the current in an inductor is dependent upon the integral of the voltage across its terminals as well as the initial current in the coil at the start of integration.

An examination of Eqs. (1-26) and (1-30) reveals an important property of inductance: *the current in an inductor cannot change abruptly in zero time.* This is made apparent from Eq. (1-26) by noting that a finite change in current in zero time calls for an infinite voltage to appear across the inductor, which is physically impossible. On the other hand, Eq. (1-30) shows that in zero time the contribution to the inductor current from the integral term is zero so that the current immediately before and after application of voltage to the inductor is the same. In this sense, then, we may look upon inductance as exhibiting the property of inertia.

Example 1-4

An inductor has a current passing through it which varies in time in the manner depicted in Fig. 1-19(a). Find the corresponding time variation of the voltage drop appearing across the inductor terminals if it is assumed that the inductance of the coil is 0.1 H.

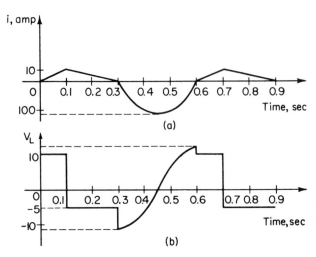

(a)

(b)

Figure 1-19 (a) Input current wave shape to an inductor; (b) corresponding voltage variation across the inductor terminal.

The solution appears in Fig. 1-19(b). Note that, in the interval from 0 to 0.1 s, $di/dt = 100$ A/s. Hence the voltage across the coil is then a constant given by

$$v_L = L\frac{di}{dt} = 0.1(100) = 10 \text{ V} \quad \text{for } 0 < t < 0.1$$

In the time range from 0.1 to 0.3 s the slope of the current curve is -50. Hence the voltage across the coil is -5 V. Finally, in the interval from 0.3 to 0.6 s the current wave slope is sinusoidal so that the corresponding voltage wave is a cosine. In drawing the cosine wave it is assumed that the maximum slope of the sine wave exceeds 100 A/s.

It is interesting to note in Fig. 1-19 that, unlike the current in an inductor, the voltage across the inductor is allowed to change discontinuously.

Energy Viewpoint

Assume that an inductor has zero initial current. Then, if a current i is made to flow through the coil across which appears the potential difference v_L, the total energy received in the time interval from 0 to t is

$$W = \int_0^t v_L i \, dt \qquad \text{joules (J)} \tag{1-31}$$

Inserting Eq. (1-26) leads to

$$W = \int_0^t \left(L \frac{di}{dt} \right) i \, dt = \int_0^i Li \, di \qquad (1\text{-}32)$$

or

$$\boxed{W = \tfrac{1}{2} Li^2} \qquad \text{J} \qquad (1\text{-}33)$$

Continuing with the assumption that the inductor has no winding resistance, Eq. (1-33) states that the inductor absorbs an amount of energy which is proportional to the inductance parameter L as well as the square of the instantaneous value of the current. Thus energy is stored by the inductor in a magnetic field. It is of finite value and retrievable. As the current is increased, so too is the stored magnetic energy. Note, however, that this energy is zero whenever the current is zero. Because the energy associated with the inductance parameter increases and decreases with the current, we can properly conclude that the inductor has the property of being capable of returning energy to the source from which it receives it.

A glance at Eq. (1-33) indicates that an alternative way of identifying the inductance parameter is in terms of the amount of energy stored in its magnetic field corresponding to its instantaneous current. Thus, in mathematical form we can write

$$\boxed{L = \frac{2W}{i^2}} \qquad \text{H} \qquad (1\text{-}34)$$

This is an energy description of the inductance parameter.

There is one final point worthy of note. It has already been demonstrated that for a potential difference to exist across the inductor terminals the current must be changing. A constant current results in a zero voltage drop across the ideal inductor. This is not true, however, about the energy absorbed and stored in the magnetic field of the inductor. Equation (1-33) readily verifies this fact. A constant current results in a fixed energy storage. Any attempt to alter this energy state is firmly resisted by the effects of the initial energy storage. This again reflects the inertial aspect of inductance.

Geometrical Viewpoint

The voltage across the terminals of an inductor may be expressed from a circuit viewpoint by Eq. (1-26). However, this same voltage may be described by Faraday's law in terms of the flux produced by the current and the number of turns N of the inductor coil. Accordingly, we may write

$$v_L = L \frac{di}{dt} = N \frac{d\phi}{dt} \qquad (1\text{-}35)$$

It then follows that

$$L = N \frac{d\phi}{di} \qquad (1\text{-}36)$$

In those cases, where the flux ϕ is directly proportional to current i for all values (i.e., for linear inductors), the last expression becomes

$$L = \frac{N\Phi}{i} \quad \frac{\text{Wb-t}}{\text{A}} \text{ or H} \tag{1-37}$$

Here the inductance parameter has a hybrid representation because it is in part expressed in terms of the circuit variable i and in part in terms of the field variable Φ. To avoid this we replace flux by its equivalent, i.e.,

$$\Phi = \frac{\text{mmf}}{\text{magnetic reluctance}} = \frac{Ni}{\mathcal{R}} \tag{1-38}$$

where *mmf* is the magnetomotive force which produces the flux Φ in the magnetic circuit having a reluctance $\mathcal{R}$. Appearing in Fig. 1-20 is an inductor consisting of N turns wound about a circular iron core. If the core is assumed to have a mean length of l meters and a cross-sectional area of A_m meters², then the magnetic reluctance can be written as

$$\mathcal{R} = \frac{l}{\mu A_m} \tag{1-39}$$

where μ is the permeability, a physical property of the magnetic material.

Figure 1-20 Linear inductor with iron core.

Upon substituting Eqs. (1-38) and (1-39) in (1-37) there results the expression for the inductance parameter of the circuit of Fig. 1-20. Thus

$$L = \frac{N^2 \mu A_m}{l} = \frac{N^2}{\mathcal{R}} \tag{1-40}$$

A study of Eq. (1-40) reveals some interesting and useful facts about the inductance parameter which are not readily available when this quantity is defined either from the circuit or the energy viewpoint. Most impressive is the fact that inductance, like resistance, is dependent upon the geometry of physical dimensions and the magnetic property of the medium. This is significant because it tells us what can be done to change the value of L. Thus, for the inductor illustrated in Fig. 1-20, the inductance parameter may be increased in any one of four ways: increasing the number of turns, using an iron core of higher permeability, reducing the length of the iron core, and increasing the cross-sectional area of the iron core.

It is interesting to note that neither the circuit viewpoint nor the energy viewpoint could tell us these things, because essentially they deal with the effects associated with a given inductor geometry. It is emphasized that all three viewpoints are needed to complete the picture of circuit parameters and to give proper perspective.

1-9 HYSTERESIS AND EDDY-CURRENT LOSSES IN FERROMAGNETIC MATERIALS

The process of magnetization and demagnetization of a ferromagnetic material in a symmetrically cyclic condition involves a storage and release of energy which is not completely reversible. As the material is magnetized during each half-cycle it is found that the amount of energy stored in the magnetic field exceeds that which is released upon demagnetization. The background for understanding this behavior was provided in Sec. 1-3. There the hysteresis loop was identified as the variation of flux density as a function of the magnetic field intensity for a ferromagnetic material in a cyclic condition. The salient feature of the hysteresis loop is the delayed reorientation of the domains in response to a cyclically varying magnetizing force. A single hysteresis loop is depicted in Fig. 1-21. The direction of the arrows on this curve indicates the manner in which B changes as H varies from zero to a positive maximum through zero to a negative maximum and back to zero again, thus completing the loop. To appreciate the meaning of the various shaded areas shown in Fig. 1-21, let us look at the units associated with the product of B and H. Thus

$$\text{units of } HB = \frac{\text{amperes}}{\text{meter}} \times \frac{\text{newtons}}{\text{ampere-meter}} = \frac{\text{newtons}}{\text{meter}^2} = \frac{\text{N}}{\text{m}^2}$$

But

$$\text{newton-meter} = \text{joule}$$

Hence

$$\text{units of } HB = \frac{\text{joules}}{\text{meter}^3} = \frac{\text{J}}{\text{m}^3}$$

which is clearly recognized as an energy density. Therefore, in dealing with areas involving B and H in connection with a hysteresis loop we are really dealing with energy densities expressed on a per cycle basis because the hysteresis loop is repeatable for each cyclic variation of H.

Area abda represents energy stored in magnetic field during positive half cycle of H

Area bdc represents energy released by magnetic field during positive half cycle

Area abca represents hysteresis loss per half cycle

Figure 1-21 Hysteresis loop and energy relationships per half-cycle.

It can be shown that the energy stored in the magnetic field during that portion of the cyclic variation of H when it increases from zero to its positive maximum value (assuming the material is already in a cyclic state) is given by

$$w_1 = \int_{B_a}^{B_b} H \, dB \qquad \text{J/m}^3 \tag{1-41}$$

when mks units are used; i.e., H must be expressed in A-t/m and B in webers per square meter. Note that the axes of Fig. 1-21 are so labeled. Moreover, during that portion of its cyclic variation when H decreases from its positive maximum value to zero (as it follows along curve bc of the hysteresis loop), energy is being released by the magnetic field and returned to the source, and this quantity can be represented as

$$w_2 = \int_{B_b}^{B_c} H \, dB \qquad \text{J/m}^3 \tag{1-42}$$

In this equation, since $B_b > B_c$, the quantity w_2 will be negative, indicating that the energy is being released rather than stored by the magnetic field.

A graphical interpretation of Eq. (1-41) leads to the result that the energy absorbed by the field when H is increasing in the positive direction can be represented by the area $abdca$. Similarly the energy released by the field as H varies from H_{max} to zero can be represented by area $bdcb$. The difference between these two energy densities represents the amount of energy which is not returned to the source but, rather, is dissipated as heat as the domains are realigned in response to the changing magnetic field intensity. This dissipation of energy is called *hysteresis loss*. Keep in mind that Fig. 1-21 depicts this energy density loss for a half-cycle variation of H. Hence area $abca$ represents the hysteresis loss per half-cycle. It certainly follows from symmetry that upon completion of the negative half-cycle variation of H an equal energy loss occurs. Therefore, as H varies over one complete cycle, the total energy loss per cubic meter is represented by the area of the hysteresis loop. More specifically this energy loss per cycle can be expressed mathematically as

$$\boxed{w_h = (\text{area of hysteresis loop})} \qquad \text{J/m}^3\text{/cycle} \tag{1-43}$$

where mks units are used for H and B.

It is frequently desirable to express the hysteresis loss of ferromagnetic materials in watts—the unit of power. A little thought about the units of w_h in Eq. (1-43) shows how this can be directly accomplished. Thus

$$w_h = \frac{\text{energy}}{\text{vol} \times \text{cycles}} = \frac{\text{power} \times \text{seconds}}{\text{vol} \times \text{cycles}} = \frac{\text{power}}{\text{vol} \times \text{cycles/second}} \tag{1-44}$$

Now let P_h = power loss, watts (W)

v = volume of ferromagnetic material

f = hertz (Hz) = frequency of variation of H

Then Eq. (1-44) becomes

$$w_h = \frac{P_h}{vf} \tag{1-45}$$

or

$$P_h = w_h v f \qquad (1\text{-}46)$$

where w_h—the energy density loss—is determined from Eq. (1-43).

To obviate the need of finding the area of the hysteresis loop in order to compute the hysteresis loss in watts from Eq. (1-46), Steinmetz obtained an empirical formula for w_h based on a large number of measurements for various ferromagnetic materials. He expressed the hysteresis power loss as

$$P_h = v f K_h B_m^n \qquad (1\text{-}47)$$

where B_m is the maximum value of the flux density and n lies in the range $1.5 \leq n \leq 2.5$ depending upon the material used. The parameter K_h also depends upon the material. Some typical values are: cast steel 0.025, silicon sheet steel 0.001, and permalloy 0.0001.

In addition to the hysteresis power loss, another important loss occurs in ferromagnetic materials that are subjected to time-varying magnetic fluxes—the *eddy-current* loss. This term is used to describe the power loss associated with the circulating currents that are found to exist in closed paths within the body of a ferromagnetic material and cause an undesirable heat loss. These circulating currents are created by the differences in magnetic potential existing throughout the body of the material owing to the action of the changing flux. If the magnetic circuit is composed of solid iron, the ensuing power loss is appreciable because the circulating currents encounter relatively little resistance. To increase significantly the resistance encountered by these eddy currents the magnetic circuit is invariably composed of very thin *laminations* (usually 14 to 25 mils thick) whenever the electromagnetic device is such that a varying flux permeates it in normal operation. This is the case with transformers and all a-c electric motors and generators. An empirical equation for the eddy-current loss is

$$P_e = K_e f^2 B_m^2 \tau^2 v \qquad \text{W} \qquad (1\text{-}48)$$

where K_e = a constant dependent upon the material
  f = frequency of variation of flux, Hz
  B_m = maximum flux density
  τ = lamination thickness
  v = total volume of the material

A comparison of this equation with Eq. (1-47) reveals that eddy-current losses vary as the square of the frequency, whereas the hysteresis loss varies directly with the frequency.

Taken together the hysteresis and eddy-current losses constitute what is frequently called the *core losses* of electromagnetic devices that involve time-varying fluxes for their operation. More than just passing attention is devoted to these losses here because, as will be seen, core losses have an important bearing on temperature rise, efficiency, and rating of electromagnetic devices.

Sec. 1-9   Hysteresis and Eddy-Current Losses in Ferromagnetic Materials   **37**

1-10 RELAYS—AN APPLICATION OF MAGNETIC FORCE

A *relay* is an electromagnetic device which can often be activated by relatively little energy, causing a movable ferromagnetic armature to open or close one or several pairs of electrical contact points located in another control circuit or in a main circuit handling large amounts of energy. Alternating- and direct-current motor starters are equipped with relays designed to ensure proper operation of motors during starting and running conditions. These devices are found in many applications in all fields of engineering, especially in situations where control of a process or machine is involved. Our objective in this section is to describe the principles that underlie the operation of these electromagnetic devices. Besides the knowledge it offers, this treatment gives the motivation for studying the theory of magnetic fields and circuits.

The derivation of a magnetic force equation is our primary interest because it shows how it is possible to do mechanical work—moving a relay armature—by abstracting energy from that stored in the magnetic field. Moreover, since the emphasis is on the principles involved, the simplifying assumptions of no saturation and no losses are imposed; i.e., linear analysis is used throughout. Accordingly, the magnetization curve of the ferromagnetic material is assumed to be a straight line as depicted in Fig. 1-22(a). (Only the curve corresponding to positive values of H is shown.) Now, by appropriately modifying the axes of the curve plotted in Fig. 1-22(a), a significant and useful thing happens. First recall that the area between the B axis and the magnetization curve represents the energy absorbed from the source and stored in the magnetic field on a per-unit-volume basis. Equation (1-41) states this result mathematically.

$$w_f = \int_{B_a}^{B_b} H \, dB \qquad \text{J/m}^3 \tag{1-49}$$

In the simplified situation of Fig. 1-22(a), Eq. (1-49) becomes merely the area of

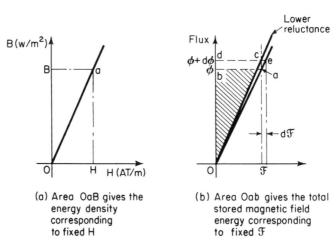

(a) Area OaB gives the energy density corresponding to fixed H

(b) Area Oab gives the total stored magnetic field energy corresponding to fixed $\mathcal{F}$

Figure 1-22   Linear magnetization curve of a relay.

triangle OaB. Thus

$$w_f = \tfrac{1}{2}BH \qquad \text{J/m}^3 \tag{1-50}$$

where H is assumed fixed at the value shown.

Moreover, since w_f is an energy density, the total energy stored in the magnetic field is found by multiplying Eq. (1-41) by the volume. Thus

$$W_f = w_f v = w_f Al \qquad \text{J} \tag{1-51}$$

where l is the length and A the cross-sectional area of the magnetic circuit. Inserting Eq. (1-50) into Eq. (1-51) yields

$$\boxed{W_f = \tfrac{1}{2}(BA)Hl = \tfrac{1}{2}\Phi\mathfrak{F}} \qquad \text{J} \tag{1-52}$$

Note that in this equation A is combined with B to identify the flux Φ and H is combined with l to identify the mmf $\mathfrak{F}$. A graphical representation of Eq. (1-52) appears in Fig. 1-22(b). It should be apparent that Fig. 1-22(b) derives from Fig. 1-22(a) by multiplying the ordinate axis by A and the abscissa axis by l, thus plotting Φ versus $\mathfrak{F}$. Then for a fixed H (or $\mathfrak{F}$) area OaB of Fig. 1-22(a) gives the energy density, whereas the corresponding area Oab of Fig. 1-22(b) gives the total energy stored in the magnetic field.

To understand how mechanical work can be done by the abstraction of energy stored in the magnetic field, consider the circuitry appearing in Fig. 1-23, which depicts the basic composition of an electromagnetic relay. It consists of an exciting coil placed on a fixed ferromagnetic core equipped with a movable element called the *relay armature*. The relay is energized from a constant voltage source through an adjustable resistor R. To begin with, consider that R is fixed at that value which makes the coil mmf equal to $\mathfrak{F}$ and produces the flux Φ as shown in Fig. 1-22(b). Then adjust R to increase the mmf by $d\mathfrak{F}$ and thereby the flux by $d\Phi$, assuming the relay armature is held fast to keep the reluctance invariant. Figure 1-22(b) shows that an additional amount of energy is absorbed from the source and stored in the magnetic field virtually equal to the area of rectangle $abdc$. Next readjust R to make $d\mathfrak{F}$ zero, and then, keeping the mmf fixed at $\mathfrak{F}$, release the armature, thus allowing it to move in the direction to decrease the air gap. Hold the armature fast again when the decreased reluctance causes the flux to increase by the amount $d\Phi$ obtained previously.

It is important to note here that, neglecting second-order effects, the new (lower-reluctance) position of the armature is also responsible for permitting the source to supply an additional amount of energy virtually equal to area $abdc$ as before but with

Confining surface

Figure 1-23 Basic composition of an electromagnetic relay.

one significant difference. Whereas, when the armature is held fixed, the additional energy supplied by the source is converted entirely to stored magnetic energy, on the other hand, when the armature is allowed to move towards a lower-reluctance position, only half of the same amount of additionally supplied energy is stored in the magnetic field. The other half is consumed in doing the mechanical work involved in moving the relay armature from the higher- to the lower-reluctance position. That the amount is one-half is readily apparent from a glance at Fig. 1-22(b) if we note that the area of triangle *Oac* is one-half the area of rectangle *abdc*. The slope of the magnetization curve is the permeability and so can be used as a measure of the reluctance. Hence the higher the slope of the magnetization curve, the higher is μ and the lower is the reluctance. For the magnetization curve in Fig. 1-22(b) to go from position *Oa* to *Oc*, it is necessary for the relay armature to be moved to a position corresponding to a smaller air gap. The mechanical work involved in accomplishing this is represented by area *Oac* in Fig. 1-22(b).

The conclusions described in the foregoing can now be expressed mathematically. It is important to keep in mind, however, that the interchange of energy between the magnetic field and the mechanical system (the relay armature) necessarily involves a *change in reluctance*. In other words, the change in energy to do mechanical work, dW_m, is equal to the change in magnetic field energy associated with a change in reluctance $d\mathcal{R}$. Thus by Eqs. (1-15) and (1-52),

$$dW_m = -\tfrac{1}{2}\Phi^2\, d\mathcal{R} \tag{1-53}$$

where the negative sign emphasizes that mechanical work is done through a *decrease* in reluctance.

An expression for the magnetic force developed on the relay armature is readily obtained by recalling that

$$F\, dx = dW_m \tag{1-54}$$

where F is the force in newtons. Inserting this expression into Eq. (1-53) then yields

$$\boxed{F = -\tfrac{1}{2}\Phi^2\, \frac{d\mathcal{R}}{dx}} \qquad \text{N} \tag{1-55}$$

Hence the magnitude of the instantaneous magnetic force is dependent upon the value of the flux as well as the rate of change of reluctance. Moreover, the direction of this force is always such as to bring about a decrease in reluctance as indicated by the minus sign.

Example 1-5

In the relay circuit of Fig. 1-23 assume the cross-sectional area of the fixed core and the relay armature to be A and the air-gap flux to be Φ. Neglecting the reluctance of the iron, find the expression for the magnetic force existing on the relay armature.

Solution We must find the rate of change of reluctance with distance along the sliding surface. Thus

$$\mathcal{R} = \frac{g}{\mu_0 A} \tag{1-56}$$

where g is the air-gap length.

$$\frac{d\mathcal{R}}{dx} = \frac{1}{\mu_0 A}\frac{dg}{dx} \tag{1-57}$$

But

$$dg = -dx \tag{1-58}$$

Hence

$$\frac{d\mathcal{R}}{dx} = -\frac{1}{\mu_0 A} \tag{1-59}$$

Inserting this last expression into Eq. (1-55) yields the desired result,

$$F = \frac{1}{2}\frac{\Phi^2}{\mu_0 A} \quad \text{N} \tag{1-60}$$

Example 1-6

The cross-sectional view of a cylindrical plunger magnet appears in Fig. 1-24. The plunger (or armature) is free to move inside a nonferromagnetic guide around which the coil is surrounded by a cylindrically shaped steel shell. The plunger is separated from the shell by an air gap of length g.

Figure 1-24 Plunger magnet for Example 1-6.

(a) Derive the expression for the magnetic force exerted on the plunger when it is in the position shown in Fig. 1-24. The length of the plunger is at least equal to that of the shell. Also it has a radius of a meters. Neglect the reluctance of the steel.

(b) Find the magnitude of the force when the mmf is 1414 A-t and the plunger magnet dimensions are

$$x = 0.025 \text{ m} \qquad h = 0.05 \text{ m}$$
$$a = 0.025 \text{ m} \qquad g = 0.00125 \text{ m}$$

Solution Before proceeding with the calculations let us review the principle that gives rise to the force. For the direction of coil current assumed, a typical flux path is that shown in Fig. 1-24. Note that the flux path must cross the air gap twice. It is particularly important to note too that the reluctance "seen" by the flux as it crosses the bottom gap is less than that seen by the same flux crossing the gap at the upper part of the plunger because the cross-sectional area of the magnetic circuit is smaller at the upper part than it is at the lower part of the plunger. As a matter of fact, at the lower part of the plunger the cross-sectional area seen by the flux is fixed for all positions of the plunger lying in the range $0 \leq x \leq h$ and specifically is equal to $2\pi ah$ (assuming g is small compared to a). Accordingly, a force is created on the plunger so directed as to decrease the reluctance. This means that the magnetic force acts to move the plunger upward.

(a) The solution of this part is obtained from Eq. (1-55), but first we need to identify the correct expression for the reluctance as seen by the flux. The reluctance of the steel is negligible. Hence the reluctance associated with the typical flux shown in Fig. 1-24 is merely the sum of the reluctances associated with each air gap. Thus

$$\mathcal{R} = \frac{g}{\mu_0 2\pi a h} + \frac{g}{\mu_0 2\pi a x} = \frac{g}{2\pi a \mu_0}\left(\frac{1}{h} + \frac{1}{x}\right) \qquad (1\text{-}61)$$

Then

$$\frac{d\mathcal{R}}{dx} = \frac{g}{2\pi a \mu_0}\left(-\frac{1}{x^2}\right) \qquad (1\text{-}62)$$

Also

$$\Phi = \frac{\mathcal{F}}{\mathcal{R}} = \frac{\mathcal{F}}{g/\mu_0 2\pi a}\frac{xh}{h+x} \qquad (1\text{-}63)$$

Inserting Eqs. (1-62) and (1-63) into Eq. (1-55) yields

$$F = \pi a \mu_0 \frac{\mathcal{F}^2 h^2}{g}\left(\frac{1}{x+h}\right)^2 = 3.94 \times 10^{-6}\frac{ah^2}{g}\mathcal{F}^2\left(\frac{1}{x+h}\right)^2 \quad \text{N} \qquad (1\text{-}64)$$

(b) Inserting the specified dimensions into Eq. (1-64) gives the magnetic force as

$$F = 70\ \text{N} = 15.74\ \text{lb}$$

PROBLEMS

1-1. A long straight wire located in air carries a current of 4 A. Assume the relative permeability of air is unity.

(a) Find the value of the magnetic field intensity at a distance of 0.5 m from the center of the wire.

(b) A second long straight wire carrying a current of 2 A is placed parallel to the first one at a distance of 0.5 m with the current flowing in the same direction. Find the direction and magnitude of the force per meter existing between the wires.

(c) Repeat part (b) for wires that are imbedded in iron having a relative permeability of 10,000 and a spacing of 0.05 m.

1-2. The wires shown in Fig. P1-2 are long, straight, and parallel and are completely embedded in iron having a relative permeability of 1000. Each wire carries a current of 10 A.

Figure P1-2

(a) Compute the magnitude and direction of the resultant force per meter on the wire in which I_2 flows.

(b) Compute the magnitude and direction of the force per meter on the wire in which I_1 flows.

(c) Repeat part (a) for the third wire.

1-3. Repeat Prob. 1-2 for current I_2 flowing opposite to I_1 and I_3.

1-4. A uniform magnetic field of 0.7 Wb/m² in the iron is applied to the configuration of Fig. P1-2.

    (a) Compute the direction and magnitude of the resultant force per meter when the magnetic field is directed perpendicularly into the plane of the paper.

    (b) What is the value of this resultant force per meter when the magnetic field is applied in the plane of the wires and directed from right to left?

    (c) Compute the resultant force per meter when the magnetic field is applied at an angle of 45° to the plane of the paper and directed into it from right to left.

    (d) What is the resultant force per meter when the magnetic field is applied at an angle of 60° to the plane of the paper and directed into it from top to bottom?

1-5. In the configuration of Fig. P1-5, the current I_1 has a value of 40 A. Find the value of I_2 that causes the magnetic field intensity at point P to disappear.

      Figure P1-5   **Figure P1-6**

1-6. A circular loop of wire of radius r meters and consisting of a single turn carries the current I as shown in Fig. P1-6. Derive the expression for the magnetic field intensity at the center.

1-7. A magnetic circuit composed of silicon sheet steel has the square construction shown in Fig. P1-7.

      Figure P1-7

    (a) Find the mmf required to produce a core flux of 25×10^{-4} Wb.

    (b) If the coil has 80 turns, how much current must be made to flow through the coil?

1-8. The magnetic circuit of Prob. 1-7 has an air gap of 0.1 cm cut in the right leg. For a coil having 100 turns find the current that must be allowed to flow in order for the core flux to be 0.0025 Wb.

1-9. In the magnetic circuit of Fig. P1-9 determine the coil mmf needed to produce a flux of 0.0014 Wb in the right leg. The thickness of the magnetic circuit is 0.04 m and is uniform throughout. Medium silicon steel is used.

Figure P1-9

1-10. Repeat Prob. 1-9 for a coil placed on the center leg.

1-11. In the magnetic circuit of Prob. 1-7 find the core flux produced by a coil current of 200 A-t.

1-12. In the magnetic circuit of Prob. 1-8 find the core flux produced by a coil current of 600 A-t.

1-13. The core shown in Fig. P1-13 has a uniform cross-sectional area of 2 in.² and a mean length of 12 in. Also, coil A has 200 turns and carries 0.5 A, coil B has 400 turns and carries 0.75 A, and coil C carries 1.00 A. How many turns must coil C have for the core flux to be 120,000 lines? The coil currents have the directions indicated in the figure. The core is made of silicon sheet steel.

Figure P1-13

1-14. In the magnetic circuit shown in Fig. P1-14 the coil F_1 is supplied with 350 A-t in the direction indicated. Find the direction and magnitude of the mmf required in coil F_2 for the air-gap flux to be 180,000 lines. The core has an effective cross-sectional area of 9 in.² and is made of silicon sheet steel. The length of the air gap is 0.05 in.

Figure P1-14

MAGNETIC THEORY AND CIRCUITS   CHAP. 1

1-15. In the magnetic circuit of Fig. P1-14 the coil F_1 is supplied with 200 A-t in the direction shown. Find the direction and magnitude of the mmf required of coil F_2 for the air-gap flux to be 90,000 lines.

1-16. Find the inductance of a coil in which:

(a) A current of 0.1 A yields an energy storage of 0.05 J.

(b) A current increases linearly from zero to 0.1 A in 0.2 s producing a voltage of 5 V.

(c) A current of 0.1 A increasing at the rate of 0.5 A/s represents a power flow of $\frac{1}{2}$ W.

1-17. The coil in the configuration of Fig. 1-20 is equipped with 100 turns. Moreover, the mean length of turn of the magnetic core is known to be 0.2 m and the cross-sectional area is 0.01 m². The value of the permeability of the iron is 10^{-3}.

(a) Find the inductance of the coil.

(b) When a d-c voltage is applied to the inductor it is found to draw a current of 0.1 A. How much energy is stored in the magnetic field?

1-18. When a d-c voltage is applied to the coil of Prob. 1-17, the current is found to vary in accordance with $i = \frac{1}{10}(1 - \epsilon^{-5t})$ expressed in amperes. Moreover, it is found that, after the elapse of 0.2 s, the voltage induced in the coil is 0.16 V. Find the inductance of the coil.

1-19. A sample of iron having a volume of 16.4 cm³ is subjected to a magnetizing force sinusoidally varying at a frequency of 400 Hz. The area of the hysteresis loop is found to be 64.5 cm² with flux density plotted in kilolines per square inch and magnetizing force in ampere-turns per inch. The scale factors used are 1 in. = 5 kilolines/in.² and 1 in. = 12 A-t/in. Find the hysteresis loss in watts.

1-20. A ring of ferromagnetic material has a rectangular cross section. The inner diameter is 7.4 in., the outer diameter is 9 in., and the thickness is 0.8 in. There is a coil of 600 turns wound on the ring. When the coil carries a current of 2.5 A, the flux produced in the ring is 1.2×10^{-3} Wb. Find the following quantities expressed in mks units:

(a) magnetomotive force; (b) magnetic field intensity; (c) flux density: (d) reluctance; (e) permeability; (f) relative permeability.

1-21. In plotting a hysteresis loop the following scales are used: 1 cm = 10 A-t/in., and 1 cm = 20 kilolines/in.² The area of the loop for a certain material is found to be 6.2 cm.² Calculate the hysteresis loss in joules per cycle for the specimen tested if the volume is 400 cm³.

1-22. The flux in a magnetic core is alternating sinusoidally at a frequency of 500 Hz. The maximum flux density is 50 kilolines/in.². The eddy-current loss then amounts to 14 W. Find the eddy-current loss in this core when the frequency is 750 Hz and the flux density is 40 kilolines/in².

1-23. The total core loss (hysteresis plus eddy current) for a specimen of magnetic sheet steel is found to be 1800 W at 60 Hz. If the flux density is kept constant and the frequency of the supply increased 50%, the total core loss is found to be 3000 W. Compute the separate hysteresis and eddy-current losses at both frequencies.

1-24. A flux ϕ penetrates the complete volume of the iron bars as shown in Fig. P1-24. When the bars are assumed separated by g meters, find the expression for the force existing between the two parallel plane faces. Neglect the reluctance of the iron.

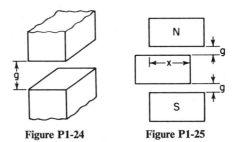

Figure P1-24   Figure P1-25

1-25. A magnetic flux ϕ penetrates a movable magnetic core which is vertically misaligned relative to the north and south poles of an electromagnet as depicted in Fig. P1-25. The depth dimension for the core and the electromagnet is b. Determine the expression for the force that acts to bring the core into vertical alignment. Express the result in terms of the air-gap flux density and the physical dimensions. Neglect the reluctance of the iron.

1-26. The magnetization curve of a relay is shown in Fig. P1-26.

Figure P1-26

(a) Find the energy stored in the field with the relay in the open position.
(b) Assume that the relay armature moves rapidly under conditions of constant flux. Compute the work done in going from the open to the closed position.
(c) Calculate the force in newtons exerted on the armature in part (b).
(d) Assume the relay armature moves slowly at constant mmf. Compute the work done in going from the open to the closed position.
(e) Does the energy of part (d) come from the original stored field energy? Explain.

1-27. In the magnetic circuit of Fig. P1-27 the cross-sectional area of the air gap is 0.0025 m² and its length is 0.05 cm. Moreover, the cast steel core has a mean length of 0.2 m and it is equipped with 1000 turns. It is desirable to operate this circuit with an air-gap density of 1 T.

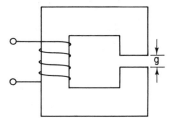

Figure P1-27

(a) Determine the coil current needed to establish this gap density.
(b) Compute the energy stored in the air gap.
(c) Find the energy stored in the cast steel section of the circuit.
(d) What is the inductance in henrys of this magnetic circuit?

CONVENTIONAL ELECTRIC MACHINES

2

Transformers

An understanding of the transformer is essential to the study of electromechanical energy conversion. Although electromechanical energy conversion involves interchange of energy between an electrical system and a mechanical system whereas the transformer involves the interchange of energy between two or more electrical systems, nonetheless the coupling device in both cases is the magnetic field and its behavior in each case is fundamentally the same. As a result we find that many of the pertinent equations and conclusions of transformer theory have equal applicability in the analysis of a-c machinery and some aspects of d-c machinery. For example, the equivalent circuit of single-phase and three-phase induction motors is found to be identical in form to that of the transformer. Other examples can be cited, as a glance through the following chapters readily reveals.

In addition to serving as a worthwhile prelude to the study of electromechanical energy conversion, an understanding of transformer theory is important in its own right because of the many useful functions the transformer performs in prominent areas of electrical engineering. In communication systems ranging in frequency from audio to radio to video, transformers are found fulfilling widely varied purposes. *Input transformers* (to connect a microphone output to the first stage of an electronic amplifier), *interstage transformers*, and *output transformers* are to be found in radio and television circuits. Transformers are also used in communication circuits as impedance transformation devices which allow maximum transfer of power from the input circuit to the coupled circuit. Telephone lines and control circuits are two more areas in which the transformer is used extensively. In electric power distribution systems the transformer makes it possible to convert electric power from a generated voltage of about 15 to 20 kV (as determined by generator design limitations) to values of 380 to 500 kV, thus permitting transmission over long distances to appropriate distribution points (e.g., urban areas) at tremendous savings in the cost of copper as well as in power losses in the transmission lines. Then at the distribution points the transformer is the means by which these dangerously high voltages are reduced to a safe level (208 to 120 V) for use in homes, offices, shops, etc. In short, the transformer is a useful device to be found in many phases of electrical engineering and therefore merits the attention given to it in this chapter.

2-1 THEORY OF OPERATION AND DEVELOPMENT OF PHASOR DIAGRAMS

In the interest of clarity and motivation the theory of operation of the transformer is developed in five steps, starting with the ideal iron-core reactor and ending with the transformer phasor diagram under load. In the process the important transformer equations are developed along with a physical explanation of how the transformer operates. Emphasis on complete understanding of the physical behavior of a device is stressed, because, once this is grasped, the theory as well as the associated mathematical descriptions readily follow.

In its simplest form a transformer consists of two coils which are mutually coupled. When the coupling is provided through a ferromagnetic ring (circular or otherwise), the transformer is called an *iron-core transformer*. When there is no ferromagnetic material but only air, the device is described as an *air-core transformer*. In this chapter attention is confined exclusively to the iron-core type. Of the two coils the one that receives electric power is called the *primary winding* whereas the other, which can deliver electric power to a suitable load circuit, is called the *secondary winding*.

Ideal Iron-Core Reactor

As the first step in the development of the theory of transformers refer to Fig. 2-1, which depicts an ideal iron-core reactor energized from a sinusoidal voltage source v_1. The reactor is ideal because the coil resistance and the iron losses are assumed to be zero and the magnetization curve is assumed to be linear.

Our interest at this point is to develop the phasor diagram that applies to the circuitry of Fig. 2-1, because thereby much is revealed about the theory. As v_1 increases

Figure 2-1 Ideal iron-core reactor: zero winding resistance and no core losses. Polarities are instantaneous.

TRANSFORMERS   CHAP. 2

in its sinusoidal variation, it causes a sinusoidal magnetizing current i_ϕ† to flow through the N_1 turns of the coil, which in turn produces a sinusoidally varying flux in time phase with the current. In other words, when the current is zero, the flux is zero and, when the current is at its positive maximum value, so too is the flux. Applying Kirchhoff's voltage law to the circuit we have

$$v_1 - e_1 = 0 \qquad (2\text{-}1)$$

where e_1 is the *voltage drop* associated with the flow of current through the coil and can be written in terms of i_ϕ as

$$e_{ab} = e_1 = L\frac{di_\phi}{dt} \qquad (2\text{-}2)$$

where L is the inductance of the coil and $i_\phi = i_{ab}$. Inserting Eq. (2-2) into Eq. (2-1) and rearranging yields

$$v_1 = e_1 = L\frac{di_\phi}{dt} = e_{ab} \qquad (2\text{-}3)$$

We also know that the magnetizing current is sinusoidal because the applied voltage is sinusoidal and the magnetization curve is assumed linear. Consequently we can write for the magnetizing current the expression

$$i_\phi = \sqrt{2}\,I_\phi \sin \omega t \qquad (2\text{-}4)$$

where I_ϕ is the rms value of the current.

Introducing Eq. (2-4) into Eq. (2-3) and performing the differentiation called for yields

$$v_1 = e_1 = \sqrt{2}\,(\omega L)I_\phi \cos \omega t = \sqrt{2}\,(\omega L)I_\phi \sin\left(\omega t + \frac{\pi}{2}\right) \qquad (2\text{-}5)$$

The maximum value of this quantity occurs when the cos ωt has the value of unity. Then

$$E_{1\,\text{max}} = \sqrt{2}\,(\omega L)I_\phi = \sqrt{2}\,E_1 \qquad (2\text{-}6)$$

where

$$E_1 \equiv \omega L I_\phi = \text{rms value of the inductive reactance drop} \qquad (2\text{-}7)$$

A comparison of Eqs. (2-4) and (2-5) shows that the magnetizing current lags the voltage appearing across the coil by 90 electrical degrees. The phasor diagram of Fig. 2-2(a) depicts this condition. Since all the quantities involved here are sinusoidal, the phasors shown represent the rms value of the quantities.

(b)   (a)   (c)

Figure 2-2 Phasor diagrams of the ideal iron-core reactor: (a) circuit viewpoint; (b) field viewpoint; (c) combined diagram.

†Lowercase letters are used to represent instantaneous values.

Another approach that can be used in arriving at the phasor diagram for the ideal iron-core reactor is worth considering because of its use later. The approach just described can be referred to as a *circuit viewpoint* because it deals solely with the electric circuit in which I_ϕ flows as well as the circuit parameter L. No direct use is made of the magnetic flux ϕ, which is a space (or field) quantity. In contrast the second approach starts with the magnetic field as it appears in Faraday's law of induction. This is one of the fundamental laws upon which the science of electrical engineering is based. Faraday's law states that the emf induced in a coil is proportional to the number of turns linking the flux as well as the time rate of change of the linking flux. In applying Faraday's law to the circuit of Fig. 2-1 the expression may be written either as a voltage drop (*a* to *b*) or as a voltage rise (*b* to *a*). By the former description we can write

$$e_{ab} = e_1 = +N\frac{d\phi}{dt} \qquad (2\text{-}8)$$

Expressed as a voltage rise, the induced emf equation becomes

$$e_{ba} = -e_1 = -N\frac{d\phi}{dt} \qquad (2\text{-}9)$$

In either case, the sign is attributable to Lenz's law. The minus sign states that the emf induced by a changing flux is always in the direction in which current would have to flow to oppose the changing flux. On the other hand, the plus sign in Eq. (2-8) denotes that the polarity of *a* is positive whenever the time rate of change of flux is positive as determined by the direction of current flow.

Equation (2-9) then represents the starting point for the second approach in analyzing the circuit of Fig. 2-1. Corresponding to the sinusoidal variation of the magnetizing current as expressed by Eq. (2-4), there is also a sinusoidal variation of flux which may be expressed as

$$\phi = \Phi_m \sin \omega t \qquad (2\text{-}10)$$

where Φ_m denotes the maximum value of the magnetic flux. Putting this expression into Eq. (2-9) and performing the required differentiation leads to

$$e_{ba} = -e_1 = -N_1\frac{d\phi}{dt} = -N_1\Phi_m\omega \cos \omega t = E_{1\,max} \sin\left(\omega t - \frac{\pi}{2}\right) \qquad (2\text{-}11)$$

where

$$E_{1\,max} \equiv N_1\Phi_m\omega \equiv \sqrt{2}\,E_1 \qquad (2\text{-}12)$$

and

$$E_1 = \text{rms value of the induced emf}$$

A comparison of Eqs. (2-10) and (2-11) makes it clear that the reaction emf $-e_1$ *lags* the changing flux that produces it by 90°. This situation is depicted in Fig. 2-2(b). Basically, the emf $-e_1$ is a voltage rise (or generated emf) which has such a direction that, if it were free to act, would cause a current to flow opposing the action of I_ϕ. To verify this solely in terms of the action-reaction law consider any two adjacent points such as a' and b' on coil N_1 in Fig. 2-1. If the flux is assumed increasing in the direction shown, then the reaction in the coil must be such that current would have to flow from point b' to a' in order to oppose the increasing flux. Recall that by the *right-hand rule* any current assumed to be flowing from b' to a' produces flux lines opposing the increasing flux. In order for this condition to prevail, it is therefore necessary for point a' to be positive with respect to b' for the instant being considered. This line of reasoning when extended to the full coil, leads to the polarity markings shown in Fig. 2-1. The interesting thing to note here is that on the basis of the flux viewpoint the emf $-e_1$ in Fig. 2-1 is treated as a reaction (or generated) voltage rise in progressing from b' to a'. On the other hand, from the circuit viewpoint the voltage e_1, and its polarity when considered in the direction of flow of the current I_ϕ, appears as a voltage drop. The voltage e_1 remains the same; it is merely the point of view that changes.

Because I_ϕ and the flux ϕ are in time phase, it is customary to combine the two viewpoints into a single phasor diagram as depicted in Fig. 2-2(c). Since $E_{ab} = E_1$ represents the voltage drop, it follows that the same induced emf viewed on the basis of a reaction to the changing flux, being a voltage rise, is equal and opposite to E_1. That is, $E_{ba} = -E_1$. Another way of describing this is to say that the terminal voltage must always contain a component equal and opposite to the voltage rise.

It should be apparent up to this point that the reaction voltage viewed from b' to a' in Fig. 2-1 does not succeed in actually establishing a reverse current to oppose the increasing flux. This never happens in the primary winding of a two-winding transformer, but it does happen in the secondary winding as is presently described.

In the work that follows in this chapter preference is given to treating the induced emf as a voltage drop. That is, in applying Faraday's law, the version described by Eq. (2-8) will be used. Due regard is given to Lenz's law in this equation by noting that, when the time rate of change of flux in the configuration of Fig. 2-1 is positive, the polarity of point a is also positive.

Practical Iron-Core Reactor

As the second step in the development of transformer theory, let us consider how the phasor diagram of Fig. 2-2(a) must be modified in order to account for the fact that a practical reactor has a winding-resistance loss as well as hysteresis and eddy-current losses. A linear magnetization curve will continue to be assumed because the effect of the nonlinearity is to cause higher harmonics of fundamental frequency to exist in the magnetizing current—which, of course, cannot be included in the phasor diagram. Only quantities of the same frequency can be shown.

To develop the phasor diagram in this case we start with the flux phasor ϕ and then draw the induced emf E_1 90° leading. This much is shown in Fig. 2-3(a). Next we must consider the location of the current. Keep in mind that the iron is in a cyclic condition, i.e., the flux is sinusoidally varying with time. Moreover, this cyclic variation takes place along a hysteresis loop of finite area because the core losses are no longer assumed zero. Now, as pointed out in Chapter 1, it is characteristic of ferromagnetic materials in a cyclic condition to have the flux *lag* behind the magnetizing current by an amount called the *hysteretic angle* γ. See Fig. 2-3(b). Since the positive direction of rotation is counterclockwise in a phasor diagram, $\bar{I}_m$ is shown

Figure 2-3 Development of the phasor diagram of a practical iron-core reactor: (a) building on the flux phasor; (b) accounting for core losses; (c) accounting for core losses and winding resistance.

leading ϕ to bear out the physical fact. Note, too, that the magnetizing current $\bar{I}_m$ is now considered made up of two fictitious components $\bar{I}_\phi$ and $\bar{I}_c$. The quantity $\bar{I}_\phi$ is the purely reactive component and is the current that would flow if the core losses were zero. The quantity I_c is that fictitious quantity which, when multiplied by the voltage E_1 produced by the changing flux, represents exactly the power needed to supply the core losses caused by the same changing flux. Thus we can write

$$I_c E_1 \equiv P_c \quad \text{W} \tag{2-13}$$

or

$$I_c \equiv \frac{P_c}{E_1} \quad \text{A} \tag{2-14}$$

where P_c denotes the sum of the hysteresis and eddy-current losses. This technique for handling the core loss of an electomagnetic device is a useful one and is frequently applied in the study of a-c machinery.

To complete the phasor diagram for the practical iron-core reactor, it is only necessary to account for the winding-resistance drop. This is readily accomplished by applying Kirchhoff's voltage law to the circuit. Hence

$$\bar{V}_1 = \bar{I}_m r_1 + \bar{E}_1 \quad \text{V} \tag{2-15}$$

where r_1 is the winding resistance in ohms. Since r_1 is a constant, it follows that the $\bar{I}_m r_1$ drop must be in phase with $\bar{I}_m$ or located parallel to $\bar{I}_m$ as depicted in Fig. 2-3(c). Note that the component of $\bar{I}_m$ that is in phase with the applied voltage $\bar{V}_1$ exceeds $\bar{I}_c$, as well it should because this component must supply not only the core losses but the winding copper losses too. Thus, the input power may be expressed as

$$P_{in} = V_1 I_m \cos \theta_0 = I_m^2 r_1 + P_c \quad \text{W} \tag{2-16}$$

where θ_0 denotes the power-factor angle.

The Two-Winding Transformer

By placing a second winding on the core of the reacor of Fig. 2-1, we obtain the simplest form of a transformer. This is depicted in Fig. 2-4. The rms value of the induced emf E_1 appearing in the primary winding ab readily follows from Eq. (2-12) as

$$E_1 = \frac{N_1 \Phi_m \omega}{\sqrt{2}} = \frac{2\pi f}{\sqrt{2}} \Phi_m N_1 = 4.44 f \Phi_m N_1 \qquad (2\text{-}17)$$

Keep in mind that this result derives directly from Faraday's law and is very important in machinery analysis. One significant result deducible from this equation is that for transformers having relatively small winding resistance (i.e., $E_1 \approx V_1$) the value of the maximum flux is determined by the applied voltage.

Figure 2-4 Two-winding transformer.

The induced emf appearing across the secondary winding terminals is produced by the same flux that causes E_1. Hence the only difference in the rms values is brought about by the difference in the number of turns. Thus we can write

$$E_2 = 4.44 f \Phi_m N_2 \qquad (2\text{-}18)$$

Dividing Eq. (2-17) by (2-18) leads to

$$\frac{E_1}{E_2} = \frac{N_1}{N_2} \equiv a = \text{ratio of transformation} \qquad (2\text{-}19)$$

Another fact worth noting is that E_1 and E_2 are in time phase because they are induced by the same changing flux.

Transformer Phasor Diagram at No-Load

When a transformer is at no-load, no secondary current flows. Figure 2-5 depicts the situation. A study of Fig. 2-5 should make it apparent that the development of

Figure 2-5 Two-winding transformer at no-load; S open.

the phasor diagram follows from Fig. 2-3(c) by making two modifications. One is to identify the secondary induced emf in the diagram. But, as already observed in the preceding step, the emf induced in the secondary winding must be in phase with the corresponding voltage E_1 of the primary winding. For convenience call this secondary induced emf E_2. Then E_2 and E_1 must both lead the flux phasor by 90° as shown in Fig. 2-6. For convenience the turns ratio a is assumed to be unity. The second modification is concerned with accounting for the fact that not all of the flux produced by the primary winding links the secondary winding. The difference between the total flux linking the primary winding and the mutual flux linking both windings is called the *primary leakage flux* and is denoted by ϕ_{l1}. Note that some of the paths drawn to represent ϕ_{l1} encircle only several turns and not all N_1 turns. As long as such closed paths encircle an mmf different from zero, such a flux path does in fact exist. This is borne out by Ampere's circuital law.

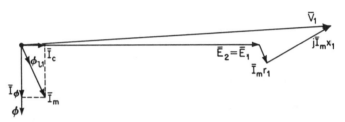

Figure 2-6   Phasor diagram of a transformer at no-load.

An important distinction exists between the mutual flux ϕ and the primary leakage flux ϕ_{l1}. It is this. The mutual flux exists wholly in iron and so involves a hysteresis loop of finite area. The leakage flux, on the other hand, always involves appreciable air paths, and, although some iron is included in the closed path, the reluctance experienced by the leakage flux is practically that of the air. Consequently the cyclic variation of the leakage flux involves no hysteresis (or lagging) effect. Hence in the phasor diagram the primary leakage flux must be placed in phase with the primary winding current as shown in Fig. 2-6. Now this leakage flux induces a voltage E_{l1} which must lead ϕ_{l1} by 90°. This emf due to leakage flux may be replaced by an equivalent *primary leakage reactance drop*. Thus

$$\boxed{I_m x_1 \equiv E_{l1}} \qquad (2\text{-}20)$$

This quantity must lead $\bar{I}_m$ by 90°. The phasor addition of this drop and the primary winding resistance drop as well as the voltage drop $\bar{E}_1$ associated with the mutual flux yields the primary applied voltage $\bar{V}_1$ as depicted in Fig. 2-6.

The quantity x_1 of Eq. (2-20) is called the *primary leakage reactance*; it is a fictitious quantity introduced as a convenience in representing the effects of the primary leakage flux.

Transformer under Load

In this last step in the development of the theory of operation of the transformer, we direct attention first to the phasor diagram representation of Kirchhoff's voltage law as it applies to the secondary circuit. For simplicity consider that the load appearing in the secondary circuit of Fig. 2-7 is purely resistive and further that the flux

Figure 2-7 Two-winding transformer under load; S closed.

has the direction indicated with the flux increasing. By Lenz's law there is emf induced in the secondary winding which instantaneously makes terminal c positive with respect to terminal d. When switch S is closed, a current $\bar{I}_2$ flows instantaneously from c through the load to d. For convenience the load resistor is assumed adjusted to cause rated secondary current to flow. $\bar{E}_2$ establishes the secondary load voltage besides accounting for other voltage drops existing in the secondary circuit. When equilibrium is established after the load switch is closed, the appropriate phasor diagram must show $\bar{V}_2$ and $\bar{I}_2$ in phase as depicted in Fig. 2-8. Since it is physically impossible for the secondary winding to occupy the same space as the primary winding, a flux is produced by the secondary current which does not link with the primary winding. This is called the *secondary leakage flux*. The voltage E_{l2} induced by this secondary leakage flux leads ϕ_{l2} by 90°. Again as a matter of convenience this secondary leakage flux voltage is replaced by a secondary leakage reactance drop. Thus

$$\boxed{E_{l2} \equiv I_2 x_2}$$

(2-21)

where x_2 is the fictitious secondary leakage reactance which represents the effects of

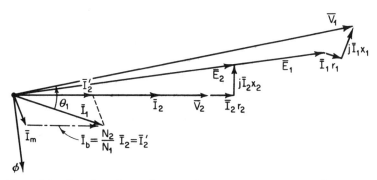

Figure 2-8 Complete phasor diagram of a transformer under load for the case where $a = N_1/N_2 > 1$.

the secondary leakage flux. Also, because leakage flux has its reluctance predominantly in air, Eq. (2-21) is a linear equation. Thus, doubling the current doubles the leakage flux, which doubles $\bar{E}_{l2}$, and by Eq. (2-21) this is represented by a doubled reactance drop. The phasor sum of the terminal voltage, the secondary winding resistance drop, and the secondary leakage reactance drop adds up to the source voltage $\bar{E}_2$ as demonstrated in Fig. 2-8.

When a phasor diagram shows $\bar{V}_2$ and $\bar{I}_2$ in phase, the implication is that power is being supplied to the load equal to $V_2 I_2$. Clearly this power must come from the primary circuit. But by what mechanism does this come about? To understand the answer to this question, consider the situation depicted in Fig. 2-7. With the flux increasing in the direction shown the secondary emf has the polarity indicated, so that when the switch is closed the action of the secondary current is to cause a decrease in the mutual flux. This tendency for the flux to decrease owing to the action of $N_2 I_2$ causes the voltage drop E_1 [or $e_1 = N_1(d\phi/dt)$] to decrease. However, since the applied voltage $\bar{V}_1$ is constant, the slight decrease in $\bar{E}_1$ creates an appreciable increase in primary current. In fact the primary current is allowed to increase to that value which, when flowing through the primary turns N_1, represents sufficient mmf to neutralize the demagnetizing action of the secondary mmf. Expressed mathematically we have

$$\bar{I}_b N_1 - N_2 \bar{I}_2 = 0 \qquad (2\text{-}22)$$

where $\bar{I}_b$ denotes the increase in primary current needed to cancel out the effect of the secondary mmf to reduce the flux. $\bar{I}_b$ is often called the *balance* current. This is above and beyond the value $\bar{I}_m$ needed to establish the operating flux. Equation (2-22) reveals that $N_1 \bar{I}_b$ must be oppositely directed to $N_2 \bar{I}_2$. In short, then, it is through the medium of the flux and its small attendant changes that power is transferred from the primary to the secondary circuit of a transformer. Of course, the total primary current is the phasor sum of $\bar{I}_b$ and $\bar{I}_m$. In this case the primary leakage flux is in phase with $\bar{I}_1$ rather than $\bar{I}_m$, which is the case at no-load.

For transformers in which the winding resistance drop and the leakage reactance drop are negligibly small, the magnetizing mmf is the same at no-load as under load. In other words, the phasor sum of the total primary mmf $N_1 \bar{I}_1$ and the secondary mmf $N_2 \bar{I}_2$ must yield $N_1 \bar{I}_m$. Thus

$$\boxed{N_1 \bar{I}_1 - N_2 \bar{I}_2 = N_1 \bar{I}_m} \qquad (2\text{-}23)$$

Dividing through by N_1 leads to

$$\bar{I}_1 - \frac{N_2}{N_1} \bar{I}_2 = \bar{I}_m \qquad (2\text{-}24)$$

Inserting Eq. (2-22) then yields

$$\bar{I}_1 - \bar{I}_b = \bar{I}_m = \bar{I}_1 - I_2' \qquad (2\text{-}25)$$

where I_2' is equivalent to I_b. These equations are readily verified in the phasor diagram of Fig. 2-8.

Finally, a comment regarding the construction features of practical transformers. Contrary to the schematic representation of Fig. 2-7 it should not be implied that the total number of primary winding turns are placed on one leg of the ferromagnetic core and all the turns of the secondary winding are placed on the other leg. Such an arrangement leads to an excessive amount of leakage flux, so that for a given source voltage the mutual flux is correspondingly smaller. The usual procedure in transformer construction is to put half the primary and secondary turns atop one another on each leg. This keeps the leakage flux within a few percent of the mutual flux. Of course still greater reduction in leakage flux can be achieved by further subdividing and sandwiching the primary and secondary turns, but this is clearly obtained at an appreciable increase in the cost of assembly.

Effect of Saturation on Magnetizing Current

All the quantities that appear in a phasor diagram such as the one depicted in Fig. 2-8 must necessarily be at the same frequency. Consequently, the magnetizing component, $\bar{I}_m$, of the current $\bar{I}_1$ has to be treated as a quantity with a sinusoidal shape when so used. However, the true shape of the actual magnetizing current is other than sinusoidal chiefly because of the influence of the magnetic core. Two factors are important in this consideration. One is the state of saturation of the magnetic iron core and the other is the combined effect of hysteresis and core losses.

Frequently, transformers function with time flux variations that place operation just above the knee of the saturation curve. It is instructive to examine the consequences of such a mode of operation. This is best accomplished by referring to Fig. 2-9. At the outset it is important to understand that the situation described in this diagram is a transitory one. We consider it in order to put into perspective the forces that create the final shape of the magnetizing current.

The voltage waveshape that is applied to the primary winding of the transformer is assumed to be a pure sinusoid. Accordingly, this voltage source is capable solely of supplying a sinusoidal magnetizing current to the primary coil of the transformer. Figure 2-9 shows this magnetizing current drawn below the magnetization curve, which in turn is depicted as possessing a fair degree of saturation but no hysteresis. The effect of hysteresis in the magnetic core material is being neglected for the moment in the interest of simplicity. The projection of the sinusoidally varying magnetizing current onto the magnetization curve produces a flat-topped flux wave as shown to the right of the magnetization curve. Moreover, this flat-topped flux wave contains primarily two harmonic components: a large fundamental component, ϕ_1, and a substantial negative third harmonic component, ϕ_3. The fundamental flux component induces a voltage, e_1, in the primary coil which is equal and opposite to the applied primary voltage, v_1, on the assumption of negligible primary winding resistance. However, the negative third harmonic flux component also induces a voltage in the primary coil, e_3. But this voltage does not encounter an equal and opposite component in the primary source voltage because it is a purely sinusoidal quantity. As a result, the coil-generated third harmonic voltage is free to act in the primary circuit to produce a negative third harmonic current which then combines with the

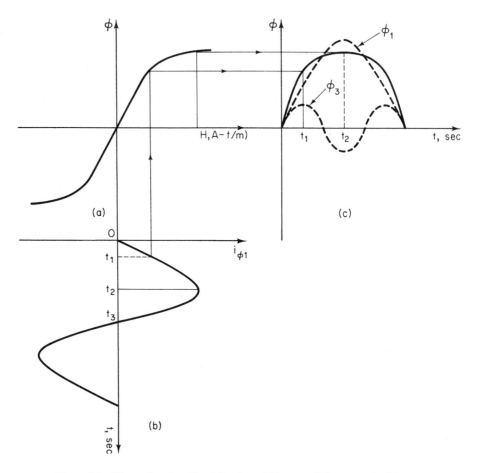

Figure 2-9 Illustrating the effect of a sinusoidal magnetizing current: (a) magnetization curve showing saturation at high magnetizing forces and no hysteresis; (b) sinusoidal magnetizing current assumed to flow from source voltage to primary coil; (c) resultant flat-topped flux wave produced by the current of (b).

positive fundamental component supplied by the source voltage. Thus, the resultant magnetizing current is the superposition of these two components. The result is a waveshape for the magnetizing current that is characteristically *peaked* as depicted in Fig. 2-10. The peaking effect is associated with the fact that the third harmonic component is made to flow in a direction opposite to the fundamental component because it originates from the primary coil.

It is interesting to observe that the action of the negative third harmonic current can also be interpreted as producing a primary-coil mmf that in turn acts to produce a negative third harmonic flux which serves to cancel the original third harmonic flux component of the flat-topped flux wave. Ultimately, of course, Kirchhoff's voltage law demands that the resultant core flux must be sinusoidal because only such a flux produces an induced coil voltage that balances the applied sinusoidal source voltage.

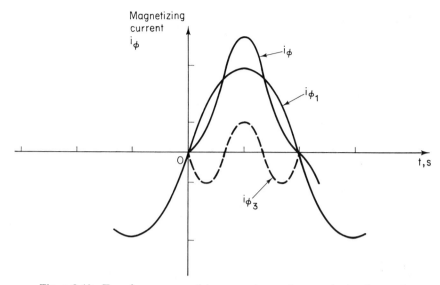

Figure 2-10 Transformer magnetizing current waveshape neglecting hysteresis and core losses. Here $i_{\phi1}$ denotes the component supplied by the source voltage and $i_{\phi3}$ is the component supplied by the transformer coil arising from saturation of the magnetic core.

The peaked nature of the resultant magnetizing current can also be demonstrated by starting with the condition that the sinusoidal source voltage can be balanced in the primary circuit only by the presence of a sinusoidal core flux. The subsequent projection of such a sinusoidal flux variation onto the saturated magnetization curve then leads directly to the peaked waveshape of the magnetization current. However, such an approach does not identify the source of the third harmonic current component, which we now know is not directly the sinusoidal source voltage.

The final waveshape of the magnetizing current is obtained by adding to the peaked curve of Fig. 2-10 a fundamental sinusoidal component (in phase with v_1 or e_1) to account for the hysteresis and eddy-current losses. Figure 2-11 shows the result.

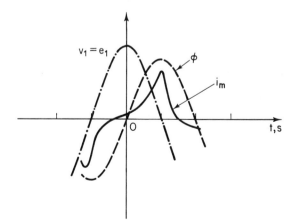

Figure 2-11 Final waveshape of magnetizing current i_m that results upon adding a sinusoidal fundamental component (in phase with the source voltage) to i_ϕ of Fig. 2-10. The in-phase component takes care of the hysteresis and eddy current losses.

2-2 THE EQUIVALENT CIRCUIT

Derivation

It is customary in analyzing devices in electrical engineering to represent the device by means of an appropriate equivalent circuit. In this way further analysis and design as well as computational accuracy is facilitated by the direct application of the techniques of electric circuit theory. This procedure is followed whenever new devices are investigated. Accordingly, equivalent circuits are derived not only for the transformer but for all a-c and d-c motors as well as important electronic devices such as the transistor. Generally, the equivalent circuit is merely a circuit interpretation of the equation(s) that describe the behavior of the device. For the case at hand there are two equations— Kirchhoff's voltage equations for the primary winding and for the secondary winding. Thus, for the primary winding we have

$$\bar{V}_1 = \bar{I}_1 r_1 + j\bar{I}_1 x_1 + \bar{E}_1 \tag{2-26}$$

and for the secondary winding

$$\bar{E}_2 = \bar{V}_2 + \bar{I}_2 r_2 + j\bar{I}_2 x_2 \tag{2-27}$$

Both equations are represented in the phasor diagram of Fig. 2-8, and the corresponding circuit interpretation appears in Fig. 2-12(a). The dot notation is used to indicate which sides of the coils instantaneously carry the same polarity. The schematic diagram of Fig. 2-12(a) can be replaced by the equivalent circuit of Fig. 2-12(b) by recognizing that $\bar{I}_1$ is composed of $\bar{I}_b$ and $\bar{I}_m$. Moreover, since by definition $\bar{I}_m$ consists of two components $\bar{I}_c$ and $\bar{I}_\phi$, it can be considered splitting into two parallel branches. One branch is purely resistive and carries the current $\bar{I}_c$, which was defined as in-phase with $\bar{E}_1$. The other branch must be purely reactive because it carries $\bar{I}_\phi$, which was defined as 90° lagging $\bar{E}_1$, as depicted in Fig. 2-3(b). The ideal transformer is included in Fig. 2-12(b) to account for the transformation of voltage and current that occurs between primary and secondary windings. Thus, with the ratio of transformation $a = 2$, the ideal transformer conveys the information that $\bar{E}_2 = \frac{1}{2}\bar{E}_1$ and that $\bar{I}_2 = 2\bar{I}_b$. The fact that the actual transformer has primary winding resistance and leakage flux and must carry a magnetizing current is taken care of by the circuitry preceding the symbol for the ideal transformer in Fig. 2-12(b).

The resistive element r_c must have such a value that, when the voltage E_1 appears across it, it permits the current I_c to flow. Thus

$$r_c \equiv \frac{E_1}{I_c} = \frac{P_c}{I_c^2} \tag{2-28}$$

The second form of this equation is obtained by replacing E_1 with the expression of Eq. (2-13). Hence, r_c may also be looked upon as that resistance which accounts for the transformer core loss. Similarly, the magnetizing reactance of the transformer may be defined as that reactance which when E_1 appears across it causes the current I_ϕ to flow. Thus

$$\boxed{x_\phi \equiv \frac{E_1}{I_\phi}} \tag{2-29}$$

Figure 2-12 Development of the exact equivalent circuit: (a) schematic diagram; (b) two-winding equivalent circuit; (c) replacing the actual secondary winding with an equivalent winding having N_1 turns; (d) final complete form.

A single-line equivalent circuit of the transformer is a desirable goal. However, as long as $E_2 \neq E_1$ this is impossible to accomplish because the potential distribution along the primary and secondary windings can never be identical. Only when $a = 1$ is this condition possible. Therefore, in the interest of achieving the desired simplicity in the equivalent circuit, let us investigate replacing the actual secondary winding with an equivalent winding having the same number of turns as the primary winding. A little thought should make it reasonable to expect that, if an equivalent secondary winding having N_1 turns is made to handle the same kilovolt-amperes, the same copper loss and leakage flux, and the same output power as the actual secondary winding, then insofar as the applied voltage source is concerned it can detect no difference between the actual secondary winding of N_2 turns and the equivalent secondary winding of N_1 turns. That is, one winding can replace the other without causing any changes in the primary quantities.

Because the equivalent secondary winding is assumed to have N_1 turns, the corresponding induced emf must be different by the factor a. We have

$$E'_2 = aE_2 = E_1 \qquad (2\text{-}30)$$

where the prime notation is used to refer to the equivalent winding having N_1 turns. For the kilovolt-amperes of the equivalent secondary winding to be the same as that of the actual secondary winding the following equation must be satisfied:

$$E'_2 I'_2 = E_2 I_2 \qquad (2\text{-}31)$$

Inserting Eq. (2-30) into the last equation shows that the current rating of the equivalent secondary winding must be

$$I'_2 = \frac{I_2}{a} \qquad (2\text{-}32)$$

Note that this current is the balance current I_b.

Since in general $a \neq 1$, it follows that the winding resistance of the equivalent secondary must be adjusted so that the copper loss is the same in both. Accordingly

$$(I'_2)^2 r'_2 = I_2^2 r_2 \qquad (2\text{-}33)$$

This equation, together with Eq. (2-32), establishes the value of the winding resistance of the equivalent secondary. Thus

$$r'_2 = \left(\frac{I_2}{I'_2}\right)^2 r_2 = a^2 r_2 \qquad (2\text{-}34)$$

Often the quantity r'_2 is described as the secondary resistance referred to the primary. The phrase "referred to the primary" is used because we are dealing with an equivalent secondary winding having the same number of turns as the primary winding.

By proceeding in a similar fashion the secondary leakage reactance referred to the primary can be expressed as

$$x'_2 = a^2 x_2 \qquad (2\text{-}35)$$

Moreover, since the kilovolt-ampere load must be the same for the two windings we have

$$V'_2 I'_2 = V_2 I_2 \qquad (2\text{-}36)$$

from which it follows that the load terminal voltage in the equivalent secondary winding is

$$V'_2 = \frac{I_2}{I'_2} V_2 = a V_2 \qquad (2\text{-}37)$$

Appearing in Fig. 2-12(c) are all of the foregoing results in addition to an ideal transformer having a unity turns ratio. Keep in mind that Fig. 2-12(c) is drawn with the equivalent secondary winding having N_1 turns, whereas Fig. 2-12(b) is drawn with the actual secondary winding. As far as the primary is concerned it cannot distinguish between the two. For the ideal transformer of Fig. 2-12(c) note that joining point a to a', b to b', and c to c' causes no disturbance because instantaneously these points are at the same potential. Accordingly, this ideal transformer may be entirely dispensed with, in which case Fig. 2-12(c) becomes Fig. 2-12(d), which is called the *complete equivalent circuit* of the transformer. This is the single-line diagram we set out to derive.

Approximate Equivalent Circuit

In most constant-voltage, fixed-frequency transformers such as are used in power and distribution applications, the magnitude of the magnetizing current I_m is only a few percent (2 to 5) of the rated winding current. This happens because high-permeability steel is used for the magnetic core and so keeps I_ϕ very small. Furthermore, steel with a few percent of silicon added is often used and this helps to reduce the core losses, thereby keeping I_c small. Consequently, little error is introduced and considerable simplification is achieved by assuming I_m so small as to be negligible. This makes the equivalent circuit of Fig. 2-12(d) take on the configuration shown in Fig. 2-13. Note that the primary current $\bar{I}_1$ is now equal to the referred secondary current $\bar{I}'_2$. Also we can now speak of an equivalent resistance referred to the primary

$$\boxed{R_{e1} \equiv r_1 + r'_2 = r_1 + a^2 r_2} \qquad (2\text{-}38)$$

and an equivalent leakage reactance referred to the primary

$$\boxed{X_{e1} \equiv x_1 + x'_2 = x_1 + a^2 x_2} \qquad (2\text{-}39)$$

The subscript 1 denotes the primary.

Figure 2-13   Approximate equivalent circuit referred to the primary.

The corresponding phasor diagram drawn for the approximate equivalent circuit appears in Fig. 2-14. Note the considerable simplification over the phasor diagram of the exact equivalent circuit which appears in Fig. 2-8.

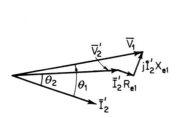

Figure 2-14 Phasor diagram of the approximate equivalent circuit for a lagging power factor angle θ_2.

Figure 2-15 Approximate equivalent circuit referred to the secondary.

The analysis of the preceding pages concerns itself with replacing the actual secondary winding by an equivalent winding having the same number of turns as the primary winding. A little thought should make it apparent that a similar procedure may be employed to draw a single-line equivalent circuit with all quantities referred to the secondary winding. In other words, the actual primary winding can be replaced by an equivalent primary winding having N_2 turns. The resulting equivalent circuit then assumes the form depicted in Fig. 2-15. It is interesting to note that, to refer the primary winding resistance r_1 to the secondary, it is necessary to divide by a^2. Thus

$$r_1' = \frac{r_1}{a^2} \tag{2-40}$$

Furthermore, the equivalent winding resistance referred to the secondary is

$$R_{e2} = r_1' + r_2 = \frac{r_1}{a^2} + r_2 \tag{2-41}$$

and the equivalent leakage reactance referred to the secondary is

$$X_{e2} = x_1' + x_2 = \frac{x_1}{a^2} + x_2 \tag{2-42}$$

Note too that the ideal transformer is included in Fig. 2-15 merely to emphasize the transformation that takes place from the primary side to the secondary side. Normally, the primary circuit is omitted entirely leaving only the single-line secondary circuit.

Meaning of Transformer Nameplate Rating

A typical transformer carries a nameplate bearing the following information: 10 kVA 2200/110 V. What are the meanings of these numbers in the terminology and analysis developed so far? To understand the answer refer to Fig. 2-16, which is really the approximate equivalent circuit drawn to include the ideal transformer in order to

Figure 2-16 Approximate equivalent circuit drawn with the ideal transformer.

bring into evidence the transformation action existing between the primary and secondary circuits. The number 110 refers to the rated secondary voltage. It is the voltage appearing across the load when rated current flows at rated kVA. We call this V_2. The 2200 number refers to the primary winding voltage rating, which is obtained by taking the rated secondary voltage and multiplying by the turns ratio a between primary and secondary. In our terminology this is V_2'. By specifying the ratio $V_2'/V_2 = 2200/110$ on the nameplate, information about the turns ratio is given. Finally, the kilovolt-ampere figure always refers to the *output* kilovolt-amperes appearing at the secondary load terminals. Of course the input kilovolt-amperes are slightly different because of the effects of R_{e1} and X_{e1}.

Transfer Function

The transfer function of the transformer may be specified with respect either to voltage or to current. The voltage transfer function is defined as the ratio of the output voltage $\bar{V}_2$ to the input voltage $\bar{V}_1$ with the load connected. Thus

$$\bar{T}_V = \frac{\bar{V}_2}{\bar{V}_1} = \frac{\bar{V}_2}{a\bar{V}_2 + \bar{I}_2'R_{e1} + j\bar{I}_2'X_{e1}} \approx \frac{1}{a} \tag{2-43}$$

Equation (2-43) indicates that when the winding resistances and leakage fluxes are negligibly small the voltage transfer function $\bar{T}_V$ is approximately the reciprocal of the ratio of transformation. When these quantities cannot be neglected, $\bar{T}_V$ will be a complex number which varies with the load.

The current transfer function $\bar{T}_I$ is defined as the output current divided by the input current. For the approximate equivalent circuit of Fig. 2-16 we have

$$\bar{T}_I = \frac{\bar{I}_2}{\bar{I}_2'} = a \tag{2-44}$$

Here the current transfer function is exactly equal to a because the magnetizing current is assumed negligible. When $\bar{I}_m$ is not negligible, $\bar{T}_I$ will be somewhat smaller than a and may be complex.

Input Impedance

The input impedance of a transformer is the impedance measured at the input terminals of the primary winding. In the case of the approximate equivalent circuit shown in Fig. 2-16 we have

$$\bar{Z}_i = \frac{\bar{V}_1}{\bar{I}_1} = R_{e1} + a^2R_L + jX_{e1} \tag{2-45}$$

A glance at this result indicates that the input impedance is dependent upon the load impedance. In the case of power applications the input impedance is of no particular concern because the internal impedance of the source is negligibly small. In communication applications, however, where $\bar{V}_1$ is obtained from a transistor, input impedance is a useful quantity because it describes the effect that the transformer has on the voltage source.

Frequency Response

One of the things we are interested in, when referring to the frequency response of a transformer, is the manner in which the voltage transfer function varies for a fixed load as the frequency of the source voltage $\bar{V}_1$ is varied. The amplitude of $\bar{V}_1$ is assumed fixed. For power transformers frequency response is of no serious interest because these units are operated at a single fixed frequency—usually 60 Hz and sometimes 50 Hz. However, in communication circuits the frequency of the source voltage is likely to vary over a wide range. For example, in the case of an output transformer which couples the power stage of an audio amplifier to a loudspeaker the frequency may vary over the entire audio range. What, then, is the effect of this varying frequency on the voltage transfer function? The answer follows from an inspection of Fig. 2-12(d). Keep in mind that inductive reactance is directly proportional to the frequency, i.e.,

$$x = \omega L = 2\pi f L \tag{2-46}$$

Hence at low frequencies the reactance x_ϕ, which normally is so high that it may be removed from the circuit, can now no longer be neglected. As a matter of fact this low magnetizing reactance can effectively shunt the fixed load impedance thereby causing a severe drop-off of output voltage as depicted in Fig. 2-18. Figure 2-17(a) shows the configuration of the equivalent circuit as it applies at very low frequencies. Note that x_1 and x_2 are so small at these frequencies that they may be omitted entirely from the diagram.

Figure 2-17 Equivalent circuit for various portions of the frequency spectrum: (a) low frequencies; (b) intermediate frequencies; (c) high frequencies.

TRANSFORMERS   CHAP. 2

In the intermediate frequency range the design of these transformers for communication systems is such that x_1 and x_2 are still quite small while x_ϕ is sufficiently large that it may be omitted. The appropriate circuitry appears in Fig. 2-17(b). Note that for a resistive load the equivalent circuit consists solely of resistive elements so that the transfer function remains constant over the band of intermediate frequencies. Refer to Fig. 2-18.

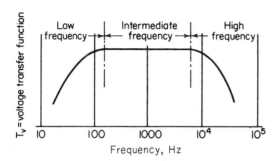

Figure 2-18   Voltage transfer function $\bar{V}_2/\bar{V}_1$ versus frequency.

At very high frequencies x_1 and x_2 are no longer negligible and therefore must be included in the circuit. However, x_ϕ may continue to be omitted. The equivalent circuit now takes the form depicted in Fig. 2-17(c). A little thought reveals that, as the frequency is allowed to get higher and higher, more and more of the source voltage $\bar{V}_1$ appears across x_1 and x_2 and less and less appears across the fixed load resistance. Therefore again a drop-off occurs in the value of the voltage transfer function as depicted in Fig. 2-18. In communication transformers a suitable frequency response such as that shown in Fig. 2-18 is very often the most important characteristic of the device.

Example 2-1

A distribution transformer with the nameplate rating of 100 kVA, 1100/220 V, 60 Hz, has a high-voltage winding resistance of 0.1 ohm (Ω) and a leakage reactance of 0.3 Ω. The low-voltage winding resistance is 0.004 Ω and the leakage reactance is 0.012 Ω. The source is applied to the high-voltage side.

(a) Find the equivalent winding resistance and reactance referred to the high-voltage side and the low-voltage side.

(b) Compute the equivalent resistance and equivalent reactance drops in volts and in percent of the rated winding voltages expressed in terms of the primary quantities.

(c) Repeat (b) for quantities referred to the low-voltage side.

(d) Calculate the *equivalent leakage impedances* of the transformer referred to the primary and secondary sides.

Solution   From the nameplate data the turns ratio is $a = 1100/220 = 5$.

(a) Hence the equivalent winding resistance referred to the primary by Eq. (2-38) is

$$R_{e1} = r_1 + a^2 r_2 = 0.1 + (5)^2(0.004) = 0.2 \ \Omega$$

Also

$$X_{e1} = x_1 + a^2 x_2 = 0.3 + 25(0.012) = 0.6 \ \Omega$$

The corresponding quantities referred to the secondary are

$$R_{e2} = \frac{0.1}{25} + 0.004 = 0.008 \, \Omega$$

$$X_{e2} = \frac{0.3}{25} + 0.012 = 0.024 \, \Omega$$

These computations reveal that $r_1 = r_2'$ and $x_1 = x_2'$. This is a very common occurrence in well-designed transformers. A well-designed transformer is one that uses a minimum amount of iron and copper for a specified output. Note, too, that the resistance of the secondary winding is considerably smaller than that of the primary winding. This is consistent with the larger current rating of the former.

(b) The primary winding current rating in this case is

$$I_2' = \frac{100 \text{ kVA}}{1.1 \text{ kV}} = 91 \text{ A}$$

Hence

$$I_2' R_{e1} = 91(0.2) = 18.2 \text{ V}$$

Expressed as a percentage of the rated voltage of the high-tension winding this is

$$\frac{I_2' R_{e1}}{V_2'} 100 = \frac{18.2}{1100} 100 = 1.65\%$$

Also

$$I_2' X_{e1} = 91(0.6) = 54.6 \text{ V}$$

and

$$\frac{I_2' X_{e1}}{V_2'} 100 = \frac{54.6}{1100} = 4.96\%$$

(c) The secondary winding current rating is

$$I_2 = aI_2' = 5(91) = 455 \text{ A}$$

Hence

$$I_2 R_{e2} = 455(0.008) = 3.64 \text{ V}$$

and

$$\frac{I_2 R_{e2}}{V_2} 100 = \frac{3.64}{220} 100 = 1.65\%$$

Note that the percentage value of the winding resistance drop on the secondary side is identical with the value on the primary side. Similarly, for the leakage reactance,

$$I_2 X_{e2} = 455(0.024) = 10.9 \text{ V}$$

$$\frac{I_2 X_{e2}}{V_2} 100 = \frac{10.9}{220} 100 = 4.96\%$$

(d) The equivalent leakage impedance referred to the primary is

$$\bar{Z}_{e1} \equiv R_{e1} + jX_{e1} = 0.2 + j0.6 = 0.634\underline{/71.6°}$$

Referred to the secondary this is

$$\bar{Z}_{e2} = \frac{\bar{Z}_{e1}}{a^2} = 0.0253\underline{/71.6°}$$

The exact equivalent circuit of the transformer has a total of six parameters as Fig. 2-12(d) shows. A knowledge of these parameters allows us to compute the performance of the transformer under all operating conditions. When the complete design data of a transformer are available, the parameters may be calculated from the dimensions and properties of the materials involved. Thus, the primary winding resistance r_1 may be found from the resistivity of copper, the total winding length, and the cross-sectional area. In a similar fashion a parameter such as the magnetizing reactance x_ϕ may be determined from the number of primary turns, the reluctance of the magnetic path, and the frequency of operation. The calculation of the leakage reactance is a bit more complicated because it involves accounting for partial flux linkages. However, formulas are available for making a reliable determination of these quantities.

A more direct and far easier way of determining the transformer parameters is from tests that involve very little power consumption, called *no-load tests*. The power consumption is merely that which is required to supply the appropriate losses involved. The primary winding (or secondary winding) resistance is readily determined by applying a small d-c voltage to the winding in the manner shown in Fig. 2-19. The voltage appearing across the winding must be large enough to cause approximately rated current to flow. Then the ratio of the voltage drop across the winding as recorded by the voltmeter, VD, divided by the current flowing through it as recorded by the ammeter, AD, yields the value of the winding resistance.

Figure 2-19   Circuitry for finding winding resistance.

Open-Circuit Test

Information about the core-loss resistor r_c and the magnetizing reactance x_ϕ is obtained from an *open-circuit test*, the circuitry of which appears in Fig. 2-20. Note that the secondary winding is open. Therefore, as Fig. 2-12(d) indicates, the input impedance consists of the primary leakage impedance and the magnetizing impedance. In this

Figure 2-20   Wiring diagram for the transformer open-circuit test.

case the ammeter and voltmeter are a-c instruments and the second letter A is used to emphasize this. Thus AA denotes a-c ammeter. The wattmeter shown in Fig. 2-20 is used to measure the power drawn by the transformer. The open-circuit test is performed by applying voltage either to the high-voltage side or the low-voltage side, depending upon which is more convenient. Thus, if the transformer of Example 2-1 is to be tested, the voltage would be applied to the low-tension side because a source of 220 V is more readily available than one of 1100 V. The core loss is the same whether 220 V is applied to the winding having the smaller number of turns or 1100 V is applied to the winding having the larger number of turns. The maximum value of the flux, upon which the core loss depends, is the same in either case as indicated by Eq. (2-17). Now for convenience assume that the instruments used in the open-circuit test yield the following readings:

$$\text{wattmeter reading} = P$$

$$\text{ammeter reading} = I_m$$

$$\text{voltmeter reading} = V_L$$

The voltmeter reading carries the subscript L to emphasize that the test is performed with the source and instruments on the low-voltage side.

Neglecting instrument losses, the wattmeter reading may be taken entirely equal to the core loss, i.e., $P_c = P$. This is because the attendant copper loss is negligibly small. Moreover, since I_m is very small, the primary leakage impedance drop may be neglected, so that for all practical purposes the induced emf is equal to the applied voltage, i.e., $E_1 = V_L$. In accordance with these simplifications the open-circuit phasor diagram is represented by Fig. 2-3(b). The no-load power-factor angle θ_0 is computed from

$$\theta_0 = \cos^{-1} \frac{P}{V_L I_m} \tag{2-47}$$

It then follows that

$$I_c = I_m \cos \theta_0 \tag{2-48}$$

and

$$I_\phi = I_m \sin \theta_0 \tag{2-49}$$

It is important to keep in mind that these currents are referred to the low-tension side. The corresponding values on the high-voltage side would be less by the factor $1/a$. The low-side core-loss resistor is

$$r_{cL} = \frac{P}{I_c^2} = \frac{P}{(I_m \cos \theta_0)^2} \tag{2-50}$$

The corresponding high-side core-loss resistor is

$$r_{cH} = a^2 r_{cL} \tag{2-51}$$

The magnetizing reactance referred to the low side follows from Eq. (2-29). Thus

$$x_{\phi L} = \frac{E_1}{I_\phi} = \frac{V_L}{I_m \sin \theta_0} \tag{2-52}$$

On the high side this becomes

$$x_{\phi H} = a^2 x_{\phi L} \qquad (2\text{-}53)$$

Although the manner of finding the magnetizing impedance is treated in the foregoing, the motivation is not so much necessity as the desire for completeness. Invariably, the use of the approximate equivalent circuit is sufficient for all but a few special applications of transformers. Accordingly, the useful information obtained from the open-circuit test is the value of the core loss, which helps to determine the efficiency of the transformer.

Short-Circuit Test

Of the six parameters of the exact equivalent circuit two remain to be determined: the primary and secondary leakage reactances x_1 and x_2. This information is obtained from a short-circuit test which involves placing a small a-c voltage on one winding and a short circuit on the other as depicted in Fig. 2-21(a). Appearing in Fig. 2-21(b) is the approximate equivalent circuit with the secondary winding short-circuited. Note that the input impedance is merely the equivalent impedance.

(a)   (b)

Figure 2-21 Transformer short-circuit test: (a) wiring diagram; (b) equivalent circuit.

When performing the short-circuit test, a reduced a-c voltage is applied to the high side usually because it is more convenient to do so. From the computations of Example 2-1 recall that the leakage impedance drop is only about 5% of the winding voltage rating. Hence 5% of 1100 V, which is 55 V, is easier and more accurate to deal with than 5% of 220 V, which is 11 V. Also, the wattmeter reading in the circuitry of Fig. 2-21(a) can be taken entirely equal to the winding copper losses. This follows from the fact that the greatly reduced voltage used in the short-circuit test makes the core loss negligibly small. Calling the wattmeter reading P_{sc} and the ammeter reading I_{sc}, we can determine the equivalent winding resistance as

$$R_{eH} = \frac{P_{sc}}{I_{sc}^2} \qquad (2\text{-}54)$$

The subscript H is used in place of L since it is known from the wiring diagram that the primary side is the high-voltage side. The equivalent resistance referred to the low side is then

$$R_{eL} = \frac{R_{eH}}{a^2} \qquad (2\text{-}55)$$

where

$$a = \frac{N_H}{N_L} = \frac{\text{high-side turns}}{\text{low-side turns}} \qquad (2\text{-}56)$$

Before the equivalent reactance can be determined, it is first necessary to find the equivalent impedance, which a glance at Fig. 2-21(b) reveals to be

$$Z_{eH} = \frac{V_{sc}}{I_{sc}} \qquad (2\text{-}57)$$

During the test the applied voltage V_{sc} is adjusted so that I_{sc} is at least equal to the rated winding current. The equivalent reactance follows from Eqs. (2-54) and (2-57). Thus

$$X_{eH} = \sqrt{Z_{eH}^2 - R_{eH}^2} \qquad (2\text{-}58)$$

It is important to note that this computation provides the *sum* of the primary and secondary leakage reactances. It gives no information about the individual breakdown between x_1 and x_2. Whenever the approximate equivalent circuit is used in analysis, such a breakdown is unnecessary. On the few occasions when the individual breakdown is needed, it is customary to assume that

$$x_1 = x_2' \qquad (2\text{-}59)$$

This statement is based on the assumption that the transformer is well-designed.

One final note. It should be apparent from the theory of transformer action that the establishment of rated current in the high-side winding requires a corresponding flow of rated current in the secondary circuit. It is for this reason that the *total* winding copper loss is measured in the short-circuit test.

Example 2-2

The following data were obtained on a 50-kVA, 2400/120-V transformer:

Open-circuit test, instruments on low side:

$$\text{wattmeter reading} = 396 \text{ W}$$
$$\text{ammeter reading} = 9.65 \text{ A}$$
$$\text{voltmeter reading} = 120 \text{ V}$$

Short-circuit test, instruments on high side:

$$\text{wattmeter reading} = 810 \text{ W}$$
$$\text{ammeter reading} = 20.8 \text{ A}$$
$$\text{voltmeter reading} = 92 \text{ V}$$

Compute the six parameters of the equivalent circuit referred to the high and low sides.

Solution From the open-circuit test

$$\theta_0 = \cos^{-1}\frac{P}{V_L I_m} = \cos^{-1}\frac{396}{120(9.65)} = 70°$$

$$\therefore \quad I_{cL} = I_m \cos\theta_0 = 9.65(0.342) = 3.3 \text{ A}$$

$$I_{\phi L} = I_m \sin\theta_0 = 9.65 \sin 70° = 9.07 \text{ A}$$

Hence

$$r_{cL} = \frac{396}{3.3^2} = 36.4 \, \Omega$$

This can also be found from

$$r_{cL} = \frac{E_1}{I_c} = \frac{V_L}{I_c} = \frac{120}{3.3} = 36.4 \, \Omega$$

On the high side this quantity becomes

$$r_{cH} = a^2 r_{cL} = 400(36.4) = 14{,}560 \, \Omega$$

The magnetizing reactance referred to the low side is

$$x_{\phi L} = \frac{E_1}{I_\phi} = \frac{V_L}{I_\phi} = \frac{120}{9.07} = 13.2 \, \Omega$$

The corresponding high-side value is

$$x_{\phi H} = a^2 x_{\phi L} = 400(13.2) = 5280 \, \Omega$$

From the data of the short-circuit test we get directly

$$Z_{eH} = \frac{92}{20.8} = 4.42 \, \Omega, \qquad \therefore \quad Z_{eL} = \frac{4.42}{400} = 0.011 \, \Omega$$

Also

$$R_{eH} = \frac{810}{(20.8)^2} = 1.87, \qquad \therefore \quad R_{eL} = \frac{1.87}{400} = 0.0047$$

and

$$X_{eH} = \sqrt{Z_{eH}^2 - R_{eH}^2} = \sqrt{4.42^2 - 1.87^2} = 4 \, \Omega$$

$$\therefore \quad X_{eL} = \frac{4}{400} = 0.01 \, \Omega$$

2-4 EFFICIENCY AND VOLTAGE REGULATION

The manner of describing the performance of a transformer depends upon the application for which it is designed. It has already been pointed out that in communication circuits the frequency response of the transformer is very often of prime importance. Another important characteristic of such transformers is to provide matching of a source to a load for maximum transfer of power. The efficiency and regulation is usually of secondary significance. This is not so, however, with power and distribution transformers, which are designed to operate under conditions of constant rms voltage and frequency. Accordingly, the material that follows has relevance chiefly to these transformers.

As is the case with other devices, the efficiency of a transformer is defined as the ratio of the useful output power to the input power. Thus

$$\eta = \frac{\text{output watts}}{\text{input watts}} \qquad (2\text{-}60)$$

Usually the efficiency is found through measurement of the losses as obtained from the no-load tests. Therefore by recognizing that

$$\text{output watts} = \text{input watts} - \Sigma \text{ losses} \qquad (2\text{-}61)$$

and inserting into Eq. (2-60), a more useful form of the expression for efficiency results. Thus

$$\boxed{\eta = 1 - \frac{\Sigma \text{ losses}}{\text{input watts}}} \qquad (2\text{-}62)$$

where

$$\Sigma \text{ losses} = \text{core loss} + \text{copper loss} = P_c + I_2^2 R_{e2} \qquad (2\text{-}63)$$

Equation (2-62) also leads to a more accurate value of the efficiency.

Power and distribution transformers very often supply electrical power to loads that are designed to operate at essentially constant voltage regardless of whether little or full-load current is being drawn. For example, as more and more light bulbs are placed across a supply line, which very likely originates from a distribution transformer, it is important that the increased current being drawn from the supply transformer does not cause a significant drop (more than 10%) in the load voltage. If this happens the illumination output from the bulbs is sharply reduced. Similar undesirable effects occur in television sets that are connected across the same supply lines. The reduced voltage means reduced picture size. An appreciable drop in line voltage with increasing load demands can also cause harmful effects in connected electrical motors such as those found in refrigerators, washing machines, etc. Continued operation at low voltage can cause these units to overheat and eventually burn out. The way to prevent the drop in supply voltage with increasing load in distribution circuits is to use a distribution transformer designed to have small leakage impedance. The figure of merit used to identify this characteristic is the voltage regulation.

The *voltage regulation* of a transformer is defined as the change in magnitude of the secondary voltage as the current changes from full-load to no-load with the primary voltage held fixed. Hence in equation form,

$$\boxed{\text{voltage regulation} = \frac{|\bar{V}_1| - |\bar{V}_2'|}{|\bar{V}_2'|} = \frac{\left|\dfrac{\bar{V}_1}{a}\right| - |\bar{V}_2|}{|\bar{V}_2|}} \qquad (2\text{-}64)$$

where the symbols denote the quantities previously defined, absolute signs are used in order to emphasize that it is the *change in magnitude* that is important. Equation (2-64) expresses the voltage regulation on a per unit basis. To convert to percentage, it is necessary to multiply by 100. The smaller the value of the voltage regulation, the better suited is the transformer for supplying power to constant-voltage loads (constant voltage is characteristic of most commercial and industrial loads).

In determining the voltage regulation it is customary to assume that $\bar{V}_1$ is adjusted to that value which allows rated voltage to appear across the load when rated current flows through it. Under these conditions the necessary magnitude of $\bar{V}_1$ is found from Kirchhoff's voltage law as it applies to the approximate equivalent circuit. Refer to Figs. 2-13 and 2-14. For the assumed lagging power-factor angle θ_2 of the secondary circuit the expression for $\bar{V}_1$ is

$$\bar{V}_1 = V_2'\underline{/0^\circ} + I_2'\underline{/-\theta_2}Z_{e1}\underline{/\theta_z} \tag{2-65}$$

where

$$\theta_z = \tan^{-1}\frac{X_{e1}}{R_{e1}} \tag{2-66}$$

Equation (2-65) is written using the referred value of rated secondary terminal voltage as the reference.

Example 2-3

For the transformer of Example 2-2(a) find the efficiency when rated kilovolt-amperes are delivered to a load having a power factor of 0.8 lagging; (b) compute the voltage regulation.

Solution (a) The losses at rated load current and rated voltage are

$$\Sigma \text{ losses} = 396 + 810 = 1206 \text{ W} = 1.2 \text{ kW}$$

Also

$$\text{output kW} = 50(0.8) = 40 \text{ kW}$$

$$\text{input kW} = 40 + 1.2 = 41.2 \text{ kW}$$

Hence

$$\eta = 1 - \frac{\Sigma \text{ losses}}{\text{input kW}} = 1 - \frac{1.2}{41.2} = 1 - 0.029 = 0.971$$

or

$$\eta = 97.1\%$$

(b) From Eq. (2-65) we have

$$\bar{V}_1 = \bar{V}_H = 2400 + 20.8\underline{/-37^\circ}\, 4.42\underline{/65^\circ}$$

where

$$\theta_z = \tan^{-1}\frac{X_{eH}}{R_{eH}} = \tan^{-1}\frac{4}{1.87} = 65^\circ$$

Therefore

$$\bar{V}_1 = \bar{V}_H = 2400 + 92\underline{/28^\circ} = 2400 + 81.1 + j43.1$$

$$\bar{V}_H \approx 2481.1$$

Note that the j-component is negligible in comparison to the in-phase component. This calculation indicates that, in order for rated voltage to appear across the load when rated current is drawn, the primary voltage $\bar{V}_1$ must be slightly different from the primary winding voltage rating of 2400 V. The percent of voltage regulation is accordingly found to be

$$\text{percent of voltage regulation} = \frac{2481.1 - 2400}{2400}100 = 3.38\%$$

2-5 MUTUAL INDUCTANCE

Analysis of the two-winding transformer by the methods of electric circuit theory involves the use of self- and mutual-inductance. So far, Kirchhoff's voltage equation applied to the primary winding has led to Eq. (2-26). In the material that follows this result will be shown to be identical with that obtained by the classical approach.

For simplicity assume the ferromagnetic material has a linear magnetization curve with no core losses. Refer to Fig. 2-22 where for the indicated current directions Kirchhoff's voltage equation becomes

$$v_1 = i_1 r_1 + L_1 \frac{di_1}{dt} - M \frac{di_2}{dt} \tag{2-67}$$

where lowercase letters for current and voltage denote instantaneous values. Here the quantity $L_1 \, di_1/dt$ denotes the voltage drop associated with the total flux produced by i_1 and linking the primary turns N_1. Similarly, the quantity $-M \, di_2/dt$ refers to the *voltage rise* associated with the flux produced by i_2 and linking the primary turns. The existence of a voltage rise in the primary circuit for the assumed direction of i_2 is easily verified. Note that the secondary mmf $N_2 i_2$, if allowed to act freely, produces a flux which threads counterclockwise in the configuration of Fig. 2-22. Then the reaction in the primary coil between two points such as a and b by Lenz's law requires that point b should be at a higher potential than a. Accordingly, this induced emf becomes a voltage rise with respect to the specified direction of i_1.

Figure 2-22 Elementary transformer for analysis by the classical approach.

Assuming that the applied primary voltage is a sinusoid, Eq. (2-67) may be written in rms quantities as

$$\bar{V}_1 = \bar{I}_1 r_1 + j\omega L_1 \bar{I}_1 - j\omega M \bar{I}_2 \tag{2-68}$$

Inserting the relationship

$$I_2 = a I_2' \tag{2-69}$$

wherein a is the ratio of transformation, we get

$$\bar{V}_1 = \bar{I}_1 r_1 + j\omega L_1 \bar{I}_1 - j\omega a M I_2' \tag{2-70}$$

But by Eq. (2-24)

$$\bar{I}_2' = \bar{I}_1 - \bar{I}_m = \bar{I}_1 - \bar{I}_\phi$$

so that

$$V_1 = I_1 r_1 + j\omega (L_1 - aM) I_1 + j\omega a M I_\phi \tag{2-71}$$

Equation (2-71) is now in a form where correspondence with Eq. (2-26) is possible after some algebraic manipulation.

TRANSFORMERS   CHAP. 2

The expression for the self-inductance of the primary winding may be written as

$$L_1 = N_1 \frac{d\phi_1}{di_1} = N_1 \frac{\Phi_1}{i_1} \tag{2-72}$$

The equivalence of the differential form with the total quantities in this last expression is allowed by the linearity condition. Recalling that the total flux linking the primary coil consists of the mutual flux Φ_m plus the primary leakage flux Φ_{l1}, Eq. (2-72) can be rewritten as

$$L_1 = N_1 \frac{\Phi_m + \Phi_{l1}}{i_1} \tag{2-73}$$

Also,

$$\Phi_m = \frac{N_1 i_1}{\mathcal{R}_m} \quad \text{and} \quad \Phi_{l1} = \frac{N_1 i_1}{\mathcal{R}_{l1}} \tag{2-74}$$

where $\mathcal{R}_m$ is the reluctance experienced by the mutual flux and consists almost exclusively of iron and $\mathcal{R}_{l1}$ is the net reluctance experienced by the leakage flux of the primary in air. Accordingly, Eq. (2-73) becomes

$$L_1 = \frac{N_1^2}{\mathcal{R}_m} + \frac{N_1^2}{\mathcal{R}_{l1}} = L_m + L_{l1} \tag{2-75}$$

where L_m denotes an inductance associated with the mutual flux and L_{l1} is the inductance associated with the primary leakage flux. It is important to note here that L_m is not the mutual inductance.

The classical definition of mutual inductance is given in this case by

$$M_{21} = N_2 \frac{d\phi_{m1}}{di_1} = N_2 \frac{\Phi_{m1}}{i_1} \tag{2-76}$$

Thus the mutual inductance is related to the number of turns of winding 2 and the change of flux with respect to the current in winding 1 which produces the mutual flux Φ_{m1}. Upon inserting $N_1 i_1 = \Phi_{m1} \mathcal{R}_m$ into Eq. (2-76), there results

$$M_{21} = \frac{N_2 N_1}{\mathcal{R}_m} \tag{2-77}$$

The expression for mutual inductance may also be found by considering a change in mutual flux corresponding to a change in secondary current i_2 linking N_1 turns. Thus

$$M_{12} = N_1 \frac{d\phi_{m2}}{di_2} = N_1 \frac{\Phi_{m2}}{i_2} \tag{2-78}$$

But

$$\Phi_{m2} \mathcal{R}_m = N_2 i_2 \tag{2-79}$$

Inserting this into Eq. (2-78) yields

$$M_{12} = \frac{N_1 N_2}{\mathcal{R}_m} = M \tag{2-80}$$

A comparison of Eq. (2-80) with Eq. (2-77) shows them to be the same, and from here on either expression will be denoted by M.

Returning to Eq. (2-71) we are now in a position to evaluate the significance of the term in parentheses. Thus, by Eqs. (2-75) and (2-77)

$$L_1 - aM = \frac{N_1^2}{\mathcal{R}_m} + \frac{N_1^2}{\mathcal{R}_{l1}} - \frac{N_1}{N_2}\frac{N_2 N_1}{\mathcal{R}_m} = \frac{N_1^2}{\mathcal{R}_{l1}} = L_{l1} \tag{2-81}$$

Clearly this quantity is the primary leakage inductance, and it represents the effect of the primary leakage flux. Of course the total quantity

$$\omega(L_1 - aM)I_1 = \omega L_{l1} I_1 = x_1 I_1 \tag{2-82}$$

is nothing more than the primary leakage reactance drop and is the counterpart of the second term of Eq. (2-26).

Let us now examine the last term of Eq. (2-71). Thus

$$\omega a M I_\phi = \omega \frac{N_1}{N_2}\frac{N_2 N_1}{\mathcal{R}_m} I_\phi = \omega N_1 \frac{N_1 I_\phi}{\mathcal{R}_m} \tag{2-83}$$

The fraction quantity on the right side of the last equation is the expression for mutual flux. Specifically, however, since it involves the reluctance and the rms value I_ϕ of the magnetizing current, this expression represents the magnitude of the maximum value of the mutual flux divided by $\sqrt{2}$. That is,

$$\frac{N_1 I_\phi}{\mathcal{R}_m} = \frac{\Phi_m}{\sqrt{2}} \tag{2-84}$$

where Φ_m denotes the maximum value of the mutual flux. Accordingly, Eq. (2-83) becomes

$$\omega a M I_\phi = \omega N_1 \frac{\Phi_m}{\sqrt{2}} = \frac{2\pi}{\sqrt{2}} f N_1 \Phi_m = E_1 \tag{2-85}$$

It therefore follows from Eq. (2-85) that the mutual inductance parameter so commonly encountered in linear circuit analysis is implied in the expression for the induced emf. Clearly both the induced emf and the mutual inductance are related by the same mutual flux.

2-6 THE EQUIVALENT CIRCUIT: COUPLED CIRCUIT VIEWPOINT

In Sec. 2-2, the equivalent circuit of the transformer is derived from the field viewpoint based on interacting mmf's and fluxes. This approach is often favored because it cultivates a better physical understanding of the mechanism that underlies the operation and behavior of this device. However, it is entirely feasible to arrive at the same equivalent circuit by applying the techniques of circuit analysis to the coupled circuit, which is what the transformer is, after all. The usual starting point in such an analysis is Eq. (2-67), which is repeated here for convenience. Thus

$$v_1 = i_1 r_1 + L_1 \frac{di_1}{dt} - M \frac{di_2}{dt} \tag{2-86}$$

Keep in mind that L_1 denotes the self-inductance of the primary winding that is associated with N_1^2; on the other hand, M represents the mutual inductance between the primary and secondary windings and involves the product of N_1 and N_2 as indi-

Figure 2-23 Schematic diagram of the two-winding transformer expressed in terms of the mutual and self-inductances.

cated by Eq. (2-80). Figure 2-23 illustrates the schematic arrangement from which it is obvious that Eq. (2-86) describes the situation that prevails in the primary circuit in a manner consistent with Kirchhoff's voltage law. For ease of handling, let us rewrite the last equation by replacing differentiation with the derivative operator s. Accordingly,

$$v_1 = i_1 r_1 + sL_1 i_1 - sMi_2 \qquad (2\text{-}87)$$

Moreover, because the secondary emf is related to the primary emf by the turns ratio a and because power is preserved assuming negligible core losses, it follows that

$$i_2 = ai_b \qquad (2\text{-}88)$$

where i_b is the required balance current needed in the primary winding to ensure power balance. In other words,

$$e_2 i_2 = \frac{e_1}{a}(ai_b) \qquad (2\text{-}89)$$

where e_1 and e_2 are, respectively, the instantaneous voltages induced in the primary and secondary windings. Inserting Eq. (2-88) into Eq. (2-87) yields

$$v_1 = i_1 r_1 + sL_1 i_1 - aMsi_b \qquad (2\text{-}90)$$

Next, we add zero to the right side of this equation in the form: $aMsi_1 - aMsi_1$. Hence we get

$$v_1 = i_1 r_1 + (L_1 - aM)si_1 + aMs(i_1 - i_b) \qquad (2\text{-}91)$$

Before proceeding with a circuit interpretation of this equation, let us repeat the foregoing analysis for the secondary circuit in Fig. 2-23.

In accordance with the dot notation, when the current i_1 takes the direction shown in Fig. 2-23, the transformer action is such that the dotted side of the secondary coil is positive and the current i_2 assumes the clockwise direction indicated. Kirchhoff's voltage law then leads to

$$M\frac{di_1}{dt} = i_2 r_2 + v_2 + L_2\frac{di_2}{dt} \qquad (2\text{-}92)$$

In this expression the voltage drops appear on the right side. The use of the derivative operator then allows us to rewrite this equation as

$$Msi_1 = i_2 r_2 + v_2 + L_2 si_2 \qquad (2\text{-}93)$$

Now multiply both sides of Eq. (2-93) by the turns ratio a to yield

$$aMsi_1 = av_2 + ai_2 r_2 + aL_2 si_2 \qquad (2\text{-}94)$$

Inserting Eq. (2-88) then leads to

$$aMsi_1 = av_2 + i_b(a^2r_2) + (a^2L_2)si_b \qquad (2\text{-}95)$$

It is now necessary to manipulate this expression in such a manner that the ensuing interpretation will be entirely consistent with Eq. (2-91), which after all is part of the same equipment with the difference that Eq. (2-91) provides a description in the primary circuit while the latter does so in the secondary circuit. Because the mutual inductance is the circuit parameter that links the primary and secondary circuits, it should play the key role in providing a linkage between Eqs. (2-91) and (2-95). A study of Eq. (2-91) suggests that the last term on the right, namely, $aMs(i_1 - i_b)$, is the term that represents this linkage. This conclusion becomes even more apparent when $(i_1 - i_b)$ is replaced by i_m as called for by Eq. (2-25) because the magnetizing current and the mutual inductance go hand in hand. Therefore, to establish this term in Eq. (2-95), it is necessary merely to add zero to the left side in the form: $aMsi_b - aMsi_b$. Upon so doing and then rearranging, there results

$$aMs(i_1 - i_b) = av_2 + i_b(a^2r_2) + (a^2L_2 - aM)si_b \qquad (2\text{-}96)$$

The circuit interpretation of Eqs. (2-91) and (2-96) taken together leads to the T configuration depicted in Fig. 2-24. Several important points are worth noting here.

Figure 2-24 Equivalent circuit of the two-winding transformer derived from the coupled-circuit viewpoint.

First, the quantity aM is really the mutual inductance referred to an equivalent winding having N_1 turns. This is apparent from

$$aM = \frac{N_1}{N_2}\left(\frac{N_1 N_2}{\mathfrak{R}_m}\right) = \frac{N_1^2}{\mathfrak{R}_m} \qquad (2\text{-}97)$$

Moreover, the quantity $(L_1 - aM)$ must necessarily represent the primary leakage flux. Recall that L_1 represents the total flux that links the primary coil which includes the leakage flux as well as the mutual flux whereas aM represents the mutual flux referred to primary turns. A similar situation prevails in the secondary part of this circuit except that all quantities are expressed in terms of an equivalence based on replacing the actual secondary turns, N_2, by the primary turns N_1. Thus, although the load voltage v_2 is derived from a secondary winding having N_2 turns, multiplication by a in Eq. (2-96) means that the level is raised to that of a winding having N_1 turns. Similarly, the self-inductance of the secondary winding, which is generally expressed as $L_2 = N_2^2/\mathfrak{R}_m$, is converted to an equivalent inductance of N_1 turns when it is multiplied by a^2 as shown in Eq. (2-96). It follows then that the quantity $(a^2L_2 - aM)$ can be considered to be an inductance that represents the secondary leakage flux. The resistance parameter too is modified in a like manner. The secondary winding resistance appears in Eq. (2-96) as well as in the circuit of Fig. 2-24 raised to a level of a^2r_2. This is consistent with the fact that the current is reduced by $1/a$, i.e., $i_b = i_2/a$, as required by the changed voltage level, av_2. The result is to keep the copper loss the same for power cannot be changed by the transformation of a winding of N_2 turns to one of N_1 turns. An ideal transformer is included in the equivalent circuit of Fig. 2-24 in order to provide the proper transition to the load where the voltage v_2 appears. Once the configuration of Fig. 2-24 is obtained, the matter of accounting for the core losses can readily be resolved by inserting a resistor in series with the mutual inductance parameter, aM, of such a value that it equals the core losses when multiplied by the square of the rms value of magnetizing current for sinusoidal sources.

2-7 THE AUTOTRANSFORMER

One of the principal features of the ordinary configurations of the transformer (see Fig. 2-22) is the electrical isolation that exists between the primary and secondary windings. Thus in the case of a distribution transformer that transforms power from 4400 V to 220 V, there does not exist the danger of the 4400 V appearing at the secondary terminals. Whenever large transformation ratios are involved, safety demands that ordinary transformers be used. However, in instances where the transformation ratio is close to unity (which minimizes the hazard), a transformer employing a single tapped winding may be used with notable advantages.

To understand some of these advantages consider the arrangement of the conventional transformer appearing in Fig. 2-25(a). Assume the transformer is ideal for simplicity. A resistive load is shown to draw 12 A at a terminal voltage of 100 V. The corresponding primary current is 10 A. An important observation to make in this arrangement is that, all along the secondary winding between points b' and c' and all along that section of the primary winding between b and c, it is possible to identify points having the same instantaneous potential. It b and b' are two such

(a)   (b)

Figure 2-25 (a) Ordinary transformer arrangement; (b) derived autotransformer arrangement. Comparison is on the basis of same output kVA.

points, they may be joined without adverse effect. Extending this procedure throughout the entire length of windings bc and $b'c'$ leads to the simplified configuration appearing in Fig. 2-25(b). This arrangement is called the *autotransformer*.

Several things are worth noting here. First, the secondary winding $b'c'$ of the ordinary transformer can be completely eliminated, and in its place an appropriate tap point, i.e., b, can be used to provide the required secondary voltage. Since the output volt-amperes remain intact in the arrangement of Fig. 2-25(b), it follows that the use of the autotransformer to provide the needed kVA at the load leads to a saving in cost. Less copper is required for the same output. As a matter of fact the saving in copper is even more pronounced than already indicated. The current that flows in the common winding of the autotransformer is always the difference between the primary and secondary currents of the ordinary transformer. In the case at hand this current is 2 A. Accordingly, the cross-sectional area of the copper of winding bc of the autotransformer can be much smaller than that of the primary of the ordinary transformer for the same design value of current density. Second, the efficiency of the autotransformer for the same output is greater. This result immediately follows from the smaller copper losses. Third, the voltage regulation of the autotransformer is superior because of the reduced resistance drop and lower leakage reactance drop. Keep in mind that one winding has been completely eliminated. Hence the resistance and leakage flux of this winding are zero. Fourth, the autotransformer is smaller in size for the same output. This too follows from the reduced copper need.

There is another way in which the autotransformer of Fig. 2-25(b) may be derived from an ordinary transformer. The second approach is predicated on the fact that it uses, completely, the primary and secondary windings of an appropriate ordinary transformer. The first approach is based on the same output for the ordinary transformer and the autotransformer. In contrast the second approach is based on the use of the same copper and iron; we should expect therefore that the autotransformer will have a greater output rating than the corresponding ordinary transformer. (The latter in this case is not the same as the ordinary transformer used in the first approach.)

A close examination of Fig. 2-25(b) reveals that the autotransformer can be derived by placing the winding voltages of two separate coils ab and bc in series aiding. The coils ab and bc can then be treated as the primary and secondary windings, respectively, of an *equivalent ordinary transformer* such as that illustrated in Fig. 2-26(a). Note that the volt-ampere rating of the equivalent ordinary transformer is 200, compared with 1200 for the derived autotransformer. In a practical situation coils ab and $b'c$ appear on the same magnetic core. Refer to Fig. 2-26(b). The series-aid connection of the coils can be achieved by first energizing one of the coils, say ab, with rated voltage, then joining one terminal of the first coil, say b, to one terminal of the second coil, say b', and finally checking that a voltmeter placed between the remaining two terminals a and c yields a reading that is the sum of the voltage ratings of the individual coils. If the voltmeter reads the difference, the connection to the second coil must be reversed. For the configuration of Fig. 2-26(b) the sense of the winding $b'c$ has been so taken that the series-aid connection results when b is joined to b'.

(a)   (b)

Figure 2-26 (a) Equivalent ordinary transformer; (b) its derived autotransformer. Comparison is on the basis of the same copper and iron.

Current and Voltage Relations for Step-Down Autotransformer

A step-down autotransformer is one in which the primary voltage is larger than the secondary voltage. By applying a source voltage to the full winding abc in Fig. 2-26(b) and placing the load across bc, operation as a step-down autotransformer results. In this configuration the applied primary voltage causes a magnetizing current to flow through the full winding ac. In view of its small magnitude compared with the rated input current, for the sake of ease and simplicity of analysis, this current will be considered equal to zero from here on.

The same basic principle that underlies the operation of the ordinary transformer is found to govern the operation of the autotransformer, i.e., the need for balancing mmf's. We use this as the starting point in calculating the current and voltage relationships for the autotransformer. To simplify the notation, use is made of the symbols depicted in Fig. 2-27. When the load switch S is closed in the secondary of the autotransformer of Fig. 2-26(b), a current $\bar{I}$ begins to flow in the common coil bc. It attempts to reduce the operating flux, but as soon as this begins to happen there

Figure 2-27 Symbolism for analysis of the step-down autotransformer.

occurs an increase in primary current $\bar{I}_H$ to such a value that the mmf in winding ab neutralizes the mmf in winding bc. That is,

$$n\bar{I}_H = N_L\bar{I} \tag{2-98}$$

or

$$(N_H - N_L)\bar{I}_H = N_L\bar{I} \tag{2-99}$$

Hence,

$$\frac{\bar{I}}{\bar{I}_H} = \frac{I}{I_H} = \frac{N_H}{N_L} - 1 = a_A - 1 \tag{2-100}$$

where

$$a_A \equiv \frac{N_H}{N_L} = \frac{N_{ac}}{N_{bc}} > 1 \tag{2-101}$$

Since the right side of Eq. (2-100) is a pure number, it follows that the currents in coils ab and cb are in-phase. The opposing action of the mmfs is derived rather from the sense of the windings.

Applying Kirchhoff's current law at point b yields

$$\bar{I}_L = \bar{I}_H + \bar{I} \tag{2-102}$$

But since $\bar{I}_H$ and $\bar{I}$ are both in time phase, it follows that Eq. (2-102) is valid also when only magnitudes are used. Thus,

$$\boxed{I_L = I_H + I} \tag{2-103}$$

Inserting $I = I_L - I_H$ into Eq. (2-100) yields the relationship between the low- and the high-side currents. Hence

$$\frac{I_L - I_H}{I_H} = a_A - 1$$

or

$$\boxed{\frac{I_L}{I_H} = a_A} \tag{2-104}$$

Finally, inserting $I_H = I_L/a_A$ into Eq. (2-100) gives

$$\boxed{\frac{I}{I_L} = \frac{a_A - 1}{a_A}} \tag{2-105}$$

The induced voltages in windings ab and bc are also in time phase because they are caused by the same flux. Letting $E_1 = E_{ac} = E_H$, $E_2 = E_{bc} = E_L$, and $E_n = E_{ab}$, we then have

$$\boxed{\frac{E_n}{E_L} = \frac{E_H - E_L}{E_L} = a_A - 1} \tag{2-106}$$

An inspection of Eqs. (2-100) and (2-106) shows that they bear out mathematically the equivalence described in Fig. 2-26. These equations indicate that the ratio of the voltages and currents in the coils ab and bc are the same as if the turns $n = N_{ab}$ formed the *primary* and the turns $N_2 = N_{bc}$ formed the *secondary* of an ordinary transformer having a ratio of transformation of $a_A - 1$. The value of a_A for the autotransformer of Fig. 2-26(b) is 1.2. Hence the equivalent ordinary transformer should have a turns ratio of $a_A - 1$, or 1/5. A glance at Fig. 2-26(a) corroborates the result.

One advantage of treating the autotransformer as an equivalent ordinary transformer is that use may be made immediately of previously established results. For example, to get the equivalent resistance of the autotransformer referred to the high side all that is necessary is to write the equivalent resistance of the equivalent ordinary transformer, treating winding ab as the primary and winding bc as the secondary. From our previous knowledge of the ordinary transformer we can then write

$$R_{eH} = r_{ab} + (a_A - 1)^2 r_{bc} = r_n + (a_A - 1)^2 r_2 \qquad (2\text{-}107)$$

where $r_n = r_{ab}$, the resistance of winding ab, and $r_2 = r_{bc}$, the resistance of winding bc. Similarly the equivalent reactance referred to the high side becomes

$$X_{eH} = x_{ab} + (a_A - 1)^2 x_{bc} = x_n + (a_A - 1)^2 x_2 \qquad (2\text{-}108)$$

The result expressed in Eq. (2-107) can be verified by writing the expression for the total copper loss when the secondary load terminals are short-circuited. If the current flowing through ab is I_H and that through bc is I, the copper loss at short circuit is then given by

$$P_{\text{Cu}} = I_H^2 r_{ab} + I^2 r_{bc} \qquad (2\text{-}109)$$

An equivalent resistance as seen from the high side may be derived from Eq. (2-109) provided I can be expressed in terms of I_H. This is readily accomplished by Eq. (2-100). Thus,

$$P_{\text{Cu}} = I_H^2 r_{ab} + (a_A - 1)^2 I_H^2 r_{bc}$$

or

$$P_{\text{Cu}} = I_H^2 [r_{ab} + (a_A - 1)^2 r_{bc}] \qquad (2\text{-}110)$$

The quantity in brackets can be interpreted as that equivalent high-side resistance which when multiplied by the high-side current squared yields the copper loss.

Power and Volt-Ampere Relations

Unlike the conventional transformer, the autotransformer has two kinds of volt-ampere power associated with it. It is the result of the direct copper connection existing between the source and the load. Refer to Fig. 2-24 to distinguish the two. Since the transformer action present in the autotransformer takes place in windings ab and bc, it follows that the number of volt-amperes in winding ab is equal to the number of volt-amperes in winding bc and specifically represents the transformed apparent power. Expressing this mathematically we have

$$\text{transformed VA} = E_{ab} I_{ab} = E_{bc} I_{bc} \qquad (2\text{-}111)$$

where VA denotes volt-amperes. If for simplicity we assume negligible winding resistance and leakage reactance, Eq. (2-111) can be written in terms of the terminal voltages. Thus

$$\text{transformed VA} = V_{ab}I_{ab} = V_{bc}I_{bc} \qquad (2\text{-}112)$$

For the values specified in Fig. 2-26 note that the transformed volt-amperes are $20(10) = 2(100) = 200$.

However, the input corresponding to the specified transformer current is $V_{ac}I_{ab} = 120(10) = 1200$ VA. This means that the number of volt-amperes that is not transormed is

$$V_{ac}I_{ab} - V_{ab}I_{ab} = (V_{ac} - V_{ab})I_{ab} = V_{bc}I_{ab} \qquad (2\text{-}113)$$

The specific value for this example is then

$$V_{bc}I_{ab} = 100(10) = 1000 \text{ VA}$$

The quantity represented by Eq. (2-113) is called the *conducted volt-amperes*. The reason for the description follows from the fact that the input current I_{ab} flows *conductively* to the load at the potential V_{bc}. Although I_{ab} starts at the potential V_{ac}, it experiences a drop V_{bc} through transformation and then proceeds to the load at the reduced potential. However, it is important to keep in mind that the reduced potential at which I_{ab} flows to the load is compensated for exactly by an increase in load current originating from the transformer action.

In the notation of Fig. 2-27 we can write for the step-down autotransformer

$$\boxed{\begin{aligned} \text{transformed VA} &= V_n I_H = V_L I & (2\text{-}114) \\ \text{conducted VA} &= V_L I_H & (2\text{-}115) \end{aligned}}$$

An interesting comparison to make at this point is the ratio of the output volt-amperes of the autotransformer to the transformed volt-amperes. The latter quantity in essence is the output of an ordinary transformer having the same current density and magnetic density ratings. Accordingly,

$$\frac{\text{output VA of autotransformer}}{\text{transformed VA}} = \frac{\text{output VA of autotransformer}}{\text{output VA of equiv. ord. transformer}}$$

$$= \frac{V_L I_L}{V_L I} = \frac{a_A}{a_A - 1} \qquad (2\text{-}116)$$

where a_A is the ratio of transformation of the autotransformer. The last term in the preceding equation follows from Eq. (2-105). Applying Eq. (2-116) to the configuration of Fig. 2-26 yields a factor of 6, which emphasizes that for many autotransformer applications (a_A close to unity) most of the power delivered to the load occurs through direct conduction and just a small portion is derived through transformation.

Phasor diagram. The phasor diagram of the step-down autotransformer can be conveniently developed by applying Kirchhoff's voltage law to the high-side circuit and making use of the equivalent ordinary trasnformer in establishing the quantity $V_{ab} = V_n$. Accordingly, in Fig. 2-25 Kirchhoff's voltage law yields

$$\bar{V}_H = \bar{V}_{ab} + \bar{V}_{bc} = \bar{V}_n + \bar{V}_L \qquad (2\text{-}117)$$

By treating winding ab as the primary of an equivalent ordinary transformer, it follows that

$$\bar{V}_{ab} = \bar{V}_n = (a_A - 1)\bar{V}_L + \bar{I}_H[r_{ab} + (a_A - 1)^2 r_{bc}] + j\bar{I}_H[x_{ab} + (a_A - 1)^2 x_{bc}]$$

(2-118)

or

$$\bar{V}_{ab} = \bar{V}_n = (a_A - 1)\bar{V}_L + \bar{I}_H(R_{eH} + jX_{eH}) = (a_A - 1)\bar{V}_L + \bar{I}_H\bar{Z}_{eH}$$ (2-119)

Inserting Eq. (2-119) into Eq. (2-117) yields

$$\bar{V}_H = (a_A - 1)\bar{V}_L + \bar{I}_H\bar{Z}_{eH} + \bar{V}_L$$

(2-120)

or more simply

$$\bar{V}_H = a_A\bar{V}_L + \bar{I}_H\bar{Z}_{eH}$$

(2-121)

Appearing in Fig. 2-28(a) is the phasor diagram representation of Eq. (2-120). The simplified version is depicted in Fig. 2-28(b).

(a)

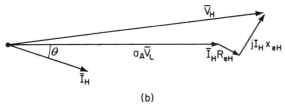

(b)

Figure 2-28 Phasor diagram of the step-down autotransformer: (a) interpretation in terms of Eq. (2-120); (b) interpretation in terms of Eq. (2-121).

Step-Up Autotransformer

The step-up autotransformer may be analyzed in a manner similar to that used for the step-down configuration. In the step-up case the source is applied to winding bc and the load is placed across the full winding ac. Refer to Fig. 2-29. The functional relationships between the input, output, and common winding currents can be derived by starting with the balancing mmf condition that must prevail in coils ab and bc. If this is done and if the ratio of transformation of the step-up case is defined as greater than unity, i.e., $a_A \equiv N_{ac}/N_{bc} = V_H/V_L$, the following results are obtained:

$$\frac{I}{I_H} = a_A - 1$$

(2-122)

$$\frac{I_L}{I_H} = a_A$$

(2-123)

Figure 2-29 Configuration and symbolism for the step-up autotransformer.

$$\frac{I}{I_L} = \frac{a_A - 1}{a_A} \tag{2-124}$$

$$I_L = I_H + I \tag{2-125}$$

and

$$E_{ab} = E_n = E_{bc}(a_A - 1) = E_L(a_A - 1) \tag{2-126}$$

A comparison of these equations with Eq. (2-100) and Eqs. (2-103) to (2-104) reveals that the results for the step-up case can be obtained from those of the step-down case by merely interchanging the roles of the currents I_L and I_H *provided that* a_A is defined the same way in both cases.

2-8 PER-UNIT CALCULATIONS

Computations in per-unit values first found wide use in transmission-line calculations often involving several generators and many transformers. One feature of the per-unit notation is that it makes it unnecessary to refer quantities on the high or low side of transformers. The per-unit value of a circuit element such as primary winding resistance is the same on either side. The bookkeeping procedure is therefore appreciably simplified by such a notation. However, the per-unit notation bears another noteworthy advantage. It reveals important information about the design and per-formance characteristics of the equipment to which it is applied. It is this feature which is emphasized in this section.

Application of the per-unit system of notation depends upon the establishment of a set of specific base quantities by which other similar quantities can be expressed. For example, all voltage drops in the low-voltage circuit of a transformer can be expressed as a ratio of the rated low-side voltage where the latter quantity is iden-tified as the base, or reference, voltage. In fact it is customary to use the *rated quan-tities* of a device as the base values whenever per-unit calculations are to be perfomed. There is nothing that restricts this selection. Rather, the choice of using the rated quantities as the base values is made solely because of the significance that can be attached to the resulting numbers. Directing attention now specifically to the two-winding transformer we identity the following base quantities:

$$\text{base voltage on high side} \quad = \text{rated voltage on high side} = V'_R$$

$$\text{base current on high side} \quad = \text{rated current on high side} = I'_R$$

$$\text{base impedance on high side} = \frac{\text{rated high-side voltage}}{\text{rated high-side current}} \quad = \frac{V'_R}{I'_R} = Z_{BH}$$

$$\text{base admittance on high side} = \frac{\text{rated high-side current}}{\text{rated high-side voltage}} \quad = \frac{I'_R}{V'_R} = Y_{BH}$$

$$\text{base output VA} \quad = \text{rated output VA} \quad = V'_R I'_R = V_R I_R$$

$$\text{base voltage on low side} \quad = \text{rated voltage on low side} \ = V_R$$

$$\text{base current on low side} \quad = \text{rated current on low side} \ = I_R$$

$$\text{base impedance on low side} = \frac{\text{rated low-side voltage}}{\text{rated low-side current}} \quad = \frac{V_R}{I_R} = Z_{BL} = \frac{Z_{BH}}{a^2}$$

$$\text{base admittance on low side} = \frac{\text{rated low-side current}}{\text{rated low-side voltage}} \quad = \frac{I_R}{V_R} = Y_{BL} = a^2 Y_{BH}$$

It is worthwhile to note that base quantities have a dimension. This is not true of the per-unit values of these quantities because by definition the per-unit value of a quantity is a ratio, or the quotient, of the actual value divided by the corresponding base value. Thus to obtain the per-unit value of the rated low-side voltage of a transformer we write

$$\text{p-u } V_R \equiv \frac{V_R \text{ (volts)}}{\text{base voltage on low side (volts)}} = \frac{V_R}{V_R} = 1 \qquad (2\text{-}127)$$

Clearly, if the rated quantities of a device are chosen to be the base values, then the per-unit values of all base quantities are equal to unity. A similar procedure is followed to find the per-unit value of other quantities. To illustrate further, let us find the per-unit value of the equivalent resistance referred to the high side of a two-winding transformer, R_{eH}. Here we write

$$\text{p-u } R_{eH} \equiv \frac{R_{eH} \text{ (ohms)}}{Z_{BH} \text{ (ohms)}} \qquad (2\text{-}128)$$

The result is a numeric value—a quantity without dimensions. However, considerable significance can be attached to this number by multiplying Eq. (2-128) by unity in the form of I'_R/I'_R. Equation (2-128) then becomes

$$\text{p-u } R_{eH} = \frac{R_{eH} I'_R}{Z_{BH} I'_R} = \frac{R_{eH} I'_R}{(V'_R/I'_R) I'_R} = \frac{I'_R R_{eH}}{V'_R} \qquad (2\text{-}129)$$

Recalling that I'_R and V'_R denote the rated values of the current and voltage, respectively, it follows from the right side of Eq. (2-129) that the per-unit value of R_{eH} gives information about the portion of the applied voltage that appears across the equivalent winding resistance as a voltage drop.

Example 2-4

A 10-kVA, 200/100-V transformer has an equivalent high-side resistance of 0.1 Ω and an equivalent high-side leakage reactance of 0.16 Ω. Determine (a) the per-unit value of R_{eH} and (b) the percent of rated voltage absorbed by the equivalent winding resistance.

Solution (a) From the transformer rating

$$I'_R = \frac{10}{0.2} = 50 \text{ A}$$

$$Z_{BH} = \frac{V'_R}{I'_R} = \frac{200}{50} = 4 \text{ }\Omega$$

Hence by Eq. (2-128)

$$\text{p-u } R_{eH} = \frac{R_{eH}}{Z_{BH}} = \frac{0.1}{4} = 0.025$$

(b) By Eq. (2-129) it follows that

$$0.025 V'_R = 0.025(200) = 5 \text{ V}$$

or 2.5% of the rated voltage appears across the equivalent winding resistance.

Still more significance can be attached to the per-unit value of R_{eH} by multiplying Eq. (2-129) once again by unity in the form of I'_R/I'_R. Thus

$$\text{p-u } R_{eH} = \frac{I'_R R_{eH}}{V'_R} \frac{I'_R}{I'_R} = \frac{I'^2_R R_{eH}}{V'_R I'_R}$$

$$= \frac{\text{total copper loss at rated value}}{\text{rated volt-amperes}} \qquad (2\text{-}130)$$

This last expression reveals that the per-unit value of R_{eH} also conveys information about the portion of the rated volt-amperes that is consumed as a copper loss at rated current. Thus for the transformer of Example 2-4 it can be said that, when rated current is delivered to the load, 2.5% of the transformer rating is dissipated as copper loss. In watts, this represents a loss of $0.025(10,000) = 250$ W. Note how much more meaningful it is to deal in the per-unit value of R_{eH} than merely to cite the absolute value.

If the equivalent resistance of a transformer is specified with reference to the low side, the per-unit value is found by using the low-side base impedance. Thus

$$\text{p-u } R_{eL} = \frac{R_{eL}}{Z_{BL}} = \frac{R_{eL}}{V_R/I_R} = \frac{I_R R_{eL}}{V_R} = \frac{I^2_R R_{eL}}{V_R I_R} \qquad (2\text{-}131)$$

A glance at this expression reveals that the per-unit value of R_{eL} carries the same interpretation as the per-unit value of R_{eH}. Moreover, the numerical values are identical for percent of voltage drop and percent of copper loss.

Computing the performance of a transformer—regulation and efficiency—can be accomplished quite conveniently in per-unit values. For example, in finding the voltage regulation for a lagging power factor, the expression for the required primary voltage we know to be

$$\bar{V}_1 = \bar{V}'_R + \bar{I}'_R \bar{Z}_{e1} = \bar{V}'_R + I'_R(\cos\theta - j\sin\theta)(R_{e1} + jX_{e1})$$

$$= V'_R \underline{/0°} + I'_R R_{e1}\cos\theta + I'_R X_{e1}\sin\theta + j(I'_R X_{e1}\cos\theta - I'_R R_{e1}\sin\theta) \qquad (2\text{-}132)$$

Dividing through both sides of this equation by V'_R yields

$$\frac{\bar{V}_1}{V'_R} = 1 + \frac{I'_R R_{e1}}{V'_R}\cos\theta + \frac{I'_R X_{e1}}{V'_R}\sin\theta + j\left(\frac{I'_R X_{e1}}{V'_R}\cos\theta - \frac{I'_R R_{e1}}{V'_R}\sin\theta\right) \qquad (2\text{-}133)$$

Keeping in mind that I'_R and V'_R denote rated quantities, we recognize the coefficients of the trigonometric terms as the per-unit values of the equivalent resistance and equivalent reactance. For ease of handling let

$$\frac{I'_R R_{e1}}{V'_R} = \text{p-u } R_{e1} \equiv \epsilon_r \qquad (2\text{-}134)$$

$$\frac{I'_R X_{e1}}{V'_R} = \text{p-u } X_{e1} \equiv \epsilon_x \qquad (2\text{-}135)$$

and

$$\frac{\bar{V}_1}{V'_R} = \text{p-u } \bar{V}_1 \qquad (2\text{-}136)$$

Equation (2-133) can then be expressed in a shorthand per-unit notation as

$$\text{p-u } \bar{V}_1 = 1 + \epsilon_r \cos\theta + \epsilon_x \sin\theta + j(\epsilon_x \cos\theta - \epsilon_r \sin\theta) \qquad (2\text{-}137)$$

The corresponding phasor diagram for this equation appears in Fig. 2-30. Since for practical transformers the per-unit values of ϵ_r and ϵ_x rarely exceed 0.1, the contribution of the j part of Eq. (2-137) in determining the per-unit value of $\bar{V}_1$ is negligibly small. Hence for all practical purposes Eq. (2-137) may be more simply expressed as

$$\text{p-u } \bar{V}_1 = 1 + \epsilon_r \cos\theta + \epsilon_x \sin\theta \qquad (2\text{-}138)$$

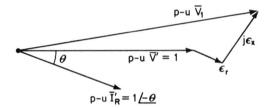

Figure 2-30   Phasor diagram of a transformer for lagging power factor, with all quantities expressed in per-unit nation.

The equation for the voltage regulation in per-unit notation then becomes

$$\epsilon = \frac{|\text{p-u } \bar{V}_1| - |\text{p-u } \bar{V}'_R|}{|\text{p-u } \bar{V}'_R|} = \frac{1 + \epsilon_r \cos\theta + \epsilon_x \sin\theta - 1}{1}$$

or

$$\boxed{\epsilon = \epsilon_r \cos\theta + \epsilon_x \sin\theta} \qquad (2\text{-}139)$$

Equation (2-139) is significant not only because it provides an accurate means of computing voltage regulation but even more because it emphasizes the factors that are responsible for its existence, i.e., the winding resistances, the leakage reactances, and the influence of the power-factor angle.

When the transformer delivers power to a leading power-factor load, θ is introduced as a negative quantity in Eq. (2-139). The expression for the per-unit voltage regulation is then found to be

$$\boxed{\epsilon = \epsilon_r \cos\theta - \epsilon_x \sin\theta} \qquad (2\text{-}140)$$

A comparison with Eq. (2-139) makes it clear that a transformer is more capable of maintaining constant load voltage in the presence of a leading power-factor than a lagging one.

The efficiency of a transformer can also be expressed in the per-unit notation. However, it is helpful first to identify the core loss as a per-unit quantity. Equation (2-28) shows the core loss to be given by $P_c = I_c^2 r_c$. If the core-loss current I_c is replaced by $E_1/r_c \approx V_R'/r_c$, the core loss may be rewritten as

$$P_c \approx \frac{(V_R')^2}{r_c} \approx (V_R')^2 g_c \tag{2-141}$$

where

$$g_c \equiv \frac{1}{r_c} = \text{core-loss conductance} \tag{2-142}$$

The core-loss conductance can then readily be expressed as a per-unit quantity by

$$\text{p-u } g_c = \frac{g_c}{Y_{BH}} = \frac{g_c}{I_R'/V_R'} = \frac{V_R' g_c}{I_R'} = \frac{(V_R')^2 g_c}{V_R' I_R'} = \frac{\text{core loss}}{\text{rated VA}} \tag{2-143}$$

Equation (2-143) indicates that the per-unit conductance representing core loss may be found either from a knowledge of the core loss itself expressed in watts or from a knowledge of the circuit parameter (g_c or r_c) associated with the core loss.

The efficiency of a transformer is affected by the presence of core losses and copper losses. Since we now know how to express both losses in the per-unit notation, let us return to the general expression for the efficiency, Eq. (2-62), and appropriately modify it so that the per-unit notation is brought into evidence. Thus

$$\eta = 1 - \frac{\text{core loss} + \text{copper loss}}{\text{rated output} + \text{core loss} + \text{copper loss}} \tag{2-144}$$

Assuming operation at rated voltage, rated current, and a power-factor of $\cos \theta$, we have

$$\eta = 1 - \frac{(V_R')^2 g_c + (I_R')^2 R_{e1}}{V_R' I_R' \cos \theta + (V_R')^2 g_c + (I_R')^2 R_{e1}} \tag{2-145}$$

Rearranging yields

$$\eta = 1 - \frac{\dfrac{(V_R')^2 g_c}{V_R' I_R'} + \dfrac{(I_R')^2 R_{e1}}{V_R' I_R'}}{\cos \theta + \dfrac{(V_R')^2 g_c}{V_R' I_R'} + \dfrac{(I_R')^2 R_{e1}}{V_R' I_R'}} \tag{2-146}$$

and by Eqs. (2-130) and (2-143) this simplifies to

$$\eta = 1 - \frac{\text{p-u } g_c + \text{p-u } R_e}{\cos \theta + \text{p-u } g_c + \text{p-u } R_e} \tag{2-147}$$

Note that the second subscript for equivalent resistance in the last equation was dropped because the per-unit value is the same irrespective of high or low side. It is important to understand that Eq. (2-147) is valid for finding the efficiency at rated load only. For any other load condition the value of per-unit R_e must be modified by the square of the per-unit value of the current. Thus to find the efficiency at load current I_2' Eq. (2-147) becomes

$$\eta = 1 - \frac{\text{p-u } g_c + (\text{p-u } R_e)(I_2'/I_R')^2}{\dfrac{I_2'}{I_R'}\cos\theta + \text{p-u } g_c + \left(\dfrac{I_2'}{I_R'}\right)^2(\text{p-u } R_e)} \qquad (2\text{-}148)$$

The per-unit value of g_c remains intact because it is assumed that operation at rated voltage continues.

Example 2-5

For the transformer of Example 2-3 compute the efficiency when rated current is delivered to a 0.8 power-factor load.

Solution

$$\text{p-u } g_c = \frac{\text{core loss}}{\text{rated volt-amperes}} = \frac{396\text{ W}}{50{,}000\text{ VA}} = 0.00792$$

$$\text{p-u } R_e = \frac{\text{copper loss at rated current}}{\text{rated volt-amperes}} = \frac{810}{50{,}000} = 0.0162$$

Inserting these values into Eq. (2-147) yields the desired result:

$$\eta = 1 - \frac{0.00792 + 0.0162}{0.8 + 0.0241} = 1 - 0.0293 = 0.9707$$

Hence

$$\eta \text{ in percent} = 97.07 \approx 97.1$$

which compares favorably with the result computed in Example 2-3.

2-9 TRANSFORMERS FOR THREE-PHASE CIRCUITS

For reasons of efficiency and considerable economy in the use of copper, the generation, transmission, and distribution of electric power takes place on a three-phase basis rather than single-phase. The process begins with the electric generator, which is designed to develop a three-phase voltage. Characteristically, the three-phase voltage system is distinguished by the fact that each phase voltage has the same maximum value but is displaced in time from the other phase quantities by 120 electrical degrees. The original energy source may be steam originating from a boiler, or water falling through an appropriate head, or the combustion of gasoline or diesel fuel. Frequently, the output voltage of the electric generator is limited by physical considerations to approximately 25 kV. The transmission of large amounts of electrical energy over considerable distances at such relatively low voltages would entail huge losses, thereby making it impractical to do so. A common solution is to introduce transformers between the generator and the transmission lines to boost the transmitting voltage to levels where the associated line losses are manageable. Often the transmitting line voltage is raised to a level of 230 kV and higher by means of step-up transformers. At the receiving end of the transmission line transformers are again needed to reduce the voltages at the substations to more reasonable levels. This is often followed by a further reduction at many places along a distribution network where the electrical energy is finally consumed. Throughout the process, transformers are used in a three-phase mode to provide the required transformations.

The three-phase mode of operation can be achieved either through use of a single three-phase transformer or by use of three single-phase transformers, which are in turn appropriately connected for three-phase operation. The single-unit three-phase transformer consists of a three-legged construction with each leg carrying the primary and secondary coils of one phase. When compared to three single-phase transformers equipped to deal with the same volt-amperes, the single three-phase unit is lighter, cheaper, requires less space, and is slightly more efficient. Some cost-conscious installations prefer them because of the smaller initial investment. However, a selection in favor of three single transformers has some notable advantages. The larger total surface area, for example, makes for better cooling and greater ease of repair. Moreover, in the initial development of a power system, it may be entirely possible to operate with just two of the transformers albeit at reduced capacity. The third unit can be added at some future date when the full capacity of the system is needed. Also, if a reserve unit is needed to meet emergency conditions, only a single smaller unit is often sufficient to have on hand as compared to the need to have a complete single-unit three-phase transformer. In addition, the latter will require more space for storage and is more difficult to move around. In view of these pros and cons, it is obvious that the choice in a specific situation will be strongly influenced by the special circumstances surrounding that particular case. Because the operation of the transformers in either mode is virtually identical, the development of the subject matter of this section is continued by dealing with the use of three single transformers. The choice is made without prejudice.

The Δ-Δ Connection

When connecting three single transformers for three-phase operation, care must be taken in order to ensure that the secondary terminals provide a balanced system of voltages. The primary coils may be arbitrarily connected. To illustrate the point, refer to the arrangement depicted in Fig. 2-31. The objective is to arrange the secondary coil connections so that the line voltages will be balanced and no circulating current takes place in the secondary coils. To aid in distinguishing the primary windings from the secondary windings, uppercase letters are used for the former and lowercase letters for the latter. Moreover, the use of common letters (whether upper- or lowercase) is to be taken to mean that the coils are associated with the same core. Thus, in Fig. 2-31(a) coil AB denotes the primary winding and coil ab denotes the secondary winding of the same single-phase transformer. The generator that supplies the voltages to the primary coils of the three transformers is assumed to furnish the balanced three-phase set that is shown in Fig. 2-31(b). Observe that this set is represented by a phasor diagram in the closed form. Note, too, that in Fig. 2-31(a), the secondary coils are left unconnected. These connections must now be made in a prescribed manner.

Begin by joining terminal b of coil ab to terminal b of coil bc. The magnitudes of the primary phase voltages are each equal to one another, i.e., $V_{AB} = V_{BC} = V_{CA} = V_{p1}$, where V_{p1} denotes primary phase voltage. Similarly, the values of the secondary phase voltages are also equal to each other based on the assumption that the transformers are identical. Thus $V_{ab} = V_{bc} = V_{ca} = V_{p2}$, where V_{p2} denotes secondary

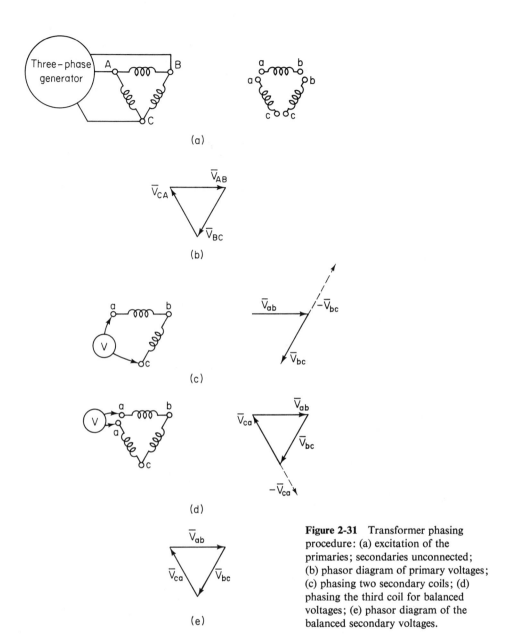

Figure 2-31 Transformer phasing procedure: (a) excitation of the primaries; secondaries unconnected; (b) phasor diagram of primary voltages; (c) phasing two secondary coils; (d) phasing the third coil for balanced voltages; (e) phasor diagram of the balanced secondary voltages.

phase voltage. It is helpful to be aware that although the magnitudes of these voltages are the same, they do differ in phase. Consequently, when a voltmeter is placed across the open terminals *ac* in the manner illustrated in Fig. 2-31(c), the reading will be one of two choices. Either it will be $\sqrt{3}\, V_{ab}$ or it will be equal in magnitude to V_{ab}. Keep in mind that the phasor voltage V_{bc} can be added to the phasor voltage V_{ab} at an angle of 60° (as shown by the dashed line in the phasor diagram of Fig. 2-31(c) or at an angle of −120° (as indicated by the solid line). These values are dictated by the relationship of V_{BC} to V_{AB} on the primaries. If the higher of the two values is found upon testing, it should be rejected and the coil connection reversed. It is important to keep in mind that in a Δ arrangement, the phase voltages are equal to the line voltages. In the last step of this *phasing procedure*, terminal c of coil ac is joined to c of coil bc, with the gap in Fig. 2-31(d) left opened except for the insertion of a voltmeter. This voltage, too, presents one of two possibilities: either the reading will be twice V_{ab} or it will be zero. The result depends upon whether the voltage of coil ac is directed along a $-60°$ direction or a direction of 120°. Clearly, that connection of coil ac must be chosen that yields a zero gap voltage. Surely, we would not want to connect the two a terminals except when the potentials of these terminals are the same. This avoids fireworks. Before leaving this matter, however, it is worthwhile to note that transformer manufacturers do use markings that lead to the proper phasing of three-phase transformers. But obviously the phasing procedure must be done by someone. Figure 2-31(e) shows the final phasor diagram as it pertains to the balanced secondary voltages.

The salient characteristics between the line and phase quantities from primary to secondary for the Δ-Δ connection may be summarized as described below. Introduce

$$a = \frac{V_{AB}}{V_{ab}} = \frac{V_{p1}}{V_{p2}}$$

where the quantities on the right side represent the primary and secondary phase voltages. Assume that a balanced three-phase load is applied to the secondary windings. Then

$$\text{Ratio of phase voltages:} \quad \frac{V_{AB}}{V_{ab}} = a$$

$$\text{Ratio of line voltages:} \quad \frac{V_{L1}}{V_{L2}} = \frac{V_{AB}}{V_{ab}} = a$$

where V_{L1} and V_{L2} represent the primary and secondary line voltages. The primary and secondary phase and line voltages are also respectively in phase with one another; i.e., $\bar{V}_{AB}$ is in phase with $\bar{V}_{ab}$, $\bar{V}_{BC}$ is in phase with $\bar{V}_{bc}$, and $\bar{V}_{CA}$ is in phase with $\bar{V}_{ca}$. A comparison of Fig. 2-31(b) with Fig. 2-31(e) makes this statement self-evident. Also,

$$\text{Ratio of phase currents:} \quad \frac{I_{AB}}{I_{ab}} = \frac{1}{a}$$

$$\text{Ratio of line currents:} \quad \frac{I_{L1}}{I_{L2}} = \frac{\sqrt{3}\,I_{p1}}{\sqrt{3}\,I_{p2}} = \frac{I_{AB}}{I_{ab}} = \frac{1}{a}$$

Here use is made of the fact that the line current in a balanced Δ-connection is $\sqrt{3}$ times the phase current.

It is important that the transformers that are used in these arrangements have equal transformation ratios in order to minimize the existence of circulating currents. Furthermore, the equivalent leakage impedances of the transformers should be as nearly identical as possible to ensure proper load sharing during operation. The use of a Δ secondary also makes only one voltage available with no possibility of including a neutral wire.

Finally, a word is in order concerning the magnetizing current in these arrangements. In the configuration of Fig. 2-31(a) a sinusoidal voltage is assumed to be applied to the primary coils of the three transformers. Because of its sinusoidal nature, the generator can deliver to the transformer a magnetizing current which is itself sinusoidal. Such a sinusoidal current, however, produces a flat-topped flux wave as a result of the saturation in the iron core of the transformer. In addition to the principal fundamental component, a flat-topped flux wave contains a large third harmonic part, which, in turn, serves to induce, by Faraday's law, a third harmonic voltage. This action takes place in each of the primary transformers. Furthermore, the third harmonic voltage induced in coil BC occurs at angle of $3(-120°) = -360°$ or in-phase with the component appearing in phase AB. Similarly, in coil CA, the phase of the third harmonic voltage is $3(120) = 360°$, so that it too is in phase with the third harmonic voltage in the other two coils. Since the sinusoidal generator contains no such component in its voltage cycle, neutralizations of the third harmonic voltages cannot take place. Accordingly, the additive effect of these third harmonic voltages is to cause a third harmonic circulating current to exist in the three transformer primaries. This current, then, serves the useful purpose of changing the magnetizing current from a sinusoidal variation to one which is peaked to the extent that is necessary to ensure a sinusoidal flux variation. The sinusoidal flux variation in turn provides a sinusoidal voltage variation at the primary terminals of the transformers, thereby satisfying Kirchhoff's voltage law in the generator-transformer primary circuit.

Δ-Y Connection

The delta-wye arrangement of transformers is commonly found at the transmitting end of transmission lines as well as at the distribution end. At the transmitting end, the Δ-Y connection offers the advantage of an increased step-up ratio beyond that associated with the individual transformers. At the distribution end the inclusion of a neutral wire allows a choice of two voltages for the customer's use. A popular combination that is available from the electric power companies to homes and commercial establishments is the 208/120-V system. The larger figure is available between any two line terminals of the Y-connection, whereas the lower figure represents the voltage from line to neutral.

The phasing procedure for a Δ-Y connection proceeds differently from that of the Δ-Δ connection insofar as the secondary connections are concerned. The primary coils in each case can be wired arbitrarily in a closed delta arrangement. Because the wye secondaries involve a neutral terminal, a modified terminal marking is used for the individual transformers, as illustrated in Fig. 2-32(a). Of course, when the primary windings are joined to form a closed circuit, the n markings may be dropped as shown in Fig. 2-32(b). However, in the phasor representation of the primary voltages that is depicted in Fig. 2-32(c) the n notation is carried along because it makes it easier to understand what is happening at the secondaries.

We begin the phasing process at the secondaries by joining n of coil na to n of coil nb. A voltmeter placed across the free terminals will indicate one of two voltages, as was the case with the Δ-Δ connection. The specific reading depends upon the relative sense of the winding. The connection causes the voltage in coil an to be added to the voltage in coil nb along a $-60°$ direction in the phasor diagram or a $120°$ direction. These are the only two choices and the one that prevails depends upon the

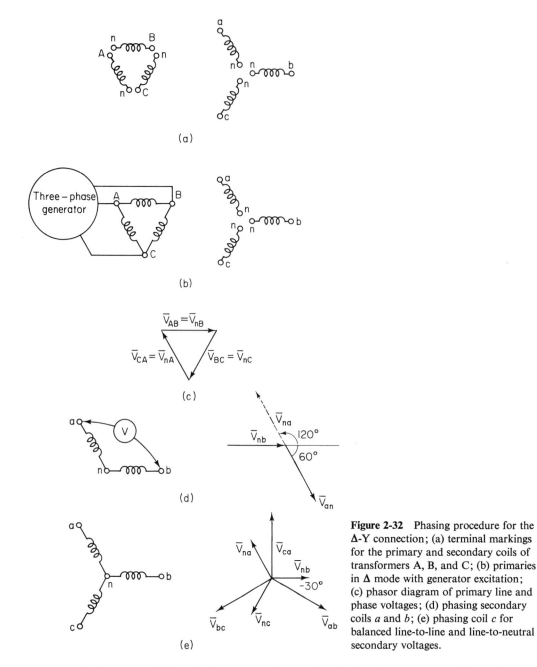

Figure 2-32 Phasing procedure for the Δ-Y connection; (a) terminal markings for the primary and secondary coils of transformers A, B, and C; (b) primaries in Δ mode with generator excitation; (c) phasor diagram of primary line and phase voltages; (d) phasing secondary coils a and b; (e) phasing coil c for balanced line-to-line and line-to-neutral secondary voltages.

relative sense of the windings when the two terminals are connected. If the voltmeter reads a value for the line voltage V_{ab} that is equal to V_{nb}, it means the addition is occurring along the 120° direction [see the dashed-line component in Fig.2-32(d)] and must be rejected. To correct this situation, it is merely necessary to reverse one of the coil connections. Now the voltmeter will yield for V_{ab} a value of $\sqrt{3} \, V_{nb}$.

TRANSFORMERS   CHAP. 2

This is the proper value for the line-to-line voltage of the wye connection. The line voltage must always be greater than the phase voltage by a factor of $\sqrt{3}$. When the third coil c is joined in the manner of Fig. 2-32(e), the correct phasing is determined when the voltages between each pair of terminals is equal in magnitude to $\sqrt{3}$ times the phase voltage. (What are the voltage readings when coil c is incorrectly phased?)

A comparison of the phasor diagram of a properly phased Δ-Y connection such as appears in Fig. 2-32(e) reveals that the line-to-line voltages at the secondaries are displaced by 30 electrical degrees from the line-to-line voltages at the primaries. Thus, for the situation depicted in Fig. 2-32, the line voltage between terminals a and b at the secondaries is $\bar{V}_{ab} = V_{ab}\underline{/-30°}$, whereas the line voltage between terminals A and B at the primaries is $\bar{V}_{AB} = V_{AB}\underline{/0°}$. Accordingly, there is a *lag* of 30 electrical degrees in this arrangement. It will be recalled that this condition did not occur in the Δ-Δ connection where it was found that the line-to-line voltages between primaries and secondaries were all in phase. The importance of this distinction lies in the conclusion that a bank of Δ-Δ connected transformers cannot be placed in parallel with a bank of Δ-Y transformers even if the voltage magnitudes correspond. The phase angle discrepancy which is inherent with these connections will result in wasteful circulating currents.

At the generator end of a transmission line, the Δ-Y connection is used to step up the generator voltage to a suitable transmission line voltage. Let the step-up voltage ratio be represented by

$$a = \frac{V_{nA}}{V_{na}} = \frac{V_{p1}}{V_{p2}} \tag{2-149}$$

where a is less than unity for a step-up arrangement and V_{p1} denotes the primary value of the phase voltage and V_{p2} is the secondary value of the phase voltage. The ratio of the line-to-line voltages at the secondaries to the line-to-line voltages at the primaries can then be expressed as

$$\frac{(V_{L\text{-}L})_Y}{(V_{L\text{-}L})_\Delta} = \frac{V_{ab}}{V_{AB}} = \frac{\sqrt{3}\,V_{p2}}{V_{p1}} \tag{2-150}$$

Upon replacing V_{p1} by its equivalence from Eq. (2-149), there results

$$\frac{(V_{L\text{-}L})_Y}{(V_{L\text{-}L})_\Delta} = \frac{V_{ab}}{V_{AB}} = \frac{\sqrt{3}\,V_{p2}}{aV_{p2}} = \frac{\sqrt{3}}{a} \tag{2-151}$$

Thus, if a is assigned a value of one-tenth, the line-to-line voltage at the secondaries will be 17.32 times greater than the line-to-line voltages at the primary. The attractive feature about this arrangement, of course, is that insulation for the secondary windings need be provided only to the extent of a factor of 10 over that of the primary voltage rather than 17.32. Furthermore, in view of the fact that output volt-amperes at the secondaries of the Y must equal the input volt-amperes at the primaries of the Δ under the assumption of negligible losses, it follows that the ratio of the line current at the secondaries $(I_L)_Y$ to the line current at the primaries $(I_L)_\Delta$

for a balanced load is the reciprocal of Eq. (2-151). That is,

$$\frac{(I_L)_Y}{(I_L)_\Delta} = \frac{I_{p2}}{\sqrt{3}\,I_{p1}} = \frac{I_{p2}}{\underbrace{\sqrt{3}\,I_{p2}}_{a}} = \frac{a}{\sqrt{3}} \tag{2-152}$$

where I_{p1} and I_{p2} denote the primary and secondary phase currents, respectively. Therefore, by raising the line voltage substantially, the corresponding transmission line currents can be markedly reduced, which in turn can serve to improve the efficiency of transmission appreciably.

The presence of a delta arrangement on the primary side, which allows a third harmonic magnetizing current component to flow, ensures that the line-to-neutral voltages on the wye side will be sinusoidal. Of course, the foregoing statement is based on the premise that the source voltage that energizes the Δ-connected primaries is itself purely sinusoidal.

Y-Y Connection

When equipped with a primary and secondary neutral, the wye-wye connection offers two advantages: (1) it provides a choice of two voltages at the secondary side; and (2) transformers need to be specified for an insulation that guards against the magnitude of the line-to-phase voltage rather than line-to-line voltage. Also, if the ratio of the primary phase voltage to the secondary phase voltage is a, this same ratio obtains for the line-to-line voltages. On the other hand, the ratio of the primary phase or line currents to the secondary phase or line currents is $1/a$. Moreover, on either side, the line voltage is phase displaced from the nearest phase voltage by a magnitude of 30 electrical degrees.

Any attempt to operate a Y-Y arrangement of transformers without the presence of a primary neutral connection will lead to difficulty and potential failure. In fact, trouble occurs even under no-load conditions. To illustrate the point, assume that the primary wye connection is energized from a sinusoidal three-phase generator. Since such a generator delivers a sinusoidal magnetizing current, the typical saturating magnetization curve of the transformer core causes the flux variation to be flat-topped. In turn, this flat flux wave contains a large third harmonic component which induces an appreciable third harmonic voltage in the transformer coils. The magnitude of this voltage depends upon the degree of saturation of the core and in most instances, this third harmonic voltage is not negligible. Recall that this same condition was noted to exist in the Δ-connected primaries. But with the Y-Y connection, there is an important difference. Whereas in the Δ-connected primaries the third harmonic voltage was presented with a closed path that allowed a circulating current to flow, this situation does not happen with the Y connection without a neutral. As a matter of fact, because the third harmonic phase voltages are each in phase with one another, they fail to make their presence felt at the line-to-line terminals. Figure 2-33 shows the phase and magnitudes of the third harmonic rms voltages from line-to-neutral in each phase. Observe that their magnitudes and phase angles are identical. To obtain the effect of these voltages between line terminals, it is necessary to perform the phasor addition

$$(\bar{V}_{ab})_3 = \bar{V}_{an_3} + \bar{V}_{nb_3} = -\bar{V}_{na_3} + \bar{V}_{nb_3} = 0$$

$$\xrightarrow{\hspace{3cm}} \quad \overline{V}_{na3} = V_{na3} \underline{/0°}$$

$$\xrightarrow{\hspace{3cm}} \quad \overline{V}_{nb3} = V_{na3} \underline{/0°}$$

$$\xrightarrow{\hspace{3cm}} \quad \overline{V}_{nc3} = V_{na3} \underline{/0°}$$

Figure 2-33 The line-to-neutral voltages caused by the third harmonic flat-topped flux wave present in a Y-Y connection without a primary netural.

which clearly adds to zero. The significance of this result in the generator-primary transformer circuit is that Kirchhoff's voltage law is satisfied despite the presence of a flat-topped (nonsinusoidal) flux wave in the transformer core. One consequence of this condition is that the resultant line-to-neutral voltage can now readily exceed the rated transformer voltage, so that harmful insulation stresses can take place. Of course, when a primary neutral connection is established, the third harmonic voltage that originates in the transformer from line to neutral is now provided with a path that allows the magnetizing current once again to become peak-shaped, thereby restoring a sinusoidal flux variation. In turn, this virtually eliminates the third harmonic voltages in the transformer coils. In those instances where the ensuing third harmonic neutral ground currents cause telephone or other related interference, it may be necessary to resort to an alternate transformer connection involving the Δ arrangement on the primary or secondary side or else to purchase transformers equipped with tertiary windings which can then be connected in the Δ mode.

Another serious shortcoming of a Y-Y connection without a primary neutral is that it makes it practically impossible to supply power to a load that is placed from line-to-neutral on the secondary side. In order for energy to be delivered to the load in such an arrangement, the generator source, which supplies the primary windings, must do so through the two unloaded primaries. However, as we learned earlier in this chapter, an unloaded transformer presents an impedance which is essentially the magnetizing reactance. This high impedance mitigates against the flow of significant currents. In this regard it is interesting to note that even if a short circuit were placed from one line-to-neutral on the secondary, very little current would flow. Because such a short circuit has the effect of placing the neutral at the potential of one line terminal, it follows that the line-to-neutral voltages appearing across each of the unloaded transformers is increased almost by a factor of $\sqrt{3}$. Such a high overvoltage can readily cause damage to the insulation. In conclusion, it ought to be obvious that a Y-Y connection should not be used without the primary neutral connected to the source.

Y-Δ Connection

The wye-delta arrangement of single-phase transformers is often found at the receiving end of a transmission line. The reason is that it provides a large step-down voltage ratio from primary line voltage to secondary line voltage. If we let a denote the ratio of the primary phase voltage (Y side) to the secondary phase voltage (Δ), it follows that the ratio of the line voltage from primary to secondary is $\sqrt{3}\,a$. Correspondingly, the ratio of the primary line current to the secondary line current is $1/\sqrt{3}\,a$.

Third harmonic voltage stresses do not exist in the wye-delta arrangement because the closed path provided by the secondary Δ connection permits the third harmonic magnetizing current to exist. In turn, this current acts to virtually eliminate the third harmonic component in the flat-topped flux wave, thus ensuring a sinusoidal flux variation. Unbalanced secondary loads, however, can present some difficulties, as explained for the Y-Y connection, without a primary neutral. To avoid problems, it is best to establish the primary neutral to the source.

PROBLEMS

2-1. A reactor coil with an iron core has 400 turns. It is connected across the 115-V, 60-Hz power line.
   (a) Neglecting the resistance voltage drop, calculate the maximum value of the operating flux.
   (b) If the flux density is not to exceed 75 kilolines/in.^{2}, what must be the cross-sectional area of the core?

2-2. A fixed sinusoidal voltage is applied to the circuit shown in Fig. P2-2. If the voltage remains connected and the shaded portion of the iron removed, state what happens to the maximum value of the flux and the maximum value of the magnetizing current. Justify your answer. Assume negligible leakage impedance and winding resistance.

Figure P2-2

2-3. Repeat Prob. 2-2 assuming the voltage applied to the core is a fixed d-c quantity. Assume d-c current is limited by an external resistor.

2-4. A transformer coil rated at 200 V and having 100 turns is equipped with a 0.5 tap. If 50 V is applied to half the number of turns, find the value of the maximum flux. The frequency is 60 Hz.

2-5. Refer to the transformer coil of Prob. 2-4.
   (a) Determine the maximum value of the operating flux when 150 V is applied across half the number of turns.
   (b) What is the required change in magnetizing current in part (a)? Neglect saturation.
   (c) Briefly describe the effect on the magnetizing current in part (b) if saturation is not neglected.
   (d) What must be the frequency of the 150-V source in part (a) if the flux is to remain unchanged?

2-6. In an open-circuit test of a 25-kVA, 2400/240-V transformer made on the low side the corrected readings of amperes, volts, and watts are, respectively, 1.6, 240, and 114. In the short-circuit test the low side is short-circuited and the current, voltage, and power to the high side are measured to be, respectively, 10.4 A, 55 V, and 360 W.
   (a) Find the core loss.
   (b) What is the full-load copper loss?
   (c) Find the efficiency for a full-load of 0.8 power factor leading.
   (d) Compute the percent voltage regulation for part (c).

2-7. A transformer with a rating of 25 kVA, 2400/240 V has the following parameters: $R_{eH} = 3.33\ \Omega$ and $Z_{eH} = 5.28\ \Omega$. The core loss at rated voltage is 114 W.

(a) Find the efficiency when the transformer delivers rated kVA at 0.8 pf lagging.

(b) What is the value of the high-side voltage in part (a)?

(c) Determine the percent voltage regulation in part (a).

2-8. A 50-kVA, 2300/230-V, 60-Hz transformer has a high-voltage winding resistance of 0.65 Ω and a low-voltage winding resistance of 0.0065 Ω. Laboratory tests showed the following results:

Open-circuit test: $V = 230$ V, $I = 5.7$ A, $P = 190$ W

Short-circuit test: $V = 41.5$ V, $I = 21.7$ A, $P = $ none used

(a) Compute the value of primary voltage needed to give rated secondary voltage when the transformer is connected *step-up* and is delivering 50 kVA at a power factor of 0.8 lagging.

(b) Compute the efficiency under the conditions of part (a).

2-9. A 100-kVA transformer was measured to have a core loss of 1250 W at rated voltage. A short-circuit test was performed at 125 % of rated current and the input power was measured to be 2875 W. Calculate the efficiency of this transformer when it delivers rated kVA at a pf of 0.9 lagging.

2-10. The following test data were taken on a 15 kVA, 2200/440-V, 60-Hz transformer:

Short-circuit test: $P = 620$ W, $I = 40$ A, $V = 25$ V

Open-circuit test: $P = 320$ W, $I = 1$ A, $V = 440$ V

(a) Calculate the voltage regulation of this transformer when it supplies full load at 0.8 pf lagging. Neglect the magnetizing current.

(b) Find the efficiency at the load condition of part (a).

2-11. The following test data were taken on a 110-kVA, 4400/440-V, 60-Hz transformer:

Short-circuit test: $P = 2000$ W, $I = 200$ A, $V = 18$ V

Open-circuit test: $P = 1200$ W, $I = 0.2$ A, $V = 4400$ V

Calculate the voltage regulation of this transformer when it supplies rated current at 0.8 pf lagging. Neglect the magnetizing current.

2-12. A 50-kVA, 2300/230-V, 60-Hz distribution transformer takes 360 W at a power factor of 0.4 with 2300 V applied to the high-voltage winding and the low-voltage winding open-circuited. If this transformer has 230 V impressed on the low side with no load on the high side, what will be the current in the low-voltage winding? Neglect saturation.

2-13. A 25-kVA, 2400/240-V transformer draws 254 W at a pf of 0.15 when 240 V is impressed on the low-voltage side with the high-voltage side open-circuited. Find the current that is drawn from the line when 2400 V is applied to the high side with the low side open-circuited.

2-14. A 10-kVA, 2400/240-V transformer draws 165 W at a pf of 0.2 when 220 V is applied to the low side with the high side open. Find the current drawn by the high side when 2400 V is applied to the high side with the low side open. Neglect saturation.

2-15. A 200/100-V, 60-Hz transformer has an impedance of $0.3 + j0.8\ \Omega$ in the 200-V winding and an impedance of $0.1 + j0.25\ \Omega$ in the 100-V winding. What are the currents on the high and low side if a short circuit occurs on the 100-V side with 200 V applied to the high side?

2-16. A 60-Hz, three-winding transformer is rated at 2300 V on the high side with a total of 300 turns. Of the two secondary windings, each designed to handle 200 kVA, one is rated at 575 V and the other at 230 V. Determine the primary current when rated current in the 230-V winding is at unity power factor and the rated current in the 575-V winding is at 0.5 pf lagging. Neglect all leakage impedance drops and magnetizing current.

2-17. A transformer is equipped with three windings. The high side (primary) is rated at 2300 V and has a total of 3000 turns. Each of the low-side secondary windings has a kVA rating of 200 and voltage ratings of 460 V and 230 V. When rated current at unity pf flows through the 460-V winding and rated current simultaneously is made to flow through the 230-V winding, the primary current is found to be $122.98\underline{/45°}$ A. Describe the type of load that appears across the 230-V winding. Neglect all leakage impedance and magnetizing current.

2-18. An "ideal" transformer has a secondary winding tapped at point b. The number of primary turns is 100 and the number of turns between a and b is 300 and between b and c is 200. The transformer supplies a resistive load connected between a and c and drawing 7.5 kW. Moreover, a load impedance of $10\underline{/45°}$ Ω is connected between a and b. The primary voltage is 1000 V. Find the primary current.

2-19. Draw a neat phasor diagram of a transformer operating at rated conditions. Assume that $N_1/N_2 = 2$ and

$$I_1 r_1 = 0.1E_1, \qquad I_2 r_2 = 0.1V_R, \qquad I_m = 0.3I'_R$$
$$I_1 x_1 = 0.2E_1, \qquad I_2 x_2 = 0.2V_R, \qquad I_c = 0.1I'_R$$

Consider the load power factor to be 0.6 leading. Use V_R as the reference phasor and show all currents and voltages drawn to scale.

2-20. A 10-kVA, 500/100-V transformer has $R_{eH} = 0.3$ Ω and $X_{eH} = 5.2$ Ω, and it is used to supply power to a load having a lagging power factor. When supplying power to this load an ammeter, wattmeter, and voltmeter placed in the high-side circuit read as follows:

$$I_1 = 20 \text{ A}, \qquad V_1 = 500 \text{ V}, \qquad \text{WM} = 8 \text{ kW}$$

For this condition calculate what a voltmeter would read if placed across the secondary load terminals. Assume the magnetizing current to be negligibly small.

2-21. A 10-kVA, 460/115-V transformer has a high-side winding resistance of 0.4 Ω and a low-side winding resistance of 0.02 Ω. The equivalent leakage reactance on the high side is 3.2 Ω. When this transformer is used to supply power to a passive load, it draws a lagging current of 21.7 A at 460 V and 8 kW. Determine the resistive and reactive components of the load impedance. Neglect the magnetizing current.

2-22. The following data are taken on a 30-kVA, 2400/240-V, 60-Hz transformer:

Short-circuit test: $V = 70$ V, $I = 18.8$ A, $P = 1050$ W
Open-circuit test: $V = 240$ V, $I = 3.0$ A, $P = 230$ W

(a) Determine the primary voltage when 12.5 A at 240 V is taken from the low-voltage side supplying a load of 0.8 pf lagging.
(b) Compute the efficiency in part (a).

2-23. A 30-kVA, 240/120-V, 60-Hz transformer has the following data:

$$r_1 = 0.14 \text{ }\Omega, \qquad r_2 = 0.035 \text{ }\Omega$$
$$x_1 = 0.22 \text{ }\Omega, \qquad x_2 = 0.055 \text{ }\Omega$$

It is desired to have the primary induced emf equal in magnitude to the primary terminal voltage when the transformer carries the full-load current. Neglect the magnetizing current. How must the transformer be loaded to achieve this result?

2-24. The two windings of a 48-kVA, 2400/240-V, 60-Hz transformer have resistances of $0.6\ \Omega$ and $0.025\ \Omega$ for the high- and low-voltage windings, respectively. This transformer requires that 238 V be impressed on the high-voltage coil in order that rated current be circulated in the short-circuited low-voltage winding.
   (a) Calculate the equivalent leakage reactance referred to the high side.
   (b) How much power is needed to circulate rated current on short circuit?
   (c) Compute the efficiency at full load when the power factor is 0.8 lagging. Assume that the core loss equals the copper loss.

2-25. A transformer coil is equipped with a 0.5 tap, has 100 turns, and is rated for 200 V. When this voltage is applied to the full winding, the magnetizing current is 1 A. If now 50 V is applied to half the number of turns, what will be the magnetizing current? Assume no saturation.

2-26. A short-circuit test is to be performed on a transformer with a rating of 50 kVA, 460/115 V. The high-side and low-side resistances are known to be $0.06\ \Omega$ and $0.004\ \Omega$, respectively. Moreover, the power factor at short-circuit conditions is 0.3. Compute the input voltage that is required to force rated current through the low-side coil at short circuit.

2-27. A coil located on an iron core has a constant 60-Hz voltage applied to it. The total inductance is found to be 10 H. A second identical coil is placed in parallel with the first one and is so wound that it produces aiding flux. What is the total inductance of the parallel combination? Neglect the winding resistance and leakage flux of each winding.

2-28. Repeat Prob. 2-27 for the case where the second identical coil is placed in series with the first coil. Assume that aiding flux is produced.

2-29. Show the derivation that allows Eq. (2-68) to be written from Eq. (2-67) for a sinusoidal forcing function and no saturation.

2-30. A two-winding transformer has a high-voltage coil equipped with 1000 turns and a low-voltage coil equipped with 100 turns. The self-inductance of the high voltage primary is known to be 20 H. Moreover, the primary winding has a leakage inductance of 1 H. The secondary leakage inductance is 0.01 H.
   (a) Find the mutual inductance in henries.
   (b) Determine the value in henries of the self-inductance of the secondary winding.

2-31. A conventional transformer with a rating of 10 kVA, 2000/1000 V is available.
   (a) Show how to arrange this transformer for use as a step-down autotransformer with a secondary voltage of 2000 V.
   (b) For a 0.8 pf lagging load, what is the maximum current that can be delivered without overloading the transformer coils?
   (c) At the load conditions of part (b), express the conducted kVA of the autotransformer as a percentage of the rating of the conventional transformer.
   (d) List the advantages which the autotransformer has over an ordinary transformer of equal kVA and voltage rating for doing the same job.

2-32. A 10-kVA, 2300/230-V transformer has an iron loss of 190 W and a copper loss of 196 W at rated voltage and rated current. The secondary winding is made up of two equal sections, each section series-connected for 230-V operation. It is desired to reconnect this conventional transformer as an autotransformer in order to raise the 2300-V supply line to 2415 V and to deliver maximum power without overheating the coils. Using *all* windings of the transformer, determine for rated current:

(a) kVA delivered to the load.

(b) kVA conducted.

(c) Efficiency for a load pf of 0.8 leading.

(d) The rating (voltage and kVA) of an ordinary transformer to do the same job. What are the advantages of using an autotransformer?

2-33. The following data were found during a short-circuit test on an autotransformer rated at 22 kVA, 440/330 V. With the low side short-circuited and voltage applied to the high side, the readings were:

$$P = 320 \text{ W}, \qquad V = 20 \text{ V}, \qquad I = 50 \text{ A}$$

(a) Find the voltage regulation of this autotransformer at full-load output and unity power factor.

(b) If the noncommon portion of the winding has a resistance of 0.064 Ω, determine the resistance in ohms of the common portion.

(c) Find the transformed and conducted power for full-load output at unity power factor.

(d) What must be the rating (kVA and voltages) of an ordinary transformer that can be used to do the same job?

2-34. The expression for the equivalent resistance of an autotransformer looking into the high-side terminals is given by Eq. (2-107). Derive the expression for the equivalent resistance of the autotransformer looking into the low-side terminals, with the high side short-circuited.

2-35. When the transformer of Prob. 2-15 is used as a step-up autotransformer (200/300), what are the line currents on the high and low sides if a short circuit occurs on the 300-V side with 200 V applied to the low side?

2-36. An autotransformer has $N_{ac} = 100$ turns and $N_{bc} = 60$ turns, where N_{ac} represents the full winding. A short-circuit test is performed by short-circuiting winding *ab* and applying a reduced voltage to winding *bc*. The equivalent impedance looking into winding *bc* is found to be: $1.5 + j4.5$.

(a) Determine the equivalent impedance looking into winding *ac* for the condition where winding *bc* is short-circuited.

(b) Find the equivalent resistance looking into *bc* when *ac* is short-circuited and voltage is applied to *bc*.

2-37. A single-phase autotransformer rated at 40 kVA supplies a load impedance of $4\underline{/36.9°}$ Ω at a terminal voltage of 200 V from a 125-V supply. All power losses and leakage reactances are negligible. Calculate the magnitudes of the currents in the common and noncommon parts of the transformer winding, taking into account the magnetizing current of 0.075 per unit.

2-38. A 25-kVA, 2400/240-V single-phase transformer is connected to act as a booster from 2400 to 2640 V. With the maximum kVA load that it is possible to supply without overloading the transformer coils, determine:

(a) kVA received and delivered.

(b) kVA transformed.

(c) kVA conducted.

Neglect all losses, magnetizing current, and leakage impedance.

2-39. Repeat Prob. 2-38 for the case where the secondary coil is reversed so that the output (load) voltage is 2160 V.

2-40. A transformer rated at 40 kVA has a total copper loss of 250 W when operating at 50% of rated current. Determine the per-unit value of the equivalent resistance.

2-41. Refer to the configuration of Fig. P2-41.

Figure P2-41

(a) Assume that the current that enters terminal a is increasing in the positive direction. Identify the instantaneous polarity markings of each terminal.

(b) The coils have the number of turns indicated. A resistive load that draws 20 A is placed across terminals c and d. A purely capacitive load drawing 40 A is placed across ef. Find the source current. Neglect all leakage impedances and magnetizing current.

(c) Disregard winding ef. Assume winding ab has a voltage rating of 400 V and a current rating of 50 A. A short-circuit test on this transformer yields the following data:

$$\text{WM} = 600 \text{ W}, \qquad \text{VM} = 50 \text{ V}, \qquad \text{AM} = 20 \text{ A}$$

Find the voltage regulation at unity pf.

2-42. The high-side and low-side winding resistances of a 100-kVA, 440/110-V single-phase transformer are 0.04 Ω and 0.0025 Ω, respectively.

(a) Find the per-unit values of the high-voltage and low-voltage winding resistances.

(b) If the core loss at rated voltage is 1200 W, find the transformer efficiency when it delivers half-rated kVA at unity power factor.

2-43. A transformer rated at 2500 kVA, 10,000/2000 V, 60 Hz is designed for maximum efficiency at 80% of rated load. The per-unit equivalent impedance of this transformer is $0.02 + j0.06$. For a resistive load and operation at maximum efficiency, find the losses and the change in voltage from 80% load (at rated voltage) to no load.

2-44. A transformer rated at 1100 kVA has a short-circuit test conducted at a current equal to 75% of rated value, and takes an amount of power equal to 7030 W. An open-circuit test is also conducted and at rated voltage the input power is measured to be 9850 W.

(a) Find the per-unit value of the equivalent resistance.

(b) Find the per-unit value of the conductance.

2-45. If the per-unit value of the equivalent resistance of the conventional transformer of Prob. 2-31 is 0.02, find the value of the equivalent resistance of the autotransformer in ohms referred to the high side.

2-46. Describe the predominate waveshape of the primary magnetizing current of a Y-Δ three-phase transformer connection employing single-phase transformers.

2-47. A three-phase transformer bank is connected with the primaries in delta and the secondaries in wye. The line voltage on the primary side is 120 V and on the secondary side 240 V. Find the ratio of primary to secondary turns for each of the single-phase transformers.

2-48. Three single-phase transformers with ratings of 2400/120 V are used to step down a distribution line voltage from 4160 V to a lower value.

  (a) Assuming the use of a Y-Δ arrangement, find the value of the secondary Δ-side voltage.

  (b) If the maximum value of the magnetic flux puts operation beyond the knee of the magnetization curve, does a third harmonic voltage exist between the line and neutral terminals on the Y-side? Explain.

2-49. Refer to the arrangement shown in Fig. P2-50. The single-phase transformers are rated for 2400/120 V. Consider that the three coils for the secondary delta connection are properly joined with regard to the fundamental induced voltage component. Assume further that the magnitude of the third harmonic induced voltage is one-third that of the fundamental.

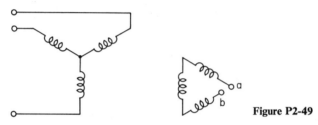

Figure P2-49

  (a) Find the gap voltage when a sinusoidal voltage of 4160 V rms is applied to the line terminals of the wye.

  (b) Calculate the line-to-neutral voltage at the Y connection.

  (c) When the gap is closed, describe the role played by the voltage found in part (a) in the operation of these transformers. Be specific.

2-50. A Y-Y connection of transformers consists of single-phase transformers which carry a rating of 69/6.9 kV. A sinusoidal transmission line voltage of 120 kV is applied to the high side. The third harmonic voltage, when it exists, is known to be one-third the value of the fundamental voltage. Neither the high nor the low side is provided with a ground neutral.

  (a) Determine the rms value of the line-to-neutral voltage on each side.

  (b) What recommendation can you make in order to eliminate the third harmonic voltage from the line-to-neutral voltage?

3

Fundamentals of Electromechanical Energy Conversion

Electromechanical energy conversion involves the interchange of energy between an electrical system and a mechanical system through the medium of a coupling magnetic field. The process is essentially reversible except for a small amount which is lost as heat energy. When the conversion takes place from electrical to mechanical form the device is called a *motor*. When mechanical energy is converted to electrical energy the device is called a *generator*. Moreover, when the electrical system is characterized by alternating current the devices are referred to as *a-c* motors and *a-c* generators, respectively. Similarly, when the electrical system is characterized by direct current the electromechanical conversion devices are called *d-c* motors and *d-c* generators.

The same fundamental principles underlie the operation of both a-c and d-c machines, and they are governed by the same basic laws. Thus, in the computation of the developed torque of an electromechanical energy-conversion device, one basic torque formula (which derives directly from Ampere's law) applies whether the machine is a-c or d-c. The ultimate forms of the torque equations appear different for the two types of machines only because the details of mechanical construction differ. In other words, starting with the same basic principles for the production of electromagnetic torque, the final forms of the torque equations differ to the extent that the mechanical details differ. These comments apply equally in generating emf in the armature winding of a machine whether it is a-c or d-c. Once again a simple basic relationship (Faraday's law) governs the voltage induced. The final forms of the voltage equations differ merely as a reflection of the differences in the construction of the machines. It is important that this is understood at the outset, because a prime objective of the treatment of the subject matter as it unfolds in this chapter is to impress upon the reader the fact that a-c machines are not fundamentally different from d-c machines. They differ merely in construction details; the underlying principles are the same.

This chapter is arranged to give still greater emphasis to this theme. Attention first is given to a basic analysis of the production of electromagnetic torque as well as the generation of voltages starting with the classical fundamental laws. Then the construction features of the various types of electric machines are examined with the purpose of indicating how the conditions for the production of torque and voltage are fulfilled and also to point out why differences are necessary. Once this background is established, the fundamental equations for developed torque and induced voltage are modified to make these expressions consistent with the particular construction features of the machines involved. In this way the equations are put in a more useful form for purposes of analysis and design. Attention is next directed in the following chapters to an analysis of each of the major types of electromechanical energy-conversion devices to be found performing innumerable tasks throughout industry and in homes, shops, and offices everywhere.

3-1 BASIC ANALYSIS OF ELECTROMAGNETIC TORQUE

Since electromechanical energy conversion involves the interchange of energy between an electrical and a mechanical system, the primary quantities involved in the mechanical system are torque and speed, while the analogous quantities in the electrical system are voltage and current respectively. Figure 3-1 represents the situation diagrammatically. *Motor action* results when the electrical system causes a current i to flow through conductors that are placed in a magnetic field. Then by Eq. (1-2) a force is produced on each conductor so that, if the conductors are located on a structure that is free to rotate, an electromagnetic torque T results, which in turn manifests itself as an angular velocity ω_m. Moreover, with the development of this motor action the revolving conductors cut through the magnetic field and thereby undergo an emf e, which is really a reaction voltage analogous to the induced emf in the primary winding of a transformer. Note that the coupling field is involved in establishing the electromagnetic torque T as well as the reaction induced emf e. In the case of *generator action* the reverse process takes place. Here the rotating member—the *rotor*—is driven by a prime mover (steam turbine, gasoline engine, etc.), causing an induced voltage e to appear across the armature winding terminals. Upon the application of an electrical load to these terminals, a current i is made to flow, delivering electrical power to the load. Of course the flow of this current through the armature conductors interacts with the magnetic field to produce a reaction torque opposing the applied torque originating from the prime mover.

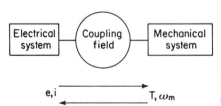

Figure 3-1 Block representation of electromechanical energy conversion.

In the interest of emphasizing further the character of this motor and generator action and its relationship to the coupling magnetic field, a mathematical description is now undertaken. Consider that we have an electromechanical energy-conversion device which consists of a stationary member—the *stator*—and the rotating member, the rotor. Refer to Fig. 3-2(a).

Assume that the stator is equipped with a coil *A-B*, which is energized to cause a constant current to flow in the direction shown by the dot-cross notation. By Flemings' right-hand rule, the flux lines that are produced by the coil ampere-turns leave the stator from the upper half, cross the air gap, penetrate the rotor iron and then once

(a)   (b)

(c)

(d)   (e)

Figure 3-2 Illustrating some construction features of an electromechanical energy conversion device: (a) stator and rotor; (b) a plan view; (c) a uniform air-gap flux density produced by coil *A-B*; (d) a full-pitch coil; (e) changing flux linkage of coil *a-b* with replacement *dx*.

again cross the air gap to form the closed path illustrated in Fig. 3-2(a). If the rotor and stator surfaces were to be cut and laid out in a flat manner, the resulting configuration would appear as shown in Fig. 3-2(b). If the reluctance of the iron is neglected, half the coil mmf is used for each crossing of the air gap. Accordingly, the value of the flux density that appears between coil sides A-B in Fig. 3-2(b) is one-half the coil mmf divided by the air-gap reluctance. Since both quantities are fixed, it follows that the flux density, too, is constant from coil side A to B and B to A. This situation is depicted in Fig. 3-2(c) in a manner that emphasizes that the flux that leaves the stator (in the upper section between coil sides A-B) behaves like a north-pole flux, while the flux that enters the stator (in the lower section between coil sides B-A) behaves like a south-pole flux.

With the uniformly distributed field now established, the focus of attention is shifted to the rotor, which is assumed to be equipped with a total of Z conductors, only two of which (a and b) are shown in Fig. 3-2(a). It is further assumed that conductor a is joined to conductor b to form a coil in the fashion illustrated in Fig. 3-2(d). Because the principle of electromechanical energy conversion is concerned with the movement of a coil relative to a magnetic field, the analysis now proceeds from the arrangement shown in Fig. 3-2(e), where only the presence of the field density B is given importance and not its source. It is helpful to observe that initially the flux that links rotor coil a-b is zero. In Fig. 3-2(e), this situation is represented by noting that coil a-b links one-half of the positive (north-pole) air-gap flux in addition to one-half of the negative (south-pole) air-gap flux for a resultant of zero. By Faraday's law, displacement of the coil by a differential distance dx to the new position a'b' causes a change in the flux linking the coil equal to twice the shaded area. Thus

$$d\phi = 2B \, dA \tag{3-1}$$

The factor 2 accounts for the two areas involved; B is the value of the flux density at the points where conductors a and b lie, and dA is the differential area through which the B-field penetrates. A little thought reveals this area to be equal to the axial length of the rotor, l, times dx. Thus

$$dA = l \, dx \tag{3-2}$$

However, if it is further assumed that dx results from a linear velocity v imparted to the conductor for the differential time dt, it follows that

$$dx = v \, dt \tag{3-3}$$

Insertion of Eqs. (3-2) and (3-3) into Eq. (3-1) then leads to

$$d\phi = 2Blv \, dt \tag{3-4}$$

The corresponding emf induced in coil ab becomes, by Faraday's law,

$$e = \frac{d\phi}{dt} = 2Blv \tag{3-5}$$

Because the coil a-b is assumed driven from left to right, the $\bar{v} \times \bar{B}$ rule indicates that the direction of the induced emf is into the paper for coil side a' and out of the paper for coil side b'. The magnetic field is considered to be directed downward on side a' and upward on side b'. Moreover, the application of the $\bar{I} \times \bar{B}$ rule discloses that a developed force exists on sides a' and b' in association with a coil current I that is directed from right to left. This produces a torque that opposes the torque that drives the coil. It is also helpful to keep in mind here that the polarity of this induced voltage is independent of the presence or absence of the current in the rotor conductors and of the polarity of this current.

The factor 2 appears in Eq. (3-5) because the coil involves two conductors. It then follows that, if the rotor is equipped with a total of Z conductors, which are all series-connected, the total induced voltage can be expressed as

$$e = ZBlv \qquad (3\text{-}6)$$

When a current i flows through conductors a and b in the configuration of Fig. 3-2(b), the direction of i and the direction of B are 90° apart. Accordingly, a force exists on each conductor which is given by Eq. (1-2). Thus

$$F_c = Bli \qquad (3\text{-}7)$$

where F_c denotes the force developed on an individual conductor. However, since there are Z conductors on the rotor surface, it follows that the total developed force is†

$$F = ZBli \qquad (3\text{-}8)$$

Moreover, if this force is made to act through a moment arm r, which is the radius of the rotor, the resulting developed torque can be expressed as

$$T = Fr = ZBlri \qquad (3\text{-}9)$$

It is important to note here that the force and torque are independent of motion and the direction of motion.

A glance at Eqs. (3-6) and (3-9) shows that both the induced voltage and the developed electromagnetic torque are dependent upon the coupling magnetic field. The factor that determines whether the induced voltage e and the developed torque T are actions or reactions depends upon whether generator or motor action is being developed.

The two mechanical quantities T and ω_m and the two electrical quantities e and i, which are coupled through the medium of the magnetic field B, are related further by the law of conservation of energy. This is readily demonstrated by dividing Eq. (3-6) by Eq. (3-9). Thus

$$\frac{e}{T} = \frac{ZBlv}{ZBlir} = \frac{\omega_m}{i} \qquad (3\text{-}10)$$

where

$$\omega_m = \frac{v}{r} \quad \text{mechanical angular velocity of the rotor} \qquad (3\text{-}11)$$

Rearranging Eq. (3-10) leads to

$$ei = T\omega_m \qquad (3\text{-}12)$$

†This equation is valid provided that the conductors are paired to form series-connected coils that have a span of one *pole pitch*. A pole pitch is equal to 180 electrical degrees.

which states that the developed electrical power is equal to the developed mechanical power. This statement is valid for generator as well as motor action.

Finally, on the basis of the foregoing discussion, it should be apparent that a conventional electromechanical energy-conversion device must involve two components. One is the *field winding*, which by definition is that part of the machine which produces the coupling field B. The other is the *armature winding*, which by definition is that part of the machine in which the "working" emf e and the current i exist.

Developed Torque for Sinusoidal B and Ni: The Case of A-C Machines

The expression for electromagnetic torque as it appears in Eq. (3-9) has limited application because B is assumed constant and all conductors are assumed to produce equal torques in the same direction—a situation which rarely occurs in practical machines. In fact in most a-c machines the field winding is intentionally designed to produce an almost sinusoidally distributed flux density, and the armature winding is similarly arranged to yield an ampere-conductor distribution which is also almost sinusoidally distributed along the periphery of the rotor structure. For this reason we shall now consider the development of an electromagnetic torque equation that applies for such cases.

The armature winding ampere-conductor distribution refers to a point-by-point plot of the current flowing through the individual conductors of the armature as depicted in Figs. 3-3(a) and (b). Note that the instantaneous value of the current existing in the conductors is assumed to be sinusoidally distributed. This situation is readily achieved in practical machines. For the 12 conductors shown in Fig. 3-3(a), 6 carry current into the paper and the other 6 carry current out of the paper, thus giving the appearance of two poles. Actually the number of poles associated with the ampere-conductor distribution of the armature winding is always equal to the number of field poles produced by the field winding.

Note too that the greater the number of conductors the more nearly sinusoidal will be the ampere-conductor distribution. In fact, with a truly large number of conductors a current sheet of sinusoidal density as shown in Fig. 3-3(c) can be used to represent the ampere-conductor distribution. Of course this means that the latter now takes on a purely sinusoidal variation. Moreover, the armature winding can be considered a single sheet of one turn having the units of ampere-turns per radian. A per-unit radian notation is used because in Fig. 3-3(c) the distribution is no longer discrete.

A study of Fig. 3-3(a) reveals that for the ampere-conductor distribution shown, the armature winding taken by itself looks like a solenoid, the mmf of which is directed towards the right. For any specified armature-winding distribution the total value of this mmf is found by applying Ampere's circuital law. The essence of Ampere's circuital law is embodied in the statement that any closed flux path can be said to exist if it encloses an mmf that is not equal to zero. Thus, if a closed flux path is taken so that it encloses just conductor 1 in Fig. 3-3(a) as it crosses the air gap twice, the associated mmf is equal to the magnitude of the current in this single conductor. If the path is extended to embrace conductors 1 and 2, the total enclosed mmf is now the sum

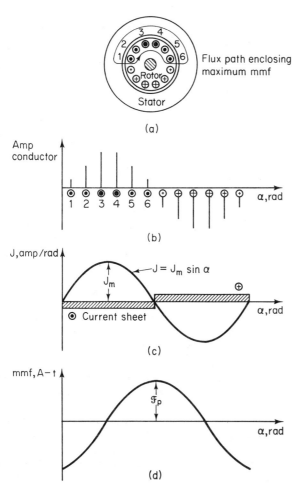

Figure 3-3 Development of the current sheet of sinusoidal density: (a) and (b) show finite ampere-conductor distribution; (c) current sheet representation; (d) corresponding mmf wave.

of the currents flowing in these two conductors. It follows then that Ampere's circuital law involves an integration (or summation) of the current sheet in order to obtain the mmf characteristic of the machine. In fact, for the closed flux path depicted in Fig. 3-3(a) as enclosing all conductors carrying current in the same direction (which is the same as enclosing the current sheet over the displacement range from 0 to π radians), the maximum mmf associated with this current sheet is given by

$$\mathcal{F} = \int_0^\pi J_m \sin \alpha \, d\alpha = 2J_m \tag{3-13}$$

where J_m denotes the peak value of the current sheet density and $\mathcal{F}$ denotes the mmf per pole pair. If the upper limit in Eq. (3-13) is left unspecified, it follows that the ensuing integration yields the negative cosine function that is illustrated in Fig. 3-3(d). Whenever such a plot is generated from Ampere's circuital law, the abscissa axis is always placed so that the areas of the mmf curve above and below this axis are the same, in recognition of the fact that the flux that leaves the stator must also be the flux that returns to the stator.

Because the flux path used in obtaining the result of Eq. (3-13) involves two air gaps, it follows that the mmf per pole (or per air-gap crossing) is

$$\boxed{\mathcal{F}_p = \frac{\mathcal{F}}{2} = J_m}$$

(3-14)

This result will be useful later in converting the basic electromagnetic torque equation to its practical forms for a-c machines.

Example 3-1

A two-pole machine has the stator construction shown in Fig. 3-4(a). Conductors 1, 2, 3, and 4 carry a current of 15 A directed out of the page. Correspondingly, conductors 5 to 8 carry the same current directed into the page.

(a) Draw the magnetomotive force as a function of the displacement along the air gap starting at point A and progressing clockwise. Use Ampere's circuital law.

(b) Find the maximum value of the mmf per pole.

(c) Identify the location of the axis of the flux produced by the mmf wave of part (a).

(d) Draw the equivalent current sheet associated with this mmf distribution.

Solution (a) Starting at point A, draw a closed path that crosses the air gap twice and encircles conductor 1 in the manner depicted in Fig. 3-4(a). By Ampere's circuital law such a flux line path does exist because the net mmf enclosed by this closed path is not zero. In a practical machine conductor 1 is connected to a matching conductor, which in the case of a full-pitch coil would be conductor 5. Hence encirclement of conductor 1 means encirclement of the turn represented by the series connection of conductors 1 and 5. Accordingly, the number of ampere-turns that act to produce such a flux line is here equal to 15 A-t. This situation is represented graphically in Figs. 3-4(b) and (c). Observe that until a flux path starting at point A is taken that reaches to enclose conductor 1, the enclosed mmf is zero. Just beyond the location of conductor 1, a step increase in mmf is assumed to occur on the assumption that thickness of the conductor is negligible. As additional flux lines, based at A are drawn, threading the space between conductors 1 and 2, no additional mmf is enclosed. Hence in Fig. 3-4(c), the mmf wave stays constant at 15 A-t until conductor 2 is reached. By drawing a closed flux path from A that just encloses conductors 1 and 2, there occurs an additional increase in the mmf enclosed by such a path. This is shown as the second step of the curve in Fig. 3-4(c). A continuation of this procedure with A as the starting point, yields the resultant mmf wave of Fig. 3-4(c). Several points are worth noting. When a flux path is drawn to enclose conductors 1 to 4, the enclosed mmf is a maximum because for such a flux path, all the conductors carry current in the same direction. Hence this spatial integration process carried on up to conductor 4 makes all positive contributions. As the process is extended beyond conductor 4, the additional contributions become negative since the newly added conductors now involve current flowing in the opposite direction. Thus if a flux path is drawn from A to include conductors 1 to 5, the net encircled mmf is now the same as when the path is made to enclose conductors 1 to 3. Of course, a flux line that encircles conductors 1 to 8 cannot in fact exist because the net mmf is zero.

(b) Keep in mind that the closed paths to find the mmf wave by spatial integration in part (a) were based on two crossings of the air gap, which is the situation that prevails in practical machines by design. The double crossing, therefore, implies that

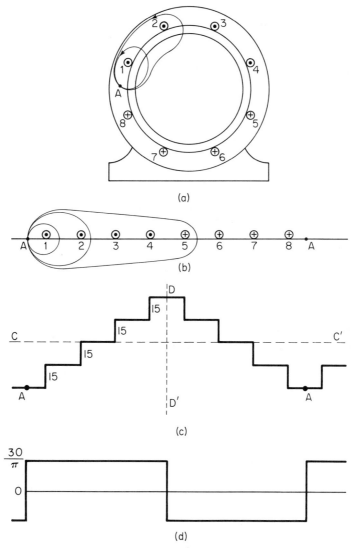

Figure 3-4 Determination of mmf distribution: (a) two-pole machine configuration; (b) plan view of the ampere-conductor distribution; (c) mmf distribution; (d) the equivalent current sheet.

the mmf associated with the closed line integrals involves a *pair* of poles. Accordingly, the mmf curve of Fig. 3-4(c) is for two poles with the mmf equally divided. This point can be stressed by drawing a line CC' in Fig. 3-4(c) that divides the area between the mmf curve and line CC' into equal positive and negative halves. In this case, the negative half represents north-pole flux and the positive half south-pole flux.

It is worthwhile to note that the spatial integration process could start at any point on the stator surface. All that is required in the end is that the resulting curve be divided equally between north- and south-pole fluxes as represented by the mmf's that produce these fluxes.

Once line CC' is properly established, it follows in this case that the maximum height of the mmf wave above (or below) line CC' is 30 A-t. Hence

$$\mathcal{F}_p = 30 \text{ A-t/pole}$$

(c) On the assumption that the presence of the iron in the closed flux paths produces negligible effects, the axis of the flux produced by the mmf wave is coincident with the axis of the mmf wave itself, which is along line DD' in Fig. 3-4(c).

(d) The equivalent current sheet is depicted in Fig. 3-4(d). The distribution is rectangular because the current carried by the conductors is the same and fixed in magnitude. Take note of the quadrature relationship that the ampere-conductor distribution (represented by the current sheet here) bears to the mmf distribution along the air gap of the machine.

We are now in a position to develop a basic torque formula as it aplies to a device that has a sinusoidally distributed flux density along the periphery of the air gap and an ampere-conductor distribution which for convenience is represented by a current-sheet density. Also, for generality, it is assumed that the current-sheet density is out of phase with the B-distribution by ψ degrees as depicted in Fig. 3-5. The procedure is really not much different from that which leads to Eq. (3-9), but several modifications are needed. We can again start with the fundamental force expression Eq. (3-7) and replace B with its sinusoidal variation $B_m \sin \alpha$. Furthermore, in place of the conductor and the current flowing through it, we now deal with a current element which is a differential strip of width $d\alpha$ of the current sheet. Accordingly, for i in Eq. (3-7) we may use

$$i = J \, d\alpha = J_m \sin(\alpha + \psi) \, d\alpha \tag{3-15}$$

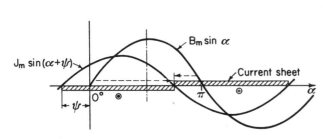

Figure 3-5 Configuration for development of the torque equation, showing the sinusoidally distributed flux density produced by the field winding and the current sheet, which represents the armature winding. The dashed arrows denote relative magnitude and the direction of the electromagnetic torque developed per pole. The magnetic field and the current sheet are stationary relative to each other.

The force appearing on the elemental current strip is then expressed as

$$F_e = Bli = (B_m \sin \alpha)l[J_m \sin(\alpha + \psi) \, d\alpha] \tag{3-16}$$

where the subscript e denotes elemental current strip. Of course the corresponding torque is obtained by multiplying by the rotor radius r. Hence

$$T_e = B_m J_m lr \sin \alpha \sin(\alpha + \psi) \, d\alpha \tag{3-17}$$

To determine the total torque developed on the current sheet beneath one pole of the field distribution, we must take the sum of the elemental torques produced by each element of the current sheet lying beneath the pole. In other words, it is necessary to integrate Eq. (3-17) over π electrical radians, which is the span of one pole. Hence

$$T_p = \sum \text{torques on current elements beneath one pole}$$

$$= B_m J_m l r \int_0^\pi \sin \alpha \sin (\alpha + \psi) \, d\alpha \qquad (3\text{-}18)$$

where T_p denotes the torque developed per pole. Clearly for a field distribution which has p poles the total resultant electromagnetic torque is

$$T = p B_m J_m l r \int_0^\pi \sin \alpha \sin (\alpha + \psi) \, d\alpha \qquad (3\text{-}19)$$

By introducing the trigonometric identity

$$\sin \alpha \sin (\alpha + \psi) = \tfrac{1}{2} \cos \psi - \tfrac{1}{2} \cos (2\alpha + \psi) \qquad (3\text{-}20)$$

into Eq. (3-19) we obtain

$$T = p B_m J_m l r \left[\int_0^\pi \tfrac{1}{2} \cos \psi \, d\alpha - \int_0^\pi \tfrac{1}{2} \cos (2\alpha + \psi) \, d\alpha \right] \qquad (3\text{-}21)$$

The second term in brackets has a value of zero, so the final expression for the developed electromagnetic torque is

$$T = \pi \frac{p}{2} J_m B_m l r \cos \psi \qquad (3\text{-}22)$$

Keep in mind that this equation is valid for sinusoidal distributions of the flux density and ampere conductors (or current sheet). Also note that for the indicated directions for the current sheet and magnetic flux, the developed torque is to the left and that its direction can be reversed when $\psi > 90°$.

An inspection of Eq. (3-22) reveals that the torque is dependent upon the space-displacement angle ψ. Thus when $\psi = 0°$ the best *field pattern* for developing torque prevails because the entire current sheet beneath a pole has a unidirectional torque. When $\psi = 45°$, as depicted in Fig. 3-5, note that a portion of the current sheet beneath a pole will produce an opposing torque, which results in a reduced torque if all other things are equal. It should not be inferred, however, that because $\psi \neq 0°$ the developed torque cannot be a maximum. In some a-c machines, operation at $\psi \neq 0°$ causes a manifold increase in J_m, which gives a considerable increase in developed torque in spite of a nonoptimum field pattern.

Often it is convenient to express Eq. (3-22) in terms of the total flux per pole, which is the flux associated with any one pole of the flux-density curve. One such pole is shown in Fig. 3-6. But before proceeding further it is necessary to distinguish between mechanical and electrical degrees. It has already been pointed out that, regardless of the number of poles in the flux-density curve, each pole spans π electrical

Figure 3-6 Relationships for computing flux per pole.

radians. Now, since there are always 2π mechanical radians in a configuration such as appears in Fig. 3-2(a), it follows that when the flux-density curve has two poles the mechanical and electrical degrees are the same. However, when the flux-density curve has four poles, then clearly the number of mechanical degrees is one-half the number of electrical degrees. For a p-pole flux-density curve the relationship is

$$\boxed{\alpha_m = \frac{2}{p}\alpha} \tag{3-23}$$

where α_m denotes mechanical degrees and α denotes electrical degrees. Therefore, in Fig. 3-6 the pole pitch expressed in meters is

$$\tau_p \equiv \text{pole pitch} = \alpha_m r = \frac{2}{p}\pi r \tag{3-24}$$

where r is the radius of the rotor in meters. Moreover, for a rotor axial length of l the area associated with a pole of the flux-density curve is

$$A_p = l\tau_p = \frac{2}{p}\pi l r \tag{3-25}$$

A further simplification in the computation leading to the expression for the flux per pole results upon recalling that the average value of the positive (or negative) half of a sine wave is

$$B_{av} = \frac{2}{\pi}B_m \tag{3-26}$$

Thus the sine wave of Fig. 3-6 may be replaced with a rectangular wave having a height of B_{av} and spanning the entire pole pitch as indicated by the broken line in Fig. 3-6. The expression for the flux per pole then follows from

$$\Phi = B_{av}A_p \tag{3-27}$$

Inserting Eqs. (3-25) and (3-26) into Eq. (3-27) yields

$$\Phi = \frac{4}{p}B_m l r \tag{3-28}$$

Again keep in mind that this result is valid for sinusoidal variations of flux density.

By means of Eq. (3-28) the equation for the developed electromagnetic torque may be expressed in terms of the flux per pole. Insertion of Eq. (3-28) into Eq. (3-22) yields

$$\boxed{T = \frac{\pi}{8}p^2\Phi J_m \cos\psi} \qquad \text{N-m} \tag{3-29}$$

A study of Eq. (3-29) points out more clearly the three conditions that must be satisfied for the development of torque in conventional electro-mechanical energy-conversion devices. There must be a field distribution represented by Φ. There must be an ampere-conductor distribution represented by J_m. Finally, there must exist a favorable space-displacement angle between the two distributions. Note that it is possible to have both a field and an ampere-conductor distribution and still the torque can be zero if the field pattern is such that equal and opposite torques are developed by the conductors, i.e., $\psi = 90°$.

It is also worthwhile to note that no really new concepts have been used to derive the torque formula of Eq. (3-29). It is merely an extension of Ampere's law.

Torque for Nonsinusoidal B and Uniform Ampere-Conductor Distribution: The Case of D-C Machines

The typical distributions are depicted in Fig. 3-7. Again for generality the current sheet that is representative of the ampere-conductor distribution is shown displaced in phase by the angle ψ. These distributions are commonly found in d-c machines,

Figure 3-7 Current-sheet and flux-density curves characteristic of d-c machines.

as is described later in the chapter. The starting point of the derivation is Ampere's law, again as expressed in Eq. (3-7). Of course modifications are needed consistent with the assumed distributions. Because the armature winding is represented by a current sheet, we deal with an elemental strip of the current sheet rather than with a particular conductor and the current it carries. Thus in Eq. (3-7) we introduce

$$i = J \, d\alpha \tag{3-30}$$

and

$$B = B_\alpha \tag{3-31}$$

where B_α denotes the nonsinusoidal variation of flux density along the air gap. An exact analytical expression for B_α is difficult to obtain, but this is not disturbing because one deals rather with the flux per pole, which is readily available. Hence the force developed by an elemental strip of the current sheet is

$$F_e = B_\alpha l J \, d\alpha \tag{3-32}$$

The corresponding torque for a p-pole machine is then

$$T_e = pJ B_\alpha lr \, d\alpha \qquad \text{N-m} \tag{3-33}$$

The total resultant torque is obtained upon integrating Eq. (3-33) over one pole pitch. Thus

$$T = pJ\left[\int_0^{\pi-\psi} B_\alpha lr \, d\alpha - \int_{\pi-\psi}^{\pi} B_\alpha lr \, d\alpha \right] \tag{3-34}$$

The second term in brackets carries the minus sign in order to account for the fact that J is negative beneath the positive pole of the B_α curve in the region from $\pi - \psi$ to π. Clearly, without an analytical expression for B_α, a closed-form solution for the torque is not readily available. However, in electro-mechanical energy-conversion devices where these distributions apply, the designer by intention makes $\psi = 0$.

Unlike the preceding case this always provides maximum torque.† Introducing $\psi = 0$ simplifies the expression for the electromagnetic torque to

$$T = pJ \int_0^\pi B_\alpha lr \, d\alpha \qquad \text{N-m} \qquad (3\text{-}35)$$

Now in the interest of avoiding B_α we can express the torque in terms of the flux per pole—a quantity which is available from other considerations. In terms of B_α the flux per pole Φ is

$$\Phi = \int_0^\pi B_\alpha \, dA_p \qquad (3\text{-}36)$$

But the area per pole can be expressed as

$$A_p = lr\alpha_m = lr\frac{2}{p}\alpha \qquad (3\text{-}37)$$

Hence Eq. (3-36) becomes

$$\Phi = \int_0^\pi B_\alpha lr \frac{2}{p} \, d\alpha = \frac{2}{p} \int_0^\pi B_\alpha lr \, d\alpha \qquad (3\text{-}38)$$

Therefore

$$\int_0^\pi B_\alpha lr \, d\alpha = \frac{p}{2}\Phi \qquad (3\text{-}39)$$

Substituting Eq. (3-39) into Eq. (3-35) yields the final form for the electro-magnetic torque developed when the distributions are as depicted in Fig. 3-7 with $\psi = 0$. Thus

$$\boxed{T = \frac{p^2}{2}J\Phi} \qquad \text{N-m} \qquad (3\text{-}40)$$

It is worthwhile to note again that electromagnetic torque is dependent upon a field distribution represented by Φ, an ampere-conductor distribution represented by J, and the angular displacement between the two distributions. Of course ψ does not appear in Eq. (3-40) because for simplicity the current sheet was assumed in phase with the flux-density curve, thereby ensuring that all parts of the current sheet experience a unidirectional torque. A glance at Fig. 3-7 should make it apparent that for $\psi = 90°$ the resultant electromagnetic torque is zero.

Example 3-2

A p-pole machine has the sinusoidal field distribution $B_p \sin(\alpha - \psi)$, where α and ψ are in electrical degrees. The armature consists of a uniform current sheet of value J, which has associated with it a triangular mmf having an amplitude per pole of F_A. The machine has an axial length l and a radius r.

(a) Determine the expression for the maximum flux/pole.

(b) Obtain the expression for the developed electromagnetic torque in terms of B_p, J, and δ.

(c) Express the torque found in part (b) in terms of F_A and the flux per pole Φ. Assume that $\delta = \pi/2$.

†See pp. 137 and 143.

(d) Is the torque found in part (b) constant at every point in the air gap? Explain.

Solution (a) The expression for the differential flux is given by

$$d\phi = B\,dA$$

where dA denotes a differential area beneath the flux density curve. This area is given by the product of the effective axial length l and the arc length, which is denoted by $r\,d\alpha_m$. Here use is made of the mechanical angle α_m because we are dealing with a physical differential arc length. See Fig. 3-8(b). By using Eq. (3-23), we can express the differential arc length in terms of electrical degrees as $(2r/p)\,d\alpha$. Accordingly, the differential flux can then be written as (for $\psi = 0$ for convenience in this part)

$$d\phi_p = \frac{2lrB}{p}\,d\alpha = \frac{2lrB_p}{p}\sin\alpha\,d\alpha$$

The total flux per pole can now be determined by integrating the last expression over one pole pitch which is equal to π electrical radians. Thus

$$\phi_p = \frac{2B_p lr}{p}\int_0^\pi \sin\alpha\,d\alpha = \frac{4B_p lr}{p}$$

Note that this result entirely agrees with Eq. (3-28).

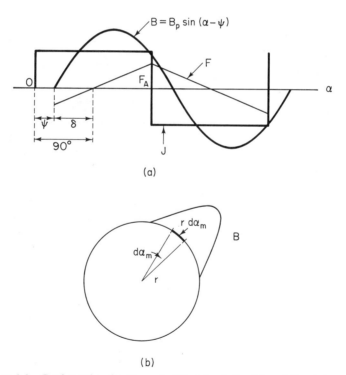

(a)

(b)

Figure 3-8 Configuration for Example 3-2: (a) relationship existing between flux density, mmf, and current sheet; (b) a view of the rotor showing one pole and the differential arc length $r\,d\alpha_m$.

(b) Let i denote the differential current element, which for a uniform current sheet of value J can be written as

$$i = J \, d\alpha$$

Associated with this differential current element under the influence of the pole flux density is a differential force $Bli = BJl \, d\alpha$. When acting on a torque arm of the machine radius r, the differential torque expression becomes

$$BJlr \, d\alpha$$

By integrating this quantity over one pole pitch (i.e., π electrical radians), the net torque produced by the interaction of the field distribution and the current sheet is determined for one pole. The total torque results upon multiplying by the number of poles. Thus the expression for the total differential torque is

$$dT = p(BJlr \, d\alpha)$$

Or the total torque produced becomes

$$T = plrJ \int_0^\pi B \, d\alpha = plrJB_p \int_0^\pi \sin(\alpha - \psi) \, d\alpha$$

Upon integrating, inserting the limits and simplifying, the result is

$$T = 2B_pJlr \cos \psi$$

Because ψ and δ are complementary angles, we get

$$\psi = 90° - \delta$$

which when introduced into the preceding equation and simplified yields

$$T = 2plrB_pJ \sin \delta$$

(c) For $\delta = \pi/2$ the last equation simplifies to

$$T = p2B_plrJ$$

Algebraic manipulation of this last expression leads to expressions which can be identified as the flux per pole and the amplitude of the mmf per pole. The flux per pole is readily identified by rearranging the last expression to read as follows:

$$T = p^2\left(\frac{4}{p}B_plr\right)\frac{J}{2} = p^2\phi_p\frac{J}{2}$$

Clearly, the quantity in parentheses is the flux per pole.

To get F_A into this formulation, it is necessary to recognize that when $\delta = \pi/2$, the total mmf per pair of poles is given by [See Fig. 3-17(a)]

$$\mathcal{F} = \pi J$$

Hence the amplitude per pole is then

$$F_A = \frac{\mathcal{F}}{2} = \frac{\pi J}{2}$$

Accordingly, by rewriting the torque equation in the further manipulated form

$$T = p^2\phi_p\frac{1}{\pi}\left(\frac{\pi J}{2}\right)$$

We get finally

$$T = \frac{p^2}{\pi}\phi_pF_A$$

(d) No. Although the current sheet is of constant value, the flux density is not. Therefore, the instantaneous value of torque varies with the value of B.

3-2 ANALYSIS OF INDUCED VOLTAGES

The electromechanical energy-conversion process involves the interaction of two fundamental quantities which are related through the law of conservation of energy and the coupling magnetic field. These are, of course, electromagnetic torque and induced voltage. The manner in which electromagnetic torque is developed is described in the preceding section. Here we consider the generation of voltage in the armature winding brought about by changing flux linkages. The polarity of this voltage depends upon the direction of the relative motion of the winding with respect to the magnetic flux and is independent of whether the machine is a motor or a generator, i.e., the direction of energy flow.

Consider the configuration shown in Fig. 3-9. The flux-density curve is assumed to be sinusoidally distributed along the air gap of the machine, and the armature winding is assumed to consist of a full-pitch coil having N turns. A full-pitch coil is one that spans π electrical radians. Just as Ampere's law serves as the starting point in the development of the torque relationships, so Faraday's law serves as the starting point in establishing the induced-voltage relationships. Equation (3-5) is not the result we are seeking, because it applies when the flux-density curve is uniform over the pole pitch. It is more generally useful to deal with a sinusoidal distribution of the flux density, as explained more fully later. Since the energy-conversion process involves relative motion between the field and the armature winding, we know that the flux linking the coil changes from a positive maximum to zero to a negative maximum and back to zero again. This cycle repeats every time the coil passes through a pair of flux-density poles. It should be clear from Fig. 3-9 that maximum flux penetrates the coil when it is in a position corresponding to $\beta = 0°$ or $\beta = 180°$ or multiples thereof. Minimum flux linkage occurs when $\beta = 90°$ or $270°$ or multiples thereof. It should be apparent too that the maximum flux that links the coil is the same as the flux per pole, which for sinusoidal distributions is given by Eq. (3-28). Moreover, as the coil moves relative to the field, it follows that the instantaneous flux linking the coil may be expressed as

$$\phi = \Phi \cos \beta = \frac{4}{p} B_m lr \cos \beta \qquad (3-41)$$

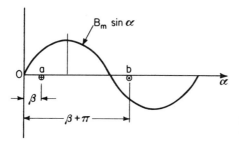

Figure 3-9 Configuration for derivation of the induced-emf equation.

The cos β factor is included because of the sinusoidal distribution. Furthermore, since β increases with time in accordance with

$$\beta = \omega t \qquad (3\text{-}42)$$

where ω is expressed in electrical radians per second, the expression for the instantaneous flux linkage becomes

$$\phi = \frac{4}{p} B_m l r \cos \omega t = \Phi \cos \omega t \qquad (3\text{-}43)$$

Then by Faraday's law the expression for the instantaneous induced voltage becomes

$$\boxed{e = -N\frac{d\phi}{dt} = \omega N \Phi \sin \omega t} \qquad (3\text{-}44)$$

It is important to keep in mind that Φ represents the *maximum flux*, which is the same as the flux per pole.

Equation (3-44) is applicable to d-c as well as a-c machines. In a later section the induced emf per phase in an a-c machine and the induced voltage appearing between brushes in a d-c machine are derived starting with Eq. (3-44). The fact that in the d-c machine the flux-density curve is nonsinusoidal, as Fig. 3-7 shows, is relatively unimportant; what is of primary importance is the fundamental component of the nonsinusoidal distribution. It is for this reason that Eq. (3-44) is derived for sinusoidal distributions. Recall that by means of the Fourier series any nonsinusoidal periodic wave may be expressed entirely in terms of sinusoids.

Example 3-3

The purpose of this example is to draw a distinction in the character of voltage that is induced in a revolving coil when it finds itself under the influence of a *vertically directed* uniform field as illustrated in Fig. 3-10(a) and a *radially directed* uniform field

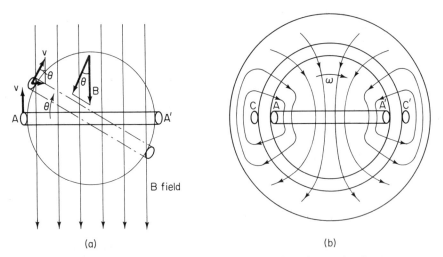

(a)   (b)

Figure 3-10 Calculation of induced voltages: (a) coil AA' operating in a vertically directed field; (b) coil AA' operating in a radially directed field.

as occurs in the configuration of Fig. 3-10(b). Assume that coil AA' is square and measures 30 cm on a side with a total of 40 turns. The coil is driven at a constant speed of 200 rad/s. The value of the uniform magnetic field in both cases is 0.1 T.

(a) Obtain an expression for the flux linkage through the coil as a function of time.

(b) Find the expression for the voltage induced in the coil as a function of time using Faraday's law.

(c) Determine the result found in part (b) by employing the Blv concept using Eq. (3-5).

(d) Now assume that coil AA is put under the influence of a radially directed field as shown in Fig. 3-10(a). Coil CC' is assumed to carry a fixed unidirectional current, thereby producing a constant field distribution over the entire periphery of the air gap. Find the coil emf in this case as a function of time.

Solution (a) Let time be counted starting with the coil in the horizontal position in Fig. 3-10(a). As time elapses, the coil, which is rotating, takes a new position displaced by $\theta = \omega t$ degrees from its reference position. In this new position, the component of the vertical field that threads the coil in a quadrature fashion is

$$B \cos \theta = B \cos \omega t$$

The corresponding amount of flux that links with the coil is described by

$$BA \cos \theta = BA \cos \omega t$$

or

$$\phi_m \cos \theta = \phi_m \cos \omega t = \phi$$

where ϕ_m denotes the maximum flux that links with the coil. But flux linkage is simply this flux multiplied by the number of turns in the coil. Thus

$$\lambda = N\phi = N\phi_m \cos \omega t = 40(0.1)(0.3)(0.3) \cos \omega t$$

$$= 0.36 \cos 200t \text{ Wb-t}$$

(b) By Faraday's law, we know that the induced emf is equal to the rate of change of flux linkage with time. Hence

$$e = -\frac{d\lambda}{dt} = -\frac{d}{dt}(0.36 \cos \omega t) = (0.36)(200) \sin 200t$$

$$= 72 \sin 200t \text{ V}$$

This result states that the emf varies sinusoidally with time.

(c) With the Blv approach it is the cutting action of the flux lines by the conductors that receives the focus. When the coil is in the horizontal position, the conductors fail to be under the influence of the B field, so the instantaneous value of the voltage is zero. The coil sides are not cutting the flux lines. It is important to note that at this position, the velocity vector of the conductor is collinear with the field. Hence the component that is in quadrature with the field lines is zero and so no cutting occurs. As time is allowed to elapse and the coil takes a position θ degrees from its horizontal reference point, the quadrature component of the velocity vector is no longer zero. Rather, it is equal to $v \sin \theta = v \sin \omega t$, as can be seen from Fig. 3-10(a). Employing only the useful part of the velocity vector then permits the induced emf to be written as

$$e = 2NBlv \sin \omega t = 2NBl\omega r \sin \omega t$$

$$= 2(40)(0.1)(0.3)(200)(0.15) \sin \omega t$$

$$= 72 \sin \omega t$$

(d) Unlike the situation that occurs in the geometry of Fig. 3-10(a), the use of a radially directed field means that the velocity vector is at all times in a position of quadrature with the *B*-field. Hence the voltage induced in this case is

$$e = 2NBlv = 2NBl\omega r = 2(40)(0.1)(0.3)(200)(0.15) = 72 \text{ V}$$

This quantity is constant in magnitude with time and specifically it is equal to the peak value found in part (c). Another point of interest associated with the geometry of Fig. 3-10(b) compared to that of Fig. 3-10(a) is that flux linkage in the case of Fig. 3-10(b) is not described by the cosine function but rather in the manner already described in Sec. 3-1.

3-3 CONSTRUCTION FEATURES OF ELECTRIC MACHINES

The energy-conversion process usually involves the presence of two important features in a given electromechanical device. These are the field winding, which produces the flux density, and the armature winding, in which the working emf is induced. In this section the salient construction features of the principal types of electric machines are described to show the location of these windings, as well as to demonstrate the general composition of such machines.

The Three-Phase Induction Motor

This is one of the most rugged and most widely used machines in industry. Its stator is composed of laminations of high-grade sheet steel. The inner surface is slotted to accommodate a three-phase winding. In Fig. 3-11(a) the three-phase winding is represented by three coils, the axes of which are 120 electrical degrees apart. Coil *aa'* represents all the coils assigned to phase *a* for one pair of poles. Similarly, coil *bb'* represents phase *b* coils, and coil *cc'* represents phase *c* coils. When one end of each phase is commonly connected, as depicted in Fig. 3-11(b), the three-phase stator winding is said to be Y-connected. Such a winding is called a three-phase winding because the voltages induced in each of the three phases by a revolving flux-density field are out of phase by 120 electrical degrees—a distinguishing characteristic of a symmetrical three-phase system.

The rotor also consists of laminations of slotted ferromagnetic material, but the rotor winding may be either the squirrel-cage type or the wound-rotor type. The

(a)   (b)

Figure 3-11 Three-phase induction motor: (a) showing a stator with three-phase winding and squirrel-cage rotor; (b) schematic representation of a three-phase Y-connected stator winding.

latter is of a form similar to that of the stator winding. The winding terminals are brought out to three slip rings as depicted in Fig. 3-12(a). This allows an external three-phase resistor to be connected to the rotor winding for the purpose of providing speed control. As a matter of fact it is the need for speed control which in large measure accounts for the use of the wound-rotor type of induction motor. Otherwise the squirrel-cage induction motor would be used. The squirrel-cage winding consists merely of a number of copper bars imbedded in the rotor slots and connected at both ends by means of copper end rings as depicted in Fig. 3-12(b). (In some of the smaller sizes aluminum is used.) The squirrel-cage construction is not only simpler and more economical than the wound-rotor type but more rugged as well. There are no slip rings or carbon brushes to be bothered with.

In normal operation a three-phase voltage is applied to the stator winding at points *a-b-c* in Fig. 3-11. As described in Sec. 4-1, magnetizing currents flow in each phase which together create a revolving magnetic field having two poles. The speed of the field is fixed by the frequency of the magnetizing currents and the number of

Figure 3-12 (a) Wound rotor for the three-phase induction motor; (b) cutaway view, showing bars in the rotor of the squirrel-cage induction motor. (Courtesy of General Electric Company.)

poles for which the stator winding is designed. Figure 3-11 shows the configuration for two poles. If the pattern a-c'-b-a'-c-b' is made to span only 180 mechanical degrees and then is repeated over the remaining 180 mechanical degrees, a machine having a four-pole field distribution results. For a p-pole machine the basic winding pattern must be repeated $p/2$ times within the circumference of the inner surface of the stator.

The revolving field produced by the stator winding cuts the rotor conductors, thereby inducing voltages. Since the rotor winding is short-circuited by the end rings, the induced voltages cause currents to flow which in turn react with the field to produce electromagnetic torque—and so motor action results.

Accordingly, on the basis of the foregoing description, it should be clear that for the three-phase induction motor the field winding is located on the stator and the armature winding on the rotor. Another point worth noting is that this machine is singly excited, i.e., electrical power is applied only to the stator winding. Current flows through the rotor winding by induction. As a consequence both the magnetizing current, which sets up the magnetic field, and the power current, which allows energy to be delivered to the shaft load, flow through the stator winding. For this reason, and in the interest of keeping the magnetizing current as small as possible in order that the power component may be correspondingly larger for a given rating, the air gap of induction motors is made as small as mechanical clearance will allow. The air-gap lengths vary from about 0.02 in. for smaller machines to 0.05 in. for machines of higher rating and speed.

Synchronous Machines

The essential construction features of the synchronous machine are depicted in Fig. 3-13. The stator consists of a stator frame, a slotted stator core, which provides a low-reluctance path for the magnetic flux, and a three-phase winding imbedded in the slots. Note that the basic two-pole pattern of Fig. 3-11(a) is repeated twice, indicating that the three-phase winding is designed for four poles. The rotor either is cylindrical and equipped with a distributed winding or else has salient poles with a coil wound on each leg as can be seen in Fig. 3-13(b). The cylindrical construction is used almost exclusively for turbogenerators which operate at high speeds. On the other hand, the salient-pole construction is used exclusively for synchronous motors operating at speeds of 1800 rpm or less.

When operated as a generator the synchronous machine receives mechanical energy from a prime mover such as a steam turbine and is driven at some fixed speed. Also, the rotor winding is energized from a d-c source, thereby furnishing a field distribution along the air gap. When the rotor is at standstill and direct current flows through the rotor winding, no voltage is induced in the stator winding because the flux is not cutting the stator coils. However, when the rotor is being driven at full speed, voltage is induced in the stator winding and upon application of a suitable load electrical energy may be delivered to it.

A little thought should make it apparent that for the synchronous machine the field winding is located on the rotor; the armature winding is located on the stator. This conclusion is valid even when the synchronous machine operates as a motor. In this mode a-c power is applied to the stator winding and d-c power is applied to the

(a)

(b)

Figure 3-13 Salient pole synchronous machine: (a) schematic diagram; (b) photograph showing field coils wrapped around salient poles. Note too the squirrel-cage winding embedded in the pole faces.

rotor winding for the purpose of energizing the field poles. Mechanical energy is then taken from the shaft. Note, too, that unlike the induction motor, the synchronous motor is a doubly fed machine. In fact it is this characteristic which enables this machine to develop a nonzero torque at only one speed—hence the name "synchronous." This matter is discussed further in Chapter 5.

Because the magnetizing current for the synchronous machine originates from a separate source (the d-c supply), the air-gap lengths are larger than those found in induction motors of comparable size and rating. However, synchronous machines are more expensive and less rugged than induction motors in the smaller horsepower ratings because the rotor must be equipped with slip rings and brushes in order to allow the direct current to be conducted to the field winding.

Direct-Current Machines

Electromechanical energy-conversion devices that are characterized by direct current are more complicated than the a-c type. In addition to a field winding and armature winding a third component is needed to serve the function of converting the induced a-c armature voltage into a direct voltage. Basically the device is a mechanical rectifier and is called a *commutator*.

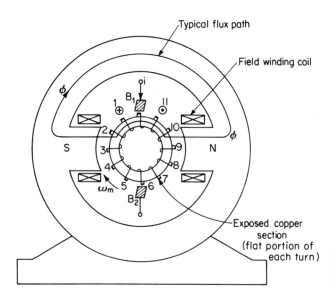

Figure 3-14 Construction features of the d-c machine, showing the Gramme-ring armature winding.

Appearing in Fig. 3-14 are the principal features of the d-c machine. The stator consists of an unlaminated ferromagnetic material equipped with a protruding structure around which coils are wrapped. The flow of direct current through the coils establishes a magnetic field distribution along the periphery of the air gap in much the same manner as occurs in the rotor of the synchronous machine. Hence in the d-c machine the field winding is located on the stator. It follows then that the armature winding is on the rotor. The rotor is composed of a laminated core, which is slotted to accommodate the armature winding. It also contains the commutator—a series of copper segments insulated from one another and arranged in cylindrical fashion as shown in Fig. 3-15. Riding on the commutator are appropriately placed carbon brushes which serve to conduct direct current to or from the armature winding depending upon whether motor or generator action is taking place.

In Fig. 3-14 the armature winding is depicted as a coil wrapped around a toroid. This is merely a schematic convenience. An actual winding is wound so that no conductors are wasted by placing them on the inner surface of the rotor core where no flux penetrates. In Fig. 3-14 those parts of the armature winding which lie directly below the brush width are assumed to have the insulation removed, i.e., the copper is exposed. This allows current to be conducted to and from the armature winding through the brush as the rotor revolves. In a practical winding each coil is made

Figure 3-15 Armature winding of a d-c machine, showing the coils connected to the commutator. (Courtesy of General Electric Company.)

accessible to the brushes by connecting the coils to individual commutator segments as indicated in Fig. 3-15 and then placing the brushes on the commutator.

For motor action direct current is made to flow through the field winding as well as the armature winding. If current is assumed to flow into brush B_1 in Fig. 3-14, then note that on the left side of the rotor for the *outside* conductors current flows into the paper while the opposite occurs for the conductors located on the outside surface of the right side of the rotor. By Eq. (1-2) a force is produced on each conductor, thereby producing a torque causing a clockwise rotation. Now the function of the commutator is to ensure that, as a conductor such as 1 in Fig. 3-14 revolves and thus goes from the left side of brush B_1 to the right side, the current flowing through it reverses, ensuring a continuous unidirectional torque for the entire armature winding. Recall that a reversed conductor current in a flux field of reversed polarity keeps the torque unidirectional. The reversal of current comes about because the commutator always allows current to be conducted in the same directions in either side of the armature winding whether or not it is rotating.

Another point of interest in Fig. 3-14 concerns the location of the brushes. By placing the brushes on a line perpendicular to the field axis all conductors contribute in producing a unidirectional torque. If, on the other hand, the brushes were placed on the same line as the field axis, then half of the conductors would produce clockwise torque and the other half counterclockwise torque, yielding a zero net torque. In the notation of Fig. 3-7 this situation would correspond to a value of ψ equal to 90°.

3-4 PRACTICAL FORMS OF TORQUE AND VOLTAGE FORMULAS

Our objective here is to modify the basic expressions for induced voltage and electro-magnetic torque to forms that are more meaningful for the specific data of a particular electromechanical energy-conversion device, whether it is alternating or direct current.

Alternating-Current Machines

If N denotes the total number of turns per phase of a three-phase winding, the instantaneous value of the emf induced in any one phase can be represented by Eq. (3-44), which is repeated here for convenience. Thus

$$e = \omega N \Phi \sin \omega t \qquad (3\text{-}45)$$

Keep in mind that Φ denotes the total flux per pole and ω represents the relative cutting speed in electrical radians per second of the winding with respect to the flux-density wave. It is related to the frequency f of the a-c device by

$$\omega = 2\pi f \qquad (3\text{-}46)$$

where f is expressed in cycles per second.

The maximum value of this a-c voltage occurs when $\sin \omega t$ has a value of unity. Hence

$$E_{\max} = \omega N \Phi \qquad \text{V} \qquad (3\text{-}47)$$

The corresponding rms value is

$$E \equiv \frac{E_{\max}}{\sqrt{2}} = \sqrt{2}\,\pi f N \Phi = 4.44 f N \Phi \qquad \text{V} \qquad (3\text{-}48)$$

A comparison of this expression with Eq. (2-18) reveals that the equations have an identical form. There is a difference, however, and it lies in the meaning of Φ. In the transformer, Φ_m is the maximum flux corresponding to the peak magnetizing current consistent with the magnitude of the applied voltage. In the a-c electromechanical energy-conversion device Φ is the maximum flux per pole that links with a coil having N turns and spanning the full pole pitch.

In any practical machine the total turns per phase are not concentrated in a single coil but rather are distributed over one-third of a pole pitch (or 60 electrical degrees for each of the three phases). In addition the individual coils that make up the total N turns are intentionally designed to span not the full pole pitch but rather only about 80 to 85% of a pole pitch. Such a coil is called a *fractional-pitch coil.* The use of a *distributed* winding employing *fractional-pitch* coils has the advantage of virtually eliminating the effects of all harmonics that may be present in the flux-density wave while only slightly reducing the fundamental component.† The reduction in the fundamental component can be represented by a winding factor denoted by K_w. Usually K_w has values ranging from 0.85 to 0.95. Accordingly, the final practical version of the induced rms voltage equation for an a-c machine is

$$\boxed{E = 4.44 f N K_w \Phi} \qquad \text{V} \qquad (3\text{-}49)$$

†See Appendix B.

There are two useful forms to which the basic expression for electromagnetic torque in a-c machines as expressed by Eq. (3-29) may be converted. In one form the ampere-conductor distribution of the armature winding is emphasized. In the other it is the conservation of energy which is stressed.

Since $\mathcal{F}$ in Eq. (3-13) is the mmf per pair of poles, it follows that J_m for sinusoidal current sheets is equal to the armature mmf per pole, denoted by $\mathcal{F}_p$. Thus

$$J_m = \mathcal{F}_p = \frac{\mathcal{F}}{2} \tag{3-50}$$

For any given a-c machine the quantity $\mathcal{F}_p$ is entirely determinable from the design data as they appear in the following equation:†

$$\mathcal{F}_p = \frac{2\sqrt{2}}{\pi} q \frac{N_2 K_{w2}}{p} I_2 = 0.9q \frac{N_2 K_{w2}}{p} I_2 \tag{3-51}$$

where q = number of phases of the armature winding
N_2 = number of turns per phase of the armature winding
K_{w2} = armature-winding factor
I_2 = armature-winding current per phase
p = number of poles

Hence the expression for electromagnetic torque as described by Eq. (3-29) becomes

$$T = \frac{\pi}{8} p^2 \Phi \mathcal{F}_p \cos \psi = \frac{\pi}{8} 0.9 p \Phi N_2 K_{w2} q I_2 \cos \psi \tag{3-52}$$

Recall too that qN_2 is the *total* number of turns on the armature surface. This can be expressed in terms of the total number of conductors, Z_2, by making use of the fact that it requires two conductors to make one turn. Thus

$$N_2 = \frac{Z_2}{2q} \tag{3-53}$$

Inserting this into Eq. (3-52) yields

$$\boxed{T = 0.177 p \Phi (Z_2 K_{w2} I_2) \cos \psi} \qquad \text{N-m} \tag{3-54}$$

The quantity in parentheses emphasizes the role played by the ampere-conductor distribution of the armature winding. Of course in order to compute the torque by this equation we need information about the space-displacement angle ψ in addition to the design data, i.e., p, Φ, Z_2, K_{w2}. Normally, ψ can be determined by the same data that lead to the determination of I_2, which is described later in the next chapter.

A second approach to obtaining a useful form of the basic torque equation is to replace Φ with Eq. (3-49) and J_m with Eqs. (3-50) and (3-51) in Eq. (3-29). Thus

$$T = \frac{\pi}{8} p^2 \left(\frac{E_2}{\sqrt{2} \pi f N_2 K_{w2}} \right) \left(2 \frac{\sqrt{2}}{\pi} q \frac{N_2 K_{w2}}{p} I_2 \right) \cos \psi$$

or

$$T = \frac{p}{4\pi f} q E_2 I_2 \cos \psi \tag{3-55}$$

†See Appendix C.

Moreover, since the mechanical angular velocity ω_m is related to the electrical angular velocity ω by

$$\omega_m = \frac{2}{p}\omega = \frac{2}{p}2\pi f = \frac{4\pi f}{p} \qquad (3\text{-}56)$$

it follows that Eq. (3-55) may be written as

$$T = \frac{1}{\omega_m}qE_2I_2 \cos\psi \qquad \text{N-m} \qquad (3\text{-}57)$$

where E_2 and I_2 are the rms induced voltage and current per phase of the armature winding.

It can be shown that the *space*-displacement angle ψ in this equation is identical with the time-displacement angle θ_2,† which is the phase angle existing between the two time phasors E_2 and I_2. Accordingly, the expression for the electromagnetic torque may be written as

$$\boxed{T = \frac{1}{\omega_m}qE_2I_2 \cos\theta_2} \qquad \text{N-m} \qquad (3\text{-}58)$$

Note the power balance, which states that the developed mechanical power is equal to the developed electrical power.

Direct-Current Machines

The useful version of the expression for the emf induced in the armature winding of a d-c machine and appearing at the brushes also derives from Eq. (3-44). The presence of the commutator, however, makes a difference in the manner in which this equation is to be treated to obtain the desired result. We are interested in the voltage as it appears at the brushes. If we follow coil 1 in Fig. 3-14 as it leaves brush B_2 and advances to brush B_1, we notice that the emf induced in this coil is always directed out of the paper beneath the south-pole flux. In fact this is true of any coil that moves from B_2 to B_1. Having established the direction of this voltage for the coils on the left side of the armature, we then note that upon tracing through the winding starting in the direction called for by the sign of the emf induced in coil 1 we terminate at brush B_2. When coil 1 rotates to a position on the right side of the armature winding, such as position 11, the direction of the induced emf is into the paper. If we start with this indicated direction and trace the armature winding, we again terminate at brush B_2. Therefore it is reasonable to conclude that brush B_1 is at a fixed *positive* polarity and brush B_2 is at a fixed *negative* polarity. In essence this means that, if we were to ride along as observers on coil 1 as it traverses through one revolution beginning at brush B_2, the fundamental component of the induced emf would exhibit the variation shown in Fig. 3-16. Note that *relative to brush B_1* the voltage induced in the coil as it moves from B_1 to B_2 appears *rectified* because the action of the commutator in conjunction with the brushes is to fix the armature winding in space in spite of its rotation. Accordingly, whether we look at a coil moving from B_2 to B_1 on the left side or from B_1 to B_2 on the

†See page 185.

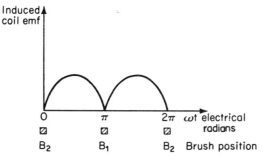

Figure 3-16 Fundamental component
of the induced emf in a coil of the
armature winding of the d-c machine.

right side, the directions of the induced emf's are such as to make B_1 assume one fixed polarity and B_2 the opposite fixed polarity.

For a single coil having N_c turns the average value of the unidirectional voltage appearing between the brushes is obtained by integrating Eq. (3-44) over π electrical radians and dividing by π. Thus

$$E_c = \frac{1}{\pi} \int_0^\pi \omega N_c \Phi \sin \omega t \, d\omega t = 4 f N_c \Phi \tag{3-59}$$

It is customary to express the frequency in terms of the speed of the armature in *rpm* which is given by the relationship†

$$\boxed{f = \frac{pn}{120}} \tag{3-60}$$

where p denotes the number of poles and n the speed in rpm. Inserting Eq. (3-60) into Eq. (3-59) and rearranging leads to

$$E_c = p\Phi \, 2N_c \frac{n}{60} = p\Phi z \frac{n}{60} \tag{3-61}$$

where z denotes the number of conductors per coil.

Equation (3-61) is the average value of the induced emf for a single coil. If many such coils are placed to cover the entire surface of the armature, the total d-c voltage appearing between the brushes can be considerably increased. If the armature winding is assumed to have a total of Z conductors and a parallel paths, then the induced emf of the armature winding appearing at the brushes becomes

$$\boxed{E_a = p\Phi \frac{Z}{a} \frac{n}{60} = \frac{pZ}{60a} \Phi n = K_E \Phi n} \tag{3-62}$$

where K_E is a winding constant defined as

$$K_E \equiv \frac{pZ}{60a} \tag{3-63}$$

---

†Rotating a coil in a two-pole field at a rate of 1 rps results in a frequency of 1 Hz. For a four-pole field each rps yields 2 Hz. For a p-pole distribution the frequency generated in hertz is

$$f = \frac{p}{2} \times \text{rps} = \frac{p}{2} \frac{\text{rpm}}{60} = \frac{pn}{120}$$

Whenever the speed is expressed in radians per second, the induced emf equation can then be expressed as

$$E_a = K_T \Phi \omega \qquad (3\text{-}62a)$$

where

$$K_T = \frac{pZ}{2\pi a} \qquad (3\text{-}63a)$$

It is worth keeping in mind that the induced-emf equation for the a-c machine and that for the d-c machine originate from the same starting point. To further emphasize this common origin let it be said that we can derive Eq. (3-62) from Eq. (3-49). It merely requires introducing the appropriate winding factor as it applies to the d-c armature winding. In fact, it is the quasi-annihilating effect of the winding factor on the harmonics that allows the derivation of Eq. (3-62) to proceed in the fundamental terms in spite of the nonsinusoidal field distribution characteristic of d-c machines. It is assumed throughout, however, that the fundamental component of flux is virtually the same as the total flux per pole.

A practical version of the electromagnetic torque formula readily follows from Eq. (3-40) if we replace the current-density sheet J with an equivalent expression involving the design data of the machine. In this connection refer to Fig. 3-17(a).

(a)   (b)

Figure 3-17 Direct-current machine: (a) armature with current sheet; (b) armature with finite ampere-conductor distribution.

Applying Ampere's circuital law to the typical flux path shown makes it apparent that the mmf per pole pair is given by

$$\mathcal{F} = \pi \, (\text{rad}) \, J\!\left(\frac{\text{A-t}}{\text{rad}}\right) \qquad (3\text{-}64)$$

Hence the mmf per pole is

$$\mathcal{F}_p = \frac{\mathcal{F}}{2} = \frac{\pi J}{2} \qquad (3\text{-}65)$$

This expression is applicable whenever the armature winding is represented by a current sheet. In a practical situation the armature winding is always represented by a finite ampere-conductor distribution as depicted in Fig. 3-17(b). In such cases the armature mmf per pole may be expressed in terms of the total current I_a entering or leaving a brush and the total conductors Z on the armature surface. Thus

$$\mathcal{F}_p = \frac{I_a}{a} \frac{Z}{2} \frac{1}{p} \qquad \text{mmf/pole, or A-t/pole} \qquad (3\text{-}66)$$

Upon equating the last two expressions we obtain the equation for J. Thus

$$J = \frac{I_a Z}{\pi p a} \qquad (3\text{-}67)$$

Insertion of Eq. (3-67) into Eq. (3-40) then yields the desired expression for the electromagnetic torque developed by the d-c machine. Hence

$$T = \frac{p^2}{2} J\Phi = \frac{pZ}{2\pi a}\Phi I_a \qquad (3\text{-}68)$$

or

$$\boxed{T = K_T \Phi I_a} \qquad \text{N-m} \qquad (3\text{-}69)$$

where

$$K_T \equiv \frac{pZ}{2\pi a} = \text{torque constant} \qquad (3\text{-}70)$$

The foregoing form of the torque formula stresses the ampere-conductor distribution taken in conjunction with the flux field. It is analogous to Eq. (3-54) for the a-c machine. Of course in Eq. (3-69) ψ does not appear because it was made zero by placing the brush axis in quadrature with the field axis. It is important to note, however, that the fundamental quantities for the production of torque are there. The equations differ only to the extent that the mechanical details of construction differ.

An alternative expression for the electromagnetic torque results upon substituting in Eq. (3-68) the expression for Φ obtained from Eq. (3-62). Thus from Eq. (3-62)

$$\Phi = \frac{60a}{pZn} E_a \qquad (3\text{-}71)$$

so that

$$T = \frac{pZ}{2\pi a} \frac{60a}{pZn} E_a I_a = \frac{60}{2\pi n} E_a I_a \qquad (3\text{-}72)$$

But

$$\omega_m = \frac{2\pi n}{60} \qquad (3\text{-}73)$$

Therefore

$$\boxed{T = \frac{1}{\omega_m} E_a I_a} \qquad (3\text{-}74)$$

A glance at this result again points out the power-balance relationship that underlies the operation of electromechanical energy-conversion devices. Equation (3-74) for the d-c machine is analogous to Eq. (3-58) for the a-c machine.

Example 3-4

An a-c machine is equipped with four poles, three phases, and 36 armature winding turns per phase. The field winding is designed to produce a flux per pole of 0.015 Wb. The armature winding is designed to be essentially resistive over its operating range and the coil span is made equal 83.3% of a pole pitch.

(a) When the motor develops a torque of 115 N-m, find the armature current in amperes.

(b) Determine the armature mmf per pole.

(c) Find the developed torque when the armature current is 15 A.

(d) In part (a) the developed power is known to be equal to 21,677 W. Calculate the armature induced voltage per phase.

(e) Compute the mechanical speed of the motor in rpm for the condition of part (a).

Solution (a) We can use Eq. (3-54) directly here to get the current I_2, which is the armature current. But first two matters need to be resolved. The first is the angle ψ. By the problem statement this quantity is $0°$ because the inductive component of the armature winding is considered to be negligible. Second, we need to determine K_{w2}. This quantity is readily calculated by invoking Eq. (B-2) of Appendix B. Thus

$$K_{w2} = \sin (0.833)\frac{\pi}{2} = \sin 75° = 0.966$$

It therefore follows that the armature current is

$$I_2 = \frac{T}{0.177p\phi Z_2(0.966)} = \frac{115}{0.177(4)(0.015)(2 \times 3 \times 36)(0.966)} = 51.9 \text{ A}$$

(b) The armature mmf per pole is found from

$$\mathcal{F}_p = J_m = 0.9q\frac{N_2K_{w2}}{p}I_2 = 0.9(3)\frac{36(0.966)}{4}51.9 = 1218.3 \text{ A-t}$$

(c) Using the second form of Eq. (3-52) we get

$$T = \frac{\pi}{8}(0.9)p\phi N_2K_{w2}qI_2 \cos \psi$$

$$= \frac{\pi}{8}(0.9)(4)(0.015)(26)(0.966)(3)(15) = 33.2 \text{ N-m}$$

(d) From Eq. (3-58)

$$3E_2I_2 = 21,677$$

$$E_2 = \frac{21,677}{3(51.9)} = 139.2 \text{ V per phase}$$

(e) By conservation of energy

$$T\omega_m = 21,677$$

$$\omega_m = \frac{21,677}{115} = 188.5 \text{ rad/s}$$

or, expressed in rpm, we write

$$n_m = \frac{30}{\pi}\omega_m = 1800 \text{ rpm}$$

where n_m denotes the mechanical speed in revolutions per minute.

3-5 THE GENERAL TORQUE EQUATION

The formulas for the electromagnetic torque that are derived in Secs. 3-1 and 3-4 apply in situations where the field distribution in one part of an electric machine interacts with the ampere-conductor distribution in the other part. Torque is produced in the machine because of the mutual action existing between the windings to bring about an alignment of their associated fields. As long as currents exist in each winding and the field pattern is proper, there exists a torque owing to this mutual coupling. Throughout the development the assumption was made that currents exist in *each* winding. However, there are machines in which it is possible to produce torque when only one part of the machine is excited. Surely this statement is not so surprising in the light of the results of Sec. 1-10, where it was shown that a force can exist whenever a flux flows in a geometrical configuration characterized by a reluctance that varies with position. Refer to Eq. (1-55). What is important to us here, however, is to recognize that the torque associated with reluctance variation cannot be computed from the results derived in Secs. 3-1 and 3-4. It seems worthwhile therefore at this point in our study to obtain a general expression for the torque developed by an electromechanical energy-conversion device—one that permits calculation of the *reluctance torque* as well as the *electromagnetic torque* of the individually excited, mutually coupled windings.

The general torque equation can be derived by analyzing the configuration shown in Fig. 3-18. A current i_1 is assumed to flow in the rotor winding producing the air-gap flux field. The current i_2 is assumed to flow in the stator winding. This current may be direct current, single-phase alternating current, or the phase current of a polyphase system applied to a polyphase stator winding; in any event irrespective of the form of i_2 it is assumed that the stator current is responsible for the ampere-conductor distribution shown. Next consider that the rotor is allowed to move by some differential angle $d\alpha$. In turn this causes a change in coupling between the rotor field distribution and the stator ampere-conductor distribution. Moreover, the induced emf's of both windings will be affected, which will influence the energy input. If the input energy originates solely with the electrical sources, it follows that corresponding to the displacement $d\alpha$ the change in input energy is described by

$$dW_e = e_1 i_1 \, dt + e_2 i_2 \, dt \tag{3-75}$$

Figure 3-18 Geometrical configuration of an electric machine used in developing the general torque equation. Current i_1 flows in the rotor winding and current i_2 flows in the stator winding; together these produce the electromagnetic torque. The nonuniform air gap gives rise to a reluctance torque.

But by Faraday's law

$$e_1 = \frac{d\lambda_1}{dt} \quad \text{and} \quad e_2 = \frac{d\lambda_2}{dt} \tag{3-76}$$

where λ_1 and λ_2 denote the total flux linkages of the rotor and stator windings respectively. Specifically, these quantities may be expressed in terms of the self- and mutual inductances provided that nonlinearities and hysteresis are neglected. In large measure this is justified by the presence of air gaps. Thus

$$\lambda_1 = L_{11}i_1 + M_{12}i_2 \tag{3-77}$$

$$\lambda_2 = M_{21}i_1 + L_{22}i_2 \tag{3-78}$$

Here L_{11} is the self-inductance of the rotor winding, L_{22} is the self-inductance of the stator winding, and M_{12} is the mutual inductance between rotor and stator windings. The positive sign is used for M_{12} because i_1 and i_2 produce fluxes in the same direction in the configuration of Fig. 3-18.

Upon substituting Eqs. (3-76) in Eq. (3-75) there results

$$dW_e = i_1 \, d\lambda_1 + i_2 \, d\lambda_2 \tag{3-79}$$

This expression emphasizes that any change in electrical input energy comes about through changes in flux linkages. Furthermore, the energy balance condition requires that this change in electrical input energy be equal to the change in field energy or the mechanical work done by the field or both. Mathematically expressed,

$$dW_e = dW_f + T \, d\alpha \tag{3-80}$$

where T denotes the total torque developed as a consequence of the rotor displacement $d\alpha$ and W_f represents the field energy associated with the currents i_1 and i_2. For the doubly excited configuration of Fig. 3-18 the expression for this field energy is given by

$$W_f = \tfrac{1}{2}L_{11}i_1^2 + \tfrac{1}{2}L_{22}i_2^2 + Mi_1i_2 \tag{3-81}$$

A word is in order at this point concerning the variational nature of the self- and mutual inductances as a function of the rotor displacement angle α. In accordance with Eq. (1-40) we can express the self-inductance of the rotor as

$$L_{11} = \frac{N_1^2}{\mathcal{R}_1} \tag{3-82}$$

where N_1 denotes the number of turns of the rotor winding and $\mathcal{R}_1$ denotes the reluctance of the flux produced by the rotor winding. It is interesting to note here that irrespective of the position taken by the rotor the reluctance $\mathcal{R}_1$ remains invariant for a symmetrical and concentrically located rotor structure. Hence for this geometry L_{11} is a constant.

However, for the stator winding, the situation is different. Although a similar expression for the stator self-inductance may be written, i.e.,

$$L_{22} = \frac{N_2^2}{\mathcal{R}_2} \tag{3-83}$$

examination of the geometry of Fig. 3-18 indicates that the reluctance $\mathcal{R}_2$ varies with the rotor position. This is readily understood by considering for a moment that i_2

is made to flow alone (i_1 assumed zero) in the stator winding to produce the ampere-conductor distribution depicted in Fig. 3-18. The stator thus behaves like a solenoid producing flux on an axis coincident with the horizontal. Clearly, then, when the rotor axis lies along this horizontal axis (i.e., $\alpha = 0$ degrees), the reluctance seen by this flux will be a minimum because the magnetic path consists of two small air gaps and a large amount of high-permeability iron. This path is called the *direct axis* of the machine and leads to a minimum reluctance $\mathcal{R}_d$. As the rotor iron is displaced towards $\alpha = \pi/2$, the reluctance increases because the stator flux is still horizontally directed but the rotor iron is advancing towards the vertical position as α increases which increases the air gap on the horizontal axis. At $\alpha = \pi/2$ the reluctance seen by the stator-produced flux is a maximum because it now consists of two very large air gaps. The rotor alignment with respect to the stator-produced flux field is now said to be in its *quadrature* axis. Call this maximum value of reluctance $\mathcal{R}_q$. As a result of the shaping of the pole pieces and the distributed nature of the stator winding,† the reluctance variation $\mathcal{R}_2$ is a cosinusoidal function of α. Since for each full revolution of the rotor structure there are two positions each of minimum and maximum reluctance, it follows that L_{22} is a double frequency function of α. The complete variation of the reluctance is depicted graphically in Fig. 3-19. The analytical expression for this reluctance is given by

$$\mathcal{R}_2 = \frac{\mathcal{R}_q + \mathcal{R}_d}{2} - \frac{\mathcal{R}_q - \mathcal{R}_d}{2} \cos 2\alpha \tag{3-84}$$

Insertion of this expression into Eq. (3-83) yields the corresponding equation that describes the variation of the stator self-inductance L_{22} with rotor position.

Reluctance

α, electrical degrees

Figure 3-19 Variations of reluctance as a function of rotor displacement, as viewed from the stator winding. To a different scale this plot also represents the variation of L_{22} versus α.

There now finally remains to be determined the manner in which the mutual inductance M_{12} varies with rotor displacement, if at all. A simple and direct way to obtain this information is to perform a straightforward test on the machine of Fig. 3-18. Apply a fixed sinusoidal voltage on the rotor winding. Remove the source V_2 from the stator winding, and replace it with a suitable rms-reading voltmeter. The voltmeter reads the voltage induced through the mutual coupling with the changing current in the rotor winding. Specifically, this induced emf is related to the rotor

†The stator winding is not concentrated as a coil but rather is distributed in slots along the entire inner periphery of the stator structure.

current by

$$e_2 = M_{12}\frac{di_1}{dt} \tag{3-85}$$

For a sinusoidal variation of i_1 the magnitude of emf induced in the stator winding, E_2, by the constant rms rotor current I_1 is

$$E_2 = \omega M_{12}I_1 \tag{3-86}$$

or

$$M_{12} = \frac{E_2}{\omega I_1} = \frac{E_{2\,max}\cos\alpha}{\omega I_1} \tag{3-87}$$

The value of E_2 in this equation is directly a cosinusoidal function of the coupling between the rotor flux field and the stator winding. A glance at Fig. 3-18 shows that at $\alpha = 0$ degrees the rotor flux has maximum linkage with the stator winding, which registers a maximum reading on the voltmeter, E_{2max}. At $\alpha = \pi/2$ the net flux linkage is zero (ideally). Hence E_2 is zero. At $\alpha = \pi$ the flux linking the stator winding is once again a maximum, but because the flux field is reversed the stator-induced emf undergoes a reversed polarity. Hence the mutual inductance is at its negative maximum value. The variation of M_{12} between its extreme values occurs in accordance with a cosinusoidal function owing to the inherent construction features of the machine. Appearing in Fig. 3-20 is the complete variation of the mutual inductance for one full revolution of the rotor. Note that, unlike the stator self-inductance, the mutual inductance has a fundamental variation with the rotor displacement. If the maximum value of the mutual inductance is called M, then

$$\boxed{M_{12} = M\cos\alpha} \tag{3-88}$$

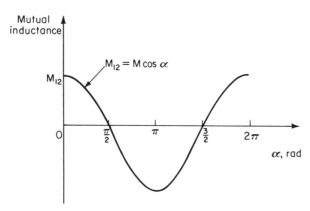

Figure 3-20 Variation of mutual inductance between the rotor and stator windings of the machine depicted in Fig. 3-18.

The foregoing results demonstrate that, for the geometrical configuration of Fig. 3-18, L_{11} is independent of rotor displacement while L_{22} and M_{12} are both functions of α. Keeping these facts in mind, we can now return to Eq. (3-79) and compute the change in the electrical energy input in terms of the self- and mutual inductances. Inserting Eqs. (3-77) and (3-78) into Eq. (3-79) leads to

$$dW_e = i_1\,d(L_{11}i_1 + M_{12}i_2) + i_2\,d(M_{12}i_1 + L_{22}i_2) \tag{3-89}$$

Since L_{11} is a constant, its differential is zero. This is not so for L_{22} and M_{12}. Accordingly, Eq. (3-89) becomes

$$dW_e = i_1 L_{11}\, di_1 + i_1 M_{12}\, di_2 + i_1 i_2\, dM_{12} + i_2 M_{12}\, di_1 + i_1 i_2\, dM_{12}$$
$$+ i_2 L_{22}\, di_2 + i_2^2\, dL_{22} \qquad (3\text{-}90)$$

Also, the change that takes place in the field energy is readily found by differentiating Eq. (3-81). Thus

$$dW_f = i_1 L_{11}\, di_1 + i_2 L_{22}\, di_2 + \tfrac{1}{2} i_2^2\, dL_{22} + M_{12} i_1\, di_2 + M_{12} i_1\, di_2$$
$$+ M_{12} i_2\, di_1 + i_1 i_2\, dM_{12} \qquad (3\text{-}91)$$

When Eqs. (3-90) and (3-91) are introduced into Eq. (3-80)—the energy balance equation—all the terms of dW_e that involve differential currents have matching terms in the expression for dW_f and so cancel out. This is by way of stating that differential changes in the currents have no bearing on the development of mechanical forces. The remaining terms in the energy balance equation are then as follows:

$$i_2^2\, dL_{22} + 2 i_1 i_2\, dM_{12} = T\, d\alpha + \tfrac{1}{2} i_2^2\, dL_{22} + i_1 i_2\, dM_{12} \qquad (3\text{-}92)$$

The left side of Eq. (3-92) represents the change in electrical input energy corresponding to the rotor displacement $d\alpha$. The last two terms on the right side denote the change in the field energy. The remainder of the input energy goes to do mechanical work as represented by $T\, d\alpha$. It follows from Eq. (3-92) that the desired general expression for the developed torque in the machine of Fig. 3-18 is

$$T = \frac{1}{2} i_2^2 \frac{dL_{22}}{d\alpha} + i_1 i_2 \frac{dM_{12}}{d\alpha} \qquad (3\text{-}93)$$

This is the expression for the instantaneous total developed torque.† It depends upon the instantaneous value of i_1 and i_2 as well as on the change of the stator self-inductance and the mutual inductance with rotor position.

A study of Eq. (3-92) leads to a useful alternative form of the general torque equation. Note that, if i_1 and i_2 are held constant, the change in field energy is exactly equal to half the change in electrical energy input. Therefore the mechanical work done is equal to the change in field energy. Expressed mathematically, this is

$$T = \left(\frac{\partial W_f}{\partial \alpha} \right)_{i_1, i_2 \text{ const}} \qquad (3\text{-}94)$$

The partial derivative is employed here to emphasize that i_1 and i_2 are held constant in this operation. Equation (3-94) can be useful in several ways. One particular application is to reveal that the positions of maximum or minimum field energy correspond to zero developed torque.

† A little thought should make it apparent that, if the machine configuration were such that L_{11} were also a function of α, the general torque equation would become

$$T = \frac{1}{2} i_1^2 \frac{dL_{11}}{d\alpha} + \frac{1}{2} i_2^2 \frac{dL_{22}}{d\alpha} + i_1 i_2 \frac{dM_{12}}{d\alpha} \qquad (3\text{-}93a)$$

Electromagnetic Torque: The Torque of Mutual Coupling

Equation (3-93) describes the total torque developed in the machine configuration of Fig. 3-18. An examination of this expression reveals some interesting facts. For example, the first term on the right side is observed to be independent of the rotor current i_1. This means that torque can be produced, even though only the stator is energized, because of the change of reluctance with rotor position. This is called the *reluctance torque* of the machine and exists by virtue of the nonuniform air gap. On the other hand, the second term is seen to depend upon the currents in the rotor and stator windings and the rate of change of the mutual coupling between these windings with position. This is usually called the *electromagnetic torque* of the energy-conversion device and represents the principal component of the total torque developed by such machines. Whenever two windings are energized and mutually coupled, there arises an electromagnetic torque between the two fields associated with the windings to bring the fields into alignment. The subject matter of Secs. 3-1 and 3-4 concerned itself with this component of the torque. The inherent formulation of the developed torque derivation therein described did not lend itself to consideration of the torque associated with a nonuniform air gap. In the interest of completeness and for better understanding, it is therefore appropriate at this point to show that the second term of Eq. (3-93) is in fact equivalent to the result derived in Sec. 3-1 and represented by Eq. (3-29).

The expression for the instantaneous electromagnetic torque component of Eq. (3-93) is valid for a two-pole configuration. For a p-pole machine geometry this becomes

$$T_t = \frac{p}{2} i_1 i_2 \frac{dM_{12}}{d\alpha} \tag{3-95}$$

where M_{12} denotes the mutual inductance between the rotor and stator windings on a *pole pair* basis. By Eqs. (2-77) and (3-88) the mutual inductance can be written as

$$M_{12} = \frac{N_1 N_2}{\mathcal{R}_{12}} \cos \alpha \tag{3-96}$$

Keep in mind that here N_1 and N_2 are the rotor and stator turns for two poles and $\mathcal{R}_{12}$ is the reluctance seen by a flux path linking two poles. Inserting Eq. (3-96) into Eq. (3-95) yields

$$T_t = \frac{p}{2} \frac{N_1 i_1}{\mathcal{R}_{12}} N_2 i_2 (-\sin \alpha) \tag{3-97}$$

To manipulate this equation for application to the a-c machine, the quantity $N_2 i_2$ (which represents the stator winding mmf for two poles) is expressed in terms of the effective turns per pole of a polyphase winding along with a cosinusoidal variation of the current i_2. That is,

$$N_2 i_2 = 2\mathcal{F}_p \cos \omega t \tag{3-98}$$

where $\mathcal{F}_p$ is the effective mmf *per pole*. [Refer to Eq. (3-51).] Moreover, the reluctance for a two-pole path may be written as

$$\mathcal{R}_{12} = \frac{2g}{\mu_0 \pi l r} \tag{3-99}$$

where g is the air-gap length at the pole center, l denotes the axial length of the machine iron, and r denotes the rotor radius. Note that πr is the pole arc length; when this quantity is multiplied by l, there results the pole area through which the air-gap flux penetrates. The reluctance of the iron portion of the magnetic circuit in Eq. (3-99) is neglected. Substituting Eqs. (3-98) and (3-99) in Eq. (3-97) gives

$$T_i = \frac{p}{2} \frac{N_1 i_1}{2g} \mu_0 \pi l r 2 \mathfrak{F}_p \cos \omega t \, (-\sin \alpha) \tag{3-100}$$

The quantity $N_1 i_1 / 2g$ is the magnetic field intensity in the air gap, and $\mu_0 N_1 i_1 / 2g = \mu_0 H_g = B_g$ is the maximum flux density in the air gap. Upon introducing this result into the last expression we obtain

$$T_i = -\frac{p}{2} B_g l r \pi 2 \mathfrak{F}_p \cos \omega t \sin \alpha \tag{3-101}$$

Furthermore, by means of Eq. (3-28) the torque can now be expressed in terms of the flux per pole. Thus

$$T_i = -\frac{p}{2} \frac{p\Phi}{4} \pi 2 \mathfrak{F}_p \sin \alpha \cos \omega t \tag{3-102}$$

Recalling that the rotor displacement angle $\alpha = \omega_r t - \delta$, where ω_r is the speed of rotation of the rotor in radians per second, the expression for the instantaneous value of the electromagnetic torque becomes

$$T_i = -\frac{\pi}{4} p^2 \Phi \mathfrak{F}_p \sin (\omega_r t - \delta) \cos \omega t \tag{3-103}$$

The average value of the electromagnetic torque is found by integrating Eq. (3-103) over one period. Thus,

$$T = \text{average } T_i = -\frac{\pi}{4} p^2 \Phi \mathfrak{F}_p \frac{1}{2\pi} \int_0^{2\pi} \sin (\omega_r t - \delta) \cos \omega t \, d\omega t$$

$$= -\frac{\pi}{4} p^2 \Phi \mathfrak{F}_p \frac{1}{\pi} \int_0^{\pi} \sin (\omega_r t - \delta) \cos \omega t \, d\omega t \tag{3-104}$$

By introducing the trigonometric identity

$$\sin (\omega_r t - \delta) \cos \omega t = \tfrac{1}{2} \sin [(\omega_r - \omega)t - \delta] + \tfrac{1}{2} \sin [(\omega_r + \omega)t - \delta] \tag{3-105}$$

into Eq. (3-104), we obtain

$$T = -\frac{p^2}{8} \Phi \mathfrak{F}_p \int_0^{\pi} \{\sin [(\omega_r - \omega)t - \delta] \, d\omega t + \sin [(\omega_r + \omega)t - \delta] \, d\omega t\} \tag{3-106}$$

A little thought shows that, whenever $\omega_r \neq \omega$ in Eq. (3-106), the average torque is zero. However, when $\omega_r = \omega$, this expression becomes

$$T = +\frac{p^2}{8} \Phi \mathfrak{F}_p \left[\sin \delta \int_0^{\pi} d\omega t + \int_0^{\pi} \sin (2\omega t - \delta) \, d\omega t \right] \tag{3-107}$$

In this case the second term in brackets is again zero but the first is not. Consequently, the average electromagnetic torque has a definite value. Specifically,

$$T = \frac{\pi}{8} p^2 \Phi \mathfrak{F}_p \sin \delta \tag{3-108}$$

The angle δ is often referred to as the *torque angle* for obvious reasons. Finally, to put Eq. (3-108) into a form that corresponds exactly with Eq. (3-29), we make use of Eq. (3-14) and the fact that δ is related to the space displacement angle ψ by†

$$\delta = \frac{\pi}{2} - \psi \qquad (3\text{-}109)$$

Accordingly, the desired result is obtained, i.e.,

$$T = \frac{\pi}{8} p^2 \Phi \mathscr{F}_p \sin \delta = \frac{\pi}{8} p^2 \Phi J_m \cos \psi \qquad (3\text{-}110)$$

Example 3-5

Two coils are mounted in a machine with a uniform air gap in the manner illustrated in Fig. 3-21. The self- and mutual inductances are given by

$$L_1 = 5\,\text{H} \qquad M = 3 \cos \alpha$$
$$L_2 = 2\,\text{H}$$

The winding resistances are negligible. Coil 2-2′ is initially placed at an angle α from coil 1-1′. Coil 1-1′ is made to carry a current of $10 \sin \omega t$, where ω is the frequency of the voltage source that energizes coil 1-1′.

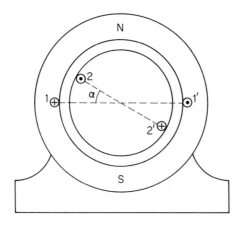

Figure 3-21 Configuration for Example 3-5.

(a) Find the expression for the torque when coil 2-2′ is supplied with a direct current of -5 A from an ideal current source. Will a displacement of the rotor occur? Explain.

(b) Now assume that the current that flows through coil 2-2′ is $-5 \sin \omega t$. Repeat part (a).

(c) At what value of α will the torque be a maximum?

†The ampere-conductor distribution of a winding is always related to the corresponding mmf wave by a displacement of $\pi/2$ electrical radians. A glance at Figs. 3-3(c) and (d) illustrates this point. It is helpful then to note that δ effectively represents the displacement angle between the axis of the field distribution of one pole polarity and the axis of the mmf wave of the armature winding of an opposite pole polarity.

Solution (a) Equation (3-93a) may be applied here. However, note that dL_1/dt and dL_2/dt are zero because L_1 and L_2 are given as constants. Hence the electromagnetic torque is attributable solely to the change in mutual inductance. Thus

$$T = i_1 i_2 \frac{dM}{d\alpha} = (10 \sin \omega t)(-5) \frac{d}{d\alpha}(3 \cos \alpha)$$

$$= 150 \sin \alpha \sin \omega t$$

Clearly, the average torque over one cycle of the source is zero and so no rotation occurs in this case. This statement is based on the assumption that the source frequency is rapid enough so that the rotor cannot make an instantaneous response.

(b) In this case we get

$$T = (10 \sin \omega t)(-5 \sin \omega t)(-3 \sin \alpha)$$

$$= 150 \sin \alpha \sin^2 \omega t$$

Introducing the trigonometric substitution

$$\sin^2 \omega t = \tfrac{1}{2} - \tfrac{1}{2} \cos 2\omega t$$

then yields

$$T = 75 \sin \alpha - 75 \cos 2\omega t$$

In this case, there does exist an average torque which is different from zero. In fact, specifically it is equal to the first term on the right side of the last equation. Of course, the second term is zero over one cycle of the source frequency.

(c) A glance at the last equation makes it obvious that the maximum torque occurs at $\alpha = 90°$. Accordingly, when coil 2-2' is energized while at position α in Fig. 3-21, the rotor will rotate in a clockwise position until it reaches a position where $\alpha = 90°$, provided that a restraining force in excess of $75 \sin \alpha$ does not appear on the rotor shaft.

The Reluctance Torque

The machine of Fig. 3-18 can be singly excited and yet produce torque. As already pointed out, this is the result of the variation of reluctance along the periphery of the air gap. It is important to note, however, that the reluctance torque can be realized in the configuration of Fig. 3-18 only if the stator is energized. When an exciting current is made to flow in the stator winding, a flux field is created in the air gap. The action of this flux is such that it seeks out the minimum reluctance path. Therefore, as long as the reluctance of the magnetic circuit, as viewed from the stator, varies for different positions of the rotor, a torque due to this reluctance variation can exist. The mathematical description of this torque is given by the first term of Eq. (3-93). Calling this quantity T_{rl}, we have

$$T_{rl} = \frac{1}{2} i_2^2 \frac{dL_{22}}{d\alpha} \tag{3-111}$$

Keep in mind that this is the expression for the instantaneous value of the reluctance torque and the second subscript serves as a reminder.

There is no reluctance torque if the machine of Fig. 3-18 is singly excited by the rotor winding. This is because, as viewed from the rotor, the reluctance of the magnetic circuit is invariant with rotor position. Accordingly, there is no tendency for the flux originating from rotor excitation to align itself with a minimum reluctance path. All rotor positions present the same reluctance path.

Let us next turn our attention to manipulating Eq. (3-111) in the interest of obtaining alternative forms that provide additional insight into the nature of the reluctance torque. Inserting the general expression for the self inductance, $L_{22} = N_2^2/\mathcal{R}_2$, into Eq. (3-111) permits us to write

$$T_{rt} = \frac{1}{2}i_2^2\frac{dL_{22}}{d\alpha} = \frac{1}{2}i_2^2\frac{d}{d\alpha}\frac{N_2^2}{\mathcal{R}_2} = \frac{1}{2}(N_2 i_2)^2\frac{d}{d\alpha}\frac{1}{\mathcal{R}_2}$$

$$= -\frac{1}{2}\left(\frac{N_2 i_2}{\mathcal{R}_2}\right)^2\frac{d\mathcal{R}_2}{d\alpha} = -\frac{1}{2}\phi^2\frac{d\mathcal{R}_2}{d\alpha} \qquad (3\text{-}112)$$

where $\mathcal{R}_2$ is given by Eq. (3-84) and ϕ denotes the instantaneous flux produced by the stator mmf $N_2 i_2$ acting in the path of reluctance $\mathcal{R}_2$. The last version of Eq. (3-112) is the rotational equivalent of the translational force described by Eq. (1-55). Note again that the negative sign indicates that the direction of the torque is such as to bring about a reduction in reluctance.

If a sinusoidal voltage of rms value V is applied to the stator winding, then

$$\phi = \Phi \cos \omega t \qquad (3\text{-}113)$$

where

$$\Phi = \frac{V}{\sqrt{2}\,\pi f N_2} = \frac{V}{4.44 f N_2} \qquad (3\text{-}114)$$

Moreover, from Eq. (3-84) we can write

$$\frac{d\mathcal{R}_2}{d\alpha} = \frac{d}{d\alpha}\left(\frac{\mathcal{R}_q + \mathcal{R}_d}{2} - \frac{\mathcal{R}_q - \mathcal{R}_d}{2}\cos 2\alpha\right)$$

$$= (\mathcal{R}_q - \mathcal{R}_d)\sin 2\alpha$$

$$= (\mathcal{R}_q - \mathcal{R}_d)\sin 2(\omega_r t - \delta) \qquad (3\text{-}115)$$

Upon substituting Eqs. (3-113) and (3-115) in Eq. (3-112), there results

$$T_{rt} = -\tfrac{1}{2}\Phi^2(\mathcal{R}_q - \mathcal{R}_d)\sin 2(\omega_r t - \delta)\cos^2 \omega t \qquad (3\text{-}116)$$

But

$$\cos^2 \omega t = \tfrac{1}{2}(1 + \cos 2\omega t)$$

Hence

$$T_{rt} = -\tfrac{1}{4}\Phi^2(\mathcal{R}_q - \mathcal{R}_d)\,[\sin 2(\omega_r t - \delta) + \sin 2(\omega_r t - \delta)\cos 2\omega t] \qquad (3\text{-}117)$$

Also,

$$\sin 2(\omega_r t - \delta)\cos 2\omega t = \tfrac{1}{2}\sin [2(\omega_r + \omega)t - 2\delta] + \tfrac{1}{2}\sin [2(\omega_r - \omega)t - 2\delta] \qquad (3\text{-}118)$$

Accordingly,

$$T_{rt} = -\tfrac{1}{4}\Phi^2(\mathcal{R}_q - \mathcal{R}_d)\,\{\sin (2\omega_r t - \delta) + \tfrac{1}{2}\sin [2(\omega_r + \omega)t - 2\delta]$$
$$+ \tfrac{1}{2}\sin [2(\omega_r - \omega)t - 2\delta]\} \qquad (3\text{-}119)$$

Equation (3-119) still represents the instantaneous value of the reluctance torque. To obtain the average value, we must integrate over one period of ωt. Thus

$$T_r \equiv \text{average } T_{rt} = \frac{1}{2\pi} \int_0^{2\pi} T_{rt} \, d\omega t = \frac{1}{\pi} \int_0^{\pi} T_{rt} \, d\omega t \qquad (3\text{-}120)$$

Specifically,

$$T_r = \frac{1}{\pi}[-\tfrac{1}{4}\Phi^2(\mathcal{R}_q - \mathcal{R}_d)]\{\int_0^{\pi} \sin(2\omega_r t - 2\delta) \, d\omega t$$

$$+ \tfrac{1}{2} \int_0^{\pi} \sin[2(\omega_r + \omega)t - 2\delta] \, d\omega t + \tfrac{1}{2} \int_0^{\pi} \sin[2(\omega_r - \omega)t - 2\delta] \, d\omega t\} \qquad (3\text{-}121)$$

Examination of the last expression reveals that the first integral term always yields a zero value. Moreover, the second and third integral terms also yield zero whenever $\omega \neq \omega_r$. If $\omega_r = \omega$, then the expression for the reluctance torque becomes

$$T_r = \frac{1}{\pi}\left[-\frac{1}{4}\Phi^2(\mathcal{R}_q - \mathcal{R}_d) \int_0^{\pi} \frac{1}{2} \sin(-2\delta) \, d\omega t \right] \qquad (3\text{-}122)$$

which leads to

$$\boxed{T_r = \tfrac{1}{8}\Phi^2(\mathcal{R}_q - \mathcal{R}_d) \sin 2\delta} \qquad (3\text{-}123)$$

Thus we see that the machine of Fig. 3-18 is capable of producing an average reluctance torque, but only when the angular speed of the rotor is equal to the angular frequency of the stator line current. Furthermore, this torque varies with an argument that is double the torque angle δ. In contrast, the electromagnetic torque, as described by Eq. (3-108), has a fundamental variation with torque angle. It follows therefore that the maximum electromagnetic torque occurs at $\delta = \pi/2$, whereas the maximum reluctance torque occurs at $\delta = 45$ degrees.

An alternate form of the average reluctance torque can be found by expressing $\mathcal{R}_d$ and $\mathcal{R}_q$ in terms of their corresponding associated reactances. We define the direct-axis reactance x_d and the quadrature-axis reactance x_q as follows:

$$x_d \equiv \omega L_d = \omega \frac{N_2^2}{\mathcal{R}_d} \qquad (3\text{-}124)$$

$$x_q \equiv \omega L_q = \omega \frac{N_2^2}{\mathcal{R}_q} \qquad (3\text{-}125)$$

where ω is the angular frequency of the stator current and N_2 denotes the stator winding turns. Inserting the values of $\mathcal{R}_d$ and $\mathcal{R}_q$ from Eqs. (3-124) and (3-125) into Eq. (3-123) yields

$$T_r = \frac{1}{4} \frac{V^2}{\omega} \frac{x_d - x_q}{x_d x_q} \sin 2\delta \qquad (3\text{-}126)$$

It is interesting to note that for a machine with a uniform air gap x_d and x_q are equal, which causes the reluctance torque to disappear.

PROBLEMS

3-1. The instantaneous voltage generated in a coil revolving in a magnetic field can be calculated either by the flux-linkage concept or by the Blv concept. With this in mind consider the following problem. A square coil 20 cm on a side has 60 turns and is so located that its axis of revolution is perpendicular to a vertically-directed uniform magnetic field in air of 0.06 Wb/m². The coil is driven at a constant speed of 150 rpm. Compute:

   (a) The maximum flux passing through the coil.
   (b) The maximum flux linkage.
   (c) The time variation of the flux linkage through the coil.
   (d) The maximum instantaneous voltage generated in the coil using both concepts referred to above. Indicate by a sketch the position of the coil at this instant.
   (e) The average value of the voltage induced in the coil over one cycle.
   (f) The voltage generated in the coil when the plane of the coil is 30° from the vertical. Compute by both methods.

3-2. A square coil of 100 turns is 10 cm on each side. It is driven at a constant speed of 300 rpm.

   (a) The coil is placed so that its axis of revolution is perpendicular to a vertically-directed *uniform* field of 0.1 Wb/m². At time $t = 0$, the coil is in a position of maximum flux linkages. Derive an expression for the instantaneous voltage generated. Sketch the voltage waveshape for one cycle.
   (b) The coil is now placed in a *radial* field. All other data remain the same as for part(a). Sketch the voltage waveshape for one cycle showing the numerical value of the maximum voltage.

3-3. In Figs. P3-3(a) and (b) the flux per pole is 0.02 Wb. The coil is revolving at 1800 rpm and consists of 2 turns.

Figure P3-3

   (a) For the configuration of Fig. P3-3(a) derive an expression for the flux linking the coil in terms of the flux per pole and α_0.
   (b) What is the instantaneous value of the coil voltage in Fig. P3-3(a)?
   (c) Compute the maximum value of the voltage induced in the arrangement of Fig. P3-3(b).
   (d) Assuming that the coil of Fig. P3-3(b) is connected to a pair of commutator segments, find the d-c value of this voltage.

3-4. An electric machine has a field distribution as shown in Fig. P3-4. The coil spans a full 180° and has N turns. The machine has two poles and an axial length l and rotor radius r.

Figure P3-4

(a) Obtain the expression for the flux linkage per pole in terms of α expressed in electrical degrees.

(b) By using Faraday's law find the expression for the induced emf in terms of the flux per pole.

(c) Use the Blv form for the induced emf and verify the result of part (b).

3-5. A two-pole field distribution has the sinusoidal shape shown in Fig. P3-5. The maximum value of the flux density is 0.4 T. A coil having full pitch, 50 turns, an axial length of 0.3 m, and a radius of 0.2 m is driven at 600 rpm.

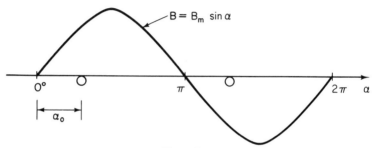

Figure P3-5

(a) Derive the expression for the flux linkage of the coil when at position α_0.

(b) Using the flux linkage approach, find the instantaneous value of the coil voltage when $\alpha_0 = 120°$.

(c) Repeat part (b) using the Blv approach.

(d) Find the maximum voltage induced in the coil and the position at which it occurs.

(e) What is the average value of the induced coil voltage?

3-6. A two-pole machine has the field distribution shown in Fig. P3-6. The coil is full pitch and has N turns, radius r, and an axial length l.

(a) Show that the flux linkage (λ) for the coil position shown is given by

$$\lambda = \frac{NB_p lr}{\pi}\left(\frac{\pi^2}{2} - 2\alpha_0^2\right) \qquad \text{where } 0 \leq \alpha_0 \leq \frac{\pi}{2}$$

(b) Assume that the coil is driven at ω_0 rad/s. Find the expression for the induced coil

Figure P3-6

voltage using the change-of-flux-linkage approach. Express the result in terms of the flux per pole and for the interval $0 \leq \omega_0 t \leq \pi/2$.

(c) Repeat part (b) using the *Blv* approach.

3-7. A two-pole machine is equipped with a total of 12 conductors uniformly distributed in slots along the inner surface of the stator periphery. The conductors are paired so that conductor 1 is series connected to conductor 7 by an appropriate end connection. Similarly, conductor 2 is joined to conductor 8 and so on until all conductors are series connected and energized by a d-c source which forces the flow of a 10-A current.

(a) Draw the mmf distribution as a function of displacement along the air gap.

(b) Determine the number of A-t per pole.

(c) Draw the equivalent current sheet of the ampere-conductor distribution.

3-8. An electric motor is designed to have a triangular field distribution as shown in Fig. P3-8. The ampere-conductor distribution is uniform at value *J*. Assume that the machine has *p* poles. A rotor radius of *r* meters and air axial length of *l* meters.

Figure P3-8

(a) Find the expression for the flux per pole.

(b) Derive the equation for the developed torque in terms of the flux per pole, ψ, and the peak mmf per pole $\mathcal{F}_p$.

3-9. A *p*-pole machine has the sinusoidal current sheet variation and constant field distribution depicted in Fig. P3-9. The machine has an axial length *l* and radius *r*.

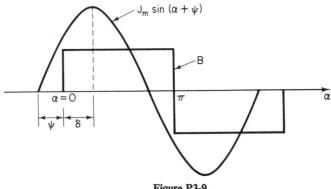

Figure P3-9

(a) Find an expression for the maximum flux per pole.

(b) Determine the expression for the developed electromagnetic torque in terms of B, J_m and δ.

(c) Is the torque constant at every point along the air gap? Explain.

3-10. Derive Eq. (3-54) by starting with the conservation of energy [i.e., Eq. (3-58)].

3-11. A 4-pole a-c induction motor is characterized by a sinusoidal rotor mmf and sinusoidal flux density. Moreover, the flux per pole is 0.02 Wb. When operating at a specified load condition, the space displacement angle is found to be 53.3°. What is the amplitude of the rotor mmf wave required to produce a torque of 40 N-m? Assume two poles.

3-12. A 60-Hz, eight-pole a-c motor has an emf of 440 V induced in its armature winding, which has 180 effective turns. When this motor develops 10 kW of power, the amplitude of the sinusoidal field mmf is equal to 800 A-t.

(a) Find the value of the resultant air-gap flux per pole.

(b) Find the angle between the flux wave and the mmf wave.

3-13. The induced emf per phase of a 60-Hz, four-pole induction motor is 120 V. The number of effective turns per phase is 100. When this motor develops a torque of 60 N-m the space displacement angle is 30°. Find the value of the total armature ampere-turns.

3-14. The three-phase armature winding of a 60-Hz, four-pole, a-c induction motor is equipped with 320 effective turns per phase. When the motor develops 10 kW it has an armature current of 40 A, a space displacement angle of 37°, and a speed of rotation of 1700 rpm.

(a) Find the amplitude of the armature mmf per pole.

(b) What is the value of the developed torque?

(c) Compute the induced emf in the armature winding per phase.

(d) Find the flux per pole.

3-15. Identify the physical location and the function(s) performed by each of the following in the d-c machine: shunt field winding, interpole winding, compensating winding, and series field winding.

3-16. Two windings are mounted in a machine with a uniform air gap as shown in Fig. P3-16. The self- and mutual inductances are

Figure P3-16

$$L_{11} = 2.2 \text{ H} \qquad M_{12} = \sqrt{2} \cos \alpha \text{ H}$$
$$L_{22} = 1.0 \text{ H}$$

The winding resistances are negligible. When winding 2-2′ is short-circuited with voltage applied to winding 1-1′, the current in winding 1-1′ is known to be $i_1 = 10\sqrt{2} \sin \omega t$, where ω is the frequency of the source.

(a) If the rotor is stationary, derive an expression for the instantaneous torque, in newton-meters, in terms of the space angle α.

(b) Compute the average torque in newton-meters when $\alpha = 45°$.

(c) If the rotor is permitted to move, will it rotate continuously or will it tend to come to rest? If the latter, at what value of α?

3-17. In the machine configuration of Fig. P3-17, the self- and mutual inductances have the following identification:

$$L_{11} = \frac{N_1^2}{\Re} \qquad M_{12} = \frac{2N_1 N_2 \alpha}{\pi \Re} \qquad \text{for } -\frac{\pi}{2} < \alpha < \frac{\pi}{2}$$

$$L_{22} = \frac{N_2^2}{\Re}$$

Figure P3-17

where N_1 denotes the stator turns and N_2 denotes the rotor turns. The machine is equipped with a uniform air gap. Through use of the rate of change of flux linkage with angular displacement, the torque is found to be

$$T = i_1 \frac{d\lambda_{12}}{d\alpha}$$

Moreover, through the use of the *Blri* relationship, the expression for developed torque is found to be $T = 2N_1 N_2 B_2 lri_2$. Finally, from the conservation of energy, the torque equation can be written as:

$$T = \frac{1}{2} i_1^2 \frac{dL_{11}}{d\alpha} + \frac{1}{2} i_2^2 \frac{dL_{22}}{d\alpha} + i_1 i_2 \frac{dM_{12}}{d\alpha}$$

Show that the torques computed by all three methods are equal.

3-18. In the configuration of Fig. P3-17 let $\alpha = \omega_0 t$. By means of the flux linkage relationship the emf induced in winding 1-1′ can be expressed by $e_1 = d\lambda_{12}/dt$. Moreover, through the use of the *Blv* relationship the emf equation becomes $e_1 = 2N_1 N_2 B_2 lv$.

Finally, the expression for the speed voltage can be written as

$$e_{1s} = i_1 \frac{dL_{11}}{dt} + i_2 \frac{dM_{12}}{dt} = \left(i_1 \frac{dL_{11}}{d\alpha} + i_2 \frac{dM_{12}}{d\alpha}\right)\omega_0$$

Show that the speed voltages computed by all three methods are equal.

3-19. In the machine configuration of Fig. P3-17 the positive directions for the currents i_1 and i_2 are defined by the dots and crosses appearing in the diagram. If both currents are of constant magnitude with $i_1(t) = I$ and $i_2(t) = -(N_1/N_2)I$, and if $\alpha = 50°$, as depicted, in what direction is the torque on the rotor? Explain. With the same currents, what should be the value of α for maximum energy to be stored in the magnetic field? Explain.

3-20. In the idealized uniform air-gap machine of Fig. P3-20, excitation of either winding produces a radial field. Coil 1 lies at radius r_1 from the axis of rotation. Coil 2 lies at radius r_2 from the axis of rotation. Given that

$$T_1 = 2N_1 B_2 lr_1 i_1$$
$$T_2 = 2N_2 B_1 lr_2 i_2$$

where B_1 is the flux density at coil 2 produced by i_1 in coil 1, and B_2 is the flux density at coil 1 produced by i_2 in coil 2.

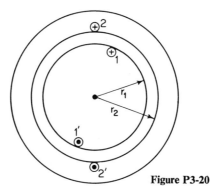

Figure P3-20

(a) Show that T_1 and T_2 have equal magnitudes.
(b) Show that T_1 and T_2 act in opposite directions.

3-21. (a) Prove the validity of Eq. (3-81) for the doubly excited configuration.
(b) Demonstrate how Eq. (3-86) follows from Eq. (3-85).

3-22. In the machine configuration of Fig. 3-18 the rotor winding is unexcited and the stator winding is energized with the direct current I_{dc}.
(a) Can a reluctance torque exist? Why?
(b) At what rotor speed, if any?
(c) Derive an expression for the reluctance torque in terms of the direct-axis self-inductance L_d and the quadrature-axis self-inductance L_q.

3-23. The machine configuration of Prob. P3-17 is doubly excited with currents i_1 and i_2. Are the positions in space at which the maximum torque occurs affected by the values of i_1 and i_2? Explain.

3-24. A reluctance torque motor has the configuration of Fig. 3-18 with a sinusoidal reluctance variation along the periphery of the air gap. The motor is energized from a 120-V,

60-Hz source. With the rotor fixed in the minimum reluctance position, a line ammeter reads 0.5 A rms. In the maximum reluctance position it reads 2 A.

(a) Determine the power developed in watts when the motor is operating with a power angle of 10 electrical degrees.

(b) What is the speed of rotation of the rotor?

(c) What is the maximum developed power in watts for this motor?

3-25. A doubly excited system with a nonuniform air gap has inductances described by

$$L_{11} = 6 + 1.5 \cos 2\alpha \qquad L_{22} = 4 + \cos 2\alpha \qquad M = 6 \cos \alpha$$

The coils are excited by d-c currents with $i_1 = 0.5$ A and $i_2 = 0.4$ A.

(a) Determine the developed torque as a function of α.

(b) Find the stored energy as a function of α.

(c) For $\alpha = 30°$ find the magnitude and direction of the torque on the rotor.

3-26. The dynamometer (moving-coil) instrument is distinguished from other measuring instruments by its linear scale. In construction the moving coil is restrained by a linear spring to have limited movement in flux field of constant value produced by a permanent magnet. Derive an expression to show that the deflection of an indicator needle attached to the moving coil is directly proportional to the coil current.

3-27. A singly excited reluctance motor has a variation of reluctance over the periphery of the air gap that is as depicted in Fig. P3-27(a) for an assumed power angle displacement δ. At the same time the variation of flux with time is described in Fig. P3-27(b).

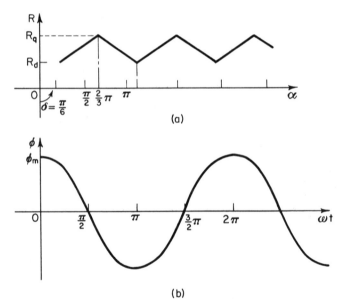

(a)

(b)

Figure P3-27

(a) Can a net torque different from zero be produced through the cooperation of the given flux and reluctance variation? Explain.

(b) Sketch the variations of the instantaneous torque for $\delta = 30°$.

(c) Compute the average value of torque associated with this instantaneous variation.

3-28. A machine configuration is such that the variation of mutual and self-inductances can be described by

$$M_{12} = 1 - \sin \alpha \quad \text{H}$$
$$L_{11} = 1 + \sin \alpha \quad \text{H}$$
$$L_{22} = 2(1 + \sin \alpha) \quad \text{H}$$

(a) Find the value and direction of the developed torque when the current in coil 1 is maintained constant at 10 amperes and that in coil 2 is fixed at -20 A. Take $\alpha = 60°$.

(b) For the condition of part (a), compute the amount of energy supplied by each electrical source.

(c) If the current of coil 1 of part (a) is changed to a sinusoidal current of 10 amperes rms and coil 2 is short-circuited, find the rms value of coil 2 current. Assume a 60-Hz, 120-V source.

(d) Determine the expression for the instantaneous torque produced in part (c).

(e) What is the average value of the torque in part (c)?

3-29. A machine is built with a two-pole single-phase winding and equipped with a d-c commutator winding as depicted in Fig. P3-29. The brush axis is displaced by 20° from the coil axis. If a single-phase voltage is applied to the stator winding, will a continuous, unidirectional torque be developed? Explain. Assume negligible armature leakage flux.

Figure P3-29

3-30. Show that the basic equation for electromagnetic torque in the form

$$T = \frac{\pi}{2} \left(\frac{p}{2} \right)^2 \Phi \mathcal{F} \sin \delta$$

is correct dimensionally.

4

Three-Phase Induction Motors

One distinguishing feature of the induction motor is that it is a *singly excited* machine. Although such machines are equipped with both a field winding and an armature winding, in normal use an energy source is connected to one winding alone, the field winding. Currents are made to flow in the armature winding by induction, which creates an ampere-conductor distribution that interacts with the field distribution to produce a net unidirectional torque. The frequency of the induced current in the conductor is affected by the speed of the rotor on which it is located; however, the relationship between the rotor speed and the frequency of the armature current is such as to yield a resulting ampere-conductor distribution that is stationary in relation to the field distribution. As a result the singly excited induction machine is capable of producing torque *at any speed below synchronous speed.*† For this reason the induction machine is placed in the class of *asynchronous machines*. In contrast, *synchronous machines* are electromechanical energy-conversion devices in which a net torque can be produced at only one‡ speed of the rotor. The distinguishing characteristic of the synchronous machine is that it is a *doubly excited* device except when it is being used as a reluctance motor.

The salient construction features of the three-phase induction motor are

---

†Synchronous speed is determined by the frequency of the source applied to the field winding and the number of poles for which the machine is designed. These quantities are related by Eq. (3-60). Thus

$$\text{synchronous speed} = \frac{120f}{p} = n_s$$

‡Theoretically, there are two rotor speeds at which a net torque different from zero can exist, but at the second speed enormous currents flow, which makes operation impractical.

164

described in Sec. 3-3. Because the induction machine is singly excited, it is necessary that both the magnetizing current and the power component of the current flow in the same lines. Moreover, because of the presence of an air gap in the magnetic circuit of the induction machine, an appreciable amount of magnetizing current is needed to establish the flux per pole demanded by the applied voltage. Usually, the value of the magnetizing current for three-phase induction motors lies between 25 and 40% of the rated current. Consequently, the induction motor is found to operate at a low power factor at light loads and at less than unity power factor in the vicinity of rated output.

We are concerned in this chapter with a description of the theory of operation and of the performance characteristics of the three-phase induction motor. Our discussion begins with an explanation of how a revolving magnetic field is obtained with a three-phase winding. After all it is this field that is the driving force behind induction motors.

4-1 THE REVOLVING MAGNETIC FIELD

The application of a three-phase voltage to the three-phase stator winding of the induction motor creates a rotating magnetic field, which by transformer action induces a working emf in the rotor winding. The rotor induced emf is called a working emf because it causes a current to flow through the armature winding conductors. This combines with the revolving flux-density wave to produce torque in accordance with Eq. (3-54). Consequently, we can view the revolving field as the key to the operation of the induction motor.

The rotating magnetic field is produced by the contributions of space-displaced phase windings carrying appropriate time-displaced currents. To understand this statement let us turn attention to Figs. 4-1 and 4-2. Appearing in Fig. 4-1 are the three-phase currents which are assumed to be flowing in phases *a*, *b*, and *c* respectively. Note that these currents are time-displaced by the equivalent of 120 electrical degrees. Depicted in Fig. 4-2 is the stator structure and the three-phase winding. Note that each phase (normally distributed over 60 electrical degrees) for convenience is represented by a single coil. Thus coil *a-a'* represents the entire phase *a* winding having its flux axis directed along the vertical. This means that whenever phase *a* carries current it produces a flux field directed along the vertical—up or down. The right-hand rule readily verifies this statement. Similarly, the flux axis of phase *b* is 120 electrical degrees displaced from phase *a*, and that of phase *c* is 120 electrical degrees

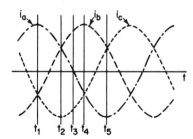

Figure 4-1 Balanced three-phase alternating currents.

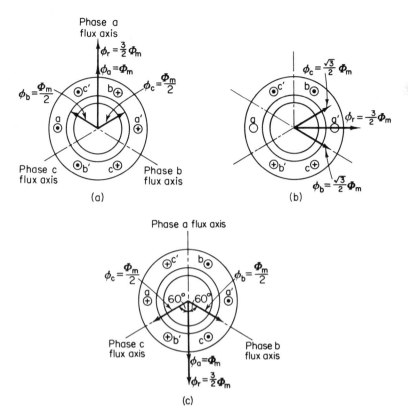

Figure 4-2 Representation of the rotating magnetic field at three different instants of time: (a) time t_1 in Fig. 4-1; (b) time t_3; (c) time t_5.

displaced from phase *b*. The unprimed letters refer to the beginning terminal of each phase.

Let us consider the determination of the magnitude and direction of the resultant flux field corresponding to time instant t_1 in Fig. 4-1. At this instant the current in phase *a* is at its positive maximum value while the currents in phases *b* and *c* are at one-half their maximum negative values. In Fig. 4-2 it is arbitrarily assumed that, when current in a given phase is positive, it flows out of the paper with respect to the unprimed conductors. Thus, since at time t_1 i_a is positive, a dot is used for conductor *a*. See Fig. 4-2(a). Of course a cross is used for *a'* because it refers to the return connection. Then by the right-hand rule it follows that phase *a* produces a flux contribution directed upward along the vertical. Moreover, the magnitude of this contribution is the maximum value because the current is a maximum. Hence $\phi_a = \Phi_m$, where Φ_m is the maximum flux per pole of phase *a*. It is important to understand that phase *a* really produces a sinusoidal flux field with the amplitude located along the axis of phase *a* as depicted in Fig. 4-3. However, in Fig. 4-2(a) this sinusoidal distribution is conveniently represented by the vector ϕ_a.

In order to determine the direction and magnitude of the field contribution of phase *b* at time t_1, we note first that the current in phase *b* is negative with respect to that in phase *a*. Hence the conductor that stands for the beginning of phase *b*

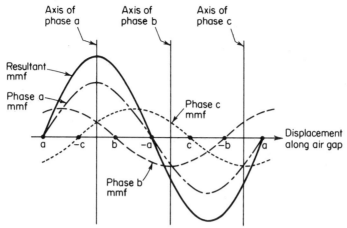

Figure 4-3 Component and resultant field distributions corresponding to t_1 in Fig. 4-1.

must be assigned a cross while b' is assigned a dot. Hence the instantaneous flux contribution of phase b is directed upward along its flux axis and the magnitude of phase b flux is one-half the maximum because the current is at one-half its maximum value. Similar reasoning leads to the result shown in Fig. 4-2(a) for phase c. A glance at the space picture corresponding to time t_1, as illustrated in Fig. 4-2(a), should make it apparent that the resultant flux per pole is directed upward and has a magnitude $\frac{3}{2}$ times the maximum flux per pole of any one phase. Figure 4-3 depicts the same results as Fig. 4-2(a) but does so in terms of sinusoidal flux waves rather than flux vectors. Keep in mind that the resultant flux vector in Fig. 4-2 shows the direction in which flux crosses the air gap. Once across the air gap, the flux is confined to the iron in the usual fashion.

Next let us investigate how the situation of Fig. 4-2(a) changes as time passes through 90 electrical degrees from t_1 to t_3 in Fig. 4-1. Here phase a current is zero, yielding no flux contribution. The current in phase b is positive and equal to $\sqrt{3}/2$ its maximum value. Phase c has the same current magnitude but is negative. Together phases b and c combine to produce a resultant flux having the same magnitude as at time t_1. See Fig. 4-2(b). It is important to note, too, that an elapse of 90 electrical degrees in time results in a rotation of the magnetic flux field of 90 electrical degrees.

A further elapse of time equivalent to an additional 90 electrical degrees leads to the situation depicted in Fig. 4-2(c). Note that again the axis of the flux field is revolved by an additional 90 electrical degrees.

On the basis of the foregoing discussion it should be apparent that the application of three-phase currents through a balanced three-phase winding gives rise to a rotating magnetic field that exhibits two characteristics: (1) it is of constant amplitude and (2) it is of constant speed. The first characteristic has already been demonstrated. The second follows from the fact that the resultant flux traverses through 2π electrical radians in space for every 2π electrical radians of variation in time for the phase currents. Hence, for a two-pole machine, where electrical and mechanical degrees are identical, each cycle of variation of current produces one complete revolution of the flux field. Hence this is a fixed relationship which is dependent upon the frequency of the currents and the number of poles for which the three-phase winding is designed. In the case where the winding is designed for four poles, it requires two cycles of variation of the current to produce one revolution of the flux field. Therefore it follows that for a p-pole machine the relationship is

$$f = \frac{p}{2} \times \text{rps} = \frac{p}{2}\frac{n}{60} \qquad (4\text{-}1)$$

where f is in cycles per second and rps denotes revolutions per second. Note that Eq. (4-1) is identical with Eq. (3-60).

An inspection of the ampere-conductor distribution of the stator winding at the various time instants reveals that the individual phases cooperate in such a fashion as to produce a solenoidal effect in the stator. Thus in Fig. 4-2(a) the directions of the currents are such that they all enter on the right side and leave on the left side. The right-hand rule indicates that the flux field is then directed upward along the vertical. In Fig. 4-2(b) the situation is similar except that now the cross and dot distribution is such that the resultant flux field is oriented horizontally toward the right. Therefore, it can be concluded that the rotating magnetic field is a consequence of the revolving mmf associated with the stator winding.

In the foregoing it is pointed out that the flow of balanced three-phase currents through a balanced three-phase winding yields a rotating field of constant amplitude and speed. If neither of these conditions is exactly satisfied, it is still possible to obtain a revolving magnetic field but it will not be of constant amplitude nor of constant linear speed. In general, for a q-phase machine a rotating field of constant amplitude and constant speed results when the following two conditions are satisfied: (1) there is a *space* displacement between balanced phase windings of $2\pi/q$ electrical degrees, and (2) the currents flowing through the phase windings are balanced and *time*-displaced by $2\pi/q$ electrical degrees. For the three-phase machine $q = 3$ and so the now familiar 120° figure is obtained. The only exception to the rule is the two-phase machine. Because the two-phase situation is a special case of the four-phase system, a value of q equal to 4 must be used.

One final point is now in order. The speed of rotation of the field as described by Eq. (4-1) is always given relative to the phase windings carrying the time-varying currents. Accordingly, if a situation arises where the winding is itself revolving, then the speed of rotation of the field relative to inertial space is different than it is relative to the winding.

Mathematical Analysis

The mmf distribution of the stator (field) winding of the three-phase induction motor is usually trapezoidal in shape, but it contains a very prominent fundamental component. In fact it is this component that is depicted in Fig. 4-3. Although some mmf harmonics do exist, we shall continue to confine our attention here to the important fundamental term.

The resultant stator mmf at any time instant is composed of the contributions of each phase. It is important to keep in mind that each phase winding makes a contribution that varies with time along a fixed space axis. An observer looking only at a single phase thus sees an alternating field. If the axis of phase a in Fig. 4-3 is taken as the zero reference point for the displacement angle α along the air gap, the alternating field created by this phase may be mathematically expressed by

$$\mathcal{F}_a = \mathcal{F}_m \cos \omega t \cos \alpha \qquad (4\text{-}2)$$

where $\mathcal{F}_m$ denotes the maximum value of the mmf per pole and is specified by Eq. (3-51) when q is set equal to unity (one phase). It is worthwhile to pause a moment to be sure that the meaning of Eq. (4-2) is understood. This equation states that an observer stationed in space, say, at $\alpha = 0°$ (i.e., on the axis of phase a) sees at time $t = 0$ the maximum value of phase a mmf. As time progresses the value of this mmf changes. Thus at $\omega t = 60°$ the observer standing at $\alpha = 0°$ now sees $\mathcal{F}_m/2$ as the phase contribution. At $\omega t = 120°$ the contribution becomes $-\mathcal{F}_m/2$, which means that the pole reverses polarity. At $\omega t = 180°$ the mmf contribution is found to be $-\mathcal{F}_m$. Accordingly, the observer placed at $\alpha = 0°$ sees the mmf contribution of phase a alternate between a positive $\mathcal{F}_m$ and a negative $\mathcal{F}_m$ as time progresses. If the point of observation is changed, Eq. (4-2) states that the observer continues to see an alternating field but now it varies between different maxima. Thus, if the observer is fixed at $\alpha = 30°$, then as time elapses phase a mmf is noted to alternate between $\pm(\sqrt{3}/2)\mathcal{F}_m$. An observer stationed at this position can never see the peak value of fundamental mmf. Of course an observer situated at $\alpha = 90°$ sees a zero value of phase a mmf for all time instants.

The expression for the mmf of phase b can be written by introducing two modifications in Eq. (4-2) relating to the time variable t and the space variable α. Recalling that a balanced three-phase voltage is applied to the balanced three-phase stator winding, it follows that the time-varying current in phase b winding is displaced by 120 electrical degrees from that in phase a as depicted in Fig. 4-1. Moreover, the axis of the winding of phase b is also displaced by 120 electrical degrees in space. Accordingly, the expression for the instantaneous mmf of phase b becomes

$$\mathcal{F}_b = \mathcal{F}_m \cos(\omega t - 120°) \cos(\alpha - 120°) \qquad (4\text{-}3)$$

where $\mathcal{F}_m$ carries the same meaning as for phase a. Proceeding in a similar fashion the mmf of phase c can be written as

$$\mathcal{F}_c = \mathcal{F}_m \cos(\omega t - 240°) \cos(\alpha - 240°) \qquad (4\text{-}4)$$

As was demonstrated in the physical analysis, the resultant mmf of the total three-phase winding is found by summing the instantaneous contributions of the alternating mmf's of each phase. Expressing it algebraically we have

$$\mathcal{F}_r = \mathcal{F}_a + \mathcal{F}_b + \mathcal{F}_c = \mathcal{F}_m[\cos \omega t \cos \alpha + \cos(\omega t - 120°)\cos(\alpha - 120°)$$
$$+ \cos(\omega t - 240°)\cos(\alpha - 240°)] \qquad (4\text{-}5)$$

Upon introducing the trigonometric identity

$$\cos x \cos y = \tfrac{1}{2}\cos(x - y) + \tfrac{1}{2}\cos(x + y) \qquad (4\text{-}6)$$

there results

$$\mathcal{F}_r = \frac{\mathcal{F}_m}{2}[\cos(\omega t - \alpha) + \cos(\omega t - \alpha) + \cos(\omega t - \alpha)$$
$$+ \cos(\omega t + \alpha - 240°) + \cos(\omega t + \alpha) + \cos(\omega t + \alpha - 120°)] \qquad (4\text{-}7)$$

Keeping in mind that this last expression describes a *space* field, it should be clear that the second, fourth, and sixth terms, being equal in amplitude and 120° apart, yield a net value of zero. Hence Eq. (4-7) simplifies to

$$\mathcal{F}_r = \tfrac{3}{2}\mathcal{F}_m \cos(\omega t - \alpha) \tag{4-8}$$

This is the equation of a revolving field of constant amplitude. Note the correspondence with the results depicted in Fig. 4-2. There it is shown that the cooperation of the three alternating fields yields a resultant field having an amplitude equal to $\tfrac{3}{2}$ the amplitude of any phase. This is consistent with the factor $\tfrac{3}{2}\mathcal{F}_m$ appearing in Eq. (4-8). To complete the correspondence we must now show that $\cos(\omega t - \alpha)$ is the mathematical expression for a traveling wave.

In this connection let us consider an observer fixed in space at a position $\alpha = 30°$ and investigate what happens as time is allowed to progress. By direct application of Eq. (4-8) for $\alpha = 30°$ we obtain:

1. At $\omega t_1 = 0°$,

$$\mathcal{F}_r = \tfrac{3}{2}\mathcal{F}_m \cos(-30°) = \tfrac{3}{2}\mathcal{F}_m \frac{\sqrt{3}}{2}$$

2. At $\omega t_2 = 30°$,

$$\mathcal{F}_r = \tfrac{3}{2}\mathcal{F}_m$$

3. At $\omega t_3 = 120°$,

$$\mathcal{F}_r = \tfrac{3}{2}\mathcal{F}_m \cos(120° - 30°) = 0$$

Depicted in Fig. 4-4 are the various positions that the constant amplitude cosinusoid must take to yield values consistent with those just computed. Note that, as time progresses by 30°, the wave position at ωt_1 must travel 30° in the positive direction of α. Only in this way can the observer fixed at the space position $\alpha = 30°$ see the

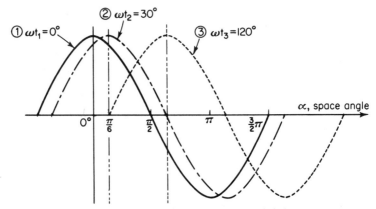

Figure 4-4 Graphical demonstration that Eq. (4-8) is the expression for a traveling wave.

peak value of the resultant wave as called for by Eq. (4-8). For an additional elapse of time corresponding to 90° the observer sees a zero value for the resultant wave. This happens when the wave shown in position 2 moves along in the positive direction of α by 90° to position 3. Therefore, Eq. (4-8) does in fact represent the equation for a traveling wave involving a trigonometric function. The same reasoning employed in the foregoing can be used to illustrate that a traveling wave of trigonometric character traveling in the negative α direction bears the form $\cos(\omega t + \alpha)$, where as before t is a time variable and α is a space variable.

4-2 THE INDUCTION MOTOR AS A TRANSFORMER

The three-phase induction motor may be compared with the transformer because it is a singly energized device which involves changing flux linkages with respect to the stator and rotor windings. In this connection assume that the rotor is of the wound type and Y-connected as illustrated in Fig. 4-5. With the rotor winding open-circuited no torque can be developed. Hence the application of a three-phase voltage to the three-phase stator winding gives rise to a rotating magnetic field which cuts both the stator and rotor windings at the line frequency f_1. The rms value of the induced emf per phase of the rotor winding is given by Eq. (3-49) as

$$E_2 = 4.44 f_1 N_2 K_{w2} \Phi \tag{4-9}$$

where the subscript 2 denotes rotor winding quantities. Note that the stator frequency f_1 is used here because the rotor is at standstill. Hence E_2 is a *line*-frequency emf. Of course the flux Φ is the flux per pole, which is mutual to the stator and rotor windings.

Applied 3-phase voltage

E_1

Stator winding

E_2

Rotor winding

Figure 4-5 Schematic representation of the three-phase wound-rotor induction motor. The dashed line indicates short-circuit links for normal operation.

A similar expression describes the rms value of the induced emf per phase occurring in the stator winding. Thus

$$E_1 = 4.44 f_1 N_1 K_{w1} \Phi \tag{4-10}$$

From Eqs. (4-9) and (4-10) we can formulate the ratio

$$\frac{E_1}{E_2} = \frac{N_1 K_{w1}}{N_2 K_{w2}} \tag{4-11}$$

Note the similarity of this expression to the voltage transformation ratio of the transformer. The difference lies in the inclusion of the winding factors of the motor, necessitated by the use of a distributed winding for the motor as contrasted to the concentrated coils used in the transformer. In essence, then, the induction motor at standstill exhibits the characteristics of a transformer wherein the stator winding is the primary and rotor winding is the secondary.

Next let us consider the behavior of the induction motor under running conditions—again with the intention of pointing out similarities to the transformer. To produce a starting torque (and subsequently a running torque) it is necessary to have a current flowing through the rotor winding. This is readily accomplished by short-circuiting the winding in the manner indicated by the broken line in Fig. 4-5. Initially the induced emf E_2 causes a rotor current per phase I_2 to flow through the short circuit, producing an ampere-conductor distribution which acts with the flux field to produce the starting torque. The sense of this torque is always to cause the rotor to travel in the same direction as the rotating field. An examination of Fig. 4-6 makes this apparent. Assume that the flux field is revolving clockwise at a speed corresponding to the applied stator frequency and the poles of the stator winding. This speed is called the *synchronous speed* and is described by Eq. (4-1). Thus

$$n_s = \frac{120 f_1}{p} \qquad \text{rpm} \qquad (4\text{-}12)$$

By the $\bar{v} \times \bar{B}$ rule the induced emf in a typical conductor lying beneath a south-pole flux is directed into the paper as indicated in Fig. 4-6. Then for this direction of current the $\bar{I}_2 \times \bar{B}$ rule reveals the torque to be directed clockwise. Therefore the rotor moves in a direction in which it tries to catch up with the stator field.

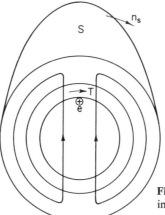

Figure 4-6 Showing the direction of induced voltage and developed torque on a typical rotor conductor.

As the rotor increases its speed, the rate at which the stator field cuts the rotor coils decreases. This reduces the resultant induced emf per phase, in turn diminishing the magnitude of the ampere-conductor distribution and yielding less torque. In fact this process continues until that rotor speed is reached which yields enough emf to produce just the current needed to develop a torque equal to the opposing torques. If there is no shaft load, the opposing torque consists chiefly of frictional losses. It is important to understand that as long as there is an opposing torque to overcome—

however small or whatever its origin—*the rotor speed can never be equal to the synchronous speed*. This is characteristic of singly excited electromechanical energy-conversion devices. Since the rotor (or secondary) winding current is produced by induction, there must always be a difference in speed between the stator field and the rotor. In other words, transformer action must always be allowed to take place between the stator (or primary) winding and the rotor (or secondary) winding.

This speed difference, or *slip*, is a very important variable for the induction motor. In terms of an equation we may write

$$\text{slip} \equiv n_s - n \quad \text{rpm} \tag{4-13}$$

where n denotes the actual rotor speed in rpm. The term slip is used because it describes what an observer riding with the stator field sees looking at the rotor—it appears to be slipping backward. A more useful form of the slip quantity results when it is expressed on a per unit basis using synchronous speed as the reference. Thus the slip in per unit is

$$\boxed{s = \frac{n_s - n}{n_s}} \tag{4-14}$$

For the conventional induction motor the values of s lie between zero and unity.

It is customary in induction-motor analysis to express rotor quantities (such as induced voltage, current, and impedance) in terms of line-frequency quantities and the slip as expressed by Eq. (4-14). For example, if the rotor is assumed to be operating at some speed $n < n_s$, then the actual emf induced in the rotor winding per phase may be expressed in terms of the line-frequency quantity E_2 as sE_2. This formulation has definite advantages, as described in the next section. In a similar fashion it is possible to express the rotor winding impedance per phase as

$$z_2 = r_2 + jsx_2 \tag{4-15}$$

where z_2 denotes the rotor phase impedance, r_2 is the rotor resistance per phase, and x_2 is the line-frequency leakage reactance per phase of the rotor winding. Of course the effective value of this reactance when the rotor operates at a speed n (or slip s) is only s times as large. Keep in mind that the frequency of the currents in the rotor is directly related to the relative speed of the stator field to the rotor winding. Accordingly, we may write

$$f_2 = \frac{p(\text{slip rpm})}{120} = \frac{p(n_s - n)}{120} \tag{4-16}$$

where f_2 is the frequency of the rotor emf and current. By means of Eq. (4-14) it is possible to rewrite Eq. (4-16) as

$$\boxed{f_2 = \frac{psn_s}{120} = s\frac{pn_s}{120} = sf_1} \tag{4-17}$$

which indicates that the rotor frequency f_2 is obtained by merely multiplying the stator line frequency by the appropriate per-unit value of the slip. For this reason f_2 is often called the *slip frequency*.

It is desirable to have an equivalent circuit of the three-phase induction motor in order to direct the analysis of operation and to facilitate the computation of performance. From the remarks made in the preceding section it should not come as a surprise that the equivalent circuit assumes a form identical to that of the exact equivalent circuit of the transformer. The derivation proceeds in a similar fashion with necessary modifications introduced to account for the fact that the secondary winding (the rotor) in this instance revolves and thereby develops mechanical power.

The Magnetizing Branch of the Equivalent Circuit

All the parameters of the equivalent circuit are expressed on a per phase basis. This applies whether the stator winding is Y- or Δ-connected. In the latter case the values refer to the equivalent Y connection. Appearing in Fig. 4-7(a) is the portion of the equivalent circuit that has reference to the stator (or primary) winding. Note that it consists of a stator phase winding resistance r_1, a stator phase winding leakage reactance x_1, and a magnetizing impedance made up of the core-loss resistor r_c and the magnetizing reactance x_ϕ. There is no difference in form between this circuit and that of the transformer. The difference lies only in the magnitude of the parameters. Thus the total magnetizing current $\bar{I}_m$ is considerably larger in the case of the induction motor because the magnetic circuit necessarily includes an air gap. Whereas in the transformer this current is about 2 to 5% of the rated current, here it is approximately 25 to 40% of the rated current depending upon the size of the motor. Moreover, the primary leakage reactance for the induction motor also is larger because of the air gap as well as because the stator and rotor windings are distributed along the periphery of the air gap rather than concentrated on a core as in the transformer. The effects of the actions that take place in the rotor (or secondary) winding must reflect themselves at the proper equivalent-voltage level at terminals a-b in Fig. 4-7(a). We next investigate the manner in which this comes about.

The Actual Rotor Circuit per Phase

For any specified load condition that calls for a particular value of slip s, the rotor current per phase may be expressed as

$$\bar{I}_2 = \frac{s\bar{E}_2}{r_2 + jsx_2} \qquad (4\text{-}18)$$

where $\bar{E}_2$ and x_2 are the standstill values. The circuit interpretation of Eq. (4-18) is depicted in Fig. 4-7(b). It illustrates that $\bar{I}_2$ is a slip-frequency current produced by the slip-frequency induced emf sE_2 acting in a rotor circuit having an impedance per phase of $r_2 + jsx_2$. In other words, this is the current that would be seen by an observer riding with the rotor winding. Furthermore, the amount of real power involved

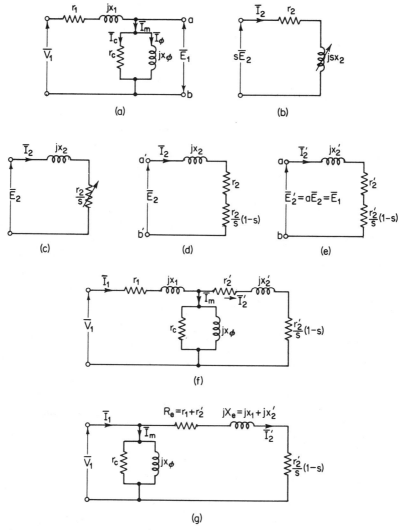

Figure 4-7 Derivation of the equivalent circuit: (a) stator winding section; (b) actual rotor circuit; (c) equivalent rotor circuit; (d) modified equivalent rotor circuit; (e) stator-referred equivalent rotor circuit; (f) exact equivalent circuit; (g) approximate equivalent circuit.

in this rotor circuit is the current squared times the real part of the rotor impedance. In fact, this power represents the rotor copper loss per phase. Hence the total rotor copper loss may be expressed as

$$P_{\text{Cu}2} = q_2 I_2^2 r_2 \qquad (4\text{-}19)$$

where q_2 denotes the number of rotor phases.

The Equivalent Rotor Circuit

By dividing both the numerator and the denominator of Eq. (4-18) by the slip s we get

$$\bar{I}_2 = \frac{\bar{E}_2}{(r_2/s) + jx_2} \qquad (4\text{-}20)$$

The corresponding circuit interpretation of this expression appears in Fig. 4-7(c). Note that the magnitude and phase angle of $\bar{I}_2$ remain unaltered by this operation. However, there is a significant difference between Eqs. (4-18) and (4-20). In the latter case $\bar{I}_2$ is considered to be produced by a line-frequency voltage $\bar{E}_2$ acting in a rotor circuit having an impedance per phase of $r_2/s + jx_2$. Hence the $\bar{I}_2$ of Eq. (4-20) is a *line-frequency* current, whereas the $\bar{I}_2$ of Eq. (4-18) is a *slip-frequency* current. It is important that this distinction be understood.

Manipulation of Eq. (4-18) by s has enabled us to go from an actual rotor circuit characterized by constant resistance and variable leakage reactance [see Fig. 4-7(b)] to one characterized by variable resistance and constant leakage reactance [see Fig. 4-7(c)]. Moreover, the real power associated with the equivalent rotor circuit of Fig. 4-7(c) is clearly

$$P = I_2^2 \frac{r_2}{s} \qquad (4\text{-}21)$$

Hence the total power for q_2 phases is

$$\boxed{P_g = q_2 I_2^2 \frac{r_2}{s}} \qquad (4\text{-}22)$$

A comparison of this expression with Eq. (4-19) indicates that the power associated with the equivalent circuit of Fig. 4-7(c) is considerably greater. For example, in a large machine a typical value of s is 0.02. Hence P_g is greater than the actual rotor copper loss by a factor of 50.

What is the meaning of this power discrepancy? The answer lies in the fact that by Eq. (4-20) $\bar{I}_2$ is a *line-frequency* current. This means that the point of reference has changed from the rotor (where slip-frequency quantities exist) to the stator (where line-frequency quantities exist). In accordance with the circuit representation of Fig. 4-7(c) the observer changes his point of reference from the rotor to the stator. This shift is significant because now, upon looking into the rotor, the observer sees not only the rotor copper loss but the mechanical power developed as well. The latter quantity is included because with respect to a stator-based observer the rotor speed is no longer zero as it is relative to a rotor-based observer. As a matter of fact Eq. (4-22) gives the total power input to the rotor. It is the power transferred across the air gap from the stator to the rotor. We can rewrite Eq. (4-22) in a manner that stresses this fact:

$$P_g = q_2 I_2^2 \frac{r_2}{s} = q_2 I_2^2 \left[r_2 + \frac{r_2}{s}(1 - s) \right] \qquad (4\text{-}23)$$

In other words, the variable resistance of Fig. 4-7(c) may be replaced by the actual rotor winding resistance r_2 and a variable resistance R_m, which represents the mechanical shaft load. That is,

$$R_m \equiv \frac{r_2}{s}(1 - s) \qquad (4\text{-}24)$$

This expression is useful in analysis because it allows any mechanical load to be represented in the equivalent circuit by a resistor. Figure 4-7(d) depicts the modified version of the rotor equivalent circuit. Finally, it should be apparent on the basis of the foregoing remarks that the rotor equivalent circuit is equivalent only insofar as the magnitude and phase angle of the rotor current per phase are concerned.

The Stator-Referred Rotor Equivalent Circuit

The voltage appearing across terminals *a-b* in Fig. 4-7(a) is a line-frequency quantity having $N_1 K_{w1}$ effective turns. The voltage appearing across terminals *a'-b'* in Fig. 4-7(d) is also a line-frequency quantity but has $N_2 K_{w2}$ effective turns. In general $E_1 \neq E_2$, so that *a'-b'* in Fig. 4-7(d) cannot be joined to *a-b* in Fig. 4-7(a) to yield a single-line equivalent circuit. To accomplish this it is necessary to replace the actual rotor winding with an equivalent winding having $N_1 K_{w1}$ effective turns as was done with the transformer. In other words, all the rotor quantities must be referred to the stator in the manner depicted in Fig. 4-7(e). The prime notation is used to denote stator-referred rotor quantities.†

The Complete Equivalent Circuit

The voltage appearing across terminals *a-b* in Fig. 4-7(e) is the same as that appearing across terminals *a-b* in Fig. 4-7(a). Hence these terminals may be joined to yield the complete equivalent circuit as it appears in Fig. 4-7(f). Note that the form is identical with that of the two-winding transformer.

The Approximate Equivalent Circuit

Considerable simplification of computation with little loss of accuracy can be achieved by moving the magnetizing branch to the machine terminals as illustrated in Fig. 4-7(g). This modification is essentially based on the assumption that $\bar{V}_1 \approx \bar{E}_1 = \bar{E}_2'$. All performance calculations will be carried out using the approximate equivalent circuit.

4-4 COMPUTATION OF PERFORMANCE

When the three-phase induction motor is running at no-load, the slip has a value very close to zero. Hence the mechanical load resistor R_m has a very large value, which in turn causes a small rotor current to flow. The corresponding electromagnetic

---

†The derivation of the reduction factors to be used in referring a rotor quantity to the stator appears in Appendix D. Both the wound rotor and squirrel-cage rotor machines are treated there.

torque, as described by Eq. (3-54), merely assumes that value which is needed to overcome the *rotational losses* consisting chiefly of friction and windage. If a mechanical load is next applied to the motor shaft, the initial reaction is for the shaft load to drop the motor speed slightly and thereby increase the slip. The increased slip subsequently causes $\bar{I}_2$ to increase to that value which, when inserted into Eq. (3-54), yields sufficient torque to provide a balance of power to the load. Thus equilibrium is established and operation proceeds at a particular value of s. In fact for each value of load horsepower requirement there is a unique value of slip. This can be inferred from the equivalent circuit, which shows that once s is specified then the power input, the rotor current, the developed torque, the power output, and the efficiency are all determined.

The use of a power-flow diagram in conjunction with the approximate equivalent circuit makes the computation of the performance of a three-phase induction motor a straightforward matter. Depicted in Fig. 4-8(a) is the power flow in statement form. Note that the loss quantities are placed on the left side of a flow point. Appearing in Fig. 4-8(b) is the same power-flow diagram but now expressed in terms of all the appropriate relationships needed to compute the performance. It should be clear that to calculate performance one must first compute the currents I_2 and I_1 from the

Figure 4-8 Power-flow diagram: (a) statement form; (b) equation form.

THREE-PHASE INDUCTION MOTORS   CHAP. 4

equivalent circuit and then make use of the pertinent relationships depicted in Fig. 4-8(b).

Example 4-1

A three-phase, four-pole, 30-hp, 220-V, 60-Hz, Y-connected induction motor draws a current of 77 A from the line source at a power factor of 0.88. At this operating condition, the motor losses are known to be the following:

Stator copper losses $= P_{cu1} = 1033$ W

Rotor copper losses $= P_{cu2} = 1299$ W

Stator core losses $= P_c = 485$ W

Rotational losses (friction, windage, and iron losses due to rotation) $= P_{rot}$ $= 540$ W

Determine: (a) the power transferred across the air gap, (b) the internally developed torque in newton-meters, (c) the slip expressed in per unit and in rpm, (d) the mechanical power developed in watts, (e) the horsepower output, (f) the motor speed in rpm and in radians per second, (g) the torque at the output shaft, (h) the torque needed to overcome the rotational losses, (i) the efficiency of operation at the stated condition.

Solution The solution to this problem provides an exercise in the application of the power flow diagram.

(a) Sufficient information is cited to permit finding the input power. Thus

$$P_i = \sqrt{3}\, V_L I_L(\text{pf}) = 3(220)77(0.88) = 25{,}820 \text{ W}$$

Figure 4-8(a) then shows that the gap power is

$$P_g = P_i - P_c - P_{cu1} = 25{,}800 - 485 - 1033 = 24{,}302 \text{ W}$$

(b) Here use is made of the important relationship

$$T\omega_s = P_g$$

where in this case

$$\omega_s = \frac{\pi}{30} n_s = \frac{\pi}{30}\left(\frac{120 f}{p}\right) = \frac{\pi}{30}(1800) = 188.5 \text{ rad/s}$$

Hence the internally developed torque is

$$T = \frac{P_g}{\omega_s} = \frac{24{,}302}{188.5} = 128.93 \text{ N-m}$$

(c) By Eq. (4-22), we have

$$s = \frac{P_{cu2}}{P_g} = \frac{1299}{24{,}302} = 0.0535 \text{ p-u}$$

Then

$$\text{slip in rpm} = s n_s = (0.0535)(1800) = 96.2 \text{ rpm}$$

(d) Again from the power flow diagram the mechanical power developed can be expressed as

$$P_d = P_g - P_{cu2} = 24{,}302 - 1299 = 23{,}003 \text{ W}$$

(e) For the output power we get

$$P_o = P_d - P_{rot} = 23{,}003 - 540 = 22{,}463 \text{ W}$$

Accordingly,

$$\mathrm{HP}_o = \frac{P_o}{746} = \frac{22{,}463}{746} = 30.1$$

(f) The actual motor speed is the synchronous speed less the slip in rpm. Thus

$$n = n_s - sn_s = 1800 - 96.2 = 1703.8 \text{ rpm}$$

(g) The expression for the output torque is

$$T_o = \frac{P_o}{\omega_m}$$

where ω_m is the actual motor speed expressed in radians per second, or

$$\omega_m = \frac{\pi}{30}(n) = \frac{\pi}{30}(1703.8) = 178.42 \text{ rad/s}$$

Therefore,

$$T_o = \frac{P_o}{\omega_m} = \frac{22{,}463}{178.42} = 125.9 \text{ N-m}$$

(h) The difference between the internally developed torque of 128.93 N-m and the output torque of 125.9 N-m is the torque required to supply the rotational losses associated with the operating speed of 1703.8 rpm. Thus

$$T_{\text{rot}} = T - T_o = 128.93 - 125.9 = 3.03 \text{ N-m}$$

It is instructive to observe that this same torque can be found from the expression that relates torque, speed, and power. Accordingly, we can also write

$$T_{\text{rot}} = \frac{P_{\text{rot}}}{\omega_m} = \frac{540}{178.42} = 3.03 \text{ N-m}$$

(i) The efficiency is

$$\eta = \frac{P_o}{P_1} = \frac{22{,}463}{25{,}820} = 0.87$$

Thus the motor operates at an efficiency of 87%.

Example 4-2

A three-phase, four-pole, 50-hp, 480-V, 60-Hz, Y-connected induction motor has the following parameters per phase:

$$r_1 = 0.10 \,\Omega, \qquad x_1 = 0.35 \,\Omega$$
$$r_2' = 0.12 \,\Omega, \qquad x_2' = 0.40 \,\Omega$$

It is known that the stator core losses amount to 1200 W and the rotational losses equal 950 W. Moreover, at no-load the motor draws a line current of 18 A at a power factor of 0.089 lagging.

When the motor operates at a slip of 2.5%, find: (a) the input line current and power factor, (b) the developed electromagnetic torque in newton-meters, (c) the horsepower output, (d) the efficiency.

Solution (a) The computations are carried out on a per-phase basis. Hence the phase voltage is $480/\sqrt{3}$ or 277.13 V, and the equivalent circuit is depicted in Fig. 4-9. The stator-referred rotor current then follows from

$$\bar{I}_2' = \frac{V_1}{r_1 + (r_2'/s) + j(x_1 + x_2')} = \frac{277.13}{4.9 + j0.75}$$
$$= 55.97\underline{/-8.7^\circ} = 55.33 - j8.47$$

Figure 4-9   Equivalent circuit for
Example 4-2.

For all practical purposes the magnetizing current may be taken equal to the no-load
current because the corresponding rotor current is negligibly small. Thus

$$\bar{I}_m = 19.64\underline{/-85°} = 1.75 - j19.58$$

Hence, the input line current is

$$\bar{I}_1 = \bar{I}_m + \bar{I}'_2 = (55.33 + 1.75) - j(19.58 + 8.47)$$
$$= 57.08 - j28.05 = 63.6\underline{/-26.2°} \qquad (4\text{-}25)$$

and

$$\text{power factor} = \cos\theta_1 = \cos 26.2° = 0.895 \text{ lagging}$$

(b) The developed torque is found from

$$T = \frac{P_g}{\omega_s} \qquad (4\text{-}26)$$

Also

$$\omega_s = \frac{2\pi n_s}{60} = \frac{2\pi(1800)}{60} = 60\pi \text{ rad/s}$$

and

$$P_g = q_1 I'^2_2\frac{r'_2}{s} = 3(55.97)^2 4.8 = 45,110 \text{ W}$$

Therefore

$$T = \frac{45,110}{60\pi} = 239.32 \text{ N-m}$$

(c) From the power-flow diagram the power output is

$$P_o = P_m - P_{\text{rot}} = P_g(1 - s) - P_{\text{rot}} = 43,982.3 - 950 = 43,032.3 \text{ W}$$

The horsepower output is therefore

$$\text{horsepower} = \frac{43,032.3}{0.746} = 57.68$$

Note that this is slightly greater than the rated horsepower of 50. Rated horse power
occurs at a slip somewhat less than 2.5%.

(d) It is more accurate to find the efficiency from the relationship

$$\boxed{\eta = 1 - \frac{\Sigma \text{ losses}}{P_i}} \qquad (4\text{-}27)$$

rather than from the output-to-input ratio. The tabulation of the losses is as follows:

$$P_c = \text{core loss} = 1200 \text{ W}$$

$$\text{stator copper loss} = q_1 I_1^2 r_1 = 3(63.6)^2 0.1 = 1213.5 \text{ W}$$

$$\text{rotor copper loss} = q_1 I_2'^2 r_2' = sP_g = 0.025(45,110) = 1127.8 \text{ W}$$

$$P_{\text{rot}} = \text{rotational loss} = 950 \text{ W}$$

$$\sum \text{losses} = 4491.3 \text{ W}$$

Also, the input power is

$$P_i = \sqrt{3}\,480(63.6)0.895 = 47,324 \text{ W}$$

Hence

$$\eta = 1 - \frac{4491.3}{47,324} = 1 - 0.095 = 0.905$$

The efficiency is 90.5%.

4-5 CORRELATION OF INDUCTION MOTOR OPERATION WITH THE BASIC TORQUE EQUATIONS

The variation of torque with speed (or slip) is an important characteristic of the three-phase induction motor. The general shape of this curve can be identified in terms of the basic torque equation [Eq. (3-110)] and knowledge of the performance computational procedure. When the motor operates at a very small slip, as at no-load, Eq. (4-18) indicates that the rotor current is very small—just enough to develop torque to supply the rotational losses. Moreover, the power-factor angle of the rotor current is practically zero. That is,

$$\theta_2 = \tan^{-1}\frac{sx_2}{r_2} \approx 0° \text{ for } s \text{ very small} \qquad (4\text{-}28)$$

Appearing in Fig. 4-10 is a cross-sectional view of a two-pole squirrel-cage induction

Figure 4-10 Cross-sectional view of two-pole squirrel-cage induction motor, depicting the flux field and corresponding induced rotor mmf distribution at very small slips.

motor showing the rotating flux field produced by the stator winding and the corresponding induced ampere-conductor distribution of the rotor. For the instant shown and the assumed counterclockwise rotation of the field the emf induced in bar 1 is a maximum and directed out of the plane of the paper. Why? Note that the emf induced in bars 2 and 10 is less than the maximum value and that induced in bars 3 and 9 at the same time instant is less still. If the value of the bar emf's are plotted as a function of the spatial displacement along the rotor periphery, a discrete distribution of sinusoidal shape results. See Fig. 3-3(b). Furthermore, because the power-factor angle is practically zero at low slips, the distribution of rotor currents is almost identical with the emf distribution. The current distribution is denoted in Fig. 4-10 by dots and crosses placed outside the rotor surface. The thicker the dot (or cross), the larger is the value of the current represented.

Let us now relate this picture to the quantities appearing in Eq. (3-110). Because there are two forms of the basic electromagnetic torque equation, two interpretations are possible. Taking the last form of Eq. (3-110) first, we note again that torque depends upon the presence of a flux field Φ, an ampere-conductor distribution (or current sheet) as implied by J_m, and a space displacement angle ψ. Examination of Fig. 4-10 discloses that for slips near zero the ampere-conductor distribution is in space phase with the field distribution, i.e., $\psi \approx 0°$. This means that the best field pattern for the development of torque exists—each bar produces a positive torque. However, in spite of this, very little net torque is developed at such low slips because the value of rotor ampere-conductor distribution is also near zero.

In the second interpretation the rotor is treated like a solenoid. In Fig. 4-10 note that the rotor current distribution of dots and crosses creates an mmf directed in quadrature with it (along the horizontal toward the right in this case). This mmf is represented for one pole by the dashed-line sine wave and is assumed to have a peak value of $\mathcal{F}_p$. Keep in mind that the magnitude of $\mathcal{F}_p$ is directly dependent upon the rotor current and is practically zero for very low values of slip. Since the torque angle is the angle between the axis of the flux field distribution and the axis of the armature (or rotor) mmf distribution, we see that its value here is almost 90°, again attesting to the existence of an optimum field pattern for the production of torque.

What happens to the field pattern as mechanical load is applied to the motor shaft? As already described in Sec. 4-4 the immediate reaction is for the motor speed to decrease slightly, which increases the slip. The increased slip produces an almost direct increase in rotor current because for values of slip below 10% the sx_2 term in Eq. (4-18) is negligibly small. When rated horsepower is delivered the slip lies between 3 and 5% for most induction motors and the value of the rotor power-factor angle lies in the vicinity of 10° (compare with Example 4-1). In spite of this, however, and for the convenience of pictorial representation we shall assume that at rated load the slip is such that Eq. (4-28) yields a rotor power-factor angle of 36°. Note that this is the number of degrees separating the rotor bars in the machine configuration of Fig. 4-10. If the field distribution is assumed to take the same instantaneous position as depicted in Fig. 4-10, then the emf distribution must be the same. This is shown with the appropriate dot-cross notation inside the rotor bars of Fig. 4-11(a). Note that the emf in bar 1 is a maximum because it is under the influence of the maximum value of the flux density. As a result of the counterclockwise rotation of

(a)

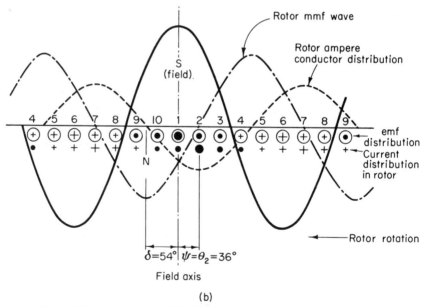

(b)

Figure 4-11 Depicting the field distribution and associated ampere-conductor and mmf waves of the rotor for $\theta_2 = \tan^{-1} (sx_2/r_2) = 36°$. Only the fundamental components are shown. The field travels from right to left.

the stator flux field we know that the emf was a maximum in bar 2 by an amount of time equal to that which the flux field takes to travel the distance between two bars, or more specifically of $\omega t = 36°$. At this point it is helpful to recall the meaning of the power-factor angle θ_2, which is the phase angle between $\bar{E}_2$ and $\bar{I}_2$. When θ_2

is a lag angle, it states that the peak value of rotor current occurs θ_2 electrical degrees in time after the peak value of the rotor induced emf has occurred. Hence for the time instant illustrated in Fig. 4-11(a) the peak outward-directed rotor current exists in bar 2. By the same reasoning outward-directed currents can be found to exist in bars 1, 10, 3, and 4, but the magnitude will be less than the peak current by the cosine of the angle that these bars are displaced from bar 2. As before, our attention is confined just to fundamental components of all distributions. The resulting rotor ampere-conductor distribution then takes the position shown in Fig. 4-11(a) and repeated for clarity's sake in Fig. 4-11(b). Note that bars 4 and 9 produce a negative (or clockwise) torque. This is the reason that the field pattern in not optimum. Nonetheless, the developed torque is much greater here than it is in Fig. 4-10 because the rotor current is many times greater. A glance at Eq. (3-54), which is the practical form of Eq. (3-110), emphasizes this point. Of the quantities appearing there, Φ is essentially fixed by the applied terminal voltage. Hence, if I_2 increases at a greater rate than cos ψ decreases, the torque increases. In normal operation the value of cos ψ rarely dips below 0.95.

These discussions revolving about the situations depicted in Figs. 4-10 and 4-11 lead to a very useful conclusion: *in the polyphase induction motor the space-displacement angle ψ is equal to the power-factor angle of the armature winding, i.e.,*

$$\psi = \theta_2 = \tan^{-1}\frac{sx_2}{r_2} \tag{4-28a}$$

One final comment is in order concerning the development of torque in the polyphase induction motor. Keep in mind that the results of Eqs. (3-54) and (3-110) are valid provided that the flux field and the ampere-conductor distributions are stationary in relation to one another. How is this condition satisfied in the situation of Fig. 4-11? To begin with we know that the speed of the stator flux field relative to the stator structure is ω_s. It therefore remains to be shown that the rotor ampere-conductor distribution has also a speed relative to the stator structure of ω_s. Since the rotor winding carries slip-frequency currents, the speed of the rotor ampere-conductor distribution relative to the rotor is at slip frequency, i.e., $s\omega_s$. However, because the rotor bars are imbedded in the rotor iron, which in turn revolves at a speed of $\omega_m = \omega_s(1-s)$ with respect to the stator structure, it follows that the total speed of rotation of the rotor ampere-conductor distribution is the sum of $s\omega_s$ and $\omega_s(1-s)$. Clearly this is ω_s. Hence the rotor ampere-conductor distribution and the flux field distribution are stationary in relation to each other. In fact this condition is fulfilled for all values of slip and is a consequence of the singly energized nature of the induction motor.

4-6 TORQUE-SPEED CHARACTERISTIC. STARTING TORQUE AND MAXIMUM DEVELOPED TORQUE

The torque-speed characteristic of the induction motor can be explained in the light of Eqs. (3-54) and (3-110). As slip is allowed to increase from nearly zero to about 10 per cent, Eq. (4-18) shows that the rotor current increases almost linearly. Moreover, for this same range of slip, ψ varies over a range of about zero to 15 degrees. This means that cos ψ remains practically invariant over the specified slip range and so torque increases almost linearly in this region.

As slip is allowed to increase still further, the rotor current continues to increase but much less rapidly than at first. The reason lies in the increasing importance of the sx_2 term of the rotor impedance. In addition, the space angle ψ now begins to increase at a rapid rate which makes the cos ψ diminish more rapidly than the current increases. Since the torque equation now involves two opposing factors, it is entirely reasonable to expect that a point is reached beyond which further increases in slip culminate in decreased developed torque. In other words, the rapidly decreasing cos ψ factor predominates over the slightly increasing I_2 factor in Eq. (3-54). As ψ increases, the field pattern for producing torque becomes less and less favorable because more and more conductors that produce negative torque are included beneath a given pole flux. A glance at Fig. 4-11 makes this statement self-evident. Accordingly, the composite torque-speed curve takes on a form similar to that shown in Fig. 4-12.

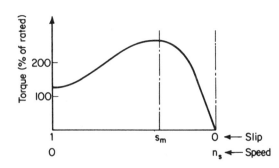

Figure 4-12 Typical torque-speed curve for a three-phase induction motor.

The *starting torque* is the torque developed when s is unity, i.e., the speed n is zero. Figure 4-12 indicates that for the case illustrated the starting torque is somewhat in excees of rated torque, which is fairly typical of such machines. The starting torque is computed in the same manner as torque is computed for any value of slip. Here it merely requires using $s = 1$. Thus the magnitude of the rotor current at standstill is

$$I_2' = \frac{V_1}{\sqrt{(r_1 + r_2')^2 + (x_1 + x_2')^2}} \qquad (4\text{-}29)$$

The corresponding gap power is then

$$P_g = q_1 I_2'^2 r_2' = \frac{q_1 V_1^2 r_2'}{(r_1 + r_2')^2 + (x_1 + x_2')^2} \qquad (4\text{-}30)$$

It is interesting to note that higher starting torques can result from increased rotor copper losses at standstill.

At unity slip the input impedance is very low so that large starting currents flow. Equation (4-29) makes this apparent. In the interest of limiting this excessive starting current, motors whose ratings exceed 3 hp are usually started at reduced voltage by means of line starters. This matter is discussed further in Sec. 4-10. Of course, starting with reduced voltage also means a reduction in the starting torque.

In fact if 50% of the rated voltage is used upon starting, then clearly by Eq. (4-30) it follows that the starting torque is only one-quarter of its full-voltage value.

Another important torque quantity of the three-phase induction motor is the maximum developed torque. This quantity is so important that it is frequently the starting point in the design of the induction motor. The maximum (or breakdown) torque is a measure of the reserve capacity of the machine. It frequently has a value of 200 to 300% of rated torque. It permits the motor to operate through momentary peak loads. However, the maximum torque cannot be delivered continuously because the excessive currents that flow would destroy the insulation.

Since the developed torque is directly proportional to the gap power, it follows that the torque is a maximum when P_g is a maximum. Also, P_g is a maximum when there is a maximum transfer of power to the equivalent circuit resistor r_2'/s. Applying the maximum-power-transfer theorem to the approximate equivalent circuit leads to the result that

$$\frac{r_2'}{s_m} = \sqrt{r_1^2 + (x_1 + x_2')^2} \tag{4-31}$$

That is, maximum power is transferred to the gap power resistor r_2'/s when this resistor is equal to the impedance looking back into the source. Accordingly the slip s_m at which the maximum torque is developed is

$$\boxed{s_m = \frac{r_2'}{\sqrt{r_1^2 + (x_1 + x_2')^2}}} \tag{4-32}$$

Note that the slip at which the maximum torque occurs may be increased by using a larger rotor resistance. Some induction motors are in fact designed so that the maximum torque is available as a starting torque, i.e., $s_m = 1$.

With the slip s_m known, the corresponding rotor current can be found and then inserted into the torque equation to yield the final form for the breakdown torque. Thus

$$\boxed{T_m = \frac{1}{\omega_s} q_1 I_2'^2 \frac{r_2'}{s_m} = \frac{1}{\omega_s} \frac{q_1 V_1^2}{2[r_1 + \sqrt{r_1^2 + (x_1 + x_2')^2}]}} \tag{4-33}$$

An examination of Eq. (4-33) reveals the interesting information that the maximum torque is independent of the rotor winding resistance. Thus increasing the rotor winding resistance increases the slip at which the breakdown torque occurs, but it leaves the magnitude of this torque unchanged. Figure 4-13 shows the effect of increasing the rotor resistance on a typical torque-speed curve.

A word of caution is appropriate at this point. Keep in mind that the results appearing in Eqs. (4-29), (4-30), (4-32), and (4-33) are predicated on the approximate equivalent circuit, which is used whenever the voltage drop across the primary leakage impedance is small compared to the stator induced emf per phase. This assumption is quite valid for normal operating conditions spanning the region from no-load to rated load. However, when interest becomes focused on maximum torque or starting torque at rated voltage, then the primary leakage voltage drop begins to

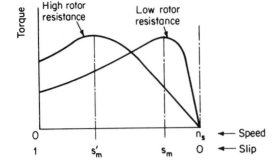

Torque

High rotor
resistance

Low rotor
resistance

0   s'_m s_m 0   ← Slip
1

n_s ← Speed

Figure 4-13 Showing the effect of increased rotor resistance on torque-speed curves.

exert some noticeable influence. Therefore, if more accurate results are required, we can base the analysis on the more exact equivalent circuit of Fig. 4-7(f) (see Example 4-4). Usually, errors of the order of 3 % are incurred when these calculations are based on the simpler, approximate equivalent circuit. Because in practical situations the extent of these errors is often less than the effects due to saturation and skin effects, which are not accounted for either, our policy here is to stay with the approximate equivalent circuit.

Example 4-3

Refer to the induction motor of Example 4-2. Assuming that operation takes place at full-rated voltage with all saturation effects neglected, determine (a) the slip at which the maximum torque is developed, (b) the current at maximum torque, and (c) the value of the maximum torque. Compare these results with those found in Example 4-2.

Solution The solution is obtained here employing the results obtained from use of the approximate equivalent circuit.

(a) Application of Eq. (4-32) yields

$$s_m = \frac{r'_2}{\sqrt{r_1^2 + (x_1 + x'_2)^2}} = \frac{0.12}{\sqrt{(0.1)^2 + (0.75)^2}} = 0.159$$

(b) The rotor current referred to the stator is

$$I'_2 = \frac{277.13}{r_1 + (r'_2/s) + j(x_1 + x'_2)} = \frac{277.13}{0.1 + (0.12/0.159) + j0.75} = \frac{277.13}{1.137\underline{/41.3°}}$$
$$= 243.74\underline{/-41.3°} \text{ A}$$

This current is about 3.83 times as large as the current when the motor delivers a little in excess of 57 hp as computed in Example 4-2 at near rated conditions.

(c) The maximum torque follows directly from Eq. (4-33). Thus

$$T_m = \frac{q_1 V_1^2}{2\omega_s} \frac{1}{r_1 + \sqrt{r_1^2 + (x_1 + x'_2)^2}} = \frac{3(277.13)^2}{2(188.5)} \frac{1}{r_1 + \sqrt{r_1^2 + (x_1 + x'_2)^2}}$$
$$= (611.5)\frac{1}{0.1 + \sqrt{(0.1)^2 + (0.75)^2}}$$
$$= 611.15(1.181) = 721.77 \text{ N-m}$$

which is approximately three times as large as the value found near rated conditions in Example 4-2.

Example 4-4

Repeat Example 4-3 using the exact equivalent circuit in place of the approximate one.

Solution In order to use the exact equivalent circuit, the value of the magnetizing reactance, x_ϕ, is needed. This is readily found from the data appearing in Example 4-2 as

$$x_\phi = \frac{V_1}{I_\phi} = \frac{277.13}{17.95} = 15.44\ \Omega$$

The exact equivalent circuit appears in Fig. 4-14(a). Observe that the core-loss resistor is omitted for simplicity; its absence causes negligible effects.

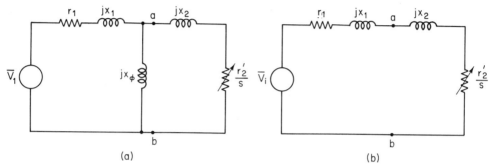

Figure 4-14 (a) Exact equivalent circuit with the core-loss resistor omitted; (b) the Thévenin equivalent of (a).

To obtain a result analogous to Eq. (4-31) by reducing the circuit in Fig. 4-14(a) to an equivalent series circuit, we invoke Thévenin's theorem. We replace the circuit to the left of terminals ab by an equivalent voltage source of value

$$\bar{V}_t = \frac{jx_\phi}{r_1 + j(x_\phi + x_1)}\bar{V}_1$$

and an internal impedance which is simply the parallel combination of $(r_1 + jx_1)$ and jx_ϕ. Thus

$$\bar{Z}_t = \frac{jx_\phi(r_1 + jx_1)}{r_1 + j(x_1 + x_\phi)} = R_t + jX_t$$

The Thévenin equivalent circuit is shown in Fig. 4-14(b). It is instructive to note that this circuit is now equivalent in form to the approximate equivalent circuit which upon analysis led to Eq. (4-31). The role of V_1 is now replaced by $\bar{V}_t$, and R_t replaces r_1, and X_t replaces x_1. Accordingly, Eqs. (4-32) and (4-33) are now applicable to the exact equivalent circuit provided that the just-mentioned replacements are made. Hence, before proceeding with the evaluation of the items specified in this example, we first determine these replacement quantities.

$$\bar{V}_t = \frac{j14.2}{0.1 + j(14.2 + 0.35)} 277.1\underline{/0^\circ} = 270.5\underline{/0.39^\circ}\ \text{V}$$

$$Z_t = \frac{j14.2(0.1 + j0.35)}{0.1 + j14.55} = 0.09 + j0.343 = R_t + jX_t$$

Observe the relatively small change that occurs in the value of R_t from r_1 and X_t from x_1. In other words, the shunting effect of the magnetizing reactance is not very great.

(a) The slip at which the maximum torque now occurs is

$$s_m = \frac{r_2'}{\sqrt{R_i^2 + (X_i + x_2')^2}} = \frac{0.12}{\sqrt{(0.09)^2 + (0.343 + 0.45)^2}} = 0.161$$

This represents a change of slightly more than 1% from the value found using the approximate equivalent circuit.

(b) The value of the current at maximum torque is

$$I_2' = \frac{V_t}{R_t + r_2' + j(X_t + x_2')} = \frac{270.5}{0.09 + 0.7453 + j(0.343 + 0.4)} = 21.9/41.6° \text{ A}$$

Here again the difference is about 1%. Of course, there are motors where the differences can be of the order of 3 or 4%. In such cases the use of the more exact equivalent circuit might appear to be justified. But it is important to note here that this amounts to quibbling because there is a more important effect taking place at the maximum torque condition which we have not accounted for. With currents as large as three times the rated current flowing at the maximum torque condition, a considerable amount of saturation occurs in the teeth of the stator and rotor. A subsequent decrease in the values of the stator and rotor leakage reactances of the order of 50% and more can easily occur. *The effect of this change is much more serious than whether the approximate or the exact version of the equivalent circuit is used.*

(c) Continuing with the original premise in the statement of the problem as set forth in Example 4-2, we get for the maximum torque

$$T_m = \frac{q_1 V_t^2}{2\omega_s[R_t + \sqrt{R_i^2 + (X_t + x_2')^2}]} = \frac{3(270.5)^2}{2(188.5)[0.09 + \sqrt{(0.09)^2 + (0.343 + 0.4)^2}]}$$

$$= (582.26)(1.193) = 694.77 \text{ N-m}$$

Here the difference is nearly 4%.

4-7 EQUIVALENT CIRCUIT PARAMETERS FROM NO-LOAD TESTS

The performance computations for a three-phase induction motor presuppose knowledge of the parameters of the equivalent circuit. This information may be available either from design data or from appropriate tests. When the design data are not available information about the magnetizing branch can be obtained from a no-load test that is performed by applying a balanced three-phase voltage to the motor which is uncoupled from its load. Often the input power and line current are measured at rated voltage and rated frequency. Sometimes this no-load test is performed with a varying voltage, which leads to the plot depicted in Fig. 4-15. Point a corresponds to operation at rated voltage. Experimental points below b are not taken because the speed will no longer be close to synchronous speed. Extrapolation of the curve to the ordinate axis gives a good indication of the friction and windage loss at normal speeds.

At no-load and rated voltage the input power is used to supply three losses: the stator copper loss, the stator core loss P_c, and the rotational losses P_{rot}. In equation form we have

$$P_0 = q_1 I_0^2 r_1 + P_c + P_{rot} \tag{4-34}$$

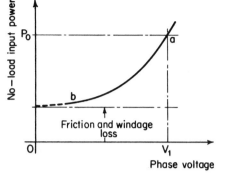

Figure 4-15 Showing the variation of no-load input power with applied voltage; ab is the curve obtained from experimental data.

where q_1 denotes the number of stator phases and r_1 is the effective stator resistance per phase. The rotational losses include the friction and windage losses as well as iron losses caused by the pulsations of flux in the stator teeth as the rotor revolves. These iron losses due to rotation are larger in machines with open slots than in those with semiopen slots. When specific information about P_c or P_{rot} is not known, it is customary to assume these quantities to be equal for they are frequently found to be so in conventional machines. Based on this assumption, the current through the core-loss resistor can then be found as follows:

$$I_c = \frac{P_c}{q_1 V_1} = \frac{(P_0 - q_1 I_0^2 r_1)}{2q_1 V_1} \tag{4-35}$$

Hence

$$\boxed{r_c = \frac{V_1}{I_c}} \tag{4-36}$$

To find the magnetizing reactance we must first obtain the reactive component of the no-load current, I_ϕ. The power-factor angle at no-load is found from

$$\theta_0 = \cos^{-1} \frac{P_0}{q_1 V_1 I_0} \tag{4-37}$$

Therefore

$$I_\phi = I_0 \sin \theta_0 \tag{4-38}$$

so that

$$\boxed{x_\phi = \frac{V_1}{I_\phi}} \tag{4-39}$$

The phasor diagram at no-load is depicted in Fig. 4-16. The angle θ_0 is usually large ($65°$ to $80°$) because of the need for a large magnetizing current to produce the required flux per pole in a magnetic circuit containing air gaps. The total in-phase component is represented by three terms to take care of the three losses appearing in Eq. (4-34). Note that the phasor sum of $\bar{I}_c$ and $\bar{I}_\phi$ yields the total magnetizing current $\bar{I}_m$ of the magnetizing branch of the equivalent circuit. The quantity $\bar{I}'_{20}$ is assumed to

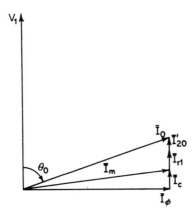

Figure 4-16 Phasor diagram of the induction motor at no-load.

be equal to $\bar{I}_c$ and represents the small rotor current that flows at no-load to supply the rotational losses. Specifically, this quantity may be written as

$$I'_{20} = \frac{P_{\text{rot}}}{q_1 V_1} = \frac{P_0 - q_1 I_0^2 r_1}{2q_1 V_1} \tag{4-40}$$

The third part of the in-phase component of the no-load current, I_{r1}, represents the stator copper loss at no-load. That is,

$$I_{r1} = \frac{I_0^2 r_1}{V_1} \tag{4-41}$$

This too can be readily computed from the measured no-load data.

Information about the winding resistances and leakage reactances is obtained from a blocked-rotor test. This test is analogous to the short-circuit test of the transformer. It requires that the rotor shall be blocked to prevent rotation and that the rotor winding shall be short-circuited in the usual fashion. Furthermore, since the slip is unity, the mechanical load resistor R_m is zero and so the input impedance of the equivalent circuit is quite low. Hence, in order to limit the rotor current in this test to reasonable values, a reduced voltage must be used—usually about 10 to 25% of rated value. Moreover, operation at such reduced voltages renders the core loss as well as the magnetizing current negligibly small. Accordingly, the equivalent circuit in this test takes on the configuration shown in Fig. 4-17.

Assume now that the following instrument readings are taken in performing a blocked-rotor test on a Y-connected three-phase wound-rotor induction motor:

P_b = total wattmeter reading, W

I_b = line current of Y connection

V_b = line voltage of Y connection

Figure 4-17 Induction motor equivalent circuit for the rotor-blocked test.

Assume too that the d-c phase winding resistances of the stator and rotor windings, $r_{1 \text{ d-c}}$ and $r_{2 \text{ d-c}}$, are made available from a simple d-c test and that the ratio of transformation, a, from stator to rotor is also available. Keep in mind that the instrument readings are consistent with operation at the rated frequency of the motor. Hence, if present, the influence of skin effect† is reflected in these readings. From the foregoing measurements it follows that the effective value of the equivalent winding resistance is given by

$$R_e \equiv r_1 + r_2' = \frac{P_b}{3 I_b^2} \qquad (4\text{-}42)$$

Here both r_1 and r_2' denote effective winding resistances per phase. To separate out the effective value of r_1 use is made of the following equation:

$$r_1 = \frac{r_{1 \text{ d-c}}}{r_{1 \text{ d-c}} + a^2 r_{2 \text{ d-c}}} R_e \qquad (4\text{-}43)$$

This last expression states that the ratio of the effective value of r_1 to the effective value of the equivalent resistance bears the same relationship as the d-c value of r_1 to the d-c value of the equivalent resistance. It is important at this point to understand that, in the equivalent circuit of the induction motor as it is used to compute performance at normal values of slip, the r_1 to be used is the effective value as determined by Eq. (4-43) but the r_2 to be used is the d-c value. The latter quantity is used because at normal slips (3 to 5%) the frequency of the rotor current is very low (2 to 3 Hz).

The equivalent phase impedance is obtained from

$$Z_e = \frac{V_b}{\sqrt{3} \, I_b} \qquad (4\text{-}44)$$

Finally, the equivalent leakage reactance is determined from

$$X_e = \sqrt{Z_e^2 - R_e^2} = x_1 + x_2' \qquad (4\text{-}45)$$

Note that, as long as computations are carried on using the approximate equivalent circuit, it is sufficient to deal directly with X_e without a further breakdown into x_1 and x_2'.

In view of the importance of working with the d-c value of r_2 in the equivalent circuit for low values of slip, how is this value found for the squirrel-cage induction motor if the design data are not available? The use of the effective value of r_2 leads to very serious error because the skin effect in the squirrel cage is very large. This is caused by the use of solid copper bars imbedded in iron. The ratio of effective to d-c value at 60 Hz is often 3 : 1. Such large ratios do not occur in the wound-rotor machine because of the use of stranded conductors. Since the phase terminals of a squirrel

---

†Skin effect refers to the tendency of time-varying current to crowd toward the surface of a conductor which reduces the effective cross-sectional area and thereby increases the resistance over the d-c value.

cage are not available, an indirect means of determining the d-c value of the rotor phase resistance must be employed. If the blocked-rotor test is performed for various frequencies at reduced voltage, the effective equivalent resistance can be found and plotted to yield the curve depicted in Fig. 4-18. Note that as the frequency is reduced the effective equivalent resistance assumes correspondingly smaller values. By extrapolating the experimental curve to the ordinate axis the d-c equivalent resistance for the squirrel-cage motor is obtained. Then the stator-referred d-c resistance per phase of the rotor is found as

$$r'_{2\,\text{d-c}} = R_{e\,\text{d-c}} - r_{1\text{d-c}} \qquad\qquad (4\text{-}46)$$

Accordingly, in the equivalent circuit for the squirrel-cage induction motor at low slips, the effective value of r_1 and the d-c value of r'_2 [Eq. (4-46)] must be used.

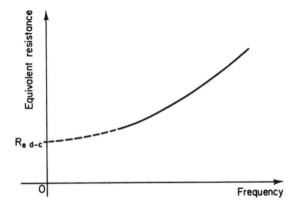

Figure 4-18 Illustrating how d-c equivalent resistance of a squirrel-cage induction motor can be found from a blocked rotor test for varying frequency.

Example 4-5

The following no-load tests are performed on a three-phase, four-pole, 60-Hz, 20-hp, 550-V, Y-connected squirrel-cage induction motor and the results are found to be:

No-load test:	550 V,	5.8 A,	754 W,	60 Hz
Blocked-rotor test at rated frequency:	123 V,	25 A,	2419 W,	60 Hz
Blocked-rotor test at low frequency:	55 V,	25 A,	2063 W,	15 Hz
D-c test on stator per phase:	15 V,	25 A		

The friction and windage losses for this motor are known to be 328 W. Determine the parameters of the approximate equivalent circuit to be used at normal operating slips (i.e., 3 to 5%).

Solution The d-c winding resistance per phase is

$$r_{1\text{dc}} = \frac{15}{25} = 0.6\ \Omega$$

Although the stator winding is a wound winding using stranded wire, some small skin effect is present. Therefore, the a-c value of the winding resistance is needed. This can be determined from the blocked-rotor test performed at rated frequency. Thus

$$R_{e_{\text{ac}}} = \frac{2419}{3(25)^2} = 1.29\ \Omega$$

From the blocked rotor test which is performed at low frequency, the skin effect is not considered to be important and so the equivalent "d-c" resistance is found to be

$$R_{e_{dc}} = \frac{2063}{3(25)^2} = 1.1 \, \Omega$$

Then, in accordance with Eq. (4-43), the a-c stator winding resistance has the value

$$r_{1_{ac}} = r_{1_{dc}} \frac{R_{e_{ac}}}{R_{e_{dc}}} = (0.6)\left(\frac{1.29}{1.1}\right) = 0.7 \, \Omega$$

This value is a bit on the pessimistic side for the reason that during the blocked-rotor test at full frequency, the skin effect that takes place in the iron-embedded bars of the rotor is greater than for the stranded wire of the stator winding. Nonetheless, this is one of the six values of the equivalent circuit. A second parameter, the "d-c" value of the rotor resistance referred to the stator, is determined by applying Eq. (4-46), which here yields

$$r'_{2_{dc}} = R_{e_{dc}} - r_{1_{dc}} = 1.1 - 0.6 = 0.5 \, \Omega$$

The use of the approximate equivalent circuit means that the two parameters x_1 and x'_2 may be evaluated and used as a lumped quantity. Because these leakage reactances are the standstill values, the data of the rotor-blocked test at rated frequency are employed. Accordingly,

$$Z_e = \frac{V_b}{\sqrt{3} \, I_b} = \frac{123}{\sqrt{3}\,(25)} = 2.84 \, \Omega$$

$$X_e = \sqrt{Z_e^2 - R_{e_{ac}}^2} = \sqrt{2.84^2 - 1.29^2} = 2.53 \, \Omega$$

The two remaining parameters are found from the no-load test data. By Eq. (4-37) the no-load power-factor angle is

$$\theta_0 = \cos^{-1}\frac{P_0}{q_1 V_1 I_0} = \cos^{-1}\frac{754}{3(550/\sqrt{3})5.8} = 82.2°$$

so that the corresponding value of the reactive component of the no-load current becomes

$$I_\phi = I_0 \sin \theta_0 = 5.8 \sin 82.2° = 5.75 \text{ A}$$

which in turn yields a magnetizing reactance of

$$x_\phi = \frac{V_1}{I_\phi} = \frac{550}{\sqrt{3}\,(5.75)} = 55.2 \, \Omega$$

Finally, the core-loss resistor is found by first calculating the value of core loss from

$$P_c = P_0 - P_{fw} - 3I_0^2 r_{1_{ac}} = 754 - 328 - 3(5.8^2)(0.7)$$
$$= 355.4 \text{ W}$$

and then noting that

$$q_1 \frac{V_1^2}{r_c} = P_c$$

Hence the value of the core-loss resistor computes to be

$$r_c = \frac{qV_1^2}{P_c} = \frac{3(550)^2}{(\sqrt{3})^2 P_c} = \frac{(550)^2}{355.4} = 851.2 \, \Omega$$

The manner of computing the performance of the polyphase induction motor by the equivalent circuit is described in Sec. 4-4. In this section we direct attention to a graphical solution of the approximate equivalent circuit of Fig. 4-7(g), not because we want to belabor unnecessarily the computational procedures for finding motor performance, but rather because the graphical solution has several notable advantages to offer. First, it makes available at a glance such important characteristics as the maximum developed torque and the maximum horsepower output. These quantities are a measure of the reserve capacity of the machine and are not so readily obtained experimentally because of the very large currents that flow at these conditions (three to four times rated current). Second, it is also possible at a glance to determine the operating point for maximum efficiency or the maximum line power factor. Third, the graphical solution offers in one representation a full panoramic view of the way the operating performance is affected by changes in machine parameters, saturation, and line voltage and frequency.

A graphical solution of the equivalent circuit of Fig. 4-7(g) essentially involves finding for the current $\bar{I}_1$ the values of slip over the range $0 \leq s \leq 1$. Since $\bar{I}_1$ is the phasor sum of the magnetizing current and the rotor current referred to the stator, the graphical solution consists of two parts. At constant applied voltage the magnetizing current remains essentially the same irrespective of the value of slip. Hence this part of the graphical solution of $\bar{I}_1$ is identical with that shown in Fig. 4-16 and repeated in the left portion of Fig. 4-20. Point P_0 is the actual no-load point where the slip assumes some very small value to supply the rotational losses. Point P_0' is the true point† for which $s = 0$, and experimentally it corresponds with the situation where the rotor of the induction motor is driven by an auxiliary machine at exactly synchronous speed. This then relieves the no-load current of the burden of supplying I_{20}'. The in-phase component of the current for operation at P_0' is just large enough to supply the no-load stator copper loss and the stator core losses. These losses are represented by the distance AD and remain constant for all values of slip.

The second part of the graphical solution involves the stator-referred rotor current. From the equivalent circuit of Fig. 4-7(g) the expression for the magnitude of this current in terms of the fixed phase voltage V_1 and the circuit parameters is

$$I_2' = \frac{V_1}{\sqrt{[r_1 + (r_2'/s)]^2 + (x_1 + x_2')^2}} \tag{4-47}$$

Also, for this portion of the circuit we can write

$$\sin \theta = \frac{x_1 + x_2'}{\sqrt{[r_1 + (r_2'/s)]^2 + (x_1 + x_2')^2}} \tag{4-48}$$

where θ is the impedance angle for this portion of the equivalent circuit. Inserting the last expression into Eq. (4-47) yields

$$\boxed{I_2' = \frac{V_1}{x_1 + x_2'} \sin \theta} \tag{4-49}$$

---

†This fine detail is introduced not so much in the interest of improving accuracy as to provide a one-to-one correspondence with the power flow diagram.

which is readily recognized as the equation of a circle in polar form. A plot of Eq. (4-49) appears in Fig. 4-19. Note that this is a circle, the center of which is displaced along the horizontal axis by $\frac{1}{2}(V_1/x_1 + x_2')$. Of course the diameter of the circle is equal to $V_1/(x_1 + x_2')$. Hence one could construct this circle once the equivalent leakage reactance is known.

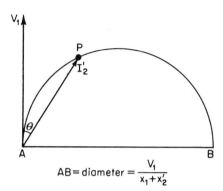

$$AB = \text{diameter} = \frac{V_1}{x_1 + x_2'}$$

Figure 4-19 Graphical representation of Eq. (4-49), showing the locus of I_2' as slip increases.

The complete graphical solution for the stator current $\bar{I}_1$ is found by combining the results depicted in Figs. 4-16 and 4-19 in the manner shown in Fig. 4-20. As values of slip increase, the locus of $\bar{I}_1$ moves along the upper semicircle in the clockwise direction.

Construction of the circle diagram proceeds as follows: First, select a suitable scale factor for current—one that yields a large enough circle to permit fairly good accuracy in making performance calculations. Call this quantity k_I and assume it

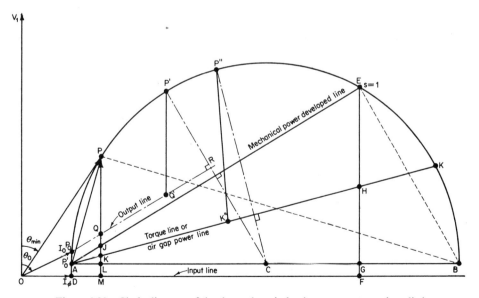

Figure 4-20 Circle diagram of the three-phase induction motor at rated applied voltage. For clarity I_0 is drawn out of proportion to the circular locus for I_2'.

bears units of amperes per centimeter. From the no-load test data mark off the distance representing the no-load current (i.e., $OP_0 = I_0/k_I$) at the power-factor angle θ_0 [see Eq. (4-37)]. Next compute I'_{20} from Eq. (4-40), and, after conversion to an appropriate length, mark it off along the vertical from P_0 to obtain P'_0. For convenience call this point A, and keep in mind that it represents one point on the circle, i.e., $s = 0$. The quantity AD represents that portion of the in-phase component of the no-load current that supplies the stator core losses and the no-load stator copper loss. Since these losses are constant, the diameter of the circle will lie on the horizontal line passing through A. Although the radius of the circle, AC, could be found by taking one-half of $V_1/(x_1 + x'_2)$ and subsequently drawing the circle, this procedure is not preferred because small errors in determining the equivalent leakage reactance reflect as large errors in the radius of the circle. A better procedure is to locate a second point on the circle, i.e., the standstill point $s = 1$. Then, with this plus the first point $s = 0$ and the knowledge that the diameter lies on the horizontal line passing through A, a more reliable determination of the location of the center of the circle is achieved. From the blocked-rotor test performed at the reduced line voltage V_b compute the power-factor angle at blocked conditions from

$$\theta_b = \cos^{-1}\frac{P_b}{\sqrt{3}\,V_b I_b} \tag{4-50}$$

where P_b denotes the total wattmeter reading and I_b denotes the line current. Keeping in mind our interest in drawing the circle diagram for rated voltage conditions, we prorate the measured blocked-rotor line current I_b to rated voltage by calculating

$$I'_b = \frac{V_L}{V_b}I_b \tag{4-51}$$

where V_L denotes the rated line voltage and I'_b denotes the value of blocked-rotor current for rated voltage. Equation (4-51) is written on the assumption that the equivalent impedance remains essentially constant. Next, starting at A, draw a line displaced from the vertical by θ_b. On this line mark off the distance $AE = I'_b/k_I$. Thus point E represents the standstill condition for operation at rated voltage. Furthermore, because AE is a chord of the circle, the construction of the perpendicular bisector of AE is found to intersect the line along which the diameter is known to lie (i.e., AB) at the center of the circle, C. Then, with C as center and AC as radius the circle is drawn.

Let us consider for the moment operation of the induction motor at rated voltage with the rotor standing still, $s = 1$. This refers to the condition that prevails when the motor is started with rated voltage applied, a custom frequently followed for ratings of 3 hp or less. The total full-voltage starting current is proportional to the distance OE or specifically expressed in amperes is equal to $(OE)k_I$. Usually this current is about 5 to 10 times the rated current. The stator-referred rotor current is given by $(AE)k_I$ and is practically equal to the stator current. Since the rotor is not yet turning, there is no mechanical output. Hence the input power must be used to supply entirely the machine losses. This input power is represented by EF, which is the in-phase component of OE. Since the construction is such that $FG = AD$, it follows that the stator core losses and the copper losses associated with the no-load current are accounted for by the length of this line. The remainder of the input power must then be used to supply the total rotor copper loss and the increased stator copper loss associated with the flow of a rotor current. The latter loss is represented by HG which is that portion of EG that satisfies the relation

$$V_1(HG)k_I = (I_b')^2 r_1 - I_0^2 r_1 \approx (I_b')^2 r_1 \qquad (4\text{-}52)$$

It is helpful to keep in mind here that I_b' is from 20 to 40 times as large as I_0. Hence

$$HG = \frac{(I_b')^2 r_1}{k_I V_1} \approx \frac{(AE)^2 k_I r_1}{V_1} \qquad (4\text{-}53)$$

Since all the quantities on the right side are known, point H is easily found. Of course, it should now be obvious that EH represents the total rotor copper loss. This line length should satisfy the equation

$$V_1(EH)k_I = [(AE)k_I]^2 r_2' \qquad (4\text{-}54)$$

or

$$EH = \frac{(AE)^2 k_I r_2'}{V_1} \qquad (4\text{-}55)$$

It is interesting to note that the line EG is split in accordance with the ratio of r_2'/r_1. This conclusion readily follows from dividing Eq. (4-55) by Eq. (4-53). Thus,

$$\frac{EH}{HG} = \frac{r_2'}{r_1} \qquad (4\text{-}56)$$

This last expression provides an alternative means of locating point H.

The length of line EH can also be used to represent the starting torque at rated voltage. From the power flow diagram of Fig. 4-8(b) we see that at $s = 1$ the gap power is given by

$$P_g = q_1(I_2')^2 r_2' = q_1[(AE)k_I]^2 r_2' = T\omega_s \qquad (4\text{-}57)$$

Upon inserting Eq. (4-54) into Eq. (4-57) the desired result is obtained:

$$T_s \equiv T = \frac{q_1 V_1 k_I}{\omega_s}(EH) = k_T(EH) \qquad (4\text{-}58)$$

where T_s is the electromagnetic torque developed at starting. Note that the starting torque depends directly upon the amount of rotor copper loss at standstill. It is also worthwhile noting that the length of line EH can be converted directly to torque expressed in newton-meters by using the torque scale factor k_T, which is expressed from Eq. (4-58) in terms of the current scale factor as

$$k_T = \frac{q_1 V_1 k_I}{\omega_s} \quad \text{N-m/cm} \qquad (4\text{-}59)$$

Attention is next turned to a consideration of the performance of the motor when it operates at a point such as P in Fig. 4-20. For convenience P is selected to correspond to operation at the maximum line power factor. This condition is imposed by drawing a line from the origin tangent to the circle. The input line current is accordingly found to be

$$I_1 = (OP)k_I \quad \text{A} \qquad (4\text{-}60)$$

and the corresponding rotor current referred to the stator becomes

$$I_2' = (AP)k_I \quad \text{A} \qquad (4\text{-}61)$$

The input power for the operating point is

$$P_i = q_1 V_1 [k_I (OP) \cos \theta_{\min}] = q_1 V_1 k_I (PM) = k_p (PM) \qquad (4\text{-}62)$$

Again note that a new scale factor—this time for power—can be obtained in terms of the scale factor for current. Thus,

$$k_p = q_1 V_1 k_I \qquad \text{W/cm} \qquad (4\text{-}63)$$

By multiplying any line length that is in phase with the V_1 direction by k_p, a quantity in watts is obtained. The significance of this quantity of course depends upon the particular distance used. Obviously, when the distance PM is used, the result is the total input power. Similarly, when LM is used, the stator core losses and the no-load stator copper loss are represented. An interesting question at this point concerns the meaning of the power that results when the length JL is used in conjunction with the power scale factor. That is, what is the significance of $P = k_p(JL)$? The answer is simply that it represents the rotor copper loss plus the stator copper loss (except for the no-load stator copper loss which is part of LM). To understand this answer note that by similar triangles we can write

$$\frac{JL}{EG} = \frac{AL}{AG} = \frac{AP \cos < PAB}{AE \cos < EAB} = \frac{AP(AP/AB)}{AE(AE/AB)} = \frac{(AP)^2}{(AE)^2} \qquad (4\text{-}64)$$

It has already been shown that at standstill the stator and rotor copper losses are represented by line EG. Also, these losses are determined by the square of the current denoted by AE. Accordingly, since JL is proportional to the square of the current AP, the quantity $k_p(JL)$ must have a similar interpretation. Thus it is equal to the rotor copper loss plus the additional stator copper loss associated with the flow of rotor current. Moreover, as was the case at standstill, this loss is divided between the stator and rotor in accordance with $JK/KL = r_2'/r_1$.

Finally, what is the meaning of the quantity $k_p(QJ)$? First, it should be understood that point Q is obtained by drawing a line parallel to line AE through P_0 and extended to about the location of the perpendicular bisector of line AE. This construction makes QJ equal to $P_0 A$. Since $P_0 A$ is a length that represents the rotational losses, it follows that $k_p(QJ)$ is equal to the rotational losses expressed in watts.

Considering that JL represents the variable copper losses with load and QJ and LM represent the fixed losses, it should be apparent that the quantity $k_p(QM)$ is equal to the sum of all losses in watts for the operating point P. The output power must then be equal to $k_p(PQ)$ watts.

It is worthwhile at this point to make some observation concerning the meaning and usefulness of several lines appearing in the circle diagram of Fig. 4-20. Because the vertical distance between any point on the circle and line OF represents the input power, OF is often called the *input line*. Similarly, the vertical distance from any point on the circle to line $P_0 R$ is a measure of the output. Hence line $P_0 R$ is called the *output line* and is useful up to the point of about maximum developed torque. Beyond this point the slip increases sufficiently to require an adjustment in the level of the rotational losses. As we move upwards along line PM from point M, it is helpful to note the correspondence with the power flow diagram of Fig. 4-8. PM represents power input, LM represents the stator core losses and the no-load stator copper loss, and KL represents the additional stator copper loss under load. Hence PK must represent the power transferred across the air gap. That is,

$$P_g = k_p(PK) \qquad (4\text{-}65)$$

where k_p is given by Eq. (4-63). For this reason line AH is called the *air-gap power line*. Moreover, because the developed torque is related to the gap power by a constant $T = (1/\omega_s)P_g$, this line is also called the *torque line*. Hence at any operating point the developed torque can be found from

$$T = k_T(PK) \qquad (4\text{-}66)$$

where k_T is given by Eq. (4-59). The removal of the rotor copper loss from the gap power yields the mechanical power developed, and this is represented by line PJ. Therefore line AE is referred to as the *mechanical power developed line*. Finally, the extraction of the rotational losses from the mechancial power leaves the output power as represented by distance PQ.

The computation of motor performance as obtained from the circle diagram can be summarized as follows:

$$P_i = \text{input power} = k_p(PM) \qquad (4\text{-}67)$$

$$P_0 = \text{output power} = k_p(PQ) \qquad (4\text{-}68)$$

$$\eta = \text{efficiency} = \frac{PQ}{PM} \qquad (4\text{-}69)$$

$$pf = \text{power factor} = \frac{MP}{OP} \qquad (4\text{-}70)$$

$$s = \text{slip in per unit} = \frac{\text{rotor copper loss}}{\text{gap power}} = \frac{JK}{PK} \qquad (4\text{-}71)$$

Note that the last three quantities result from the ratio of line lengths that are readily measured from the circle diagram.

To obtain the operating point corresponding with that at which the motor develops its maximum output horsepower, we must locate that point along the arc AE for which PQ is a maximum. A little thought reveals that this point is easily found by drawing a line from the center perpendicular to the output line and intersecting the circle at P'. Then

$$\text{maximum } P_0 = k_p(P'Q') \qquad (4\text{-}72)$$

and

$$\text{maximum } HP_0 = \frac{k_p(P'Q')}{746} = k_{HP}(P'Q') \qquad (4\text{-}73)$$

In a similar fashion the maximum torque developed by the motor is found by constructing a line from the center perpendicular to the torque line and intersecting the circle at P''. Then

$$T_{\max} = k_T(P''K'') \qquad (4\text{-}74)$$

Both Eqs. (4-73) and (4-74) provide information about the motor that is not readily obtained experimentally.

What are some of the principal assumptions involved in the use of the circle diagram for determining motor performance? The use of the approximate equivalent circuit rather than the more exact form of Fig. 4-7(f) is certainly one factor.

Another is the assumption that the leakage reactances remain constant. Recall that the equivalent leakage reactance is found in the blocked-rotor test for I_b in the vicinity of rated current. This current is then prorated to rated voltage by calculation to locate point E in Fig. 4-20. However, if I'_b were found experimentally for rated voltage, the excessive magnitude of I'_b would cause saturation of the iron of the rotor and stator teeth which would cause a reduction of the leakage reactances. This reduction could mean a noticeable change in the circle diameter. Although the prorated procedure for finding point E is to be preferred because it causes no error in the normal operating range, it should be realized that some error will occur when computing maximum horsepower and maximum torque since much larger currents are associated with these operating points. Another assumption concerns the manner in which the stator copper loss is computed. A review of the outlined procedure shows that the stator copper loss for any operating point is found by adding to the no-load stator copper loss the additional loss associated with the stator-referred rotor current. The correct procedure is to use the total primary current. This situation is a direct consequence of using the approximate equivalent circuit. However, it is assumed that one method is virtually equivalent to the other. The difference is generally inconsequential because even for an error as much as 20% in a quantity which may be only 3% of the output power the net effect is 0.6 of 1%. This certainly is less than the accuracy of the construction of the circle diagram itself.

4-9 SPEED CONTROL

In most industrial applications the essentially constant speed characteristic of the induction motor is a desirable one. However, there are some applications (e.g., conveyors, hoists, and elevators) where speed control capability is a transcendent factor. Therefore in this section we explore the extent to which the induction motor lends itself to adjustment of speeds under various load conditions.

The squirrel-cage induction motor is examined first because it is simpler owing to the closed nature of its rotor winding, which eliminates all possibility of control from that end. Any control that is possible must take place through manipulations at the stator. Here three possibilities present themselves, two of which are evident from Eq. (4-12). The first is to *change the line frequency*. As the line frequency is raised or lowered the synchronous speed is correspondingly raised or lowered, which provides satisfactory control. However, the serious shortcoming of this procedure is that a variable frequency source is not available generally. Moreover, if such a source is to be provided locally, a motor-generator set is needed equipped with appropriate controls and power rated at a value exceeding the rating of the induction motor it is to control. Furthermore, if maximum iron-flux densities are not to be exceeded as the frequency is dropped, corresponding decreases in the applied voltage must be effected. Consequently, this method of control rarely, if ever, is used because of these uneconomical aspects.

A second method of control is to *change the number of poles*. Recall that the number of poles is determined by the winding layout. It is possible to arrange the winding in appropriately paired sections for each phase. When the two sections, e.g., are connected in a way to ensure symmetrical orientations of current flow, a four-pole machine results. However, if the current through the second section is made to flow in the reversed direction by throwing a switch, a two-pole machine results. Accordingly, the synchronous speed can be changed by a factor of 2. Such a machine is called a *multispeed motor*. The motor can operate at one of two speeds, but unfortunately once a particular speed is selected there is no further control available. Speed control in this instance occurs in a discrete fashion. For a line frequency of 60 Hz such a machine allows operation near either 3600 or 1800 rpm. If speed control is desired between 1800 and 3600 rpm, it cannot be achieved with the multispeed squirrel-cage induction motor.

The third method of speed control involves *reducing the applied line voltage*. Operation at reduced voltages means that the ordinate values of torque on the torque-speed curve assume new values in accordance with the square of the reduced voltage ratio. Thus, if the squirrel-cage motor is operated at 70.7% of rated voltage, all points on the new torque-speed curve have half the value of the original curve. The maximum torque is reduced by one-half; the starting torque is also reduced to one-half its original value, etc. Figure 4-21 depicts the distinctions over the complete speed range. Also shown are the torque-speed curves of a constant-torque load so characteristic of conveyors and the variable load torque characteristic of fans and other propeller type loads. It is assumed that at rated voltage the induction motor delivers rated torque to each load type. The operating point is denoted by a, where the slip is the vicinity of about 5%. A reduction of 30% in the applied voltage causes the operating point for the constant-torque load to move from a to b and from a to c for the variable-torque load. Note that this scheme may allow the slip to double or perhaps triple. A little thought indicates that this is not an appreciable amount of speed adjustment—and certainly not for the price paid. Speed control by voltage

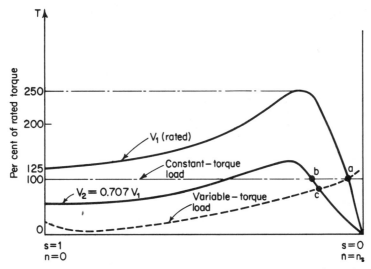

Figure 4-21 Illustrating how some degree of speed control is obtained by reducing the applied voltage.

control has serious disadvantages which account for its infrequent use, especially for polyphase motors. First, it is expensive. Variable three-phase voltage is obtained efficiently through the employment of a motor-generator set or a three-phase variac having ratings at least as large as the motor being controlled. Second, when the motor is operating at reduced voltage, its reserve capacity is dangerously reduced. For the situation illustrated in Fig. 4-21 note that the reserve capacity for the constant-torque load is 150% of rated torque at full voltage and only 125% at 70.7% of rated voltage. Third, if an attempt were made to operate the motor over the full range at the reduced voltage, a glance at Fig. 4-21 would make it clear that only the variable-torque load can be so operated. At reduced voltage the motor's starting torque of 62.5% is insufficient to move the constant-torque load.

In conclusion therefore it can be stated that speed adjustment of the squirrel-cage induction motor is difficult to achieve and expensive besides. This motor is best suited for constant-speed loads.

Speed control is easier to achieve in the wound-rotor induction motor than in the squirrel-cage motor because of the availability of the rotor terminals. Of course the adjustment of speed by the three methods already outlined apply with equal validity to the wound-rotor machine, but these are dismissed for the reasons previously cited.

Three additional methods by which the speed can be controlled include the following: increasing the total rotor reactance per phase, x_2'; increasing the total rotor resistance per phase, r_2'; and injecting suitable voltages at the rotor terminals. The use of an external three-phase reactor connected to the rotor terminals is rarely, if ever, used because of its very adverse effect on the line power factor. In addition the size of the reactor needs to be large because of the effect of the slip term. By far the most common method of adjusting the speed of a three-phase induction motor is to use an external three-phase resitor connected to the rotor terminals. The larger the value of this external resistance per phase, the lower the speed is. The effect is illustrated in Fig. 4-22. Recall from Eq. (4-32) that, as resistance is added to the rotor winding, the slip at which the maximum torque is developed increases. When this is combined with, say, the constant-torque characteristic of the load, we see that

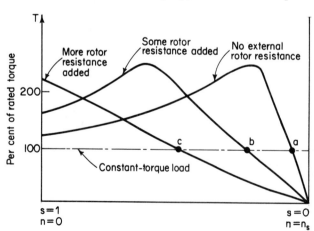

Figure 4-22 Illustrating speed control by external rotor resistance.

the same torque can be delivered over a considerable range of speed. It is customary with this method to obtain speed control to about 50% below synchronous speed. Beyond this point the efficiency becomes quite poor, being less than 50%. In fact this is the price paid for attaining speed control. When rated torque is delivered to a load at a slip of 50% rather than the usual 5% the power delivered to the load is less by about 45 per cent. This difference of 45% of rated power is consumed as heat in the external resistors and so is unrecoverable.

From the nature of the torque-speed curve we know that at a specified slip s there occurs a developed torque T. How do we determine the amount of resistance to be added to the rotor winding per phase so that this same developed torque prevails at an increased slip s'? The circle diagram is useful here in leading directly to the desired result. Refer to Fig. 4-23. Recall that, at standstill with the rotor short-circuited, line EG is divided into two parts in accordance with the ratio of r'_2/r_1. If resistance is now assumed to be added to the rotor winding by an amount of R'_a ohms per phase referred to the stator, then this ratio must change. The effect is that the mechanical power developed line (AE) is moved counterclockwise to the new position AE', where now

$$\frac{E'H'}{H'G'} = \frac{R'_a + r'_2}{r_1} \tag{4-75}$$

In essence the standstill point E has been moved closer to A. This conclusion is also consistent with the fact that with increased rotor resistance the rotor-blocked current is smaller in magnitude and more in phase with V_1 than it is without external rotor resistance. The position of the air-gap power line AH is not affected because r_1 has not been disturbed. Inspection of Fig. 4-23 discloses that for the operating point P the insertion of external rotor resistance causes a change in the total rotor copper loss

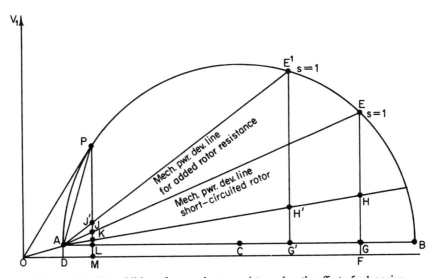

Figure 4-23 The addition of external rotor resistance has the effect of advancing counterclockwise the mechanical power-developed line.

from an amount represented by JK to an increased amount represented by $J'K$. It is important to note that for an assumed constant-torque load neither the rotor current nor the line current changes. The condition of constant developed torque demands that PK should remain fixed, and so P cannot take a new position on the circle. What does happen is that the mechanical power developed decreases by the amount JJ', which represents the power dissipated in the external resistors. When the motor operates with the rotor short-circuited at point P, the expression for the slip is

$$s = \frac{JK}{PK} = \frac{q_1[(AP)k_I]^2 r_2'}{P_g} \qquad (4\text{-}76)$$

where k_I is the current scale factor. Also, when the motor operates at the same point P but with external resistance added, the slip is increased and takes on the value

$$s' = \frac{J'K}{PK} = \frac{q_1[(AP)k_I]^2 (R_a' + r_2')}{P_g} \qquad (4\text{-}77)$$

The quantity P_g is the same in Eqs. (4-76) and (4-77) in order to ensure the same developed torque $T = (1/\omega_s)P_g$. Upon equating the expressions for P_g from each of these equations, the desired result is obtained, i.e.,

$$R_a' = \frac{s' - s}{s} r_2' \qquad (4\text{-}78)$$

If the actual resistance R_a to be added per phase at the rotor terminals is desired without having it referred to the stator winding, the expression to use is simply

$$\boxed{R_a = \frac{s' - s}{s} r_2} \qquad (4\text{-}79)$$

To illustrate the use of this equation, let us find the amount of resistance to be added to the rotor winding per phase to yield the maximum torque at starting. Assume that this torque normally occurs at a 10% slip with the rotor short-circuited. In this case $s' = 1$ and $s = 0.1$; hence

$$R_a = \frac{1 - 0.1}{0.1} r_2 = 9r_2 \qquad (4\text{-}80)$$

Thus, by adding an external resistance equal to nine times the actual rotor winding resistance per phase, the starting torque is equal to the maximum torque.

The achievement of speed control by injecting a three-phase voltage into the rotor winding can be understood by studying the sequence depicted in Fig. 4-24. For simplicity it is assumed that rated torque is to be delivered to a constant-torque load at varying speeds. Moreover, the rotor leakage reactance is assumed to be negligibly small. This means that the space-displacement angle ψ of Eq. (4-28) is zero so that the basic torque equation is $T = K\Phi I_2$. Since Φ is fixed by the constant applied voltage, it follows that to maintain constant torque at various speeds it is necessary merely to keep I_2 constant. Depicted in Fig. 4-24(a) is the situation existing in the rotor at normal operation, i.e., with the rotor short-circuited and rated motor torque delivered to the load. Note that the resultant slip voltage $s\bar{E}_2$ need only be large

Figure 4-24 Speed control by the rotor injected-voltage method: (a) Normal operation, with rated motor torque delivered to constant-torque load; (b) V_i injected opposed to sE_2; (c) V_i injected in-phase with sE_2 of part (a) but less in magnitude; (d) V_i injected in-phase with original sE_2 and equal to it; (e) V_i injected in-phase with original sE_2 but greater in magnitude.

enough to overcome the rotor winding resistance drop. Now assume that a voltage of rms value $\bar{V}_i$ per phase is injected into the rotor in such a direction as to oppose the original $s\bar{E}_2$ voltage. The immediate effect of $\bar{V}_i$ is to cause the rotor current to drop, which in turn causes the developed torque to decrease. Then, since the motor can no longer supply rated torque to the load, the speed decreases. As the speed decreases, the slip emf increases and thereby gradually overcomes the influence of $\bar{V}_i$ on the rotor current. Equilibrium is again established when the resultant rotor voltage per phase is once more equal to $\bar{I}_2 r_2$. This happens when algebraically

$$s'E_2 = sE_2 + V_i \tag{4-81}$$

where s' denotes the new increased slip. The condition is depicted in Fig. 4-24(b). The net effect of the injected voltage is to bring about an increase in slip in a manner not unlike that achieved with external rotor resistance. As a matter of fact the corresponding value of external rotor resistance that yields the same slip can be computed from

$$I_2 R_a = V_i \qquad (4\text{-}82)$$

The basic difference between the two methods, however, lies in the fact that, whereas with the resistance method power is wasted as heat, the injected voltage source can either usefully consume the power associated with the drop in speed or else return it to the line source.

What happens to the induction motor speed when the injected voltage is introduced in phase with the original slip emf but smaller in magnitude? The condition is depicted in Fig. 4-24(c). The immediate effect of $\bar{V}_i$ is to cause $\bar{I}_2$ to increase beyond the rated value, which causes an accelerating torque that *raises* the motor speed beyond the normal value. Equilibrium is then established when $s'\bar{E}_2 + \bar{V}_i = \bar{I}_2 r_2$. It is interesting to note by way of comparison that this is a situation that cannot be realized with external rotor resistance. The use of external rotor resistance allows speed control at increased slips—never at reduced slips.

An interesting situation occurs when the injected voltage is made just large enough to establish the rotor current entirely by itself. Refer to Fig. 4-24(d). The motor is thus relieved of having to play any part in producing the rotor current. Accordingly, the slip becomes zero and the motor *operates at synchronous speed*. This should not cause too much surprise because in point of fact the motor is no longer singly but, rather, doubly excited and so has every right to behave like a synchronous motor. Depicted in Fig. 4-24(e) is the situation where $\bar{V}_i$ is made larger than is needed to produce the required rated rotor current. Consequently, the motor will accelerate above synchronous speed and cause operation to take place at negative slips.

What price is paid for speed control by the injected rotor voltage method? To begin with, an auxiliary device is needed to generate the injected voltage. An important characteristic of the unit is that it must make available at the rotor terminals of the induction motor a voltage for injection that *always has the correct slip frequency;* otherwise, addition to the motor slip-frequency emf cannot be achieved. Moreover, the size and rating of the auxiliary unit must be consistent with the degree of control to be affected. Thus, if the speed is to be controlled to 50% below synchronous speed, its rating must be at least equal to one-half that of the motor being controlled. However, with the injected voltage source, speed control to the extent of 50% is possible above as well as below synchronous speed so that the total range over which control is possible is 3 : 1 as compared with 2 : 1 for the external rotor resistance method. The Schrage motor is an induction motor that employs injected armature voltages to obtain speed control. This motor has been manufactured by the General Electric Company for many years to supply up to 50 hp. GE calls it the "brush-shift" motor.

Example 4-6

A 15-hp, 60-Hz, four-pole, 480-V, three-phase, wound-rotor, Y-connected induction motor has a line frequency rotor induced emf per phase of 141 V, a rotor leakage reactance per phase of 0.27 Ω, and a rotor winding resistance of 0.16 Ω per phase. When the motor operates with the rotor winding short-circuited, it develops a torque of 62.4 N-m at a slip of 3.5% and carries a rotor current of 31.1 A. Determine the magnitude and phase of the injected voltage to be introduced at the terminals of the rotor winding (with the short circuit removed) so that the same torque is developed at a slip which is 40% above synchronous speed. The rotor leakage reactance at this high slip is not to be neglected.

Solution We know from the basic torque equation [see Eq. (3-54)] of a-c machines that the condition of constant torque requires that

$$I_2 \cos \psi_2 = I_2^* \cos \psi_2^*$$

where the * notation denotes operation at the new slip condition. The angle ψ of course denotes the power factor angle at the rotor winding at the particular operating slip. For the case at hand, we have

$$\psi_2 = \tan^{-1} \frac{s x_2}{r_2} = \tan^{-1} \frac{(0.035)(0.27)}{0.16} = 3.4°$$

and

$$\psi_2^* = \tan^{-1} \frac{s^* x_2}{r_2} = \tan^{-1} \frac{(0.4)(0.27)}{0.16} = 34°$$

Accordingly the rotor current at the new slip condition is

$$I_2^* = I_2 \frac{\cos \psi_2}{\cos \psi_2^*} = 31.1 \left(\frac{\cos 3.4°}{\cos 34°} \right) = 37.5 \text{ A}$$

When the motor operates with the rotor short-circuited, the resultant emf that produces the current of 31.1 A has a magnitude of

$$E_R = I_2 |r + j s x_2| = 31.1 |0.16 + j(0.035)(0.27)| = 4.94 \text{ V}$$

This quantity is shown in Fig. 4-25(a) placed along a vertical line in quadrature with the flux phasor. Actually, the phasor $\bar{E}_R$ should be drawn lagging the flux phasor by an angle of $(90° + 3.4°)$, but this is so close to a 90° lag that it is so indicated. However,

Figure 4-25 Diagram for Example 4-6: (a) showing relationships between slip-induced emfs and currents at two values of slip; (b) determination of the required injected voltage to achieve speed control above synchronous speed.

when the motor operates at a high slip, the rotor winding phase angle can be appreciable as already demonstrated ($\psi_2^* = 34°$). Although the resultant rotor emf at slip s^* must still bear a quadrature relationship to the flux that produces it, the magnitude now is no longer just dependent upon the winding resistance. The leakage reactance also makes a contribution. Thus the magnitude of the resultant rotor winding induced emf now computes to be

$$E_R^* = I_2^* |r_2 + js^* x_2| = 37.5|0.16 + j(0.4)(0.27)| = 7.24 \text{ V}$$

This quantity is also depicted in Fig. 4-25(a) together with its phasor components. Observe that the projection of I_2^* on the vertical is the same for both slip conditions, thus ensuring constant torque.

Keep in mind that by design the injected voltage is introduced always in quadrature with the flux phasor for the achievement of speed control. Because in this case we want speed control above synchronous speed, the injected voltage must be introduced 90° *lagging* the flux phasor in the manner illustrated in Fig.4-25(b). The quantities E_R^*, $s^* E_2$, and V_i are all collinear and the relationship between them expressed arithmetically is simply

$$V_i - s^* E_2 = E_R^*$$

$$V_i = E_R^* + s^* E_2 = 7.24 + (0.4)(141) = 63.64 \text{ V}$$

4-10 THE ELECTRONIC METHOD OF SPEED CONTROL

Reference to Eq. (4-12) makes it obvious that an attractive means of speed control is the availability of an adjustable frequency source. It is mentioned in the preceding section that one method of obtaining such a source is to use a dc motor-ac generator set. Unfortunately, this solution projects its own shortcomings such as the need for a suitable d-c supply, the need to purchase two additional machines, a more restrictive limit on the range of speed control as well as reduced efficiency. A better solution lies in the use of an *inverter* which can be made to change a d-c source into a variable frequency source. Because d-c power is not readily available, it becomes necessary to generate it through the use of suitable rectifier circuits. A block diagram of this *electronic scheme* of speed control of a squirrel-cage induction motor is shown in Fig. 4-26. It is significant to note that this is one of the few methods that is available to achieve speed control for the highly rugged squirrel-cage induction motor.

Figure 4-26 Block diagram to illustrate the principal equipment needed in the electronic speed control of a squirrel-cage three-phase motor.

The three-phase rectifier in Fig. 4-26 serves the purpose of converting the commonly available three-phase voltage to a d-c source. Usually, the output of the rectifier contains higher harmonies of the fundamental frequency of the a-c source and these are conveniently removed by an appropriate filter. Figure 4-27(a) illustrates a typical three-phase half-wave rectifier the output voltage of which is drawn

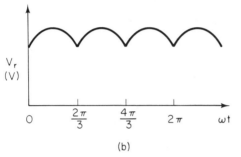

Figure 4-27 (a) A three-phase, half-wave rectifier using semiconductor diodes; (b) output waveshape of the rectifier.

in Fig. 4-27(b). By passing this wave shape through the *LC* filter, the output voltage becomes essentially d-c. It then becomes the task of the inverter to generate a new three-phase voltage source which in general exhibits the properties of variable frequency, adjustable voltage, and even adjustable phase. Simultaneous adjustment of the output voltage of the inverter with frequency is needed in order to prevent operation at values of flux per pole which deviate sharply from the rated value of the controlled motor. This can be achieved by designing the speed control system so that it keeps the ratio of the inverter output voltage to the controlled frequency a constant. By Eq. (4-9) it then becomes clear that by this technique the flux per pole can be preserved over the operating range. If this precaution were not to be observed, then at low frequency severe saturation of the magnetic circuit could easily occur, while at high frequencies there would be a diminution of the motor's reserve capacity.

The circuitry of Fig. 4-27(a) shows the use of diodes in the rectifying circuit. It is useful to note that such diodes are available in voltage ratings up to 5000 V and current ratings of 7500 A. The output voltage can be made adjustable by replacing the diodes with silicon-controlled rectifiers (SCRs). By placing a suitable voltage pulse on the gate terminal of the SCR, the instant of firing of the diode section can be delayed, thereby reducing the value of the rectified output. Although the voltage ratings of the SCRs match those of the power diodes, the current ratings are about half as much.

A schematic diagram of a typical electronic speed control system employing the foregoing notions is depicted in Fig. 4-28. The components shown here are those called

Figure 4-28 Schematic diagram for electronic speed control of a three-phase squirrel-cage induction motor. The silicon controlled rectifiers (SCR) are marked S1 through S6.

for by the block diagram of Fig. 4-26 but modified to include the control circuit for the silicon-controlled rectifiers. Moreover, the details of the inverter are also indicated. The switches marked S1 through S6 are the SCRs and each is appropriately gated by signals obtained from the gate control section. It is the purpose of the gate signal to sequence the ON and OFF times of each SCR so that the output voltage appearing at terminals a, b, c will exhibit the characteristics of a three-phase voltage set. The speed control voltage E_s serves two functions. First, it controls the rectifying action of the three-phase rectifier in a fashion that allows the magnitude of the d-c output to be directly proportional to E_s; and second, the speed control voltage also determines the output frequency of a voltage-controlled oscillator (VCO) in a direct manner. Expressed mathematically, we can write

$$V_{dc} = k_1 E_s \qquad (4\text{-}83)$$

and

$$f_o = k_2 E_s \qquad (4\text{-}84)$$

where V_{dc} denotes the output voltage of the three-phase rectifier plus filter and f_o is the output frequency of the oscillator. The quantities k_1 and k_2 are appropriate scale factors. In fact, it is this frequency which the inverter imposes at the terminals of the squirrel-cage induction motor. The diodes that are placed in parallel with the SCRs in a reversed orientation are needed to provide a path for reactive currents when the corresponding SCR is gated off. They also are useful in allowing energy to flow back to the source whenever the induction motor is made to operate as a generator.

We now undertake a description of how the inverter takes the d-c source and manipulates it to produce a three-phase a-c voltage set. The gate control section of the system plays a crucial role in the process, for it provides the gating signals, which in turn establish the proper gating sequence as well as the duration of the gating (i.e., the portion of the frequency cycle each SCR is allowed to stay conductive). The switching sequence used in this explanation is shown in Fig. 4-29. One period of the frequency f_o is shown and it is divided into six equal parts of $\pi/3$ radians. In the interest of realizing a maximum d-c voltage each SCR is gated to be in the ON mode for close to π radians. As a practical matter, the ON-time must be kept just under π radians in order subsequently to allow S4 to be gated ON at exactly π radians. A glance at Fig. 4-28 makes clear the importance of gating S1 to the OFF state before S4 is turned on. Failure to do so places a short circuit across the d-c supply.

The switching sequence appearing in Fig. 4-29 is determined as follows. Start with S1 and draw a line over the first three 60° periods, thus indicating that S1 is to be kept in a gated ON state for the first 180 electrical degrees. Next, because we are dealing with three-phase, start the conduction period of S2 at a position 120 electrical degrees later than the start of S1. Similarly, start the conduction period of S3 at a position 120 electrical degrees later than the start of S2. This pattern is then repeated for the second set of SCRs. Accordingly, the ON-time for S2 is made to begin at π radians and to endure for 180°. The conduction period for S5 is then made to begin 120° later than for S4 and to endure for 180°. Finally, conduction for S6 is programmed to commence 120° after S4 is placed into conduction and again allowed to stay conductive for almost π radians. By examining the first 60° column of Fig. 4-29, we can properly conclude that over this period the SCRs that are gated on are S1,

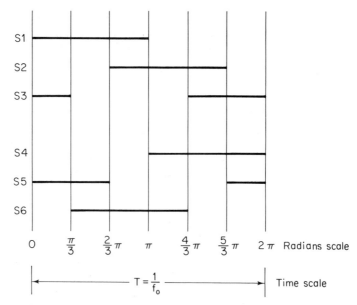

Figure 4-29 Switching sequence diagram for the SCRs in Fig. 4-28.

S3, and S5. The remaining three SCRs are simultaneously and intentionally gated to the OFF state. During the period that a specific SCR is programmed to be in the ON state, it is important to maintain the gate signal so that the switch continues to stay closed whether or not current is flowing at the moment. Upon repeating the procedure for each 60° period, the following table is obtained that describes the switching sequence.

Interval	SCRs in ON state
0–60°	S1, S5, S3
60–120°	S1, S5, S6
120–180°	S1, S2, S6
180–240°	S4, S2, S6
240–300°	S4, S2, S3
300–360°	S4, S5, S3

Once the switching sequence is established, it is a routine matter to identify the line-to-line voltages as they appear at the terminals of the squirrel-cage motor, which is assumed to have a Y-connected stator winding. The inverter output appears at terminals a, b, c. During the first 60° interval, switches S1, S3, and S5 are in the ON state. Because S1 and S3 are conductive, terminals *a* and *b* are connected to the + side of the d-c source and through the action of S5 terminal c is joined to the negative side. This arrangement places phases *a* and *b* of the induction motor in parallel with each other and the combination in series with phase *c*. Refer to the schematic diagram for the 60° interval in line (a) of Table 4-1, which illustrates the distribution of the d-c input voltage to the inverter during the first three 60° intervals of the operating fre-

TABLE 4-1 DISTRIBUTION OF THE D-C INPUT VOLTAGE DURING THE FIRST THREE 60° INTERVALS OF THE OPERATING FREQUENCY

Interval	ON-gated SCRs	Voltage distribution on induction motor phases	Corresponding line voltages
(a) 0–60°	S1, S3, S5	$\frac{1}{3}V_{dc}$ (phase a), $\frac{1}{3}V_{dc}$ (phase c), $\frac{2}{3}V_{dc}$ (phase b)	$V_{ab} = V_{dc}$ $V_{bc} = -V_{dc}$ $V_{ca} = 0$
(b) 60–120°	S1, S5, S6	$\frac{2}{3}V_{dc}$ (phase a), $\frac{1}{3}V_{dc}$ (phase c), $\frac{1}{3}V_{dc}$ (phase b)	$V_{ab} = V_{dc}$ $V_{bc} = 0$ $V_{ca} = -V_{dc}$
(c) 120–180°	S1, S2, S6	$\frac{1}{3}V_{dc}$ (phase a), $\frac{2}{3}V_{dc}$ (phase c), $\frac{1}{3}V_{dc}$ (phase b)	$V_{ab} = 0$ $V_{bc} = V_{dc}$ $V_{ca} = -V_{dc}$

quency determined by the voltage-controlled oscillator in response to the speed control voltage E_s. For balanced phase windings it follows that the distribution of the d-c source voltage puts $V_{dc}/3$ across the two parallel-connected phases (i.e., coils an and bn) and $2/3V_{dc}$ across phase c. A simple application of Kirchhoff's voltage law then shows that the line voltages are $V_{ab} = V_{dc}$, $V_{bc} = -V_{dc}$, and $V_{ca} = 0$.

During the second 60° interval, the foregoing table shows that switches S1, S5, and S6 are in the ON state. In this case, only terminal a is joined to the positive side of the d-c source while b and c are both connected to the negative side. The effect is to put phases b and c in parallel and the combination in series with phase a in the manner illustrated in line (b) of Table 4-1. While reading these diagrams, it is helpful to keep in mind that terminals marked with the same polarity are joined together for the specified interval. Now the distribution of the d-c voltage puts two-thirds across the winding of phase a and one-third across each of the coils b and c with the polarity indicated. The result is that the line voltages during this second period are described by $V_{ab} = V_{ac}$, $V_{bc} = 0$, and $V_{ca} = -V_{dc}$.

The situation that prevails in the third 60° interval leads to the results shown in line (c) of Table 4-1. Here the coils of phase a and b are in parallel. The corresponding line voltages become $V_{ab} = 0$, $V_{bc} = V_{dc}$, and $V_{ca} = -V_{dc}$. A continuation of this procedure through to the sixth 60° interval leads to the results depicted in Fig. 4-30.

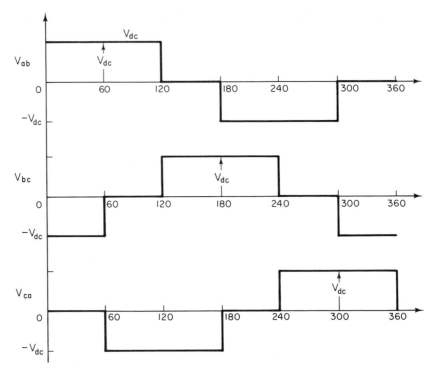

Figure 4-30 Variations of the inverter output line-to-line voltages corresponding to the switching sequence of Fig. 4-29 for one full cycle of the controlled frequency f_0.

An examination of these waveshapes reveals that the line voltages do exhibit the characteristics of a balanced three-phase voltage system. They are of equal magnitude and the individual line voltages are displaced in time by the required 120 electrical degrees. Of course, these waveshapes are not sinusoidal but they do contain a large fundamental component which is the one that serves to drive the induction motor.

The peak value of this fundamental component is readily found by applying Fourier series analysis to the rectangular waveshapes that represent the line-to-line voltages. For convenience, this basic waveform is redrawn in Fig. 4-31 with the zero position relocated in the interest of taking advantage of the quarter-wave symmetry contained therein. Then from Fourier series analysis the amplitude of the fundamental component is

$$V_{1p} = \frac{1}{\pi} \int_0^{2\pi} V_{dc} \cos \omega t \, d(\omega t) = \frac{4}{\pi} \int_0^{} V_{dc} \cos \omega t \, d(\omega t)$$

or

$$V_{1p} = \frac{2\sqrt{3}}{\pi} V_{dc} \tag{4-85}$$

The corresponding rms value of this voltage is then

$$V_1 = \frac{V_{1p}}{\sqrt{2}} = \frac{\sqrt{6}}{\pi} V_{dc} = 0.78 V_{dc} \tag{4-86}$$

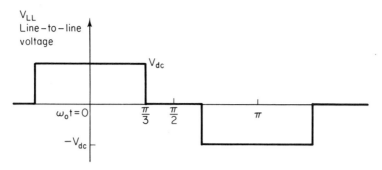

Figure 4-31 Variation of each line-to-line voltage with zero time placed at the center of the rectangular wave.

The rectangular nature of the line voltages means, of course, that harmonics do exist in these voltages. However, the harmonic situation is not as bad as it may at first appear. For example, there are no even harmonics in these line voltages because of the mirror symmetry. Furthermore, although a third harmonic voltage exists in the line-to-neutral voltages, they do not appear in the line-to-line voltages because cancellation occurs when adding the phase voltages to obtain the line voltages. Hence the first important harmonic becomes the fifth. But Fourier analysis tells us that the magnitude of the fifth harmonic is only one-fifth that of the fundamental. When this is coupled with the fact that the frequency of the fifth harmonic is five times that of the fundamental frequency and that the induction motor presents an inductive load, which exerts a substantial filtering effect, the total effect of this harmonic is not serious. The other higher harmonics cause even less disturbance. The influence of the harmonic content of the line voltages can be further reduced by employing a switching arrangement that yields a more nearly sinusoidal waveshape for the line voltages (see Prob. 4-66). Unfortunately, this usually carries with it the penalty of reduced magnitude for the rms output voltage.

As a final point, let us examine the variation, if any, that takes place in the flux per pole of the induction motor over its operating range. Insertion of Eq. (4-83) into Eq. (4-86) gives the rms value of the voltage applied to the induction motor as

$$V_1 = 0.78V_{dc} = 0.78k_1E_s = k_0E_s \qquad (4\text{-}87)$$

where k_0 denotes a new constant equal to $0.78k_1$. Except at very low frequencies, where machine inductance plays a reduced role in determining motor performance, Eq. (4-87), when divided by $\sqrt{3}$, also expresses the induced emf that occurs per phase in the induction motor as it operates at various frequencies in response to the speed control voltage E_s. Hence we may write

$$\frac{V_1}{\sqrt{3}} \approx E_1 = 4.44NK_wf_0\phi = \frac{k_0}{\sqrt{3}}E_s$$

or

$$\phi = \frac{k_0E_s}{\sqrt{3}(4.44)NK_wf_0} = \frac{k_0}{K}\frac{E_s}{f_0} \qquad (4\text{-}88)$$

where $K = (\sqrt{3})4.44NK_W$, which is a constant. But the commanded frequency is established by the voltage-controlled oscillator in accordance with the relationship appearing in Eq. (4-84). Upon introducing this expression for f_0 into Eq.(4-88), there results

$$\phi = \frac{k_0 E_s}{K k_2 E_s} = \frac{k_0}{K k_2} = \text{a constant} \qquad (4\text{-}89)$$

Therefore, the speed control offered by the electronic system depicted in Fig. 4-28 is achieved in a manner that ensures operation at essentially constant flux.

4-11 RATINGS AND APPLICATIONS OF THREE-PHASE INDUCTION MOTORS

Now that the theory of operation, the characteristics, and the performance of the three-phase induction motor are understood, we can study their standard ratings and typical applications. Of course, before it is possible to specify a particular motor for a given application, the characteristics of the load must be known. These include such items as horsepower requirement, starting torque, acceleration capability, speed variation, duty cycle, and the environment in which the motor is to operate. Once this information is available, it is often possible to select a general-purpose motor to do the job satisfactorily. Table 4-2 is a list of such motors which are readily available and standardized in accordance with generally accepted criteria established by the National Electrical Manufacturers Association (NEMA). The table is essentially self-explanatory.

A brief description of the salient features of the common classes of squirrel-cage induction motors follows. These distinctions arise from the differences in the construction details of the rotor slots which accommodate the squirrel-cage.

NEMA Class A: The General-Purpose Motor

The rotor stampings for this machine usually have semiclosed slots of medium depth in order to limit the skin effect. See Fig. 4-32(a). The low-resistance cage that is placed in these slots leads to a performance characterized by high efficiency and high power factor under rated load conditions. The disadvantage of such a construction is that at rated voltage the motor draws from the line source very large starting currents in a ratio of five to eight times rated current. The limitations imposed by the electric power companies in restricting the size of the motor starting current often means that motors rated in excess of 7.5 hp cannot be started directly across the supply lines. Instead, these motors must be energized through a voltage-reducing device such as an autotransformer (also called a starting compensator). When this procedure becomes necessary, the starting torque as well as the accelerating torque available to a shaft-connected load will be reduced. Recall that the starting torque is directly proportional to the square of the applied voltage. Hence a 20% reduction in voltage applied to the motor at starting yields a 36% reduction in the starting torque.

In sizes and circumstances where line starting is tolerable, however, this motor will be found to be the least costly of the group and also the least demanding in maintenance. Moreover, the class A motor is designed to have a generous reserve

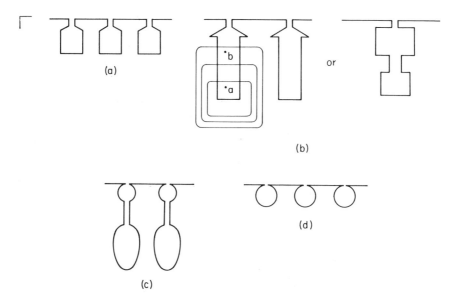

Figure 4-32 Construction details of the rotor slots of squirrel-cage induction motors: (a) class A; (b) class B; (c) class C; (d) class D.

capacity, as indicated by its value of maximum developed torque, which typically is close to 250% of rated torque. This puts the motor in a good position to manage successfully temporary overloads that are likely to be encountered in general-purpose applications. Figure 4-33 depicts the shape of its torque-speed curve.

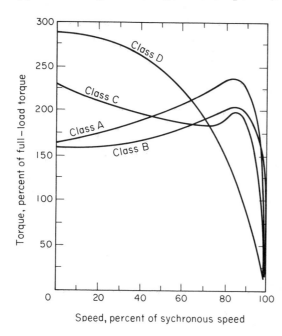

Figure 4-33 Typical torque-slip curves for various classes of squirrel-cage induction motors.

TABLE 4-2 CHARACTERISTICS AND APPLICATIONS OF POLYPHASE 60-CYCLE A-C MOTORS

Type classification	hp range	Starting torque[a] (%)	Maximum torque[a] (%)	Starting current[a] (%)
General-purpose, normal torque and starting current, NEMA class A	0.5–200	Poles–Torque 2–150 4–150 6–135 8–125 10–120 12–115 14–110 16–105	Up to 250 but not less than 200	500–1000
General-purpose, normal torque, low starting current, NEMA class B	0.5–200	Same as above or larger	About the same as class A but may be less	About 500–550, less than average of class A
High torque, low starting current, NEMA class C	1–200	200–250	Usually a little less than class A but not less than 200	About same as class B
High torque, medium and high slip, NEMA class D	0.5–150	Medium slip 350 High slip 250–315	Usually same as standstill torque	Medium slip 400–800, high slip 300–500
Low starting torque, either normal starting current, NEMA class E, or low starting current, NEMA class F	40–200	Low, not less than 50	Low, but not less than 150	Normal 500–1000, low 300–500
Wound-rotor	0.5–5000	Up to 300	200–250	Depends upon external rotor resistance but may be as low as 150

[a]Figures are given in per cent of rated full-load values.
Source: By permission from M. Liwschitz-Garik and C. C. Whipple, *Electric Machinery*, vol. II (Princeton, N.J.: D. Van Nostrand Co., Inc., 1946).

Slip (%)	Power factor (%)	Efficiency (%)	Typical applications
Low, 3–5	High, 87–89	High, 87–89	Constant-speed loads where excessive starting torque is not needed and where high starting current is tolerated. Fans, blowers, centrifugal pumps, most machinery tools, woodworking tools, line shafting. Lowest in cost. May require reduced voltage starter. Not to be subjected to sustained overloads, because of heating. Has high maximum torque
3–5	A little lower than class A	87–89	Same as class A—advantage over class A is lower starting current, but power factor slightly less
3–7	Less than class* A	82–84	Constant-speed loads requiring fairly high starting torque and lower starting current. Conveyors, compressors, crushers, agitators, reciprocating pumps. Maximum torque at standstill
Medium 7–11; high, 12–16	Low	Low	Medium slip. Highest starting torque of all squirrel-cage motors. Used for high-inertia loads such as shears, punch presses, die stamping, bulldozers, boilers. Has very high average accelerating torque. High slip used for elevators, hoists, etc., on intermittent loads
1 to 3$\frac{1}{2}$	About same as class A or class B	About same as class A or class B	Direct-connected loads of low inertia requiring low starting torque, such as fans and centrifugal pumps. Has high efficiency and low slip
3–50	High, with rotor shorted same as class A	High, with rotor shorted the same as class A, but low when used with rotor resistor for speed control	For high-starting-torque loads where very low starting current is required or where torque must be applied very gradually and where some speed control (50%) is needed. Fans, pumps, conveyors, hoists, cranes, compressors. Motor with speed control more expensive and may require more maintenance.

NEMA Class B : Low Starting Current, Normal Starting Torque

The rotor laminations in the class B motor may be either of two types, as illustrated in Fig. 4-32(b). The slot is shaped to accommodate either a single deep bar or two bars of slightly different sizes with the upper one of larger cross-sectional area. At line frequency the leakage flux produced by a current flowing through the deep bar assumes the pattern depicted in Fig. 4-32(b). As this flux changes with the changing bar current, the associated emf induced at the deep end of the bar (point *a*) is greater than that produced at the upper end of the bar (point *b*). Accordingly, current tends to flow more in the upper portion of the bar than in the lower portion, thus reducing the effective cross-sectional area of the conductor for current flow. Of course, for a nonvarying current flow (d-c) the counter emf is zero, so the entire cross-sectional area of the bar is effective in supporting current flow. As the frequency of the bar current increases there is a greater squeezing of the current toward the upper portion of the bar. Figure 4-34 depicts the effects of this situation (the skin effect ratio) for a typical solid deep bar as a function of the slip. Observe that at standstill the a-c resistance is 3 times greater than the d-c resistance. In essence, then, the class B motor at starting exhibits a high rotor resistance, which in turn yields ample starting torque at reduced starting current. The latter is about 75 % of the value of the class A motor used in application. On the debit side for the class B motor, however, it is necessary to list a lower maximum torque, slightly lower power factor, and a somewhat greater cost.

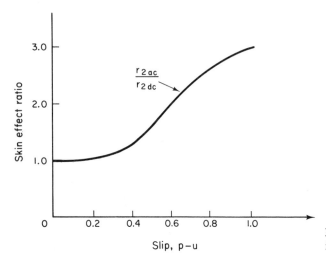

Figure 4-34 Illustrating the skin effect in a deep rotor bar.

NEMA Class C : Low Starting Current; High Starting Torque

The rotor slot construction detail for the class C motor appears in Fig. 4-32(c). A double squirrel-cage is always used. At standstill the impedance of the lower cage is much greater than the impedance of the upper cage. The upper cage is characterized by high resistance and low leakage reactance while the lower cage exhibits high leakage reactance and low resistance. Accordingly, at starting, the upper cage plays the dominant role; and because it has high rotor resistance, the advantages of a high starting torque at low starting current are realized. Then, as the motor reaches its normal operating speeds (where the cage current frequency is very low), control passes on to the lower cage, which displays a much lower resistance than the upper cage, thus improving efficiency during normal operation. Of course, the high leakage reactance of the lower cage is diminished in importance by operation at low slips. However, both efficiency and power factor for the class C motor will be less than for either classes A or B. A glance at the torque-speed curve for the class C motor shows that it can provide more accelerating torque than either the A or B motors. In fact, a distinguishing characteristic of this motor is that it develops a starting torque that is larger than its pull-out torque.

The availability of the class C motor has eliminated the need to specify the wound-rotor induction motor in some applications.

NEMA Class D: Highest Starting Torque; Lowest Starting Current; High Slip

A single squirrel-cage is used for the class D motor. A slot of shallow depth and small cross-sectional area is used. Refer to Fig. 4-32(d). Bars of higher resistivity are also often employed. The objective is to obtain the high resistance needed to obtain very high starting and accelerating torques, as revealed by the typical torque-speed characteristic shown in Fig. 4-33. Of course, the use of a single high-resistance cage means that the efficiency will be lower and the slip higher than the preceding classes. However, the high slip characteristic is a very useful one in applications where this motor excels, such as for punch presses, hoists, cranes, etc. When equipped with a flywheel in these applications, the drop in speed that occurs during the work portion of the duty cycle enables the flywheel to release some of its stored energy, thus permitting a decrease in the peak level of power the motor draws from the line source.

The horsepower rating of an a-c motor is limited by the allowable temperature rise under operating conditions. Although copper, aluminum, and iron as individual elements can withstand very high temperature increases, this does not apply to the insulating materials that are used for the stator winding in both the wound-rotor and the squirrel-cage induction motors as well as for the rotor winding of the wound-rotor motor. Insulating materials deteriorate rapidly whenever the maximum temperatures they are designed to withstand are exceeded. Consequently, the capability of an insulating material to withstand high temperature rises plays a crucial role in establishing the physical dimensions of a motor with a specific horsepower rating. When a motor is equipped with a superior insulating material, the motor size can be made smaller for a given horsepower because it is possible to work the motor harder magnetically (i.e., operate at higher flux densities in the iron) as well as electrically (i.e., operate at higher current densities in the copper or aluminum conductors).

There are three classes of NEMA-rated insulating materials. Class B insulation is composed of such materials as mica, glass fibers, and asbestos in a manner that allows a maximum temperature rise of 80°C when operation occurs with fan cooling. Class F insulation includes similar elements in addition to some synthetic substances all designed to provide an allowable temperature rise of 100°C. The class H category refers to a silicone-type insulation which permits an even higher temperature rise of 125°C.

If an electric motor is operated at overload for a sustained period of time, the life of the motor can be markedly reduced. This fact is graphically demonstrated in Fig. 4-35 for an insulating material of the class B type. An examination of the plot reveals an interesting bit of information: if a motor equipped with this insulation is operated continuously at an overload condition that causes an increased temperature rise of 8 to 10°C over the rated value of the insulation, the life of the motor is approximately halved. Moreover, because of the linear character of this semilogarithmic plot, the statement is true for each additional increase in operating temperature of 10°C.

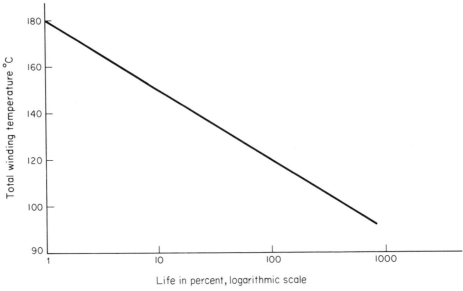

Life in percent, logarithmic scale

Figure 4-35 Illustrating the effect of winding temperature on motor life.

4-12 CONTROLLERS FOR THREE-PHASE INDUCTION MOTORS

After the right motor is selected for a given application, the next step is to select the appropriate controller for the motor. The primary functions of a controller are to furnish proper starting, stopping, and reversing without damage or inconvenience to the motor, to other connected loads, or to the power system. However, the controller fulfills other useful purposes as well, especially the following:

1. It limits the starting torque. Some connected shaft loads may be damaged if excessive torque is applied upon starting. For example, fan blades can be sheared off or gears having excessive backlash can be stripped. The controller supplies reduced voltage at the start and as the speed picks up the voltage is increased in steps to its full value.

2. It limits the starting current. Most motors above 3 hp cannot be started directly across the three-phase line because of the excessive starting current that flows. Recall that at unity slip the current is limited only by the leakage impedance, which is usually quite a small quantity, especially in the larger motor sizes. A large starting current can be annoying because it causes light to flicker and may even cause other connected motors to stall. Reduced-voltage starting readily eliminates these annoyances.

3. It provides overload protection. All general-purpose motors are designed to deliver full-load power continuously without overheating. However, if for some reason the motor is made to deliver, say, 150% of its rated output continuously, it will proceed to accommodate the demand and burn itself up in the process. The horsepower rating of the motor is based on the allowable temperature rise that can be tolerated by the insulation used for the field and armature windings. The losses produce the heat that raises the temperature. As long as these losses do not exceed the rated values, there is no danger to the motor, but, if they are allowed to become excessive, damage will result. There is nothing inherent in the motor that will keep the temperature rise within safe limits. Accordingly, it is also the function of the controller to provide this protection. Overload protection is achieved by the use of an appropriate time-delay relay which is sensitive to the heat produced by the motor line currents.

4. It furnishes undervoltage protection. Operation at reduced voltage can be harmful to the motor, especially when the load demands rated power. If the line voltage falls below some preset limit, the motor is automatically disconnected from the three-phase line source by the controller.

Controllers for electric motors are of two types—manual and magnetic. We shall consider only the magnetic type, which has many advantages over the manual type. It is easier to operate. It provides undervoltage protection. It can be remotely operated from one or several different places. Moreover, the magnetic controller is automatic and reliable whereas the manual controller requires a trained operator, especially when a sequence of operations is called for in a given application. The one disadvantage is the greater initial cost of the magnetic type.

Appearing in Fig. 4-36 is the schematic diagram of a magnetic full-voltage starter for a three-phase induction motor. The operation is simple. When the start button is pressed, the relay coil M is energized. This moves the relay armature to its closed position, thereby closing the main contactors M, which in turn apply full voltage to the motor. When the relay armature moves to its energized position, it also closes an auxiliary contactor M_a which serves as an electrical interlock, allowing the operator to release the start button without de-energizing the main relay. Of course contactors M are much larger in size than M_a. The former set must have a current

Figure 4-36 Full-voltage magnetic starter.

rating that enables it to handle the starting motor current. The latter needs to accommodate just the exciting current of the relay coil. Figure 4-36 also shows that the motor line current flows through two overload heater elements. If the temperature rise becomes excessive, the heater element causes the overload contacts in the control circuit to open. In controller diagrams it is important to remember that all contactors are shown in their de-energized state. Thus the symbol ⎯⊣⊢⎯ means that the contactors M are open when the coil is not energized. Similarly, the symbol ⎯⊣⊬⎯ means that these contactors are closed in the de-energized state.

Undervoltage protection is inherent in the magnetic starter of Fig. 4-36. This comes about as a result of designing the coil M so that, if the coil voltage drops below a specified minimum, the relay armature can no longer be held in the closed position.

A full-voltage magnetic starter equipped with the control circuitry to permit reversing is illustrated in schematic form in Fig. 4-37. A three-phase induction motor is reversed by crossing two of the three line leads going to the motor terminals. In this connection note the criss-cross of two of the R contactors. Pressing the forward (FWD) button energizes coil F which in turn closes the main contactors F as well as the interlock F_a. This allows the motor to reach its forward operating speed. To reverse the motor the REV button is pushed. This does two things. One, it de-energizes

Figure 4-37 Full-voltage starter equipped with reversing control.

coil F, thus opening the F contactors. Two, it energizes the R relay coil, thus closing the R contactors which apply a reversed-phase sequence to the motor causing it to attain full speed in the reverse direction. Putting the REV switch in the F_a interlock circuit is a safety measure which prevents having both the R and F contactors closed at the same time.

An illustration of a reduced-voltage magnetic controller using limiting resistors in the line circuit appears in Fig. 4-38. This unit is frequently referred to as a three-step acceleration starter because the line resistors are removed in three steps. Pushing the start button energizes coil M and closes contactors M, thus applying a three-phase voltage to the motor through the full resistors. In addition to contactors M and M_a coil M is also equipped with a time-delayed contactor T_M. This contactor is so designed that it does not close until a preset time *after* the armature of coil M is closed.

Figure 4-38 Three-step reduced-voltage starter for a three-phase induction motor.

The delay is usually obtained through a mechanical escapement of some sort which is actuated by the relay armature. Of course the time delay is needed to permit the motor to accelerate to a speed corresponding to the reduced applied voltage. After the elapse of the preset time, the T_M contacts close, energizing coil $1A$, which in turn closes contactors $1A$, short-circuiting the first part of the series resistor. Coil $1A$ is also equipped with a time-delay contactor T_{1A}, which is designed to allow the motor to accelerate to a higher speed before it closes. When contactors $1A$ do close, coils $2A$ are energized. This immediately closes contactors $2A$, shorting out the second section of the line resistor. Then, after still another time delay, contactor T_{2A} closes, applying an excitation voltage to coils $3A$. With the closing of contactors $3A$ the full line voltage is applied to the motor. In this manner the motor is brought up to speed in a "soft," smooth fashion without drawing excessive starting current or developing large starting torques.

PROBLEMS

4-1. A three-phase armature winding is shown in Fig. P4-1. Sinusoidal currents having an amplitude of 100 A flow in the three phases. Each coil consists of three turns. Carefully sketch to scale the actual mmf distribution along the air gap for a span of two poles for the indicated time instants t_1 and t_2.

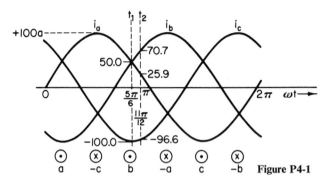

Figure P4-1

4-2. A 60-Hz polyphase induction motor runs at a speed of 873 rpm at full-load. What is the synchronous speed? Find the frequency of the rotor currents.

4-3. A 50-Hz polyphase induction motor runs at a speed of 1475 rpm at full load. What is the synchronous speed of the motor? How many poles is it designed for?

4-4. State the general conditions under which it is possible for m alternating fields (with axes fixed in space) to yield a revolving field constant in amplitude and traveling at constant speed.

4-5. Determine the magnitude and direction of the resultant flux field in the machine configuration of Fig. 4-2 corresponding to time instants t_2 and t_4 in Fig. 4-1.

4-6. In the time variation of the phase currents depicted in Fig. 4-1 assume that the amplitude of phase b current is one-half that of phases a and c but that each is displaced by 120° from the other. Find the magnitude and direction of the resultant flux for the time instants t_1, t_3, and t_5, and compare with the results shown in Fig. 4-2.

4-7. An unbalanced three-phase mmf flows through three coils that are space displaced by 120° as shown in Fig. 4-2. The nature of the unbalance is such that phase b lags behind phase a by 90° instead of 120°. However, each phase has the same amplitude. Investigate the character of the resultant flux field corresponding to various instants of time.

4-8. Demonstrate that a resultant mmf that is described by an equation of the form

$$\mathcal{F}_r = \mathcal{F} \cos(\omega t + \alpha)$$

represents a wave that travels in the negative α direction. Keep in mind that t is a time variable and α is a space variable.

4-9. A balanced two-phase voltage is applied to identical coils, the axes of which are 90° apart. Derive the expression for the resultant mmf in terms of the mmf amplitude of either coil ($\mathcal{F}_m$), the angular frequency ω of the exciting currents, and the space variable α. Assume that each coil produces a sinusoidal field distribution.

4-10. A balanced, three-phase, 60-Hz voltage is applied to a three-phase, four-pole induction motor. When the motor delivers rated output horsepower, the slip is found to be 0.05. Determine the following:

(a) The speed of the revolving field relative to the stator structure, which accommodates the exciting winding.

(b) The frequency of the rotor currents.

(c) The speed of the rotor mmf relative to the rotor structure.

(d) The speed of the rotor mmf relative to the stator structure.

(e) The speed of the rotor mmf relative to the stator field distribution.

(f) Are the conditions right for the development of a net unidirectional torque? Explain.

4-11. Repeat Prob. 4-10 for the case where the rotor structure is blocked, thus preventing rotation in spite of the application of a balanced three-phase voltage to the stator.

4-12. Repeat Prob. 4-11 for the case where the slip has a value of 0.03.

4-13. Determine the no-load speed of a six-pole, wound-rotor, three-phase induction motor, the stator of which is connected to a 60-Hz line and the rotor of which is connected to a 25-Hz line, when:

(a) The stator field and the rotor field revolve in the same direction.

(b) The stator field and the rotor field revolve in opposite directions.

4-14. Determine the speed of operation of a four-pole, wound-rotor, three-phase induction motor the stator of which is connected to a 60-Hz line and the rotor of which is connected to a 50-Hz line when:

(a) The stator field and the rotor field revolve in the same direction.

(b) The stator field and the rotor field revolve in opposite directions.

4-15. Answer whether or not the following statements are true or false. When the statement is false, provide the correct answer.

(a) In a three-phase induction motor the rotor ampere-conductor distribution is stationary with respect to the stator magnetic field distribution at *all* speeds of the rotor.

(b) The load on the shaft of a three-phase induction motor is increased. This causes the speed of the rotor mmf relative to the stator mmf to change, thus permitting an increased input current to provide power balance.

4-16. The shaft output of a three-phase, 60-Hz induction motor is 75 kW. The friction and windage losses are 900 W, the stator core loss is 4200 W, and the stator copper loss is 2700 W. If the slip is 3.75%, what is the percent efficiency at this output?

4-17. Refer to Prob. 4-16. The rotor winding resistance referred to the stator is known to be 0.1 Ω. Determine the value of the stator-referred rotor current for the specified operating conditions.

4-18. A 15-hp, 220-V, three-phase, 60-Hz, 6-pole, Y-connected induction motor has the following parameters per phase: $r_1 = 0.128\ \Omega$, $r_2' = 0.0935\ \Omega$, $x_1 + x_2' = 0.496\ \Omega$, $r_c = 183\ \Omega$, $x_\phi = 8\ \Omega$. The rotational losses are equal to the stator hysteresis and eddy-current losses. For a slip of 3%, find:

(a) The line current and power factor.

(b) The horsepower output.

(c) The starting torque.

4-19. A 10-hp, 220-V, Y-connected, four-pole, 60-Hz, three-phase induction motor draws a line current of 26.2 A at a pf of 0.78 lag when it operates at a slip of 5%. Rotational losses amount to 250 W. The motor is known to have the following parameters expressed in ohms per phase:

$$r_1 = 0.3 \qquad x_1 = x_2' = 1.25 \qquad r_c = 150 \qquad x_\phi = 18$$

(a) Calculate the rotor current per phase referred to the stator.

(b) Find the value of output horsepower.

(c) Determine the efficiency.

(d) Find the developed torque in N-m.

(e) Calculate the starting line current at rated line voltage.

4-20. A three-phase, 60-Hz, six-pole, wound-rotor induction motor draws 10 kW when driving its normal load. It draws 700 W when the load is disconnected. The rotor and stator copper losses under normal load are 295 W and 310 W, respectively. Calculate the efficiency, the speed, and the shaft torque of this motor under normal load conditions. Assume equal rotational and core losses and negligible no-load copper losses.

4-21. A three-phase induction motor has a Y-connected rotor winding. At standstill the rotor induced emf per phase is 100 V rms. The resistance per phase is 0.3 Ω, and the leakage reactance is 1.0 Ω per phase.

(a) With the rotor blocked, what is the rms value of the rotor current? What is the power factor of the rotor circuit?

(b) When the motor is running at a slip of 0.06, what is the rms value of the rotor current? What is the power factor of the rotor circuit?

(c) Compute the value of the developed power in part (b).

4-22. A three-phase, 12-pole, 60-Hz, 2200-V induction motor runs at no-load with rated voltage and frequency impressed and draws a line current of 20 A and an input power of 14 kW. The stator is Y-connected, and its resistance per phase is 0.4 Ω. The rotor resistance r_2' is 0.2 Ω per phase. Also, $x_1 + x_2' = 2.0 \Omega$ per phase. The motor runs at a slip of 2% when it is delivering power to a load. For this condition compute:

(a) The developed torque.

(b) The input line current and power factor.

4-23. A three-phase, 440-V, 60-Hz, Y-connected, eight-pole, 100-hp induction motor has the following parameters expressed per phase:

$$r_1 = 0.06\ \Omega, \qquad x_1 = x_2' = 0.26\ \Omega$$
$$r_2' = 0.048\ \Omega \qquad r_c = 107.5\ \Omega$$
$$x_\phi = 8.47\ \Omega$$

The rotational losses are 1600 W. Using the approximate equivalent circuit determine for $s = 0.03$:

(a) The input line current and power factor.

(b) The efficiency.

4-24. A three-phase, 335-hp, 2000-V, six-pole, 60-Hz, Y-connected squirrel-cage induction motor has the following parameters per phase that are applicable at normal slips:

$$r_1 = 0.2\ \Omega \qquad x_1 = x_2' = 0.707\ \Omega$$
$$r_2' = 0.203\ \Omega \qquad r_c = 450\ \Omega$$
$$x_\phi = 77\ \Omega$$

The rotational losses are 4100 W. Using the approximate equivalent circuit, compute for a slip of 1.5%:

(a) Line power factor and current.

(b) Developed torque.

(c) Efficiency.

4-25. A six-pole, three-phase, 40-hp, 60-Hz, induction motor has an input when loaded of 35 kW, 51 A, 440 V and a speed of 1152 rpm. When uncoupled from the load, the readings are found to be: 440 V, 21.3 A, 2.3 kW, and 1199 rpm. The resistance measured between terminals for the stator winding is 0.25 Ω for a Y-connection. The stator core losses and the rotational losses are known to be equal. Determine:

(a) The power factor of the motor when loaded.
(b) The motor efficiency when loaded.
(c) The horsepower rating of the load.

4-26. Answer each part briefly:

(a) What is the effect of doubling the air gap of an induction motor on the magnitude of the magnetizing current and of the maximum value of the flux per pole? Neglect the effect of the leakage impedance.

(b) Describe the effect of a reduced leakage reactance on the maximum developed torque, the power factor at full-load, and the starting current of a three-phase induction motor.

4-27. The manufacturers' specifications concerning the ratings of induction motors invariably list the maximum (or breakdown) torque as a percentage of rated torque. Its value is usually 200%. Explain the significance of choosing so high a value.

4-28. In a given situation only 220-V, 30-Hz, three-phase service is available. A plant manager has an opportunity to buy cheaply a 10-hp, 440-V, three-phase, 60-Hz, four-pole squirrel-cage motor, which is to be used to supply power to a constant torque load. The load torque corresponds to that of the 10-hp motor.

(a) Can the motor be used? Justify your answer.
(b) If so, what would be its rating?
(c) How would the efficiency of operation be affected?
(d) Find the approximate change in starting torque.

4-29. Would you recommend use of a 115-V, 60-Hz, three-phase induction motor on a 460-V, 400-Hz system to deliver power to a constant torque load? Explain.

4-30. When a 10-hp, 220-V, 60-Hz, three-phase induction motor operates at 220 V, 50 Hz delivering the same constant rated torque, what change occurs in the rotor current?

4-31. In what way, if any, does a reduced applied voltage affect the breakdown torque of a three-phase induction motor? In what way, if any, does an increased frequency affect the breakdown torque?

4-32. A 115-V, three-phase, 60-Hz induction motor is used on a 115-V, 50-Hz source. Assuming that the delivered torque is to be the same when operating at either frequency, determine:

(a) The change in operating flux.
(b) The change in rotor current.
(c) The change in synchronous speed.
(d) Would you permit this motor to be used continuously on the 50-Hz source? Explain.

4-33. Refer to the motor of Prob. 4-19.

(a) Compute the slip at which maximum torque occurs.
(b) Determine the value of the maximum torque.

4-34. Refer to the motor of Prob. 4-19.

(a) For the operating condition described determine the missing parameter of the exact equivalent circuit: namely, the stator-referred rotor resistance per phase.

(b) Using the exact equivalent circuit obtain the speed at which maximum torque occurs.

(c) Find the stator-referred rotor current at maximum torque. How does this differ from the input stator current?

(d) Compute the value of the maximum torque.

4-35. A three-phase, Y-connected, 440-V, 200-hp induction motor has the following blocked-rotor data: $P_b = 10$ kW, $I_b = 250$ A, $V_b = 65$ V, and $r_1 = 0.02$ Ω. Find the value of the rotor resistance referred to the stator.

4-36. A six-pole, 60-Hz, three-phase, Y-connected wound-rotor (three phases also) induction motor has a standstill induced rotor voltage of 130 V per phase. At short circuit with the rotor blocked, this voltage produces a current of 80 A at a power factor of 0.3 lagging. At full-load the motor runs at a slip of 9%. Find the full-load developed torque.

4-37. A 500-hp, three-phase, 2200-V, 25-Hz, 12-pole, Y-connected wound-rotor induction motor has the following parameters: $r_1 = 0.225$ Ω, $r_2' = 0.235$ Ω, $x_1 + x_2' = 1.43$ Ω, $x_\phi = 31.8$ Ω, $r_c = 780$ Ω. A no-load and a blocked-rotor test are performed on this machine.

(a) With rated voltage applied in the no-load test, compute the readings of the line ammeters as well as the total wattmeter reading.

(b) In the blocked-rotor test the applied voltage is adjusted so that 228-A line current is made to flow in each phase. Calculate the reading of the line voltmeter and the total wattmeter reading.

4-38. A three-phase, 2000-V, Y-connected wound-rotor induction motor has the following no-load and blocked-rotor test data:

No-load: 2000 V,   15.3 A,   10.1 kW

Blocked-rotor: 440 V,   170.0 A,   36.4 kW

The resistance of the stator winding is 0.22 Ω per phase. The rotational losses are equal to 2 kW. Calculate all the necessary data for the approximate equivalent circuit at a slip of 2%, and draw the circuit showing all parameter values.

4-39. For the machine of Prob. 4-37 compute the following:

(a) The slip at which maximum torque occurs.

(b) The input line current and power factor at the condition of maximum torque.

(c) The value of the maximum torque.

4-40. Refer to the motor of Prob. 4-37, and find the value of resistance that must be externally connected per phase to the rotor winding in order that the maximum torque be developed at starting. What is the value of this torque?

4-41. A three-phase, 440-V, 50-hp, 60-Hz, Y-connected induction motor yields the following no-load test data:

D-c test on stator per phase: 5 V,   50 A

Blocked-rotor test at rated frequency: 106 V,   65 A,   3423 W,   60 Hz

Blocked-rotor test at low frequency: 42 V,   65 A,   2910 W,   15 Hz

No-load test: 440 V,   16.3 A,   2157 W,   60 Hz

The friction and windage losses for this motor are 834 W. Compute the parameters of the approximate equivalent circuit to be used at normal operating slips.

4-42. From the no-load test on a 10-hp, four-pole, 230-V, 60-Hz, three-phase, Y-connected induction motor with rated voltage applied, the no-load current was found to be 9.2 A and the corresponding input power 670 W. Also, with 57 volts applied in the blocked-rotor test, it was found that the motor took 30 A and 950 W from the line. The stator winding resistance was measured to be $0.15 \, \Omega$ per phase. When this motor is coupled to its mechanical load, the input to the motor is found to be 9150 W at 28 A and a power factor of 0.82. The stator core losses are equal to the rotational losses.

(a) Compute the rotor current referred to the stator.

(b) Find the developed torque.

(c) What is the value of the slip?

(d) At what efficiency is the motor operating?

4-43. A Y-connected, wound-rotor induction motor is found to have the following parameters per phase:

$$r_1 = 0.26 \, \Omega, \qquad x_1 = 0.6 \, \Omega, \qquad r_c = 143 \, \Omega$$
$$r_2' = 0.40 \, \Omega, \qquad x_2' = 1.4 \, \Omega, \qquad x_\phi = 22.2 \, \Omega$$

(a) If rated voltage of 220 V is applied to the motor at no-load, compute the line current and power factor. Assume the stator core losses are equal to the rotational losses.

(b) If the rotor is short-circuited and blocked and a line voltage of 60 V is applied, find the power input in watts.

4-44. A three-phase induction motor has the following data: 20 hp rated, 220 V, Y-connected, four-poles, 60 Hz. Stator effective resistance is $0.118 \, \Omega$ per phase; rotor effective resistance referred to the stator is $0.102 \, \Omega$ per phase. Equivalent stator referred reactance is $0.208 \, \Omega$ per phase. Total core loss is 408 W, of which one-half is for the rotor. Friction and windage is 400 W.

(a) What is the starting torque at full voltage?

(b) If the motor is driven at 1800 rpm against the revolving field, what is the brake effect obtained in watts at the shaft? Assume that rotor resistance and core loss remain unchanged.

4-45. The approximate circle diagram of a three-phase, 220-V, 60-Hz, four-pole induction motor is given in Fig. P4-45, drawn to a scale of 12 A/in. Determine:

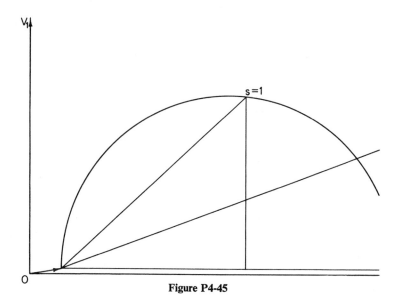

Figure P4-45

(a) The ratio of stator referred rotor resistance to stator resistance.

(b) The equivalent leakage reactance referred to the stator.

(c) The total rotor resistance needed so that the starting torque will be 31.2 lb-ft.

4-46. For the motor of Prob. 4-45 determine the following:

(a) The maximum horsepower output.

(b) The maximum developed torque and the corresponding line current.

(c) The developed torque at maximum efficiency and the corresponding line current.

(d) Compare the results of parts (b) and (c) and comment.

(e) The starting torque.

4-47. List the complete laboratory data that is necessary for the construction of the circle diagram of a polyphase induction motor.

4-48. A three-phase, Y-connected, 440-V induction motor has the following blocked-rotor data: 150 V line, 57 A, and 2800 W. Moreover, the stator resistance per phase is 0.18 Ω and the wattless component of the no-load data is 18.4 A. The stator core losses are 573 W.

(a) Determine the length of the diameter of the circle in amperes, and draw the circle.

(b) Construct the torque line.

(c) Compute the performance corresponding to maximum efficiency. Assume rotational losses are 600 W.

4-49. A three-phase wound-rotor induction motor delivers fixed torque to a load. The rotor leakage reactance is negligible. For this load condition the rotor resistance drop is found to be 20 V. Determine the rotor injected voltage needed to produce twice this torque at 10 times the slip.

4-50. A 625-V (line voltage), three-phase, Y-connected wound-rotor induction motor has a ratio of transformation from stator to rotor of 3.0. At a specified load condition with the rotor short-circuited, the motor operates with a slip emf (i.e., sE_2) of 6 V per phase.

Find the value of the injected voltage needed to increase the slip to a value of 40%. Assume that the torque varies in a manner proportional to the square root of the slip. Neglect rotor leakage reactance and rotational losses.

4-51. A four-pole, three-phase, wound-rotor induction motor is running at 1700 rpm with the rotor short-circuited and rated voltage at rated frequency applied to the stator winding. The rotor slip emf (sE_2) is 10 V per phase and the rotor current is 40 A per phase. An injected voltage of slip frequency and constant rms value of 80 V per phase is then introduced. This voltage is in phase opposition to the normal slip emf. Rotor leakage reactance is negligible.

(a) What is the new motor speed if the torque required by the load is constant as the speed varies?

(b) Repeat part (a), assuming that the torque varies directly with the speed.

4-52. A three-phase wound-rotor induction motor delivers rated torque and draws a line current of 100 A at a pf of 0.707. The effective turns ratio from stator to rotor is 1.41 and the rotor resistance per phase is 0.1 Ω. The rotor is wound for three phases. Assume that the stator core loss, stator winding resistance, and stator and rotor leakage reactances are all negligible.

(a) Calculate the value of rotor injected voltage needed to increase the line pf to 90% lagging.

(b) Find the power supplied by the source of injected voltage.

4-53. A 60-Hz, six-pole, three-phase induction motor has its speed adjusted by a rotor injected voltage. When it develops rated torque, the motor slip is 5%, and the normal slip frequency voltage is 20 V. The rotor resistance per phase is 0.1 ohm and the rotor leakage reactance is negligible. An injected voltage of 100 V per phase is inserted into the rotor in phase with the original sE_2 voltage. If the torque demand remains unchanged, determine:
(a) The new value of slip.
(b) The new operating speed in rpm.
(c) The power supplied by the injected voltage source.

4-54. A three-phase, 1000-hp, 2200-V, 25-Hz, 12-pole, wound-rotor induction motor has a ratio of transformation from stator to rotor of 22 to 15. Both stator and rotor windings are Y-connected. At a slip of 2% the motor develops its rated torque and draws from the line a current of 180 A at a pf of 0.866. Consider the quantities r_1, x_1, x_2, and r_c negligible. The motor is to develop the same torque at a slip of of 10%, while simultaneously operating at unity pf, by means of a rotor injected voltage. Determine the magnitude and phase of the injected voltage relative to the original slip emf.

4-55. A three-phase, Y-connected, wound-rotor induction motor delivers rated torque at a slip of 4%, and draws a line current of 100 A at a pf of 80%. The effective turns ratio from stator to rotor is unity, and the rotor is wound for three phases. The rotor resistance per phase is 0.1 Ω, and the rotor standstill emf is 200 V per phase. For simplicity assume that $r_1 = x_1 = x_2' = 0$.
(a) Find the rotor injected voltage needed to give the same torque at a slip of 20% below synchronous speed.
(b) Calculate the rotor injected voltage required to make the line pf unity.

4-56. A three-phase induction motor with slip rings delivers power to a constant torque load. The rotor current is 40 A per phase; the rotor resistance is 0.4 Ω per phase. In order to bring the pf to unity, a voltage is injected into the rotor whose magnitude in this case is equal to the normal slip emf induced in the rotor at this particular slip. The speed of the rotor is now to be decreased so that the slip is five times greater than before. The torque remains fixed. What must now be the magnitude and phase of the injected voltage to produce unity pf?

4-57. A three-phase, wound-rotor induction motor delivers rated torque at a slip of 2%. At this slip the value of the rotor induced emf is 10 V. The rotor resistance per phase is 0.2 Ω per phase. If it is desirable to produce *twice* rated torque at the same slip of 2%, determine:
(a) The magnitude, phase, and frequency of the voltage that must be injected into the rotor.
(b) The amount of power that the auxiliary source must supply to the induction motor.
Note: At the low slip of 2% the rotor leakage reactance may be neglected.

4-58. A 1000-hp, three-phase, 2200-V, 25-Hz, 12-pole, wound-rotor induction motor has the following parameters per phase:

$$r_1 = 0.093 \ \Omega, \qquad x_1 = 0.295 \ \Omega$$

$$r_2 = 0.072 \ \Omega, \qquad x_2 = 0.15 \ \Omega$$

stator to rotor effective turns ratio: 22/15.

When operating at a slip of 1.72% this machine develops a torque of 15,000 lb-ft, corresponding to a rotor current per phase of 205 A. The friction and windage losses total 6 kW.

(a) Determine the magnitude and phase of the minimum value of injected rotor voltage needed to change the slip to 50% above synchronous speed at constant torque. Do not neglect rotor leakage reactance.

(b) Determine the mechanical power output of the rotor in part (a).

(c) Determine the rotor electrical power "output" in part (a).

(d) Find the total rotor copper loss. How is this loss supplied?

4-59. In the configuration of Fig. P4-59 balanced three-phase currents of 20 Hz variation are fed to the brushes of the d-c winding. The rotor rotates at a frequency of 35 Hz. What is the frequency of the emf induced at the brushes? Why?

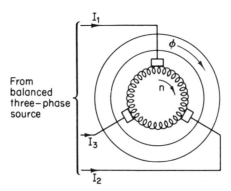

Figure P4-59

4-60. The resistance measured between each pair of slip rings of a three-phase, 60-Hz, 300-hp, 16-pole induction motor is 0.04 Ω. With the rings short-circuited, the full load slip is 0.025, and it may be assumed that the slip-torque curve is a straight line from no-load to full-load. The motor drives a fan which requires 300 hp at the full-load speed of the motor. The torque required to drive the fan varies as the square of the speed. What value of resistance should be connected in series with each slip ring so that the fan will run at 300 rpm?

4-61. A three-phase, 60-Hz, four-pole, wound-rotor induction motor has a rated speed of 1746 rpm. The actual rotor resistance per phase is 0.8 Ω. What starting resistance must be inserted in the rotor circuit so that rated torque will be developed at starting? Neglect any change of rotor resistance with frequency.

4-62. For the polyphase induction motor, sketch the variation of torque as a function of horsepower output over the speed range from zero to synchronous speed. *Hint:* Apply the circle diagram.

4-63. Continue the analysis displayed in Table 4-1 by completing the tabulation for the remaining three 60° intervals of the cycle.

4-64. Draw the line-to-neutral voltage variations for each phase over one complete cycle for the speed control system of Fig. 4-28 corresponding to the switching sequence displayed in Fig. 4-29.

4-65. In the interest of generating a more nearly sinusoidal waveform for the line voltages in the speed control system of Fig. 4-28, the gated ON interval for each SCR is specified as 120°. Draw the switching sequence diagram for the inverter.

4-66. (a) Use the results of Prob. 4-65 and sketch the variation of each line-to-neutral voltage over one cycle of the control frequency.

(b) Sketch the variations of the line-to-line voltages over one complete cycle of the control frequency.

(c) Determine the expression for the rms value of the line-to-line voltages and compare it with the result that appears in Eq. (4-86).

5

Three-Phase
Synchronous Machines

The synchronous generator is universally used by the electric power industry for supplying three-phase as well as single-phase power to its customers. The single-phase power that is brought to homes, shops, and offices originates from one phase of the three-phase system. Moreover, the assignment of commercial load circuits to each phase is made in an effort to keep the phases balanced.

The basic construction features of these machines are described in Sec. 3-3 and illustrated in Fig. 3-13. It is worth noting here that synchronous generators are classified in two types. The first is the *low-speed* (engine- or water-driven) type, which is characterized physically by having salient poles, a large diameter, and small axial length. The second is the *turbogenerator*, which uses the steam turbine as the prime mover. The usual salient-pole rotor construction is abandoned in these high-speed generators in favor of the cylindrical (or smooth) rotor because the protruding-pole construction gives rise to dangerously high mechanical stresses. For 60-Hz three-phase power the two-pole generator must operate at 3600 rpm. Moreover, since turbogenerators are invariably designed for two poles, it follows that these machines are further characterized by a small diameter and long axial length.

5-1 GENERATION OF A THREE-PHASE VOLTAGE

How is a three-phase voltage generated? This can be readily understood from a study of Fig. 5-1(a). Depicted here is a two-pole rotor, the field winding of which is assumed to be energized from a d-c source to create the pole flux. It is further assumed that the pole pieces are shaped to produce a sinusoidal flux field. Appearing in the stator is a

238

Figure 5-1 Induced-voltage distribution in the stator winding of a synchronous generator: (a) phase a at a positive maximum (time t_1 in Fig. 5-2); (b) phase b at a negative maximum (time t_2).

balanced three-phase winding with the axis of each phase displaced by 120°. The complete winding of each phase is represented in Fig. 5-1 by a single coil. Now consider that the rotor is driven by the prime mover in a counterclockwise direction at synchronous speed. Applying the $\bar{v} \times \bar{B}$ rule for the field direction shown reveals that the instantaneous voltage induced in coil sides a, b', c' is directed out of the paper while in coil sides a', b, c it is directed into the paper. Furthermore, since coil sides a and a' are located beneath the maximum value of the flux-density wave, the induced emf in phase a is at its maximum value. The corresponding induced emf in phase b, as identified by coil side b, is seen to be of opposite sign and of smaller magnitude. In fact, since coil sides b and b' are both displaced from the maximum flux-density position by 60°, it follows that this instantaneous voltage is at one-half (cos 60°) its maximum value. A similar reasoning yields the same result for phase c.

Next consider that the rotor has advanced by 60° in the counterclockwise direction. This puts the amplitude of the north-pole flux directly beneath b', indicating that the induced emf is now a negative maximum in phase b. Also, coil c-c' now finds itself under the influence of flux of reversed polarity as indicated in Fig. 5-1(b). Hence the direction of the emf in c is now out of the paper and into the paper for c'. The instantaneous values of the three-phase voltages for the first and second time instants are depicted in Fig. 5-2 and are identified as t_1 and t_2, respectively.

By repeating the foregoing procedure many times and plotting the results for each instant of time we obtain the complete curves of Fig. 5-2. Note that the induced-emf variation for each phase is identical with the other two except for time displace-

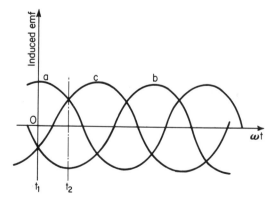

Figure 5-2 Time history of three-phase induced emfs for the generator of Fig. 5-1.

Sec. 5-1   Generation of a Three-Phase Voltage

239

ments of 120° and 240°, respectively. As a matter of fact this variation is a direct consequence of having the beginning of each phase space-displaced by 120 electrical degrees. Physical connection of the ends of each phase (points a', b', c') gives a Y-connected stator winding.

5-2 LINEAR ANALYSIS BY THE GENERAL METHOD

The basic principles and characteristics of the cylindrical-pole synchronous generator can be readily described by developing the phasor diagram as was done for the transformer. However, there exists an important distinction that must be understood right at the start. The transformer (as also the three-phase induction motor) is essentially a constant-flux device, i.e., the amplitude of the resultant operating air-gap flux is fixed by the applied a-c voltage. In contrast the resultant air-gap flux in the synchronous machine depends upon the magnitude of the exciting current in the field winding (located on the rotor structure) as well as on the value of the armature (or load) current.† Recall that for the synchronous machine the armature winding is the stator winding. Consequently, when analyzing the synchronous generator, it becomes necessary to introduce the magnetization curve that applies for its magnetic circuit. If the design data are available, the magnetization curve can be computed by following the procedure outlined in Chapter 1. Otherwise, if the machine is available in the laboratory, the magnetization curve can be found by performing a simple *open-circuit test*. The wiring diagram for such a test is shown in Fig. 5-3. The rotor

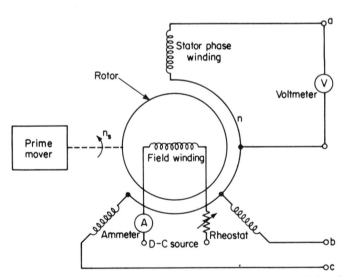

Figure 5-3 Wiring diagram for determining the no-load characteristic of the synchronous generator.

†Here we are treating the synchronous machine as an isolated unit. If it were assumed to be connected to an infinite bus system (i.e., in parallel with other synchronous generators many times its size), the air-gap flux would be invariant.

is driven at synchronous speed throughout the test by the prime mover. The d-c field current is varied by means of the rheostat and at each setting the corresponding line-to-neutral voltage is measured. A plot of these data points is depicted as curve (a) in Fig. 5-4 and is often called the *no-load characteristic*. The straight-line portion of this curve is called the *air-gap characteristic*. In this section our interest is confined to *linear* analysis so that for our immediate purposes the magnetization curve to be used will be curve (b) of Fig 5-4.

Figure 5-4 No-load characteristics: (a) nonlinear open-circuit characteristic; (b) linear air-gap characteristic; (c) short-circuit characteristic.

Another useful characteristic in connection with synchronous generator analysis is the *short-circuit characteristic*. This is found by short-circuiting terminals *a*, *b*, and *c* in the diagram of Fig. 5-3, driving the rotor at synchronous speed, and measuring the short-circuit line currents for various values of field current. A plot of the short-circuit line current versus field current leads to curve (c) of Fig. 5-4. This characteristic is linear because the amount of air-gap flux needed to produce even as much as twice rated armature current is so small as to cause little or no saturation of the magnetic-circuit iron. Only enough flux is needed to produce an emf which overcomes the leakage impedance drop.

Before proceeding with a description of the phasor diagram, a few remarks are in order at this point concerning the inclusion in one phasor diagram of both the time-varying quantities (such as voltage and current) and the mmf space sinusoids.

Figure 5-5 depicts a two-pole turbogenerator. The field winding is located on the rotor and is distributed in suitably placed slots. When a direct current is made to flow through this winding, an ampere-conductor distribution is produced. The illustration shows the current moving into the conductors on the right side, and flowing out of the conductors on the left side. The corresponding mmf produced by this field winding has a sinusoidal distribution in space with its axis directed along the vertical. Because the maximum value of the flux density lies along this vertical too, it follows that the voltage induced in coil e_f-e_f' is a maximum value. Moreover, application of the $\bar{v} \times \bar{B}$ rule indicates that this emf is directed out of the paper for the top conductor and into the paper for the bottom conductor. Although it is assumed that the stator is equipped with a three-phase winding, essentially coil e_f-e_f' can be considered to represent the emf induced in this entire winding for the instant shown. Accordingly, if one were to look at the emf induced in each coil side of the entire three-phase winding, a sinusoidal emf distribution would result, with its peak lying along the vertical. As a matter of convenience, therefore, it can be said that insofar as the armature induced emf is concerned, the entire three-phase winding can be represented by the single coil e_f-e_f'. Note that the axis of this coil lies along the horizontal. In general, the axis of the coil that represents the emf distribution in the three-phase winding is always 90 electrical degrees displaced from the axis of the field mmf $\mathfrak{F}_F$. As a result of this quadrature relationship it will be customary for us in the phasor diagrams which follow to locate the field mmf $\mathfrak{F}_F$ at a position 90 degrees leading the direction of the excitation emf phasor. The excitation emf is the voltage induced by the field winding flux alone. It is helpful to note in Fig. 5-5 that when the emf in coil e_f-e_f' is a maximum, the flux linkage is zero.

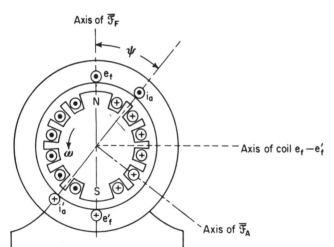

Axis of $\overline{\mathfrak{F}}_F$

Axis of coil e_f — e_f'

Axis of $\overline{\mathfrak{F}}_A$

Figure 5-5 Diagram illustrating the relationship between mmf space sinusoids and the associated time-varying quantities of excitation voltage and armature current. The rotor is assumed to be driven counterclockwise at ω radians per second.

A similar convention is adopted concerning the phasor representation of the armature current and the corresponding mmf associated with this current. For example, in Fig. 5-5, coil i_a-i_a' is used to denote the sinusoidal ampere-conductor distribution of the entire three-phase armature winding for a phase lag relative to the excitation emf of ψ degrees. Note that the axis of this coil coincides with the location

THREE-PHASE SYNCHRONOUS MACHINES   CHAP. 5

of the peak value of the corresponding sinusoidal armature mmf $\mathcal{F}_A$ and is space displaced from the field mmf axis $\mathcal{F}_F$ by $(90° + \psi)$. Another point worthy of note is that as the phase angle between armature current and excitation emf changes, a corresponding change occurs in the axis of $\mathcal{F}_A$. This fact is noted in subsequent phasor diagrams by placing the armature mmf coincident with the phasor for armature current.

In synchronous machine analysis it is important to note that the two space sinusoids $\overline{\mathcal{F}}_F$ and $\overline{\mathcal{F}}_A$, having the same frequency of rotation and the same number of poles, can be added to yield a resultant space sinusoid denoted by $\overline{\mathcal{F}}_R$, which in turn is responsible for generating the resultant induced emf in the armature winding $\bar{E}_r$. As a convenience, the result of this addition is also represented in the phasor diagram in spite of the fact that these mmfs are space sinusoids and not time-varying sinusoids. This procedure is permissible because space sinusoids can also be summed by phasor methods.

One final point. Application of the force rule $(\vec{I} \times \vec{B})$ discloses that an average torque is produced on the armature (stator) winding to cause counterclockwise rotation. However, because the stator is immobile, there occurs instead a clockwise torque on the rotor. The latter represents a counter-torque to the prime mover of the turbogenerator.

The development of the phasor diagram is done on a per phase basis. Assume that the armature winding resistance r_a and leakage reactance x_l per phase are known along with the air-gap characteristic. Assume too that the synchronous generator is delivering rated armature current to a lagging power-factor load at rated terminal voltage. Call the rated terminal voltage per phase V_t and the load power-factor angle θ. Let it be desired that the value of the field current corresponding to this load condition should be found. Before proceeding, it is helpful to keep in mind the factors that must be reckoned with in establishing the rated terminal voltage when rated current flows. It is shown in Sec. 4-1 that the existence of three-phase currents in a three-phase winding leads to a revolving mmf $\overline{\mathcal{F}}_A$. It follows then that the resultant mmf $\overline{\mathcal{F}}_R$ is made up of the vector sum of the field mmf $\overline{\mathcal{F}}_F$ and the armature mmf $\overline{\mathcal{F}}_A$. In equation form we have

$$\overline{\mathcal{F}}_F + \overline{\mathcal{F}}_A = \overline{\mathcal{F}}_R \tag{5-1}$$

In turn, this resultant mmf is responsible for the resultant air-gap flux Φ_r that induces the actual armature winding voltage $\bar{E}_r$. The terminal voltage is then obtained by subtracting the armature leakage impedance drop per phase from $\bar{E}_r$.

We shall now develop the phasor diagram by starting with the known terminal conditions: rated armature current I_a lagging the rated terminal voltage V_t by θ degrees. Refer to Fig. 5-6. The induced armature winding voltage per phase is obtained as the phasor sum of $\bar{V}_t$ and the armature leakage impedance drop. Thus, employing the horizontal axis as the zero reference line for angle measurements, we can compute

$$\bar{E}_r = \bar{V}_t + \bar{I}_a(r_a + jx_l)$$

or

$$E_r\underline{/\gamma} = V_t\underline{/0°} + I_a\underline{/-\theta}\,(r_a + jx_l) \tag{5-2}$$

The quantity R that produces E_r is found by entering the air-gap characteristic. Note that R is expressed in field amperes rather than field ampere-turns $\overline{\mathcal{F}}_R$. This is irrelevant here in view of our objective, which is to find the required field current for the specified load condition.

The complete expression for the field current representing the resultant flux is then

$$\bar{R} = R\underline{/90°} + \gamma \qquad (5\text{-}3)$$

where γ is determined from the right side of Eq. (5-2). It is convenient now to write the counterpart of Eq. (5-1) in field amperes, which after all is the quantity that usually appears as the abscissa quantity in the experimentally established curves of Fig. 5-4. Thus

$$\bar{F} + \bar{A} = \bar{R} \qquad (5\text{-}4)$$

where $\bar{A}$ denotes the armature mmf expressed in equivalent field amperes.

A glance at Eq. (5-4) shows that with $\bar{R}$ available $\bar{F}$ can be determined once $\bar{A}$ is known. There are two ways to determine $\bar{A}$, either from the design data or from appropriate no-load tests. The latter approach is discussed in Sec. 5-4. When the design data are available, Eq. (3-51) may be used to represent the armature mmf per pole. Keep in mind that this equation is derived (see Appendix C) as the amplitude of the fundamental component of the mmf distribution of the armature winding. Also, the distribution of the field winding on the cylindrical-rotor surface is such that it leads practically to a sinusoidal mmf wave having a peak mmf per pole of $N_f I_f$, where N_f is the number of turns per pole. Accordingly, since the armature mmf and the field mmf waves are both of the same shape, it follows that the actual armature mmf as described by Eq. (3-51) can be replaced by an equivalent field winding current A. Thus

$$A = \frac{1}{N_f} 0.9 q \frac{N_2 K_{w2}}{p} I_a \qquad \text{equivalent field amperes} \qquad (5\text{-}5)$$

where q = number of phases

p = number of poles

$N_2 K_{w2}$ = effective armature winding turns per phase

I_a = armature current per phase.

Equation (5-5) indicates that A and I_a differ by a fixed relationship as already described. Hence the phasor expression for the equivalent field current representing the armature mmf for lagging power-factor load is

$$\bar{A} = A\underline{/-\theta} \qquad (5\text{-}6)$$

where A is described by Eq. (5-5). In the phasor diagram of Fig. 5-6 note that $\bar{A}$ and $\bar{I}_a$ are shown on the same line as a matter of convenience.

With $\bar{R}$ and $\bar{A}$ determined, the required field current is found to be

$$\bar{F} = \bar{R} - \bar{A} = R\underline{/90°} + \gamma - A\underline{/\theta} \qquad (5\text{-}7)$$

The quantity $\bar{E}_f$ appearing in Fig. 5-6 and shown 90° behind $\bar{F}$ is the voltage produced by $\bar{F}$ at the armature terminals with the load removed (i.e., $I_a = 0$) and no saturation of the iron. It is usually called the excitation voltage. The foregoing procedure for finding the field current is referred to as the *general method.*

The positive direction of rotation for the phasor quantities and the space

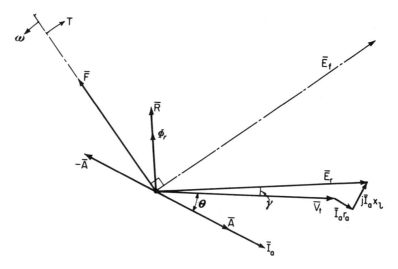

Figure 5-6 Phasor diagram of a synchronous generator for a lagging power-factor load. The general method of analysis is indicated.

sinusoids depicted in Fig. 5-6 is counterclockwise and is indicated by the direction of the arrow marked ω. It was also previously pointed out that for generator action a counter torque exists on the rotor. This is indicated by the arrow marked T in Fig. 5-6. It is useful to note here that in generator action the torque on the field structure is always such as to bring the field axis in alignment with the resultant flux.

Example 5-1

A 10,000-kVA, three-phase, Y-connected, two-pole, 60-Hz, 13,800-V, line-to-line turbogenerator has the air-gap characteristic depicted in Fig. 5-4. The armature winding resistance is 0.07 Ω per phase and the armature winding leakage reactance per phase is 1.9 Ω. Also, the armature mmf at rated current is known to be equal to 155 equivalent field amperes. Find the field current that produces rated terminal voltage when rated armature current is delivered to a 0.8 lagging power-factor three-phase balanced load.

Solution The rated phase voltage is

$$V_t = \frac{13,800}{\sqrt{3}} = 7967.4 \text{ V} \tag{5-8}$$

and the rated armature current is

$$I_a = \frac{10,000}{\sqrt{3}\,13.8} = 418.4 \text{ A} \tag{5-9}$$

The actual armature induced emf is found from Eq. (5-2) to be

$$\bar{E}_r = E_r\underline{/\gamma} = 7967.4\underline{/0°} + 4178.4\underline{/-36.8°}(0.07 + j1.9) \tag{5-10}$$

Note that r_a is small enough to be neglected. Thus,

$$\bar{E}_r = 7967.4 + 4178.4(0.8 - j0.6)j1.9$$

$$= 7967.4 + 477 + j636$$

$$= 8444.4\underline{/4.3°} \tag{5-11}$$

The corresponding value of R from the air-gap characteristic is 157 A. Hence by Eq. (5-3)

$$\bar{R} = 157\underline{/90°} + 4.3° = 157\underline{/94.3°} \qquad (5\text{-}12)$$

Equation (5-7) then yields the desired result:

$$\bar{F} = \bar{R} - \bar{A} = 157\underline{/94.3°} - 155\underline{/-36.8°}$$
$$= -11.7 + j156.6 - 124 + j92.8$$
$$= -135.79 + j249.1 = 284\underline{/119°} \text{ A} \qquad (5\text{-}13)$$

The excitation voltage produced by this field current is found on the air-gap line to be

$$\bar{E}_f = 15,300\underline{/29°} \text{ V} \qquad (5\text{-}14)$$

This quantity is almost twice the rated terminal voltage. It is a highly exaggerated result because of the assumption of no saturation. Practical and useful results are obtained when this assumption is dropped as described in Sec. 5-4.

5-3 LINEAR ANALYSIS BY THE SYNCHRONOUS REACTANCE METHOD. THE EQUIVALENT CIRCUIT

In any treatment of the behavior of synchronous machines it is necessary to take due account of the effects produced by the current that flows in the armature winding—frequently referred to as *armature reaction*. In the preceding section this was accomplished by working directly with the armature winding mmf expressed in equivalent field amperes. A second approach is possible by treating the armature reaction in terms of a suitable flux and its associated induced voltage. Since the assumption of no saturation has been made, it is to be expected that either viewpoint will lead to the same result. As a matter of fact this is precisely the reason for imposing the linearity condition. In this way the introduction of new concepts can readily be correlated with preceding ideas with consistency.

By means of superposition the effect of the rotating armature mmf can be treated individually. Accordingly, the set of revolving mmf poles created by the three-phase armature current in the three-phase winding can be considered to produce an armature reaction flux Φ_{ar}. Of course this flux bears a fixed relationship to its mmf $\bar{A}$. Moreover, it revolves with respect to the stationary armature winding at a speed corresponding with 60 Hz. Hence Φ_{ar} can be said to induce a voltage (rise) of value E_{ar} which can be placed 90° behind Φ_{ar} as depicted in Fig. 5-7(a). In situations of this kind it is customary to replace such an armature flux voltage by a reactance drop that involves the associated current (I_a in this case) directly. Recall that this same technique was used to replace the effect of leakage flux in the transformer as well as armature leakage flux occurring around the slots of the stator windings. Expressed mathematically this is

$$\boxed{\bar{E}_{ar} \equiv -j\bar{I}_a x_\phi} \qquad (5\text{-}15)$$

where x_ϕ is the *armature reaction reactance*. The $-j$ factor accounts for the 90° lag

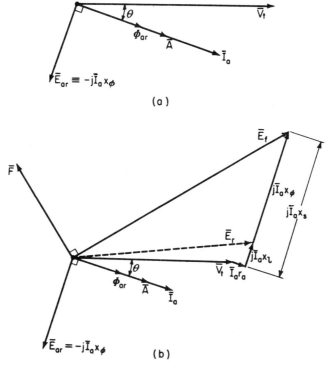

Figure 5-7 Illustrating the replacement of the effect of armature mmf by an equivalent induced emf E_{ar}.

existing between $\bar{I}_a$ and $\bar{E}_{ar}$. Furthermore, the linearity condition enables a constant value of x_ϕ to replace the effect of armature flux. Thus a 50% increase in armature reaction flux brought about by a 50% increase in armature current causes E_{ar} to increase to 1.5 times its previous value. But the presence of I_a on the right side of Eq. (5-15) automatically accounts for this. If I_a and Φ_{ar} were not assumed to be linearly related, this correspondence would not prevail. Finally, it should be apparent that, when the synchronous generator operates at a field current $\bar{F}$, the associated excitation voltage must contain a component equal and opposite to that specified by Eq. (5-15).

Based on this background the complete phasor diagram of the effect of armature mmf treated in terms of its equivalent reactance drop can be developed. We begin by assuming rated armature current flows to a balanced three-phase lagging power-factor load at rated terminal voltage. As with the general method, the armature leakage impedance drop can be added to $\bar{V}_t$ to yield $\bar{E}_r$. Refer to Fig. 5-7(b). However, at this point, instead of dealing with mmf's, the analysis continues in voltages. Thus to $\bar{E}_r$ is added an armature reaction reactance drop that is equal and opposite to $\bar{E}_{ar}$. The resultant is the excitation voltage. An important observation to make in Fig. 5-7(b) is that the leakage reactance drop and the armature reaction reactance drop add along the same line, both being perpendicular to $\bar{I}_a$. This allows the two quantities to be added algebraically, which leads to a total reactance called the *synchronous reactance* and denoted by x_s. In equation form

$$x_s \equiv x_l + x_\phi \tag{5-16}$$

where x_l is the armature winding leakage reactance per phase and x_ϕ is defined by Eq. (5-15).

A study of Fig. 5-7(b) discloses that with the use of the synchronous reactance the required field excitation for any specified load condition can be found in two steps. The first step involves finding $\bar{E}_f$ directly from the equation

$$\bar{E}_f = \bar{V}_t + \bar{I}_a(r_a + jx_s) \approx V_t\underline{/0°} + I_a\underline{/-\theta}\ jx_s \tag{5-17}$$

In step two the air-gap characteristic is entered at the level of E_f and the corresponding value of F is recorded. This procedure is called the *synchronous reactance method*. It is useful to keep in mind that the synchronous reactance is a fictitious quantity that replaces the effect of the armature winding leakage flux and the armature winding rotating mmf.

A circuit interpretation of Eq. (5-17) leads directly to the equivalent circuit of the synchronous generator and is depicted in Fig. 5-8. This is an immediate consequence of introducing the concept of synchronous reactance.

Figure 5-8 The per-phase equivalent circuit of the synchronous generator lies to the left of the load terminal.

How is the synchronous reactance determined? If the design data are available so that the effective turns of the stator as well as the reluctance of the magnetic circuit can be computed, then an expression like Eq. (3-124) can be used. Otherwise x_s can be found from the air-gap characteristic and the short-circuit characteristic, i.e., through the use of curves (b) and (c) of Fig. 5-4. To understand how this comes about let us examine the phasor diagram of the synchronous generator under short-circuit conditions with the field current adjusted to the value that causes rated armature current to flow. The armature winding mmf remains the same, but the terminal voltage is zero. Accordingly, the phasor diagram of Fig. 5-7(b) reduces to that shown in Fig. 5-9(a). Note that the field current is used almost entirely to neutralize the effect of the armature winding mmf. The remainder, R_{sc}, is only large enough to overcome the leakage impedance drop, which for all practical purposes is equal to the leakage reactance drop. Note too that the armature current in short-circuit conditions lags the excitation voltage by nearly 90°. For obvious reasons at such low power factors the quantities $\bar{R}$, $\bar{A}$, and $\bar{F}$ are essentially collinear. However, the important thing to note in Fig. 5-9(a) is that the excitation voltage at short circuit is practically equal to the synchronous impedance drop. This observation provides the key for the determination of x_s. If F_{sc} is maintained and the short circuit is removed, the open-circuit voltage produced by this field current is obviously E_{fsc}. Therefore, with F_{sc}

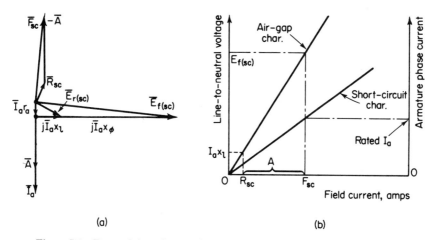

Figure 5-9 Determining the synchronous reactance from no-load tests: (a) phasor diagram at short-circuit; (b) no-load characteristics.

known, the air-gap characteristic can be used in the manner illustrated in Fig. 5-9(b) to obtain E_{fsc}. Similarly, by entering the short-circuit characteristic the corresponding current produced by F_{sc} can be found. The synchronous reactance is then the ratio of these two quantities. Thus

$$x_s = \frac{E_{fsc}}{I_a} \qquad (5\text{-}18)$$

Keep in mind that this is the *unsaturated* value of synchronous reactance because we are still working with the air-gap line.

Example 5-2

Determine the following information for the machine of Example 5-1: (a) the value of the unsaturated synchronous reactance and (b) the field current needed to establish rated terminal voltage when rated current is delivered to a balanced three-phase load of 0.8 power factor lagging, using the synchronous reactance method.

Solution (a) A glance at Fig. 5-4 indicates that a field current of $F_{sc} = 170$ A is needed to force rated current of 418.4 A through the armature winding at short-circuit. Furthermore, the excitation voltage corresponding to 170 field amperes is $E_{fsc} = 9200$ V. Hence

$$x_s = \frac{E_{fsc}}{I_{a \text{ (rated)}}} = \frac{9200}{418.4} = 22\ \Omega \qquad (5\text{-}19)$$

(b) By Eq. (5-17) we have

$$\bar{E}_f = V_t \underline{/0^\circ} + j\bar{I}_a x_s = 7967.4 \underline{/0^\circ} + 418.4(0.8 - j0.6)j22$$

$$= 7967.4 + 5523 + j7364 = 13,490 + j7364$$

$$= 15,369 \underline{/28.6^\circ} \qquad (5\text{-}20)$$

Then from the air-gap characteristic

$$F = 285\ \text{A} \qquad (5\text{-}21)$$

A comparison of these results with Eqs. (5-13) and (5-14), which are obtained by the general method, shows them to be almost identical. Such correspondence is entirely in keeping with the linearity assumption.

5-4 NONLINEAR MACHINE ANALYSIS

The power output capability of electric machines is directly dependent upon the degree to which the material is worked magnetically and electrically. The higher the operating flux density in the iron parts of the machine and the current density in the copper parts are, the greater is the output. Consequently, in most electric machines the level of flux density chosen puts operation in the saturated or nonlinear region. If analytical procedures are to be used to predict machine performance, then the effect of the nonlinearity must be appropriately accounted for. Otherwise realistic results will be almost impossible to attain. This is a point to keep well in mind when sophisticated procedures are employed to predict machine behavior, formulated in terms of inductance parameters based on linearity.

The obvious way to account for operation of the machine in its nonlinear region is to work directly with the nonlinear characteristic usually represented in a graphic form such as curve (a) in Fig. 5-4. This curve is the actual *open-circuit characteristic* and may be either computed from the design data by employing the procedures discussed in Chapter 1 or found experimentally as described in Sec. 5-2. It should not be inferred that the mere use of the actual nonlinear characteristic guarantees realistic results. That this need not be so will be illustrated presently. Attention is now turned to the application of the general method and the synchronous reactance method to find the necessary excitation currents for synchronous generators by utilizing the actual, nonlinear open-circuit characteristic instead of the linear air-gap characteristic. It is worthwhile to note that information about the range of the excitation currents is necessary for the proper design of a voltage regulator which will act to maintain constant terminal voltage in the presence of changing load demands.

The General Method

Again, let us find the excitation needed to produce rated terminal voltage when rated current flows to a balanced lagging power-factor load. The phasor diagram of Fig. 5-6 applies with undiminished validity. The actual armature winding emf induced by the resultant flux is computed as in the linear case. Thus

$$\bar{E}_r = V_t \underline{/0^\circ} + \bar{I}_a(r_a + jx_l) = 7966.4 + 418.4\underline{/-36.8^\circ}(0.07 + j1.9)$$

$$= 8444.4\underline{/4.3^\circ} \text{ V} \tag{5-22}$$

The value of the resultant mmf expressed in field amperes that produces this result is found by entering the nonlinear open-circuit characteristic (Fig. 5-4) at 8444.4 volts and reading the associated abscissa value. This yields

$$R = 222 \text{ A}$$

so that the complete phasor expression becomes

$$\bar{R} = 222\underline{/94.3^\circ} \text{ A} \tag{5-23}$$

A comparison with the result obtained through linear analysis indicates a wide discrepancy (222 versus 157 A), which attests to the fact that the machine is operating at a fairly high level of saturation.

The armature reaction mmf is again treated in terms of an equivalent field current of 155 A. Application of Eq. (5-7) then provides the required value of the field current:

$$\bar{F} = \bar{R} - \bar{A} = 222\underline{/94.3°} - 155\underline{/-36.8°}$$

$$= -140.6 + j314$$

$$= 344\underline{/114°} \text{ A} \tag{5-24}$$

Eq. (5-13) shows that the result obtained by linear analysis is considerably lower and, if used for the design of a voltage regulator, would obviously lead to an inadequate design.

The field excitation voltage that corresponds to the field current of 344 amperes is found on the curve of the open-circuit characteristic to be

$$E_f = 9560 \text{ V} \tag{5-25}$$

This value stands in sharp contrast to the figure of 15,300 found in the linear case. The latter quantity is highly unrealistic, and it is easy to appreciate why by noting in Fig. 5-4 the wide divergence existing between the open-circuit and the air-gap characteristics for field current in the vicinity of 300 A.

Equation (5-25) identifies the value of the terminal voltage that occurs if the excitation current is maintained at 344 A and the load is disconnected. In other words the no-load terminal voltage is 9560 V. Accordingly, the voltage regulation of this machine in percent is found to be

$$\text{voltage regulation} = \frac{E_f - V_t}{V_t} 100 = \frac{9560 - 7960}{7960} 100 = 20.1\% \tag{5-26}$$

where E_f is the excitation voltage found on the magnetization curve. This figure is not only realistic but typical of such machines. Calculation of the voltage regulation for the linear case yields a meaningless result.

The Saturated Synchronous Reactance Method

It is shown in Sec. 5-3 that the use of the unsaturated synchronous reactance method in finding excitation currents leads to unrealistic solutions for reasons therein described. However, it is possible to alter the method of computation in a way that is consistent with the state of saturation that the machine is in and thereby obtain results that are comparable with those of the general method.

Frequently, the design of synchronous machines is such that when rated power is delivered at rated voltage the operating point on the magnetization curve is located somewhere along the knee of the curve. Accordingly, the machine can be described as operating in a partially saturated state. A good indication of the degree of saturation of the magnetic circuit is given by the resultant flux Φ_r or its associated emf E_r. Thus, if a synchronous generator is assumed to be delivering rated power at rated voltage and a specified power factor, the actual armature induced emf $\bar{E}_r$ can be found as illustrated by Eq. (5-22) and then located on the magnetization curve. This is illustrated in Fig. 5-10, where E_r is represented by point a on the saturation curve and is produced by the field current R. It is worthwhile to note here that if there were no saturation of the iron, the voltage E_r would be induced by a smaller excitation R_{ag}. In essence, then, it can be said that the degree of saturation prevailing under these conditions is represented by the ratio of these two quantities. That is, the reluctance of the magnetic circuit has been increased by a saturation factor k where,

$$k = \frac{R}{R_{ag}} = \frac{E_{r(ag)}}{E_r} \qquad (5\text{-}27)$$

The second form of this equation follows from the geometry of similar triangles in Fig. 5-10.

How would this synchronous generator perform if it were now assumed that the degree of saturation remained fixed at Φ_r and the armature current were reduced to zero while the field current remained constant? A study of Fig. 5-10 discloses that the operating point would move from a to c along a linearized magnetization curve that corresponds to the resultant flux. Point c is found as the point of intersection of a line drawn from the origin through a and the ordinate line placed at the constant field excitation (F). The no-load voltage that corresponds to c is identified as E'_f and

Figure 5-10 Illustrating the saturated synchronous reactance method.

represents the no-load terminal voltage that would be produced if the air-gap flux remained invariant as the armature current diminished to zero. Since this does not actually happen, clearly E'_f is entirely a fictitious quantity. On the other hand, the field current F (that is associated with E'_f) is not fictitious, but real. In fact, F is the quantity that we are seeking, and the foregoing construction has been introduced simply to facilitate the determination of F. The determination immediately follows from the similar triangles OaR and OcF. Thus

$$F = \frac{E'_f}{E_r} R \tag{5-28}$$

where these quantities are defined as illustrated in Fig. 5-10.

In applying Eq. (5-28), the quantity E_r is found by means of Eq. (5-22); R is found from the actual magnetization curve. E'_f is then computed by the synchronous reactance method in the manner illustrated in Example 5-2 but with the necessary modification that that value of synchronous reactance be used that corresponds to the degree of saturation represented by the resultant flux. If x_s were entirely associated with the iron structure of the machine, the saturated synchronous reactance could be found by dividing the unsaturated x_s by the saturation factor k. However, as indicated by Eq. (5-16), the synchronous reactance also includes the effect of leakage reactance, which remains essentially unaffected by the changes in the degree of saturation of the main iron paths of the machine. Accordingly, the saturation factor must be applied to x_ϕ only in Eq. (5-16). This leads to the following expression for the saturated synchronous reactance x_{ss}:

$$x_{ss} = x_l + \frac{x_s - x_l}{k} \tag{5-29}$$

where x_l denotes leakage reactance, k is the saturation factor given by Eq. (5-27), and x_s is the unsaturated synchronous reactance given by Eq. (5-18). The following example illustrates the effectiveness of this method.

Example 5-3

For the machine of Example 5-1, find the field current that produces rated terminal voltage when rated armature current is delivered to a 0.8 lagging pf three-phase balanced load. Use the saturated synchronous reactance method.

Solution The values of E_r and R are given by Eqs. (5-22) and (5-23), which are repeated here for convenience.

$$E_r = 8444.4 \text{ V} \quad \text{and} \quad R = 222 \text{ A}$$

The saturation factor is found to be

$$k = \frac{R}{R_{ag}} = \frac{222}{157} = 1.41$$

Hence by Eq. (5-29) the saturated synchronous reactance is

$$x_{ss} = x_l + \frac{x_s - x_l}{k} = 1.9 + \frac{22 - 1.9}{1.41} = 16.2 \ \Omega$$

Accordingly, we get

$$\bar{E}'_f = V_t\underline{/0°} + \bar{I}_a(jx_{ss})$$
$$= 7967.4 + j418.4(0.8 - j0.6)16.2$$
$$= 7967.4 + 4042 + j5423$$
$$= 12,009 + j5423 = 13,177\underline{/24.3°}$$

Finally, by Eq. (5-28) the required excitation current is found to be

$$F = \frac{13,177}{8444.4}(222) = 346 \text{ A}$$

which is identical to the result found by the general method.

When it is desirable to compute voltage regulation by the saturated synchronous reactance method, a glance at Fig. 5-10 reveals that the no-load terminal voltage to be used in Eq. (5-26) is to be found on the actual magnetization curve corresponding to the value of F computed by Eq. (5-28). This is denoted by point b in Fig. 5-10.

5-5 EXPERIMENTAL DETERMINATION OF LEAKAGE REACTANCE x_l AND ARMATURE REACTION MMF A

Nonlinear machine analysis by the general method presupposes knowledge of the armature winding leakage reactance as well as the armature winding mmf expressed in equivalent field amperes. When the complete design data of a machine are available, these quantities can be readily calculated. However, in the absence of such data other means must be sought to furnish the information.

As a review, let us return to Fig. 5-9, which depicts the situation in the synchronous generator at short-circuit with rated current flowing. Recall that the armature current flows at almost zero power factor and so the quantities $\bar{F}_{sc}$, $\bar{R}_{sc}$, and $\bar{A}$ are collinear. In Fig. 5-9(b), F_{sc} is a measured quantity and is accordingly available. If the winding data are known, the quantity A can be computed from Eq. (5-5) and then algebraically subtracted from F_{sc} to yield R_{sc}. Then on the air-gap line corresponding to R_{sc}, the leakage reactance drop is found. Strictly speaking, this ordinate quantity is $E_{r(sc)}$, but because $x_l > r_a$ little error is made by taking $E_{r(sc)} = I_a x_l$. Since $E_{r(sc)}$ and I_a are then both known, x_l can be found. Unfortunately, this determination of x_l is dependent upon knowledge of A which is generally not available and must also be found.

On the basis of the foregoing it appears worthwhile to explore further the operation of the synchronous generator at zero power factor. The wiring diagram for such a test is shown is Fig. 5-11. The zero power-factor load consists of purely inductive coils arranged in a Y-connection. Initially the coils are set for zero reactance and the field current is adjusted to yield rated current. This corresponds with point F_{sc} in Fig. 5-9(b). The coil reactance is then increased, which causes a drop in current. The field current is then increased to reestablish the rated value of the armature current. Now the terminal voltage is no longer zero but, rather, equal to $I_a X_L$, where X_L is the load reactance. By continuing this procedure, rated terminal voltage can be made to appear across the zero power-factor load with rated armature current

Figure 5-11 Wiring diagram for obtaining the zero power-factor characteristic.

flowing. A plot of terminal voltage versus field current yields the *zero power-factor characteristic*.

If the machine were entirely linear (i.e., with no saturation occurring), the zero power-factor characteristic could be readily drawn from a knowledge of F_{sc} alone. The construction should be obvious from an inspection of Fig. 5-12. The zero power-factor characteristic is generated merely by displacing triangle Oac so that Oa lies on the air-gap characteristic. Note that to produce rated terminal voltage at zero power factor, the total field current must be F_z. By subtracting A from F_z, R is obtained corresponding to which is the armature induced voltage E_r on the air-gap line. The vertical distance between E_r and V_t (line $a'b'$ in Fig. 5-12) continues to be the leakage reactance drop because at zero power factor E_r and V_t are also essentially collinear.

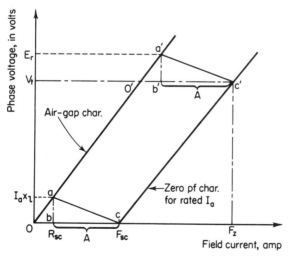

Figure 5-12 Illustrating how the zero power-factor characteristic can be derived from the short-circuit data for conditions of no saturation.

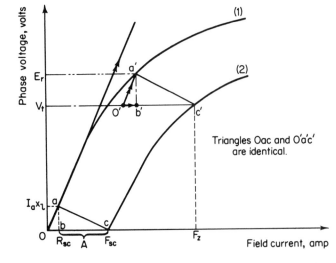

Triangles Oac and O'a'c'
are identical.

Figure 5-13 Graphical determination
of leakage reactance x_l and armature
mmf expressed in field amperes.
Triangle $a'b'c'$ is known as the *Potier*
triangle.

Let us next consider what happens to the situation depicted in Fig. 5-12 when
the actual nonlinear characteristics are used. Refer to Fig. 5-13. Curve 1 denotes the
open-circuit characteristic and curve 2 denotes the zero power-factor characteristic
for rated armature current. A little thought should make it clear that the zero power-
factor curve cannot be derived from a knowledge of triangle Oac as was so for the
linear case. Hence this characteristic must be found experimentally (or at least points
c and c'). A very interesting and useful thing happens if we attempt to construct triangle $O'a'c'$ of Fig. 5-12 on the characteristics of Fig. 5-13. To accomplish this, it
is necessary to mark off line $O'c'$ equal to Oc. Then at O' a line is drawn parallel to
the air-gap line until it intersects the open-circuit characteristic at point a'. Thus the
triangle is simulated and, what is more, as a result of this construction the two items
being sought are immediately available. Line $a'b'$ represents the leakage reactance
drop and can easily be measured in volts. Then

$$x_l = \frac{a'b' \text{ (volts)}}{I_a \text{ (rated)}} \tag{5-30}$$

This quantity is called the *Potier reactance*. Moreover, the armature winding mmf
expressed in equivalent field amperes is immediately available upon measuring line
$b'c'$ against the abscissa axis. Thus

$$A = b'c' \quad \text{in field amperes} \tag{5-31}$$

5-6 THEORY FOR SALIENT-POLE MACHINES

In contrast to the cylindrical-pole synchronous machine, which has a uniform air gap,
the salient-pole machine has a highly nonuniform air gap because of the use of a
protruding-pole structure. A glance at Fig. 3-13 makes this obvious. A flux path that
includes two protruding poles has two small gaps to cross. This path is indicated in

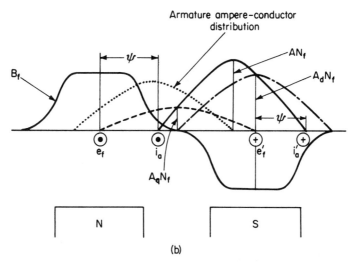

Figure 5-14 (a) Identification of direct- and quadrature-axis flux paths; (b) illustrating how the armature mmf, AN_f, is resolved into two components: one acting in the direct axis, A_dN_f; the other acting in the quadrature axis, A_qN_f.

Fig. 5-14 by ϕ_d and is called the *direct axis*. It is the path of minimum reluctance. On the other hand the path denoted by ϕ_q in Fig. 5-14 involves two large air gaps and is the path of maximum reluctance. This is called the *quadrature axis*. In normal operation the armature (or stator) winding mmf is distributed with its peak value located somewhere between the direct and quadrature axes. Accordingly it produces a significant effect in each axis—but in a different way because of the considerable difference in the reluctance variation of each axis. Consequently, the cylindrical-rotor theory developed in the preceding sections cannot be used for the salient-pole machine with satisfactory results. A modified theory is required that takes due account of this difference. In this section such a theory is developed for the salient-pole machine in both a two-reactance method (analogous to the synchronous reactance method for smooth-rotor machines) and a method based on mmfs.

Two-Reactance Method for Salient-Pole Machines

To provide background, let us consider first the manner of locating the position of the armature winding mmf relative to the rotor field distribution when current flows. Call the angle between the excitation voltage and the armature current ψ. Again refer to Fig. 5-14(a). For the indicated direction of the rotor flux and speed the induced emf in those conductors under the influence of the north-pole flux is directed outward; and in those inductors under the influence of the south-pole flux, it is directed into the plane of the paper. Coil e_f-e_f' is a single coil that corresponds to the location of the amplitude of the fundamental sinusoidal emf distribution that takes place in the entire three-phase winding for the instant shown. The resultant distribution of the currents in the same three-phase winding for a lagging power-factor load is represented by coil i_a-i_a', which is located behind the emf distribution by the specified phase angle ψ. Note that when expressed in electrical degrees ψ denotes a space angle as well as a time angle. Appearing in Fig. 5-14(b) is the developed representation of Fig. 5-14(a). The armature ampere-conductor distribution is shown dotted and lagging behind the field flux distribution by ψ. On the other hand the armature mmf, which in synchronous machines is being denoted in terms of equivalent field mmf, is represented by AN_f, where A is the equivalent field current as described by Eq. (5-5). It is always desirable to work in equivalent field current so that the abscissa axis of the open-circuit characteristic may be used directly in accounting for armature reaction mmf. Again note that the armature mmf is 90 degrees behind the ampere-conductor distribution. Accordingly, the armature mmf curve AN_f takes the position shown in Fig. 5-14(b). Clearly, the location of this wave is dependent upon the phase angle between E_f and I_a, which in turn is dependent upon the power factor of the load and the amount of power delivered (see Sec. 5-7).

It is helpful to consider for a moment the general effects caused by various locations of the armature mmf wave. If the load conditions are such that ψ becomes equal to 90°, the armature mmf wave lies entirely in the direct axis and directly opposes the field flux. In other words, the armature mmf has a *demagnetizing* effect on the field flux. When ψ assumes a value of 0 degrees (as nearly happens for a unity power-factor load), then clearly the armature wave lies entirely in the cross axis and its effect is to cause a distortion of the field flux. That is, the armature mmf wave now has a *cross-magnetizing* effect on the field flux. When the armature current leads the excitation voltage (as happens with leading power-factor loads) by, say, 90°, the effect of the armature mmf is to produce *magnetization* of the field flux. Because of the nature of industrial and domestic loads, the armature current usually lags behind the excitation voltage so that the situation depicted in Fig. 5-14 is the customary one. This means that generally the armature mmf wave is so located that it produces simultaneously an effect on both the direct and the cross axes.

In view of the fact that the direct axis is influenced by iron saturation as mmf is added or subtracted in this axis whereas iron saturation is of little consequence in the cross axis, it is imperative that for any specified ψ the armature mmf wave be composed of two components—one that acts in the direct axis and the other that acts in the quadrature axis. The amplitude of the direct-axis component is denoted by $A_d N_f$ in Fig. 5-14. A study of Fig. 5-14(b) reveals that

$$\bar{A}_d = \bar{A} \sin \psi \quad \text{A} \tag{5-32}$$

where $\bar{A}_d$ is the direct-axis armature mmf expressed in equivalent field amperes. Similarly, the component of the quadrature-axis mmf can be expressed as

$$\bar{A}_q = \bar{A} \cos \psi \quad \text{A} \tag{5-33}$$

Appearing in Fig. 5-15 are the respective flux-density wave forms produced by the direct-axis and quadrature-axis armature mmf components. The direct-axis mmf produces a quasi-truncated sinusoid, whereas the quadrature-axis mmf produces a saddle-shape flux-density wave because of the high reluctance of the interpolar space. The effects are indeed quite different and accordingly must be treated separately.

Figure 5-15 Flux-density wave forms produced by the direct- and quadrature-axis mmfs.

The key to the reactance method of treating the salient-pole synchronous machine lies in Eqs. (5-32) and (5-33). In accordance with these equations the armature current can be considered composed of a direct-axis and a quadrature-axis current. That is,

$$\bar{I}_d = \bar{I}_a \sin \psi \tag{5-34}$$

$$\bar{I}_q = \bar{I}_a \cos \psi \tag{5-35}$$

Hence, if we assume for the moment that $\bar{E}_f$, $\bar{I}_a$, and ψ are known, then $\bar{I}_d$ and $\bar{I}_q$, as well as their associated mmfs $\bar{A}_d$ and $\bar{A}_q$, will be represented as depicted in Fig. 5-16. Note that the direct axis is that along which the field mmf $\bar{F}$ acts. By superposition each mmf is then considered to produce its own induced emf in the armature winding. Thus by following the same procedure outlined in the synchronous reactance method for the cylindrical rotor machine, through superposition we can associate with $\bar{A}_d$ an emf which is replaced by the reactance drop in the direct axis, i.e.,

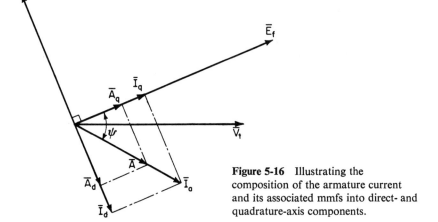

Figure 5-16 Illustrating the composition of the armature current and its associated mmfs into direct- and quadrature-axis components.

$j\bar{I}_d x_{ad}$. The j term provides the necessary 90-degree lead that must exist between a current and its related reactance drop. Here x_{ad} *is the armature reaction reactance in the direct axis.* In a similar fashion the effect of $\bar{A}_q$ is replaced by a reactance drop $+j\bar{I}_q x_{aq}$ leading $\bar{A}_q$ (or $\bar{I}_q$) by 90°. The quantity x_{aq} *is called the armature reaction reactance in the quadrature axis.* A single armature reaction reactance cannot be used here because of the wide difference in the flux-density curves associated with the two axes. The value of x_{aq} is always less than x_{ad} since the emf induced by a given mmf acting on each axis is always smaller for the quadrature axis owing to its higher reluctance.

We next describe the development of the phasor diagram for the salient-pole synchronous machine, assuming the following data to be available: rated terminal voltage per phase, V_t; rated armature current per phase, I_a; load power-factor angle, θ; armature winding leakage reactance per phase, x_l; and the armature reaction reactance in each axis expressed per phase, x_{ad} and x_{aq}. For simplicity the armature winding resistance is assumed negligible. Refer to Fig. 5-17. Add $j\bar{I}_a x_l$ to $\bar{V}_t$ to get the induced emf in the armature winding caused by the resultant gap flux. This is quantity Ob. At b mark off the quantity $+j\bar{I}_q x_{aq}$ leading $\bar{I}_q$ by 90° and terminating at point d. The excitation voltage $\bar{E}_f$ lies along line Od. Finally, the magnitude of $\bar{E}_f$ is found by adding the armature reaction reactance drop in the direct axis, i.e., $j\bar{I}_d x_{ad}$, along line Od starting at point d. The quantity Of measured on a suitable voltage scale yields the value of the excitation voltage.

Generally, unless the design data are furnished, the quantities x_{aq} and x_{ad} are not available. Even an appropriate no-load test does not provide information about the armature reactances directly. Rather, it is information about the combined effects of the leakage reactance with the armature reaction reactances that becomes directly available. In this connection then let us study the phasor diagram of Fig. 5-17 in more detail. First note that

$$\angle\, cbd = \psi \qquad (5\text{-}36)$$

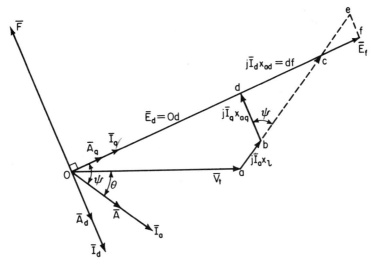

Figure 5-17 Development of a phasor diagram for the salient-pole synchronous machine, by the armature-reaction reactances.

It then follows that

$$bc = \frac{bd}{\cos \psi} = \frac{I_q x_{aq}}{\cos \psi} = \frac{I_a \cos \psi x_{aq}}{\cos \psi} = I_a x_{aq} \qquad (5\text{-}37)$$

In addition,

$$ac = ab + bc = I_a(x_l + x_{aq}) \qquad (5\text{-}38)$$

Therefore we can define

$$\boxed{x_q \equiv x_l + x_{aq}} \qquad (5\text{-}39)$$

where x_q is the *quadrature-axis synchronous reactance*. Note the correspondence with Eq. (5-16) for the cylindrical rotor machine. Equation (5-38) furnishes an alternative way of locating the direction along which the $\bar{E}_f$ phasor lies. It merely requires that the quantity $j\bar{I}_a x_q$ be added to $\bar{V}_t$ to locate point c in Fig. 5-17.

By proceeding in a similar fashion a synchronous reactance associated with the direct axis can be defined. From the geometry of Fig. 5-17 we can write that

$$be = \frac{I_d x_{ad}}{\sin \psi} = \frac{I_a \sin \psi x_{ad}}{\sin \psi} = I_a x_{ad} \qquad (5\text{-}40)$$

Also,

$$ae = ab + be = I_a(x_l + x_{ad}) \qquad (5\text{-}41)$$

Therefore

$$\boxed{x_d \equiv x_l + x_{ad}} \qquad (5\text{-}42)$$

where x_d denotes the *direct-axis synchronous reactance*. This quantity is comparable to x_s for the smooth-rotor machine. Fairly accurate values for x_q and x_d can be obtained from a no-load test called the *slip test*, which is described in Appendix E. Our discussions hereinafter proceed on the assumption that information about x_q and x_d is available either from the design data or from an appropriate no-load test such as the slip test.

The determination of $\bar{E}_f$ (and thence the corresponding field current $\bar{F}$) by the use of x_d and x_q has caused the method to be called the *two-reactance method*. Depicted in Fig. 5-18 is the phasor diagram expressed solely in terms of synchronous reactance drops. The solution procedure then involves just three steps as outlined below:

1. Locate the direction of $\bar{E}_f$ from

$$\psi = \tan^{-1}\frac{V_t \sin\theta + I_a x_q}{V_t \cos\theta} \tag{5-43}$$

2. Compute the phasor expressions for direct and quadrature currents.

$$\bar{I}_q = (I_a \cos\psi)\underline{/\Delta} \tag{5-44}$$

$$\bar{I}_d = (I_a \sin\psi)\underline{/\Delta - 90°} \tag{5-45}$$

where

$$\Delta \equiv \psi - \theta \tag{5-46}$$

and the angles are all positive for lagging power factor.

3. Finally, as can be seen from Fig. 5-18,

$$\bar{E}_f = \bar{V}_t + j\bar{I}_q x_q + j\bar{I}_d x_d \tag{5-47}$$

The field current that produces this excitation voltage is then found from the open-circuit characteristic.

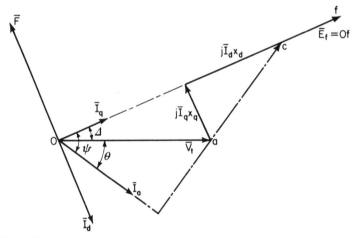

Figure 5-18 Phasor diagram of the salient-pole machine in terms of the synchronous reactances in the direct and quadrature axes.

The same limitations that apply to the synchronous reactance method apply to the two-reactance method. To obtain reliable results the direct-axis synchronous reactance must be found corresponding to the proper flux conditions in the machine. Of course, x_q presents no such problem because it is associated with large air paths where saturation is of little or no consequence. Surely, whenever there is a choice, the preferred way of dealing with the effect of armature reaction in the direct axis is to work directly with the open-circuit characteristic. This is possible by employing the general method for salient-pole machines which is described next.

General Method for Salient-Pole Machines

The phasor diagram to be used is the one depicted in Fig. 5-17. The procedure is as follows:

1. Find ψ by means of Eq. (5-43).
2. Compute the emf induced by the resultant flux in the direct axis. This is the quantity Od and shall now be called E_d. Its magnitude is given by

$$E_d = V_t \cos(\psi - \theta) + I_a x_l \sin \psi \qquad (5\text{-}48)$$

3. Corresponding to E_d on the open-circuit characteristic, find the resultant mmf expressed in field amperes. Call this R_d.
4. Find A by means of Eq. (5-31) and compute

$$A_d = A \sin \psi$$

5. Then for ψ lagging the required field current is found from the algebraic sum of steps 3 and 4. Thus

$$F = R_d + A \sin \psi \qquad (5\text{-}49)$$

It is interesting to note that by computing E_d in the manner described by Eq. (5-48), there is no need to have knowledge of x_{aq}. Of course the magnitude of $\bar{E}_d$ is the same as that which would be obtained from the expression

$$\bar{E}_d = \bar{V}_t + jI_q x_l + jI_q x_{aq} \qquad (5\text{-}50)$$

Finally, note that steps 4 and 5 of the foregoing procedure is the mmf equivalent of adding the reactance drop $I_d x_{ad}$ to E_d to obtain the excitation voltage. The advantage of the mmf version is that one automatically accounts for the correct level of saturation by working directly with the nonlinear open-circuit characteristic. This method yields highly realistic results.

Example 5-4

A three-phase, 15-kVA, 220-V, Y-connected, 60-Hz, six-pole salient-pole synchronous generator has the open-circuit, short-circuit, and zero power factor characteristics depicted in Fig. P5-10. The armature winding resistance is negligible. It is also known that for this type machine the synchronous reactance in the quadrature axis is 60% of the value associated with the direct axis. Moreover, the leakage reactance is known to have a value of 0.9 Ω per phase.

(a) Identify the base quantities to be used in calculating the performance employing the per-unit notation.

(b) Determine the per-unit values of the leakage reactance and the direct-axis synchronous reactance per phase as derived from the graphical data.

(c) Determine the excitation emf when this machine delivers rated kVA to a 0.8 pf lagging load using the two reactance method. Make all calculations in terms of per-unit quantities.

(d) Find the excitation emf for the conditions of part (c) using the general method for the salient pole machine.

Solution (a) It is customary to use the rated phase voltage as the base voltage and the rated phase current as the base current. Thus in this case

$$V_{\text{base}} = V_{\text{phase}} = \frac{V_L}{\sqrt{3}} = \frac{220}{\sqrt{3}} = 127 \text{ V/phase}$$

$$I_{\text{base}} = I_{\text{phase}} = \frac{15,000}{\sqrt{3}\,(220)} = 39.4 \text{ A/phase}$$

Correspondingly, the base impedance becomes

$$Z_{\text{base}} = \frac{V_{\text{base}}}{I_{\text{base}}} = \frac{127}{39.4} = 3.22 \text{ }\Omega/\text{phase}$$

A useful, alternative form for the calculation of the base impedance can be obtained by writing

$$Z_{\text{base}} = \frac{V_{\text{phase}}}{I_{\text{phase}}} = \frac{V_L/\sqrt{3}}{\text{VA}/\sqrt{3}\,V_L} = \frac{V_L^2}{\text{VA}}$$

where V_L is the line-to-line voltage (irrespective of whether or not the stator winding is Y- or Δ-connected) and VA denotes the volt-ampere rating of the generator. Observe that with this formulation the $\sqrt{3}$ is not a factor in the evaluation. Applied to the case at hand, we get

$$Z_{\text{base}} = \frac{(220)^2}{15,000} = 3.23 \text{ }\Omega/\text{phase}$$

(b) The per-unit value of the leakage reactance per phase readily follows from the definition of a per-unit value. Thus

$$\text{p-u } x_l = \frac{x_l(\Omega)}{Z_B(\Omega)} = \frac{0.9}{3.22} = 0.28$$

A convenient way to find the per-unit value of the direct-axis synchronous reactance when graphical data is available is to find the *short-circuit ratio* (SCR). By definition the SCR is the armature current, expressed in per unit, which flows in a short-circuit test corresponding to that value of field excitation that produces rated voltage in an open-circuit test. Algebraically stated, we have

$$\text{SCR} = \text{p-u } I_{\text{sc}}$$

where the right side is the short-circuit current caused by the field current that on open circuit yields rated voltage. Applying this definition to the graphical data, we see that to produce a rated voltage of 127 V on the open-circuit characteristic a field current of 4.15 A is needed. However, this same field current causes a short-circuit current of 41 A, which expressed as a per-unit quantity is p-u $I_{\text{sc}} = 41/39.4 = 1.041$ = SCR. Consequently, the per-unit value of x_d becomes simply

$$\text{p-u } x_d = \frac{\text{p-u } V_R}{\text{p-u } I_{\text{sc}}} = \frac{1}{\text{p-u } I_{\text{sc}}} = \frac{1}{\text{SCR}} = \frac{1}{1.041} = 0.961$$

(c) Delivering rated kVA means that both the voltage and current are at their rated values which on a per-unit basis means values of unity. The solution procedure for this part is outlined by Eqs. (5-43) through (5-47). For a 0.8 pf the power factor angle is

$$\theta = \cos^{-1} 0.8 = 36.9°$$

Also, the p-u value of the synchronous reactance in the quadrature axis is

$$\text{p-u } x_q = (0.6)(\text{p-u } x_d) = 0.577$$

Hence

$$\psi = \tan^{-1}\frac{(1) \sin 36.9° + (1)(0.577)}{(1) \cos 36.9°} = \tan^{-1}\frac{0.6 + 0.577}{0.8} = 55.8°$$

$$\text{p-u } \bar{I}_q = (\text{p-u } \bar{I}_a \cos \psi)/\underline{\Delta} = \cos 55.8°/\underline{\Delta} = 0.562/\underline{\Delta}$$

where

$$\Delta = \psi - \theta = 55.8° - 36.9° = 18.9°$$

$$\text{p-u } \bar{I}_d = (\text{p-u } I_a \sin \psi)/\underline{-90° + \Delta} = 0.827/\underline{-71.1°}$$

Therefore,

$$\text{p-u } \bar{E}_f = \bar{V}_t + j\bar{I}_q x_q + j\bar{I}_d x_d = 1/\underline{0°} + (0.562)/\underline{18.9°}(0.577)/\underline{90°} + (0.827)/\underline{-71.1°}$$

$$= 1/\underline{0°} + (0.562)/\underline{18.9°}(0.577)/\underline{90°} + (0.827)/\underline{-71.1°}(0.961)/\underline{90°}$$

$$= 1/\underline{0°} - 0.105 + j0.307 + 0.752 + j0.257 = 1.647 + j0.564$$

$$= 1.74/\underline{18.9°}$$

In accordance with this result it follows that the voltage regulation of this machine is a high 74%. This value is highly pessimistic, which we already know is characteristic of the synchronous reactance method. A more realistic result can be obtained by using the general method, which is done in part (d).

(d) The general method also begins with the evaluation of ψ. The emf induced by the resultant flux in the air gap of the direct axis is then found with Eq. (5-48). Thus

$$\text{p-u } E_d = V_t \cos (\psi - \theta) + I_a x_l \sin \psi$$

$$= (1) \cos (18.9°) + (1)(0.28) \sin 18.9° = 0.946 + 0.091 = 1.037$$

The corresponding voltage value is

$$E_d = 1.037(127) = 131.7 \text{ V}$$

and the resultant field current, R_d, associated with this voltage as obtained from the open-circuit characteristic is found to be

$$R_d = 4.6 \text{ A}$$

Also, by construction of the Potier triangle the armature reaction mmf expressed in terms of equivalent field amperes is found to be $A = 2.96$ A. Accordingly, the resultant field current that produces the excitation emf is determined by Eq. (5-49) to be

$$F = R_d + A \sin \psi = 4.6 + 2.96 \sin 55.8° = 7.05 \text{ A}$$

Entering the open-circuit characteristic for this value of field current yields an excitation voltage of

$$E_f = 166 \text{ V/phase}$$

The per-unit value is

$$\text{p-u } E_f = \frac{166}{127} = 1.31$$

Thus a much more realistic value of 31% voltage regulation is now obtained.

5-7 POWER

The expression for power in an electromechanical energy-conversion device is always a matter of primary importance. It was shown in Chapter 3 that, once the developed torque is computed, the developed power can be found from

$$P_m = T\omega_m \tag{5-51}$$

where ω_m denotes the mechanical speed of rotation expressed in radians per second. However, it is frequently useful to have an equation for power that is expressed in terms of appropriate machine parameters, voltages, and phase angles because of the insight it gives regarding machine performance. Such an equation can be derived from the phasor diagram for the cylindrical and also the salient-pole machine.

Cylindrical-Rotor Machine

Use is made of the phasor diagram of Fig. 5-7(b), which is repeated in Fig. 5-19. Since the effect of armature winding resistance is very small, it is neglected here. Consequently, the expression for the developed power can be written in one of two ways, i.e.,

$$P_m = qV_tI_a \cos \theta = qE_fI_a \cos \psi \tag{5-52}$$

This statement follows from Fig. 5-19 upon noting that the projection of $\bar{E}_f$ on the $\bar{I}_a$ phasor is identical with the projection of $\bar{V}_t$ on $\bar{I}_a$. Moreover, from the geometry of Fig. 5-19 we see that the quantity ab may also be expressed in two ways. Thus

$$ab = E_f \sin \Delta = I_a x_s \cos \theta \tag{5-53}$$

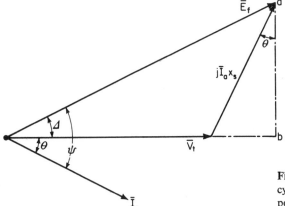

Figure 5-19 Phasor diagram of the cylindrical-rotor machine for lagging power-factor.

from which it follows that

$$I_a \cos \theta = \frac{E_f}{x_s} \sin \Delta \qquad (5\text{-}54)$$

Insertion of Eq. (5-54) into Eq. (5-52) yields the desired result. Hence the power developed in the synchronous generator where armature winding resistance is negligible is given by

$$P_m = q \frac{V_t E_f}{x_s} \sin \Delta \qquad (5\text{-}55)$$

where q denotes the number of phases.

Because of the dependence of the power upon the angle Δ, this angle has come to be called the *power angle*. Equation (5-55) indicates that, if the power angle is zero, the synchronous machine cannot develop useful power. It further reveals that the developed power is a sinusoidal function of the power angle and that it has its maximum value at Δ equal to 90°. Figure 5-20 depicts the variation P_m as a function of positive and negative values of Δ. As described in Sec. 5-11, negative values of power angle refer to the operation of the synchronous machine as a motor. It is also worthwhile to note the correspondence of Eq. (5-55) with Eq. (3-110). Both the developed electromagnetic torque and the developed power are sinusoidal functions of an appropriate power angle.† Upon multiplying Eq. (3-110) by ω_m, a value of developed power should be obtained which is the same as that found from Eq. (5-55). Clearly, such correspondence is critically dependent upon the use of a reliably good value of synchronous reactance in Eq. (5-55). It is helpful to keep in mind that the calculation of power by $T\omega_m$ or by Eq. (5-51) is always more reliable than by Eq.

Figure 5-20 Graphical representation of developed power as a function of the power angle for a cylindrical-rotor synchronous machine.

†Take care not to confuse the torque angle δ of Eq. (3-110) with the power angle Δ of Eq. (5-55). In the former, δ is defined as the angle between the axis of the flux-density wave produced by the field winding and the axis of the armature winding mmf $\mathcal{F}$ (which is 90° displaced from the armature ampere-conductor distribution). In the latter case, Δ is defined as the time angle between the excitation voltage (produced by the field flux) and the terminal voltage (produced by the resultant flux minus the leakage flux).

(5-55). However, the latter expression is particularly useful in furnishing insight into machine performance. A reading of Sec. 5-15 illustrates the point.

Salient-Pole Machine

The derivation of the expression for the developed power of the salient-pole machine proceeds in a similar fashion. However, in this case use is made of the phasor diagram based on the two-reactance theory. The diagram is repeated for convenience in Fig. 5-21. Again, the phasor diagram as it applies to the synchronous generator (i.e., positive power angle) is used, but it should be understood that the results apply equally for operation as a synchronous motor. Moreover, the need to use a reliably correct value of x_d is imperative if accurate values of power are to be computed by the derived expression. On the other hand, if the use of this expression is merely to expand one's insight into the machine, then the need is not at all important.

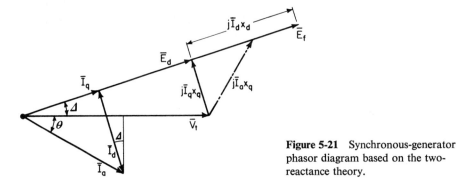

Figure 5-21 Synchronous-generator phasor diagram based on the two-reactance theory.

Because of the assumption of negligible armature resistance, the power output and the power developed are equal. Accordingly we may write

$$P_m = qV_tI_a \cos \theta \qquad (5\text{-}56)$$

A study of Fig. 5-21 discloses that the component of armature current that is in phase with the terminal voltage is related to I_d and I_q by

$$I_a \cos \theta = I_q \cos \Delta + I_d \sin \Delta \qquad (5\text{-}57)$$

Moreover, an expression for I_q can be found by noting that by the geometry of Fig. 5-21

$$I_qx_q = V_t \sin \Delta \qquad (5\text{-}58)$$

from which

$$I_q = \frac{V_t}{x_q} \sin \Delta \qquad (5\text{-}59)$$

In a similar fashion an expression for I_d can be found expressed in terms of voltage and the machine parameters. Thus,

$$E_f = V_t \cos \Delta + I_dx_d \qquad (5\text{-}60)$$

Therefore,

$$I_d = \frac{E_f - V_t \cos \Delta}{x_d} \tag{5-61}$$

Upon substituting Eqs. (5-59) and (5-61) in Eq. (5-57), there results

$$I_a \cos \theta = \frac{V_t}{x_q} \sin \Delta \cos \Delta + \frac{E_f}{x_d} \sin \Delta - \frac{V_t}{x_d} \sin \Delta \cos \Delta \tag{5-62}$$

The expression for the developed power then becomes

$$P_m = q\frac{V_t E_f}{x_d} \sin \Delta + q\frac{V_t^2}{x_d x_q}(x_d - x_q) \sin \Delta \cos \Delta \tag{5-63}$$

Introducing the trigonometric identity

$$\sin 2\Delta = 2 \sin \Delta \cos \Delta \tag{5-64}$$

into the last equation yields the desired result. Hence,

$$\boxed{P_m = q\frac{V_t E_f}{x_d} \sin \Delta + q\frac{V_t^2}{2x_d x_q}(x_d - x_q) \sin 2\Delta} \tag{5-65}$$

A plot of this equation appears in Fig. 5-22 for motor as well as generator operation. Curve (a) is a plot of the first term of Eq. (5-65) and is identical with the result obtained for the cylindrical-rotor machine. It represents the power associated with the electromagnetic torque produced by the mutual coupling between the rotor field winding (indicated by E_f) and the stator armature winding (indicated by V_t). Curve (b) is a graphical representation of the second term of Eq. (5-65). It is independent of excitation voltage which discloses that this term exists even when the

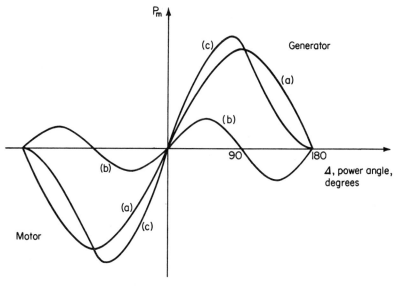

Figure 5-22 Developed power as a function of the power angle for the synchronous salient-pole machine.

field current is zero. Of course it is assumed that the machine is connected to an infinite bus system, otherwise the second term would be meaningless. When connected to the infinite bus, an armature current can flow without field excitation. Accordingly, a revolving mmf exists and it attempts to align itself with the minimum reluctance path as explained in Chapter 3. Since it arises because of the difference in reluctance between the direct and quadrature axes, it is called the *reluctance power* and has a value usually in the vicinity of 20 to 25% of the rating of the machine. Moreover, it involves double the power angle and thereby causes a peaking of the resultant developed power curve as indicated by curve (c) in Fig. 5-22. The effect of the reluctance power term in the expression for developed power is not only to increase the maximum value of developed power but also to cause it to occur at a value of Δ less than 90°. Finally, it is interesting to note that, when Eq. (5-65) is applied to the cylindrical-rotor machine, the expression reduces to that of Eq. (5-55) because x_d and x_q are equal.

5-8 PARALLEL OPERATION OF SYNCHRONOUS GENERATORS

The electrical needs of industry, commercial establishments, and individual consumers are supplied almost exclusively by synchronous generators. In many large regions of the country, there are virtually hundreds of such generators operating in parallel to meet these needs. It is the purpose of this section to describe some of the principles that govern the steady-state behavior of alternators that operate in parallel.

The Need to Synchronize

Initially we focus on the parallel operation of two isolated synchronous generators of equal ratings. Assume that the first generator G_1 is already connected to a three-phase resistive load as indicated in Fig. 5-23 by the closed load switch and that it is operating at rated frequency and rated voltage. It is now desirable to place the second alternator in parallel with the first in order that it might help share the current load as well as to prepare for the future increased power requirements of this load.

If severe shocks to both generators are to be avoided, a definite procedure,

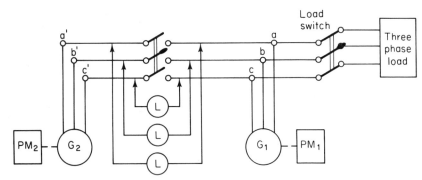

Figure 5-23 Synchronizing two synchronous generators (alternators) for parallel operation.

called the *synchronizing procedure*, must be followed before the switch (or circuit breaker) that parallels the two machines can be closed. The process is started by adjusting the speed of the prime mover of generator G_2 to correspond exactly to that of generator G_1. The field current of G_2 is then adjusted to yield the same nominal voltage between lines a', b', and c'. A suitable bank of lamps† is placed in series with lines a'-a, b'-b, c'-c, as shown in Fig. 5-23. These lamps will light up if either the voltage or the frequency of G_2 differs slightly from those of G_1. When the two frequencies are identical but the magnitude of the line voltages are slightly different, the lamps will be on steadily. A subsequent adjustment of the field current of G_2 can then be used to cause the lamps to darken at which time the synchronizing switch can be closed, thereby placing both machines in parallel without incident. On the other hand, if the two sets of three-phase voltages are equal in magnitude but a bit different in frequency, the lamps will again register light but now they will flicker at a rate equal to the difference frequency. At one point in the difference cycle, the voltage across the lamps will reach twice the line voltage; a half-cycle later it will be zero. An adjustment in the prime mover speed can serve to reduce this difference frequency to a very small value. Then, as a dark period is approached, the line switch can be closed with little or no disturbance.

Effect of Field Excitation (Control of Reactive A-C Power)

Once G_2 is placed in parallel with G_1, in the manner described, and no further changes are made, the second machine assumes a state of simply idling on the line. If the assumed fixed load demand is to be equally shared by G_1 and G_2, it is necessary to make appropriate adjustments on both prime movers in a fashion to be described presently. For the moment, however, let us assume that each generator is delivering half the load power at rated frequency and rated load. Now we investigate the effect of varying the field excitation of each machine that is operating in parallel. This initial state of affairs is depicted in Fig. 5-24, corresponding to the unprimed quantities. This diagram is drawn for the case of negligible armature (stator) winding resistance. Moreover, the excitation current is initially assumed to be set at that value which yields unity power factor for each machine. This is illustrated by placing I_1 and I_2, the respective G_1 and G_2 stator currents, in phase with the voltage phasor. It is instructive to note here that synchronous generators operating in parallel have a unique field excitation that yields unity power factor at a fixed load share.

In order to better understand the sequence of events that unfolds with variation of excitation, we begin by first increasing the field current of G_1. In turn, this causes an increase in the field excitation phasor E_{f1} to a larger value E'_{f1}. The new position of phasor E'_{f1} in Fig. 5-24 must be such that its tip lies on the horizontal locus line in order to ensure that the vertical distance between this line and the line along which V lies remains a constant. This condition is demanded by Eq. (5-55), which for constant power requires that $E_{f1} \sin \Delta_1 = E'_{f1} \sin \Delta'_1$. In addition, the operating power angle of G_1 must assume a reduced value in the presence of the augmented E'_{f1}.

What happens to this power system if no subsequent adjustment is made on

†The equivalent commercial version of these lamps is called a *synchroscope*.

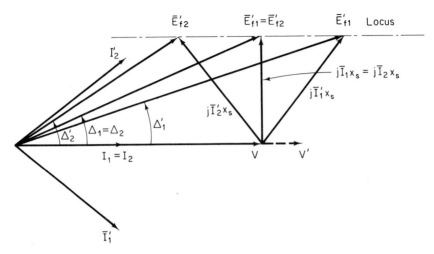

Figure 5-24 Phasor diagram drawn to illustrate the effect of field excitation on two synchronous generators operating in parallel. Real power remains fixed.

the second generator? The phasor diagram for the new condition in G_1 immediately reveals that G_1 can no longer operate at unity power factor. A reaction is, therefore, set in motion to reduce the effect of the increased excitation. In a generator overexcitation calls for a lagging reactive current component because it produces a demagnetizing effect. But because the load is assumed to be resistive, this generator lagging current component cannot be accommodated by the load. Therefore, it must exist as a circulating current between the two stator windings and as such must necessarily take on the character of a leading current in G_2. Since a leading current in a generator produces a magnetizing effect on the air gap flux, it follows that the net effect of increasing only one field current is to cause the terminal voltage of the parallel set to increase to V' located midway between the horizontal projections of E_{f1} and E_{f1}' onto the V line.

On the basis of the foregoing analysis, it should be apparent that to maintain operation at the original value of terminal voltage (after the field current of G_1 is increased) it becomes necessary to *decrease* the field current of G_2 by a corresponding amount. The required condition is depicted in Fig. 5-24 by the prime notation on the quantities of G_2. Here again note that the phasor voltage differences between the new excitation voltages and the terminal voltage call for a reactive current flow—lagging in the overexcited machine G_1 and leading in the underexcited machine G_2. Accordingly, we are led to the conclusion that the effect of changing the field excitation of generators operating in parallel is merely to change the amount of reactive voltamperes associated with each machine. There is no change in the amount of real power delivered.

Effect of Prime Mover Power (Control of Real Power)

Attention is next directed to a study of altering the power delivered to the shaft of each of two synchronous generators operating in parallel with fixed field excitations. The starting point is again taken to be that corresponding to which each generator delivers the same electrical power to the load at that excitation that puts the power factor of each machine at unity. It is important to understand at this juncture that the electrical energy developed by each generator is mechanically derived from the individual prime movers. These prime movers may be steam turbines, gasoline engines, waterwheels, or any practical primary energy source. The speed-power characteristics of these prime movers are decisive in establishing the manner in which generators operating in parallel will share a common load. Moreover, these characteristics assume a drooping bias in the fashion illustrated in Fig. 5-25. Because G_1 and G_2 are assumed to be identical machines, their speed-torque characteristics are also assumed to be the same. They are depicted as mirror images in Fig. 5-25 for reasons of clarity. At the initial operating point, the prime movers are revolving at a speed that yields the frequency required at the electrical load. At this frequency the prime mover of G_1 delivers power P_1 while the prime mover of G_2 delivers power $P_2 = P_1$.

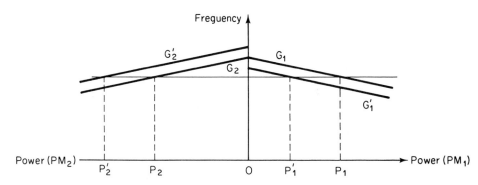

Figure 5-25 Speed-power characteristics of the prime movers of generators G_1 and G_2 of Fig. 5-23.

Suppose now that by suitable adjustment of the prime mover of G_1 the power delivered to G_1 is reduced. Simultaneously, in the interest of maintaining a constant frequency, let the input power to G_2 from its prime mover be increased in order to continue to operate at constant frequency. These prime mover adjustments are represented in Fig. 5-25 by a displacement of the speed-power characteristics as illustrated. Observe that to reduce the share of load supplied by generator G_1 its speed-power characteristic is displaced downward from position G_1 to G_1'. Similarly, the characteristic of the prime mover of machine two is raised in reference to its initial position. The result is that, at the maintained frequency, G_1 is now assuming a smaller share of the load $P_1' < P_1$, while G_2 takes on a greater share $P_2' > P_2$. The sum of the primed and unprimed powers is constant, however.

The effects of these prime mover adjustments on the electrical quantities of the individual generators are shown in Fig. 5-26. The reduced share of the electrical load supplied by G_1 is accompanied by a decrease in its power angle from Δ_1 to Δ_1' as demanded by Eq. (5-55). The phasor voltage difference between E_{f1} and V is correspondingly diminished and this in turn produces a smaller stator current $\bar{I}_1'$ for G_1. Two points are worth noting about this new value of G_1 current. First, it has a smaller in-phase component, which is consistent with the fact that it is delivering

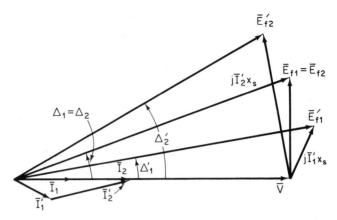

Figure 5-26 Phasor diagrams illustrating the effect of changing the prime mover power at constant excitation for two generators operating in parallel.

a smaller share of the electrical power taken by the load. Second, the power factor of this current is no longer unity. In fact, it is a lagging power factor. Corresponding changes take place in generator G_2. Here the power angle increases from Δ_2 to Δ_2', thus producing a larger phasor voltage difference which acts to establish an increased value of stator current in G_2 as represented by $\bar{I}_2'$. This current exhibits a greater in-phase component because it delivers a greater load share. In addition, it too now possesses a reactive current which is a leading component. Since the load is resistive, the reactive current of both generators constitute a common circulating current. Hence, if it is of a lagging nature in one generator, it must be leading in the other. Once again, we have an illustration of the fact that at each load condition there is a unique excitation for operation at unity power factor.

Connection to an Infinite Bus

For the sake of argument, let G_1 in Fig. 5-23 represent many large synchronous generators connected to a common power grid. This interconnection of generators frequently involves the use of solid copper bars and has come to be called a *bus*. Appearing at such a bus are characteristics of the electric power system that exhibit virtually constant voltage and frequency. It has come to be called an *infinite bus* because no single generator on the interconnected system can by itself in normal circumstances influence either of these quantities to any noticeable extent.

The synchronizing procedure needed to connect G_2 to the infinite bus is the same as for the interconnection of two isolated generators as described previously. Assume that this has been done and that G_2 is now idling on the power grid line. Any increase in the excitation of G_2 produces a lagging current with a power factor very close to zero. This reactive current serves to neutralize the effect of the increased field mmf in order that the terminal voltage stay invariant as demanded by the infinite bus. Actually, this reactive generator current serves to relieve the other generators on the power grid of some of their reactive volt-ampere burden. Correspondingly, a decrease in excitation causes G_2 to draw a leading current from the bus system. This current serves to produce a magnetizing effect of sufficient quantity to ensure an air-gap flux that is consistent with the bus voltage. A return of the excitation to a value of 100% (i.e., $E_f = V$), puts the machine once again in the idling state.

Let the prime mover of G_2 now be adjusted to raise its speed-power characteristic. This has the effect of increasing the input power to G_2, which in turn responds by delivering a corresponding amount to the power grid. As the speed of the prime mover of G_2 is momentarily increased to raise its speed-power characteristic, it does make an effort to raise the frequency of the entire power system. Of course, it fails to do so because it is no match for the enormous total capacity already on the grid to which G_2 would need to supply synchronizing power in order to bring about a frequency change. Instead, G_2 remains content with simply relieving the system of a tiny fraction of the delivered electrical power.

5-9 TRANSIENT AND SUBTRANSIENT REACTANCES

The notion of replacing the effect of the armature mmf by a machine parameter called the synchronous reactance is introduced in Sec. 5-3. In fact, it is this reactance which limits the current when a short-circuit appears across the armature terminals of a synchronous generator at normal excitation in the steady state. The situation then is not unlike the conditions that prevail during a short-circuit test. The difference, of course, is that the corresponding short-circuit stator winding current will be several times larger than the rated value, when the short-circuit is imposed at full excitation. It is important to note in these remarks that the emphasis is on the short-circuit current flow at *steady state*. The clear implication is that between the instant that a short circuit appears at the generator terminals and the actual settlement into steady state, a transient period occurs. It is the purpose of this section to characterize the behavior of the generator during this transient period and even to determine, to the extent possible, parameters to represent this characterization. Although the steady-state short-circuit currents can be quite large at normal excitation, the transient currents can be expected to be even larger. Because it is important to interrupt such transient currents before serious damage is inflicted on the equipment, a reliable characterization becomes a necessity.

To make it easier to understand the factors that come into play in establishing the behavior of the synchronous generator during the transient period, we focus initially on the relationship between the field winding of a turbo-generator and one phase of the stator winding. Refer to Fig. 5-27. Here the horizontally placed coil on the rotor represents the entire field winding which is assumed to be carrying the steady-state d-c current I_f. Similarly, the coil *a-a* on the stator is meant to represent the net action of phase *a* of the stator winding. For the purpose of this discussion we assume that the flux linkage between the field winding and coil *a-a* is zero at the instant a three-phase symmetrical short circuit occurs at the stator terminals. This is represented in the diagram by the quadrature orientation of the two coils. We know from our study of Chapter 3 that the instantaneous value of the emf in the stator phase is a maximum at this position. On the assumption that the armature winding resistance is negligible, it follows that the instantaneous value of the armature current is zero at the moment of the short circuit. Recall that the armature circuit is entirely

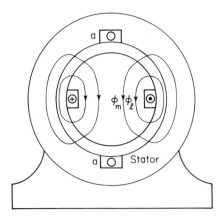

Figure 5-27 Diagram showing the position of zero flux linkage between the field coil and the stator phase coil *a-a*.

inductive once short circuit occurs. Hence the armature phase *a* current must be at its zero value, when the voltage is passing through its peak value.

Our interest next is to describe the fundamental principle that dictates the actions that occur following the short circuit. But first it is helpful to write the expressions for the flux linking the individual coils immediately before short circuit takes place. Proceeding on the assumption that all mmf's are required to force flux to cross the air gap, the total flux that links the field coil is

$$\lambda_f = N_f(\phi_m + \phi_l) = I_f(L_m + L_f) = L_m I_f(1 + l_f) \tag{5-66}$$

where λ_f = flux leakage of the field coil
  ϕ_m = flux that links the stator and rotor
  ϕ_l = flux that links only the rotor coil
  L_m = inductance associated with the mutual flux ϕ_m
  L_f = *leakage* inductance associated with the leakage flux ϕ_l
  N_f = number of turns of the field coil

and

$$l_f = \frac{L_l}{L_m} = \text{leakage coefficient of the field winding} \tag{5-67}$$

The quantity $(L_m + L_f)$ is nothing more than the self-inductance of the field winding. Correspondingly, the flux that links phase coil *a-a* is zero both for the reason that its phase current is zero and also because the linkage with the field flux is zero at this instant. Thus

$$\lambda_a = 0 \tag{5-68}$$

where λ_a denotes the flux linking the armature winding turns.

It is a fact of physics that stored energy, such as the magnetic energy represented by the field flux, cannot change instantaneously in the presence of finite forces. Now observe that a linkage of the field flux with phase coil *a-a* begins to occur immediately following the short circuit because the rotor is spinning at synchronous speed ω_s. Hence in the differential time Δt after the short circuit, there is an angular displacement of

$$\alpha = \omega_s(\Delta t) \tag{5-69}$$

This causes an armature current i_a to flow in coil a-a, which in turn produces an additional current i_f to flow in the field coil to counteract the action of i_a in attempting to change the field coil linkage. The new expression for the armature winding flux linkage now becomes

$$\lambda_a = i_a(L_m + L_a) + (i_f + I_f)L_m \sin \alpha$$
$$= i_a L_m(1 + l_a) + (i_f + I_f)L_m \sin \alpha \tag{5-70}$$

where

$$l_a = \frac{L_a}{L_m} = \text{leakage coefficient of the armature phase coil } a\text{-}a \tag{5-71}$$

and L_a denotes the inductance associated with the leakage flux around the stator coil a-a caused by the flow of current i_a. The first term on the right side of Eq. (5-70) represents the total flux produced by i_a in coil a-a in the mutual path and the leakage path, whereas the second term denotes the linkage of the flux produced by the total field current acting through the mutual flux path at the displacement angle α. In a similar manner the expression for the net flux linkage of the field winding at position α is

$$\lambda_f = (i_f + I_f)L_m(1 + l_f) + i_a L_m \sin \alpha \tag{5-72}$$

The first term on the right side is the total flux produced by the total field current. The second term represents the coupling with the field winding of the flux produced by the armature current i_a acting in the mutual flux path at angle α.

The significant point at this stage is to recognize that the value of the flux linkages at time Δt must be the same as that which existed at time $t = 0^+$ immediately following the short circuit. Thus Eq. (5-70) is equated to Eq. (5-68) and Eq. (5-72) is equated to Eq. (5-66). Accordingly, we have

$$i_a L_m(1 + l_a) + (i_f + I_f)L_m \sin \alpha = 0 \tag{5-73}$$

and

$$(i_f + I_f)L_m(1 + l_f) + i_a L_m \sin \alpha = I_f L_m(1 + l_f) \tag{5-74}$$

The simultaneous solution of these equations yields expressions for the armature phase current and the increased field current as a function of α and the machine parameters. The results are

$$i_a = I_f \frac{(1 + l_f) \sin \alpha}{\sin^2 \alpha - (1 + l_a)(1 + l_f)} \tag{5-75}$$

$$i_f = -I_f \frac{\sin^2 \alpha}{\sin^2 \alpha - (1 + l_a)(1 + l_f)} \tag{5-76}$$

It is interesting to note here that as α is allowed to increase through several cycles, a sinusoidal variation of i_a develops. On the other hand, the concomitant change in the field current remains unidirectional.

Frequently the matter of chief interest is the peak value of the stator phase current which clearly results when $\alpha = \pi/2$. Upon insertion into Eq. (5-75), this peak is found to be

$$i_{a\,\text{max}} = I_f \frac{1 + l_f}{1 - (1 + l_a + l_f + l_a l_f)} = -I_f \frac{1 + l_f}{l_a + (1 + l_a)l_f} \tag{5-77}$$

Keeping in mind that the leakage coefficients are relatively small quantities, it then follows that the maximum phase current is a sizable multiple of the d-c current existing before the short circuit.

The result appearing in the last equation is useful for another important reason. By appropriate manipulation of the right side it is possible to arrive at an equivalent circuit which in fact determines this current. We begin by rewriting the right side of Eq. (5-77) as

$$i_{a\,max} = -I_f \frac{1 + \omega L_f/\omega L_m}{(\omega L_a/\omega L_m) + (1 + \omega L_a/\omega L_m)(\omega L_f/\omega L_m)}$$

$$= -I_f \omega L_m \frac{\omega L_m + \omega L_f}{(\omega L_a)(\omega L_m) + (\omega L_m)(\omega L_f) + (\omega L_a)(\omega L_f)} \qquad (5\text{-}78)$$

Furthermore, the following identifications can be made for use in this last expression:

$\omega L_f \equiv x_f$ = leakage reactance of the field winding
$\omega L_a \equiv x_l$ = leakage reactance per phase of stator winding
$\omega L_m \equiv x_\phi$ = reactance associated with the armature mmf acting in the mutual flux path $\qquad (5\text{-}79)$
$I_f \omega L_m \equiv E_a$ = armature induced emf associated with the mutual flux path

Inserting these relationships into Eq. (5-78) accordingly yields

$$i_{a\,max} = -E_a \frac{x_\phi + x_f}{x_l x_\phi + x_\phi x_f + x_l x_f}$$

$$= -E_a \frac{(x_\phi + x_f)}{x_l(x_\phi + x_f)x_\phi x_f}$$

$$= -\frac{E_a}{x_l + [x_\phi x_f/(x_\phi + x_f)]}$$

$$= -\frac{E_a}{x_s'} \qquad (5\text{-}80)$$

where

$$\boxed{x_s' = x_l + \frac{x_\phi x_f}{x_\phi + x_f}} \qquad (5\text{-}81)$$

A circuit interpretation of Eq. (5-81) is shown in Fig. 5-28. For obvious reasons, this reactance has come to be known as the *transient reactance* of the synchronous generator. It is instructive to note that besides the armature reaction reactance, the transient reactance is dependent upon the *leakage* reactances of both the stator winding per phase and the field winding. Consequently, the transient reactance has

Figure 5-28 Equivalent circuit for the transient reactance. Here x_l and x_f are leakage reactances of the armature and field windings respectively.

THREE-PHASE SYNCHRONOUS MACHINES   CHAP. 5

a value that is very close to the sum of the two leakage reactances. This explains why the current during the transient state can reach extremely large values.

A comment is now in order concerning the exact appropriateness of the circuit of Fig. 5-28. The absence of a dissipative element (i.e., a resistor) implies that whatever transient currents are called for by Eqs. (5-75) and (5-76) would continue indefinitely. Experience shows, however, that the transient current does diminish exponentially to the steady-state value, which is essentially determined by the synchronous reactance in a manner similar to that illustrated in Fig. 5-29. The exponential decrease of the short-circuit current is dictated primarily by the time constant of the field winding, which in addition to the field leakage inductance also involves the field winding resistance. These transients are known to endure from one to several seconds which means they can persist from less than 100 cycles to almost 200 cycles of the line frequency. The situation shown in Fig. 5-29 assumes that the rotor is not equipped with a squirrel-cage damper winding.

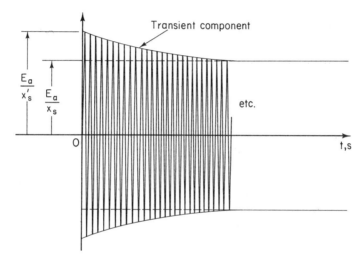

Figure 5-29 Illustrating the transient period to show the effect of the rotor field winding time constant on the phase current of coil *a-a* in Fig. 5-26 when a symmetrical short circuit occurs. Although not so depicted, there may be a hundred or more cycles occurring during this transient period. Here the rotor is assumed to have no damper winding.

In many salient-pole synchronous generators, however, the rotor field structure is equipped with a damper winding. Under these circumstances, when current begins to flow in coil *a-a* immediately following the short circuit, the shorted squirrel-cage damper winding is interlinked with the field winding and the stator phase winding. As short-circuit current begins to build up in coil *a-a* of Fig. 5-27, it causes an increase in the field current by i_f and the interlinkage in turn causes current to emerge in the damper winding. This damper winding current acts in a direction to aid the field winding current, thus causing a still greater increase in the stator phase coil in its attempt to satisfy the boundary condition that calls for the flux linkage to remain fixed at the value prevailing at the moment the short circuit at the stator terminals occurs. In essence, from the viewpoint of the stator current the damper winding acts in *parallel* with the field winding. Accordingly, an analysis similar to that which led to Eq. (5-81) reveals that the stator winding equivalent circuit during a symmetrical short circuit on a machine equipped with a damper winding in addition to a field winding assumes the configuration depicted in Fig. 5-30. The reactance appearing at the stator terminals in this case is

$$x_s'' = x_l + \frac{x_\phi x_f x_d}{x_f x_d + x_\phi x_d + x_\phi x_f} \tag{5-82}$$

and is called the *subtransient synchronous reactance*. Here x_d denotes the *leakage* reactance of the damper winding per phase.

Figure 5-30 Equivalent circuit for the subtransient synchronous reactance.

Although a resistive element does not appear in this equivalent circuit either, the actual circuit experienced by the short-circuit armature current does in fact encounter the effects of dissipative elements. The design of the damper winding in these machines is such that the ratio of the winding resistance to the leakage inductance of the damper winding (r_d/L_d) is much larger than the corresponding quantity for the field winding. In other words, the damper winding time constant is much smaller than the field winding time constant. Consequently, the transients involved in connection with the squirrel-cage damper winding endure for a very short time (usually a few cycles), which is the reason why these transients have come to be called the *subtransients*. In the next 100 or more cycles the field winding exerts the dominant influence. Finally, following the expiration of the field winding transients, the sustained short-circuit current attains a value consistent with the synchronous reactance parameter x_s. This complete sequence of events is illustrated in Fig. 5-31.

A striking aspect of the response appearing in Fig. 5-31 is the symmetry of the transient and sustained short-circuit current over the entire time history. This is characteristic of the behavior of the short-circuit current in a stator phase whenever the symmetrical three-phase short circuit occurs at a condition of zero flux linkage between the field winding and the given stator phase winding. Let us suppose, however, that the short circuit occurs at the instant when the coupling of the field winding with the stator phase winding is a maximum. Because the induced emf at this instant is zero, the steady-state value of the current in the short circuit mode should be at its instantaneous maximum. Of course, the flux condition just before short circuit must remain the same immediately after the short circuit since the energy contained in the magnetic field cannot be allowed to change instantaneously. Hence reactions must take place both in the stator phase coil and the field winding to preseve this initial maximum flux condition. The composition of the incremental

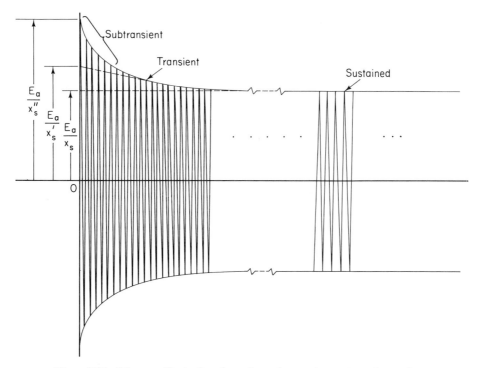

Figure 5-31 Diagram illustrating the subtransient and transient phases that precede the sustained short circuit current for a symmetrical short circuit occurring in a stator phase at the moment of zero flux linkage with the field winding. The rotor carries a damper squirrel-cage winding.

field and armature currents in this case is more involved than in the zero-flux linkage case. In addition to the previously described a-c current that flows in the stator coil, there must also be included a d-c component which is needed in the inductive armature circuit to ensure that the armature current starts from a value of zero current just before the short circuit occurs. Because the a-c value of this current in the first cycle is E_a/x_s' and because initially it must be zero on the assumption that the armature circuit is open-circuited just prior to the occurrence of the short circuit, it therefore follows that the tendency for the flux to *want* to change creates an initial value of d-c component of armature current which must be equal and opposite to the quantity E_a/x_s'. One apparent consequence of this situation is that a half-cycle later in time the armature current reaches a value of nearly twice this magnitude, i.e., $i_{a\,max} = 2E_a/x_s'$. Refer to Fig. 5-32 for a graphical description of the situation. The decay rate of this d-c component is determined by the resistance and equivalent inductance that is seen at the stator terminals. It is instructive to note that the presence of this d-c component in the stator phase coil gives rise to an additional current component in the field winding which assumes a damped sinusoidal character. This increase in field current is above and beyond the component that was encountered when the short circuit took place at the instant of zero flux linkage [cf. Eq. (5-76)].

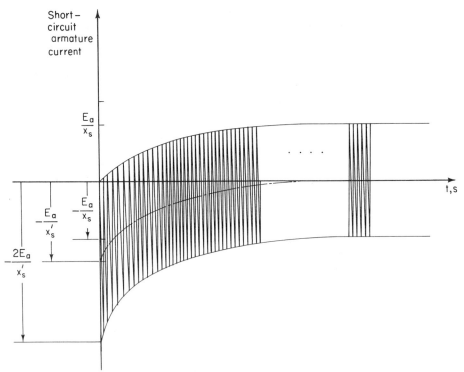

Figure 5-32 Variation of short-circuit current in a stator phase when the short circuit occurs at an instant corresponding to maximum flux linkage between the field coil and the stator phase coil. No damper winding present.

Example 5-5

A synchronous generator is operating as an isolated unit in an idle state with the field excitation set at that value which permits rated terminal voltage to appear at its line terminals. The per-unit values of the machine parameters are: $x_s = 0.9$, $x_s' = 0.35$, and $x_s'' = 0.20$. A three-phase short circuit occurs across the armature terminals at the instant when the coupling of the field winding with the stator phase winding a is a maximum.

(a) Find the value of the d-c component of the armature current in phase a during the first cycle of the short circuit. Express the result in per unit.

(b) Determine in per unit the maximum current that flows in phase a during this period.

(c) The subtransient components of the armature winding currents disappear in a few cycles. On the other hand, the transient components persist for much longer. Calculate the value of the armature current in phase a immediately following the subtransient period. Assume that the d-c component is barely diminished in phase a.

(d) What is the value of the sustained short-circuit current in phase a after all transients have decayed to a negligible level?

Solution (a) The voltage behind the subtransient reactance in the case where no load current is being delivered is simply the terminal voltage. The d-c component is essentially equal and exactly opposite to the maximum current produced by this voltage acting on the subtransient reactance. Thus,

$$i_{a\,d\text{-}c} = \sqrt{2}\,\frac{\text{p-u }V}{x_s''} = \frac{\sqrt{2}}{0.2} = 7.07 \text{ p-u}$$

(b) The maximum value of the short circuit current can be taken to be

$$\text{p-u } I_a'' = 2\,(7.07) = 14.14 \text{ p-u}$$

A glance at Fig. 5-31 should make this result obvious. However, as a practical matter, there is a slight decrease in this current because of the time constant associated with the stator winding during the subtransient state.

(c) Here use is made of the transient reactance, thereby leading to a value of short-circuit current in phase a which is

$$\text{p-u } I_a' = 7.07 + \sqrt{2}\left(\frac{1}{0.35}\right) = 11.11 \text{ p-u}$$

(d) At this stage in the short-circuit condition the operative reactance is solely the synchronous reactance, so that so long as the field winding remains energized at the prefault value, the steady-state short-circuit current becomes

$$\text{p-u } I_a = \frac{\text{p-u }V}{x_s} = \frac{1}{0.9} = 1.11 \text{ p-u rms}$$

5-10 DYNAMICAL BEHAVIOR OF THE PRIME MOVER–GENERATOR SYSTEM

In Sec. 5-8 it was learned that a synchronous generator operating in parallel with other synchronous generators can be made to assume a greater or lesser share of the total electrical load being delivered from the common bus system by making appropriate adjustments in the speed-power characteristics of the prime mover. This adjustment establishes the amount of mechanical power that is delivered to the generator via the shaft connection. In turn, the synchronous generator delivers electrical energy which is equal to the prime mover shaft input minus appropriate losses. By Eq. (5-55) the associated mechanical power developed in this manner calls for a specific value of generator power angle for a fixed field excitation. Our interest in this section is to describe the dynamical behavior of this mechanically linked prime mover–generator system as changes are made to alter a quiescent operating point. For example, we are concerned with the manner in which the power angle seeks a new equilibrium value consistent with a change in mechanical input power at constant excitation.

Because the synchronous machine is a doubly excited device, it can produce a net unidirectional torque only if it operates at a single fixed speed—the synchronous speed. Unlike the induction machine, which is singly excited, the synchronous machine cannot adapt to changing load conditions by changing its speed; rather, the synchronous machine responds by changing its power angle. Accordingly, it is frequently referred to as a position-sensitive device, as contrasted to the induction motor, which is a speed-sensitive device. As a result of this difference in sensitivity, the synchronous machine can be prone to severe oscillations as load conditions change because the differential equation that describes its dynamical behavior is second order. On the other hand, machines that are speed sensitive often have their dynamics described by a first-order differential equation.

The occurrence of a second-order differential equation for the synchronous machine of course implies the presence of two independent energy-storing elements. What are they? An examination of the prime mover and generator combination readily reveals the combined angular inertia of the two units as one of the energy-storing elements. Clearly, energy is stored in the spinning mass and it is specifically represented by the expression $\frac{1}{2}J\omega_m^2$, where J is the total inertia of the system and ω_m denotes the angular velocity of the generator. The second energy-storing element is the shaft stiffness, which necessarily plays a role because of the position-sensitivity feature of the synchronous machine. This quantity is analogous to the spring constant in a mass-spring translational system.

When the synchronous machine is operating as a generator and the input power from the prime mover is increased, the immediate reaction is that the speed changes *momentarily* in order to allow the power angle to change to a new equilibrium position. A description of the transition of the power angle from the quiescent value to the new equilibrium value is obtained from a solution to the appropriate differential equation. By the law of conservation of energy we can write

$$T_{\text{sh}} = J\alpha_m + T_d + T_s \qquad (5\text{-}83)$$

where T_{sh} denotes the shaft torque in N-m delivered by the prime mover, T_d represents the torque developed in the damper squirrel-cage winding with which the rotor is assumed to be equipped, T_s denotes the counter electromagnetic torque which changes with the power angle in accordance with Eq. (5-55), and α_m is the mechanical acceleration of the rotating mass. It is helpful to keep in mind that the two terms, $J\alpha_m$ and T_d, on the right side of the last equation exist only during the brief transient period when the power angle undergoes alteration in response to a change in energy coming from the prime mover.

It is now desirable to modify the formulation of Eq. (5-83) in order to bring the power angle Δ of the synchronous generator explicitly into evidence. In order to do this in a convincing manner, let us examine the sequence of events that takes place as the synchronous generator takes on load. We start with the machine operating at no load immediately following synchronization to the bus line. Ideally, at this load, the power angle that separates the field excitation phasor $\bar{E}_f$ and the terminal bus voltage phasor $\bar{V}_t$ is zero. Keep in mind that in the phasor diagram representation these two phasors are assumed rotating at synchronous frequency but are stationary relative to each other. If we call the displacement angle of a point on the rotor field structure with respect to a fixed external reference point θ, then clearly while in synchronism with the bus system this angle increases with time t in accordance with

$$\theta = \omega_s t \qquad (5\text{-}84)$$

where ω_s is the synchronous angular velocity of the bus system and θ is measured in electrical degrees. Now assume that the prime mover of this local generator is adjusted to increase the shaft power input to the generator. The increased shaft power causes a momentary increase in the angular velocity of the rotor field structure which may be written more generally as

$$\omega = \frac{d\theta}{dt} \qquad (5\text{-}85)$$

It is important to understand the distinction between Eqs. (5-84) and (5-85). The former describes the total displacement with respect to a fixed external reference point while the generator is synchronized at no load. The latter is a general expression for the angular velocity of the rotor of the synchronous generator at all times. Moreover, in a phasor diagram such as Fig. 5-19, it is helpful to associate the placement of the field excitation phasor $\bar{E}_f$ with the angular velocity ω of Eq. (5-85) and the bus voltage phasor $\bar{V}_t$ with the synchronous angular velocity ω_s. Accordingly, when ω experiences a momentary increase above its synchronous value in response to the increased input power, the difference in the angular velocity of phasor $\bar{E}_f$ and of phasor $\bar{V}_t$ causes $\bar{E}_f$ to advance in phase relative to $\bar{V}_t$. The amount of the increase can be expressed mathematically as

$$\Delta = \int_0^{t_1} (\omega - \omega_s)\, dt \qquad (5\text{-}86)$$

where t_1 denotes the time period during which the generator is temporarily out of synchronism with the bus system. Of course, the longer t_1 is, the greater will Δ become. An equilibrium condition, however, is rather quickly achieved because with the increased power angle the generator delivers electrical power to the line in the manner dictated by Eq. (5-55). Consequently, the momentary change in ω in Eq. (5-86) yields that Δ which is consistent with the increased shaft power input coming from the prime mover. Once synchronism is re-established, the expression for the total angular displacement of the rotor structure relative to a fixed reference then becomes

$$\theta = \omega_s t + \Delta \qquad (5\text{-}87)$$

Observe that this last expression differs from Eq. (5-84) because it represents θ under a load condition other than zero. The instantaneous value of the angular velocity of the generator is available upon differentiating Eq. (5-87). Thus,

$$\omega = \frac{d\theta}{dt} = \omega_s + \frac{d\Delta}{dt} \qquad (5\text{-}88)$$

This expression also reveals that the angular velocity of the rotor structure (i.e., $d\theta/dt$) once more becomes ω_s, once balance between the mechanical input power and the electrical output power is achieved and Δ ceases to change any further (i.e., $d\Delta/dt = 0$).

Use can now be made of Eq. (5-88) to get an expression for the acceleration of the inertia of the prime mover and generator combination. Any further increase in shaft input power to the generator is accompanied by a momentary change in its angular velocity, which in turn is the acceleration. The mathematical expression for this acceleration readily follows upon differentiation of Eq. (5-88). Thus

$$\frac{d\omega}{dt} = \frac{d^2\theta}{dt^2} = \frac{d^2\Delta}{dt^2} \qquad (5\text{-}89)$$

All the angular quantities in this equation are expressed in electrical radians. However, the acceleration quantity α_m appearing in Eq. (5-83) is in terms of mechanical degrees.

Accordingly, if the synchronous generator has p poles, the last equation can be re-written as

$$\frac{p}{2}\frac{d^2\theta_m}{dt^2} = \frac{d^2\Delta}{dt^2}$$ (5-90)

where θ_m denotes the displacement angle of the rotor structure measured with respect to fixed point expressed in mechanical degrees. Rearranging yields

$$\frac{d^2\theta_m}{dt^2} = \frac{2}{p}\frac{d^2\Delta}{dt^2} = \alpha_m$$ (5-91)

which is the expression for the acceleration of the rotor structure. It is expressed in terms of the power angle measured in electrical degrees. Inserting this last result into Eq. (5-83) now permits the basic relationship for describing the transient stability of the prime mover–generator systems to be written as

$$T_{sh} = \frac{2}{p}J\frac{d^2\Delta}{dt^2} + T_d + T_s$$ (5-92)

A distinguishing characteristic of the operation of a synchronous machine is that it must exhibit a unique value of the power angle for a given developed power and field excitation. It is this feature in conjunction with two energy-storing elements that can give rise to a substantial oscillation as the generator seeks out the proper power angle during the transient state. A proper description of this behavior is dependent upon expressing both the damping torque, T_d, and the synchronizing torque, T_s, in terms of this crucial power angle variable. It has already been stated that the first two terms on the right side of Eq. (5-92) are nonzero only when the synchronous generator is momentarily out of synchronism, i.e., during the transient state. However, once steady state is reached, the left side of Eq. (5-92) must have a balancing term on the right side. Clearly, this must be the role of T_s. In fact, T_s is nothing more than the electrical version of the mechanical input T_{sh}. Described mathematically, we have

$$T_s\omega_s = q\frac{V_tE_f}{x_s}\sin\Delta$$

or

$$T_s = q\frac{V_tE_f}{\omega_s x_s}\sin\Delta \qquad \text{N-m}$$ (5-93)

This is properly called the synchronizing torque because it represents the electrical equivalent of the mechanical torque input from the prime mover at steady state. When $T_s = T_{sh}$, the transients have effectively decayed to zero and the generator is once again synchronized to the bus system.

The damping torque attributable to the squirrel-cage winding placed in the field pole structure involves a different mathematical focus of the power angle from either of the two terms on the right side of Eq. (5-92). It is useful to recall that currents can be induced in the squirrel-cage winding only when the rotor is out of synchronism with the frequency of the bus system. But this in turn means that the power angle is changing with time. Moreover, the greater the rate of change of the power angle with time, the larger will be the induced voltage that creates the squirrel-cage winding current. The action of this current is always to produce a torque that opposes the torque that produced it. Thus the squirrel-cage produces a reaction to help restore the generator to its equilibrium condition of synchronism with the bus system. Accordingly, one can represent the torque T_d analytically as

$$T_d = k_d \frac{d\Delta}{dt} \tag{5-94}$$

where k_d is an appropriate design constant of the squirrel-cage winding that correctly describes this damping effect. The units of k_d are N-m per electrical degrees per second.

The insertion of Eqs. (5-93) and (5-94) into Eq. (5-92) then yields the non-homogeneous differential equation involving the single important dependent variable Δ. Thus

$$T_{sh} = \frac{2}{p} J \frac{d^2\Delta}{dt^2} + k_d \frac{d\Delta}{dt} + q \frac{V_t E_f}{\omega_s x_s} \sin \Delta \tag{5-95}$$

The quantity J in this last equation is usually computed from knowledge of the weight and the radius of gyration of the prime mover–generator rotor structure. Customarily, the units of this calculation are kg-m² or N-m/rad/s². But because Δ is expressed in electrical degrees rather than radians, it is necessary to multiply J, so expressed, by $180°/\pi$ for consistency. Thus, Eq. (5-95) becomes

$$T_{sh} = \frac{2\pi}{180p} J \frac{d^2\Delta}{dt^2} + k_d \frac{d\Delta}{dt} + q \frac{V_t E_f}{\omega_s x_s} \sin \Delta \tag{5-96}$$

when J is used in units of kg-m² or N-m/rad/s².

This last equation can be written more simply as

$$T_{sh} = k_J \ddot{\Delta} + k_d \dot{\Delta} + k_s \sin \Delta \tag{5-97}$$

by noting that

$$k_J = \frac{\pi}{90p} J = \text{inertial torque per unit acceleration}$$
$$k_s = q \frac{V_t E_f}{\omega_s x_s} = \text{synchronizing torque coefficient} \tag{5-98}$$

and k_d is as previously defined. Also, the dot and double dot notation are introduced to represent the first and second derivatives with respect to time. A moment's study of Eq. (5-97) readily discloses that this is a nonlinear differential equation and therefore a bit more of a nuisance to solve than linear differential equations. However, the recourse to a computer easily removes the difficulty (see Sec. 11-9). An alternative procedure is to linearize Eq. (5-97) by replacing the $\sin \Delta$ by Δ expressed in radians. This is generally permissible for values of Δ up to 15 electrical degrees. Because our interest here is more with a qualitative rather than the precise quantitative description of the transient behavior of the generator in response to load changes, we continue the analysis with the linearized version of the differential equation which accordingly reads as

$$T_{sh} = k_J \ddot{\Delta} + k_d \dot{\Delta} + k_s \Delta \tag{5-99}$$

and where it is now to be understood that k_s is expressed in N-m per degree. This allows Δ to be expressed in degrees throughout.

The dynamical behavior of the system described by this differential equation can be completely described by examining the roots of the characteristic equation. There is no need here to obtain a formal complete solution of the nonhomogeneous differential equation. By Laplace transforming the homogeneous equation for zero initial conditions the characteristic equation in normalized form is found to be

$$s^2 + \frac{k_d}{k_J}s + \frac{k_s}{k_J} = 0 \tag{5-100}$$

where s is the Laplace transform variable. Once the constants k_J, k_d, and k_s are known for a given generator, the nature of the ensuing oscillatory response following a load change is routinely determined through the use of two "figures of merit" ζ and ω_n as they appear in the universal version of Eq. (5-100), which is

$$s^2 + 2\zeta\omega_n s + \omega_n^2 = 0 \tag{5-101}$$

By matching the coefficients of this expression with those of Eq. (5-100) the values of the system's damping ratio ζ and its natural frequency ω_n can then be found. The quantity ζ is useful in that it yields information about the maximum percent overshoot that occurs in the power angle during the transient state as the load is moved from one nonzero operating point to another nonzero operating point. The relationship of the maximum percent overshoot versus ζ for step commands appears in Fig. 5-33. The second piece of characterizing information that completes the description of the transient behavior is obtained from knowledge of the system's natural frequency ω_n, the value of which can be shown to be dependent upon the parameters of the two energy storing elements. Knowledge of ω_n gives information about the settling time, i.e., the time it takes for the transients to fall within an acceptable tolerance band of error. The expression used is†

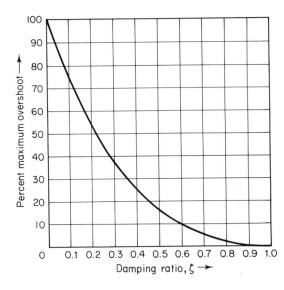

Figure 5-33 Percent maximum overshoot versus damping ratio for a linear second-order system.

†For a derivation, see V. Del Toro, *Principles of Electrical Engineering*, 2nd ed. (Englewood Cliffs, N.J.: Prentice-Hall, Inc., 1972), pp. 634–642.

$$t_s = \text{settling time} = \frac{5}{\zeta\omega_n} \qquad (5\text{-}102)$$

The use of 5 on the right side of this equation implies a tolerance band error of 1%. In other words, the new value of power angle will be within 99% of its final value in a time determined by Eq. (5-102).

Example 5-6

A six-pole, 60-Hz synchronous generator with constant excitation is delivering a total of 553 kW of electrical power to a bus system at a power angle of 10 electrical degrees. The inertia acceleration coefficient for the combined prime mover–generator is known to be 1.6 N-m/electrical degree/s². Moreover, the damping torque coefficient is given by 11.2 N-m/electrical degree/s.

The input power from the prime mover is increased so that the generator increases its share of the bus electrical load to 1000 kW.

(a) Determine the maximum value of the power angle as it moves to its new steady-state value.

(b) How long does it take for steady state to be reached expressed in terms of a 1% tolerance band?

Assume no losses and linear analysis.

Solution   (a) The initial value of the power angle is $\Delta_i = 10°$. The final value for the linearized analysis is readily found to be

$$\Delta_f = \frac{1000 \text{ kW}}{553 \text{ kW}}\Delta_i = \frac{1000}{553}(10) = 18.1°$$

The description of the transient response of the generator in moving from an initial steady-state value of 10° to a new steady value of 18.1° is described by examining the characteristic equation. The problem statement gives $k_j = 1.6$ and $k_d = 11.2$. The third parameter can be found from the first steady-state condition. At a power of 553 kW, the corresponding synchronizing torque is

$$T_s = \frac{553,000}{\omega_s} = \frac{553,000}{\frac{120(60)}{6}\left(\frac{\pi}{30}\right)} = 4399.4 \text{ N-m}$$

But

$$k_s\Delta = T_s$$

$$k_s = \frac{T_s}{\Delta_i} = \frac{4399.4}{10} = 440 \frac{\text{N-m}}{\text{degree}}$$

Inserting these three parameters into Eq. (5-100) yields

$$s^2 + \frac{11.2}{1.6}s + \frac{440}{1.6} = 0$$

or

$$s^2 + 7s + 275 = 0$$

The natural frequency is thus determined to be

$$\omega_n = \sqrt{275} = 16.58 \text{ rad/s}$$

Expressed in cycles per second this is

$$f_n = \frac{16.58}{2} = 2.64 \text{ Hz}$$

The actual damped frequency of oscillation of the transient response will be just slightly less than this figure. The exact value is expressed by

$$\omega_d = \omega_n\sqrt{1 - \zeta^2}$$

The damping ratio ζ is found from

$$2\zeta\omega_n = 7$$

$$\zeta = 0.21$$

which leads to

$$\omega_d = 16.58\sqrt{1 - (0.21)^2} = 16.25 \text{ rad/s}$$

or

$$f_d = 2.58 \text{ Hz}$$

With the value of ζ available, the maximum percent overshoot is obtained by entering Fig. 5-33. This yields 52%. It is important to understand that this overshoot is 52% of the newly commanded value, which is the difference between the final and initial value of the power angle. Accordingly, the maximum value of power angle attained by the generator in this transient state is

$$\Delta_{\max} = \Delta_i + 1.52(\Delta_f - \Delta_i)$$

$$= 10 + 1.52(18.1 = 10) = 22.31°$$

A graphical representation of the transient behavior is sketched in Fig. 5-34.

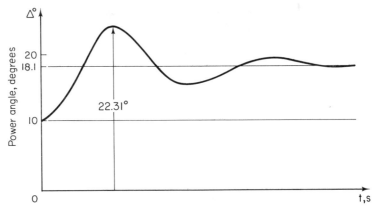

Figure 5-34 Transient response of the prime mover–synchronous generator system in response to an increased load demand.

As a point of interest it is useful to note that if the generator were operating at a much higher initial level of power angle, the oscillatory nature of the transient response in lightly damped systems can act to bring the generator close to losing its synchronism. This occurs when the power angle exceeds 90°. In fact, use of the nonlinear differential equation makes the situation even worse because a larger power angle is needed for a given increase in load power.

(b) The time it takes for the transient to decay is readily found from Eq. (5-102).

Thus,

$$t_s = \frac{5}{\zeta \omega_n} = \frac{5}{3.5} = 1.43 \text{ s}$$

Thus the duration is not very long.

5-11 THE SYNCHRONOUS MOTOR

The significant and distinguishing feature of synchronous motors in contrast to induction motors is that they are doubly excited. Electrical energy is supplied to both the field and the armature windings. When this is done torque can be developed at only one† speed—the *synchronous speed*. At any other speed the average torque is zero. The synchronous speed refers to that rotor speed at which the rotor flux field and the armature ampere-conductor distribution (or the armature flux field) are stationary with respect to each other. To illustrate this point consider the conventional synchronous motor equipped with two poles. It has a d-c voltage applied to the rotor winding and a three-phase a-c voltage applied to the stator winding which is usually 60 Hz. The three-phase voltage produces in the stator an ampere-conductor distribution which revolves at a rate of 60 Hz. The d-c (or zero-frequency) currents in the rotor set up a two-pole flux field which is stationary so long as the rotor is not turning. Accordingly, we have a situation in which there exists a pair of revolving armature poles and a pair of stationary rotor poles. A little thought should make it clear that for a period of half a cycle (or $\frac{1}{120}$ s) a positive torque is developed by the cooperation of the field flux and the revolving ampere-conductor distribution. However, in the next half cycle this torque is reversed so that the average torque is zero. This is the reason that a synchronous motor per se has no starting torque.

In order to develop a continuous torque, then, it is necessary not only to have a flux field and an appropriately displaced ampere-conductor distribution, *but the two quantities must be stationary with respect to each other*. In singly fed machines such as the induction motor this comes about automatically. The relative speed between the stator field and the rotor winding induces rotor currents at those slip frequencies which yield a revolving rotor mmf. Superposing this upon the rotor speed gives a revolving rotor mmf, which in turn is stationary relative to the stator field. In the synchronous motor this condition cannot take place automatically because of the separate excitation used for the field and armature windings. In the light of these comments it is reasonable to expect that, if an induction motor is doubly energized, it too will behave as a synchronous motor. Refer to Sec. 4-8.

On the basis of the foregoing comments it follows that to produce a continuous nonzero torque it is first necessary to bring the d-c excited rotor to synchronous speed by means of an auxiliary device. Sometimes the auxiliary device takes the form of a small d-c motor mounted on the rotor shaft. Most often, however, use is made of a squirrel-cage winding similar to that used in induction motors and imbedded in the pole faces. By means of this winding (also called the *amortisseur winding*) the syn-

†There is a second speed, too, but it is only of academic interest because of the excessive currents that flow.

chronous motor is brought up to almost synchronous speed. Then, if the field winding is energized at the right moment, a positive torque will be developed for a sufficiently long period to allow the armature poles to pull the rotor poles into synchronism.

A general expression may be written to identify that frequency of rotation required of the rotor so that the stator and rotor fields are stationary with respect to each other. Let

f_1 = frequency of the currents through the stator winding

f_2 = frequency of the currents through the rotor winding

f_r = frequency of rotation of the rotor structure

Then the defining equation is

$$f_1 = f_2 + f_r \qquad \text{(5-103)}$$

In the conventional synchronous motor the frequency of the rotor current f_2 is zero, since it is direct current. Hence for an f_1 of 60 Hz the rotor must revolve at 60 Hz in order for nonzero torque to be developed. On the other hand, note that, if the rotor winding were energized with 20-Hz current in place of direct current, then for an f_1 of 60 Hz the synchronous speed would be 40 Hz.

5-12 SYNCHRONOUS MOTOR PHASOR DIAGRAM AND EQUIVALENT CIRCUIT

When the synchronous machine makes the transition from generator to motor action, there occurs a reversal of power flow. Thus, instead of current flowing *out of* the armature winding to a suitable electrical load, the current flows *into* the armature winding originating from a suitable electical source. Since this transition occurs with the direction of rotation and the direction of field current flow unaltered from the generator case, it follows that the relationship between F and its associated excitation voltage E_f remains invariant. An observer looking only at this relationship certainly could not distinguish whether operation is as a motor or generator. One way to make this distinction, however, is to determine whether the armature current flows in a direction that is essentially opposed to that of the excitation voltage. If it is, the machine is behaving as a motor. In this condition the phasor diagram of the synchronous motor can then be shown to differ from that of the generator in two respects. The first involves the voltage equation for the stator circuit. In the generator, $\bar{E}_f$ (the excitation voltage) played the role of a source voltage and the terminal voltage $\bar{V}_t$ was dependent upon the magnitude of $\bar{E}_f$ and the synchronous impedance drop. In the motor the roles are interchanged. Now $\bar{V}_t$ is the source voltage applied to the synchronous motor armature winding and $\bar{E}_f$ is a reaction or counter emf which is internally generated. It is assumed that the terminal voltage originates from an infinite bus system† and so remains invariant. Applying Kirchhoff's voltage law to the synchronous motor for negligible armature resistance leads to

$$\boxed{V_t\underline{/0^\circ} = \bar{E}_f + j\bar{I}_a x_s} \qquad \text{(5-104)}$$

---

†A power system of tremendous capacity compared with the rating of the synchronous motor.

Clearly, this equation states that the applied stator voltage is equal to the sum of the drops. The excitation voltage is treated as a reaction voltage drop in much the same way as $\bar{E}_1$ was treated in the case of the transformer.

The circuit interpretation of Eq. (5-104) leads to the equivalent circuit as it applies to the synchronous motor. This appears in Fig. 5-35.

The second point of view in which the phasor diagram of the motor differs from that of the generator involves the angle Δ. Physically, for the generator, the phase of $\bar{E}_f$ is ahead of $\bar{V}_t$ in time because of the driving action of the prime mover. Keep in mind that $\bar{E}_f$ may be considered associated with the rotor field axis. A similar line of reasoning pertains for the motor. At no-load, Δ is zero and so the axis of a field pole and the resultant flux axis associated with the terminal voltage are coincident. As shaft load is applied, however, the rotor falls slightly behind its no-load position and in this way causes Δ to increase but in a sense opposite to that for the generator.

Figure 5-35 The equivalent circuit of the synchronous motor appears to the right of terminals *ab*.

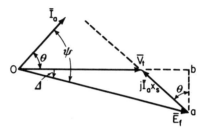

Figure 5-36 Phasor diagram of the synchronous motor.

On the basis of these modifications the phasor diagram of the synchronous motor becomes that shown in Fig. 5-36. Note that $\bar{E}_f$ now lags $\bar{V}_t$ and that it is the phasor sum of $\bar{E}_f$ and $j\bar{I}_a x_s$ which equals $\bar{V}_t$, as called for by Eq. (5-104). Another interesting point about this diagram is revealed upon comparison with Fig. 5-19. In both cases the magnitude of the excitation voltage $\bar{E}_f$ exceeds that of the terminal voltage $\bar{V}_t$. This is referred to as a condition of *overexcitation* (i.e., $|E_f| > |V_t|$). Note, however, that overexcitation in the synchronous generator is accompanied by a current of *lagging* power factor, whereas in the synchronous motor overexcitation is accompanied by a current of *leading* power factor. When the magnitude of $\bar{E}_f$ is equal to $\bar{V}_t$ the condition is referred to as *100%* excitation. A situation where $|E_f| < |V_t|$ is described as *underexcitation*.

5-13 *CORRELATION OF SYNCHRONOUS MOTOR OPERATION WITH THE BASIC TORQUE EQUATION*

It is the purpose of this section to outline in some detail the manner in which the factors appearing in the basic torque equation [Eq. (3-110)] are manifested in synchronous motor operation for several power factor conditions. The doubly excited feature of the synchronous motor complicates the situation somewhat because field winding conditions of overexcitation or underexcitation influence the reactive component of the armature current to become leading or lagging, respectively. See Sec. 5-15. Attention is directed first to the case where ψ equals $0°$, i.e., the armature current is in phase with the excitation voltage (and treated as a voltage drop). Refer to Fig. 5-37(a). Our concern is with the cylindrical-rotor machine in spite of the use of a salient-pole structure; the latter is employed merely to facilitate the graphical representation. For the indicated polarity of the rotor field flux and direction of rotation, the location of the amplitude of the emf distribution of the entire three-phase winding is represented by coil e_f-e_f' having the instantaneous directions shown. When the machine is operating as a motor, the armature current flows *opposed* to the polarity of the e_f shown because e_f is represented here as a *voltage rise* with respect to the

(a)

(b)

(c)

Figure 5-37 Space and time relationships in the synchronous motor for $x = 0°$: (a) space diagram depicting the axis of ampere-conductor distribution; (b) corresponding phasor diagram; (c) detailed representation of (a). Field motion is toward the left.

THREE-PHASE SYNCHRONOUS MACHINES   CHAP. 5

field flux that produces it. However, when viewed with respect to the current,† the $\bar{E}_f$ phasor is treated as a *voltage drop* and is so represented in Fig. 5-37(b). The coil i_a–i_a' in Fig. 5-37(a) represents the location of the amplitude of the assumed sinusoidal ampere-conductor distribution. Note that the armature current flows inward for the upper conductor and outward for the lower conductor because i_a is specified as in-phase with the excitation voltage expressed as a voltage drop and hence must be opposite to the direction of e_f treated as a voltage rise. The armature mmf (represented by $\bar{A}$) is therefore directed from right to left and cooperates with $\bar{F}$ to produce the resultant $\bar{R}$. Note that the $\bar{R}$ in Fig. 5-37(a) is consistent with that shown in Fig. 5-37(b). For simplicity the armature leakage reactance is assumed to be negligible so that the armature induced emf $\bar{E}_r$ may be taken equal to the terminal voltage $\bar{V}_t$. It is for this reason that $\bar{R}$ is drawn perpendicular to $\bar{V}_t$ in Fig. 5-37(b). Then to be consistent with this the amplitude of the distribution of the terminal voltage in the armature winding can be represented in the space picture of Fig. 5-37(a) by coil v_t–v_t'. It is worthwhile to note at this point that by means of such a representation a physical interpretation of the power angle Δ is obtained.

Also worthy of note is the fact that application of the $\bar{I} \times \bar{B}$ rule indicates a clockwise torque produced on the armature (stator) winding, while the torque on the rotor is counterclockwise. For the motor, therefore, the direction of torque and the direction of rotation are the same. In Fig. 5-37(b) note again that the direction of the torque on the rotor is such that it tends to align the field axis with the axis of the resultant flux, as was the case for generator action.

The information contained in Figs. 5-37(a) and (b) can also be represented in more detail by employing the developed form depicted in Fig. 5-37(c). The fundamental component of the field flux produced by $\bar{F}$ is denoted by ϕ_f. The ampere-conductor distribution is denoted by the dashed line. The armature flux produced by the armature mmf associated with the ampere-conductor distribution is represented by ϕ_a; this ϕ_a wave is properly displaced 90° behind the ampere-conductor distribution and in-phase with the armature mmf wave. Of course the resultant flux is produced by summing ϕ_f and ϕ_a and is permitted by the linearity assumption. This is denoted by ϕ_r. Keep in mind that for a constant terminal voltage the amplitude of the ϕ_r wave remains invariant. A study of these distributions makes the correspondence with the basic torque equation self-evident. Because $\psi = 0°$, each conductor beneath a given field pole carries current in the same direction and thereby makes a contribution to the developed torque. This viewpoint is consistent with that version of the basic torque equation described by

$$T = \frac{\pi}{8}p^2\Phi J_m \cos \psi = \frac{\pi}{8}p^2\Phi J_m \tag{5-105}$$

where Φ is the flux produced by the field winding and J_m denotes the amplitude of the ampere-conductor distribution.

The second viewpoint involves the use of the *torque angle* δ. Recall that δ was defined in Fig. 3-18 as the angle between the axis of the field pole and the axis of the armature mmf pole. Moreover, δ is always the complement of angle ψ. A glance

†See Sec. 2-1.

at Fig. 5-37(c) discloses that δ is 90° for this case so that the second form of the basic torque equation becomes

$$T = \frac{\pi}{8} p^2 \Phi \mathfrak{F}_p \sin \delta = \frac{\pi}{8} p^2 \Phi \mathfrak{F}_p \qquad (5\text{-}106)$$

where $\mathfrak{F}_p$ denotes the armature mmf per pole.

Figure 5-38 depicts the situation that exists in the synchronous motor when $\psi = 60°$ leading. Note that the field pattern for producing torque is not optimum because now some of the armature conductors find themselves under a pole flux of the wrong polarity. This is accounted for in Eq. (5-105) by the need to use $\cos \psi = \cos 60°$ rather than unity as in the preceding case. When the situation is described in terms of the formulation of Eq. (5-106), the required value of δ is obviously 30°. It is worthwhile to note in Fig. 5-38(c) that the orientation of the armature mmf (or

(a)

(b)

(c)

Figure 5-38 Space and time relationships in the synchronous motor for $\psi = 60°$ leading: (a) space diagram; (b) phasor diagram; (c) detailed form of (a).

its associated flux) with the field flux is such as to produce a net demagnetization. This is entirely consistent with the vector relationship shown between $\bar{F}$ and $\bar{A}$ in Fig. 5-38(a).

5-14 COMPUTATION OF SYNCHRONOUS MOTOR PERFORMANCE

The analysis of the performance of a synchronous motor is readily determined by the phasor diagram, the power-flow diagram, and the expression for the mechanical power developed. The usefulness of the power-flow diagram has already been demonstrated in connection with the induction motor. Hence it is presented in Fig. 5-39 without much additional comment. There is one step less in this flow graph because the a-c source does not have to supply the rotor winding copper loss since the rotor is traveling at synchronous speed. This is not to say that the rotor winding copper losses do not exist. They do, of course, but the power is supplied from the separate d-c source. Figure 5-39 represents the flow of power from the a-c source to the shaft of the synchronous motor.

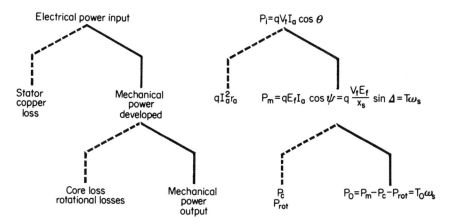

Figure 5-39 Power-flow diagram of the synchronous motor; a-c power only.

The phasor diagram and the expression for the mechanical power developed in terms of the power angle have already been discussed and so are in need of no further elaboration here.

What is the mechanism by which the synchronous motor becomes aware of the presence of a shaft load, and how is the electric energy source made aware of this so that it proceeds to provide energy balance? To answer this question we start first with a study of the conditions prevailing at no-load. As the power-flow diagram indicates, the only mechanical power needed is that which is required to supply the rotational and core losses. This calls for a very small value of Δ, as depicted in Fig. 5-40. The machine is assumed to be overexcited since $\bar{E}_f > \bar{V}_t$. Note that Δ_0 is just large enough so that the in-phase component of $\bar{I}_a$ is sufficient to supply the losses. Consider next that a large mechanical load is suddenly applied to the motor shaft.

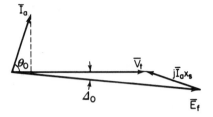

Figure 5-40 Phasor diagram of a synchronous motor at no-load and overexcited.

The first reaction is to cause a momentary drop in speed. In turn this appreciably increases the power angle and thereby causes a phasor voltage difference to exist between $\bar{V}_t$ and the excitation voltage $\bar{E}_f$. The result is the flow of an increased armature current at a very much improved power factor compared with no-load. As a matter of fact the speed changes momentarily by a sufficient amount to allow the power angle to assume that value which enables the armature current and input power factor to assume those values which permit the input power to balance the power required at the load plus the losses. Figure 5-36 is typical of what the final phasor diagram looks like. Note that the power component of the current is considerably increased over the no-load case. The increase in power angle from no-load to the load case can be measured by means of any stroboscopic instrument. The stroboscope furnishes a convincing demonstration of the physical character of the power angle.

The phasor diagram for an overexcited salient pole synchronous motor is depicted in Fig. 5-41. The motor is assumed to be connected to an infinite bus and operating at a leading power factor angle θ. A comparison of this diagram with Fig. 5-18, which illustrates the phasor diagram of an overexcited synchronous generator, reveals a considerable similarity. There are differences, however. One is that overexcitation produces opposite effects regarding the power factor; it is leading for the motor and lagging for the generator. Another is that for the generator the terminal voltage plus the reactance drops yields the excitation voltage whereas for the motor it is the excitation voltage summed with the reactance drops that equates to the terminal voltage.

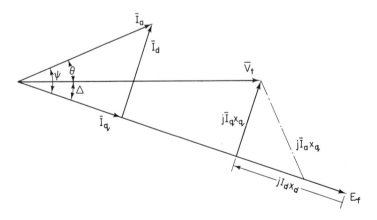

Figure 5-41 Phasor diagram for an overexcited salient pole synchronous motor.

Because the geometry of the phasor diagram is the same for Fig. 5-41 as for Fig. 5-18, the expression for the determination of the angle ψ, which locates the direction along which the excitation phasor lies, is identical to that of Eq. (5-43), which is repeated here for convenience. Thus

$$\psi = \tan^{-1} \frac{V_t \sin \theta + I_a x_q}{V_t \cos \theta}$$

The angle θ is to be used as an absolute quantity because the expression derives from geometrical considerations. Also, although the expressions for the magnitudes of the direct-axis and quadrature-axis currents are the same in both cases, the angles of these complex quantities are reversed in sign. Accordingly, we have

$$\bar{I}_q = (I_a \cos \psi) /\!\!-\Delta^\circ$$
$$I_d = (I_a \sin \psi) /\!90^\circ - \Delta^\circ \qquad (5\text{-}107)$$

where in absolute terms the power angle takes on the value

$$\Delta = \psi - \theta$$

Finally, with $\bar{V}_t$ taking on the role of the applied line voltage, Kirchhoff's voltage law applied on a per phase basis to the stator winding yields

$$\bar{V}_t = \bar{E}_f + j\bar{I}_d x_d + j\bar{I}_q x_q \qquad (5\text{-}108)$$

from which the expression for the excitation voltage readily follows. Hence

$$\bar{E}_f = \bar{V}_t - j\bar{I}_d x_d - j\bar{I}_q x_q \qquad (5\text{-}109)$$

Example 5-7

A 2300-V, three-phase, 60-Hz, Y-connected, cylindrical-rotor synchronous motor has a synchronous reactance of 11 Ω per phase. When it delivers 200 hp the efficiency is found to be 90% exclusive of field loss, and the power angle is 15 electrical degrees as measured by a stroboscope. Neglect ohmic resistance and determine (a) the induced excitation voltage per phase, E_f, (b) the line current I_a, and (c) the power factor.

Solution (a) The power input and the mechanical power developed are the same. Hence

$$3\frac{V_t E_f}{x_s} \sin \Delta = \frac{200(746)}{0.9}$$

Inserting $V_t = 2300/\sqrt{3}$, $\Delta = 15^\circ$, and $x_s = 11$ yields

$$E_f = 1790 \text{ V/phase}$$

(b) The armature current follows directly from Eq. (5-104). Thus

$$\bar{I}_a = \frac{1330/\!0^\circ - 1790/\!-15^\circ}{11/\!90^\circ} = 55.4/\!41.4^\circ$$

Thus since the angle of $\bar{I}_a$ is positive, the power factor is leading.

(c) The line power factor is merely

$$\text{power factor} = \cos 41.4^\circ = 0.755 \text{ leading}$$

Example 5-8

Consider that the synchronous machine of Example 5-4 is now operated as an over-excited synchronous motor drawing rated current at rated voltage from the three-phase line source at a leading pf of 0.8.

(a) Determine the excitation voltage of the synchronous motor at this operating point.

(b) Obtain the value of the mechanical power developed in watts as it derives from the presence of the field excitation.

(c) Find the reluctance power in watts.

(d) Compute the total mechanical power developed and compare it with the input power drawn from the line.

(e) Calculate the developed motor torque.

Solution (a) Angle ψ is found to be the same as the value found for the overexcited generator. Thus

$$\psi = \tan^{-1} \frac{V_t \sin \theta + I_a x_q}{V_t \cos \theta} = \tan^{-1} \frac{\sin 36.9° + 0.577}{0.8} = 55.8°$$

Also,

$$\bar{I}_d = (I_a \sin \psi)\underline{/90° - \Delta°} = 0.827\underline{/71.1°}$$
$$\bar{I}_q = (I_a \cos \psi)\underline{/-\Delta°} = 0.562\underline{/-18.9°}$$

where Δ is found from

$$\Delta = \psi - \theta$$

Finally, from Eq. (5-109) we get

$$
\begin{aligned}
\text{p-u } \bar{E}_f &= \text{p-u } \bar{V}_t - j(\text{p-u } \bar{I}_d)x_d - j(\text{p-u } \bar{I}_q)x_q \\
&= 1\underline{/0°} - (0.827\underline{/71.1°})(0.961\underline{/90°}) - (0.562\underline{/-18.9°})(0.577\underline{/90°}) \\
&= 1 - (-0.752 + j0.257) - (0.105 + j0.307) \\
&= 1.74\underline{/-18.9°}
\end{aligned}
$$

It should not be surprising that this result is identical to that found in Example 5-4 for the overexcited synchronous generator except for the sign of the power angle.

(b) The developed mechanical power that originates from the excitation of the field winding is described by the first term on the right side of Eq. (5-65). Thus, evaluated in terms of per-unit quantities, we get

$$\text{p-u } P_{mf} = \frac{(\text{p-u } V_t)(\text{p-u } E_f)}{\text{p-u } x_d} \sin \Delta = \frac{1(1.74)}{0.961} = \sin 18.9° = 0.5865$$

To obtain this result in watts it is simply necessary to multiply by the base kVA which is the rating of the motor, namely, 15 kVA. Thus,

$$P_{mf} = (0.5865)(15,000) = 8797 \text{ W}$$

(c) The expression for the reluctance power is the second term of Eq. (5-65). Calling this part P_{mr}, we evaluate

$$\text{p-u } P_{mr} = \frac{(\text{p-u } V_t)^2}{2(\text{p-u } x_d)(\text{p-u } x_q)}(\text{p-u } x_d - \text{p-u } x_q) \sin 2\Delta$$

$$= \frac{1^2}{2(0.961)(0.577)} = (0.961 - 0.577) \sin 37.8°$$

$$= 0.2122$$

and expressed in watts this becomes

$$P_{mr} = (0.2122)(15,000) = 3183 \text{ W}$$

(d) The total mechanical power developed by this motor is the sum of the excitation power and the reluctance power. On a per-unit basis this value is

$$\text{p-u } P_m = \text{p-u } P_{mf} + \text{p-u } P_{mr} = 0.5865 + 0.2122 = 0.7987$$

This compares favorably with the per-unit input power, which is

$$\text{p-u } P_i = (\text{p-u kVA}) \cos \theta = 1(\cos 36.9) = 0.8$$

Expressed in watts this input power is

$$P_i = (0.8)(15,000) = 12,000 \text{ W}$$

The corresponding expression derived from the total mechanical power developed is

$$P_m = (0.7987)(15,000) = 11,985 \text{ W}$$

(e) As a six-pole machine operating on a 60-Hz power supply, the synchronous motor shaft speed is

$$n_s = \frac{120(60)}{6} = 1200 \text{ rpm}$$

Or, expressed in radians per second,

$$\omega_s = 1200\left(\frac{\pi}{30}\right) = 125.7 \text{ rad/s}$$

Hence the torque expressed in newton-meters is

$$T = \frac{P_m}{\omega_s} = \frac{11,985}{125.7} = 95.4 \text{ N-m}$$

which in English units is

$$T = (0.7376)95.4 = 70.35 \text{ lb-ft}$$

5-15 POWER-FACTOR CONTROL

For a fixed mechanical power developed (or load) it is possible to adjust the reactive component of the current drawn from the line by varying the d-c field current. This feature is achievable in the synchronous motor precisely because it is a doubly excited machine. Thus, although operation at a constant applied voltage demands a fixed resultant flux, both the d-c source and the a-c source may cooperate in establishing this resultant flux. If the field winding current is made excessively large, then clearly the resultant air-gap voltage in the motor tends to be larger than that demanded by the applied voltage. Accordingly, a reaction occurs which causes the armature current to assume such a power-factor angle that the armature mmf exerts that amount of demagnetizing effect which is needed to restore the required resultant flux. Similarly, if the field is underexcited, then the resultant gap flux tends to be too small. This also creates a phasor voltage difference between the line voltage and the motor excitation voltage which acts to cause the armature current to flow at that power-factor angle which enables it to magnetize the air gap to the extent needed to provide the necessary resultant flux.

Whether the reactive component of current must be leading or lagging readily follows from an investigation of the phasor diagrams depicted in Fig. 5-42 for various

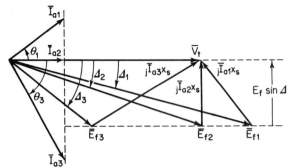

Figure 5-42 Showing the effect of varying excitation on power-factor.

values of excitation voltage. The diagram applies to an assumed constant mechanical power developed. Hence as $\bar{E}_f$ is changed, the sin Δ must change correspondingly to keep E_f sin Δ invariant. In Fig. 5-42 this means that, as the excitation is varied, the locus of the tip of the $\bar{E}_f$ phasor is the broken line. Furthermore, the in-phase component of the current in each case must be the same. In the first case, where the excitation voltage is $\bar{E}_{f1}$, the field current is producing too much flux. This creates a reaction between the motor and the source which calls for a leading current of such a magnitude that it provides an amount and direction of synchronous reactance drop that when added to $\bar{E}_{f1}$ yields the fixed terminal voltage $\bar{V}_t$. Physically, what is happening is that a leading reactive current is made to flow, which acts to demagnetize the flux field to the extent needed. When the excitation is reduced to $\bar{E}_{f2}$, there is no excess flux produced by the field winding. Consequently, the a-c line current contains no reactive component. It merely has the value of the in-phase component needed to supply power to the load. At the excitation corresponding to $\bar{E}_{f3}$ the machine is greatly underexcited. To compensate for this a reaction occurs which allows the line to deliver a large lagging reactive current, which helps to establish the value of air-gap flux demanded by the terminal voltage. Note, too, that as the excitation is decreased the power angle must increase for a fixed mechanical power developed.

If a plot is made of the armature current as the excitation is varied for fixed mechanical power developed, the current is observed to be large for underexcitation and overexcitation and passing through a minimum at some intermediate point. The plot in fact resembles a V shape as demonstrated in Fig. 5-43.

The ability of the synchronous motor to draw leading current when overexcited can be used to improve the power factor at the input lines to an industrial establishment that makes heavy use of induction motors and other equipment drawing

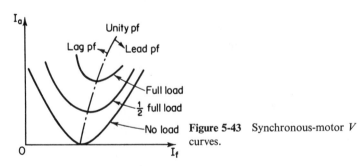

Figure 5-43 Synchronous-motor V curves.

power at a lagging power factor. Many electric power companies charge increased power rates when power is bought at a poor lagging power factor. Over the years such increased power rates can result in an appreciable expenditure of money. In such instances the installation of a synchronous motor operated at overexcitation can more than pay for itself by improving the overall input power factor to the point that the penalty clause no longer applies.

5-16 SYNCHRONOUS MOTOR TRANSIENTS CAUSED BY LOAD CHANGES

A three-phase synchronous machine that has its stator winding synchronized to a three-phase bus sytem can be made to operate either as a generator or a motor. If a primary source of power, such as a steam turbine, is connected to the shaft of the synchronous machine and mechanical power is derived from the turbine, then the synchronous machine performs as a synchronous generator. A description of the transient behavior of this mode of operation for changes in mechanical input is the subject matter of Sec. 5-10. On the other hand, if a passive mechanical load is attached to the motor shaft in the manner illustrated in Fig. 5-44, the synchronous machine performs as a motor that draws electrical power from the three-phase bus system and delivers it in mechanical form to the load attached to the motor shaft. Our interest here is to examine the nature of the transient response that results as the power requirements at the load change.

Figure 5-44 Synchronous machine operating in the motor mode.

Our experience with the analysis of linear systems in the past has surely taught us that dynamical behavior is determined solely by the system parameters. We have reason to suspect therefore that the form of the describing differential equation for motor action should not differ from result that was previously obtained to describe the transients associated with generator action. If the machine is the same and the attached shaft inertia is unchanged, then the resulting transient behavior for motor action exactly corresponds to that for generator action. To demonstrate that this is in fact the case, let us examine the specific conditions that prevail for synchronous motor operation and proceed to derive the differential equation that describes its behavior. It is helpful in this connection to keep in mind that the power angle for motor action is negative, i.e., the field excitation voltage lags the bus terminal voltage as illustrated in Fig. 5-36. The total displacement angle θ of the rotor structure relative to a fixed external point expressed in electrical degrees is

$$\theta = \omega_s t - \Delta \tag{5-110}$$

where ω_s denotes the synchronous angular velocity of the bus system. A comparison of this expression with Eq. (5-87) for the generator shows that the difference lies in the sign attached to the power angle. It is positive for the generator and negative for the motor. Now, if the demand for power by the attached mechanical load is increased, a momentary *drop* in speed occurs in order to allow the power angle to increase so that at constant excitation the phasor voltage difference between $\bar{V}_t$ and $\bar{E}_f$ is sufficiently augmented to allow a new current with a larger in-phase component to flow. This in turn represents an increased electrical input power which serves to meet the increased demands at the load shaft. The corresponding change in the power angle needed to accommodate the increased shaft power is expressed mathematically as

$$\text{change in } \Delta = \int_0^{t_1} (\omega_s - \omega)\, dt \tag{5-111}$$

where t_1 is the time needed to provide the change in power angle that provides energy balance between the electrical system and the mechanical load. The angular velocity of the rotor of the synchronous motor during this transient state can be expressed as

$$\omega = \frac{d\theta}{dt} = \omega_s - \frac{d\Delta}{dt} \tag{5-112}$$

which is simply the derivative of Eq. (5-110). The corresponding acceleration of the shaft inertia associated with this change in angular velocity is clearly obtained by differentiating a second time. Thus,

$$\frac{d^2\theta}{dt^2} = -\frac{d^2\Delta}{dt^2} \tag{5-113}$$

Application of the law of conservation of energy to the system of Fig. 5-44 allows us to write

$$\text{motor developed torque} = \Sigma \text{ opposing torques}$$

This equation applies at all times including the transient state. Proceeding on the assumption that the synchronous motor is equipped with a squirrel-cage winding, the last equation may be written in some detail as

$$k_s \sin \Delta = J\alpha_m + T_d + T_L \tag{5-114}$$

where k_s is the synchronizing torque coefficient as defined by Eq. (5-98), α_m is the mechanical acceleration of the attached shaft inertia, T_d is the counter torque exerted by the squirrel-cage winding, and T_L denotes the new, increased shaft-applied load torque.

Because α_m implies mechanical degrees and θ in Eq. (5-113) involves electrical degrees, it follows that an adjustment is needed which includes the number of poles of the synchronous motor. Thus,

$$\alpha_m = \frac{2}{p}\frac{d^2\theta_m}{dt^2} = -\frac{2}{p}\left(\frac{d^2\Delta}{dt^2}\right) \tag{5-115}$$

Moreover, as pointed out in the transient analysis of generator action, the damping torque of the squirrel-cage winding is related to the rate of change of the power angle with time. However, the difference for motor action is that the power angle is negative, so we can write

$$T_d = k_d\left(-\frac{d\Delta}{dt}\right) \tag{5-116}$$

where k_d has the same definition as for the generator mode, namely, it expresses the number of newton-meters of torque developed by the squirrel-cage winding per electrical degree change in power angle per second. Upon inserting Eqs. (5-115) and (5-116) into Eq. (5-114), there results

$$k_s \sin \Delta = -\frac{2}{p} J \frac{d^2\Delta}{dt^2} - k_d \frac{d\Delta}{dt} + T_L \qquad (5\text{-}117)$$

Or, by rearranging and linearizing through replacement of sin Δ by Δ, we get

$$k_J \ddot{\Delta} + k_d \dot{\Delta} + k_s \Delta = T_L \qquad (5\text{-}118)$$

where k_J has the same identity as Eq. (5-98).

A comparison of Eq. (5-118) with Eq. (5-99) for the generator mode of operation of the synchronous machine reveals the expressions to be identical provided that T_L in the case of the motor is made to take the role of the input shaft torque T_{sh} in the case of generator operation. Since the characteristic equation is now the same as for the motor, the resulting description of the dynamical behavior of the synchronous motor to load changes is identical to that described in Example 5-6. The sole adjustment that is required is in the sign of the power angle. A plot similar to that shown in Fig. 5-34 would be marked with negative angles rather than positive ones.

A final comment is in order. If the effective inertia associated with motor action driving a particular load is appreciably different than the inertia of the attached prime mover rotor in generator action, then clearly coefficient k_J will be different. In turn, this changes the coefficient of the first derivative in the normalized characteristic equation as well as the constant term. The resulting dynamic behavior can accordingly be significantly different.

5-17 SYNCHRONOUS MOTOR APPLICATIONS

Synchronous motors are rarely used below 50 hp in the medium-speed range because of their much higher initial cost compared with induction motors. In addition they require a d-c excitation source, and the starting and control devices are usually more expensive—especially where automatic operation is required. However, synchronous motors do offer some very definite advantages. These include constant-speed operation, power-factor control, and high operating efficiency. Furthermore, there is a horsepower and speed range at which the disadvantage of higher initial cost vanishes, even putting the synchronous motor at advantage. This is demonstrated in Fig. 5-45. Where low speeds and high horsepower are involved, the induction motor is no longer cheaper because it must use large amounts of iron in order not to exceed a gap flux density of 0.7 T. In the synchronous machine, on the other hand, a value twice this figure is permissible because of the separate excitation.

Appearing in Table 5-1 are some of the more important characteristics of the synchronous motor along with some typical applications. In addition to the area of application listed there, these motors are also quite prominently found in the following applications, which are characterized by operation at low speeds and high horsepower: large, low-head pumps; flour-mill line shafts; rubber mills and mixers; crushers; shippers; pulp grinders and jordans and refiners used in the papermaking industry.

Figure 5-45 Indicating the general areas of application of synchronous and induction motors.

TABLE 5-1 SYNCHRONOUS MOTOR CHARACTERISTICS AND APPLICATIONS

Type designation	Synchronous, high speed, above 500 rpm	Synchronous, low speed, below 500 rpm
Starting torque (% of normal)	Up to 120	Low 40
Pull-in torque (% of normal)	100–125	30
Pull-out torque	Up to 200	Up to 180
Starting current	500–700	200–350
Slip	Zero	Zero
Power factor	High, but varies with load and with excitation	High, but varies with excitation
Efficiency, %	Highest of all motors, 92–96	Highest of all motors, 92–96
Typical applications	Fans, blowers, d-c generators, line shafts, centrifugal pumps and compressors, reciprocating pumps and compressors. Useful for power-factor correction. Constant speed. Frequency changers	Lower-speed direct-connected loads such as reciprocating compressors when started unloaded, d-c generators, rolling mills, band mills, ball mills, pumps. Useful for power-factor control. Constant speed. Flywheel used for pulsating loads

Source: By permission from M. Liwschitz-Garik and C. C. Whipple, *Electric Machinery*, vol. II (Princeton, N.J.: D. Van Nostrand Co., Inc., 1946).

5-18 *FIELD EXCITATION BY RECTIFIED SOURCES*

A distinguishing feature of the conventional synchronous motor is its need for d-c as well as an a-c electrical supply. Because the a-c source is the one that is frequently available, special provision must be made for a d-c source. In years past, one solution was to equip the synchronous motor with an auxiliary self-excited d-c generator attached to the motor shaft. Once the synchronous motor reached nearly synchronous speed through the induction-motor action of its squirrel-cage winding, the d-c generator field rheostat was adjusted to furnish the proper excitation for the synchronous motor, which in turn pulled the synchronous motor into synchronism with the line voltage. In applications of large synchronous motors today, the exciter generator is conveniently replaced by a d-c source obtained by electronic methods employing diodes and silicon-controlled rectifiers (SCRs), which are fully capable of handling a large range of currents over a wide range of voltages.

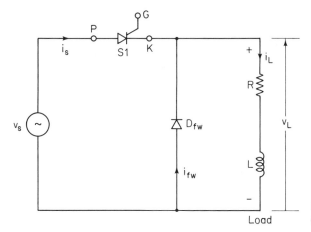

Figure 5-46 Single-phase, half-wave controlled rectifier circuit.

A typical controlled electronic rectifier circuit is shown in Fig. 5-46. This circuit is a single-phase, half-wave controlled rectifier which can be used to adjust the field current in the field winding of a synchronous motor. The field winding is represented by the resistance R and the inductance L and is called the load. Two electronic elements are used: the diode D_{fw} that serves to shunt the load and the silicon-controlled rectifier S1 which behaves as a controllable switch. Of course the source voltage is the frequently available a-c supply. The manner in which this circuit performs the conversion from a fixed a-c supply to an adjustable d-c source can be understood by referring to Fig. 5-47. It is helpful to recall that the SCR conducts when two conditions are satisfied at its three terminals. The first is that the voltage from plate (P) to cathode (K) be positive. The second is that the gate-cathode voltage be set to allow conduction to take place during the time the plate-cathode voltage is positive. By proper adjustment of this gate voltage it is possible to permit conduction to occur over the entire time the plate-cathode voltage is positive or during no part of this time. Figure 5-47(a) shows two cycles of the a-c source voltage. In Fig. 5-47(b) the rectified version of the a-c source is shown. Observe that a positive potential appears across S1 (the SCR) only during the positive portion of the a-c input voltage variation. During the negative portion of the cycle, a reversed (negative) voltage appears across the SCR and so conduction is not possible. This accounts for the zero level of voltage from $\omega t = \pi$ to $\omega t = 2\pi$. In the first half-cycle, however, S1 is made to conduct at the angle θ, which is determined by the voltage on the gate terminal. This angle is often referred to as the *firing angle* of the SCR and is adjustable between 0 and π

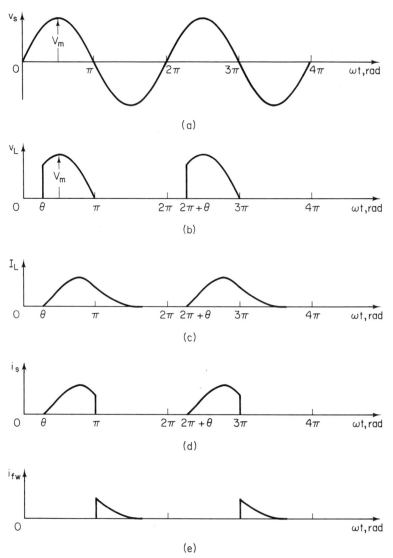

Figure 5-47 Time variations of the electrical quantities appearing in Fig. 5-46: (a) source voltage; (b) rectified voltage across the R-L load; (c) the R-L load current variation; (d) source current; (e) current in the free-wheeling shunt diode. The firing angle of the SCR (S1) is θ.

radians in this case. When firing of the thyristor (SCR) is made to occur at $\theta = 0°$, the voltage developed across the load is a maximum. The associated average value of this voltage can easily be shown to be V_m/π where V_m denotes the peak value of the a-c supply voltage.

The variation of the load current produced by the rectified voltage of Fig. 5-47(b) is depicted in Fig. 5-47(c). It is particularly interesting here to note that action of the inductance causes the current to exist beyond the time when the thyristor ceases to conduct. Since the current at $\omega t = \pi$ is not zero and there exists an amount of energy stored in the magnetic field of the inductor consistent with this nonzero current, a path must be provided to permit this current to continue to flow. As soon as the SCR cuts off at $\omega t = \pi$, the load current tries to change instantaneously. Then, by Lenz's law, an internally generated emf is induced in the inductor of a polarity that acts to sustain the original direction of the current. This action puts a positive potential across the shunt diode, D_{fw}, which makes it conductive thus providing a path for the load current. Current decay subsequently proceeds at a rate determined by the field winding time constant. The shunt diode is called a *free-wheeling* diode for the reason that it freely responds to conditions that call for an auxiliary current path whenever inductive elements are involved.

The variations with time of the source current and the current in the free-wheeling diode are depicted in Figs. 5-47(d) and (e) respectively. The abrupt change in the source current at $\omega t = \pi$ is allowed on the assumption that the a-c source has negligible internal inductance, which is reasonable. The magnitude of the step current that appears in i_{fw} of course matches the cutoff value that is found in the variation of i_s.

The half-wave controlled rectifer has the disadvantage of relatively low currents because of the absence of any contributions from the source when its potential is undergoing a negative excursion. This situation, however, is nicely resolved by employing the circuitry illustrated in Fig. 5-48, which is called a full-wave controlled rectifier. By

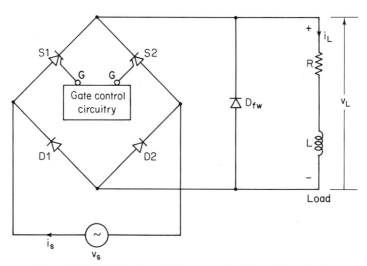

Figure 5-48   Single-phase, full-wave controlled rectifier circuit.

adding an additional SCR and diode along with appropriate modifications in the gate control circuitry, conduction to the load can be achieved in a controlled fashion during each successive half-cycle of the source frequency. For an explanation of the operation of this circuit refer to Fig. 5-49. During the first positive half-cycle of the source voltage, a positive potential appears across the plate-cathode terminals of thyristor S1. In Fig. 5-48 the left side of the source is placed positive while the cathode of D2 is

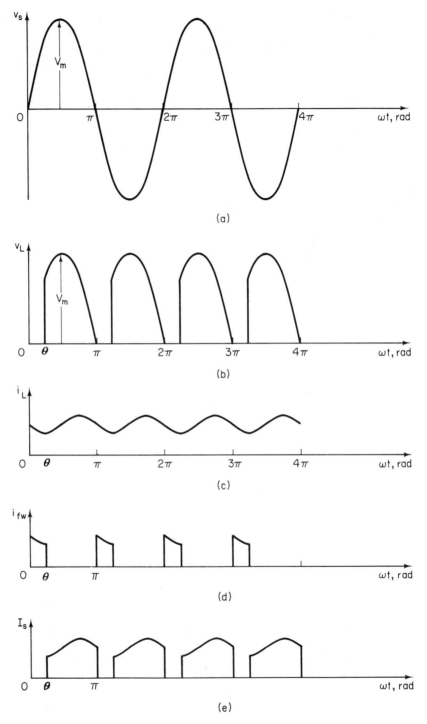

Figure 5-49 Time variations of the electrical quantities found in Fig. 5-48: (a) source voltage; (b) rectified load voltage; (c) resultant load current variation in steady state; (d) current in the free-wheeling diode; (e) the source current. The SCR firing angle is set to θ.

placed negative. When the gate terminal is pulsed at θ, S1 conducts and thereby applies the source voltage to the R-L load by completing the circuit through diode D2. In other words, when the source voltage is positive, there exists a series path connecting the source to the load through thyristor S1 and diode D2. As soon as the source goes negative just beyond π radians in Fig. 5-49(a), S1 goes off as does D2. But now the right side of the source in Fig. 5-48 is positive and the left side is negative, which means that a positive potential appears across both S2 and D1. Again a closed path becomes established between the source and the load upon the arrival of the gate pulse at $\omega t = \pi + \theta$, but now it is thyristor S2 and diode D1 that are effective in providing the connection. Observe too that the sequencing of the switching operation is such as to maintain the polarity of the voltage across the load the same as it was in the preceding half-cycle. See Fig. 5-49(b). This ensures a unidirectional current flow. Because a contribution occurs for each half-cycle of the source voltage, the average voltage that is available to the load doubles over that of the half-wave rectifier to a maximum value of $(2/\pi)V_m$. Moreover, control is available from zero to this maximum through the action of the gate control circuit, which varies θ from π radians to zero.

The corresponding variation of the current in the load is depicted in Fig. 5-49(c). This graph shows the variation after steady state is reached following a suitable transient response. Note the appreciable improvement over the half-wave circuit. A glance at Fig. 5-49(d) reveals the essential contribution of the free-wheeling diodes which is to provide a current path over the time interval between the start of any half-cycle of the source voltage and the point of firing of the thyristors. Figure 5-49(e) illustrates the variations of the source current during the time that the SCRs are in the ON state. These current pulses are supplied alternately through thyristors S1 and S2 to the load from the source.

The foregoing analysis illustrates the improvement that occurs when a full-wave rectifier circuit is used in place of a half-wave rectifier circuit. However, a further improvement is possible when a *three-phase* full-wave arrangement is used in place of the *single-phase* full-wave circuit. Such a circuit is depicted in Fig. 5-50. It is useful to note that a total of six thyristors (SCRs) is needed with this arrangement. The availability of a rectifier circuit that uses a three-phase source is attractive for the

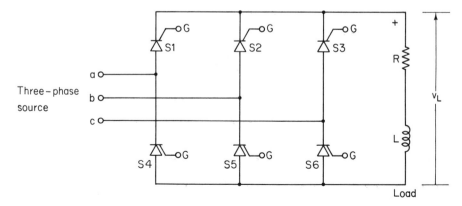

Figure 5-50 Three-phase, full-wave controlled rectifier circuit.

reason that the three-phase source is already needed to drive the three-phase synchronous motor and, accordingly, is assumed to be available. By the proper sequential pulsing of the gate terminals of the thyristors in the circuitry of Fig. 5-50 it is possible to vary the average voltage across the field winding over a range that varies from zero to $(3/\pi)V_m$, where V_m denotes the maximum value of the a-c voltage between lines.

The success of this circuit depends upon a biasing scheme that permits only two thyristors to be in a conducting state at any time while simultaneously ensuring that the remaining SCRs are in a state of reversed bias. A typical switching sequence for this circuit is illustrated in Fig. 5-51 as it applies to a line voltage sequence of a, b, c. Each thyristor is scheduled for a maximum ON period of 120 electrical degrees for each cycle of the source voltage. The gating pulse is designed to allow a full variation over this interval. Starting with the begining of a cycle, S1 is forward-biased to be in the ON state potentially over the first 120 electrical degrees while S2 and S3 are scheduled potentially to be in the ON state for each of the next 120° intervals respectively. Then, beginning with the start of the next half-cycle at π radians, a similar schedule is established for thyristors S4, S5, and S6. Accordingly, S4 is arranged to have a total possible ON period of 120° extending from π to $\frac{5}{3}\pi$ radians. Similarly, S5 is forward-biased for a possible ON period extending from $\frac{5}{3}\pi$ radians right on into the first $\pi/3$ radians of the next cycle. As a matter of convenience this second 60° portion is shown reflected into the first cycle in Fig. 5-51. Thyristor S6 is subsequently scheduled for a potential ON period covering the next 120 electrical degrees. Reflected back into the first cycle, S6 is scheduled to occupy the period from $\pi/3$ to π radians.

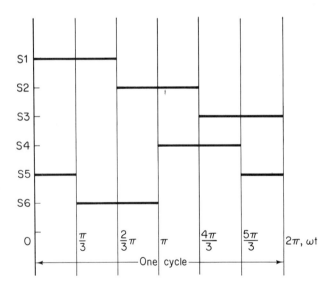

Figure 5-51 Switching sequence diagram for the thyristors of Fig. 5-50. The solid lines indicate the specific portions of a cycle when each thyristor is forward biased for conduction to occur subject to the action of the gate signal.

As a result of the switching sequence schedule depicted in Fig. 5-51 for the line voltage sequence a, b, c, it follows that both thyristors S1 and S2 are forward-biased for conduction during the first $\pi/3$ radians of a cycle that begins with v_{ab} at $(\sqrt{3}/2)V_m$ and moving to its peak while v_{ac} is simultaneously passing through zero and moving positively. The specific point in this interval at which conduction actually occurs is determined by the gate pulsing signals. Similarly, in the next 60° interval, it is thyristors S1 and S6 that are involved in the conduction process. In the third 60° interval the responsibility for conduction passes over to thyristors S2 and S6. This process continues for the remainder of the cycle and then repeats itself for each cycle. The result is an effective control of the field winding d-c current at entirely acceptable levels.

Brushless Excitation for Synchronous Generators

Synchronous generators employed by the public utility companies can have ratings of 1000 MVA and more (where MVA denotes megavolt-amperes). To equip such machines with d-c power to be conducted to the rotating field winding through slip rings presents a significant maintenance problem as well as the loss of a substantial amount of power consumed as a heat loss in the carbon brushes used to conduct the large d-c currents to the slip rings. An innovative application of the use of the circuitry of Fig. 5-50 can be used to eliminate these problems. The arrangement is illustrated in Fig. 5-52.

Figure 5-52 Schematic diagram illustrating how brushless excitation of a large synchronous generator is obtained.

The object of this scheme is to make available a d-c source which is revolving at the same speed as the field winding of the main generator. A direct connection to the terminals of the field winding is then possible thereby eliminating the need for brushes. Attached to the shaft of the main generator is an exciter generator the size of which is sufficient to provide the power required by the field winding to maintain the level of d-c current needed by the synchronous generator. Also attached to the shaft is the rectifier of Fig. 5-50. A small d-c current (in comparison to that required by the synchronous generator) is supplied to the field winding of the exciter generator. An auxiliary d-c source derived from a static rectifier can be used because the currents involved here are manageable. When the prime mover of the synchronous generator reaches synchronous speed, the a-c exciter generator develops a balanced three-phase voltage. Application of this voltage set to the three-phase full-wave rectifier attached to the generator shaft produces a d-c voltage source at terminals F1–F2 which revolves at the same velocity as the generator field winding itself. Hence, a firm, terminal connection is possible and slip rings and brushes for the main generator are no longer necessary.

PROBLEMS

5-1. A synchronous generator has a short-circuit characteristic such that rated armature current is obtained by 0.75 per-unit excitation. Rated voltage on the air-gap line is obtained by 1.25 per-unit excitation. What per-unit excitation is required to supply a load at rated voltage, rated kVA, and 80% pf lagging?

5-2. Repeat Prob. 5-1 for the case where the load power factor is 80% leading.

5-3. A synchronous generator has the following rating:

$$\text{180-kVA, 440-V, 300-rpm, 60-Hz, three-phase, Y-connection,}$$

$$r_l \approx 0 \quad \text{and} \quad x_l = 0.296 \, \Omega$$

The air-gap line is described by the equation $E = 17I_f$, expressed per phase. The short-circuit characteristic is described by $I_{sc} = 10.75I_f$. Compute the field current needed to provide rated voltage when rated current is delivered to a load at a pf of 0.8 *leading*. Use the general method linear analysis. Assume that $I_A \simeq 17.9$ equivalent field amperes.

5-4. Repeat Prob. 5-3 for the case where the pf of the load is 0.8 lagging.

5-5. When the machine of Prob. 5-3 delivers rated armature current to a unity power factor load, the field winding current is measured to be 30 A. Moreover, the armature current is found to be lagging the excitation voltage by 50 electrical degrees.
(a) Compute the resultant voltage induced in the armature winding by the air-gap flux.
(b) What is the value of the terminal voltage per phase?
(c) Find the power delivered to the load.

5-6. Repeat Prob. 5-5 for the case where the pf of the load is 0.8 lagging.

5-7. A 5-kVA, 220-V, 60-Hz, six-pole, Y-connected synchronous generator has a three-phase winding with 135 effective turns per phase. Moreover, its field winding is equipped with 200 turns. The leakage reactance per phase is 0.78 Ω and the armature winding resistance is negligible.
(a) Compute the field current required to establish rated voltage across the terminals of a unity power factor load that draws rated generator armature current.
(b) Determine the field current needed to provide rated terminal voltage to a load that draws 125% of rated current at 0.8 lagging pf.

5-8. A 5-kVA, 220-V, six-pole, 60-Hz, Y-connected synchronous generator has a three-phase winding with 135 effective turns per phase and an armature winding leakage reactance of 0.8 Ω per phase. The equation for the air-gap line is $E = 25I_f$, expressed in induced volts per phase. The generator has 200 field turns. When this machine delivers rated armature current to its connected load, like poles of the armature mmf are known to lag behind its field poles by 150 electrical degrees. Assume negligible saturation.
(a) Find the value of the terminal voltage per phase.

(b) What is the pf of the load?

(c) How much power does the load take?

5-9. The machine of Prob. 5-3 has a cylindrical rotor.

(a) Find the value of the unsaturated synchronous reactance.

(b) What is the value of the armature reaction reactance?

(c) By means of the synchronous reactance method find the field current needed to yield rated terminal voltage at rated current for a 0.8 lagging pf load.

(d) Prepare a carefully drawn phasor diagram.

5-10. A three-phase, 15-kVA, 220-V, Y-connected, six-pole synchronous generator has a cylindrical rotor and negligible armature resistance. The open-circuit, short-circuit, and full-load zero pf characteristics are shown in Fig. P5-10.

Figure P5-10

(a) Determine the armature leakage reactance per phase.

(b) Compute the voltage regulation at unity pf by the general method.

(c) Find the voltage regulation at pf $= 1$ by the use of the saturated synchronous reactance.

(d) Compare the results of parts (b) and (c) and comment.

5-11. The number of field turns per pole for the machine of Prob. 5-10 is 400. Determine the number of effective armature turns per pole.

5-12. At no-load, the field current of the machine of Prob. 5-10 is adjusted to 5.8 A and kept at this value. A resistive load is then applied to the generator until rated current flows through the armature winding. At this condition the angle between the no-load excitation voltage E_f and the full-load terminal voltage is measured to be 46 electrical degrees. Find the value of the terminal voltage per phase by the general method.

5-13. A cylindrical pole synchronous generator has the following rating:

$$200\text{-kVA},\ 440\text{-V},\ 300\text{-rpm},\ 60\text{-Hz, three-phase, Y-connection, } r_a \approx 0$$

The open-circuit and rated current zero pf characteristics are shown in Fig. P5-13.

Figure P5-13

Compute the field current needed to provide rated voltage when rated current is delivered to a load at a pf of 0.8 lagging. Use the general method.

5-14. A 40,000-kVA, 14,000-V, three-phase, Y-connected alternator has negligible armature resistance and a leakage reactance of 1 Ω per phase. Other pertinent data are as follows:

Short-circuit characteristic:	$I_a = 7I_f$
Air-gap line, volts per phase:	$E = 33I_f$
Open-circuit characteristic volts per phase:	$E = \dfrac{21,300I_f}{430 + I_f}$

The equation for the open-circuit characteristic is not valid for values of I_f about the origin. For a constant terminal voltage of 14,000 V (line) compute the variation of field current from no-load to full-load kVA at 80% lagging pf. Use the general method for nonlinear analysis.

5-15. A 2100-kVA, 6600-V, Y-connected, 25-Hz, three-phase alternator is operated at synchronous speed with the armature short-circuited. The excitation is adjusted to produce full-load current in the armature. Without changing the field current, the armature circuit is opened and the terminal voltage is found to rise to 2900 V. Armature resistance is negligible.
   (a) Determine the voltage regulation for a 0.8 lagging pf by the synchronous reactance method.
   (b) Repeat for a pf of 0.8 leading.

5-16. A three-phase, Y-connected synchronous generator is rated at 200 kVA and 2300 V. It is supplying rated current at rated terminal voltage to a 0.8 lagging pf load. When the load is completely disconnected (without any change of field excitation), the terminal voltage rises to 3800 V.
   (a) Draw a phasor diagram, illustrating the foregoing load condition. Mark each phasor by its symbol and its numerical value.
   (b) Calculate the synchronous reactance of this generator. Assume r_a negligible.

5-17. A 50-kVA, 550-V, 25-Hz, single-phase alternator has an effective resistance of 0.25 Ω and a synchronous reactance of 3.2 Ω. Determine the percent regulation for rated load at 0.8 lagging pf.

5-18. A cylindrical-pole synchronous generator supplies rated kVA at a leading pf of 80%. Its per-unit synchronous reactance is 1.0, and the armature resistance may be assumed negligible. The base impedance for such a machine is defined as the ratio of the rated phase voltage to rated phase current. Compute the per-unit value of the excitation voltage and the associated power angle.

5-19. A synchronous generator operates at rated voltage and rated current supplying power to a 0.866 leading pf load. The per-unit value of excitation voltage is 1.732. The base voltage is the rated value of the phase voltage. Armature resistance is negligible. Find the per-unit value of the synchronous reactance.

5-20. An 1800-rpm, 60-Hz, 50,000-kVA, 13,800-V, synchronous generator has the following test data available:

Open-circuit test:

| I_f (A): | 175 | 200 | 225 | 250 | 275 |
| V (line to line) kV: | 12 | 13.0 | 13.8 | 14.5 | 15.1 |

Short-circuit test:
   $I_f = 193$ A, $I_a = 2090$ A

Zero pf characteristic:
$$I_f = 468 \text{ A}, \qquad I_a = 2090 \text{ A}, \qquad V = 13.8 \text{ kV}$$
Compute the field current needed to maintain rated terminal voltage for rated kVA at 80% pf.

5-21. A 5-kVA, 220-V, Y-connected, three-phase, salient-pole synchronous generator is used to supply power to a unity pf load. By means of a slip test, the direct-axis synchronous reactance is found to be 12 Ω and the quadrature-axis synchronous reactance 7 Ω. Assume that rated current is delivered to the load at rated voltage and that armature resistance is negligible.
    (a) Compute the corresponding value of the excitation voltage.
    (b) What is the value of the internal phase displacement angle ψ?
    (c) What is the value of the power angle?
    (d) Draw a carefully labeled phasor diagram.

5-22. Repeat Prob. 5-21 for the case where the load pf is 0.8 lagging.

5-23. Repeat Prob. 5-21 for the case where the load pf is 0.8 leading.

5-24. Assume that the characteristics depicted in Fig. P5-10 apply to the machine of Prob. 5-21. Moreover, assume that rated kVA is delivered to a unity pf load at rated voltage.
    (a) By the general method for salient-pole machines compute the value of excitation voltage E_f and compare with the result of Prob. 5-21(a).
    (b) What is the field current?
    (c) Draw a carefully labeled phasor diagram for this load condition.

5-25. Repeat Prob. 5-24 for the case where the load pf is 0.8 leading.

5-26. Repeat Prob. 5-24 for the case where the load pf is 0.8 lagging.

5-27. A six-pole, 60-Hz, Y-connected synchronous generator is rated at 15-kVA, 220-V, three-phase. A slip test is performed at a fixed reduced voltage of 96 V (line to line). The maximum and minimum readings of the ammeter used in the test are 20 A and 12 A, respectively. Compute the value of the excitation voltage when this machine delivers rated current to a 0.8 lagging pf load. Neglect armature resistance.

5-28. What physical significance can be attached to the synchronous reactances x_d and x_q as they relate to the salient-pole synchronous machine? Which reactance is larger and why?

5-29. A three-phase, four-pole, 60-Hz cylindrical-rotor synchronous machine is rated at 10,000 kW, 5000 V, and 0.8 pf lead. The synchronous reactance is 0.8 p-u. The normal open-circuit field current is 200 A. Neglect saturation and armature resistance. Assume a constant field excitation. Calculate each of the following when the machine is taking rated power from the line at rated current and voltage: **(a)** reactive volt-amperes, **(b)** field current, and **(c)** power angle.

5-30. Answer whether or not the following statements are true or false. When the statement is false, provide the correct answer. Assume a Y-connection and 0.8 pf lead.
    (a) In a synchronous machine the relative position of the field mmf axis and the armature mmf axis is solely dependent upon the power factor of the load.
    (b) The load on a synchronous motor is increased. This causes the speed of the armature mmf to drop to a lower value at steady state so that more input current can flow to supply the need for increased shaft power.

5-31. Refer to the schematic diagram of Fig. 5-23. Describe the reactions that occur during the synchronizing procedure when the speeds of the two generators are identical and the switch is closed at an instant when the voltages of the two generators are not exactly equal.

5-32. Repeat Prob. 5-31 for the case where the voltages are exactly the same but the frequencies of the two generators are slightly different.

5-33. Two generators of equal rating are operating in parallel supplying power to a lagging pf load for a condition as depicted in Fig. P5-33. Describe the adjustments to be made so that each generator is made to operate at the same power factor of the load. Is there any advantage to the new operating condition? Explain.

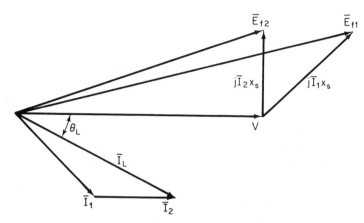

Figure P5-33

5-34. For the situation depicted in Fig. P5-33, describe the adjustments that must be made in order to permit both generators to operate at unity power factor when the load is unequally shared. What happens to the power angles of the individual generators?

5-35. A three-phase, 1732-V (line to line), Y-connected, cylindrical-pole synchronous motor has $r_a = 0$ and $x_s = 10\ \Omega$ per phase. The friction and windage plus the core losses amount to 9 kW. The motor delivers an output of 390 hp. The greatest excitation voltage that may be obtained is 2500 V per phase.
  (a) Calculate the magnitude and pf of the armature current for maximum excitation at the specified load.
  (b) Compute the smallest excitation for which the motor will remain in synchronism for the given power output.

5-36. A 2300-V, three-phase, 60-Hz, Y-connected, cylindrical-rotor synchronous motor has a synchronous reactance of 11 Ω per phase. When it delivers 200 hp, the efficiency is 90% and the power angle is 15 electrical degrees. Neglect r_a and determine: **(a)** the excitation emf per phase; **(b)** the line current; **(c)** the power factor.

5-37. A synchronous motor is operated at half-load. An increase in its field current causes a decrease in armature current. Does the armature current lead or lag the terminal voltage? Explain.

5-38. A three-phase, Y-connected synchronous motor is operating at 80% leading pf. The synchronous reactance is 2.9 Ω per phase and the armature winding resistance is negligible. The armature current is 20 A per phase. The applied line voltage is 440 V.
  (a) Find the excitation voltage and the power angle.
  (b) It is desired to increase the armature current to 40 A per phase and maintain the pf at 80% leading. Show clearly in a phasor diagram how the field excitation must be changed. Can this be accomplished if the shaft load remains fixed? Explain.

5-39. A Y-connected, three-phase, 60-Hz, 13,500-V synchronous motor has an armature resistance of 1.52 Ω per phase and a synchronous reactance of 37.4 Ω per phase. When the motor delivers 2000 hp, the efficiency is 96% and the field current is so adjusted that the motor takes a leading current of 85 A.
  (a) At what pf is the motor operating?
  (b) Calculate the excitation emf.
  (c) Find the mechanical power developed.
  (d) If the load is removed, describe (do not calculate) how the magnitude and pf of the resulting armature current compare with the original.

5-40. A 1250-hp, three-phase, cylindrical-pole, synchronous motor receives constant power of 800 kW at 11,000 V. Armature winding resistance is negligible and the synchronous reactance per phase is 50 Ω. The armature is Y-connected. The motor has a rated full-load current of 52 A. If the armature current shall not exceed 135% of this value, determine the range over which the excitation emf can be varied through adjustment of the field current.

5-41. A synchronous machine is supplied from a constant-voltage source. At no-load the motor armature current is found to be negligible when the excitation is 1.0 per unit. The per-unit motor constants are $x_d = 1.0$ and $x_q = 0.6$.
  (a) If the machine loses synchronism when the angle between the quadrature axis and the terminal voltage phasor direction is 60 electrical degrees, what is the per-unit excitation at pullout?
  (b) What is the load on the machine at pullout? Assume the same excitations as part (a).

5-42. A synchronous machine operating at rated voltage and rated current shows a leading pf of 86.6%. The per-unit excitation voltage is 1.732. The per-unit synchronous reactance is 1.00.
  (a) Does this condition represent motor or generator operation? Explain.
  (b) Compute the power angle.

5-43. As a synchronous motor is loaded from zero to full-load, can the pf ever become leading if the field excitation is maintained at 75%? Explain.

5-44. A synchronous motor delivers rated power to a load. Is there a lower limit on the excitation current beyond which it may not be reduced? Explain.

5-45. A 440-V, three-phase, Y-connected synchronous motor has a synchronous reactance of 6.06 Ω per phase. The armature resistance is negligible, and the induced excitation emf per phase is 200 V. Moreover, the power angle between $\bar{V}_t$ and $\bar{E}_f$ is 36.4 electrical degrees.
  (a) Calculate the line current and pf.
  (b) What values of excitation emf and power angle are necessary to make the pf unity for the same input?

5-46. A synchronous motor at no-load is connected to an infinite bus system. The field circuit is accidentally opened. Explain what happens to the armature current.

5-47. An eight-pole synchronous motor draws 45 kW from the 208-V, 60-Hz, three-phase power system at a pf of 0.8 lagging. The motor is Y-connected and has a synchronous reactance of 0.6 Ω per phase. Armature resistance is negligible. Without any further manipulations on the motor, what is the highest possible value of its steady-state torque?

5-48. Answer whether the following statements are true or false. When the statement is false, describe why.

(a) Increasing the air-gap of a machine reduces the synchronous reactance and raises the steady-state power limit of the machine.

(b) For a synchronous motor, the developed electromagnetic torque is in a direction opposite to the direction of rotation.

(c) In a synchronous machine, the space-phase relation between the field mmf wave and armature-reaction-mmf wave is determined by the pf of the load.

(d) The sum of the armature mmfs in the three phases of the stator winding is zero at any given instant.

(e) The sum of the main field flux Φ_f and the armature reaction flux Φ_{ar} always add to give the resultant flux Φ_r.

(f) A synchronous motor operating at leading pf is underexcited.

(g) The high currents flowing during a short circuit cause a synchronous machine to operate under saturated conditions.

(h) The load on a synchronous motor is increased. This causes the main field mmf axis to drop further behind the air-gap mmf axis, so as to increase the torque angle.

(i) An overexcited synchronous motor is a generator of lagging kvars. This means that it supplies lagging kvars to the system to which it is connected.

(j) For large machines the armature resistance and leakage reactance are about 10% on the machine rating and the synchronous reactance is about 100%.

(k) The equivalent circuit developed for the synchronous machine assumes that saturation is negligible and that the air gap of the machine is uniform.

(l) The electromagnetic torque developed in a machine is proportional to the magnitudes of the main-field and air-gap-mmf waves and the sine of the angle between them.

(m) By loss of synchronism we mean that the field and air-gap mmf waves are no longer stationary with respect to each other.

5-49. A synchronous motor is fed from an infinite bus with 1.00 per-unit voltage, through a line containing inductance and no resistance. With a per-unit excitation of 0.9, the motor current is negligible at no-load. However, when the motor is loaded without changing excitation, it falls out of step at a load of 0.81 per unit of rated output. Full-load current is produced by 0.75 per-unit excitation during a short-circuit test.
(a) What is the per-unit inductive reactance of the line?
(b) What is the current under the above conditions?

5-50. A six-pole *synchronous motor* (cylindrical rotor) has a per-unit synchronous reactance of 0.75, the resistance being negligibly small. Its excitation emf is adjusted to a per-unit value of 1.50. As the motor is loaded, the rotor is observed to drop back 10 degrees from its initial space position corresponding to zero load. Losses are negligible. For rated voltage impressed on the motor, compute:
(a) Line current in per-unit values.
(b) Power factor, lagging or leading.

5-51. A synchronous motor with salient poles is operating at no-load with the per-unit rated terminal voltage $V_t = 1$ and per-unit excitation voltage of $E_f = 0.65$. The motor is found to draw an armature current of 0.5 per unit. Assuming that the ratio $x_d/x_q = 1.75$, determine the per-unit values of the direct-axis and quadrature-axis synchronous reactances.

5-52. The direct- and quadrature-axis synchronous reactances of a salient-pole synchronous generator are $x_d = 1.00$ per unit, and $x_q = 0.50$ per unit. The stator resistance and the rotational losses are both negligible. The generator is connected to an infinite bus of rated voltage and frequency. The generator field excitation is adjusted so that at no-load the line current is zero. The power input to the generator from its prime mover is then slowly increased until the generator is pulled out of synchronism. What is the per-unit line current and the pf when the machine is on the verge of pulling out? Prepare a neat, clear phasor diagram illustrating your calculations.

5-53. Derive an expression for the armature and field winding current increases, i_a and i_f, respectively, for the case where the field coil in Fig. 5-27 is completely linked with the stator phase coil a-a. What are the maximum values of these currents? Sketch the general shape of the field winding current following the short circuit.

5-54. A synchronous generator is operating as an isolated unit delivering rated kVA to a unity pf load at rated voltage. The per-unit values of the machine parameters are $x_s = 0.9$, $x_s' = 0.35$, and $x_s'' = 0.20$. A three-phase short circuit occurs across the armature terminals at the instant when the coupling of the field winding with stator phase winding a is a maximum.

(a) Find the value of the d-c component of the armature current in phase a in per unit immediately following the short circuit.

(b) Determine the maximum current that flows in per unit immediately after the fault.

(c) The subtransient components of the armature current disappear in a matter of a few cycles. Calculate the approximate value of the armature current immediately following the subtransient period. Assume that the d-c component of current is barely diminished in phase a.

(d) Compute the value of the sustained short-circuit current after all transients have decayed to a negligible level.

5-55. A synchronous generator is operating as an isolated unit in an idle state with the field current set to yield 125% excitation. The per-unit values of the machine parameters are $x_s = 0.85$ p-u, $x_s' = 0.30$ p-u, and $x_s'' = 0.15$ p-u. A three-phase short circuit occurs at the armature terminals when the coupling of the field coil with phase a of the stator winding is a minimum.

(a) Find the maximum value of phase a armature current during the first few cycles following the short circuit.

(b) What is the value of this short-circuit current during the transient phase?

(c) Determine the value of the sustained short-circuit current.

5-56. A three-phase synchronous generator has the following machine parameters: $x_l = 0.07$ p-u, $x_\phi = 0.8$ p-u, $x_f = 0.4$ p-u, and $x_d = 0.215$ p-u. Compute the transient and subtransient reactances.

5-57. A three-phase synchronous generator has an armature leakage reactance per phase of 0.08 per unit and a mutual reactance per phase of 0.85 per unit. When idling at 100% excitation, a three-phase symmetrical short circuit occurs at the stator terminals producing a phase a current of 2.857 per unit when the linkage between phase a and the field winding is a minimum. The machine is not equipped with a damper winding. Find the per-unit value of the field winding leakage reactance.

5-58. A three-phase synchronous generator equipped with a squirrel-cage damper winding has a leakage armature winding reactance of 0.1 per unit, a mutual reactance of 0.9 per unit and a field winding leakage reactance of 0.35 per unit. A three-phase short circuit occurs when stator phase a has zero flux linkage with the field coil which is energized for 110% excitation. The ensuing maximum short-circuit current is found to be 4.4 per unit. Determine the leakage reactance of the damper winding in per unit.

5-59. A three-phase synchronous generator is found to have the following machine parameters expressed in per unit: $x_1 = 0.06$; $x_f = 0.17$. Moreover, the transient reactance is known to be 0.2 per unit.

(a) Find the value of the mutual reactance per phase.

(b) Determine the maximum value of the transient short-circuit current.

(c) Compute the sustained short-circuit current.

Assume 100% excitation and an isolated unit operating at no load. The machine is not equipped with a damper winding.

5-60. A three-phase, four-pole, 60-Hz, 13,800-V, 100-mVA synchronous generator operates at no-load idling on an infinite bus system. The damping torque parameter of the squirrel-cage winding is 295 N-m/electrical degree. The inertia torque parameter expressed in N-m/electrical degree/s² is 40. Moreover, at rated conditions and fixed excitation the power angle is 12 electrical degrees. Adjustments are made in the prime mover to permit the synchronous generator to deliver 20 mW to the bus system. Describe the ensuing dynamic response of the prime mover–generator system by determining the following quantities:

(a) The steady-state value of the power angle expressed in electrical degrees.

(b) The maximum value of the power angle during the transient state.

(c) The frequency of oscillation of the mechanical system in cycles per second.

(d) The duration time of the transients.

Assume linear operation and neglect all losses. The value of $x_s = 1.9\ \Omega$.

5-61. A three-phase, six-pole, 100-kVA synchronous generator has its excitation set so that it delivers rated output at a power angle of 15 electrical degrees. The damping coefficient of the squirrel cage is known to be 1.4 N-m/electrical degree and the inertia constant is 0.3 N-m/electrical degree/s². Describe the dynamic behavior of this system when it is suddenly made to deliver power to the bus system by computing

(a) The maximum percent overshoot of the power angle about its steady-state value.

(b) The settling time for a 1% tolerance band.

(c) The frequency of oscillation of the mechanical system.

Assume linear analysis and no losses.

5-62. A three-phase, 2300-V, 60-Hz, 18-pole, 250-hp synchronous motor is running idly on an infinite bus system. At rated horsepower this motor operates at a power angle of 14 electrical degrees. The damping coefficient of its squirrel-cage winding is 3.1 N-m/electrical degree. The inertia coefficient of the motor and its connected mechanical load is known to be 0.5 N-m/electrical degree/s². Describe the dynamic behavior of this system when the mechanical load is suddenly applied to the shaft of the synchronous motor. Assume linear analysis and no losses.

5-63. For the machine described in Example 5-8 determine the following.

(a) The power angle at which maximum power is developed.

(b) The maximum value of the developed power expressed in per unit.

5-64. The synchronous motor of Example 5-8 is operated with a change in field current that causes the power factor to change to 0.8 lagging from 0.8 leading. Determine the new value of the excitation voltage.

5-65. A cylindrical-pole synchronous motor is to be used to improve the power factor of a load of 1500 kW at a power factor of 0.85 lagging. The synchronous motor is to operate at a 0.75 leading power factor. What must be the kVA rating of the synchronous motor if it is required to raise the total power factor to 0.95 lagging?

5-66. The source voltage appearing in Fig. 5-48 is assumed to be 120 V rms and 60 Hz.

(a) Determine the value in volts of the average load voltage when the firing angle of the SCR is set for $\pi/4$ radian.

(b) Let the resistor in this circuit be 10 Ω and the inductor 5 H. Compute the average value of the load current for the firing angle of part (a).

6

Direct-Current
Machines

The d-c machine is a highly versatile machine. It can provide high starting torques as well as high accelerating and decelerating torques. It is capable of quick reversals, and speed control over a range of 4:1 is achieved with relative ease in comparison with all other electromechanical energy-conversion devices. These are features that are responsible for its use in the really tough jobs in industry, such as are found in steel mills. Unfortunately, the need for a mechanical rectifier (in the form of a commutator) to convert the a-c emf that is induced in each armature coil to a unidirectional voltage makes it one of the least rugged of electric machines as well as more expensive.

The principles underlying basic torque production and induced voltages in the d-c machine are outlined in detail in Chapter 3. Here we investigate the operational characteristics of the various types of d-c machines, beginning with a description of how the direct voltage is obtained. The influence of the armature winding mmf on machine behavior is explored with particular emphasis on its effect on commutation and its external characteristics. The subject of commutation occupies a position of preeminence in the study of d-c machines because without good commutation the machine is rendered almost useless. The motor speed-torque curves are analyzed and also the various methods of speed control. Finally, the procedure for finding machine performance through the use of the governing equations, the equivalent circuit, the power-flow diagram, and the magnetization curve is outlined and illustrated.

6-1 GENERATION OF UNIDIRECTIONAL VOLTAGES

It is helpful at the start of our study of d-c machines to understand how a direct voltage can be produced at the armature output terminals. Appearing in Fig. 6-1(a) is a plot of the flux density produced by the field winding as a function of displacement

(a)

(b)

(c)

Figure 6-1 Generation of unidirectional voltage: (a) shape of air-gap flux density produced by field winding; (b) emf generated is a single-coil armature for four instants of time; (c) plot of E_a as a function of time, with specific time instants of part (b) shown.

along the periphery of the rotor. The flat-topped portion of the plot is attributable to the constant air gap between the rotor surface and the faces of the pole pieces. A sharp falloff occurs in the interpolar space because of the effect of the large air gaps there. Figure 6-1(b) depicts a cross-sectional view of a two-pole machine having an armature winding consisting of a simple coil, a–a'. Coil side a is joined to coil side a' by a back connection that is not shown. The front ends of each coil side are joined permanently to copper segments as indicated. Note that both coil sides are effective in voltage production with this arrangement, unlike the Gramme-ring winding. The copper segments are part of the rotor structure, thereby rotating with it. Any emf that is induced in the coil a–a' appears at these copper segments. Placed in contact with the copper segments and fixed in space are two carbon brushes B_1 and B_2. These are used to collect the voltage induced in the armature winding and to make it available to the external circuit. The coil in Fig. 6-1(b) is shown in four different positions relative to the flux field. The rotor is assumed revolving at speed n in the counterclockwise direction. At time instant 1, the emf induced in the coil is zero because each coil side finds itself at a point of zero flux density. At instant 2, there is an emf induced in each coil side and its magnitude is proportional to the value of flux density, as obtained from Fig. 6-1(a), as well as to the velocity and length of the coil sides. The direction of the induced voltage in coil side a is such that it makes the polarity of brush B_1 *positive*. As the rotor revolves an additional 90 degrees, the situation illustrated in 3 of Fig. 6-1(b) is found to prevail. Again note the zero voltage value. At instant 4, coil side a is under the influence of south-pole flux and so has an emf of reversed polarity induced. However, note that its attached commutator segment is now in contact with brush B_2, which keeps this brush at a negative polarity. A complete plot of the voltage appearing at the armature terminals is depicted in Fig. 6-1(c) with the four time instants of Fig. 6-1(b) specifically indicated. It is important to note that, although the voltage in each coil side alternates for each revolution, the voltage appearing at the brushes is unidirectional because of the effect of the segmented commutator.†

Although the use of a single coil in conjunction with the commutator furnishes a unidirectional voltage, the resulting wave shape of Fig. 6-1(c) is unsatisfactory because the magnitude is not constant over the full period. A considerable improvement can be achieved by increasing and distributing the number of armature coils. Figure 6-2(a) depicts in developed form a two-pole machine having eight slots on the rotor equipped with two coil sides per slot, yielding a total of eight coils. This representation is typical of the situation found in practical machines. The lower part of the figure illustrates the manner in which the various coil sides are joined to give a summed quantity. The winding layout shown is referred to as a *wave winding* for obvious reasons.‡ As a general rule the sides of a given coil are made to span one pole pitch (i.e., 180 electrical degrees). Moreover, in the double layer winding, such as the one illustrated here, each coil is so placed that one side occupies a lower position and the other side an upper position. For example, coil sides 1 and 10 are joined by the back-end connection to constitute one coil. Note that the span is 180° and that coil side 10 occupies the lower position in slot 5, while coil side 1 occupies the upper position in slot 1.

A study of Fig. 6-2(a) discloses that the armature winding completely closes on itself and that in particular the wave winding has two armature paths with respect to the brushes. The arrowheads attached to each coil side indicate the direction in which the induced emf causes current to flow when a load is placed across the brushes. In the plan view of Fig. 6-2(a) the direction of the emf for the direction of rotation shown is downward beneath a north pole and upward beneath a south pole. For clarity's sake the south pole is repeated once; on the left side the upper coil sides are specifically illustrated, while on the right side the coil sides located in the lower

---

†For better understanding the distinction with the a-c generator is worth making. The arrangement of Fig. 6-1(b) becomes an a-c generator when each coil side is connected to a closed copper ring and each brush is made to ride on one ring. Then the positive and negative variations of induced emfs are made to appear at the brushes and therefore the armature terminals.

‡Another frequently used arrangement is the lap winding.

Figure 6-2 Generation of a d-c voltage: (a) winding layout for a simple machine; (b) schematic representation of two armature paths existing between the brushes at the time t_0 illustrated in (a); (c) total induced armature voltage as a function of time; complete cycle is shown for each of four coils.

portions of the slots are shown. Also brush B_2 is assumed to be resting on commutator segment 4, and simultaneously brush B_1 is assumed to be in contact with commutator segment 8. Examination of the circuitry at B_2 discloses that current converges at this brush from two directions—from coil side 2 and coil side 7. Further investigation reveals that coil side 2 is associated with a path that consists of four coils: 8–15, 6–13, 4–11, and 2–9. These coils are drawn with solid lines in Fig. 6-2(a) and are denoted by small series-connected d-c sources in Fig. 6-2(b). Similarly coil side 7 is associated with the remaining four coils: 7–16, 5–14, 3–12, and 1–10. These coils are represented in Fig. 6-2(a) with broken lines and in Fig. 6-2(b) by series-connected d-c sources.

If attention is directed solely at the fundamental component of the flux-density curve of Fig. 6-1(a), it should be clear that for the time instant being considered the values of the coil voltages are not the same. For example, the instantaneous values of coils 3–12, 4–11, 5–14, and 6–13 are displaced from their positions of maximum value by 22.5° while the remaining coils are away by 67.5°. A partial time history of these coil voltages for various positions of the armature relative to the field distribution is depicted in Fig. 6-2(c). The instant represented in Fig. 6-2(a) is identified as t_0. Note that each coil voltage over any full cycle appears as a rectified wave because of the action of the commutator. Furthermore, only the four coils on one side of the armature winding are depicted. The total induced voltage appearing between the brushes B_1 and B_2 at time t_0 is the sum of the instantaneous coil voltages (E_{a0}), i.e., $E_{a0} = 2E_m$ cos 22.5 + $2E_m$ cos 67.5, where E_m is the peak voltage induced in the coil. If the conditions leading to Fig. 6-2(b) are analyzed for a time instant t_1—22.5° later—then coils 5–14 and 6–13 will experience their peak values, while the emf induced in coils 2–9 and 1–10 will be zero. The instantaneous values of the induced emfs in the remaining coils will be E_m cos 45°. The total contribution at time t_1 is therefore $E_{a1} = E_m + 2E_m$ cos 45° + 0. This total is identified in Fig. 6-2(c) and is less than the value occurring at time t_0.

A glance at Fig. 6-2(c) makes it obvious that an almost constant voltage now appears between the brushes in contrast to the situation of Fig. 6-1(c). The small ripple is attributable to the small number of slots per pole. In practical machines this number is very large so that the ripple is hardly detectable even with sensitive instruments.

6-2 DIRECT-CURRENT GENERATOR TYPES

The d-c machine functions as a generator when mechanical energy is supplied to the rotor and an electrical load is connected across the armature terminals. In order to supply electrical energy to the load, however, a magnetic field must first be established in the air gap. The field is necessary because it serves as the coupling device permitting the transfer of energy from the mechanical to the electrical system. There are two ways in which the field winding may be energized to produce the magnetic field. One method is to excite the field separately from an auxiliary source as depicted in the schematic diagram of Fig. 6-3. But clearly this scheme is disadvantageous because of the need of another d-c source. After all, the purpose of the d-c generator is to make available such a source. Therefore, invariably d-c generators are excited by the second method which involves a process of self-excitation. The wiring diagram appears in

Figure 6-3 Schematic diagram of a separately excited d-c generator.

Figure 6-4 Schematic diagram of a self-excited shunt generator.

Fig. 6-4, and the arrangement is called the *self-excited shunt generator*. The word shunt is used because the field winding appears in parallel with the armature winding. That is, the two windings form a shunt connection.

To understand how the self-excitation process takes place we must shart with the *magnetization curve* of the machine. Sometimes this is called the *saturation curve*. Strictly speaking, the magnetization curve represents a plot of air-gap flux versus field winding mmf. However, in the d-c generator where the winding constant K_E is known and the speed n is fixed, the magnetization curve has come to represent a plot of the open-circuit induced armature voltage as a function of the field winding current. With K_E and n fixed, Eq. (3-62) shows that ϕ and E_a differ only by a constant factor. Figure 6-5 depicts a typical magnetization curve, valid for a constant speed of rotation of the armature. It is especially important to note in this plot that even with zero field current an emf is induced in the armature of value Oa. This voltage is due entirely to residual magnetism, which is present because of the previous excitation history of the magnetic-circuit iron. The linear curve appearing on the same set of axes is the *field-resistance line*. It is a plot of the current caused by the voltage applied to the series combination of the field winding and the active portion of the field rheostat. Clearly, then, the slope of the linear curve is equal to the sum of the field-winding resistance R_f and the active rheostat resistance R_{rh}. The voltage Oa due to residual magnetism

Figure 6-5 Illustrating the build-up procedure of a self-excited shunt generator.

appears across the field circuit and causes a field current Ob to flow. But in accordance with the magnetization curve this field current aids the residual flux and thereby produces a larger induced emf of value bc. In turn this increased emf causes an even larger field current, which creates more flux for a larger emf, etc. This process of voltage build-up continues until the induced emf produces just enough field current to sustain it. This corresponds to point f in Fig. 6-5. Note that in order for the build-up process to take place three conditions must be satisfied: (1) There must be a residual flux.† (2) The field winding mmf must act to aid this residual flux. (3) The total field-circuit resistance must be less than the critical value. The *critical field resistance* is that value which makes the resistance line coincide with the linear portion of the saturation curve.

In addition to the shunt generator there are the other generator types—the compound generator and the series generator. A *compound generator* is a shunt generator equipped with a *series winding*. The series winding is a coil of comparatively few turns wound on the same magnetic axis as the field winding and connected in series with the armature winding. Consult Fig. 6-6. Because the series field winding must be capable of carrying the full armature current, its cross-sectional area is much greater than that used in the shunt field winding. The purpose of the series field is to provide additional air-gap flux as increased armature current flows, in order to neutralize the armature winding resistance drop as well as the voltage drops occurring in the feeder wires leading to the load. In such cases the generator is usually referred to as a *cumulatively* compounded generator because the series field *aids* the shunt field flux. If the series field connection were reversed, its flux would oppose the shunt field flux in which case the configuration is referred to as a *differentially* compounded generator.

By imposing the appropriate constraint on the connection diagram depicted in Fig. 6-6(a), we can identify any one of the three modes of operation of the d-c generator. Thus in addition to the armature winding we have the following:

Compound generator: shunt and series field windings

Shunt generator: shunt field winding

Series generator: series field winding

Since the series generator is rarely used except for special applications, all further treatment of generators is confined to the shunt and compound modes.

6-3 DEMAGNETIZING EFFECT OF THE ARMATURE WINDING MMF

The armature winding mmf produces two adverse effects: it causes a net reduction in the field flux, and it makes it more difficult for the armature current in the coils to commutate. We turn our attention to the first of these here; the problem of commutation is treated in the next section.

---

†When the iron is in its original virgin state, the residual flux is negligible. In such cases the field winding must receive an initial excitation to give the machine a usable level of residual flux. This process is called *flashing the field*.

(a)

(b)

Figure 6-6 Cumulatively compounded generator: (a) schematic diagram; (b) detailed diagram depicting the location of series and shunt windings on the pole structure.

A glance at Fig. 6-6(b) shows that the flow of current in the armature winding produces an ampere-conductor distribution that makes the armature behave like a solenoid. The associated mmf is directed downward along the brush axis and in a position of quadrature to the field axis. If there were no saturation of the iron present, the effect of this cross-armature mmf would be merely to cause a distortion of the flux. However, in most practical machines operation occurs around the knee of the magnetization curve and as a result a net demagnetization takes place. To understand why this happens, consider the situation illustrated in Fig. 6-7 which is that of a shunt generator delivering current to a load (the latter not shown). The armature mmf wave is assumed to be triangular. This is a valid assumption whenever the number of surface armature conductors is large; otherwise it would be a stepped trapezoid. The flux-density wave produced by this armature mmf is saddle-shaped because of the high

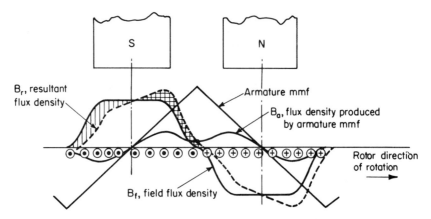

Figure 6-7 Effect of armature mmf on the air-gap flux density. Demagnetization occurs if the cross-hatched area to the left of the pole axis is greater than the double-cross-hatched area to the right of the pole axis.

reluctance of the interpolar space. The resultant air-gap flux density is the sum of the B_f and the B_a curves. It should be noted that the armature mmf acts to increase the flux in the trailing half of the pole, while it causes a diminution of flux in the leading half. If the increase of flux in the trailing half is equal to the decrease in the leading half, no net reduction in field flux takes place. If the increase in the trailing half is less than the decrease in the leading half, demagnetization occurs. In Fig. 6-7 the cross-hatched area represents the amount of demagnetization that occurs in the leading half of the pole, the double cross-hatched area denotes the amount of magnetization caused by the cross-magnetizing armature mmf. When a net demagnetization of the field flux results, the former area exceeds the latter. Because machines are operated in a partially saturated state, there often exists some degree of demagnetization.

A measure of the amount of this demagnetization expressed in equivalent field amperes can be obtained from a knowledge of the magnetization curve and appropriate design data. The procedure is graphically illustrated in Fig. 6-8. The air-gap

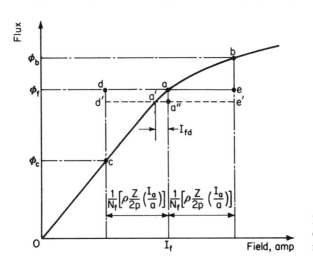

Figure 6-8 Illustrating the demagnetizing mmf by means of the magnetizing curve.

DIRECT-CURRENT MACHINES   CHAP. 6

flux per pole produced by the field winding acting alone is denoted by Φ_f. Let ρ be defined as the ratio of the pole face width, b_p, to the width of a pole pitch, τ_p. That is,

$$\rho \equiv \frac{b_p}{\tau_p} \tag{6-1}$$

Then the amount of armature mmf acting at either pole tip is

$$\text{armature mmf at pole tip} = \rho \frac{Z}{2p} \frac{I_a}{a} \tag{6-2}$$

where Z denotes the total armature conductors, p is the number of poles and a denotes the number of armature paths. The factor 2 converts conductors to turns. To express Eq. (6-2) in equivalent field amperes, it is necessary to divide by the number of turns per pole of the field winding, N_f. This quantity is shown on the abscissa axis of Fig. 6-8 with some exaggerations for the sake of clarity. The value of the resultant air-gap flux at the center of the pole is Φ_f. However, at the trailing pole tip the magnetizing action of the armature mmf raises this quantity to Φ_b (as is also indicated in Fig. 6-7), whereas at the leading pole tip it drops to Φ_c. Area acd is a measure of the amount of demagnetization that occurs in the leading half of a pole; area abe is a measure of the amount of magnetization that occurs in the trailing half of a pole. For machines that exhibit a measurable degree of saturation area acd is greater than area abe. Hence demagnetization occurs. The value of this demagnetization is found by displacing line dae to that position $d'a'e'$ such that area $a'd'c$ equals area $a'be'$. The quantity aa' projected on the field current axis yields the desired result. Thus

$$I_{fd} \equiv a'a'' \quad \text{measured in field amperes}$$

where I_{fd} is used to denote the demagnetizing effect of the armature mmf in equivalent field amperes.

6-4 COMMUTATION

Good commutation is indispensable for the satisfactory operation of d-c machines. This matter is so important that it takes precedence in design considerations over the problem of heat dissipation. Without proper commutation, deterioration of the carbon brushes and the copper commutator segments can rapidly set in. This causes severe sparking and thereby renders the machine useless. Even such an important characteristic as the maximum torque is limited by commutation rather than by heating.

Commutation is concerned with providing a suitable transition of the armature current in a coil from a value $+I_a/a$ to a value of $-I_a/a$ as it passes beneath a brush from one pole to an adjacent pole. A glance at Fig. 6-9(a) makes it evident that when coil 1–10 passes beneath the brushes and occupies the position now being occupied by coil 15–8, the current in the coil will have reversed completely as indicated by the cross-dot notation. Commutation of the current from coil 1–10 begins to take place when brush B_1 makes contact with commutator segment 1 as shown in Fig. 6-9(b). Note that when this happens coil 1–10 is actually short-circuited by brush B_1 through segments 8 and 1. All commutator segments are insulated from one another by a suitable material. A similar situation occurs at brush B_2 where coil 2–9 is being short-circuited. Because of this and in the interest of preventing large circulating currents

(a)

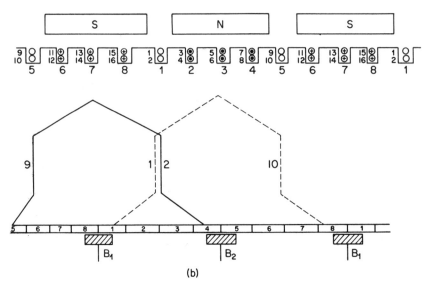

(b)

Figure 6-9 Illustrating the commutation problem: (a) coil 1–10 must undergo a complete current reversal as it advances counterclockwise to the left side of the brush axis; (b) for the position shown, coil 1–10 is short-circuited through segments 8 and 1 by B_1, and coil 2–9 is short-circuited through segments 4 and 5 by B_2.

in these coils, which in turn can cause severe sparking as the circulating currents are interrupted upon leaving the brush, the coils undergoing commutation should be free of voltage sources. The time instant depicted in Fig. 6-9(b) is 22.5° later than that shown in Fig. 6-2(a). Hence coils 1–10 and 2–9 find themselves in the interpolar space where the flux density produced by the field mmf is zero. Accordingly, no Blv voltage is induced in these coils from this source. However, the presence of the armature mmf causes the resultant flux density to have a finite value in the interpolar space as is evident from Fig. 6-7. Since the polarity of this flux is the same as that of the main field pole which the short-circuited coils are leaving, an emf is induced that acts to maintain current flow in the same direction as exists before commutation begins. Therefore this induced emf due to the armature mmf acts to hinder commutation; it becomes more difficult for the coil current to reverse by the time it leaves the brush.

Types of Commutation

There are two types of commutation possible: resistance commutation and voltage commutation. Moreover, the latter may be divided into the two classes of under-commutation and overcommutation. The distinction between these types is depicted by the various commutation curves of Fig. 6-10.

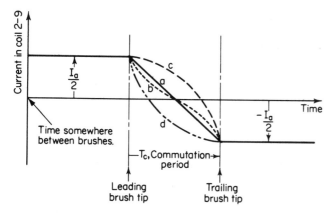

Figure 6-10 Commutation curve of coil 2–9 in Fig. 6-9. During the commutation period the curve is a plot of the circulating current in the short-circuited coil undergoing commutation.

Curve a represents *linear* commutation, and it occurs only when the brush contact resistance is the exclusive factor influencing commutation. All induced emf's from whatever sources must add to zero and the coil resistance must be negligible compared with the brush contact resistance. Under these conditions the current in the coil changes uniformly during the commutation period T_c. The mechanism of commutation by this method is illustrated in Fig. 6-11. Figure 6-11(a) depicts the situation just before

Figure 6-11 Illustrating linear commutation in coil 2–9: (a) conditions just before commutation begins; (b) conditions at a point one-quarter through the commutation period; (c) conditions at a point halfway through the commutation period.

commutation of the current in coil 2–9 is about to occur. Brush B_2 is resting on commutator segment 4, and of the total armature current I_a flowing through this brush half comes from one armature path involving coil 16–7 and the remaining half comes from the second path involving coil 2–9. The direction of rotation of the armature with respect to the brush is towards the left. Depicted in Fig. 6-11(b) is the situation where 25% of the brush is in contact with commutator segment 5. The contact resistance of that part of the brush surface in contact with segment 5 is three times as large as the resistance of the brush in contact with segment 4. Hence division of the current coming from the two armature paths takes place in accordance with this ratio. Thus the current $I_a/2$ coming from the right side divides at junction a in the ratio of 3:1 so that the current passing to B_2 from segment 5 from this source is $\frac{1}{4}(I_a/2)$ or $I_a/8$. The portion passing to B_2 through segment 4 is $\frac{3}{4}(I_a/2)$ or $\frac{3}{8}I_a$. Similarly, the current $I_a/2$ originating from the armature path on the left side splits, so that $\frac{3}{4}(I_a/2)$ or $\frac{3}{8}I_a$ flows to B_2 through segment 4 and $\frac{1}{4}(I_a/2)$ or $\frac{1}{8}I_a$ flows from b to a to segment 4 and thence to B_2. The net current that flows from segment 4 to B_2 is therefore $\frac{3}{8}I_a + \frac{3}{8}I_a$ or $\frac{3}{4}I_a$, whereas the net current to B_2 from segment 5 is $(I_a/8) + (I_a/8)$ or $I_a/4$. The resultant current in coil 2–9 flows from a to b and is equal to $\frac{3}{8}I_a - \frac{1}{8}I_a$ or $I_a/4$. Consequently, one-quarter through the period of commutation the current in coil 2–9 has changed from $I_a/2$ to $I_a/4$. Keep in mind that the commutation curve is a plot of the current in the short-circuited coil as a function of time. Appearing in Fig. 6-11(c) is the distribution of currents that prevail at a time midway through the commutation period. Note that now the current in the short-circuited coil is zero.

Curve b in Fig. 6-10 represents resistance commutation. It is the curve that results when the resistance of the short-circuited coil is not negligible compared with the brush contact resistance. Note that at the leading brush tip the slope of curve b is steeper than that of curve a, which indicates that the coil resistance enhances the current reversal. This is apparent from an examination of Fig. 6-11(b) where it is seen that at junction a the presence of coil resistance means that more current flows to segment 5. However, at the trailing brush tip (or during the second half of the commutation period) the coil resistance serves to retard the current reversal. The current density, which is represented by the slope of the commutation curve, is therefore higher at both the leading and trailing brush edges for resistance commutation.

The commutation curve denoted as c in Fig. 6-10 is described as *under commutation*. It is a form of voltage commutation in which current reversal has been delayed by the action of the emf of self-induction in the short-circuited coil or of the Blv voltage induced in the short-circuited coil by the flux produced by the armature mmf or of a combination of both. A decelerated commutation of this kind is characterized by high current density at the trailing brush tip, which can promote a deterioration of the brush material and also bring about undesirable chemical changes in the copper segments of the commutator.

A scheme that is frequently used to neutralize this emf of self-induction and the effect of the armature mmf is the inclusion of small poles located on the brush axis and energized by a coil carrying the armature current. Because of their location and function, these poles are called *interpoles* or *commutating poles*. The use of armature-current excitation is dictated by the fact that both the emf of self-induction in the short-circuited coil and the flux in the brush axis produced by the armature mmf are proportional to armature current. For generator action the polarity of the interpole must be the same as that of the field pole into which the coil is moving. Figure 6-12 depicts the situation for a two-pole machine. It is helpful to compare the resultant flux-density curve here with that appearing in Fig. 6-7. Note that the value of the flux density on the interpolar axis in Fig. 6-12 is opposite to that occurring in Fig. 6-7. If the magnitude of the flux density produced by the interpole is adjusted properly, it can be made to neutralize the emf of self-induction completely. Of course the effect of the armature mmf has already been neutralized in the representation of Fig. 6-12 because the resultant flux density is shown to be of opposite polarity. With such an adjustment the type of commutation that prevails is linear commutation, provided that the coil resistance compared to the brush contact resistance is inconsequential. Otherwise resistance commutation takes place.

Figure 6-12 Influence of interpoles on the resultant flux-density wave of a d-c machine.

A considerable improvement in the commutation process can be achieved by increasing the interpole flux to the point at which, in addition to canceling the effects of the emf of self-induction and the armature mmf, there is induced a voltage equal to twice the brush voltage drop, $2I_aR_b$. Under these conditions that special condition of *overcommutation* results that leads to a value of zero current density at the trailing edge of the brush. This situation is denoted by curve d in Fig. 6-10. Note that the current in the coil undergoing commutation has already reached its value of $-I_a/2$ before the coil leaves the brush. This results in no sparking at the trailing edge of the brush and represents the most favorable condition for good commutation.

The relationship between the contact voltage drop at a brush and the current density is shown in Fig. 6-13. For average carbon brushes the contact drop tends to remain essentially constant at about 1 volt. This is consistent with the fact that the conduction in the space between the copper segments and the carbon brushes occurs in accordance with arc phenomenon.

When d-c machines are used in applications that involve severe duty cycles (e.g., steel mills), swiftly changing loads and high overloads can cause breakdown of the air space between commutator segments. This breakdown can spread rapidly over the entire commutator, thus causing damage to the commutator as well as a short

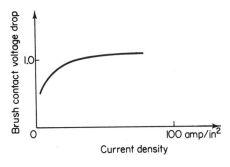

Figure 6-13 Illustrating the arc-type conduction characteristics of carbon brushes.

circuit on the d-c supply lines. Unfortunately, this situation can occur in spite of the presence of interpoles.

The most effective way to prevent this flashover condition is to neutralize the armature reaction that causes the severe peaking of the air-gap flux density that leads to breakdown. This is best achieved by the use of a *compensating winding* embedded in slots distributed along the pole faces. By allowing armature current to flow through this pole-face winding with a polarity opposite to that of the armature winding and a suitable number of ampere-turns, there can result complete cancellation of the armature reaction. The chief disadvantage of this method of eliminating or reducing the commutation problem is the high expense of pole-face windings.

6-5 COMPUTATION OF GENERATOR PERFORMANCE

Appearing in Fig. 6-14 is the equivalent circuit of the compound generator. The armature winding is replaced by a source voltage having the induced emf E_a and a resistance R_a, which represents the armature circuit resistance.† The series field is replaced by its resistance R_s, and the same for the shunt field.

The governing equations for determining the performance are the following:

$$E_a = \frac{pZ}{60a}\Phi n = K_E \Phi n \tag{6-3}$$

$$T = \frac{pZ}{2\pi a}\Phi I_a = K_T \Phi I_a \tag{6-4}$$

$$V_t = E_a - I_a(R_a + R_s) \tag{6-5}$$

$$I_a = I_L + I_f \tag{6-6}$$

These include the two basic relationships for induced voltage and electromagnetic torque as developed in Sec. 3-4. Equations (6-5) and (6-6) are merely statements of Kirchhoff's voltage and current laws as they apply to the equivalent circuit of Fig. 6-14. For the compound generator the expression for the air-gap flux must include the effect

†The armature-circuit resistance includes the armature winding resistance, interpole winding resistance, compensating winding resistance, plus the effect of the voltage drop in the carbon brushes.

Figure 6-14 Equivalent circuit of the compound d-c generator.

of the series field as well as the shunt field. Thus†

$$\Phi = \Phi_{sh} + \Phi_s \tag{6-7}$$

where Φ_{sh} denotes the flux produced by the shunt field winding and Φ_s denotes the flux caused by the series field winding. For the shunt generator, of course, Φ_s is zero in Eq. (6-7) and so too is R_s in Eq. (6-5).

The power-flow diagram for the d-c generator is depicted in Fig. 6-15. Note the similarity it bears to that of the synchronous generator. The field winding loss is included in the power flow directly because it is assumed that the generator is self-excited. Of course, if the field winding is separately excited, the field losses are not supplied from the prime mover and so must be handled separately. Moreover, note that the field losses are represented in terms of the product of the field terminal voltage and the field current. This ensures that the field winding rheostat losses are included as well.

Figure 6-15 Power-flow diagram of the d-c generator.

The quantity P_{rot} that appears in the power flow diagram represents the iron losses that arise from rotation of the armature (rotor) structure as well as friction and windage losses. The iron losses associated with rotation include hysteresis and eddy current losses. Although the slotted rotor structure, which accommodates the armature winding, is composed of laminations, eddy currents are caused to flow in the individual laminations as the rotor iron is driven alternately beneath poles of N polarity and S polarity. This very same alternation of the flux in the iron is responsible for the hysteresis losses represented by the generated hysteresis loops. Both losses are functions of the maximum flux density and the speed. Moreover, for machines with open slots, there is an additional component of P_{rot} that accounts for the losses incurred in the pole faces because of the ripple that is made to appear in the field distribution curve by the slot effect. The major part of the frictional losses is contributed by the friction of the brushes on the commutator segments. Also included but to a lesser degree are

†Algebraic summation is permitted only when saturation is negligible.

the machine bearing losses. Typically, the total of these losses represents 3 to 5% of the machine rating. However, it is useful to keep in mind that the per-unit value of all machine losses diminishes as the machine rating increases.

The armature winding copper losses are essentially losses that occur under load conditions. As such these losses account for all armature winding copper losses, including the skin effect that takes place in the individual coils of the armature winding, the electrical loss in the carbon brushes, (which is frequently taken to be $2I_a$ on the assumption that the voltage in the brushes is about 2 V), and the short-circuit loss in those coils that undergo commutation. For convenience all of these are assumed to be represented in the *armature winding circuit resistance*. Thus the parameter R_a is not to be treated as merely the d-c armature winding resistance.

The per-unit value of the armature circuit resistance indicates the ratio of the armature circuit losses expressed in units of power to the rated power output of the generator. If we denote the rated armature current as the base current, I_B, the rated armature voltage as the base voltage V_B, and the rated output power as the base power P_B, then

$$\text{p-u } R_a = \frac{I_B^2 R_a}{V_B I_B} = \frac{I_B R_a}{V_B} = \frac{R_a}{V_B/I_B} = \frac{R_a}{R_B}$$

where R_B is the base resistance and is determined as the ratio of the base voltage to the base current. Thus a value of p-u $R_a = 0.05$ means that 5% of the rated output of the generator is consumed as losses in the armature circuit when rated current is being delivered. This per-unit value takes on larger values for generators of small ratings. Similarly, for large generators the p-u R_a diminishes.

Example 6-1

The magnetization curve of a 10-kW, 250-V, d-c self-excited shunt generator driven at 1000 rpm is shown in Fig. 6-16. Each vertical division represents 20 V and each horizontal unit represents 0.2 A. The armature circuit resistance is 0.15 Ω and the field current is 1.64 A when the terminal voltage is 250 V. Also, the rotational losses are known to be equal to 540 W. Find at rated load (a) the armature induced emf, (b) the developed torque, (c) the efficiency. Assume constant speed operation.

Solution   (a) With the generator delivering rated load current it follows that

$$I_L = \frac{10,000}{250} \text{W} = 40 \text{ A}$$

Hence from Eq. (6-6) the armature current is

$$I_a = I_L + I_f = 40 + 1.64 = 41.64 \text{ A}$$

The induced armature voltage then follows from Eq. (6-5):

$$E_a = V_t + I_a R_a = 250 + 41.64(0.15) = 256.25 \text{ V} \tag{6-8}$$

(b) To determine the developed torque it is first necessary to compute the electromagnetic power $E_a I_a$. Thus

$$E_a I_a = 256.25(41.64) = 10,670 \text{ W}$$

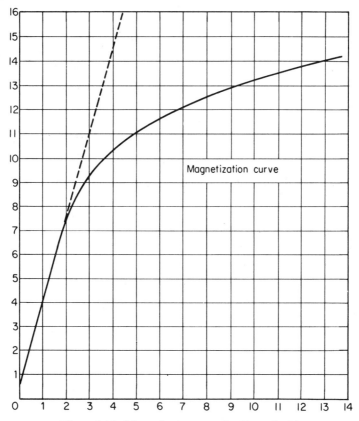

Figure 6-16 Magnetization curve for Example 6-1.

The power-flow diagram reveals that this power may also be found from

$$E_a I_a = P_0 + I_a^2 R_a + V_t I_f = 10 \text{ kW} + (41.64)^2 0.15 + 250(1.64)$$
$$= 10 \text{ kW} + 260 + 410 = 10,670 \text{ W} \qquad (6\text{-}9)$$

which checks with the preceding calculation. Hence the developed torque is

$$T = \frac{E_a I_a}{2\pi n/60} = \frac{10,670}{2\pi(1000)/60} = 101.9 \text{ N-m} \qquad (6\text{-}10)$$

(c) The efficiency is found from Eq. (4-27), which for the d-c generator takes the form

$$\eta = 1 - \frac{\sum \text{losses}}{E_a I_a + P_{\text{rot}}} \qquad (6\text{-}11)$$

In this case

$$\sum \text{losses} = P_{\text{rot}} + I_a^2 R_a + V_t I_f$$
$$= 540 + 260 + 410 = 1210 \text{ W}$$

Hence

$$\eta = 1 - \frac{1210}{10,670 + 540} = 1 - \frac{1210}{11,210} = 1 - 0.108 = 0.892$$

The efficiency is therefore 89.2%.

6-6 EXTERNAL GENERATOR CHARACTERISTIC. NONLINEAR ANALYSIS

The external characteristic of a generator is a description of the manner in which the terminal voltage varies with armature current. A plot of the external characteristic of a self-excited shunt generator is shown in the right half of Fig. 6-17. The drop in the curve is attributable to three factors: the effect of the armature circuit resistance drop, the demagnetizing action of the armature winding mmf, and the decrease in field current caused by the resulting drop in terminal voltage associated with the first two factors. It is assumed that the generator is driven at constant speed.

The precise way in which these factors combine to yield a stable operating point can be understood in the graphical representation appearing on the left side of Fig. 6-17. Assume that the following information is available: armature circuit resistance, field circuit resistance, magnetization curve, and the equivalent field current that represents the demagnetization caused by the armature mmf (i.e., I_{fd}). The last item is available either from the design data or an appropriate test. The intersection of the field resistance line with the nonlinear magnetization curve shows that at no-load the equilibrium field current is I_{f0} corresponding to the no-load terminal voltage V_0. When the load condition is such that the armature current is I_{a1}, the external characteristic shows that the terminal voltage has the value V_1. Extension of a horizontal line at the V_1 level results in a point of intersection with the field resistance line identified as O'. The projection of O' onto the field axis discloses that the current flowing in the field circuit is now I_{f1}. A comparison with the value at no-load makes it clear that a *reduction* occurs equal to

$$\Delta I_f = I_{f0} - I_{f1} \tag{6-12}$$

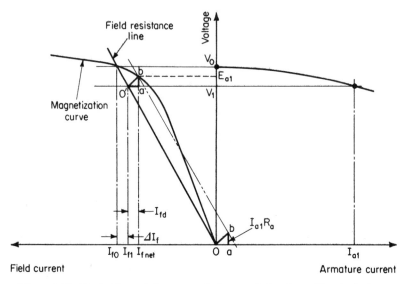

Figure 6-17 Derivation of the external characteristic of a self-excited shunt generator.

DIRECT-CURRENT MACHINES   CHAP. 6

To find the net field current, which in turn is responsible for producing the air-gap flux that provides the induced armature voltage E_a, it is necessary to subtract from I_{f1} the demagnetization effect I_{fd} of the armature mmf. Thus

$$I_{f \text{ net}} = I_{f1} - I_{fd} \qquad (6\text{-}13)$$

Or expressing this more generally in terms of the field current existing at no load, we can write

$$I_{f \text{ net}} = I_{f0} - \Delta I_f - I_{fd} \qquad (6\text{-}14)$$

The induced armature voltage for the specified load condition E_{a1} is then found on the nonlinear open-circuit characteristic corresponding to $I_{f \text{ net}}$. This is represented in Fig. 6-17 by point b. Upon subtracting from E_{a1} the armature circuit resistance drop $I_{a1}R_a$, the terminal voltage V_1 results. The $I_{a1}R_a$ drop is denoted by line ab in Fig. 6-17. It should be noted that there is only one position in the vicinity of the upper portion of the magnetization curve where triangle $O'ab$ will properly fit between the field resistance line and the magnetization curve. In fact it is this position that defines the equilibrium point that is being described.

On the basis of the foregoing discussion it can be seen that the external characteristic can be derived from the magnetization curve, the field resistance line, and knowledge of I_{fd} and R_a. To find the terminal voltage corresponding to any specified value of I_a, first construct the triangle Oab at the origin as shown in Fig. 6-17. Side ab is equal to I_aR_a and side Oa is equal to the amount of demagnetization associated with the specified I_a. Then draw a line through point b parallel to the field resistance line and intersecting the magnetization curve in two places. Triangle Oab may then be translated to position $O'ab$ taking care to keep O' on the field resistance line and b on the magnetization curve. The terminal voltage is the ordinate value corresponding to point a. By repeating this procedure for various values of armature current the complete characteristic is obtained.

A similar procedure may be used to describe the external characteristic of the self-excited cumulatively compounded generator. Very often the contribution of the series field is responsible for a rising external characteristic† such as the one depicted on the right side of Fig. 6-18. This characteristic can also be explained in terms of the same information that is used for the shunt generator, but in addition information about the series field contribution is required, usually expressed in terms of an equivalent field current. The last quantity is readily written as

$$I_{fs} \equiv \frac{N_s}{N_f} I_s \qquad (6\text{-}15)$$

where I_{fs} is the field current equivalent of the series field mmf, N_s denotes the number of turns of the series field per pole, N_f denotes the number of shunt field turns per pole, and I_s is the current flowing through the series field winding. Often I_s is equal to the armature current, but sometimes a portion of the armature current is diverted through a low resistance shunt (called a *diverter*) placed across the series field winding in order to limit its contribution.

At no-load the operating point is on the magnetization curve. The no-load

---

†Often referred to as *overcompounding*.

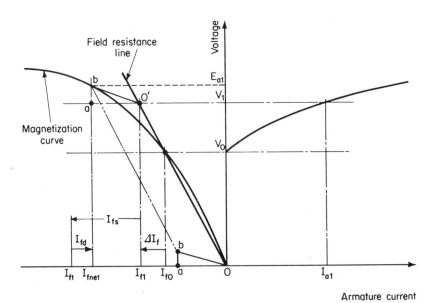

Figure 6-18 Derivation of the external characteristic of a cumulatively compounded generator.

terminal voltage is identified as V_0 and its corresponding field current as I_{f0}. It is helpful to think of this point as identical with the no-load point of the self-excited shunt generator. Next consider that load current is allowed to flow so that the armature current becomes I_{a1}, corresponding to which the terminal voltage becomes V_1 as shown in Fig. 6-18. With the assumption that (V_1, I_{a1}) is the new equilibrium point, it follows that the field current is *greater* by the amount ΔI_f which yields a total current in the field winding of value $I_{f1} = I_{f0} + \Delta I_f$. The plus sign for ΔI_f always appears whenever the external characteristic is a rising one. The current I_{f1} is associated with point O' on the field resistance line.

To the shunt field current must now be added the effect of the series field as expressed by Eq. (6-15). This yields a total main axis excitation expressed by

$$I_{ft} = I_{f1} + \frac{N_s}{N_f}I_s = I_{f1} + I_{fs} \tag{6-16}$$

The net field current, which is responsible for the armature induced emf, is found by subtracting the demagnetizing effect of the armature mmf. Thus

$$I_{f\,net} = I_{ft} - I_{fd} \tag{6-17}$$

Corresponding to $I_{f\,net}$ is the induced armature voltage E_{a1}. The terminal voltage is then found by removing the armature circuit resistance drop from E_{1a}, which yields V_1. The $I_{a1}(R_a + R_s)$ drop is represented by line *ab* in Fig. 6-18.

The external characteristic can be derived by following a procedure similar to that described for the shunt generator. Construct triangle *Oab* at the origin. Line *ab* is equal to the armature circuit resistance drop for the specified armature current. Line Oa is equal to the resultant of the magnetizing effect of the series field and the demagnetizing effect of the armature mmf, i.e.,

$$Oa = I_{fs} - I_{fd} \tag{6-18}$$

Then through b draw a line parallel to the field resistance line intersecting the magnetization curve at point b. Dropping down from b by the $I_{a1}(R_a + R_s)$ drop gives the desired terminal voltage V_1. A comparison with the procedure employed for the shunt generator discloses that the triangle Oab is placed on the left side of the origin rather than the right side. This is dictated by the fact that the base of the triangle for the cumulatively compounded generator represents a net magnetization rather than a net demagnetization.

6-7 DIRECT-CURRENT MOTOR ANALYSIS

A d-c motor is a d-c generator with the power flow reversed. In the d-c motor electrical energy is converted to mechanical form. Also, as is the case of the generator, there are three types of d-c motors: the *shunt* motor, the *cumulatively compounded* motor, and the *series* motor. The compound motor is prefixed with the word cumulative in order to stress that the connections to the series field winding are such as to ensure that the series field flux *aids* the shunt field flux. The series motor, unlike the series generator, finds wide application, especially for traction-type loads. Hence due attention is given to this machine in the treatment that follows.

The performance of the d-c motor operating in any one of its three modes can conveniently be described in terms of an equivalent circuit, a set of performance equations, a power-flow diagram, and the magnetization curve. The equivalent circuit is depicted in Fig. 6-19. It is worthwhile to note that now the armature induced voltage is treated as a reaction or counter emf. Observe that the defined direction of I_a is reversed from the case for a d-c generator. The generated voltage is caused by the relative motion of the rotor with respect to the magnetic field, and its polarity is independent of the direction of armature current. By imposing constraints similar to those applied to the d-c generator, we obtain the correct equivalent circuit for the desired mode of operation. For example, for a series motor the appropriate equivalent circuit results upon removing R_f from the circuitry of Fig. 6-19.

Figure 6-19 Equivalent circuit of the d-c motor.

The set of equations needed to compute the performance is listed below.

$$E_a = K_E \Phi n \qquad (6\text{-}19)$$

$$T = K_T \Phi I_a \qquad (6\text{-}20)$$

$$V_t = E_a + I_a(R_a + R_s) \qquad (6\text{-}21)$$

$$I_L = I_f + I_a \qquad (6\text{-}22)$$

The first two equations are identical with those used in generator analysis. Note, however, that the next two equations are modified to account for the fact that for the motor V_t is the applied or source voltage and as such must be equal to the sum of the voltage drops. Similarly the line current is equal to the *sum* rather than the *difference* of the armature current and field currents.

The power-flow diagram, depicting the reversed flow from that which occurs in the generator, is illustrated in Fig. 6-20. The electrical power input $V_t L_L$ originating from the line supplies the field power needed to establish the flux field as well as the armature circuit copper loss needed to maintain the flow of I_a. This current flowing through the armature conductors imbedded in the flux field causes torque to be developed. The law of conservation of energy then demands that the electromagnetic power $E_a I_a$ be equal to $T\omega_m$, where ω_m is the steady-state operating speed. Removal of the rotational losses from the developed mechanical power yields the mechanical output power. The use of the foregoing tools in computing motor behavior is illustrated by the following examples.

Figure 6-20 Power-flow diagram of the d-c motor.

Example 6-2

A 20-hp 230-V 1150-rpm shunt motor has four poles, four parallel armature paths, and 882 armature conductors. The armature-circuit resistance is 0.188 Ω. At rated speed and rated output the armature current is 73 A and the field current is 1.6 A. Calculate: (a) the electromagnetic torque, (b) the flux per pole, (c) the rotational losses, (d) the efficiency, (e) the shaft load.

Solution (a) The torque can be computed from Eq. (6-10), but we first need E_a. Hence

$$E_a = V_t - I_a R_a = 230 - 73(0.188) = 230 - 13.7 = 216.3 \text{ V}$$

Also

$$\omega_m = \frac{2\pi n}{60} = \frac{2\pi(1150)}{60} = 120 \text{ rad/s}$$

Therefore

$$T = \frac{E_a I_a}{\omega_m} = \frac{216.3(73)}{120} = 132 \text{ N-m}$$

(b) $E_a = K_E \Phi n = \dfrac{p}{60} \dfrac{Z}{a} \Phi n = \dfrac{4}{60} \dfrac{882}{4} \Phi(1150)$

$$\therefore \quad \Phi = \frac{216.3(60)}{882(1150)} = 0.0128 \text{ Wb}$$

(c) From the power-flow diagram

$$P_{\text{rot}} = P_m - P_0 = E_a I_a - 20(746) = 15{,}790 - 14{,}920 = 870 \text{ W}$$

(d) First find the sum of the losses. Thus

$$\Sigma \text{ losses} = P_{\text{rot}} + I_a^2 R_a + V_t I_f = 870 + (73)^2(0.188) + 230(1.6)$$
$$= 870 + 1002 + 368 = 2240 \text{ W}$$

Hence

$$\eta = 1 - \frac{\Sigma \text{ losses}}{E_a I_a + I_a^2 R_a + V_t I_f} = 1 - \frac{2240}{17{,}160} = 0.869$$

The efficiency is 86.9%.

(e) $T_0 = \dfrac{1}{\omega_m} P_0 = \dfrac{1}{120}(20 \times 746) = 124 \text{ N-m}$

The difference between this torque and the electromagnetic torque is the amount needed to overcome the rotational losses.

Example 6-3

The shaft load on the motor of Example 6-2 remains fixed, but the field flux is reduced to 80% of its value by means of the field rheostat. Determine the new operating speed.

Solution Information about the speed is available from Eq. (6-19). However, in turn, knowledge of flux and armature induced emf E_a is needed. By the statement of the problem the new flux Φ' is related to the original flux by

$$\Phi' = 0.8\Phi$$

To obtain information about E_a, we must determine the change in I_a, if any. By the constant-torque condition we have

$$K_T \Phi I_a = K_T \Phi' I_a'$$

Hence

$$I_a' = \frac{\Phi}{\Phi'} I_a = \frac{1}{0.8} 73 = 91.3 \text{ A}$$

Consequently

$$E_a' = V_t - I_a' R_a = 230 - 91.3(0.188) = 212.8 \text{ V}$$

Returning to Eq. (6-19) we can now formulate the ratio

$$\frac{E_a'}{E_a} = \frac{K_E \Phi' n'}{K_E \Phi n}$$

from which the expression for the new operating speed becomes

$$n' = \frac{E'_a}{E_a}\frac{\Phi}{\Phi'}n = \frac{212.8}{216.3}\frac{1}{0.8}1150 = 1414 \text{ rpm}$$

6-8 MOTOR SPEED-TORQUE CHARACTERISTICS.
SPEED CONTROL

How does the d-c motor react to the application of a shaft load? What is the mechanism by which the d-c motor adapts itself to supply to the load the power it demands? The answers to these questions can be obtained by reasoning in terms of the performance equations that appear on page 346. Initially our remarks are confined to the shunt motor, but a similar line of reasoning applies for the others. For our purposes the two pertinent equations are those for torque and current. Thus

and

$$T = K_T\Phi I_a \tag{6-20}$$

$$I_a = \frac{V_t - K_E\Phi n}{R_a} \tag{6-23}$$

Note that the last expression results from replacing E_a with Eq. (6-19) in Eq. (6-21). With no shaft load applied, the only torque needed is that which overcomes the rotational losses. Since the shunt motor operates at essentially constant flux, Eq. (6-20) indicates that only a small armature current is required compared with its rated value to furnish these losses. Equation (6-23) reveals the manner in which the armature current is made to assume just the right value. In this expression V_t, R_a, K_E, and Φ are fixed in value. Therefore the speed is the critical variable. If, for the moment, it is assumed that the speed has too low a value, then the numerator of Eq. (6-23) takes on an excessive value and in turn makes I_a larger than required. At this point the motor reacts to correct the situation. The excessive armature current produces a developed torque which exceeds the opposing torques of friction and windage. In fact this excess serves as an accelerating torque, which then proceeds to increase the speed to that level which corresponds to the equilibrium value of armature current. In other words, the acceleration torque becomes zero only when the speed is at that value which by Eq. (6-23) yields just the right I_a needed to overcome the rotational losses.

Consider next that a load demanding rated torque is suddenly applied to the motor shaft. Clearly, because the developed torque at this instant is only sufficient to overcome friction and windage and not the load torque, the first reaction is for the motor to lose speed. In this way, as Eq. (6-23) reveals, the armature current can be increased so that in turn the electromagnetic torque can increase. As a matter of fact the applied load torque causes the motor to assume that value of speed which yields a current sufficient to produce a developed torque to overcome the applied shaft torque and the frictional torque. Power balance is thereby achieved, because an equilibrium condition is reached where the electromagnetic power E_aI_a is equal to the mechanical power developed, $T\omega_m$.

A comparison of the d-c motor with the three-phase induction motor indicates that both are *speed-sensitive* devices in response to applied shaft loads. An essential difference, however, is that for the three-phase induction motor developed torque is adversely influenced by the power-factor angle of the armature current. Of course no analogous situation prevails in the case of the d-c motor.

On the basis of the foregoing discussion it should be apparent that the speed-torque curve of d-c motors is an important characteristic. Appearing in Fig. 6-21 are the general shapes of the speed-torque characteristics as they apply for the shunt, cumulatively compounded, and series motors. For the sake of comparison the curves are drawn through a common point of rated torque and speed. An understanding of why the curves take the shapes and relative positions depicted in Fig. 6-21 readily follows from an examination of Eq. (6-19), which involves the speed. For the shunt motor the speed equation can be written as

$$n = \frac{E_a}{K_E\Phi_{sh}} = \frac{V_t - I_aR_a}{K_E\Phi_{sh}} \tag{6-24}$$

Figure 6-21 Typical speed-torque curves of d-c motors.

The only variables involved are the speed n and the armature current I_a and its effect I_{fd}. At rated output torque the armature current is at its rated value and so, too, is the speed. As the load torque is removed, the armature current becomes correspondingly smaller, making the numerator term of Eq. (6-24) larger. This results in higher speeds. The extent to which the speed increases depends upon how large the armature circuit resistance drop is in comparison with the terminal voltage and the value of I_{fd}. It is usually around 5 to 10%. Accordingly, we can expect the percent change in speed of the shunt motor to be about the same magnitude. This change in speed is identified by a figure of merit called the *speed regulation*. It is defined as follows:

$$\text{percent of speed regulation} = \frac{\text{no-load speed} - \text{full-load speed}}{\text{full-load speed}} \times 100 \tag{6-25}$$

The speed equation as it applies to the cumulatively compounded motor takes the form

$$n = \frac{V_t - I_a(R_a + R_s)}{K_E(\Phi_{sh} + \Phi_s)} \tag{6-26}$$

where Φ_s denotes the increase in flux produced by the series field mmf. (Note that this is not a linear addition except when saturation is negligible.)

A comparison with the analogous expression for the shunt motor bears out two differences. One, the numerator term includes the voltage drop in the series field winding besides that in the armature winding. Two, the denominator term is increased to account for the effect of the series field flux Φ_s. Starting at rated torque and speed, Eq. (6-26) makes it clear that as load torque is decreased to zero there is an increase in the numerator term which is necessarily greater than it is for the shunt motor. At the same time, moreover, the denominator term decreases because Φ_s reduces to zero as the torque goes to zero. Both effects act to bring about an increase in speed. Therefore the speed regulation of the cumulatively compounded motor is greater than for the shunt motor. Figure 6-21 presents this information graphically.

The situation regarding the speed-torque characteristic of the series motor is significantly different because of the absence of a shunt-field winding. Keep in mind that the establishment of a flux field in the series motor comes about solely as a result of the flow of armature current through the series field winding. In this connection, then, the speed equation for the series motor becomes

$$n = \frac{E_a}{K_E \Phi_s} = \frac{V_t - I_a(R_a + R_s)}{K_E' I_a} \tag{6-27}$$

where K_E' denotes a new proportionality factor which permits Φ_s to be replaced by the armature current I_a. When rated torque is being developed the current is at its rated value. The flux field is therefore abundant. However, as load torque is removed less armature current flows. Now since I_a appears in the denominator of the speed equation, it is easy to see that the speed will increase greatly. In fact, if the load were to be disconnected from the motor shaft, dangerously high speeds would result because of the small armature current that flows. The centrifugal forces at these high speeds can easily damage the armature winding. For this reason a series motor should never have its load uncoupled.

Because the armature current is directly related to the air-gap flux in the series motor, Eq. (6-19) for the developed torque may be modified to read as

$$T = K_T \Phi I_a = K_T' I_a^2 \tag{6-28}$$

Thus the developed torque for the series motor is a function of the square of the armature current. This stands in contrast to the linear relationship of torque to armature current in the shunt motor. Of course in the compound motor an intermediate relationship is achieved. It is interesting to note, too, that as the series motor reacts to develop greater torques, the speed drops correspondingly. It is this capability which makes the series motor so well-suited to traction-type loads.

Speed Control

One of the attractive features the d-c motor offers over all other types is the relative ease with which speed control can be achieved. The various schemes available for speed control can be deduced from Eq. (6-24), which is repeated here with one modification:

$$n = \frac{V_t - I_a(R_a + R_e)}{K_E \Phi} \tag{6-29}$$

The modification involves the inclusion of an external armature circuit resistance R_e. Inspection of Eq. (6-29) reveals that the speed can be controlled by adjusting any one of the three factors appearing on the right side of the equation: V_t, R_e, or Φ. The simplest to adjust is Φ. A field rheostat is used such as that shown in Fig. 6-19. If the field rheostat resistance is increased, the air-gap flux is diminished, yielding higher operating speeds. General-purpose shunt motors are designed to provide a 200% increase in rated speed by this method of speed control. However, because of the weakened flux field the permissible torque that can be delivered at the higher speed is correspondingly reduced in order to prevent excessive armature current.

A second method of speed adjustment involves the use of an external resistor R_e connected in the armature circuit as illustrated in Fig. 6-22. The size and cost of this resistor are considerably greater than those of the field rheostat because R_e must be capable of handling the full armature current. Equation (6-29) indicates that the larger R_e is made, the greater will be the speed change. Frequently the external resistor is selected to furnish as much as a 50% drop in speed from the rated value. The chief disadvantage of this method of control is the poor efficiency of operation. For example, a 50% drop in speed is achieved by having approximately half of the terminal voltage V_t appear across R_e. Accordingly, almost 50% of the line input power is dissipated in the form of heat in the resistor R_e. Nonetheless, armature circuit resistance control is often used—especially for series motors.

Figure 6-22 Speed adjustment of a shunt motor by an external armature-circuit resistance.

A third and final method of speed control involves adjustment of the applied terminal voltage. This scheme is the most desirable from the viewpoint of flexibility and high operating efficiency. But it is also the most expensive because it requires its own d-c supply. It means purchasing a motor-generator set with a capacity at least equal to that of the motor to be controlled. Such expense is not generally justified except in situations where the superior performance achievable with this scheme is indispensable, as is the case in steel mill applications. Armature terminal voltage control is referred to as the *Ward-Leonard* system. See Chapter 11 for further details.

Example 6-4

In an industrial application the need arises to provide a constant torque of 105 lb-ft over a speed range of 250 to 1000 rpm. All losses are to be considered negligible for simplicity.

(a) Determine the hp rating and speed rating of the motor to accomplish this task when the control of speed is to be obtained through adjustment of the armature voltage.

(b) Repeat part (a) for the condition where speed control is to be obtained from adjustment of the field current.

(c) Assume that the motors in parts (a) and (b) are designed to operate at 220 V. Find the value of the armature current of both motors when they operate at 1000 rpm delivering the specified torque.

(d) Plot the variation of armature current and armature voltage of both motors over the specified speed range.

(e) Compute the value of the factor $K_T\phi$ that appears in Eq. (3-69) [or $K_E\phi$ that appears in Eq. (3-62)] for both motors at a speed of 1000 rpm. Also, plot the variation of this parameter for each motor over the speed range.

(f) Show a plot of the manner in which the delivered hp varies over the specified speed range.

(g) What is the maximum hp that can be delivered by each motor at an operating speed of 250 rpm?

Solution (a) When armature voltage control is used, the base speed is taken to be the maximum value of the specified speed range. In this case, it is 1000 rpm. Since the motor must deliver a torque of 105 lb-ft = 105(1.356) N-m [see Fig. 6-23(a)], it follows that the required hp rating is determined to be

$$hp = \frac{(105)(1.356)\left(1000 \times \frac{\pi}{30}\right)}{746} = \frac{(105)(1000)}{5252} = 20$$

Thus the rating for motor A (for armature control) is

$$A: 20 \text{ hp, } 1000 \text{ rpm}$$

(b) When speed is varied by field current control, the maximum speed of the specified speed range is attained corresponding to minimum field current (i.e., the minimum flux condition). The minimum speed of 250 rpm is then obtained as the field current is advanced to its maximum setting. Because the base speed of a motor is defined as the speed that corresponds to maximum flux at rated armature voltage, it follows that the base speed in this case must be 250 rpm. The required hp rating of this motor can be calculated at the maximum speed value of the speed range. Because the torque is to be constant over the speed range, which includes 1000 rpm, the required hp is the same as for machine A of part (a), namely

$$hp = \frac{105(1000)}{5252} = 20$$

Hence the complete specification for motor F (for field control) is

$$F: 20 \text{ hp, } 250 \text{ rpm}$$

A comment is in order at this point concerning motors A and F. First, however, it is useful to cite a general expression for the rating P of electrical machines that puts the emphasis on physical dimensions. This is

$$P = kRD^2Ln \tag{6-30}$$

where D is the rotor diameter, L the effective axial core, n the speed in rpm, and R is an appropriate design parameter. In essence the last expression states that the power rating of a machine is a function of the product of its volume and speed. Accordingly, the greater the speed, the smaller will be the volume. Therefore, motor F in this application will be larger, heavier and so more expensive than motor A. This greater volume of motor F is entirely consistent with the fact that at the lower speeds it must be capable of accommodating values of flux that are much larger than at higher speeds.

(c) Because losses are being neglected, we can set input power equal to output power. Thus,

$$I_a(220) = 20(746)$$

or

$$I_a = 67.8 \approx 68 \text{ A}$$

This armature current is the same for both motors at 1000 rpm, the high end of the speed range.

(d) The variation of the armature current for the motor (A) operated with adjustable armature voltage is depicted in Fig. 6-23(b) as curve A. This motor operates with constant field current at all speeds. Hence, its air-gap flux is also constant. The basic torque expression of Eq. (3-69) thus can be written as $T = K_T\phi I_a = KI_a$. Consequently, a constant torque demands a constant armature current.

The situation is different for the motor (F) that employs field control. In this case a drop in speed is the result of an increase in flux. At a constant torque condition this flux increase means a corresponding decrease in armature current. The variation of this current over the speed range is illustrated as curve F in Fig. 6-23(b).

Appearing in Fig. 6-23(c) are the plots of the armature voltage with speed. Motor F operates at a constant armature voltage. The armature voltage variation of motor A, however, must change with the speed. In fact, the speeds below 1000 rpm are achieved precisely because of the reduction in armature voltage consistent with the equation for the counter emf of the armature which, by the assumption of zero losses, becomes

$$V = K_E\phi n = K_E'n = E_a$$

Hence, if rated voltage corresponds to operation at 1000 rpm, then operation at 250 rpm is obtained when the armature voltage is reduced to 55 V.

(e) Motor A operates at a constant flux and a constant armature current. The quantity $K_T\phi$ readily follows from Eq. (3-69) as

$$(K_T\phi)I_a = T$$

$$K_T\phi(68) = 105(1.356) = 142.4 \text{ N-m}$$

$$K_T\phi = 2.1 \text{ N-m/A}$$

Of course, this quantity stays fixed throughout the speed range as indicated by curve A in Fig. 6-23(d).

In the case of motor F part (c) shows that the armature current is also 68 A at 1000 rpm. But because the lower speeds in this motor are the result of increasing flux values, it follows that the quantity $K_T\phi$ for motor F increases with decreasing speed as illustrated by curve F of Fig. 6-23(d). It is instructive to note that for constant torque the variations of $(K_T\phi)$ and armature current for motor F are complementary. The increase in $K_T\phi$ is accompanied by a corresponding decrease in armature current.

In a motor with negligible losses by the law of conservation of energy we can write

$$T\omega_m = E_aI_a$$

$$K_T\phi I_a\omega_m = K_E\phi\omega_m I_a$$

or

$$K_T\phi = K_E\phi \qquad (6\text{-}31)$$

Therefore, the variation of $K_T\phi$ which bears units of N-m/A is identical to the parameter $K_E\phi$, which bears units of V/rad/s. As an illustration of the use of this result, we can find the armature voltage of motor A at a speed of 250 rpm ($= 26.18$ rad/s) by

(a)

(b)

(c)

(d)

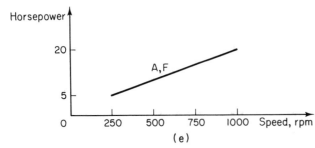

(e)

Figure 6-23 Performance feature of two motors used to deliver constant torque over a 4:1 speed range: (a) constant torque characteristic needed by load; (b) armature current variation at constant torque; (c) armature voltage variation at constant torque; (d) flux factor, $k\phi$, variation; (e) horsepower variation. A, motor with armature voltage control; F, motor with field current control.

simply multiplying the parameter 2.1 by the speed expressed in radians per second. Thus,

$$2.1(16.18) = 55 \text{ V}$$

(f) The plot for both motors is shown in Fig. 6-23(e). Of course, the hp changes in direct proportion to the speed.

(g) The maximum hp for motor A is 5 because the motor is working as hard electrically and magnetically as it has been designed to work. It is at full capacity electrically since it is drawing rated armature current. It is also at full capacity magnetically since it is operating at its fixed excitation value.

The situation, however, is different with motor F. Although this motor has rated armature voltage applied, it is drawing only 17 A at 250 rpm. Accordingly, this motor could be called on to deliver at 250 rpm an increased torque to the level of 4(105) = 425 lb-ft, corresponding to a fourfold increase in armature current thus raising it to its rated value of 68 A. The corresponding delivered hp is then 20, which is four times larger than that of motor A.

6-9 PERMANENT MAGNET (PM) D-C MOTORS

In order for a d-c motor to develop torque and thereby attain an operating speed, it is necessary that there exist a flux density distribution along the air gap of the machine. We already know that in conventional d-c motors this field distribution is established by the action of a d-c current that flows through the field winding which wraps around the salient poles of the stator structure. However, circumstances do arise in the application of d-c motors where machines so equipped offer no particular advantage and, in fact, may turn out to be impractical, inconvenient, and/or more expensive than alternative ways of furnishing the machine with flux density. A particularly striking example of this situation is the application of d-c motors in countless motor vehicles throughout the world, doing such tasks as driving windshield wipers, fan motors, windows, air conditioners, radio antennas, and the like. The selection of the d-c motor is clearly a natural choice because of the availability of a d-c source in the form of a rechargeable battery. The motor size needed to accomplish most of these jobs is very small and in most cases is a small fraction of a horsepower. Accordingly, the use of permanent magnets (PM) in place of field windings offers the advantages of lower cost and simpler construction. It also means higher efficiency because of the absence of the field winding copper losses but in these small sizes this is generally not a significant factor.

Permanent magnet d-c motors have also found an application niche in situations requiring integral horsepower sizes exceeding even 100 hp. These motors can be found in systems where motor speed control below base speed only is required and is achieved by armature voltage control through electronic methods like those described in Sec. 6-10. Of course, in applications where control of the motor speed must extend beyond base speed, adjustment of the field flux is necessary and a wound-field machine must be specified because the PM d-c motor does not offer this flexibility.

The material used to construct a PM d-c motor may be one of three types: Alnico magnets, ceramic (ferrite) magnets, or rare earth magnets. These materials have

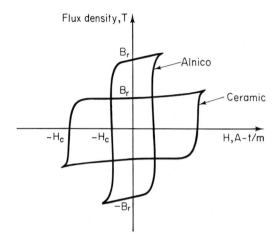

Flux density, T

B_r

—Alnico

B_r

— Ceramic

$-H_c$ $-H_c$ H, A-t/m

$-B_r$

Figure 6-24 General shapes of the hysteresis loops of two permanent magnet materials: (a) alnico; (b) ceramic (ferrite).

the distinguishing feature of very high residual flux density and high coercive force. A typical hysteresis loop is shown in Fig. 6-24. Alnico has a residual flux density of the order of 1 T, which is comparable to the flux densities found in wound-field machines, but it falls mostly on the low end of the range. Consequently, for a given hp rating the Alnico-type PM d-c motor has a slightly larger armature than its wound-field counterpart. The ceramic magnet has a residual flux density that is usually less than half that of Alnico and it is also the least expensive of the PM materials. As a result ceramic magnets are used for the construction of PM d-c motors in the fractional hp range while Alnico magnets are used invariably for motors in the integral hp range.

There are certain disadvantages associated with the use of strong permanent magnets in d-c motors especially when the ratings are in the integral hp range. Because the magnetic field is active at all times even when the motor is not being used, the motor enclosure must be more carefully designed to guard against attracting foreign matter that could be harmful to the motor. A more serious drawback concerns demagnetization of the permanent magnets. This can occur especially if the armature current should reach excessive values as might happen under fault conditions such as an armature short circuit. Equipping the motor with a compensating winding is one effective way to avoid this problem but of course this increases the cost of the motor.

Demagnetization is a problem that is inherently characteristic of permanent magnets. To better understand this statement, let us examine the situation illustrated in Fig. 6-25. Depicted in Fig. 6-25(a) is a permanent magnet attached to a very high permeability magnetic material to form a closed magnetic circuit. If the magnetic material attached to the PM is assumed to offer zero reluctance, the flux density that then penetrates the magnetic circuit is B_r, the residual flux density of the PM. Consider next the situation that is shown in Fig. 6-25(b), which is more in keeping with what is encountered in the actual magnetic circuit of the machine because of the presence of the two air gaps. The insertion of the air gaps into the magnetic circuit of Fig. 6-25(a) necessarily leads to a level of flux density in the magnetic circuit of Fig. 6-25(b) which is less than that found in Fig. 6-25(a). The presence of the air gaps increases the reluctance of the magnetic circuit and the effect may be interpreted in terms of an equivalent demagnetizing field intensity (H_d) associated with the insertion of the air gaps.

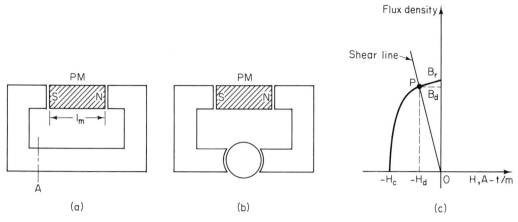

Figure 6-25 Application of the PM to magnetic circuits: (a) PM in a magnetic circuit of zero reluctance; (b) PM in a magnetic circuit with two air gaps; (c) locating the operating flux density value in the magnetic circuit of (b).

Expressed mathematically we can write

$$-H_d l_m = 2\mathfrak{F}_g \qquad (6\text{-}32)$$

where $\mathfrak{F}_g$ is the mmf required to force the flux density of the PM to cross one air gap, l_m is the effective length of the PM and H_d is the equivalent demagnetizing field intensity associated with the air gaps. The quantity $H_d l_m$ is that magnetomotive force which would have to be applied on the PM in a direction opposite [hence the negative sign in Eq. (6-32)] to the residual flux to produce the *same* effect as the insertion of the air gaps. The gap mmf is

$$\mathfrak{F}_g = \phi_g \mathfrak{R}_g \qquad (6\text{-}33)$$

where ϕ_g is now the gap flux associated with a partially demagnetized flux density $B_d \, (< B_r)$ and the cross-sectional area A of the magnetic circuit. Thus

$$\phi_g = B_d A \qquad (6\text{-}34)$$

Hence

$$\mathfrak{F}_g = \phi_g \mathfrak{R}_g = B_d A \left(\frac{g}{\mu_0 A}\right) = \frac{B_d g}{\mu_0} \qquad (6\text{-}35)$$

where μ_0 denotes the permeability of air.

Introducing the last expression into Eq. (6-32) and rearranging terms leads to

$$B_d = -\left(\frac{2\mu_0 l_m}{g}\right) H_d \qquad (6\text{-}36)$$

This is clearly the equation of a straight line that passes through the origin and exhibits a negative slope given by the coefficient of H_d. The negative slope plots the line on the demagnetizing side of the B-H curve of the permanent magnet. The intersection of this line with the hysteresis loop of the PM material yields the new operating flux density B_d when the configuration in which the PM magnet is placed changes from that depicted in Fig. 6-25(a) to that of Fig. 6-25(b). Figure 6-25(c) illustrates the result. The straight line of Eq. (6-36) is called the *shear line* of the magnetic circuit in which the PM acts.

It is instructive to note that when the gap length is allowed to approach zero, the slope of the shear line becomes infinite. This means that the operating flux density returns to the residual value.

In any practical PM d-c motor the operating point is represented by P in Fig. 6-25(c) which means that the actual flux density is less than the residual flux density that is associated with the *B-H* curve of the PM material. If an Alnico PM motor is subjected to a severe armature current, a further demagnetization takes place from the operating point P. Upon subsequent removal of this demagnetizing force, the effect of a minor hysteresis loop prevents the operating point from returning to point P. Instead the operating flux density assumes a somewhat smaller level. It is useful to note, however, that the foregoing situation does not duplicate itself when the PM material is ceramic. For permanent magnets constructed of ceramic, the operating point P is virtually re-established upon removal of the demagnetization associated with an excessive armature current.

6-10 SPEED CONTROL BY ELECTRONIC MEANS

The development of reliable thyristors (SCRs) with high current and voltage ratings has made the speed control of d-c motors by armature voltage adjustment more viable. Keep in mind that adjustment of the armature voltage offers the advantage of speed control below base speed as well as operation at high efficiency coupled with good speed regulation. Moreover, the control of speed can be achieved whether the motor is operating under load conditions or no load.

Single-Phase, Half-Wave Thyristor Drive

Figure 6-26 shows the arrangement that is used when a single-phase source is available. It is frequently used for motors which are equipped with permanent magnet fields and rated for 1 hp or less. Because a single thyristor is used in this circuit, control of the firing point, θ_f, by the gate signal can occur anywhere between 0 and π radians. Illustrated in Fig. 6-27 is a typical situation that can prevail for a firing angle of $\theta_f = 90°$ during steady-state conditions. Time in this diagram is assumed to be measured after steady state is reached. Moreover, the load torque to be supplied by the motor is assumed to be at such a level as to require an average armature current of I_a projected over a full cycle. To produce this average armature current, however, it is necessary for the actual variation of the armature current to have such an instantaneous variation that integration over the conduction period yields an average value of I_a. The conduction period is the time during which the thyristor is conducting and generally this is represented by

$$\theta_c = \theta_f - \theta_e \tag{6-37}$$

where θ_f is the angle in the positive excursion of the supply voltage where the gate signal fires the thyristor and θ_e is the extinction angle of the thyristor.

The extinction angle is determined as that point in the cycle where the actual armature current i_a returns to zero, which in turn depends upon the relationship of the instantaneous value of the supply voltage v_i and the variation of the armature counter

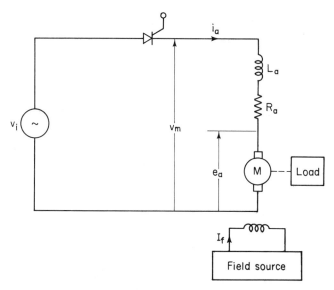

Figure 6-26 Armature voltage control of a d-c motor through use of a single-phase, half-wave controlled rectifier. The motor may have either a permanent magnetic field or a separate source of excitation.

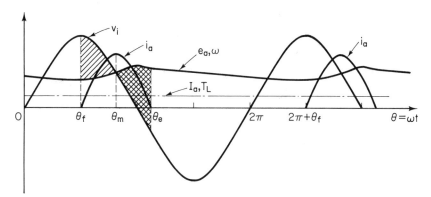

Figure 6-27 Waveshapes of the various electrical quantities and motor speed and load torque associated with the system of Fig. 6-26. The thyristor firing angle is illustrated for $\theta_f = 90°$.

emf e_a. This e_a must always assume such a position relative to v_i that it produces a time variation in i_a which yields a value of the average armature current that allows a torque to be developed to meet the needs of the load. It is instructive to note here that the instantaneous armature current i_a undergoes an increase in value from θ_f to θ_m, where θ_m corresponds to the time instant where $v_i = e_a$. In this interval the voltage across the armature winding inductance in Fig. 6-26 is positive. Consequently, an increase in current from zero takes place which can be described by a form of Faraday's law as

$$\Delta i_a = \frac{1}{L_a} \int_{\theta_f}^{\theta_m} (v_i - e_a) \, dt \tag{6-38}$$

where L_a denotes the armature winding inductance. Beyond θ_m the quantity $(v_i - e_a)$ becomes negative. By Eq. (6-38) the effect of these negative volt-seconds, appearing across the armature inductance, is to cause the armature current to diminish. In fact, the extinction angle is that point in the cycle where the sum of the negative volt-seconds (i.e., the double-crosshatched area in Fig. 6-27) is equal to the sum of the positive volt-seconds (i.e., the single-crosshatched area).

When the load calls for an increased developed torque by the motor, the motor responds by reducing its speed (and therefore e_a) by an amount that allows a sufficient increase in positive volt-seconds to yield a greater instantaneous current flow, i_a. Correspondingly, a larger average value of I_a is made to exist over a full cycle thereby meeting the increased torque demand of the load. The variation with time of e_a and the motor speed, ω_m, are represented in Fig. 6-27 by the same curve. It is understood, of course, that the curve is drawn to a different scale. The same applies to the curves marked I_a and T_L. In the interval from θ_f to θ_e observe that the motor is receiving a pulse of energy from the source which manifests itself as a slight increase in speed. The energy stored in the rotating mass, consisting of the rotor of the motor and its attached mechanical load (i.e., the total inertia J), is thereby increased. However, because there is no armature current flow in the interval from θ_e to $(2\pi + \theta_f)$, the motor finds itself coasting along. During this period, energy is delivered to the load by extracting it from the kinetic energy stored in the rotating mass. The result is a dip in speed. When the motor is called on to deliver full load torque at low speeds, the speed dip will be greater than when it operates at speeds close to its base value.

Although small d-c motors do operate successfully in the circuitry of Fig. 6-26, it is not without notable shortcomings. For example, the need to meet load torque requirements in terms of an armature current averaged over a full cycle of the supply frequency means that large pulses of armature current must flow during the conduction period. This leads to much higher heating losses. Unless special arrangements are made for cooling, it may be necessary to operate the motor at a reduced horsepower rating. Another disadvantage is related to the fact that an average d-c current is made to flow through the a-c source. If this source should be a single-phase transformer, then it is possible for the transformer to operate in a state of partial saturation. These problems are essentially avoided by operating from a three-phase source.

Three-Phase, Half-Wave Thyristor Motor Drive Circuit

The circuit diagram for this mode of operation is depicted in Fig. 6-28. A three-phase voltage source is assumed to be available with an accessible neutral. It is the line-to-neutral voltages which appear across the series combination of the individual thyristors and the armature winding circuit. The motor field is presumed to be originating from a permanent magnet or from a separate d-c source which itself may involve rectification of an a-c supply. The application of the three-phase voltage set of Fig. 6-29 in the drive circuit of Fig. 6-28 causes the thyristors S_a, S_b, and S_c to fire in sequence over a potential conduction period of 120°. The actual conduction period depends on when the firing gate signal is applied during this 120° interval. A glance at Fig. 6-29 should make it apparent that between $\theta = 30°$ and $\theta = 150°$ a positive potential appears across thyristor S_a. Accordingly, it can act as a switch (hence the notation S for the

Figure 6-28 Schematic diagram of the three-phase, half-wave controlled rectifier drive circuit for the d-c motor.

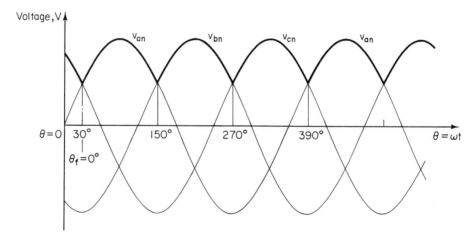

Figure 6-29 Waveforms of the balanced three-phase source. The heavy lines indicate the 120° intervals associated with each phase voltage during which thyristor firing can be initiated.

thyristor) that serves to put a portion or all of v_{an} across the armature terminals of the motor. Simultaneously, during this period, the voltage appearing across the other two thyristors is negative and so S_b and S_c are kept in a cutoff state. Similarly, in the interval from $\theta = 150°$ to $\theta = 270°$, thyristor S_b is prepared for firing because here it now possesses a positive forward voltage while thyristors S_a and S_c have negative voltages on their anode-cathode terminals. In the interval from 270° to 390° thyristor S_c is ready for conduction. This sequenced firing of the thyristors results in a rectification of the three-phase line-to-neutral voltages which has a controllable average value that is determined by the point at which the thyristors are allowed to conduct through action of the gate signals. It is this nonzero average voltage that drives the d-c motor.

Let us now obtain an expression that relates this controllable average voltage to the firing angles of the thyristors. Call θ_f the thyristor firing angle and for convenience let $\theta_f = 0°$ correspond to $\theta = \omega t = 30°$ in Fig. 6-29. Because θ_f can be assigned any value from 30° to 150° (or $\pi/6$ to $\pi/6 + \frac{2}{3}\pi$), it follows that the expression for the average voltage over the first conduction period corresponding to firing at angle θ_f is

$$V_m = V_{dc} = \frac{1}{\frac{2}{3}\pi} \int_{\theta_f + \pi/6}^{\theta_f + (5/6)\pi} V_p \sin (\omega t) \, d(\omega t) \qquad (6\text{-}39)$$

where V_p denotes the peak value of the line to neutral voltage and V_m refers to the motor armature voltage. It is important to note in this equation that the upper limit of the integral is written in recognition of the fact that the thyristor stays in the ON state for a period of 120° (i.e., $\frac{2}{3}\pi$) once it is allowed to initiate conduction. Performing the integration and inserting the limits, the desired result is found to be

$$V_{dc} = V_m = 0.827 V_p \cos \theta_f \qquad (6\text{-}40)$$

This equation clearly reveals the controllable aspect of the rectified a-c source voltage. A setting of the thyristor gate circuit to yield $\theta_f = 0°$ means that a maximum d-c voltage appears across the armature terminals. At the other extreme a setting of $\theta_f = 90°$ reduces this voltage to zero. Accordingly, there is at our disposal a conveniently adjustable armature voltage source to permit speed control below the base speed of the motor.

An illustration of the speed control capability of the drive circuit of Fig. 6-28 for three values of the thyristor firing angle is shown in Fig. 6-30. Part (a) of this diagram depicts the situation in the drive circuit when the motor is at no load and $\theta_f = 0°$. Thyristor S_a initiates conduction at $\theta = 30°$ and so applies the value of v_{an} to the armature winding of the motor over the period from $\theta = 30°$ to $\theta = 150°$. A similar situation prevails when phases b and c become most positive. The average value of the rectified source is at its maximum. If the armature winding resistance drop is assumed to be zero, then E_a becomes equal to this value and ω_m takes on a corresponding value consistent with the state of its field excitation. Based on the assumption that the maximum rectified voltage is the same as the rated d-c motor voltage and the flux field is at its maximum setting, the resulting speed is the base speed. In Fig. 6-30(b) the firing angle is delayed 60°. Consequently, the initial value of v_{an} that appears on the armature circuit is its peak value along with the variation of v_{an} that occurs over the next 120°. In other words, that section of v_{an} is placed across the armature circuit that corresponds to curve def. The average value of this section is $0.413V_p$ and is consistent with a firing angle of 60°. Of course, this process is repeated for each third of the cycle involving in succession thyristors S_b and S_c. Now the speed is at a level one-half the base speed again based on the assumption of negligible armature winding resistance drop and thyristor drop. Figure 6-30(c) depicts the situation as it exists for $\theta_f = 90°$. Thyristor S_a initiates conduction at $\theta = 120°$ (i.e., $\theta = 30° + \theta_f = 30° + 90°$) and continues in the ON state until $\theta = 240°$. Accordingly, the portion of v_{an} that appears across the motor armature winding is the section represented by $d'e'f'$, which already has an average value of zero. Hence the motor speed becomes zero.

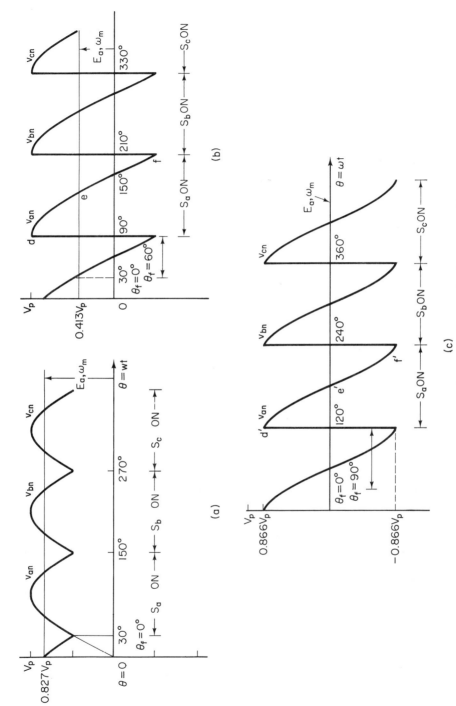

Figure 6-30 Waveshapes of the rectified three-phase source for three values of the thyristor firing angle. Also shown are the average values of the armature induced emf E_a and the motor speed ω_m: (a) full-firing of the thyristors for $\theta_f = 0°$; (b) $\theta_f = 60°$; (c) $\theta_f = 90°$.

363

When the d-c motor is operating at a subbase speed under no-load conditions, what is the effect of applying load? The demand for increased torque to be delivered to the attached mechanical load is met by a decrease in speed just as it is in the conventional d-c motor. Although the average armature voltage and current are derived from rectification, Eqs. (6-19) to (6-21) continue to be valid provided that the terminal average values are used. As a result in those instances where the armature winding voltage drop is moderate or small, relatively small changes in the speed accompany the changes in load. This matter is illustrated in the following example.

Example 6-5

A d-c motor, which is rated at 550 V and 60 hp, draws 30 A when it delivers rated torque at speed of 800 rpm. The field is set at a fixed value that gives the motor a torque and speed parameter of 5.94 N-m/A (or V/rad/s). The motor is supplied from a three-phase, 60-Hz, 1100 line-to-line source through a half-wave thyristor drive circuit. The armature circuit resistance is 1 Ω. Assume that the thyristors operate with negligible voltage drop.

(a) Determine the required firing angle when the motor operates at rated load conditions.

(b) Compute the change in speed that occurs when the load torque is reduced to one-half the rated value. What is the percent speed change?

(c) Find the firing angle that is needed when rated torque is to be delivered at a speed of 500 rpm.

Solution (a) Kirchhoff's voltage law employing average values appearing at the motor terminals allows us to write

$$V_{dc} = V_m = E_a + I_a R_a$$

$$0.827 V_p \cos \theta_f = (5.94)\omega_m + 30(1)$$

Here

$$\omega_m = 800\left(\frac{\pi}{30}\right) = 83.78 \text{ rad/s}$$

$$V_P = \sqrt{2}\left(\frac{1100}{\sqrt{3}}\right) = 898.1 \text{ V}$$

Thus

$$742.73 \cos \theta_f = 497.65 + 30 = 527.65$$

$$\theta_f = 44.73°$$

(b) With the load torque reduced to one-half, the required average armature current is also halved. Since the firing angle is maintained, we have

$$527.65 = 5.94\omega_m + 15$$

or

$$\omega_m = 86.3 \text{ rad/s}$$

Expressed in revolutions per minute, the speed is

$$n_m = \omega_m\left(\frac{30}{\pi}\right) = 86.3\left(\frac{30}{\pi}\right) = 824.1 \text{ rpm}$$

The change in speed from the rated value expressed in percent is

$$\frac{824.1 - 800}{800} \times 100 = 3.02\%$$

(c) We proceed as in part (a). However, in this part the speed is

$$\omega_m = \frac{5}{8}(83.78) = 52.36 \text{ rad/s}$$

Hence

$$742.73 \cos \theta_f = (5.94)(52.36) + 30 = 341 \text{ V}$$

$$\theta_f = 62.7°$$

Chopper Drives

There is an area of application of d-c motors where d-c energy is available, often in the form of batteries, and where operation below base speed is extremely important. This is the case for example with forklifts, golf carts, electrically powered trucks, and even rapid-transit cars. Of course, as already pointed out in the preceding section, one workable scheme is to use a series resistor. But we already know that this solution carries with it such serious shortcomings as poor efficiency, poor speed regulation, and high cost. A popular method of providing armature voltage control today in such applications is to use the thyristor in a pulse width modulation mode or a pulse frequency modulation mode. In the former the d-c source voltage is switched on and off for a variable part of a cycle in the interest of adjusting the level of average voltage applied to the motor armature winding. In the latter a pulse of voltage of fixed width, which is derived from the d-c source, is repeated more or fewer times over a given time period. A typical schematic arrangement of the circuitry is shown in Fig. 6-31. Because the principles are the same, attention is directed here to just one of the schemes, namely, pulse width modulation of the d-c source voltage.

Figure 6-31 Chopper drive circuit for armature voltage control of a d-c series motor.

The control circuitry that is connected to the thyristor has a twofold purpose. First, it serves to turn the thyristor on over a variable part of a period T_p as illustrated in Figs. 6-32(a) and (b). In the time interval from 0 to T_0 the gate signal puts the thyristor into a conductive state thereby passing the battery voltage to the armature winding. Second, at time T_0, a reverse voltage is instantaneously (measured in the

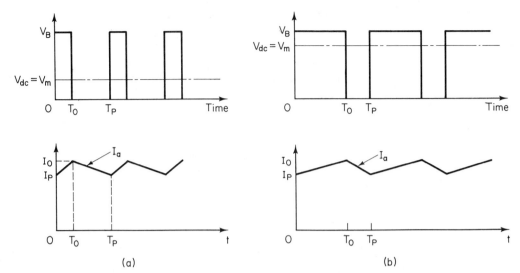

Figure 6-32 Voltage and current waveshapes appearing at the armature terminals of the motor in Fig. 6-31: (a) thyristor fired for low-speed operation; (b) thyristor fired for high-speed operation.

tens of microseconds) applied to the thyristor thereby cutting it off. At the end of the period represented by T_P, the process is repeated. In essence then the control circuitry acts to "chop" the d-c supply voltage before it is applied to the motor armature winding. These drive circuits are called chopper drives for obvious reasons. The average value of the d-c voltage that is transferred to the motor is quite simply given by

$$V_{dc} = V_m = \frac{T_0}{T_p} V_B \qquad (6\text{-}41)$$

where the meaning of the quantities are self-evident from their identities in Fig. 6-32.

The frequency at which the thyristor is pulsed is critical to the design of the control circuitry if large peaks in the armature current are to be avoided. In a properly designed chopper drive circuit, the variation of the armature current looks very much like that illustrated in Fig. 6-32 during steady-state operation. For equilibrium to exist the increase in current during time T_0 is equal to the decrease in current during the time $(T_P - T_0)$. In terms of the quantity of volt-seconds applied to the armature inductance, we can write as the increase in current in period T_0

$$\Delta I_a = \frac{1}{L_a} \int_0^{T_0} [V_B - (V_m - I_a R_a)] \, dt = \frac{[V_B - (V_m - I_a R_a)]T_0}{L_a} \qquad (6\text{-}42)$$

since the integrand is essentially constant. The motor voltage V_m is given by Eq. (6-41). Of course, the effect of this increased current is to augment the magnetic energy stored in the armature field. Then, following cutoff of the thyristor at time T_0, the energy is released to the motor through the circuit established by the free-wheeling diode D_{fw}. The rate of decay of the armature current during the time $(T_P - T_0)$ is determined by the time constant of the armature circuit which is often very large compared to the cycle time of the pulses. For this reason the decay is illustrated as a straight line.

DIRECT-CURRENT MACHINES   CHAP. 6

Because of the important role of the armature inductance in storing sufficient magnetic energy, chopper drives are frequently used with series motors. The series motor has larger armature inductance than the shunt motor, although perfectly acceptable performance can be had when the latter machine is used with an external series inductor.

Example 6-6

Refer to the d-c motor described in Example 6-5 and assume that it is to be operated with a chopper drive from a 500-V d-c source.

(a) Find the motor speed when the thyristor is made to operate with a ratio of ON-time T_0 to cycle time T_P of 0.4. Assume that the motor delivers rated torque.

(b) Determine the required pulse frequency of the thyristors that serves to limit the total change of armature current during the ON period to 5 A. Assume an armature winding inductance of 0.1 H.

(c) Repeat parts (a) and (b) for $T_0/T_P = 0.7$.

Solution (a) From Eq. (6-41) the average value of the armature terminal voltage is

$$V_m = \frac{T_0}{T_P} V_B = 0.5(550) = 220 \text{ V}$$

The corresponding value of the induced armature voltage then becomes

$$E_a = V_m - I_a R_a = 220 - (30)(1) = 190 \text{ V}$$

Therefore

$$\omega_m = \frac{190}{5.94} = 32 \text{ rad/s} \quad \text{or} \quad 305.4 \text{ rpm}$$

(b) The time T_0 is found directly from Eq. (6-42). Thus,

$$T_0 = \frac{L_a(\Delta I_a)}{V_B - E_a} = \frac{0.1(5)}{550 - 190} = 1.4 \times 10^{-3} \text{ s}$$

Accordingly,

$$T_P = \frac{T_0}{0.4} = \frac{1.4}{0.4} \times 10^{-3} = 3.5 \times 10^{-3} \text{ s}$$

and this leads to the following required pulses per second:

$$f = \frac{1}{T_P} = \frac{10^3}{3.5} = 286 \text{ Hz}$$

(c) Now Eq. (6-41) yields

$$V_m = 0.7(550) = 385 \text{ V}$$

$$E_a = 385 - 30 = 355 \text{ V}$$

$$\omega_m = \frac{355}{5.94} = 64.81 \text{ rad/s or } 618.9 \text{ rpm}$$

Also, from Eq. (6-42),

$$T_0 = \frac{(0.1)5}{195} = 2.56 \times 10^{-3}$$

$$T_P = \frac{2.56 \times 10^{-3}}{0.7} = 3.66 \times 10^{-3}$$

$$f = \frac{1}{T_P} = 273 \text{ Hz}$$

6-11 APPLICATIONS OF D-C MOTORS

The d-c motor is often called upon to do the really tough jobs in industry because of its high degree of flexibility and ease of control. These features cannot easily be matched by other electromechanical energy-conversion devices. The d-c motor offers a wide range of control of speed and torque as well as excellent acceleration and deceleration. For example, by the insertion of an appropriate armature circuit resistance, rated torque can be obtained at starting with no more than rated current flowing. Also, by special design of the shunt-field winding speed adjustments over a range of 4:1 above rated speed are readily obtainable. If this is then combined with armature-voltage control, the range of speed adjustment spreads to 8:1. In some electronic control devices that are used to provide the d-c energy to the field and armature circuits, a speed range of 40:1 is possible. The size of the motor being controlled, however, is limited.

TABLE 6-1 CHARACTERISTICS AND APPLICATIONS OF D-C MOTORS

Type	Starting torque (%)	Max. running torque, momentary (%)
Shunt, constant speed	Medium—usually limited to less than 250 by a starting resistor but may be increased	Usually limited to about 200 by commutation
Shunt, adjustable speed	Same as above	Same as above
Compound	High—up to 450, depending upon degree of compounding	Higher than shunt—up to 350
Series	Very high—up to 500	Up to 400

Table 6-1 lists some of the salient characteristics and typical applications of the three types of d-c motors. It is interesting to note that the maximum torque in the case of the d-c motor is limited by commutation and not, as with all other motor types, by heating. Commutation refers to the passage of current from the brushes to the commutator and thence to the armature winding itself. The passage from the brushes to the commutator is an arc discharge. Moreover, as a coil leaves a brush the current is interrupted, which causes sparking. If the armature current is allowed to become excessive, the sparking can become so severe as to cause flashover between brushes. This renders the motor useless.

Another point of interest in the table is the considerably higher starting torque of the compound motor in comparison with the shunt motor. This feature is attributable to the contribution of the series field winding. The same comment is valid as regards the maximum running torque. In each case, of course, the limit for the armature current is the same.

Speed-regulation or characteristic (%)	Speed control (%)	Typical application and general remarks
5–10	Increase up to 200 by field control; decrease by armature-voltage control	Essentially for constant-speed applications requiring medium starting torque. May be used for adjustable speed not greater than 2:1 range. For centrifugal pumps, fans, blowers, conveyors, woodworking machines, machine tools, printing presses
10–15	6:1 range by field control, lowered below base speed by armature-voltage control	Same as above, for applications requiring adjustable speed control, either constant torque or constant output
Varying, depending on degree of compounding—up to 25–30	Not usually used but may be up to 125 by field control	For drives requiring high starting torque and only fairly constant speed; pulsating loads with flywheel action. For plunger pumps, shears, conveyors, crushers, bending rolls, punch presses, hoists
Widely variable, high at no-load	By series rheostat	For drives requiring very high starting torque and where adjustable, varying speed is satisfactory. This motor is sometimes called the traction motor. Loads must be positively connected, not belted. For hoists, cranes, bridges, car dumpers. To prevent overspeed, lightest load should not be much less than 15 to 20% of full-load torque

Source: By permission from M. Liwschitz-Garik and C. C. Whipple, *Electric Machinery*, vol. I (Princeton, N.J.: D. Van Nostrand Co., Inc., 1946).

6-12 STARTERS AND CONTROLLERS FOR D-C MOTORS

The limitations imposed by commutation as well as voltage-dip restrictions on the source as set forth by the electric utility company make it necessary to use a starter or controller on all d-c machines whose ratings exceed 2 hp. A glance at Eq. (6-23) discloses that at starting ($n = 0$) the armature current is limited solely by the armature circuit resistance. Hence, if full terminal voltage is applied, excessive armature currents will flow. This is especially so where large machines are involved because the armature resistance gets smaller as the rating increases. In addition to limiting the armature current, controllers fulfill other useful functions as described in Sec. 4-10.

Depicted in Fig. 6-33 is a simple line starter which is used for small d-c motors. The operation is straightforward. Pushing the start button energizes the main coil which then closes the main contactors M and the interlock M_a. Note that, when the main contactors close, voltage is applied to the armature winding through the starting resistor R and simultaneously to the field winding. This arrangement prevents "shock" starting because some time is needed before the field flux is fully established. Note, too, that this starter has no provisions for removing the starting resistor once the motor has attained its operating speed. In small motors this is of little consequence and it makes for an inexpensive starter. However, the starter resistor in this case does serve another purpose. When the motor is stopped, the field winding is disconnected from the line. The energy stored in the magnetic field then discharges through the starting resistor, preventing possible damage to the field winding.

Figure 6-33 Line starter for a low-horsepower d-c motor.

Three types of magnetic controllers are in use today for controlling the starting current, the starting torque, and the acceleration characteristics of d-c motors. One type is the *current-limit* controller, which works on the principle of keeping the current during the starting period between specified minimum and maximum limits. It is not too commonly used because its success depends upon a tricky electrical interlock arrangement which must be kept in excellent operating condition at all times. No further consideration is given to this type. The second type is the *counter-emf* type which is illustrated in Fig. 6-34. Two kinds of relays are used in this controller. One is a light, fast-acting unit called an accelerating relay (AR). The other is the strong, heavy-duty type previously discussed. The accelerating relays appearing across the armature circuit in Fig. 6-34 are voltage-sensitive devices. They are designed to close when the voltage across the coil exceeds a preset value. For the controller under discussion the

Figure 6-34 Counter-emf magnetic controller for a d-c motor.

accelerating relay 1AR is usually adjusted to "pick up" at 50% of line voltage and 2AR is adjusted to close at 80% of line voltage.

Pressing the start button energizes coil M, which then closes interlock M_a and the main contactors M. Since the armature is initially stationary, the accelerating relays are de-energized so that both steps of the starting resistor are in the circuit. As the armature gains speed and develops an induced emf exceeding 50% of line voltage, coil 1AR snaps closed, closing contactors 1AR. In turn, coil 1A is energized, closing contactors 1A, which short out the first section of the starting resistor. This then applies increased voltage to the armature, which furnishes further acceleration. When the armature induced emf exceeds 80% of line voltage, accelerating relay 2AR closes. This excites coil 2A, which shorts out the second section of the starting resistor. The motor then accelerates to its full-voltage operating speed.

The controller of Fig. 6-34 is also equipped with a resistor that shunts the armature terminals. It is called a *dynamic breaking resistor* because it serves to bring the motor to a quick halt by dissipating energy obtained from the rotating armature inertia. The diagram of Fig. 6-34 shows that a path is established for the dynamic breaking resistor when the stop button is closed and coil M becomes de-energized. This action closes the normally closed contactor M which is placed in series with the dynamic breaking resistor. Since the motor is still turning at a high speed with stored kinetic energy in its inertia and because the shunt field remains energized, the armature behaves as a generator which in turn delivers current to the dynamic breaking resistor. The subsequent expenditure of energy as heat in the resistor acts quickly to drain the stored kinetic energy of the armature, thus reducing the speed to zero rapidly.

The counter-emf controller has the advantage of providing a contactor closing sequence which adjusts itself automatically to varying load conditions. Furthermore, this is accomplished in a manner that maintains uniform accelerating current and torque peaks. There can be no question about the desirability of such starting performance; however, there is one disadvantage. The contactor closing sequence is based on the assumption that the motor will start on the first step. If it fails to do so, all subsequent operations cannot take place. Furthermore, the starting resistor is in danger of burning up. To avoid such occurrences general-purpose d-c motors are most often equipped with *definite time-limit controllers.* Figure 6-35 shows the schematic diagram of such a controller. After a relay coil is energized, the corresponding contactors do not close until the elapse of a preset time delay. The time delay is achieved either by means of magnetic flux decay, a pneumatic device, or a mechanical escapement.

Figure 6-35 Definite time-limit controller for a d-c motor.

Pressing the start button energizes coil M and closes interlock M_a and the main contactors M, thereby applying voltage to the armature winding through the starting resistor. A definite time after the armature of relay coil M closes, contactors T_M in the control circuit close, regardless of whether the rotor is turning or not. This energizes coil 1A, which closes contactors 1A and thus shorts out the first section of the starting resistor. At a preset time delay after coil 1A is energized, contactors T_{1A} close. Coil 2A becomes energized, removing the entire starting resistor. The motor then assumes its normal operating speed.

In the definite time-limit controller the preset time intervals between the closing of contactors are adjusted to obtain smooth acceleration and uniform current peaks for average load conditions. If a heavy starting condition occurs and the motor fails to start on the first step, the first accelerating (time-delay) contactors close anyway. This allows an increased starting torque to be developed. Thus the motor is made to "work harder" if it does not start on the first step. Accordingly, whenever a controller must be selected for a general-purpose motor, it is wiser to prescribe the definite time-limit type. The reason is that as a rule for general-purpose applications the starting conditions are not well known.

PROBLEMS

6-1. The rotor of a two-pole d-c machine has 12 slots. The wave-wound, full-pitch armature winding has two coil sides per slot. There are 12 commutator segments. Show a plan view of the winding layout. Clearly identify the position of the brushes on the commutator and relative to the poles. Also show the direction of the induced emf in the coil sides for a counterclockwise rotation of the rotor.

6-2. The rotor of a two-pole d-c machine carries a wave-wound armature winding having two coil sides per slot in its 12 slots. The winding is full-pitch.
  (a) Draw the schematic diagram that shows the two armature paths existing between the two brushes. Represent each suitably marked coil voltage by small d-c sources with properly identified polarities. Consider that instant when slots 1 and 7 are located exactly between the poles.
  (b) Directing attention solely to the fundamental component of the flux density in the air-gap, compute the voltage value appearing at the brushes.

6-3. Repeat Prob. 6-2 for the case where slot 1 lies to the left of the interpolar axis by 15°.

6-4. For the machine of Prob. 6-2 draw the variation of the armature induced voltage measured at the brushes as a function of $\omega_r t$ where ω_r is the angular rotor velocity. Dealing only with the fundamental component of the flux density curve, show the variation of each coil voltage as well as that of the total. Compute the maximum and minimum values of the coil and total voltages.

6-5. Repeat Prob. 6-1 for a d-c machine with the following specifications: two-poles, six slots, full-pitch, wave-wound armature winding having two coil sides per slot.

6-6. For the machine of Prob. 6-1 sketch the variation of the emf induced in each coil voltage as well as between brushes as a function of $\omega_r t$, where ω_r is the rotor angular velocity. Direct attention solely to the fundamental component of the air-gap flux density. Indicate the maximum and minimum values of coil and brush voltages.

6-7. Explain your answer to each part:
  (a) Can a separately excited d-c generator operate below the knee of its magnetization curve?
  (b) Can a d-c shunt generator operate below the knee of its magnetization curve?

6-8. The magnetization curve for a d-c shunt generator driven at a constant speed of 1000 rpm is shown in Fig. 6-16. Each vertical division represents 20 V; each horizontal division represents 0.2 A.
  (a) Compute the critical field resistance.
  (b) What voltage is induced by the residual flux of this machine?
  (c) What must be the resistance of the field circuit in order that the no-load terminal voltage be 240 V at a speed of 1000 rpm?
  (d) Determine the field current produced by the residual flux voltage when the field circuit resistance has the value found in part (c).
  (e) At what speed must the generator be driven in order that it would fail to build up when operating with the field circuit resistance of part (c)?

6-9. A d-c shunt generator has a magnetization curve given by Fig. 6-16, where each ordinate unit is made equal to 20 V and each abscissa unit is set equal to 0.2 A and the speed of rotation is 1000 rpm. The field circuit resistance is 156 Ω. Determine the voltage induced between brushes when the generator is operated at a reduced speed of 800 rpm.

6-10. It is desired to reverse the terminal voltage polarity of a generator that has been operating properly as a cumulative compound generator. The machine is stopped. Residual magnetism is reversed by temporarily disconnecting the shunt field and separately exciting it with reversed current. The connections are then restored exactly as they were before.

(a) Does the terminal voltage build up? Explain.

(b) If answer to part (a) is yes, will the compounding be cumulative or differential? Explain.

6-11. A two-pole d-c machine is equipped with a wave winding having 120 armature conductors. The pole faces span 70% of a pole pitch and the field winding has 1050 turns. This machine operates with a fixed field current of 1.6 A. The magnetization curve is depicted in Fig. 6-16 when each ordinate unit represents 0.1 Wb and each abscissa unit represents 0.2 A. At rated load the armature current is 40 A.

(a) Determine the amount of demagnetization of the field flux caused by the armature mmf.

(b) What is the value of the air-gap flux under load?

(c) What is the value of flux at the leading pole tip?

6-12. In a lap-wound armature winding there are as many brushes as there are poles. Hence, for a four-pole d-c machine it is customary to use four brushes, with each pair of brushes being responsible for conducting one-half of the total armature current. Show the distribution of current in the coils undergoing commutation for two instances: just before undergoing commutation and five-eighths through the commutation period. Assume resistance commutation throughout.

6-13. The no-load characteristic of a 10-kW, 250-V, d-c self-excited shunt generator driven at 1000 rpm is shown in Fig. 6-16. Each vertical unit denotes 20 V, and each horizontal unit represents 0.2 A. The armature circuit resistance is 0.3 Ω, and the field circuit resistance is set at 157 Ω. Determine:

(a) The critical resistance of the shunt field circuit.

(b) The change in voltage from no-load to full-load. Neglect armature demagnetization.

(c) The shunt field current under rated load conditions.

6-14. Repeat Prob. 6-13(b) and (c) for the case where the armature demagnetizing effect I_{fd} is known to be 0.2 A expressed in terms of equivalent field current.

6-15. The machine of Prob. 6-13 operates with a field winding resistance of 143 Ω.

(a) What is the no-load voltage?

(b) When the generator delivers power to a specified load, the terminal voltage is found to drop to 220 V and the corresponding demagnetization effect of the armature mmf is known to be 0.25 equivalent field ampere. Determine the armature winding current.

(c) What is the value of the field current at the load condition of part (b)?

(d) Find the armature winding emf under load.

6-16. The no-load characteristic of a 5-kW, 125-V shunt generator driven at 1000 rpm is shown in Fig. 6-16. Each ordinate unit represents 10 V, and each abscissa unit represents 0.1 A. The armature circuit resistance is 0.2 Ω, and the field circuit resistance is 157 Ω. When the generator delivers rated armature current, the armature induced emf is 116 V.

(a) Find the terminal voltage.

(b) What is the value of the demagnetization of the armature mmf expressed in equivalent field amperes?

(c) What is the field current?

(d) How much power is delivered to the load?

6-17. The no-load characteristic of a d-c shunt generator driven at a speed of 800 rpm is approximated in its useful range by the equation

$$E = \frac{300 I_f}{2 + I_f}$$

where E is armature induced emf and I_f is the field current. The armature circuit resistance is 0.1 Ω. The field winding resistance is 20 Ω. The demagnetizing effect of armature reaction may be neglected.

(a) The terminal voltage is to be 225 V when the armature current is 150 A. The generator is driven at 800 rpm. Find the resistance of the field rheostat for this condition.

(b) Change the field rheostat setting to 10 Ω. Let the load be disconnected and the speed reduced to 720 rpm. Find the terminal voltage.

(c) Return to the conditions of part (a). Assume the rotational losses to be 2 kW. Find the output power, the efficiency, and the shaft input torque from the prime mover of the generator.

6-18. The no-load characteristic of a d-c machine in its useful operating range and driven at 1500 rpm may be approximated by

$$E_a = \frac{400 I_f}{3 + I_f} \qquad \text{or} \qquad I_f = \frac{3 E_a}{400 - E_a}$$

This machine is operated as a d-c shunt generator whose field circuit resistance is 50 Ω.

(a) Compute the no-load terminal voltage when the generator is driven at 1500 rpm, indicating any necessary assumptions.

(b) What will be the terminal voltage at no-load when the speed is reduced to 750 rpm? Account for the large difference.

6-19. The magnetization curve of a d-c shunt generator is represented in Fig. 6-16, provided that each ordinate unit is multiplied by 40 V and each abscissa unit by 0.5 A. The armature circuit resistance is 0.4 Ω, and the field winding resistance is 80 Ω.

(a) This machine is made to deliver an armature current of 60 A. Find the resistance of the field rheostat (in series with the field winding) so that the terminal voltage will be 520 V at a speed of 1000 rpm.

(b) With the field rheostat set as in part (a) find the terminal voltage at no-load and 1000 rpm.

6-20. A cumulative-compound generator driven at 1000 rpm has the magnetization curve shown in Fig. 6-16, provided that each unit of the ordinate axis is multiplied by 20 V and each abscissa unit is multiplied by 1 A. The armature circuit resistance is 0.06 Ω. In addition:

$$\text{Series turns per pole } (N_s) = 20$$
$$\text{Shunt turns per pole } (N_f) = 1000$$
$$\text{Armature-reaction demagnetizing effect at}$$
$$I_a \text{ of } 100 \text{ A} = 600 \text{ A-t per pole}$$

The field circuit resistance is 45 Ω.

(a) Determine the no-load terminal voltage.

(b) Find the terminal voltage when the load is such that it causes 100 A to flow through the armature winding.

(c) Find the new value of the field current.

(d) What is the value of the armature induced emf at the specified load.

6-21. Repeat Prob. 6-20 for the case where the field circuit resistance is reduced to 31.5 Ω and the number of series field turns per pole is increased to 40. The armature current remains at 100 A.

6-22. A cumulative-compound generator has the magnetization curve shown in Fig. 6-16, where each ordinate unit is 20 V and each abscissa unit is 1 A. Determine the required field circuit resistance that allows the generator to operate at a terminal voltage of 240 V when the armature current is 100 A. In addition:

Armature circuit resistance drop = 10 V
Series field turns per pole = 25
Shunt field turns per pole = 1000
Armature demagnetizing A-t per pole = 800

6-23. A 10-hp, 230-V shunt motor has an armature circuit resistance of 0.5 Ω and a field resistance of 115 Ω. At no-load and rated voltage the speed is 1200 rpm and the armature current is 2 A. If load is applied, the speed drops to 1100 rpm. Determine:
(a) The armature current and the line current.
(b) The developed torque.
(c) The horsepower output assuming the rotational losses are 500 W.
Neglect armature reaction.

6-24. A 230-V, 50-hp, d-c shunt motor delivers power to a load drawing an armature current of 200 A and running at a speed of 1100 rpm. The magnetization curve is given by Fig. 6-16, where each vertical unit represents 20 V and each horizontal unit represents 2 A. Neglect armature reaction. Also $R_a = 0.02$ Ω.
(a) Find the value of the armature induced emf at this load condition.
(b) Compute the motor field current.
(c) Compute the value of the load torque. The rotational losses are 600 W.
(d) At what efficiency is the motor operating?
(e) At what percentage of rated power is it operating?

6-25. Refer to Prob. 6-24, and assume the load is reduced so that an armature current of 75 A flows.
(a) Find the new value of speed.
(b) What is the new horsepower being delivered to the load?

6-26. The magnetization curve of Fig. 6-16 applies to a 230-V, 15-hp motor, where each ordinate unit represents 20 V and each abscissa unit represents 0.2 A. At no-load, the field circuit draws a current of 1.9 A from the line, while the armature current is negligibly small. The speed at no-load is 1000 rpm. A mechanical load is then applied to the motor shaft. The measured armature current is found to be 40 A, causing an armature circuit resistance drop of 16 V. Moreover, the demagnetizing effect of the armature mmf is found to be 0.3 equivalent field ampere.
(a) Compute the load speed in rpm, neglecting the demagnetizing effect of the armature reaction.
(b) Find the speed for the specified load when the demagnetizing effect is accounted for. (*Hint:* The procedure is similar to that outlined for generator behavior, except that now the voltage equation for the motor is used in place of that for the generator. Also, keep in mind that the magnetization curve is valid only for a single speed, so that adjustments are needed.)
(c) Compute the percentage of change in speed from no-load to full-load for parts (a) and (b); compare and explain the difference.

6-27. A 230-V, 15-hp, d-c shunt motor has the magnetization curve shown in Fig. 6-16, where each ordinate unit represents 20 V and each abscissa unit represents 0.2 A. This curve is valid for operation at 1000 rpm.
(a) Rated voltage is applied to this motor, and at no-load the speed is found to be

900 rpm. Negligible armature current flows at this condition. Find the total field circuit resistance for this point of operation.

(b) Load is applied to this motor, corresponding to an armature current of 40 A. If the armature circuit resistance drop is 16 V and the demagnetizing effect of the armature mmf is 0.3 equivalent field ampere, determine the new speed in rpm.

6-28. A 115-V, 15-hp, d-c cumulative-compound motor has the magnetization curve shown in Fig. 6-16, where each ordinate unit represents 10 V and each abscissa unit represents 1 A. The motor speed at no-load and rated voltage applied is 1000 rpm.

(a) Assuming negligible armature current at no-load, find the field circuit resistance for this operating point.

(b) The motor delivers power consistent with an armature current of 100 A, which is also made to flow through the series field winding. The magnetization effect of the series field is represented by 3 equivalent field amperes. The demagnetizing effect of the armature winding is denoted by 0.7 equivalent field ampere. The armature circuit resistance drop (including the series field) is 6 V. Find the speed under load.

6-29. A 20-hp, 230-V, 1150-rpm, 4-pole, d-c shunt motor has a total of 620 conductors arranged in two parallel paths and yielding an armature circuit resistance of 0.2 Ω. When it delivers rated power at rated speed, the motor draws a line current of 74.8 A and a field current of 3 A. Compute:

(a) The flux per pole.

(b) The developed torque.

(c) The rotational losses.

(d) The total losses expressed as a percentage of the rated power.

Assume negligible armature reaction.

6-30. A 230-V, 10-hp, d-c series motor draws a line current of 36 A when delivering rated power at its rated speed of 1200 rpm. The armature circuit resistance is 0.2 Ω, and the series field winding resistance is 0.1 Ω. The magnetization curve may be considered linear. Effects of armature reaction may be neglected.

(a) Find the speed of this motor when it draws a line current of 20 A.

(b) What is the developed torque at the new conditions?

(c) How does this torque compare with the original value? Why?

6-31. A 250-V, 50-hp, 1000-rpm, d-c shunt motor drives a load that requires a constant torque regardless of the speed of operation. The armature circuit resistance is 0.04 Ω. When this motor delivers rated power, the armature current is 160 A.

(a) If the flux is reduced to 70% of its original value, find the new value of armature current.

(b) What is the new speed?

6-32. When a 250-V, 50-hp, 1000-rpm, d-c shunt motor is used to supply rated output power to a constant-torque load, it draws an armature current of 160 A. The armature circuit has a resistance of 0.04 Ω, and the rotational losses are equal to 2 kW. An external resistance of 0.5 Ω is inserted in series with the armature winding. For this condition compute:

(a) The speed.

(b) The developed power.

(c) The efficiency assuming the field loss is 1.6 kW.

Armature reaction is negligible.

6-33. A d-c shunt motor has the magnetization curve shown in Fig. 6-16, where one vertical unit represents 20 V and one horizontal unit represents 2 field amperes. The armature circuit resistance is 0.05 Ω. At a specified load and a speed of 1000 rpm, the field current is found to be 12 A when a terminal voltage of 245 V is applied to the motor. The rotational losses are 2.5 kW, and armature reaction is negligible.

(a) Find the armature current.

(b) Compute the developed torque.

(c) What is the efficiency?

6-34. Answer whether the following statements are true or false. When the statement is false, provide the correct answer.

(a) The emf induced in a single coil of the armature of a d-c machine is necessarily unidirectional.

(b) Putting a suitable limit on the amount of heat dissipated is the single most important design factor for all electric machines.

(c) A d-c motor responds to the need for increased power at its output shaft by changing its ampere-conductor distribution relative to its field distribution.

6-35. A 230-V, d-c shunt-wound motor has an armature circuit resistance of 0.5 Ω. At full-load the armature winding draws 40 A and the speed is measured at 1100 rpm corresponding to a field circuit resistance of 115 Ω.

(a) Find the developed motor torque in newton-meters.

(b) The field circuit resistance is increased to 144 Ω. Find the new operating speed assuming the developed torque remains constant in demand to load requirements. Neglect saturation.

(c) Calculate the efficiency in part (b). Assume that rotational losses amount to 600 W.

6-36. A 230-V, d-c shunt motor is used as an adjustable speed drive over a range of zero to 2400 rpm. Speeds from zero to 1600 rpm are obtained by adjusting the armature terminal voltage from zero to 230 V, with the field current kept fixed at full field value. Speeds from 1600 rpm to 2400 rpm are obtained by increasing the field current, with the armature terminal voltage maintained at 230 V. Ignore machine losses and armature reaction effects.

(a) The torque required by the load remains constant over the entire speed range. Show the general form of the curve for armature current versus speed, over the full speed range.

(b) Instead of keeping the load torque constant, suppose that the armature current is not to exceed a specified value. Show the general form of the curve for allowable load torque versus speed, over the entire speed range.

6-37. A customer is interested in buying a d-c shunt motor to supply a load requiring a constant torque of 525 lb-ft continuously, over a speed range of 500 to 2000 rpm.

(a) List three ways that this job can be done.

(b) For each method of part (a), specify the base speed and horsepower rating of the motor.

(c) List the outstanding advantage associated with each method of control.

6-38. Two adjustable-speed, d-c shunt motors have maximum speeds of 2000 rpm and minimum speeds of 500 rpm. Motor A drives a load requiring constant horsepower over the speed range. Motor B drives one requiring constant torque. Neglect all losses and armature reaction.

(a) If speed adjustment is obtained by field control and if the horsepower outputs are equal at 500 rpm and the armature currents are each 100 A, compute the armature currents at 2000 rpm.

(b) Specify the nameplate horsepower and speed ratings of the individual motors to be purchased to do these jobs. Assume that at 500 rpm the torque for both loads is the same, and equal to 1050 lb-ft.

6-39. Figure P6-39 shows a d-c series motor, with a rheostat in parallel with the field winding. The resistance of this rheostat is equal to two-thirds of the field winding resistance. The motor is operating in the steady state, with the rheostat switch open. The terminal voltage is 250 V, the armature current 40 A, the speed 700 rpm. When the load torque is increased by 50% and the rheostat switch is closed, find the new steady-state values of speed and armature current. Given values:

Armature resistance 0.2 Ω, series field resistance 0.15 Ω.

Neglect saturation, armature reaction, and rotational power losses.

Figure P6-39

6-40. A d-c series motor runs at 300 rpm, and draws a current of 75 A from the 500-V line. The total resistance of the armature and field circuits is 0.4 Ω. Find the new steady-state values of armature current and speed if the line voltage is increased from 500 to 600 V. Assume that the developed torque is to remain constant.

6-41. Read the following sentence carefully, and correct any wrong statement that you may find in it:

A d-c shunt motor operates under load in the steady state. At the time $t = 0$, the field rheostat resistance is reduced; this increases the flux and thereby the torque. So the motor accelerates and reaches a new steady-state speed which is higher than it was before.

6-42. A d-c shunt motor draws 10.0 A from its 220-V supply line. The resistance of its armature circuit (including brushes, brush contacts and interpoles) is $R_a = 1\ \Omega$. The field winding resistance is $R_f = 300\ \Omega$. (There is no field rheostat.) The rotational losses (hysteresis, eddy currents, friction and windage) are 10% of the output power. Find the efficiency, as accurately as the given information and the use of a calculator permit.

6-43. A d-c shunt motor is rated at 3 hp, 115 V, 1000 rpm. At rated operating conditions, the efficiency is 82.5%. The armature resistance (including brush contact) is 0.44 Ω, the field circuit resistance is 145 Ω.

(a) Calculate the value of the induced voltage E_a at rated operating conditions.

(b) If this motor is tested at *no-load*, with the supply voltage adjusted to the value calculated under (a) and the speed kept at 1000 rpm (so that the *rotational losses* remain the same as at full load), how much is the armature current under this condition? (*Note:* The $I_a r_a$ drop at no-load can be neglected.)

6-44. Two identical d-c machines are tested in "opposition" as shown in Fig. P6-44 (armatures in parallel, shafts coupled, losses supplied from an electrical source). The letters G and M stand for *generator* and *motor*.

Figure P6-44

(a) Which machine has the higher armature copper losses?
(b) Which machine has the higher field copper losses?
(c) Which machine has the higher core losses?
To each answer, add a brief explanation.

6-45. The magnetization curve of Fig. 6-16 shows the no-load characteristic for a d-c shunt generator at 900 rpm, where each abscissa unit represents 0.5 A and each ordinate unit represents 20 V. The armature circuit resistance is 0.2 Ω, the field circuit resistance (without field rheostat) 30 Ω. The effect of armature reaction is equivalent to four demagnetizing ampere-turns per pole for every ampere of armature current. The field winding has 1250 turns per pole.

(a) Find the resistance of the field rheostat (in series with the field winding) for the machine operating as a generator driven at 900 rpm and carrying an armature current of 125 A at a terminal voltage of 220 V.
(b) The field rheostat remains as for part (a). The load is now disconnected. The speed remains unchanged. What is the new value of the terminal voltage?
(c) Find the field rheostat resistance for a no-load voltage of 420 V at a speed of 1800 rpm.

6-46. Depicted in Fig. P6-46 is the reversing controller circuitry for a d-c shunt motor. Iden-

Figure P6-46

tify the unmarked armature contacts and explain the reversing operation. Explain, too, why the auxiliary interlock contacts F_a and R_a are placed in series with the Rev and Fwd switches.

6-47. A current-limit type controller for a d-c shunt motor is shown in Fig. P6-47. The accelerating relays (AR) respond to armature current. These relays have a light-weight armature so that they easily pick up and open their contacts whenever the armature current exceeds the rated value. By means of a step-by-step procedure, describe the operation of this controller. Be careful to identify the proper closing and opening sequence whenever two or more contactors appear in series.

Figure P6-47

6-48. In the three-phase half-wave controlled rectifier circuit of Fig. 6-28, sketch the waveform of the source voltage that appears across the armature circuit when the firing angle of the SCRs is set at 45°. What is the average value of this voltage?

6-49. Repeat Prob. 6-48 for the case where the firing angle of the SCRs is set at 75°.

6-50. A 25-hp, 250-V, d-c motor is driven from a three-phase, 550-V source through a half-wave thyristor drive circuit as displayed in Fig. 6-28. The armature circuit resistance is 0.1 Ω and the developed torque parameter of this motor is known from design data to be 6.25 N-m/A. The thyristor voltage drops and losses are negligible.

(a) When the firing angle of the SCRs is set at 30° and the motor is operating at no load, find the speed in rpm.

(b) Rated load torque of 575 N-m is applied to the motor shaft. Determine the new operating speed in rpm.

(c) Calculate the speed regulation of this electronically controlled d-c motor.

6-51. It is desirable that the motor of Prob. 6-50 be made to deliver its rated torque of 575 N-m at a speed of 250 rpm.

(a) Find the required firing angle of the SCRs.

(b) Compute the speed regulation.

6-52. The motor of Prob. 6-50 is driven by the chopper circuitry of Fig. 6-31 from a 250-V d-c source. The thyristor control circuitry is arranged to provide a ratio of ON-time to cycle time of 0.5.

(a) Find the operating speed when rated torque is delivered by the motor.

(b) Determine the no-load speed and the speed regulation for the specified switching ratio.

(c) Repeat parts (a) and (b) for $T_0/T_p = 1.0$.

6-53. The motor of Prob. 6-50 is driven by a chopper drive of a 250-V source where the ratio of ON-time to cycle time is 0.5. Determine the required pulse frequency of the thyristors in order that the change in armature current during the ON period be limited to 5 A when operating at rated load. The armature inductance is 0.05 H.

6-54. Repeat Prob. 6-53 for the case where the ratio of ON-time to cycle time is 0.8.

SPECIAL
ELECTRIC
MACHINES

7

Two-Phase Servomotors

The two-phase servomotor has gained widespread use as an output actuator in feedback control systems. A high rotor resistance is one of its distinguishing design characteristics. This feature ensures a negative slope for the torque-speed character-istic over its entire operating range and thereby furnishes the motor with positive damping for good stability. The a-c servomotor often operates with unbalanced two-phase voltages. A fixed reference voltage is applied to its main winding and a varying voltage is applied to its control winding. The control winding is frequently fed from the output terminals of a servoamplifier with a signal that is usually displaced 90° from that on the main winding.

The analytical treatment of the two-phase servomotor supplied with unbalanced voltages can be conveniently accomplished by the use of symmetrical components. This technique is described in the first section of this chapter for the more general case of an unbalanced four-phase system and is subsequently applied to the two-phase case. The symmetrical components approach allows the unbalanced voltages to be replaced by two or more sets of balanced voltages. In this way the unbalanced machine is treated in terms of theoretical considerations that have already been fully exposed. (The computation of the performance of polyphase motors is described in Chapter 4.)

7-1 SYMMETRICAL COMPONENTS

The method of replacing an unbalanced four-phase voltage system with balanced voltage sets is described first chiefly because a more complete formulation of the problem can be made. If the unbalanced two-phase voltage system were treated directly, there would be no need to deal with zero-sequence voltages or balanced single-phase voltage sets so that some generality would thereby be lost. In Fig. 7-1(a) is shown an unbalanced four-phase voltage set. In a balanced system the phase voltages would be equal in amplitude and exactly 90° displaced from one another. Neither condition is satisfied by the set depicted in Fig. 7-1(a). Figure 7-1(b) shows that the phasor sum is not zero but adds, rather, to a resultant quantity which is identified as $4\bar{V}_0$. The quantity $\bar{V}_0$ is called the *zero-sequence* voltage and the factor 4 states that this unbalance can be assumed to have been brought about by the existence of an equal unbalance in each phase voltage. It is called zero-sequence because the $\bar{V}_0$ in any one phase voltage is *in-phase* with the $\bar{V}_0$ quantity of the other phase voltages.

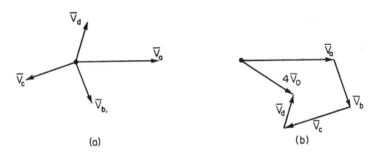

Figure 7-1 Unbalanced four-phase voltage system: (a) star notation; (b) identifying the zero-sequence component $\bar{V}_0$.

For any given unbalanced system the zero-sequence component can be readily found from

$$\bar{V}_0 = \tfrac{1}{4}(\bar{V}_a + \bar{V}_b + \bar{V}_c + \bar{V}_d) \tag{7-1}$$

The removal of $\bar{V}_0$ from each of the phase voltages results in a four-phase voltage system that adds up to zero. Thus

$$(\bar{V}_a - \bar{V}_0) + (\bar{V}_b - \bar{V}_0) + (\bar{V}_c - \bar{V}_0) + (\bar{V}_d - \bar{V}_0) = 0 \tag{7-2}$$

For convenience let

$$\begin{aligned} \bar{V}_A &\equiv \bar{V}_a - \bar{V}_0 \\ \bar{V}_B &\equiv \bar{V}_b - \bar{V}_0 \\ \bar{V}_C &\equiv \bar{V}_c - \bar{V}_0 \\ \bar{V}_D &\equiv \bar{V}_d - \bar{V}_0 \end{aligned} \tag{7-3}$$

Equation (7-2) then becomes

$$\bar{V}_A + \bar{V}_B + \bar{V}_C + \bar{V}_D = 0 \tag{7-4}$$

The foregoing is graphically represented in Fig. 7-2. It is important to keep in mind that, although the phasor sum of $\bar{V}_A$, $\bar{V}_B$, $\bar{V}_C$, and $\bar{V}_D$ is zero, these quantities still represent an unbalanced four-phase system in the sense that the voltages are unequal in magnitude and have unequal phase-displacement angles.

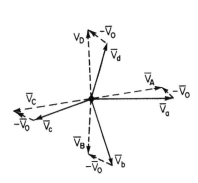

Figure 7-2 Removing the zero-sequence voltage $\bar{V}_0$ from each phase voltage.

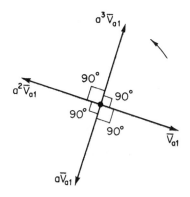

Figure 7-3 Balanced positive-sequence four-phase voltage set.

Attention is next directed to the conditions that must be imposed in order generally to represent the unbalanced voltage system of Eq. (7-4). Simplicity is served if we can succeed in representing this unbalanced system by groups of balanced voltages. Specifically, because we are dealing with a four-phase system, it is desirable to make this representation in terms of *balanced four-phase systems* to the extent possible. In this connection then a good starting point is to assume that each of the phase voltages of the system of Eq. (7-4) contains an appropriate component of a *balanced positive sequence set* as illustrated in Fig. 7-3. The quantity $\bar{V}_{a1}$ is assumed to be a component of $\bar{V}_A$. The subscript 1 is used to denote the *positive sequence*. Note that in a positive sequence set of four-phase balanced voltages, the phase voltages are equal in magnitude and 90 degrees displaced from one another and the rotation is counterclockwise (i.e., positive). The quantity $a\bar{V}_{a1}$ is the corresponding component of $\bar{V}_B$ where the symbol a denotes the rotational operator defined by

$$a = 1\underline{/-90°} \qquad (7\text{-}5)$$

Similarly, $a^2\bar{V}_{a1}$ is the positive sequence component of $\bar{V}_C$ and $a^3\bar{V}_{a1}$ is the positive sequence component of $\bar{V}_D$. Observe that $a^2 = -1$ and $a^3 = 1\underline{/90°}$.

What can we now add to the positive-sequence four-phase voltage set of Fig. 7-3 to ensure at least some unbalance in the resulting phase voltages? The inclusion of a second set of balanced positive-sequence voltages is fruitless, as a glance at Fig. 7-4 reveals. The result of adding two positive-sequence voltage sets still yields a balanced four-phase system. It therefore seems worthwhile at this point to investigate the effect of adding instead a balanced *negative-sequence* four-phase voltage set to the system of Fig. 7-3. Now the result is that illustrated in Fig. 7-5. Note that the selection of the opposite sequence has the effect of unbalancing two pairs of voltages while still preserving a balanced system, i.e., yielding a zero resultant. The quantity $\bar{V}_{a2}$ is the negative-sequence component of $\bar{V}_A$. The subscript 2 is used to denote the negative sequence. Note too that for the negative-sequence set it is $a^3\bar{V}_{a2}$ that is a component of $\bar{V}_B$, and $a\bar{V}_{a2}$ is a component of $\bar{V}_D$. This marks a further step in achieving our goal of representing an unbalanced voltage system in terms of suitable balanced sets.

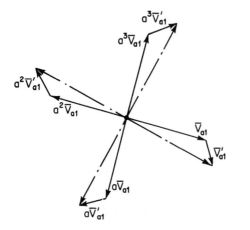

Figure 7-4 Illustrating how the addition of two positive-sequence voltage sets yields a balanced resultant set.

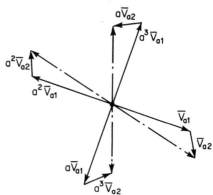

Figure 7-5 Addition of positive- and negative-sequence balanced four-phase voltage sets.

It is now desirable to add a third set of voltages to each phase of the system depicted in Fig. 7-5 but in such a way as to cause each of the phase voltages (and not just pairs) to be different in phase as well as in magnitude. In this connection let us explore the effect of adding a balanced single-phase set, keeping in mind that the resultant of the four components is to yield zero in order to preserve balance. Accordingly, if we assume that a third component $\bar{V}_{a3}$ is added to $\bar{V}_{a1}$ and $\bar{V}_{a2}$ to yield $\bar{V}_A$, then an equal and opposite quantity, $a^2\bar{V}_{a3}$, is assumed added to $a\bar{V}_{a1}$ and $a^3\bar{V}_{a2}$ to yield $\bar{V}_B$. Similarly, the third component of $\bar{V}_C$ is taken to be $\bar{V}_{a3}$, while the third component of $\bar{V}_D$ is taken to be $a^2\bar{V}_{a3}$. The results of these additions are depicted in Fig. 7-6. Note that this single-phase reversed-phase set differs from the zero-sequence set in that the former set taken together adds to zero. An inspection of Fig. 7-6 discloses that, with the use of the indicated balanced voltage sets, it becomes possible to represent any four-phase system of unbalanced voltages. The reason is that with such a composition there exists sufficient flexibility to make each phase

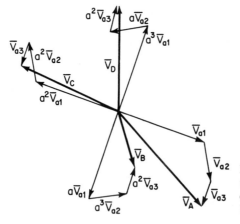

Figure 7-6 Illustrating how an unbalanced set of four-phase voltages can be represented generally by balanced sets.

voltage different in angle and magnitude from the others. However, our interest really focuses on the reverse situation, i.e., finding the balanced voltage components in each phase when an unbalanced system is specified. The foregoing treatment was described as presented in order to emphasize better how each component affects the total voltage.

The unbalanced phase voltages of the four-phase system can now be expressed mathematically as follows:

$$\bar{V}_A = \bar{V}_{a1} + \bar{V}_{a2} + \bar{V}_{a3} \tag{7-6}$$
$$\bar{V}_B = a\bar{V}_{a1} + a^3\bar{V}_{a2} + a^2\bar{V}_{a3} \tag{7-7}$$
$$\bar{V}_C = a^2\bar{V}_{a1} + a^2\bar{V}_{a2} + \bar{V}_{a3} \tag{7-8}$$
$$\bar{V}_D = a^3\bar{V}_{a1} + a\bar{V}_{a2} + a^2\bar{V}_{a3} \tag{7-9}$$

The first term on the right side of each equation represents a component of the balanced positive-sequence set, the second term represents a component of the balanced negative-sequence set, and the third term represeresents a component of the balanced single-phase set. Since each component set is a balanced set, together they yield a resultant that is zero. This is not surprising in view of the fact that the unbalanced component was previously removed. [Refer to Eq. (7-3).]

If Eq. (7-3) is inserted into Eqs. (7-6) to (7-9), expressions for the original unbalanced voltages in terms of the appropriate components are obtained. Thus

$$\bar{V}_a = \bar{V}_{a1} + \bar{V}_{a2} + \bar{V}_{a3} + \bar{V}_0 \tag{7-10}$$
$$\bar{V}_b = a\bar{V}_{a1} + a^3\bar{V}_{a2} + a^2\bar{V}_{a3} + \bar{V}_0 \tag{7-11}$$
$$\bar{V}_c = a^2\bar{V}_{a1} + a^2\bar{V}_{a2} + \bar{V}_{a3} + \bar{V}_0 \tag{7-12}$$
$$\bar{V}_d = a^3\bar{V}_{a1} + a\bar{V}_{a2} + a^2\bar{V}_{a3} + \bar{V}_0 \tag{7-13}$$

The manner of finding the zero-sequence component $\bar{V}_0$ in terms of the given unbalanced four-phase system is indicated by Eq. (7-1). An appropriate matter of concern at this point is consideration of how the remaining three balanced components can be determined in terms of the given unbalanced system. Note that, if the positive-sequence component $\bar{V}_{a1}$ of phase a is determined, then the corresponding components in each of the remaining phases are also known because they differ by a rotation factor only. A similar situation obtains for the negative-sequence and the balanced single-phase components.

To find the positive-sequence component, Eqs. (7-10) through (7-13) must be manipulated in such a way as to isolate $\bar{V}_{a1}$. A study of these equations reveals that this can be accomplished by multiplying Eq. (7-11) by a^3, Eq. (7-12) by a^2, and Eq. (7-13) by a and then summing each of these equations with Eq. (7-10). Thus

$$\bar{V}_a = \bar{V}_{a1} + \bar{V}_{a2} + \bar{V}_{a3} + \bar{V}_0 \tag{7-14}$$
$$a^3\bar{V}_b = \bar{V}_{a1} + a^2\bar{V}_{a2} + a\bar{V}_{a3} + a^3\bar{V}_0 \tag{7-15}$$
$$a^2\bar{V}_c = \bar{V}_{a1} + \bar{V}_{a2} + a^2\bar{V}_{a3} + a^2\bar{V}_0 \tag{7-16}$$
$$a\bar{V}_d = \bar{V}_{a1} + a^2\bar{V}_{a2} + a^3\bar{V}_{a3} + a\bar{V}_0 \tag{7-17}$$

$$\overline{\bar{V}_a + a^3\bar{V}_b + a^2\bar{V}_c + a\bar{V}_d = 4\bar{V}_{a1} + 0 + 0 + 0} \tag{7-18}$$

Keep in mind that the quantities $(1 + a + a^2 + a^3)$ and $(1 + a^2)$ are identically equal to zero. From the last equation it follows that the value of the positive-sequence component of voltage is

$$\boxed{\bar{V}_{a1} = \tfrac{1}{4}(\bar{V}_a + j\bar{V}_b - \bar{V}_c - j\bar{V}_d)} \tag{7-19}$$

since $a^3 = +j$, $a^2 = -1$, and $a = -j$.

Applying a similar procedure to the negative-sequence and the balanced single-phase set leads to the following results:

$$\boxed{\bar{V}_{a2} = \tfrac{1}{4}(\bar{V}_a - j\bar{V}_b - \bar{V}_c + j\bar{V}_d)} \tag{7-20}$$

and

$$\boxed{\bar{V}_{a3} = \tfrac{1}{4}(\bar{V}_a - \bar{V}_b + \bar{V}_c - \bar{V}_d)} \tag{7-21}$$

Therefore by means of Eqs. (7-1), (7-19), (7-20), and (7-21) the values of balanced voltage sets can be found to replace an unbalanced four-phase system of voltages.

Example 7-1

The phase voltages of an unbalanced four-phase system are given by

$$\bar{V}_a = 120\underline{/0°} = 120 + j0 \tag{7-22}$$
$$\bar{V}_b = 75\underline{/-60°} = 37.5 - j65 \tag{7-23}$$
$$\bar{V}_c = 100\underline{/-150°} = -86.6 - j50 \tag{7-24}$$
$$\bar{V}_d = 50\underline{/80°} = 8.7 + j49 \tag{7-25}$$

(a) Find the zero-sequence, positive-sequence, negative-sequence voltages and the balanced single-phase set.

(b) Draw the phasor diagram showing the manner by which the individual phase components add to yield the given unbalanced voltages.

Solution (a) By Eq. (7-1) the zero-sequence component for each phase is found to be
$$\bar{V}_0 = \tfrac{1}{4}(\bar{V}_a + \bar{V}_b + \bar{V}_c + \bar{V}_d)$$
$$= \tfrac{1}{4}(79.6 - j66) = 19.9 - j16.5 = 25.8\underline{/-39.8°} \tag{7-26}$$

The positive-sequence component as obtained from Eq. (7-19) is
$$\bar{V}_{a1} = \tfrac{1}{4}(\bar{V}_a + j\bar{V}_b - \bar{V}_c - j\bar{V}_d)$$
$$= \tfrac{1}{4}(120 + j0 + j37.5 + 65 + 86.6 + j50 - j8.7 + 49)$$
$$= 80.2 + j19.7 = 82.6\underline{/13.8°} \tag{7-27}$$

The negative-sequence component is determined from Eq. (7-20) to be
$$\bar{V}_{a2} = \tfrac{1}{4}(\bar{V}_a - j\bar{V}_b - \bar{V}_c + j\bar{V}_d)$$
$$= \tfrac{1}{4}(120 + j0 - j37.5 - 65 + 86.6 + j50 + j8.7 - 49)$$
$$= 23.2 + j5.3 = 23.7\underline{/12.9°} \tag{7-28}$$

Finally, the $\bar{V}_{a3}$ component is computed to be

$$\bar{V}_{a3} = \tfrac{1}{4}(\bar{V}_a - \bar{V}_b + \bar{V}_c - \bar{V}_d)$$
$$= \tfrac{1}{4}(120 + j0 - 37.5 + j65 - 86.6 - j50 - 8.7 - j49)$$
$$= -3.2 - j8.5 = 8.9\underline{/-107.8^\circ} \tag{7-29}$$

(b) The required phasor diagram appears in Fig. 7-7.

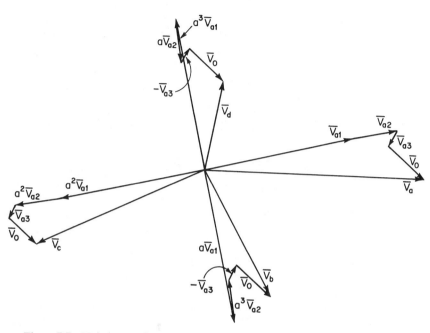

Figure 7-7 Unbalanced four-phase voltage system of Example 7-1, and the balanced voltage components.

7-2 SYMMETRICAL COMPONENTS APPLIED TO TWO-PHASE SERVOMOTOR

The unbalanced two-phase voltage system can be treated as a special case of the unbalanced symmetrical four-phase system. The unbalanced symmetrical four-phase system is one in which the voltages $\bar{V}_c$ and $\bar{V}_d$ are respectively equal to $-\bar{V}_a$ and $-\bar{V}_b$. Thus in the phasor diagram $\bar{V}_c$ and $\bar{V}_d$ are mirror images of $\bar{V}_a$ and $\bar{V}_b$. Appearing in Fig. 7-8(a) is the schematic representation of an unbalanced two-phase system. Keep in mind that for a balanced two-phase system $|\bar{V}_a| = |\bar{V}_b|$; also these two quantities are exactly out of phase by 90 degrees. If either condition is not satisfied, the system is unbalanced. Depicted in Fig. 7-8(b) is the schematic diagram of the corresponding symmetrical four-phase system. The use of any two successive phases yields a two-phase system.

What are the balanced components that can now be used to represent an unbalanced two-phase system of voltages? Do we need as many components as were

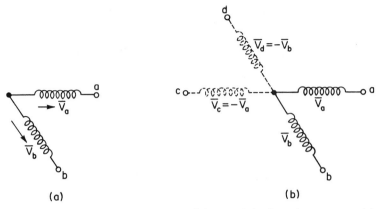

Figure 7-8 Two-phase system as a special case of the four-phase system: (a) unbalanced two-phase system; (b) unbalanced symmetrical four-phase system.

required to represent the unbalanced four-phase system? The answers to these questions follow immediately by treating the two-phase system in terms of the symmetrical four-phase system and then applying Eqs. (7-1) and (7-19) to (7-21). Introducing $\bar{V}_c = -\bar{V}_a$ and $\bar{V}_d = -\bar{V}_b$ into Eq. (7-1) leads to the conclusion that the zero-sequence voltage is zero. This is entirely consistent with the fact that, as a symmetrical system, the resultant of the four-phase voltages must be identically equal to zero. Also, the insertion of these same quantities into Eq. (7-19) discloses that the value of the positive-sequence component of phase a is given by

$$\bar{V}_{a1} = \tfrac{1}{4}(\bar{V}_a + j\bar{V}_b + \bar{V}_a + j\bar{V}_b)$$

or

$$\boxed{\bar{V}_{a1} = \tfrac{1}{2}(\bar{V}_a + j\bar{V}_b)} \qquad (7\text{-}30)$$

Similarly, by Eq. (7-20) the negative-sequence component is found to be

$$\bar{V}_{a2} = \tfrac{1}{4}(\bar{V}_a - j\bar{V}_b + \bar{V}_a - j\bar{V}_b)$$

or

$$\boxed{\bar{V}_{a2} = \tfrac{1}{2}(\bar{V}_a - j\bar{V}_b)} \qquad (7\text{-}31)$$

The value of $\bar{V}_{a3}$ is determined from Eq. (7-21). Thus

$$\bar{V}_{a3} = \tfrac{1}{4}(\bar{V}_a - \bar{V}_b - \bar{V}_a - \bar{V}_b) = 0 \qquad (7\text{-}32)$$

On the basis of the foregoing analysis we reach the conclusion that *any unbalanced two-phase system can be completely represented by a two-phase balanced positive-sequence component and a two-phase balanced negative-sequence component.* The unbalanced phase voltages can then be expressed simply as

$$\boxed{\begin{aligned} \bar{V}_a &= \bar{V}_{a1} + \bar{V}_{a2} \qquad\qquad (7\text{-}33)\\ \bar{V}_b &= a\bar{V}_{a1} + a^3\bar{V}_{a2} = -j\bar{V}_{a1} + j\bar{V}_{a2} \quad (7\text{-}34) \end{aligned}}$$

where $\bar{V}_{a1}$ and $\bar{V}_{a2}$ are given by Eqs. (7-30) and (7-31).

Example 7-2

An unbalanced two-phase voltage system is described by

$$\bar{V}_a = 120\underline{/0^\circ} = 120 + j0 \qquad (7\text{-}35)$$

$$\bar{V}_b = 75\underline{/-60^\circ} = 37.5 - j65 \qquad (7\text{-}36)$$

Find the positive- and negative-sequence components that can be used to represent the unbalanced voltage system.

Solution The positive-sequence component of phase a is found from Eq. (7-30) to be

$$\bar{V}_{a1} = \tfrac{1}{2}(\bar{V}_a + j\bar{V}_b) = \tfrac{1}{2}(120 + j0 + j37.5 + j65)$$

$$= 92.5 + j18.8 = 94.4\underline{/11.5^\circ} \qquad (7\text{-}37)$$

Figure 7-9(a) depicts a simple graphical procedure for finding $\bar{V}_{a1}$. It first requires that $\bar{V}_b$ be rotated by 90° in the positive direction to yield $j\bar{V}_b$, then finding the phasor sum of $\bar{V}_a$ and $j\bar{V}_b$, and finally taking one-half of this resultant. The desired quantity is indicated by OA.

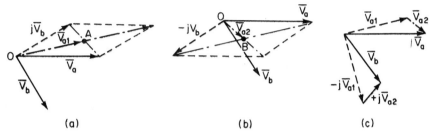

(a) (b) (c)

Figure 7-9 Phasor diagrams for Example 7-2: (a) graphical determination of the positive-sequence component; (b) graphical determination of the negative-sequence component; (c) unbalanced voltages with balanced components.

The analytical determination of the negative-sequence component follows from Eq. (7-31). Thus

$$\bar{V}_{a2} = \tfrac{1}{2}(\bar{V}_a - j\bar{V}_b) = \tfrac{1}{2}(120 - j0 - j37.5 - 65)$$

$$= 27.5 - j18.8 = 33.3\underline{/-34.4^\circ} \qquad (7\text{-}38)$$

The graphical determination of $\bar{V}_{a1}$ is illustrated in Fig. 7-9(b). It first requires that $\bar{V}_b$ be rotated in the negative direction by -90° to yield $-j\bar{V}_b$, then adding $\bar{V}_a$ and $-j\bar{V}_b$, and finally taking one-half of this sum. The desired quantity is identified as OB and is equivalent to the result expressed in Eq. (7-38).

The complete expressions for the phase voltages in terms of the positive- and negative-sequence components are then

$$\bar{V}_a = \bar{V}_{a1} + \bar{V}_{a2} = 94.4\underline{/11.5^\circ} + 33.3\underline{/-34.4^\circ} \qquad (7\text{-}39)$$

$$\bar{V}_b = a\bar{V}_{a1} + a^3\bar{V}_{a2} = -j\bar{V}_{a1} + j\bar{V}_{a2} = 94.4\underline{/78.5^\circ} + 33.3\underline{/55.6^\circ} \qquad (7\text{-}40)$$

The graphical representation of the last two equations appears in Fig. 7-9(c).

The results embodied in Eqs. (7-33) and (7-34) lead to a very useful and interesting physical interpretation of the analysis of the unbalanced two-phase servomotor. Each phase voltage consists of a component of a balanced two-phase set, one positive and the other negative. The positive-sequence components in each phase, $\bar{V}_{a1}$ and $a\bar{V}_{a1}$, can be considered to cooperate in creating a revolving magnetic field that interacts with the rotor winding to produce a positive-sequence torque T_1 At the same time the negative-sequence components in each phase, $\bar{V}_{a2}$ and $a^3\bar{V}_{a2}$, can be considered to combine in creating a revolving magnetic field of reversed direction that interacts with the rotor winding to produce a negative-sequence torque T_2. Of course the rotor responds to the resultant torque, which is the difference between T_1 and T_2. A schematic representation of this situation is depicted in Fig. 7-10. Note that for convenience the rotor is treated as consisting of two rotors that are mechanically coupled. The positive-sequence voltage set—$\bar{V}_{a1}$ and $a\bar{V}_{a1}$—produces T_1, and the negative-sequence set produces T_2. All the known techniques of balanced operation can now be used to determine T_1 and also T_2, and the results can then be superposed to obtain the resultant. Unbalanced operation is thereby treated entirely in terms of the techniques of balanced operation. Accordingly, no new theory is needed to handle the two-phase servomotor beyond this point.

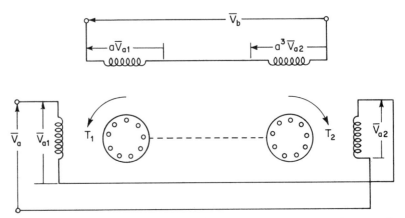

Figure 7-10 Representation of the unbalanced two-phase servomotor in terms of positive- and negative-sequence voltages.

7-3 EQUIVALENT CIRCUIT AND PERFORMANCE BASED ON SYMMETRICAL COMPONENTS

The equivalent circuit of the unbalanced two-phase servomotor actually consists of two parts, one representing the motor response to the positive-sequence voltage set and the other representing the response to the negative-sequence set. Because each voltage sequence is a balanced set, the equivalent circuit configuration expressed on a per phase basis is similar to that shown in Fig. 4-7(f). The circuit is repeated in Fig. 7-11(a). The applied stator voltage for this case is the positive-sequence voltage $\bar{V}_{a1}$. If the main winding (phase a) of the servomotor is assumed identical with the control winding (phase b), then the equivalent circuit for each winding is identical. In this chapter we restrict attention to servomotors that have identical stator phase windings.

Figure 7-11 Two-part equivalent circuit of the two-phase servomotor with unbalanced applied voltages: (a) equivalent circuit per phase of the balanced positive-sequence set; (b) the per phase equivalent circuit of the negative-sequence balanced set.

However, in those instances where this condition is not met, the use of a suitable transformer can be made to bring about the balanced condition.

The parameters appearing in the equivalent circuit of Fig. 7-11(a) have the same meanings as described in Chapter 4. The quantity r_1 denotes the stator winding resistance per phase. For the servomotor with balanced windings this represents the winding resistance of either phase. The total magnetizing reactance per phase is denoted by x_ϕ, and the total leakage reactances per phase of the stator and rotor windings are respectively denoted by x_1 and x_2'. The prime notation indicates that the last quantity is referred to the stator winding. The total rotor resistance per phase is identified as r_2'. Use of the quantity r_2'/s in the equivalent circuit implies that the rotor travels in the direction of the positive-sequence rotating field. Finally, note that a core loss resistor is not included. Hence the assumption throughout is that core loss is negligible.

The equivalent circuit of the servomotor in response to the balanced negative-sequence set is depicted in Fig. 7-11(b). Each circuit parameter has the same meaning as it has for the equivalent circuit of the positive-sequence set. By superposition the negative-sequence voltage set is assumed to be applied to the same stator windings and producing its own reactions in the same rotor winding. This is the meaning that is intended in the schematic diagram of Fig. 7-10. There is really only one pair of stator coils and one rotor, but the effect is the same as if there were two identical pairs of stator coils and two identical rotors mechanically coupled. A comparison of Fig. 7-11(b) with 7-11(a) shows differences occurring in two places. One, the stator applied voltage now has the value $\bar{V}_{a2}$ instead of $\bar{V}_{a1}$. Two, the slip of the actual rotor rotation with respect to the revolving field of the negative-sequence voltage set is $2 - s$ instead of s. Keep in mind that even with unbalanced voltages the rotor can revolve at only *one* speed. The slip of the rotor with respect to the positive-sequence field is then clearly

$$s = \frac{n_s - n}{n_s} \qquad (7\text{-}41)$$

Because the negative-sequence voltage set establishes a field that revolves in the opposite direction to that of the positive sequence, it follows that the slip of the rotor with respect to the negative-sequence field, s_2 is given by

$$s_2 = \frac{n_s - (-n)}{n_s} = 1 + \frac{n}{n_s} \tag{7-42}$$

In the interest of dealing solely in terms of s—the slip with respect to the positive-sequence field—use is made of Eq. (7-41) in Eq. (7-42) to yield

$$s_2 = 1 + \frac{n}{n_s} = 2 - s \tag{7-43}$$

Therefore in all future work that involves the use of positive- and negative-sequence fields, the rotor slip with respect to the negative-sequence field will always be denoted by $2 - s$.

The computation of performance can be readily determined through the use of our knowledge of the operation of a polyphase induction motor in response to balanced applied voltages. Let us find the expression for the resultant torque and the individual phase currents when the servomotor operates at a slip s with unbalanced stator voltages. First we find the behavior of the motor in response to the positive-sequence voltages. This is easily accomplished by means of the equivalent circuit of Fig. 7-11(a). In a typical problem $\bar{V}_{a1}$, s, and the circuit parameters are specified. As the initial step the input impedance occurring with the positive-sequence voltage is found. Calling this impedance Z_1 we have

$$\bar{Z}_1 = \bar{z}_1 + \bar{Z}_{p1} = r_1 + jx_1 + \frac{jx_\phi[(r_2'/s) + jx_2']}{(r_2'/s) + j(x_\phi + x_2')} \tag{7-44}$$

where Z_{p1} obviously denotes the impedance of the parallel arrangement of the circuit elements. The second subscript, 1, identifies the positive sequence. With the positive-sequence input impedance known, the positive-sequence component of the current in phase a is given by

$$\bar{I}_{a1} = \frac{\bar{V}_{a1}}{\bar{Z}_1} \tag{7-45}$$

The rotor current produced by the action of the positive-sequence rotating field follows from

$$\bar{I}'_{21} = \frac{\bar{I}_{a1}\bar{Z}_{p1}}{(r_2'/s) + jx_2'} \tag{7-46}$$

where the second subscript denotes the positive sequence. A little thought discloses that an alternative expression for $\bar{I}'_{21}$ can be had by applying the current divider theorem for the parallel circuit. However, Eq. (7-46) is usually preferred because the quantities on the right side are frequently available in a convenient form.

The gap power associated with the positive-sequence rotating field, P_{g1}, is then found from

$$P_{g1} = q(I'_{21})^2 \frac{r_2'}{s} = 2(I'_{21})^2 \frac{r_2'}{s} \tag{7-47}$$

The corresponding developed torque is then

$$T_1 = \frac{P_{g1}}{\omega} \quad \text{N-m} \tag{7-48}$$

where ω is the synchronous angular frequency expressed in radians per second.

The performance of the motor in response to the negative-sequence set is found in a similar fashion. Thus the negative-sequence input impedance can be expressed by

$$\bar{Z}_2 = \bar{z}_1 + \bar{Z}_{p2} = r_1 + jx_1 + \frac{jx_\phi[(r_2'/2 - s) + jx_2']}{(r_2'/2 - s) + j(x_\phi + x_2')} \tag{7-49}$$

where $\bar{Z}_{p2}$ is the impedance of the parallel arrangement appearing in the circuit of Fig. 7-11(b). The negative-sequence component of the current in stator phase a becomes

$$\bar{I}_{a2} = \frac{\bar{V}_{a2}}{\bar{Z}_2} \tag{7-50}$$

The associated rotor current takes on the value

$$\bar{I}'_{22} = \frac{\bar{I}_{a2}\bar{Z}_{p2}}{(r_2'/2 - s) + jx_2'} \tag{7-51}$$

so that the expression for the power transferred across the air gap by the negative-sequence flux field becomes

$$P_{g2} = q(I'_{22})^2 \frac{r_2'}{2 - s} = 2(I'_{22})^2 \frac{r_2'}{2 - s} \tag{7-52}$$

and

$$T_2 = \frac{P_{g2}}{\omega} \tag{7-53}$$

The resultant torque produced by the simultaneous effects of the positive- and negative-sequence voltages expressed in newton-meters is

$$\boxed{T = T_1 - T_2 = \frac{1}{\omega}(P_{g1} - P_{g2})} \tag{7-54}$$

For servomotors used in an instrument type of servomechanism the value of torque in newton-meters is a very small quantity. It is customary in such instances to refer rather to the torque in terms of its proportional quantity, power. Hence it is not uncommon for control engineers to refer to a servomotor torque as "so many units of synchronous watts." We too adopt this practice.

The resultant line current in phase a is

$$\boxed{\bar{I}_a = \bar{I}_{a1} + \bar{I}_{a2}} \tag{7-55}$$

and that in phase b is similarly obtained as

$$\boxed{\bar{I}_b = -j\bar{I}_{a1} + j\bar{I}_{a2}} \tag{7-56}$$

Information about the power factor of each stator phase current is implied in the polar expression of the phase currents. In normal applications of the servomotor, efficiency is not very important and so this matter receives no further attention here.

Example 7-3

A 5-W, 60-Hz, 120-V, two-pole, two-phase servomotor has the following parameters:

$$r_1 = 285 \ \Omega, \qquad r'_2 = 850 \ \Omega$$

$$x_1 = 60 \ \Omega, \qquad x'_2 = 60 \ \Omega$$

$$x_\phi = 995 \ \Omega$$

When it is operating at a slip of 0.6, determine (a) the resultant torque in synchronous watts and (b) the stator phase currents. Assume the servomotor operates with the following unbalanced two-phase voltage: $\bar{V}_a = 120\underline{/0°}$ and $\bar{V}_b = 75\underline{/-60°}$.

Solution (a) The positive- and negative-sequence voltages for the specified unbalanced voltages are computed in Example 7-2 to be as follows:

$$\bar{V}_{a1} = 94.4\underline{/11.5°} \quad \text{and} \quad \bar{V}_{a2} = 33.3\underline{/-34.4°}$$

We first compute the motor response to the positive-sequence set. Thus

$$\bar{Z}_{p1} = \frac{j995[(850/0.6) + j60]}{(850/0.6) + j1055} \approx \frac{j955(1415)}{1760\underline{/36.8°}} = 800\underline{/53.2°} \tag{7-57}$$

and

$$\bar{Z}_1 = \bar{z}_1 + \bar{Z}_{p1} = 285 + j60 + 481 + j644 = 1038\underline{/42.6°} \tag{7-58}$$

Hence

$$\bar{I}_{a1} = \frac{\bar{V}_{a1}}{\bar{Z}_1} = \frac{94.4\underline{/11.5°}}{1038\underline{/42.6°}} = 0.091\underline{/-31.1°} \tag{7-59}$$

and

$$\bar{I}'_{21} = \frac{\bar{I}_{a1}\bar{Z}_{p1}}{1415 + j60} = \frac{0.091\underline{/-31.1°}(800\underline{/53.2°})}{1415} = 0.0514\underline{/22.1°} \tag{7-60}$$

The positive-sequence gap power then becomes

$$P_{g1} = 2(0.0514)^2 1415 = 2(3.7) = 7.4 \text{ W} \tag{7-61}$$

The response to the negative-sequence voltage set is computed to be as follows:

$$\bar{Z}_{p2} = \frac{j995[(850/1.4) + j60]}{(850/1.4) + j(995 + 60)} \approx \frac{j995(606)}{1218\underline{/60.1°}} = 496\underline{/29.9°} \tag{7-62}$$

$$\bar{Z}_2 = \bar{z}_1 + \bar{Z}_{p2} = 285 + j60 + 431 + j248 = 779\underline{/23.3°} \tag{7-63}$$

$$\bar{I}_{a2} = \frac{\bar{V}_{a2}}{\bar{Z}_2} = \frac{3.3\underline{/-34.4°}}{779\underline{/23.3°}} = 0.042\underline{/-57.7°} \tag{7-64}$$

$$\bar{I}'_{22} = \frac{\bar{I}_{a2}\bar{Z}_{a2}}{606 + j60} = \frac{0.042\underline{/-57.7°}(496\underline{/29.9°})}{606} = 0.034\underline{/-33.5°} \tag{7-65}$$

$$P_{g2} = 2(0.0344)^2 \frac{r'_2}{2 - s} = 2(0.0344)^2(606) = 2(0.716) = 1.43 \text{ W} \tag{7-66}$$

Hence the resultant gap power is

$$P_g = P_{g1} - P_{g2} = 7.4 - 1.43 = 5.97 \text{ synchronous watts} \tag{7-67}$$

(b) By Eq. (7-55) the expression for the current in stator phase a is

$$\bar{I}_a = \bar{I}_{a1} + \bar{I}_{a2} = 0.09\underline{/-31.1°} + 0.042\underline{/-57.7°}$$

$$= 0.0778 - j0.047 + 0.0224 - j0.0355 \qquad (7\text{-}68)$$

$$= 0.1002 - j0.0825 = 0.13\underline{/-39.5°}$$

Similarly by Eq. (7-56) the phase current in the control winding b is

$$\bar{I}_b = -j\bar{I}_{a1} + j\bar{I}_{a2} = -j0.0778 - 0.047 + j0.0224 + 0.0355$$

$$= -0.0115 - j0.0554 = 0.0566\underline{/-101.7°} \qquad (7\text{-}69)$$

Attention is next turned to the derivation of an oftentimes useful equation for computing the resultant torque. The insertion of Eqs. (7-45) and (7-46) into Eq. (7-47) leads to the following expression for the gap power of the positive-sequence field

$$P_{g1} = 2\left(\frac{V_{a1}}{Z_1}\right)^2\left[\frac{Z_{p1}}{(r_2'/s) + jx_2}\right]^2\left(\frac{r_2'}{s}\right) \qquad (7\text{-}70)$$

Let us now determine the expression for the gap power when the motor is operated with balanced two-phase voltages, i.e., $V_a = V_b$ and each displaced from the other by 90 degrees. In this instance the phase current is given by

$$\bar{I}_a = \frac{\bar{V}_a}{\bar{Z}_1} \qquad (7\text{-}71)$$

where $\bar{Z}_1$ is described by Eq. (7-44). Moreover the rotor current expression is

$$\bar{I}_2' = \frac{\bar{I}_a \bar{Z}_{p1}}{(r_2'/s) + jx_2'} \qquad (7\text{-}72)$$

where $\bar{Z}_{p1}$ is specified as before. The expression for the total gap power for balanced operation, P_g, becomes therefore

$$P_g = 2(I_2')^2\frac{r_2'}{s} = 2\left(\frac{V_a}{Z_1}\right)^2\left(\frac{Z_{p1}}{(r_2'/s) + jx_2'}\right)^2\left(\frac{r_2'}{s}\right) \qquad (7\text{-}73)$$

By formulating the ratio of Eq. (7-70) to Eq. (7-73) there results

$$P_{g1} = \left(\frac{V_{a1}}{V_a}\right)^2 P_g \qquad (7\text{-}74)$$

By proceeding in a similar fashion it can be shown that

$$P_{g2} = \left(\frac{V_{a2}}{V_a}\right)^2 P_g' \qquad (7\text{-}75)$$

where P_g' denotes the *gap power computed for balanced two-phase operation at a slip of $2 - s$*. The resultant gap power is accordingly expressed as

$$P_{gr} = P_{g1} - P_{g2} = \left(\frac{V_{a1}}{V_a}\right)^2 P_g - \left(\frac{V_{a2}}{V_a}\right)^2 P_g' \qquad (7\text{-}76)$$

Correspondingly, the resultant torque can then be written as

$$T = \frac{1}{\omega}\left[\left(\frac{V_{a1}}{V_a}\right)^2 P_g - \left(\frac{V_{a2}}{V_a}\right)^2 P_g'\right] \qquad (7\text{-}77)$$

or

$$T = \left(\frac{V_{a1}}{V_a}\right)^2 T_{B1} - \left(\frac{V_{a2}}{V_a}\right)^2 T_{B2} \qquad (7\text{-}78)$$

where $T_{B1} = P_g/\omega$ and denotes the developed torque computed at slip s for balanced two-phase operation at rated voltage, and $T_{B2} = P'_g/\omega$ and denotes the developed torque computed at slip $2 - s$ for balanced two-phase operation at rated voltage.

The usefulness of Eq. (7-78) lies in the fact that the resultant torque at any slip and for any condition of unbalanced two-phase voltages may be computed in terms of the torque for full-voltage balanced operation. This information is usually supplied by the motor manufacturer. Hence the torque-speed characteristic for various values of the control winding voltage is readily determined from Eq. (7-78) without resorting to the equivalent-circuit computational procedure. This matter is further elaborated in the next section.

Example 7-4

Compute the resultant torque of the servomotor of Example 7-3 by employing Eq. (7-78).

Solution Under conditions of balanced two-phase operation the phase current becomes

$$\bar{I}_a = \frac{\bar{V}_a}{\bar{Z}_1} = \frac{120}{1038}\underline{/-42.6°} = 0.116\underline{/-42.6°} \qquad (7\text{-}79)$$

The corresponding rotor current is

$$\bar{I}'_2 = \frac{\bar{I}_a \bar{Z}_{p1}}{(r'_2/s) + jx'_2} \approx \frac{0.116\underline{/-42.6°}(800\underline{/53.2°})}{1415} = 0.0655\underline{/10.6°}$$

$$\therefore \quad P_g = 2(I'_2)^2\frac{r'_2}{s} = 2(0.0655)^2(1415) = 12.1 \text{ W}$$

Therefore the gap power of the positive-sequence set becomes by Eq. (7-61)

$$P_{g1} = \left(\frac{V_{a1}}{V_a}\right)^2 P_g = \left(\frac{94.4}{120}\right)^2(12.1) = 7.45 \qquad (7\text{-}80)$$

A comparison with Eq. (7-61) indicates favorable agreement.

To find P'_g, the motor is assumed to be operating at a slip of $2 - s = 1.4$ with balanced two-phase voltages applied. Then

$$\bar{I}_a = \frac{\bar{V}_a}{\bar{Z}_2} = \frac{120}{779\underline{/23.3°}} = 0.152\underline{/-23.3°} \qquad (7\text{-}81)$$

and

$$\bar{I}'_2 = \frac{\bar{I}_a \bar{Z}_{p2}}{(r'_2/2 - s) + jx'_2} \approx \frac{0.152\underline{/-23.3°}(496\underline{/29.9°})}{606} = 0.1245\underline{/6.6°}$$

$$P'_g = 2(0.1245)^2\frac{r'_2}{2 - s} = 2(0.0155)606 = 18.8 \text{ W} \qquad (7\text{-}82)$$

Hence by Eq. (7-75) the gap power of the negative-sequence field is

$$P_{g2} = \left(\frac{V_{a2}}{V_a}\right)^2 P'_g = \left(\frac{33.3}{120}\right)^2(18.8) = 1.45 \qquad (7\text{-}83)$$

which compares favorably with Eq. (7-66).

The resultant torque is therefore

$$T = \frac{1}{\omega}(7.45 - 1.45) = \frac{1}{375}(6) = 0.0159 \text{ N-m} \tag{7-84}$$

7-4 SERVOMOTOR TORQUE-SPEED CURVES

The torque-speed curve of a conventional two-phase induction motor supplied by a balanced two-phase voltage has a shape similar to that of the three-phase induction motor. This characteristic is depicted as curve a in Fig. 7-12. The use of such a motor in control systems is intolerable because of the positive slope that prevails over most of the operating speed range. The positive slope represents negative damping in the control system, which in turn can lead to a condition of instability. Therefore for control systems applications the motor must be modified in a way that ensures positive damping over the full speed range. A convenient way to achieve this result is to design the rotor with very high rotor resistance. The torque-speed characteristic then assumes the shape shown by curve b of Fig. 7-12. It is sketched over a range of slip from zero to 2 because of the pertinence of that portion of the curve lying between $s = 1$ and $s = 2$ when negative-sequence voltages exist in unbalanced operation. Whenever the two-phase induction motor includes the high rotor resistance design feature, it is referred to as a two-phase *servomotor*.

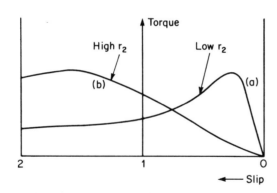

Figure 7-12 Torque-speed curves: (a) for the balanced two-phase induction motor with low rotor resistance; (b) for the same motor with high rotor resistance.

In many applications of the servomotor in feedback control systems, phase a is energized with fixed rated voltage (often called the *reference voltage*), while phase b is energized with a varying control voltage that is usually obtained from a preceding amplifier stage. Moreover, the arrangement in these configurations is such that the control voltage is frequently adjusted to be exactly 90° out of phase with the voltage applied to phase a. In these instances unbalanced operation is really an unbalance in voltage magnitude only. If we proceed under the assumption that the reference voltage $\bar{V}_a$ and the control voltage $\bar{V}_b$ are always 90° apart in phase and if we define the quantity ρ such that

$$\rho \equiv \frac{V_b}{V_a} \tag{7-85}$$

then the phasor expression for the control voltage becomes

$$\bar{V}_b = -j\rho\bar{V}_a \tag{7-86}$$

The $-j$ factor accounts for the 90° lag between the two stator phase voltages, and p expresses the magnitude of V_b as a per unit ratio of the reference voltage. Upon inserting Eq. (7-86) into Eq. (7-30), the expression for the positive-sequence voltage becomes

$$\bar{V}_{a1} = \tfrac{1}{2}(\bar{V}_a + j\bar{V}_b) = \frac{\bar{V}_a}{2}(1 + p) \qquad (7\text{-}87)$$

Similarly

$$\bar{V}_{a2} = \tfrac{1}{2}(\bar{V}_a - j\bar{V}_b) = \frac{\bar{V}_a}{2}(1 - p) \qquad (7\text{-}88)$$

Figure 7-13 Torque-speed curves of the two-phase servomotor for various values of p.

Appearing in Fig. 7-13 is a set of actual torque-speed curves for a two-phase servomotor corresponding to various values of the control voltage expressed as a fraction (p) of the reference voltage. The curve marked 115 V (or $p = 1$) represents the characteristic for balanced operation. However, note that, as the control voltage is reduced, the associated torque-speed characteristic becomes severely restricted. If information about the torque-speed characteristic of the servomotor is supplied by the manufacturer for balanced rated operation over the full range of slip from zero to 2, then the torque-speed characteristic for any value of p can be easily derived by means of Eqs. (7-87), (7-88), and (7-78). The procedure is illustrated in Fig. 7-14.

Consider that the servomotor operates with the control voltage fixed at one-half the value of the reference voltage, i.e., $p = \tfrac{1}{2}$. Then by Eq. (7-87) the positive-sequence voltage has a magnitude of

$$\bar{V}_{a1} = \frac{\bar{V}_a}{2}(1 + p) = \frac{\bar{V}_a}{2}(1 + \tfrac{1}{2}) = \tfrac{3}{4}\bar{V}_a \qquad (7\text{-}89)$$

TWO-PHASE SERVOMOTORS CHAP. 7

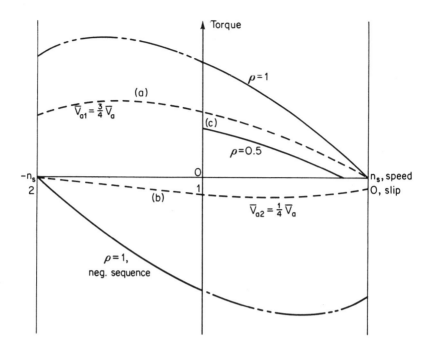

Figure 7-14 Graphical determination of the servomotor torque-speed curves for $\rho = 0.5$.

and the negative-sequence voltage is found by Eq. (7-88) to be

$$\bar{V}_{a2} = \frac{\bar{V}_a}{2}(1 - \rho) = \frac{\bar{V}_a}{2}(1 - \tfrac{1}{2}) = \tfrac{1}{4}\bar{V}_a \qquad (7\text{-}90)$$

Inserting these results into Eq. (7-78) leads to the following expression for the resultant torque:

$$T_r = \left(\frac{V_{a1}}{V_a}\right)^2 T_{B1} - \left(\frac{V_{a2}}{V_a}\right)^2 T_{B2} = \tfrac{9}{16} T_{B1} - \tfrac{1}{16} T_{B2} \qquad (7\text{-}91)$$

The quantity $\frac{9}{16}T_{B1}$ is readily obtained over the entire speed range by taking nine-sixteenths of the ordinate value of the characteristic for full-voltage balanced operation at all values of slip. This result is identified as curve a in Fig. 7-14. In like manner the quantity $\frac{1}{16}T_{B2}$ is easily determined over the entire speed range by taking one-sixteenth of the value of the characteristic supplied by the manufacturer for full-voltage balanced operation. This is represented by curve b in Fig. 7-14 and is drawn below the abscissa axis because the negative-sequence torque is opposed to the positive-sequence torque. The resultant torque-speed characteristic over the useful speed range is found by adding curves a and b, which leads to curve c. So long as the servomotor is operated with a voltage $\bar{V}_b = -j0.5\bar{V}_a$, curve c describes its operating behavior. By repeating the foregoing procedure the resultant torque-speed characteristic for other values of ρ can be determined. It is interesting to note that these characteristics are obtainable without the necessity of making repeated calculations in conjunction with the equivalent circuit as outlined in Example 7-3. However, the alternative procedure described here does require having the torque-speed characteristic for balanced operation at full voltage over the slip range from 0 to 2.

Another useful characteristic of the servomotor is a description of the developed torque as a function of the control voltage with speed as a parameter. These curves, too, can be derived with the use of Eq. (7-78) and the availability of the torque-speed characteristic for balanced operation at rated voltage. Appearing in Fig. 7-15 are the torque-speed characteristics corresponding to various values of ρ for both the positive- and negative-sequence voltages. The curves for $\rho = 0$ are identical but merely reversed in position consistent with their association with the positive- and negative-sequence voltage sets. As ρ is increased toward unity the positive-sequence contribution becomes increasingly larger than it is for $\rho = 0$, while the corresponding negative-sequence contribution becomes increasingly smaller than it is for $\rho = 0$. To obtain a plot of resultant torque as a function of control voltage expressed as a ratio of the rated reference voltage at zero speed, it is merely necessary to subtract the negative-sequence contribution from the positive-sequence torque. Thus at zero speed (or $s = 1$) the resultant torque is $a_1 - a_2 = 0$ for $\rho = 0$. For $\rho = 0.2$ the resultant torque is $b_1 - b_2$, and this quantity is greater than zero. The net torque for $\rho = 0.4$ is $c_1 - c_2$, and it should be apparent that this quantity is greater than $b_1 - b_2$. The

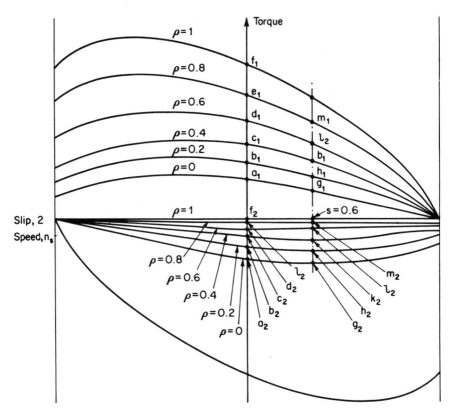

Figure 7-15 Graphical construction for deriving the curves of Fig. 7-16. The abscissa axis is slip or speed.

TWO-PHASE SERVOMOTORS   CHAP. 7

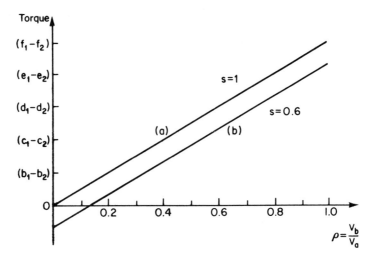

Figure 7-16 Variation of resultant torque versus control voltage with slip as a parameter.

complete plot is depicted in Fig. 7-16 as curve (*a*). Repeating the procedure at a second value of slip leads to curve (*b*). Note the region of negative torque. More is said about this matter in the next chapter.

PROBLEMS

7-1. An unbalanced four-phase system is described by the following voltage set:

$$\bar{V}_a = 110\underline{/15°}$$
$$\bar{V}_b = 110\underline{/-50°}$$
$$\bar{V}_c = 110\underline{/-135°}$$
$$\bar{V}_d = 110\underline{/90°}$$

   (a) What is the value of the zero-sequence voltage?
   (b) Find the phasor expression of the positive-sequence voltage set, and draw the corresponding phasor diagram.
   (c) Repeat part (b) for the negative-sequence set.
   (d) Repeat part (b) for the balanced single-phase set.
   (e) Draw the complete phasor diagram, showing how the individual components combine to yield the given unbalanced voltage set.

7-2. A four-phase voltage system is described by the following phasors:

$$\bar{V}_a = 110\underline{/15°}$$
$$\bar{V}_b = 60\underline{/-50°}$$
$$\bar{V}_c = 140\underline{/-135°}$$
$$\bar{V}_d = 40\underline{/90°}$$

   (a) Determine that set of balanced components that can replace the unbalanced system.

(b) Draw the complete phasor diagram of this four-phase system, showing all components.

7-3. Show the derivation that leads to Eq. (7-20).

7-4. Show the derivation that leads to Eq. (7-21).

7-5. The positive-sequence component of phase a of a four-phase voltage system is known to be equal to $60\underline{/-30^\circ}$ V.

(a) What is the phasor expression of the positive-sequence component of phase c? Assume the phase rotation is a-b-c-d.

(b) What is the positive-sequence component of phase b?

(c) Draw the complete phasor diagram for the positive-sequence set.

7-6. Under what conditions is it possible for an unbalanced four-phase voltage system to yield a resultant voltage that is zero?

7-7. One phase of a four-phase voltage system has the following components:

$$\bar{V}_{a1} = 60\underline{/-60^\circ}$$
$$\bar{V}_{a2} = 30\underline{/-90^\circ}$$
$$\bar{V}_{a3} = 20\underline{/-110^\circ}$$
$$\bar{V}_{a0} = 0$$

(a) Find the phasor expression of phase b voltage.

(b) Is this a balanced four-phase system? Explain.

7-8. A four-phase system has the positive-, negative- and zero-sequence voltages specified in Prob. 7-7. Draw to scale the complete phasor diagrams of the resulting four-phase system. A graphical solution is permissible.

7-9. A four-phase voltage system is described as follows:

$$\bar{V}_a = 100\underline{/0^\circ}$$
$$\bar{V}_b = 50\underline{/-90^\circ}$$
$$\bar{V}_c = 100\underline{/-180^\circ}$$
$$\bar{V}_d = 50\underline{/90^\circ}$$

(a) Is there need for a balanced single-phase set when replacing this voltage system with balanced sets? Explain.

(b) Find the positive-sequence voltage set.

(c) Find the negative-sequence voltage set.

7-10. The positive-sequence component of phase a of a four-phase system is $75\underline{/-30^\circ}$, and the negative-sequence component is $25\underline{/-30^\circ}$. The zero-sequence and the balanced single-phase sets are both zero. Determine the expressions for the phase voltages of the original four-phase system.

7-11. A four-phase voltage system is given by:

$$\bar{V}_a = 100\underline{/0^\circ}$$
$$\bar{V}_b = 60\underline{/-90^\circ}$$
$$\bar{V}_c = 60\underline{/-180^\circ}$$
$$\bar{V}_d = 60\underline{/90^\circ}$$

(a) Determine the phasor expressions of that four-phase system which has a zero resultant.

(b) Find the positive-sequence voltage set for the given system.

(c) Is the positive-sequence set the same for the system of part (a)? Explain.

(d) Find the negative-sequence voltage set.

(e) Compute the balanced single-phase set.

(f) Draw the phasor diagram for phase b, clearly identifying each component.

7-12. Compute the zero-, positive- and negative-sequence sets as well as the balanced single-phase set that can be used to replace the following four-phase system:

$$\bar{V}_a = 100/\underline{0°}$$
$$\bar{V}_b = 60/\underline{-65°}$$
$$\bar{V}_c = 100/\underline{-180°}$$
$$\bar{V}_d = 60/\underline{115°}$$

What generalization can you draw concerning the component sets that make up the original voltage system?

7-13. Find the unbalanced two-phase system that is represented by a positive-sequence component of $95.2/\underline{14.6°}$ and a negative-sequence component of $31/\underline{40.6°}$.

7-14. An unbalanced two-phase voltage system is described by:

$$\bar{V}_a = 120/\underline{0°}$$
$$\bar{V}_b = 80/\underline{-53°}$$

(a) Find the positive-sequence voltage set of this system.

(b) What is the negative-sequence set?

(c) Draw the complete phasor diagram of this two-phase system, showing each positive- and negative-sequence component.

7-15. Repeat Prob. 7-14 for the case where $\bar{V}_a = 120/\underline{0°}$ and $\bar{V}_b = 80/\underline{-90°}$.

7-16. A two-phase, four-pole servomotor is energized by an auxiliary winding voltage that is 90° out of phase with the main winding voltage. The main winding voltage rating at 60 Hz is 75 V. The motor parameters are as follows:

$$r_1 = 50\ \Omega, \qquad r'_2 = 100\ \Omega$$
$$x_1 = 120\ \Omega, \qquad x_\phi = 100\ \Omega$$

The rotor leakage reactance is negligible. Compute the developed torque at a slip of 0.4 when the auxiliary voltage is 37.5 V.

7-17. Appropriate no-load tests on a two-phase servomotor rated at 115-V, 60-Hz, two-poles, reveal the following parameters:

$$r_1 = 302\ \Omega, \qquad r'_2 = 1380\ \Omega$$
$$x_1 = 385\ \Omega, \qquad x'_2 = 385\ \Omega$$
$$x_\phi = 695\ \Omega$$

This servomotor is operated with $\bar{V}_a = 115/\underline{0°}$ and $\bar{V}_b = 80/\underline{-90°}$, and a rotor slip of 0.25.

(a) Draw the appropriate equivalent circuit for the positive- and negative-sequence voltage sets. Show the proper applied voltage in each case.

(b) Compute the developed torque, expressed in synchronous watts and newton-meters.

(c) What is the rms value of the current that flows through the control winding (phase b)?

7-18. Refer to the servomotor of Prob. 7-17 and assume that it continues to operate at a slip of 25%.

(a) Compute the torque developed by this motor when it operates as a balanced two-phase motor, i.e., $\bar{V}_a = 115\underline{/0°}$ and $\bar{V}_b = 115\underline{/-90°}$.

(b) Compute the developed torque when the motor operates as a balanced two-phase motor at a slip of 1.75 per unit.

(c) By employing the results of parts (a) and (b), compute the torque developed by this motor when $\bar{V}_a = 115\underline{/0°}$ and $\bar{V}_b = 80\underline{/-90°}$.

7-19. Refer to the servomotor of Prob. 7-17. When this motor operates at a slip of 0.5, the positive-sequence current is found to be $80.5\underline{/-51.1°}$ ma and the negative-sequence current is $38.5\underline{/-98.4°}$ ma.

(a) Determine the voltages applied to the main and control windings.

(b) Find the control winding current.

7-20. A two-phase servomotor has $115\underline{/0°}$ V applied to the main winding. At a particular point of operation corresponding to a control winding voltage of $75\underline{/-80°}$, the positive- and negative-sequence impedances are, respectively, $220\underline{/67°}$ and $175\underline{/57°}$.

(a) Determine the positive- and negative-sequence components of the main and control winding currents.

(b) Find the total values of the main and control winding currents.

7-21. The servomotor of Prob. 7-17 is operated with the main and control winding voltages always in quadrature.

(a) Compute the stall torque when the control winding voltage equals the main winding voltage, i.e., $\rho = 1$.

(b) Determine the stall torque for $\rho = 0.5$.

(c) Assuming the no-load speed is 3400 rpm, calculate the ratio of stall torque to speed, in radians per second, for $\rho = 1$.

(d) Of what usefulness is the quantity computed in part (c)?

7-22. A two-phase, two-pole, a-c servomotor equipped with a drag-cup rotor has the following parameters at 60 Hz:

$$r_1 = 360\ \Omega, \qquad r_2' = 260\ \Omega$$
$$x_1 = 50\ \Omega, \qquad x_2' = 50\ \Omega$$
$$x_\phi = 890\ \Omega$$

The main winding is identical to the control winding, and both are rated at 115 V. The source of the control winding voltage is arranged to provide a quadrature voltage at all times. Moreover, the internal impedance of this source is negligibly small.

(a) Determine the developed torque in synchronous watts for $\rho = 0.8$ and $s = 0.3$.

(b) How much power is supplied to the main winding in part (a)?

(c) What is the power in the control winding in part (a)?

7-23. (a) For the servomotor of Prob. 7-22 compute the value of the developed torque corresponding to operation with $\rho = 0.1$ and $s = 0.05$.

(b) Repeat this calculation for the servomotor of Prob. 7-17. Compare and comment.

7-24. Refer to the servomotor of Prob. 7-22.

(a) Determine the developed torque in synchronous watts for $\rho = 1$ and $s = 0.3$.

(b) Repeat part (a) for $\rho = 1$ and $s = 1.7$.

(c) Find the developed torque for $\rho = 0.7$ and a rotor slip of 30%.

7-25. For the operating condition specified in Prob. 7-17, compute the ratio of the positive-sequence flux wave to the negative-sequence flux wave. Neglect saturation.

7-26. For the operating condition specified in Prob. 7-22(a) determine the ratio of the positive-sequence flux wave to the negative-sequence flux wave. What significance can you draw from this result?

Single-Phase
Induction Motors

By far the vast majority of single-phase induction motors are built in the fractional-horsepower range. Single-phase motors are found in countless applications doing all sorts of jobs in homes, shops, offices, and on the farm. An inventory of the appliances in the average home in which single-phase motors are used would probably number close to a dozen. An indication of the volume of such motors can be had from the fact that the sum total of all fractional-horsepower motors in use today far exceeds the total of integral-horsepower motors of all types.

This chapter more appropriately belongs in Part I, since the single-phase motor is not customarily considered a special machine. It is included here, however, because the analysis of performance can be very easily handled as a special case of the symmetrical components theory developed in the preceding chapter. Moreover, the background of single-phase motor theory is needed in order to facilitate the understanding of the single-phasing problem associated with servomotors.

8-1 HOW THE ROTATING FIELD IS OBTAINED

In its pure and simple form the single-phase motor usually consists of a distributed stator winding (not unlike one phase of a three-phase motor) and a squirrel-cage rotor. The a-c supply voltage is applied to the stator winding, which in turn creates a field distribution. Since there is a single coil carrying an alternating current, a little thought reveals that the air-gap flux is characterized by being fixed in space and alternating in magnitude. If hysteresis is neglected, the flux is a maximum when the current is instantaneously a maximum and it is zero when the current is zero. Such an arrangement gives the single-phase motor no starting torque. To understand this in terms of the concepts discussed in Sec. 3-1 refer to Fig. 8-1. As a matter of convenience the distributed single-phase winding is represented by a coil wrapped around protruding pole pieces; some motors do in fact use this configuration. Assume that instantaneously the flux-density wave is increasing in the upward direction as shown. Then by transformer action a voltage is induced in the rotor having that distribution which enables the corresponding rotor mmf to oppose the changing flux. To accomplish this, current flows out of the right-side conductors and into the left-side conductors as illustrated in Fig. 8-1. Note that this resulting ampere-conductor distribution corresponds to a space phase angle of $\psi = 90°$. By Eq. (3-54) the net torque is therefore zero. Of course what this means is that beneath each pole piece there are as many conductors producing clockwise torque as there are producing counterclockwise torque. This condition, however, prevails only at standstill. If by some means the rotor is started in either direction, it will develop a nonzero net torque in that direction and thereby cause the motor to achieve normal speed. The problem therefore is to modify the configuration of Fig. 8-1 in such a way that it imparts to the rotor a nonzero starting torque.

Stator winding

Figure 8-1 Simplified diagram of the single-phase motor.

The answer to this problem lies in so modifying the motor that it closely approaches the conditions prevailing in the *two-phase* induction motor. In accordance with the general results developed in Sec. 4-1 we know that to obtain a revolving field of constant amplitude and constant linear velocity in a two-phase induction motor two conditions must be satisfied. One, there must exist two coils (or windings) whose axes are space-displaced by 90 electrical degrees. Two, the currents flowing through these coils must be time-displaced by 90 electrical degrees and they must have such magnitudes that the mmf's are equal. If the currents are less than 90° apart in time but greater than 0°, a rotating field will still be developed but the locus of the resultant flux vector will be an ellipse rather than a circle. Hence in such a case the linear velocity of the field varies from one point in time to another. Also, if the currents are 90° apart but the mmf's of the two coils are unequal, an elliptical locus for the rotating field again results. Finally, if the currents are neither 90° time-displaced nor of a magnitude to furnish equal mmf's, a rotating magnetic field will continue to be developed but now the locus will be more elliptical than in the previous cases. However, the important aspect of all this is that a revolving field can be so obtained even if its amplitude is not constant during its time history, and satisfactory performance can be achieved with such a revolving field. Of course such performance items as power factor and efficiency will be poorer than for the ideal case, but this is not too serious because the motors are of relatively small power.

Appearing in Fig. 8-2 is the schematic diagram which shows the modifications needed to give the single-phase motor a starting torque. A second winding called the *auxiliary* winding is placed in the stator with its axis in quadrature with that of the main winding. Usually the main winding is made to occupy two-thirds of the stator slots and the auxiliary winding is placed in the remaining one-third. In this way the space-displacement condition is met exactly. The time displacement of the currents through the two windings is obtained at least partially by designing the auxiliary winding for high resistance and low leakage reactance. This is in contrast to the main winding, which has low resistance and higher leakage reactance. Figure 8-3 depicts the time displacement existing between the auxiliary winding current $\bar{I}_A$

Figure 8-2 Schematic diagram of the resistance split-phase motor.

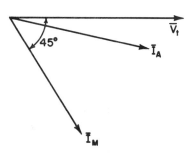

Figure 8-3 Phasor diagram showing the time-phase displacement between the auxiliary and main winding currents at standstill.

and the main winding current $\bar{I}_M$ at standstill. Frequently in motors of this design the $\bar{I}_A$ and $\bar{I}_M$ phasors are displaced by about 45° in time. Thus with the arrangement of Fig. 8-2 a revolving field results and so the motor achieves normal speed. Because of the high-resistance character of the auxiliary winding, this motor is called the *resistance-start split-phase* induction motor. Also, the auxiliary winding used in these motors has a short time power rating and therefore must be removed from the line once the operating speed is reached. To do this a cut-out switch is placed in the auxiliary winding circuit which, by centrifugal action, removes the auxiliary winding from the line when the motor speed exceeds 75 per cent of synchronous speed.

8-2 EQUIVALENT CIRCUIT BY SYMMETRICAL COMPONENTS

The single-phase induction motor can be treated as that special case of a two-phase induction motor where only one stator phase winding is used. In the notation employed in the preceding chapter only stator phase a† is provided in the machine;

†Lowercase a is not to be confused with the capital A used in Sec. 8-1. Phase a here is identical with the main winding referred to in Sec. 8-1.

phase *b* is eliminated entirely. Accordingly, it is reasonable to expect that, in the expressions for the positive- and negative-sequence currents as represented by Eqs. (7-55) and (7-56), the phase *b* current $\bar{I}_b$ is identically equal to zero. Therefore

$$\bar{I}_{a1} = \frac{\bar{I}_a}{2} \tag{8-1}$$

$$\bar{I}_{a2} = \frac{\bar{I}_a}{2} \tag{8-2}$$

which leads to the obvious conclusion that for the single-phase induction motor the positive- and negative-sequence currents are the *same*. This result is not surprising in view of the fact that the motor is equipped with only a single stator coil.

Figure 8-4(a) depicts the equivalent circuits corresponding to the positive- and negative-sequence magnetic fields as derived in Sec. 7-3. However, because the positive- and negative-sequence currents are the same, *a series connection of these individual equivalent circuits is now permissible.* The result is shown in Fig. 8-4(b) and it is worthwhile to note that now both the current and voltage equations for the stator winding can be represented by a single circuit in terms of the appropriate

(a)   (b)

(c)

Figure 8-4 Equivalent circuit of the single-phase induction motor, based on symmetrical components: (a) the equivalent circuits associated with the positive- and negative-sequence fields; (b) series connection representation that is permitted because the positive- and negative-sequence currents are always the same; (c) simplified representation in terms of the positive-sequence input impedance $\bar{Z}_1$ and the negative-sequence input impedance $\bar{Z}_2$. Core loss is neglected throughout.

positive- and negative-sequence components. The resultant stator current is given by

$$\bar{I}_a = \bar{I}_{a1} + \bar{I}_{a2} \tag{8-3}$$

where $\bar{I}_{a1}$ and $\bar{I}_{a2}$ are expressed by Eqs. (8-1) and (8-2). The applied single-phase stator voltage continues to be expressed generally as

$$\bar{V}_a = \bar{V}_{a1} + \bar{V}_{a2} \tag{8-4}$$

but the positive- and negative-sequence voltage components can no longer be validly determined by Eqs. (7-30) and (7-31). This happens because a definite, second-phase voltage does not exist to *impose* specific positive- and negative-sequence voltages on the machine. Nor is it correct to set $\bar{V}_b$ equal to zero and thereby expect to obtain useful results from these equations. An examination of Fig. 8-4(b) discloses that the values of $\bar{V}_{a1}$ and $\bar{V}_{a2}$ depend upon the slip at which the motor is operating. Recall from Chapter 7 that the input impedance $\bar{Z}_1$ of the polyphase induction motor to the positive sequence is expressed by Eq. (7-44) and the input impedance $\bar{Z}_2$ to the negative sequence is given by Eq. (7-49). A simplified equivalent circuit of the single-phase induction motor can then be represented as shown in Fig. 8-4(c). *Since $\bar{Z}_1$ and $\bar{Z}_2$ are permitted to be series connected*, it follows that the positive- and negative-sequence voltages for a given $\bar{V}_a$ depend solely upon $\bar{Z}_1$ and $\bar{Z}_2$. In fact, in accordance with the voltage-divider theorem the appropriate expressions become

$$\bar{V}_{a1} = \frac{\bar{Z}_1}{\bar{Z}_1 + \bar{Z}_2} \bar{V}_a \tag{8-5}$$

$$\bar{V}_{a2} = \frac{\bar{Z}_2}{\bar{Z}_1 + \bar{Z}_2} \bar{V}_a \tag{8-6}$$

Keep in mind that $\bar{Z}_1$ and $\bar{Z}_2$ are unequal whenever the slip is other than unity. The difference in $\bar{Z}_1$ and $\bar{Z}_2$ is consistent with the use of r_2'/s for the positive sequence and $r_2'/(2-s)$ for the negative sequence.

As the speed of the single-phase induction motor changes, so too do the values of the $\bar{V}_{a1}$ and $\bar{V}_{a2}$. This behavior stands in sharp contrast to the manner in which the unbalanced two-phase motor operates. In the latter, once the unbalanced voltages $\bar{V}_a$ and $\bar{V}_b$ are applied and maintained, the corresponding positive- and negative-sequence voltage components as determined from Eqs. (7-30) and (7-31) remain invariant and thus yield a single torque-speed curve that applies over the full operating range. As the operating point changes, say, in response to load conditions, the $\bar{V}_{a1}$ and $\bar{V}_{a2}$ components do not correspondingly change. Rather, the point on the derived† torque-speed curve changes.

Physical Conception of Single-Phase Motor Operation

There may be some question in the mind of the reader at this point concerning the propriety of representing the single-phase induction motor in terms of positive- and negative-sequence voltage sets when in fact there exists only a single-phase voltage applied to a single-phase stator winding. One might argue that in the two-phase

†By the technique described in Sec. 7-4.

motor subjected to unbalanced voltages a revolving field does in fact exist in the original machine and so representation in terms of two balanced but oppositely rotating fields reflects an acceptable degree of consistency. In contrast, the single-phase voltage produces in the machine an alternating field, the axis of which is fixed in space by the location of the axis of the stator winding. Fortunately, this representation is not as far-fetched as it may initially seem, simply because every alternating field may be replaced by two oppositely revolving fields having half the amplitude of the alternating field and traveling at the same frequency as the stationary field alternates. We can turn to mathematics to illustrate this equivalence in precise terms.

As described in Sec. 4-1 the equation for an alternating magnetic field whose axis is fixed in space is given by

$$b(\alpha) = B_m \sin \omega t \cos \alpha \qquad (8\text{-}7)$$

where B_m denotes the amplitude of the sinusoidally distributed air-gap flux density produced by a properly distributed stator winding carrying an alternating current of frequency ω, and α denotes a space-displacement angle measured from the axis of the stator (or field) winding. If the trigonometric identity

$$\sin x \cos y = \tfrac{1}{2} \sin (x - y) + \tfrac{1}{2} \sin (x + y) \qquad (8\text{-}8)$$

is introduced into Eq. (8-7), the expression for the magnetic field becomes alternatively

$$b(\alpha) = \frac{B_m}{2} \sin (\omega t - \alpha) + \frac{B_m}{2} \sin (\omega t + \alpha) \qquad (8\text{-}9)$$

The first term on the right side of Eq. (8-9) is the equation of a revolving field that moves in the positive α direction (see Sec. 4-1) and has an amplitude equal to one-half that of the original alternating field. The second term is the equation of a revolving field whose amplitude is also one-half that of the alternating field and whose direction of rotation is in the negative α direction. The field that moves in the positive α direction is sometimes referred to as the *forwardly rotating field*. In our description it is called the positive-sequence field and is assumed to be produced by $\bar{V}_{a1}$ and its balanced two-phase counterpart $-j\bar{V}_{a1}$. Similarly, the backwardly rotating field corresponds to the negative-sequence field. By definition the positive direction is that direction in which the single-phase motor is initially started.

The absence of a starting torque in a single-phase motor that is not equipped with an auxiliary starting device may now be understood in terms of effects produced by the positive- and negative-sequence fields. At standstill $s = 1$, and so $\bar{Z}_1 = \bar{Z}_2$. Then by Eqs. (8-5) and (8-6), $\bar{V}_{a1} = \bar{V}_{a2}$, and under these conditions a glance at Fig. 8-4(b) indicates that the magnetizing current associated with the positive sequence $\bar{I}_{\phi1}$ is equal to that of the negative sequence $\bar{I}_{\phi2}$. This is another way of saying that the amplitudes of the two revolving fields in Eq. (8-9) are the same and thereby yield no net revolving field to produce a starting torque. The two revolving fields in the motor produce effects that are equal and opposite.

If the motor is given an initial starting torque and made to operate at a slip s lying in value between zero and unity, then $\bar{Z}_1$ is greater than $\bar{Z}_2$ and $\bar{I}_{\phi1} > \bar{I}_{\phi2}$. In fact in the normal range where $0 \leq s \leq 0.15$, the resistance $r'_2/s > r'_2/(2 - s)$ and so $\bar{I}_{\phi1} > \bar{I}_{\phi2}$. Consequently, the forwardly rotating field represented by the vector

$\Phi_f = \Phi_1$ in Fig. 8-5 is appreciably greater than the amplitude of the backwardly rotating field $\Phi_b = \Phi_2$, and so a resultant field flux is produced that lies off the stator winding axis. As time continues to elapse in Fig. 8-5, Φ_2 moves clockwise while Φ_1 moves counterclockwise. Note that, when $\omega t = \pi/2$, Φ_1 and Φ_2 lie along the horizontal and thereby yield a resultant field that lies to the left on the horizontal and is equal to $\Phi_1 - \Phi_2$. This quantity in fact denotes the minor axis of the elliptical locus of the resultant field that is produced as the angular frequency changes through one complete cycle. At standstill when $s = 1$ and $\Phi_1 = \Phi_2$, the resultant flux quantity always lies along the vertical and thus yields an alternating field rather than a revolving one.

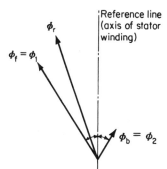

Figure 8-5 Illustrating how a resultant revolving field is obtained when the motor operates at a slip $\neq 1$.

Simplified Approximate Equivalent Circuit

The equivalent circuit appearing in Fig. 8-4(b) is approximate because core loss has been neglected. Of course due account can be taken of this quantity by merely placing a core loss resistor in parallel with the magnetizing reactance of the positive-sequence part of the circuit. It is usually not necessary to include the core loss resistor for the negative-sequence section. Why?

Several simplifications of the equivalent circuit of Fig. 8-4(b) can be introduced. The first involves combining the stator impedance drop into a single total value which is permissible for the single-phase induction motor because the positive- and negative-sequence currents are the same. Thus,

$$\text{total impedance drop} = \bar{I}_{a1}(r_1 + jx_1) + \bar{I}_{a2}(r_1 + jx_1)$$

$$= \frac{\bar{I}_a}{2}(r_1 + jx_1) + \frac{\bar{I}_a}{2}(r_1 + jx_1) = \bar{I}_a(r_1 + jx_1) \qquad (8\text{-}10)$$

The second simplification comes about upon noticing that the voltage drop across $\bar{Z}_{p1}$ [the parallel combination of jx_ϕ and $(r_2'/s + jx_2')$] and $\bar{Z}_{p2}$ can be written as

$$\frac{\bar{I}_a}{2}\bar{Z}_{p1} = \bar{I}_a\frac{\bar{Z}_{p1}}{2} \qquad (8\text{-}11)$$

and

$$\frac{\bar{I}_a}{2}\bar{Z}_{p2} = \bar{I}_a\frac{\bar{Z}_{p2}}{2} \qquad (8\text{-}12)$$

Whenever the stator impedance drop is expressed by the result of Eq. (8-10), the preferred way of expressing the voltage drops appearing across the parallel elements of the positive- and negative-sequence circuits is by the right side of the last two equations. With the formulations represented by Eqs. (8-10), (8-11), and (8-12), the equivalent circuit of Fig. 8-4(b) can be replaced by the simpler version depicted in Fig. 8-6. Note that it is now the total stator current $\bar{I}_a$ that flows in the series arrangement of the circuit elements, $r_1 + jx_1$, $\bar{Z}_{p1}/2$, and $\bar{Z}_{p2}/2$ in response to the driving force represented by $\bar{V}_a$. The manipulation involved in Eqs. (8-11) and (8-12) calls for the use of one-half the values of the actual rotor winding resistance and leakage reactance per phase as well as one-half the actual value of the magnetizing reactance.

Figure 8-6 Simplified approximate equivalent circuit of the single-phase induction motor based on symmetrical components.

Figure 8-7 A modified simplified equivalent circuit applies for $0 \leq s \leq 0.15$.

Once the single-phase induction motor is started and achieves normal speed, the slip frequently lies in the range $0 \leq s \leq 0.15$. Moreover, for conventional single-phase motors $x_\phi > |r_2' + jx_2'|$. Accordingly, very little error is made if the magnetizing reactance is entirely eliminated from the impedance $\bar{Z}_{p2}/2$. Also, $r_2'/[2(2-s)] \approx r_2'/4$ and, since this resistor carries the same current as r_1, the former may be combined with the latter. Similarly, $jx_2'/2$ may be combined with jx_1. Imposing these assumptions on the circuit of Fig. 8-6 leads to a still simpler form as depicted in Fig. 8-7.

The form of the configuration of Fig. 8-7 is like that of the per phase equivalent circuit of a balanced polyphase induction motor. A considerable and useful simplification has clearly been made possible in a situation that at first appeared quite complex.

8-3 THE TORQUE-SPEED CHARACTERISTIC

We have seen that the curve relating torque as a function of speed is a very important characteristic of electromechanical energy-conversion devices operating as motors. The single-phase motor is no exception. It is to be expected, however, that the resultant torque-speed that applies over the entire operating range is not as straight-forward to obtain as it was for the two-phase induction motor subjected to unbalanced two-phase voltages. The difference lies in the fact that with the latter machine the positive- and negative-sequence voltage components remain fixed for all values of slip [compare Eqs. (7-30) and (7-31)], whereas for the single-phase motor these voltage components change in value with slip [refer to Eqs. (8-5) and (8-6)]. Accordingly, when the new resultant torque-speed curve is found by applying Eq. (7-78), care must be taken to compute T_1 and T_2 at a given slip by first calculating those values of $\bar{V}_{a1}$ and $\bar{V}_{a2}$ that are valid for the slip concerned.

To emphasize this point further, refer to the plot appearing in Fig. 8-8(a). It shows the plot of the positive-sequence torque-speed curve T_1 corresponding to $\bar{V}_{a1} = \bar{V}_a/2$, and the negative-sequence torque-speed curve T_2 corresponding to $\bar{V}_{a2} = \bar{V}_a/2$. When this plot is made to apply to the single-phase induction motor, the only point on the resultant curve shown that has any validity is the one for $s = 1$. It indicates that the resultant torque is zero—a familiar result by now. When the motor is assumed to be operating at a normal slip, say s_n, it is incorrect to identify the corresponding torque as T'_n in Fig. 8-8(a), because $\bar{V}_{a1}$ and therefore T_1 are much larger than the values prevailing at standstill. Rather, the correct value of the resultant torque at slip s_n is determined by first finding the T_1 and T_2 quantities using the appropriate values of $\bar{V}_{a1}$ and $\bar{V}_{a2}$ and then subtracting to obtain T_n as illustrated in Fig. 8-8(b). The true resultant torque-speed curve of the single-phase induction motor assumes a form like that depicted in Fig. 8-8(c). If the torque-speed characteristic of the motor for balanced two-phase operation is known over the entire speed range, then the curve of Fig. 8-8(c) can be readily computed by a direct application of Eqs.

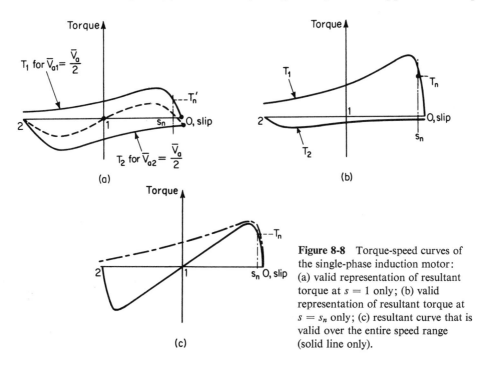

(a)

(b)

(c)

Figure 8-8 Torque-speed curves of the single-phase induction motor: (a) valid representation of resultant torque at $s = 1$ only; (b) valid representation of resultant torque at $s = s_n$ only; (c) resultant curve that is valid over the entire speed range (solid line only).

(8-5), (8-6), and (7-78). It must be remembered that for the resultant torque computed at each slip by Eq. (7-78), it is necessary first to find the appropriate values of $\bar{V}_{a1}$ and $\bar{V}_{a2}$ by Eqs. (8-5) and (8-6).

The broken-line curve of Fig. 8-8(c) denotes the torque-speed curve of the two-phase motor operated under balanced conditions. A comparison with single-phase operation reveals two interesting facts: (1) operation as a single-phase motor at small slips is quite acceptable in spite of increased rotor losses; (2) the single-phase motor will run in the direction in which it is started.

Effect of Rotor Resistance on the Torque-Speed Curve

In the case of the polyphase induction motor it was shown that neither the shape nor the maximum value of the torque-speed curve was influenced by the value of rotor resistance. Refer to Eq. (4-33). However, this is not so for the single-phase induction motor and the reason is obvious from a glance at the equivalent circuit of Fig. 8-7. Recall that the slip at which the maximum torque is developed corresponds to the conditions for which maximum power is delivered to the resistor $r_2'/2s$. The slip at which this occurs is found by setting $r_2'/2s$ equal to the magnitude of the impedance looking back into the equivalent circuit from the terminals of this resistance. Clearly, the ensuing value for s_n will be found to depend upon r_2'. Accordingly, the resultant torque too is so influenced by r_2', and the manner in which this occurs is depicted in Fig. 8-9. Note that it not only causes a reduction in the peak developed torque but it brings about a flattening of the curve as well. This behavior of the torque-speed curve restricts the use of external rotor resistance for controlling the speed of the single-phase induction motor.

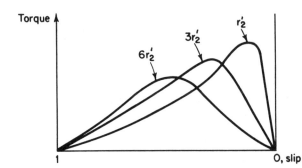

Figure 8-9 Showing the effect of rotor resistance on the torque-speed curve of a single-phase induction motor.

8-4 PERFORMANCE ANALYSIS

The calculation of the performance of the single-phase induction motor can be readily determined through the use of the equivalent circuit and the appropriate power-flow diagram. The approximate equivalent circuit of Fig. 8-7 may be used whenever it is desired to compute the behavior of the motor in the slip range of $0 \leq s \leq 0.15$. Otherwise use is made of the circuit depicted in Fig. 8-6.

The flow of power from the line to the shaft in the single-phase motor is a bit more complicated than it is in the balanced polyphase motor because of the need

$$P_i = V_a I_a \cos \theta$$

$I_a^2 r_1$
P_c

Gap | Power

$$P_{g2} = (I_{22}')^2 \frac{r_2'}{2(2-s)}$$

$$P_{g1} = (I_{21}')^2 \frac{r_2'}{2s}$$

$$(I_{21}')^2 \frac{r_2'}{2}$$

$$P_{m1} = P_{g1}(1-s)$$

$$P_{g2} + (1-s) P_{g2}$$

P_{rot}

P_0

Total rotor copper loss
of negative sequence

Σ losses

Figure 8-10 Power-flow diagram of the single-phase induction motor.

to work with the positive- and negative-sequence fields. The complete power-flow diagram appears in Fig. 8-10. The electrical input power is applied to the motor at the stator winding. After removing a small part of this power to supply the stator winding copper loss and the core loss produced by the resultant field, the remainder is the power transferred across the air gap. However, note that the total gap power is now expressed in two parts, one associated with the positive-sequence field, P_{g1}, and the other associated with the negative-sequence field, P_{g2}. Because the rotor travels in the same direction as the positive-sequence field, the negative-sequence gap power is necessarily consumed entirely as rotor copper loss. It is worthwhile to note too that this gap power P_{g2} only partially supplies the total copper loss associated with the rotor resistance of the negative-sequence impedance. From the equivalent circuit of Fig. 8-6 it is seen that the negative-sequence gap power is the power associated with $r_2'/2(2-s)$ and is expressed by

$$P_{g2} = (I_{22}')^2 \frac{r_2'}{2(2-s)} \tag{8-13}$$

Upon rearrangement of this equation, it follows that the expression for the total copper loss in the rotor resistance of the negative sequence becomes

$$(I_{22}')^2 \frac{r_2'}{2} = (2-s)P_{g2} = P_{g2} + (1-s)P_{g2} \tag{8-14}$$

The right side of Eq. (8-14) is written in two parts to emphasize that a portion of this copper loss, P_{g2}, is supplied directly from the line through the medium of the negative-sequence field, while the remaining portion, $(1-s) P_{g2}$, is supplied mechanically by the rotor in response to the positive-sequence field. At normal operating slips the positive-sequence gap power is many times greater than that of the negative-sequence. Upon subtracting the copper loss in the rotor winding associated with the positive-sequence [i.e., $(I_{21}')^2 (r_2'/2)$ from P_{g1}], there results the useful mechanical developed power P_{m1}. Specifically,

$$P_{m1} = P_{g1}(1 - s) \qquad (8\text{-}15)$$

and this quantity represents the force that drives the load. Further examination of the power-flow diagram indicates, however, that before useful power can be delivered to the load the mechanical power developed by the positive-sequence field must first supply the rotational losses as well as the power needed to drive the rotor against the opposing action of the negative-sequence field. This latter quantity is already denoted in Eq. (8-14) as $(1 - s)P_{g2}$.

Example 8-1

A $\frac{1}{4}$-hp, 110-V, 60-Hz, four-pole single-phase induction motor has a rotational loss of 15 W at normal speeds. The equivalent circuit parameters are known to be as follows:

$$r_1 = 1.3 \ \Omega, \qquad r'_2 = 3.2 \ \Omega$$
$$x_1 = 2.5 \ \Omega, \qquad x'_2 = 2.2 \ \Omega$$
$$x_\phi = 48 \ \Omega$$

Determine the performance (i.e., line current, line power factor, horsepower output, and efficiency) of this motor when it operates at a slip of 4%.

Solution The approximate equivalent circuit of Fig. 8-7 is used. It is repeated in Fig. 8-11 with all parameter values shown. The impedance of the parallel circuit is given by

$$\frac{\bar{Z}_{p1}}{2} = \frac{j\frac{x_\phi}{2}\left(\frac{r'_2}{2s} + j\frac{x'_2}{2}\right)}{\frac{r'_2}{2s} + j\left(\frac{x_\phi}{2} + \frac{x'_2}{2}\right)} = \frac{j24(40 + j1.1)}{40 + j25.1} = 10.8 + j17.2 = 20.4\underline{/57.9^\circ} \qquad (8\text{-}16)$$

The total input impedance is therefore

$$\bar{Z}_T = \left(r_1 + \frac{r'_2}{4}\right) + j\left(x_1 + \frac{x'_2}{2}\right) + \frac{\bar{Z}_{p1}}{2} = 2.1 + j3.6 + 10.8 + j17.2$$

$$= 12.9 + j20.8 = 24.4\underline{/58.4^\circ} \qquad (8\text{-}17)$$

Accordingly, the line current is

$$\bar{I}_a = \frac{\bar{V}_a}{\bar{Z}_T} = \frac{110}{24.4\underline{/58.4^\circ}} = 4.51\underline{/-58.4^\circ} \ \text{A} \qquad (8\text{-}18)$$

and the line power factor is

$$\text{power factor} = \cos 58.4^\circ = 0.525 \ \text{lagging} \qquad (8\text{-}19)$$

Figure 8-11 Equivalent circuit for Example 8-1.

SINGLE-PHASE INDUCTION MOTORS   CHAP. 8

Such a low value for the line power factor is typical of single-phase motor operation. The input power is found to be

$$P_i = V_a I_a \cos \theta = 110(4.51)0.525 = 261 \text{ W} \tag{8-20}$$

The positive-sequence gap power is found from

$$P_{g1} = (I_a)^2 \frac{1}{2} R_{p1} = (4.51)^2 10.8 = 220 \text{ W} \tag{8-21}$$

Also, the negative-sequence gap power is determined to be

$$P_{g2} = (I'_{22})^2 \frac{r'_2}{4} \approx I_a^2 \frac{r'_2}{4} = (4.51)^2 0.8 = 16.3 \text{ W} \tag{8-22}$$

From the power-flow diagram the expression for the output power is

$$P_0 = P_{m1} - P_{\text{rot}} - (1-s)P_{g2}$$
$$= P_{g1} - sP_{g1} - P_{\text{rot}} - (1-s)P_{g2} \tag{8-23}$$

Inserting values

$$P_0 = 220 - (0.04)220 - 15 - (0.96)16.3$$
$$= 220 - 8.8 - 15 - 15.6$$
$$= 182.4 \text{ W} \tag{8-24}$$

The horsepower delivered to the load is therefore

$$\text{horsepower} = \frac{P_0}{746} = \frac{182.4}{746} = 0.245 \tag{8-25}$$

This result indicates that the motor is delivering almost rated output power.
The efficiency is found to be

$$\eta = \frac{P_0}{P_i} = \frac{182.4}{261} = 0.70, \text{ or } 70\% \tag{8-26}$$

By repeating the calculations described in the foregoing example for various values of slip, the performance characteristics of the motor can be determined over the full operating range. Curves for a typical $\frac{1}{4}$-hp, 1725-rpm motor are depicted in Fig. 8-12.

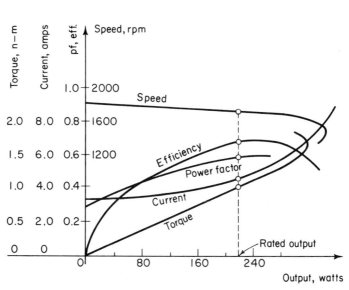

Figure 8-12 Typical performance characteristics for a single-phase induction motor, rated at $\frac{1}{4}$ hp, 60 Hz, 4 poles.

8-5 THE APPROXIMATE DETERMINATION OF THE EQUIVALENT CIRCUIT PARAMETERS

The data obtained in a blocked-rotor test and a no-load test can be used to furnish information about the parameters appearing in the equivalent circuit. The blocked-rotor test data are used to reveal information about the leakage reactances and the stator-referred rotor winding resistance. The determination of these parameters is approximate because the associated procedure involves simplifying assumptions. One assumption is to consider that for $s = 1$ the impedance of $x_\phi/2$ in the equivalent circuit of Fig. 8-6 is so large compared with $(r_2'/2) + j(x_2'/2)$ that for all practical purposes it may be eliminated entirely from the equivalent circuit. Accordingly, the equivalent circuit of Fig. 8-6 reduces to the form depicted in Fig. 8-13.

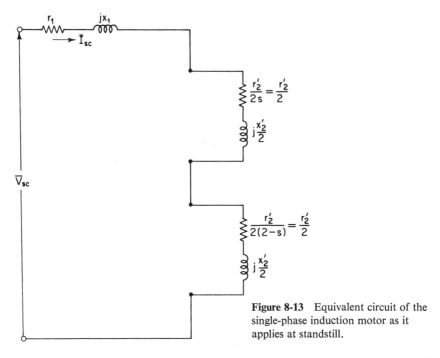

Figure 8-13 Equivalent circuit of the single-phase induction motor as it applies at standstill.

If we let the increased values of voltage, current, and power obtained in the rotor-blocked test be denoted respectively by V_{sc}, I_{sc}, and P_{sc}, it then follows that the equivalent impedance referred to the stator coil is

$$Z_e = \frac{V_{sc}}{I_{sc}} \tag{8-27}$$

Moreover, the equivalent resistance component of Eq. (8-27) is readily determined from the wattmeter and ammeter readings. Thus

$$R_e \equiv r_1 + r_2' = \frac{P_{sc}}{I_{sc}^2} \tag{8-28}$$

To determine r_2' from the result of Eq. (8-28), it is necessary to introduce a second important assumption, i.e.,

$$r_1 = r_2' \qquad (8\text{-}29)$$

This assumption is valid for all well-designed machines, i.e., those machines for which a minimum amount of iron and copper is used for a specified output rating. Unless otherwise stated, the assumption stated in Eq. (8-29) is always made in analyzing the single-phase motor. Therefore, the rotor winding resistance term appearing in the equivalent circuit can be identified as

$$\frac{r_2'}{2} = \frac{R_e}{4} \qquad (8\text{-}30)$$

where R_e is determined by Eq. (8-28).

The value of the equivalent leakage reactance is obtained from

$$X_e = \sqrt{Z_e^2 - R_e^2} \qquad (8\text{-}31)$$

where Z_e and R_e are found from Eqs. (8-27) and (8-28). The stator-referred rotor winding leakage reactance is then found by assuming that

$$x_1 = x_2' \qquad (8\text{-}32)$$

and this leads directly to

$$\frac{x_2'}{2} = \frac{X_e}{4} \qquad (8\text{-}33)$$

The sole parameter of the equivalent circuit that remains to be determined is the magnetizing reactance x_ϕ. Its value can be established from the no-load test data and knowledge of x_1, x_2', and r_2' as derived from the rotor-blocked test. The procedure is illustrated by referring to the equivalent circuit of Fig. 8-14 which is derived from Fig. 8-6 upon imposing simplifying assumptions. Recall that at no-load the slip is very close to zero; it then follows that the quantity $r_2'/2s$ represents a very high resistance and so may be omitted without much error. Therefore, at no-load $\bar{Z}_{p1}/2$ is assumed to be equal simply to $jx_\phi/2$. Similarly, the impedance $\bar{Z}_{p2}/2$ associated with

Figure 8-14 Equivalent circuit of the single-phase induction motor at no-load ($s \approx 0$).

the negative-sequence field can be simplified to the low impedance represented by $(r_2'/4) + j(x_2'/2)$—which is many times smaller than $jx_\phi/2$. Certainly this assumption is even more valid than the corresponding assumption used in the rotor-blocked test.

Determination of x_ϕ essentially requires finding the voltage across a and b in Fig. 8-14. By letting the instrument readings for voltage, current, and power in the no-load test be denoted respectively by V_n, I_n, and P_n, the power factor angle of the current can be found from

$$\cos \theta_n = \frac{P_n}{V_n I_n} \tag{8-34}$$

Then,

$$\bar{V}_{ab} = \bar{V}_n - I_n\underline{/-\theta_n}\left[\left(r_1 + \frac{r_2'}{4}\right) + j\left(x_1 + \frac{x_2'}{2}\right)\right] \tag{8-35}$$

Since all the quantities on the right side of this equation are known, the magnetizing reactance is readily computed as

$$\frac{x_\phi}{2} = \frac{V_{ab}}{I_n} \tag{8-36}$$

The foregoing procedure can be somewhat simplified by using an approximate expression for the voltage across a and b that uses magnitudes instead of phasors. Thus

$$V_{ab} \approx V_n - I_n\left(x_1 + \frac{x_2'}{2}\right) \tag{8-37}$$

The validity of this assumption is based on the fact that the single-phase induction motor operates at very poor power factor at no-load. A glance at Fig. 8-15 indicates that the three quantities involved in Eq. (8-37) are practically collinear. When Eq. (8-37) is used to determine V_{ab}, there is no need to use Eq. (8-34); x_ϕ is found directly through the application of the last two equations only.

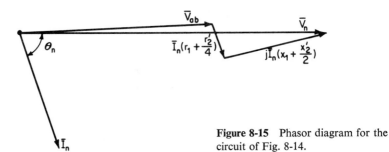

Figure 8-15 Phasor diagram for the circuit of Fig. 8-14.

Example 8-2

Laboratory tests on a single-phase induction motor yield the following data:

Blocked-rotor test: $V_{sc} = 110$ V, $I_{sc} = 14.8$ A, $P_{sc} = 1130$ W
No-load test: $V_n = 110$ V, $I_n = 2.8$ A, $P_n = 60$ W
Rotational losses: 17 W

Determine the five parameters of the equivalent circuit and the value of the core loss in watts.

Solution From Eq. (8-28) the equivalent resistance is

$$R_e = r_1 + r_2' = \frac{P_{sc}}{I_{sc}^2} = \frac{1130}{24.8^2} = 5.15 \ \Omega$$

Hence,

$$r_1 = r_2' = 2.65 \ \Omega$$

and

$$\frac{r_2'}{2} = 1.33 \ \Omega$$

Also,

$$Z_e = \frac{V_{sc}}{I_{sc}} = \frac{110}{14.8} = 7.44 \ \Omega$$
$$X_e = \sqrt{7.44^2 - 5.15^2} = 5.36 \ \Omega$$

Therefore,

$$x_1 = x_2' = \frac{5.36}{2} = 2.68 \ \Omega$$

and

$$\frac{x_2'}{2} = 1.34 \ \Omega$$

Moreover, by Eq. (8-37) the voltage across the magnetizing reactance at no-load is

$$V_{ab} = V_n - I_n\left(x_1 + \frac{x_2'}{2}\right) = 110 - 2.8(2.68 + 1.34) = 110 - 11.2 = 98.8 \ \text{V}$$

Hence,

$$\frac{x_\phi}{2} = \frac{V_{ab}}{I_n} = \frac{98.8}{2.8} = 35.3 \ \Omega$$

The core loss is found from

$$P_c = P_n - I_n^2\left(r_1 + \frac{r_2'}{4}\right) - P_{\text{rot}}$$
$$= 60 - 2.8^2(3.17) - 17 = 18.2 \ \text{W} \qquad (8\text{-}38)$$

The equivalent circuit showing the computed parameters appears in Fig. 8-16.

Figure 8-16 Equivalent circuit showing parameters computed in Example 8-2.

8-6 TYPES OF SINGLE-PHASE MOTORS

Many different types of single-phase motors have been developed primarily for two reasons. One, the torque requirements of the appliances and applications in which they are used vary widely. Two, it is desirable to use the lowest-priced motor that will drive a given load satisfactorily. For example, a high-torque version of the split-phase induction motor just discussed is designed almost exclusively for washing-machine applications. This motor is available in a single speed rating and in just two horsepower ratings. This, coupled with the large volume of sales, enables it to be the least expensive motor in its category. See line two of Table 8-1 for more details.

The Capacitor-Start Induction-Run Motor

The chief difference between the various types of single-phase motors lies in the method used to start them. In the case of this motor a capacitor is placed in the auxiliary winding circuit so selected that it brings about a 90-degree time displacement between $\bar{I}_A$ and $\bar{I}_M$. See Fig. 8-17. The result is a much larger starting torque than is achievable with resistance split-phase starting. This motor is widely used for general-purpose applications.

Figure 8-17   Capacitor-start induction-run motor: (a) schematic diagram; (b) phasor diagram at starting.

The Capacitor-Start Capacitor-Run Motor

As Fig. 8-18 shows, two capacitors are used in the auxiliary circuit of this motor. By keeping one capacitor in during normal operation, improved performance is obtained because the motor then behaves more like the balanced two-phase motor. The improved performance is manifested in less noise and higher efficiency and power

Figure 8-18   Capacitor-start capacitor-run motor.

Figure 8-19   Permanent-split capacitor motor.

factor. A second capacitor is needed at starting because the reactive component of the input impedance of the auxiliary circuit is considerably different at standstill than at full speed.

The Permanent-Split Capacitor Motor

In this motor a single capacitor is used both for starting and for running. To take advantage of improved running performance, the value of the capacitor used is that needed at full speed. Consequently, starting torque must be sacrificed. The schematic wiring diagram appears in Fig. 8-19.

The Shaded-Pole Motor

The shaded-pole motor is very extensively used in applications that require $\frac{1}{20}$ hp or less. The construction is extremely rugged, as Fig. 8-20 reveals. There is little that can go wrong with this motor aside from overheating. Note that it contains no cut-out switch, which could be a source of trouble, nor does it have an auxiliary winding, which could burn up—especially if the cut-out switch become faulty. The shaded-pole motor consists essentially of copper wound on iron and not much more.

Figure 8-20 Construction features of the shaded-pole motor.

The manner of obtaining a moving-flux field, however, is different in this motor than it is in those considered previously. Here use is made of a copper ring imbedded in each salient pole piece. As the gap flux is changing in response to the alternating coil current, transformer action causes currents to be induced in the copper rings, always so directed as to oppose the changing flux. The net effect of this action is that, for one portion of the cycle, the ring currents cause the flux to concentrate in that part of the pole piece which is free of the ring. At a subsequent portion of the cycle the ring currents act to crowd the gap flux through that part of the pole piece around which the ring is wrapped. In this fashion the air-gap flux undergoes a sweeping motion across the pole face. It appears to be moving from the unshaded to the shaded (or ring) portion of the pole. The sweeping action of the flux occurs periodically, thereby producing a starting as well as a running torque. The starting torque is normally 50 per cent of rated value. The breakdown torque is also relatively low.

Type designation	Starting torque (% of normal)	Approx. comparative price (%)	Breakdown torque (% of normal)	Starting current at 115 V
General-purpose split-phase motor	90–200 Medium	85	185–250 Medium	23 1/4 hp
High-torque split-phase motor	200–275 High	65	Up to 350	32 High 1/4 hp
Permanent-split capacitor motor	60–75 Low	155	Up to 225	Medium
Permanent-split capacitor motor	Up to 200 Normal	155	260	
Capacitor-start general-purpose motor	Up to 435 Very high	100	Up to 400	
Capacitor-start capacitor-run motor	380 High	190	Up to 260	
Shaded-pole motor	50	—	150	

Source: By permission from M. Liwschitz-Garik and C. C. Whipple, *Electric Machinery*, vol. II (Princeton, N.J.: D. Van Nostrand Co., Inc., 1946).

8-7 CHARACTERISTICS AND TYPICAL APPLICATIONS

The principal performance features as well as typical applications of single-phase a-c motors appear in Table 8-1. Note that for applications involving less than $\frac{1}{20}$ hp the shaded-pole motor is invariably used. On the other hand, for applications involving $\frac{1}{20}$ hp to $\frac{3}{4}$ hp the choice of the motor depends upon such factors as the starting and breakdown torque and even quietness of operation. Where a minimum of noise is desirable and low starting torque is adequate, such as for driving fans and blowers, then the motor to choose is the permanent-split type. If quietness is to be combined with high starting torque, as might be the case where a compressor drive is needed, then the suitable choice would be the capacitor-start capacitor-run motor. Of course,

Power factor (%)	Efficiency (%)	Horsepower range	Application and general remarks
56–65	62–67	1/20 to 3/4	Fans, blowers, office appliances, food-preparation machines. Low- or medium-starting torque, low-inertia loads. Continuous-operation loads. May be reversed.
50–62	46–61	1/6 to 1/3	Washing machines, sump pumps, home workshops, oil burners. Medium- to high-starting torque loads. May be reversed.
80–95	55–65	1/20 to 3/4	Direct-connected fans, blowers, centrifugal pumps. Low-starting torque loads. Not for belt drives. May be reversed.
80–95	55–65	1/6 to 3/4	Belt-driven or direct-drive fans, blowers, centrifugal pumps, oil burners. Moderate-starting torque loads. May be reversed.
80–95	55–65	1/8 to 3/4	Dual voltage. Compressors, stokers, conveyors, pumps. Belt-driven loads with high static friction. May be reversed.
80–95	55–65	1/8 to 3/4	Compressors, stokers, conveyors, pumps. High-torque loads. High power factor. Speed may be regulated.
30–40	30–40	1/300 to 1/20	Fans, toys, hair dryers, unit heaters. Desk fans. Low-starting torque loads.

in circumstances where the compressor is located in a noisy environment, clearly the choice should be the capacitor-start induction-run motor because it is less expensive.

PROBLEMS

8-1. Two field coils are space displaced by 45 electrical degrees. Moreover, sinusoidal currents that are time displaced by 45 electrical degrees flow through these windings. Will a revolving flux result? Explain.

8-2. Two coils are placed in a low reluctance magnetic circuit with their axes in quadrature —the first along the vertical and the second along the horizontal. Sinusoidal currents that are time displaced by 90 electrical degrees and of the same frequency ω are made to flow through the coils. Moreover, the mmf of the first coil is twice that of the second.

    (a) Assuming that at time ωt_0 the mmf of the first coil is at its positive peak value, determine the relative magnitude and direction of the resultant mmf.

    (b) Repeat part (a) for a time ωt_1 that is 45° later than ωt_0.

    (c) Repeat part (a) for a time ωt_2 that is 90° later than ωt_0.

    (d) What conclusion can you draw from parts (a), (b), and (c)?

8-3. The two coils of Prob. 8-2 are now energized with sinusoidal currents that are 45 electrical degrees out of phase but yield equal mmfs.

    (a) Assuming that at time ωt_0 the mmf of the first coil is at its positive peak value, find the relative magnitude and direction of the resultant mmf.

    (b) Repeat part (a) for a time ωt_1 that is 45° later than ωt_0.

    (c) Repeat part (a) for a time ωt_2 that is 90° later than ωt_0.

    (d) What conclusion can you draw from parts (a), (b), and (c)?

8-4. In Fig. P8-4, I, II, and III denote the three coils of a balanced three-phase winding which are spaced displaced by 120° from one another. If a single-phase voltage is applied to the three-phase, squirrel-cage motor in the manner shown, will the motor operate? Justify your answer.

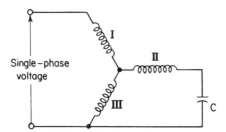

                                                  Figure P8-4

8-5. The stator windings of a two-phase, two-pole induction motor are in space quadrature. When the effective values of the phase currents are $\bar{I}_a = 10\underline{/0°}$ and $\bar{I}_b = 10\underline{/90°}$ A, the resulting stator mmf wave rotates clockwise, and its peak value is 2000 A-t per pole. If the phase currents are changed to $\bar{I}_a = 5\underline{/0°}$ and $\bar{I}_b = 10\underline{/53.1°}$, find the peak values of the clockwise and counterclockwise rotating mmf waves.

8-6. The phase currents in the stator windings of a two-phase induction motor are $\bar{I}_a = 8\underline{/0°}$ and $\bar{I}_b = 8\underline{/-90°}$ A. Together they produce an mmf wave rotating counterclockwise and having a peak value of 1000 A-t per pole. These currents are then changed to the following: $\bar{I}_a = 10\underline{/0°}$ and $\bar{I}_b = 6\underline{/+36.9°}$. Find the peak values of the resulting clockwise and counterclockwise rotating mmf waves.

8-7. At standstill, the currents in the main and auxiliary windings of a single-phase induction motor are given by $\bar{I}_M = 10\underline{/0°}$ and $\bar{I}_A = 5\underline{/60°}$. There are five turns per pole on the auxiliary winding for every four turns on the main winding. Moreover, these windings are in space quadrature. What should be the magnitude and phase of the current in the auxiliary winding to produce an mmf wave of constant amplitude and velocity?

8-8. The armature mmf of a machine is described by the equation

$$\mathcal{F}_a = K_1 \cos \omega t \cos \alpha + K_2 \cos \omega t \sin \alpha$$

Is this a single-phase or a polyphase machine? Explain your answer.

8-9. The results of a load test on a $\frac{1}{4}$-hp, 115-V, 60-Hz, four-pole, single-phase induction motor are as follows:

$$
\begin{aligned}
\text{applied voltage} &= 115 \text{ V} \\
\text{main winding current} &= 3.7 \text{ A} \\
\text{power input} &= 270 \text{ W} \\
\text{stator core loss} &= 10 \text{ W} \\
\text{slip} &= 90 \text{ rpm}
\end{aligned}
$$

The stator winding resistance is 2 Ω and the stator referred rotor resistance is 1.6 Ω. The starting winding was disconnected during this test. Calculate the power output during the test. Assume that the rotational losses are equal to the stator core losses.

8-10. A 110-V, 60-Hz, single-phase induction motor has the following parameters:

$$
r_1 = 1.86 \text{ } \Omega, \qquad r_2' = 1.78 \text{ } \Omega
$$
$$
x_1 = 2.56 \text{ } \Omega, \qquad x_2' = 1.28 \text{ } \Omega
$$
$$
x_\phi = 25 \text{ } \Omega
$$

At a slip of 5% find:
(a) Input line current and power factor.
(b) Horsepower output.
(c) The efficiency.

8-11. A 115-V, $\frac{1}{4}$-hp, 60-Hz, single-phase induction motor has the following design parameters:

$$
r_1 = 1.48 \text{ } \Omega, \qquad r_2' = 1.68 \text{ } \Omega
$$
$$
x_1 = 3.22 \text{ } \Omega, \qquad x_2' = 1.6 \text{ } \Omega
$$
$$
x_\phi = 25 \text{ } \Omega
$$

Also, stator core loss is negligible.
(a) Determine the operating performance of this motor when it runs at a slip of 10%.
(b) Would you permit continuous operation of this motor under the conditions of part (a)?

8-12. The rotor of the machine described in the preceding problem is blocked and rated voltage is applied. Compute the readings of a line wattmeter and ammeter. Make approximations where proper.

8-13. A $\frac{1}{2}$-hp, 115-V, four-pole, single-phase induction motor presents the following test data:

At no load: 115 V, 4.81 A, 105 W

With rotor blocked: 115 V, 27.2 A, 1308 W

The stator winding resistance is 0.83 Ω. The stator core losses are equal to the rotational losses. Determine the parameters of the equivalent circuit.

8-14. The following no-load and blocked rotor test data are available for a single-phase induction motor:

Blocked rotor: $I_{sc} = 6.6$ A, $V_{sc} = 53$ V, $P_{sc} = 210$ W

No-load: $I_n = 3.85$ A, $V_n = 208$ V, $P_n = 160$ W

The stator resistance is 1.48 Ω and the rotational losses are 60 W.

(a) Determine the parameters of the equivalent circuit.

(b) What is the value of the core loss?

(c) Compute the value of resistance that can be used to represent this core loss in the equivalent circuit. Show its location.

8-15. A $\frac{1}{4}$-hp, 110-V, 60-Hz, four-pole, single-phase induction motor has the following parameters:

$$r_1 = 1.5\,\Omega, \qquad r_2' = 1.4\,\Omega$$
$$x_1 = 2.4\,\Omega, \qquad x_2' = 1.2\,\Omega$$
$$x_\phi = 40\,\Omega$$

The stator core loss is 20 W.

(a) Calculate the developed torque at a slip of 5%.

(b) If the rotational losses are equal to 20 W, what is the shaft torque?

8-16. A $\frac{1}{2}$-hp, 220-V, 60-Hz, four-pole, single-phase induction motor has the following equivalent circuit parameters:

$$r_1 = 2.8\,\Omega, \qquad r_2' = 2.4\,\Omega$$
$$x_1 = 5.5\,\Omega, \qquad x_2' = 2.8\,\Omega$$
$$x_\phi = 90\,\Omega$$

The core loss is 30 W and the rotational losses are 46 W. Determine for $s = 0.05$:

(a) Line current and line power factor.

(b) Horsepower output.

(c) Efficiency.

8-17. The following data apply to a $\frac{1}{4}$-hp, 110-V, 60-Hz, four-pole, single-phase induction motor.

No-load test: $\qquad V_n = 110$ V, $I_n = 2.73$ A, $P_n = 56$ W

Blocked rotor test: $\quad V_{sc} = 110$ V, $I_{sc} = 16.6$ A, $P_{sc} = 1260$ W

The resistance of the main winding is 1.5 Ω. Assume the rotational losses are equal to the stator core loss. On a carefully drawn equivalent circuit, show the value of each parameter for a slip of 5%.

8-18. A two-phase, 60-Hz, squirrel-cage induction motor has a blocked resistance per phase of 6 Ω and a reactance of 8 Ω. When connected as a normal two-phase motor, its starting torque is 200%. It is desired to use this motor on a single-phase circuit.

(a) A resistance of 10 Ω is inserted in series with one phase and the machine is operated as a split-phase motor. Determine the developed torque at standstill. Consider that the torque is proportional to the product of the phase currents as well as the sine of the phase angle between them.

(b) A capacitor is to be used in place of the resistor. Calculate the microfarad rating of this capacitor so that the phase angle between the currents is 90°.

(c) Determine the starting torque for the condition of part (b).

9

Stepper Motors

The explosive growth of the computer industry in recent years has meant an enormous growth too for stepper motors because these motors provide the driving force in many computer peripheral devices. They can be found, for example, driving the paper feed mechanism in line printers and printing terminals. Stepper motors are also used exclusively in floppy disk drives where they provide precise positioning of the magnetic head on the disks. The x- and y-coordinate pens in expensive plotters are driven by stepper motors. Here they offer good dynamic performance while eliminating the need for a heavy maintenance schedule associated with alternative drive schemes that employ gears, slide wires, and brushes. The stepper motor is especially suited for such applications because essentially it is a device which serves to convert input information in *digital form* to an output that is mechanical. It thereby provides a natural interface with the digital computer.

The stepper motor, however, can be found performing countless tasks outside the computer industry as well. In many commercial, military, and medical applications, the stepper motors perform such functions as mixing, cutting, stirring, metering, blending, and purging. They provide many supporting roles in the manufacture of packaged foodstuffs, commercial end products, and even the production of science fiction movies. The advantages offered by the stepper motor in these applications can easily be exploited with the use of two pieces of auxiliary equipment: the microprocessor and controlled switches in the form of transistors.

Although the stepper motor is driven by a properly sequenced set of digital signals, the motor itself exhibits the characteristics of a synchronous motor. In one of its most popular forms, it acts as a doubly excited machine, which in turn means it produces a steady-state torque at one speed.

9-1 CONSTRUCTION FEATURES

The construction of the stepper motor is quite simple. It consists of a slotted stator equipped with two or more individual coils and a rotor structure that carries no winding. The classification of the stepper motor is determined by how the rotor is designed. If it is provided with a permanent magnet attached to its shaft, it is called a permanent magnet (PM) stepper motor. If the permanent magnet is not included, it is simply classified as a reluctance-type stepper motor. Attention here is directed first to the PM stepper motor. Of course, the presence of the permanent magnet furnishes the motor with the equivalent of a constant d-c excitation. Thus, when one or more of the stator coils are energized, the machine behaves as a synchronous motor.

The basic construction details of the PM stepper motor are illustrated in Fig. 9-1(a) for a four-pole stator and a five-pole rotor structure. Observe that the action of the permanent magnet on the particular orientation of the rotor structure is to magnetize each of the poles (or slots) at one end of the rotor with a N-pole polarity and each of the poles at the other end with a S-pole polarity. Moreover, the N-pole set of rotor poles is arranged to be displaced from the S-pole set of rotor poles by one-half of a pole-pitch and is readily evident by a comparison of Figs. 9-1(b) and (c). The symmetry that exists between the stator poles and each set of rotor poles makes it apparent that each rotor pole set behaves in an identical fashion. For example, if coil A-A' is assumed to be energized to yield a N-pole polarity at A and a S-pole polarity at A', then the relationship between N of the stator and S1, S2 and S5 of the rotor in Fig. 9-1(b) corresponds exactly to that of S of the stator and N1, N2, and N5 of the rotor in Fig. 9-1(c). A similar statement can be made for the remaining sets of poles.

9-2 METHOD OF OPERATION

The manner in which the PM stepper motor can be used to perform precise positioning is explained by examining the sequence of diagrams that appear in Fig. 9-2. The starting point is represented in Fig. 9-2(a) with coil A-A' energized to carry positive d-c current so that the upper stator pole is given a N-pole polarity and the corresponding lower stator pole is made to behave with a S-pole polarity. Coil B-B' is assumed to be de-energized. The rotor maintains this orientation as long as the stator coil currents remain unchanged. At this position, the net torque developed by the stepper motor is zero. This conclusion is readily apparent in this case from symmetry considerations. For example, at the S-pole end of the rotor it is seen that S1 is in alignment with the N-pole of the stator. The torque produced by the action of the field distributions of the stator pole and the rotor pole S1 is dependent on the sine of the angular displacement of these two flux fields. With exact alignment the angle is clearly zero and so too is the torque between these two fields. Proceeding on the basis of superposition, we note that a torque of attraction exists between the stator N-pole field and the field associated with S5, which acts to cause counterclockwise (ccw) rotation. The magnitude of this torque is proportional to the sin 72°, which is the

Figure 9-1 Diagram illustrating the construction features of the PM stepper motor: (a) axial view showing the permanent magnet attached to the rotor shaft; (b) cross-sectional view at *a-a* showing rotor poles of S-pole polarity; (c) cross-sectional view at *b-b* showing poles of N-pole polarity.

(a)

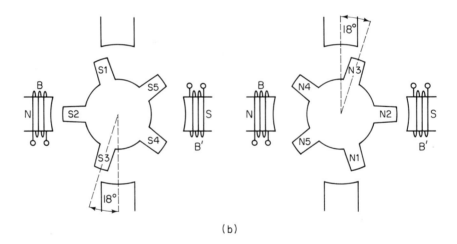

(b)

Figure 9-2 Illustrating the indexing of the rotor in a ccw direction for a full cycle of coil *A-A'*/coil *B-B'* excitation in sequence through positive and negative values. Observe that the orientation between rotor and stator in (e) exactly corresponds to the starting point in (a) except for rotation. (a) Coil *A-A'* energized for a N-S orientation; coil *B-B'* de-energized. Comment: starting point. (b) Coil *A-A'* de-energized; coil *B-B'* energized for a N-S orientation. Result: Rotor advances ccw 18° from its position in (a). (c) Coil *A-A'* energized with reversed current for a S-N orientation; coil *B-B'* de-energized. Result: Rotor advances ccw an additional 18° from its position in (b). (d) Coil *A-A'* de-energized; coil *B-B'* energized with reverse current for a S-N orientation. Result: Rotor advances ccw an additional 18° from its position in (c). (e) Coil *A-A'* energized with positive current for a N-S orientation; coil *B-B'* de-energized. Result: Rotor advances an additional 18° for a total of 72° or one rotor pole pitch corresponding to one cycle of ± excitation of stator coils.

Figure 9-2 (*cont.*)

rotor pole-pitch angle. However, the action between the N-pole stator field and S2 is also one of attraction but it is exactly equal and opposite to that produced between the stator N-pole and S5. A similar result is obtained upon examining the situation that exists under the S-pole stator field. It is useful to note here that a difference in detail does exist in that the torques involved are of a repulsive nature rather than attractive as occurs beneath the N-pole stator field for this orientation. Because the repulsive force has its maximum effect when a rotor flux field is in exact alignment with a stator flux field, the cosine of the displacement angle between the axes of two flux fields of opposite angle must be used. Accordingly, there is a torque existing between the stator S-pole and the rotor S3 pole which is proportional to the strength of these fields as well as the cos 36°, which acts to produce a clockwise rotation. But the action between S4 and the stator S-pole is exactly equal and opposite, thus yielding a resultant of zero torque produced by conditions beneath the stator S-pole. The net torque is zero and so the rotor remains in a position of equilibrium.

Let us now examine the effect of alternately energizing the two quadrature stator coils in sequence through positive and negative values of d-c excitation. The first step in the sequence is represented in Fig. 9-2(b) by placing coil *A-A'* in a de-energized state and coil *B-B'* in a fully energized state. The rotor is shown in the final steady-state position that results as long as coil *B-B'* is energized with positive current (i.e., so that the left pole becomes a N-pole and the right pole assumes a S-pole polarity). The net rotor displacement for this switch in coil excitation in this case is 18°. Once the rotor assumes the position depicted in Fig. 9-2(b), the orientation of the rotor poles relative to the newly located stator poles is identical to that appearing in Fig. 9-2(a) and so the rotor is again in equilibrium. It is instructive, however, to examine the torque conditions that develop when coil *A-A'* is de-energized and coil *B-B'* is energized while the rotor is still in the position corresponding to Fig. 9-2(a). We proceed on the assumption that the currents in these coils change instantaneously, i.e., the time constants of the windings are very small compared to the mechanical time constants involved. This is not an unrealistic assumption. Because the magnitude of the flux field produced by a given coil excitation in the fixed geometry of a given machine is determinable and since the permanent magnet flux field of each rotor pole is also calculable, we can represent the maximum torque produced by these fields as T_m. When the rotor orientation is such that the torque developed is other than maximum, the appropriate sine or cosine function is introduced depending upon whether the torque is one of attraction or one of repulsion. Thus with coil *B-B'* energized and the rotor still in the position depicted in Fig. 9-2(a), the expression for the torque employing superposition is

$$T = T_m \left[\sin \frac{\angle S2}{N} + \sin \frac{\angle S1}{N} - \sin \frac{\angle S3}{N} + \cos \frac{\angle S5}{S} - \cos \frac{\angle S4}{S} \right] \quad (9\text{-}1)$$

where the plus sign is used if the torque acts to produce ccw movement and the negative sign is used if the movement would be cw. Inserting the proper angles corresponding to the stator pole orientation of Fig. 9-2(b) and the rotor position of Fig. 9-2(a), Eq. (9-1) becomes

$$T = T_m(\sin 18° + \sin 90° - \sin 54° + \cos 18° - \cos 54°)$$
$$= T_m(0.309 + 1.0 - 809 + 0.951 - 0.588)$$
$$= 0.863T_m \quad (9\text{-}2)$$

Because this is a positive torque, the rotor in Fig. 9-2(a) moves to the new equilibrium position shown in Fig. 9-2(b).

Consider next that coil *B-B'* is switched off and coil *A-A'* is again energized but this time with current flowing in the opposite direction. The upper stator pole assumes a S-pole polarity and the lower stator pole takes on a N-pole polarity. Accordingly, the new stator flux configuration [in Fig. 9-2(c)] exerts a torque on the rotor orientation of Fig. 9-2(b) which is again described by Eq. (9-2). The ensuing torque displaces the rotor by an additional 18° counterclockwise in order to reach the new equilibrium position shown in Fig. 9-2(c). In the subsequent step in the sequence, the current in coil *A-A'* is switched off while the current in coil *B-B'* is switched on but in the reversed direction. The stator poles then assume the polarity depicted in Fig. 9-2(d), which in turn produces a torque on the rotor to yield a further 18° displacement in the ccw direction. In other words, the rotor moves to the new equilibrium position that finds S4 in alignment with the N-pole associated with coil *B-B'*.

It is instructive to observe here that once a sequence is initiated and maintained through the proper positive and negative alternate excitation of the stator coils, the action of the torque produced by the subsequent coil excitation is to displace the rotor in the established direction. This is done in a manner to ensure that the rotor pole having the least angular alignment with a stator pole of opposite polarity is brought into exact alignment with that pole. Accordingly, when coil *A-A'* is next energized, the result is to move S5 into alignment with the upper stator N-pole. The new equilibrium position is shown in Fig. 9-2(e). Observe that this last orientation is identical to the first one except that the entire rotor structure has been indexed through a total displacement equal to one rotor pole pitch, or 72° in this case. Note too that for a *two*-coil (or two-phase) stator with *two* states (positive and negative), it requires *four* switching operations to complete a switching cycle. Thus, this motor is made to *step* four times during each cycle of excitation which in turn produces a net displacement of one rotor pole pitch. Or, stated differently, each switch of the stator excitation produces a rotor displacement of one-fourth a rotor pole pitch. The *stepping* action exhibited by this motor in response to alternate sequenced excitation of stator coils is the reason it is called a *stepper motor*.

The direction of rotation of the stepper motor can be conveniently reversed by merely reversing the switching sequence. Thus, if in Fig. 9-2(b) the excitation of coil *B-B'* were reversed, then the action of the developed torque would be to displace S5 in Fig. 9-2(a) into a position of exact alignment with the N-pole flux which would now be associated with coil *B'*. The result is a clockwise displacement of 18° in contrast to the counterclockwise displacement that occurs when coil *B* is made to produce a N-pole polarity.

9-3 DRIVE AMPLIFIERS AND TRANSLATOR LOGIC

The foregoing description of the method of operation of the stepper motor makes it plain that the stepping action of the motor is dependent on a specific switching sequence that serves to energize and de-energize the stator coils. A practical scheme that achieves this objective is the one of a drive amplifier consisting of transistors that are switched on and off sequentially in accordance with the signals that originate from an appropriate translator logic control circuit. A block diagram of the

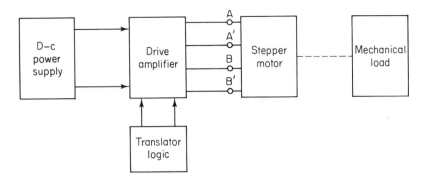

Figure 9-3 Block diagram of the stepper motor and its switched supply.

arrangement is shown in Fig. 9-3. The drive amplifier provides the d-c excitation to coils *A-A'* and *B-B'* through switched circuitry that is controlled by the translator logic device. Appearing in Fig. 9-4 is a detailed description of the composition of one type of drive amplifier, namely, the *bipolar* type. A total of four switching transistors is needed for each phase coil whenever a single power supply is used and current reversal in the phase coils is desired. The term bipolar is used for such a circuit because the circuitry permits either end of the phase coil to be connected to the positive side of the power supply. There is a drive circuit available that employs two power supplies in which case it becomes necessary to use only two transistors per phase coil. In addition to the eight transistors, the drive amplifier of Fig. 9-4 also includes eight

Figure 9-4 Diagram illustrating the details of a bipolar drive amplifier.

diodes. Of course, the diodes are required in order to provide paths for the inductive coil currents to flow whenever the transistors are switched off by logic control circuitry. For example, if transistors Q1 and Q2 are ON during the first quarter of a switching cycle thus causing current to flow in the phase coil from A to A', the transition into the next quarter of the cycle calls for Q1 and Q2 to be switched OFF. Because of the magnetic storage associated with the current in coil A-A', the current cannot be reduced to zero instantaneously. But, as it attempts to do so, a large emf is induced in the coil with terminal A' positive and A negative. This induced emf puts a positive voltage across diodes D3 and D4 thus furnishing a path for the switched-off current in coil A-A'. The current decays rapidly because of the small time constant of the discharging circuit.

How is the logic schedule of the translator control circuit determined? This is best explained by developing the logic required to perform the operations of the stepper motor described in connection with Fig. 9-2. In this case, a complete switching cycle required that the stator coils be energized in accordance with the schedule in the table below for a counterclockwise rotation.

Coil A-A'	Coil B-B'
ON+	OFF
OFF	ON+
ON−	OFF
OFF	ON−

Here ON+ is distinguished from ON− by a current reversal in the coil. The logic schedule that achieves this switching sequence for the drive amplifier of Fig. 9-4 expressed in binary language where 1 denotes ON and 0 denotes OFF is as follows:

Coil A-A'				Coil B-B'			
Q1	Q2	Q3	Q4	Q5	Q6	Q7	Q8
1	1	0	0	0	0	0	0
0	0	0	0	1	1	0	0
0	0	1	1	0	0	0	0
0	0	0	0	0	0	1	1

Each repetition of this switching cycle provides an advancement of the stepper motor equal to one rotor pole pitch. Hence in this case five cycles of this switching sequence are needed in order to complete one full revolution of the rotor.

A very popular circuit for the drive amplifier of stepper motors is the one illustrated in Fig. 9-5. This circuit clearly has the advantage of fewer electronic components and simpler bias circuitry for the switching transistors. Each phase coil of the stepper motor that is used with these drives is equipped with a center tap and the switching logic for the associated transistors arranged so that current flows only in one section at a time. Reversal of the magnetic pole polarity is achieved by directing

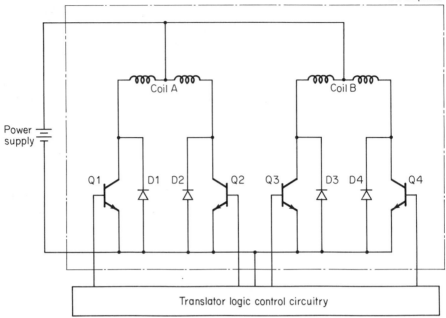

Power
supply

Coil A

Coil B

Q1   D1   D2   Q2   Q3   D3   D4   Q4

Translator logic control circuitry

Figure 9-5 Drive amplifier of the bifilar, unipolar type. Each phase coil of the stepper motor comes equipped with a center tap.

the current either to one section or to the other. The disadvantage of such a scheme is that the copper is not made to work continuously in producing useful output. This arrangement is called the bifilar unipolar drive. It is referred to as unipolar because the ends of both coil sections stay connected to the negative side of the power supply during operation unlike the bipolar case where the connection alternates between the high and low side of the source.

The logic schedule to be furnished by the translator control circuitry for each full switching cycle of the operation of a stepper motor such as the one illustrated in Fig. 9-2 is the following:

Coil A		Coil B	
Q1	Q2	Q3	Q4
1	0	0	0
0	0	1	0
0	1	0	0
0	0	0	1

It is assumed here that coil A is equipped with a center tap by using the point of connection of coil A and coil A' in Fig. 9-1. A similar arrangement is assumed to be used for coil B.

The diodes D1 through D4 appearing in the circuitry of the drive amplifier shown in Fig. 9-5 are for the purpose of maintaining a current path following the cutoff of a particular transistor switch. For example, when current is flowing through the left-half section of coil A in Fig. 9-5 and Q1 is subsequently cut off, a circuit is established by transformer action in the second section of coil A which allows the current in the first section to drop to zero as the current in the second section instantaneously rises to the level of the current in the first section. The latter current then flows through a path that involves the power supply and diode D2. Diode D2 is made conductive since a positive voltage appears across it as a result of the transformer voltage induced in the right-half section of coil A. The transfer of current from the left section to the right section of coil A is allowed to occur instantaneously because there is *no change* in magnetic energy in this process. The mmf of both coils act on the same magnetic circuit.

9-4 HALF-STEPPING AND THE REQUIRED SWITCHING SEQUENCE

The description of the operation of the stepper motor so far has been limited to situations where either one stator coil or the other was allowed to be energized at any one time. What would be the consequence if a switching schedule is devised which permits both stator coils to be excited simultaneously at certain periods in a cycle? To examine this matter, let us return to the configuration depicted in Fig. 9-2(a). Recall that a full 18° ccw displacement resulted from switching coil A-A' off and coil B-B' on. However, suppose that coil B-B' is switched on while maintaining the excitation on coil A-A'. In other words, coil A-A' is not de-energized when excitation is applied to coil B-B'. Intuition, based on the symmetry prevalent in this situation, leads us to suspect that the motor takes a *half-step*, i.e., it advances in the ccw direction by one-eighth of a rotor pole-pitch. An examination of the resultant torque on the rotor structure at this displacement angle reveals the motor to be in equilibrium. This is readily verified by writing the expression for the net torque on the rotor corresponding to this condition. The situation is illustrated in Fig. 9-6 for a half-step angle of 9°. Note that all four stator poles are energized with the polarities indicated. Applying superposition, the expression for the net torque can be written as

$$T_R = T_m \left[-\sin \left\langle {{N_A} \atop {S1}} \right. - \sin \left\langle {{N_A} \atop {S2}} \right. + \sin \left\langle {{N_A} \atop {S5}} \right. - \cos \left\langle {{S_{A'}} \atop {S3}} \right. + \cos \left\langle {{S_{A'}} \atop {S4}} \right. \right.$$

$$\left. + \sin \left\langle {{N_B} \atop {S2}} \right. + \sin \left\langle {{N_B} \atop {S1}} \right. - \sin \left\langle {{N_B} \atop {S3}} \right. + \cos \left\langle {{S_{B'}} \atop {S5}} \right. - \cos \left\langle {{S_{B'}} \atop {S4}} \right. \right]$$

where the first line represents the torque produced by the stator poles created by coil A-A' interacting with the S-pole section of the rotor and the terms on the second line represent the torque contributions associated with the stator poles of coil B-B' and the S-pole section of rotor. Upon inserting the angular displacements existing between the axes of the indicated poles, we get

$$T_R = T_m(-\sin 9° - \sin 81° + \sin 63° - \cos 27° + \cos 45°$$
$$+ \sin 9° + \sin 81° - \sin 63° + \cos 27° - \cos 45°)$$
$$= 0$$

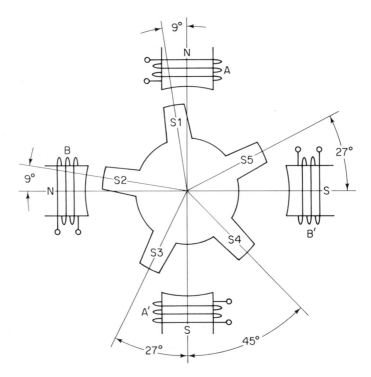

Figure 9-6 Half-step equilibrium position that occurs when the rotor is initially at the position illustrated in Fig. 9-2(a) and both coils *A-A'* and *B-B'* are energized to produce the field orientation shown here. The result is a 9° displacement of the rotor as indicated here.

Hence, this half-step displacement of the rotor represents an equilibrium position.

If coil *A-A'* is now de-energized while the excitation of coil *B-B'* is maintained, the motor steps an additional 9°, thus yielding the originally expected 18° that results when one coil is switched off and the other is switched on. An extension of this analysis through a complete switching cycle executed by half-step increments leads to the following switching sequence for counterclockwise rotation:

Coil *A-A'*	Coil *B-B'*	
ON+	OFF	
ON+	ON+	
OFF	ON+	
ON−	ON+	One complete
ON−	OFF	switching cycle
ON−	ON−	
OFF	ON−	
ON+	ON−	
ON+	OFF	
.	.	
.	.	
.	.	

A full switching cycle now includes a total of eight switching states, which yield eight-half steps, to give a total displacement of one rotor pole pitch. The switching schedule required to half-step the motor of Fig. 9-1 arranged to operate in the bifilar mode and driven by the unipolar drive amplifier of Fig. 9-5 is detailed in the following table:

Coil A		Coil B	
Q1	Q2	Q3	Q4
1	0	0	0
1	0	1	0
0	0	1	0
0	1	1	0
0	1	0	0
0	1	0	1
0	0	0	1
1	0	0	1

The advantage of half-stepping of course is that it improves the precision of the stepper motor in applications where more precise positioning is demanded.

9-5 THE RELUCTANCE-TYPE STEPPER MOTOR

The reluctance-type stepper motor is not equipped with a permanent magnet on the rotor. It develops an indexing torque in response to the sequenced d-c excitation of stator coils by virtue of the large difference in magnetic reluctance that exists between a direct-axis path and a quadrature-axis path. The resulting stationary flux fields produced by the d-c current in selected stator coils always exert a force which acts to minimize the reluctance of the flux path in accordance with the characterization of Eq. (1-55). A typical configuration of the reluctance-type stepper motor is illustrated in Fig. 9-7. The stator in this case is designed for eight poles and is equipped with four phase coils each of two sections. The rotor is slotted to produce the effect of six poles corresponding to which is a pole pitch of 60°. The rotor is at a position of equilibrium which occurs when coil A-A' (i.e., phase A) is energized and the remaining phases are left unexcited. Observe that the constant flux field produced by phase A encounters a path of minimum reluctance by passing through rotor poles 1 and 4 which are aligned with the stator pole structures associated with phase A.

For the combination of stator and rotor poles that is designed into this motor, it is instructive to note that the axes of both rotor poles 2 and 5 are 15° displaced from the axes of the two stator poles associated with phase B. Hence, if phase A is de-energized and phase B is energized to produce a N-pole polarity at coil B and a S-pole polarity at coil B', a reluctance torque is developed that acts to align the axes of rotor poles 2 and 5 with the axes of the stator poles at B-B'. The rotor advance is 15° ccw which in this machine too is one-quarter of a rotor pole pitch. Successive

Figure 9-7 Construction details of a reluctance-type stepper motor equipped with four stator phases (i.e., eight poles) and six rotor poles.

excitations of phases C, D, and A yield a total advance in the ccw direction equal to a rotor pole pitch. A sixfold repetition of this switching sequence for the excitation of the stator phases results in one revolution of the motor shaft.

The equilibrium position of the rotor in Fig. 9-7 also prepares the motor to move in a clockwise direction. However, it now requires that coil D' be energized for a N-pole polarity and coil D for a S-pole polarity following the removal of excitation on phase A. A reluctance torque now develops that pulls rotor teeth 6 and 3 into alignment with the poles produced by coil D'-D. Clearly this results in a clockwise rotation.

The design of the drive amplifiers and the translator logic for the reluctance-type stepper motor differ in some details from those reported for the PM stepper motor. This is necessary in order to account for differences in the design aspects of the two types of motors. The principles, however, remain the same and are easily transferable from one motor type to the other.

9-6 RATINGS AND OTHER CHARACTERISTICS

On the basis of the material that is related in the preceding pages concerning the stepper motor, it should come as no surprise that these motors are used in circumstances where the need for large horsepower is not a factor. A motor's output power is dependent upon the extent to which the copper is worked electrically and the iron is worked magnetically. The stepper motor ranks low on both counts. To achieve the characteristics which make the stepper motor unique, it is necessary that during each complete switching cycle each of the phase coils spend time in a de-energized state and that some parts of the magnetic flux paths be essentially free of field flux. Consequently, stepper motors are found to range in output power from about 1 W to a maximum size of 3 hp. In terms of physical dimensions, the largest machines are built with diameters up to seven inches while the smallest ones are constructed in a pancake form with diameters as small as 1 in.

Many step sizes are also available with stepper motors. They can be purchased with step sizes as small as 0.72° or as large as 90°. The most common step sizes are 1.8°, 7.5°, and 15°. To procure a step size of 1.8° in a PM type, it is necessary to design the stator for 40 poles and the rotor for 50 poles (or slots).

The shape of the torque-speed curve of the stepper motor takes on the general form depicted in Fig. 9-8. The speeds indicated in this particular motor are achieved as a result of programming the switching sequence for appropriate repetition. For example, in order to realize one revolution per second, it is necessary for the trans-lators' logic to switch a 1.8° step motor at the rate of 200 steps per second. Any attempt to operate this motor at a rate of five revolutions per second fails because the torque drops to zero. This decrease in torque is attributable to the effects of the speed volt-ages induced in the phase coils at the higher speeds as well as to the inductance of these coils. Compensation circuits are available which can double the useful speed range but this is at the cost of increased circuit complexity for the drive amplifiers.

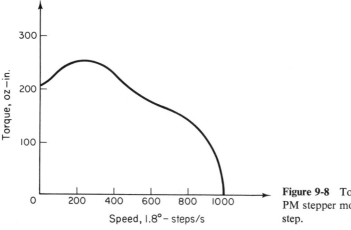

Figure 9-8 Torque-speed curve of a PM stepper motor operated at full step.

PROBLEMS

9-1. Consider the situation corresponding to the rotor position depicted in Fig. 9-2(b). Assume that coil A-A' is energized so that the magnetic material on which coil A' is wrapped is of a N-pole polarity and that of coil A is of a S-pole polarity.

   (a) Write the expression for the torque produced on the rotor by the S-pole section of the rotor. Assume sinusoidal flux distributions throughout.

(b) How many degrees of rotation occur before a position of equilibrium is reached?

9-2. Repeat Prob. 9-1 for the conditions that exist at the N-pole section of the rotor.

9-3. Let T_m denote the maximum torque produced by the cooperation of a single PM rotor pole with the stator pole in the configuration of Fig. 9-2. Calculate the net torque produced on the rotor shaft for the condition represented by Fig. 9-2(c).

9-4. The schematic diagram of a four-phase, two-pole PM stepper motor is shown in Fig. P9-4. The phase coils are excited in sequence by means of a translator logic circuit.

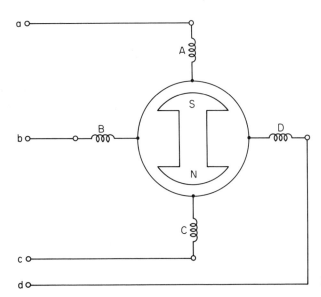

Figure P9-4

(a) Write the logic schedule for full-stepping of this motor.

(b) What is the displacement angle of the full step?

(c) Write the logic schedule for operating this motor in the half-step mode.

(d) Make a sketch of the stator poles and the relative rotor position when coils B and C are simultaneously energized.

(e) What is the total rotor displacement in part (d) measured from the position depicted in Fig. P9-4?

9-5. The excitation signals used to drive the coils in the motor of Fig. P9-4 originate from digital signals that drive the supply amplifier. A continuous sequencing of these signals will cause the stepper motor to run at a constant speed (thus behaving as a synchronous motor). Determine the frequency rate of these digital signals if the speed of rotation is to be 360 rpm. Assume a half-stepping pattern.

9-6. Determine the frequency rate at which the digital signals that drive the amplifier of the motor of Fig. 9-2 must be set in the half-step mode if the motor is to run at a speed of 360 rpm.

9-7. The drive amplifier of Fig. 9-4 is used to drive the stepper motor of Fig. 9-2 in the full-step mode. Write the logic schedule for switching the transistors in a manner such that it yields a *clockwise* direction of rotation.

9-8. A PM stepper motor is designed to provide a full step size of 7.5°.

(a) Determine the number of rotor poles.

(b) How many stator poles are needed?

9-9. A PM stepper motor is designed to provide a half-step size of 0.9°.

    (a) Determine the rotor pole pitch in degrees.

    (b) Find the number of stator poles.

9-10. A PM stepper motor is designed to provide a full-step size of 15°. Determine the required number of stator and rotor poles.

9-11. Draw the diagram that illustrates the orientation of the stator poles with the N-pole section of the rotor for the half-step case illustrated in Fig. 9-6.

10

Synchros

Synchros are a-c electromagnetic devices that find wide applications in servomechanisms.† However, they can also be used as indicating instruments for registering mechanical displacements remotely. In this chapter attention is first directed to a description of the construction features of the synchros as well as an explanation of their theory of operation. This is then followed by a discussion of the types of errors that are found in such devices which include static errors, dynamic errors, and residual voltages.

10-1 CONSTRUCTION FEATURES

There are four basic types of synchros: the control transmitter (denoted by cx), the control transformer (cT), the control differential (cD), and the control receiver (cR). The essential construction characteristics of the control transmitter are represented in Fig. 10-1(a). It consists of a stator and a rotor. The inner surface of the stator is slotted to accommodate a balanced three-phase winding, which is usually of the concentric-coil type. The rotor is of the dumbbell construction with many turns of wire wrapped around the stem. An a-c voltage is applied to the rotor winding through a pair of slip rings, which causes an excitation current to flow and produce an alternating flux field with respect to the rotor structure. By definition the electrical zero position of the rotor corresponds to maximum coupling with the turns of winding 2 of the stator. Moreover, since there are two rotor positions for maximum coupling, a

†A servomechanism is an error-actuated feedback control system in which the controlled variable is mechanical.

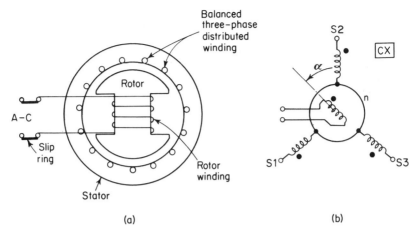

Figure 10-1 Synchro control transmitter: (a) construction features; (b) schematic diagram.

unique electrical zero is identified by matching dots as shown in Fig. 10-1(b). Figure 10-1(b) depicts the rotor axis displaced from the electrical zero by the angle α. Note the similarity in construction to the three-phase synchronous generator; the synchro can be aptly described as a miniature generator. It is important, however, to understand the distinction in the way the two machines are used. The synchronous generator is excited with a d-c voltage and driven at constant speed, and produces a three-phase voltage at the stator terminals. In contrast, the synchro is excited with an a-c voltage and is often merely displaced by finite amounts from the electrical zero, and thus induces *single-phase* voltages through transformer action in the stator winding. Sometimes, in control system applications, the synchros are driven at low tracking speeds.

The construction details of the synchro control transformer differ somewhat from those of the control transmitter. One important difference is that the air gap is uniform owing to the rotor's cylindrical or umbrella-like construction. Refer to Fig. 10-2(a) and (b). This feature is included to keep the magnetizing current drawn by the control transformer to a minimum. Another difference is the way in which the electrical zero is defined. For the control transformer it is defined as that position of the rotor which makes the coupling with winding 2 of the stator zero. This particular configuration is depicted in Fig. 10-2(c). The stator winding is also a balanced three-phase winding; however, it differs from the transmitter in that the transformer has a higher impedance per phase. This latter feature permits several control transformers to be fed from a single transmitter.

The differential synchro has a balanced three-phase distributed winding in both the stator and the rotor. Moreover, the rotor is cylindrically shaped as shown in Fig. 10-3(a). Although three-phase windings are involved, it is important to keep in mind that these units deal solely with single-phase voltages.

The basic structure of the control receiver is the same as that of the control transmitter. It has a balanced three-phase stator winding and a salient-pole rotor. The one detail in which the receiver differs from the transmitter is the inclusion of a mechanical viscous damper on its shaft. In normal use both the rotor and stator

(a)   (b)

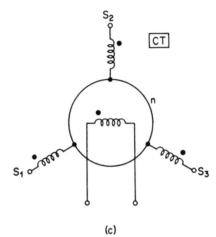

(c)

Figure 10-2 Control transformer:
(a) construction features of cylindrical
rotor type; (b) construction features of
umbrella rotor type; (c) schematic
diagram with rotor winding at
electrical zero.

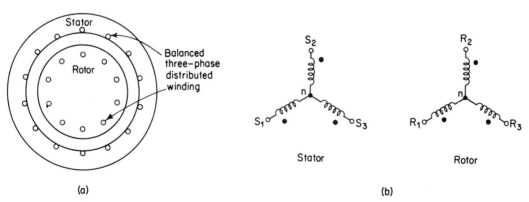

(a)   (b)

Figure 10-3 Differential synchro: (a) construction features; (b) schematic
representation.

windings are excited with single-phase currents. Accordingly, the field distribution of the rotor interacts with the ampere-conductor distribution of the stator to produce torque and hence rotation. The purpose of the damper is to permit the receiver rotor to respond to the changing stator ampere-conductor distribution without causing the rotor to overshoot its mark. If the overshoot is large, average torque may be produced such that the receiver runs as a single-phase motor.

10-2 VOLTAGE RELATIONS

A description of how the stator phase voltages vary with rotor displacement is useful in understanding the various applications of synchros. In this connection refer to the schematic diagram of the control transmitter shown in Fig. 10-4. Assume that the rotor winding is excited with a single-phase a-c voltage and that the rotor is displaced from its electrical zero position by the angle α and held fast. By transformer action single-phase voltages having the same frequency as the rotor applied voltage will be induced in each stator phase. The value of the induced voltage depends upon the coupling of each phase with the rotor winding. To describe these relationships mathematically let

$$a \equiv \frac{\text{effective stator turns}}{\text{effective rotor turns}} \qquad (10\text{-}1)$$

$$\bar{E}_r = \text{rms value of the induced emf in the rotor winding} \qquad (10\text{-}2)$$

and

$$\bar{E} = a\bar{E}_r \qquad (10\text{-}3)$$

The last expression denotes the maximum value of rms voltage that is induced in any balanced stator phase.

The rms value of the emf induced in winding 2 of the stator when the rotor is at the angle α is described by

$$\bar{E}_{2n} = \bar{E} \cos \alpha \qquad (10\text{-}4)$$

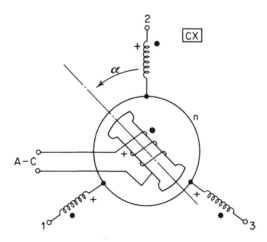

Figure 10-4 Circuit for deriving the stator phase and line voltage variations as a function of rotor displacement α.

Note that, when α is zero, the rotor is completely linked with winding 2 and so yields the maximum rms voltage $\bar{E}$. This situation is entirely consistent with the definition of the electrical zero for the control transmitter. The expression for the instantaneous value of this voltage can be written as

$$e_{2n}(t) = \sqrt{2}\,E \sin \omega t \cos \alpha \qquad (10\text{-}5)$$

where ω is the frequency of the applied rotor voltage. A study of the last expression should make it apparent that the $\sin \omega t$ term accounts for the transformer action that takes place in the synchro and that $\cos \alpha$ describes the coupling between the rotor winding and winding 2 of the stator. The use of a properly distributed and pitched stator winding that yields a quasisinusoidally distributed field is the reason for the cosine function in Eq. (10-4). If the winding were distributed differently, a different functional relationship would be needed.† However, throughout this chapter, we assume the use of sinusoidally distributed windings only.

The value of the rms voltage induced in winding 1 is given by

$$\bar{E}_{1n} = \bar{E} \cos (\alpha - 120°) \qquad (10\text{-}6)$$

The argument is modified by $-120°$ in order to account for the fact that the axis of winding 1 is located $120°$ ahead of the axis of winding 2. The corresponding expression for the instantaneous voltage then becomes

$$e_{1n}(t) = \sqrt{2}\,E \sin \omega t \cos (\alpha - 120°) \qquad (10\text{-}7)$$

Note that the $\sin \omega t$ factor remains the same as in Eq. (10-5), which emphasizes that this induced emf is of the same frequency and of the same phase as that occurring in winding 2. Only the amplitudes differ because of the difference in coupling.

Finally the emf induced in winding 3 can be expressed as

$$\bar{E}_{3n} = \bar{E} \cos (\alpha + 120°) \qquad (10\text{-}8)$$

The presence of $120°$ in the argument accounts for the fact that winding 3 is located behind the axis of winding 2 by this amount. The expression for the instantaneous voltage in winding 3 is accordingly

$$e_{3n}(t) = \sqrt{2}\,E \sin \omega t \cos (\alpha + 120°) \qquad (10\text{-}9)$$

Equations (10-4) to (10-9) describe the stator *phase* voltages. Information about the associated line voltages is readily obtained from the phase voltages. Thus the induced voltage appearing between terminals 1 and 2 in Fig. 10-4 can be written as

$$\bar{E}_{12} = \bar{E}_{1n} + \bar{E}_{n2} = \bar{E}_{1n} - \bar{E}_{2n} \qquad (10\text{-}10)$$

Inserting Eqs. (10-4) and (10-6) into Eq. (10-10) yields

$$\bar{E}_{12} = \bar{E} \cos \alpha \cos 120° + \bar{E} \sin \alpha \sin 120° - \bar{E} \cos \alpha$$

$$= \left(-\frac{3}{2} \cos \alpha + \frac{\sqrt{3}}{2} \sin \alpha \right) \bar{E} \qquad (10\text{-}11)$$

Or combining the sine and cosine functions into an equivalent cosine function leads to

$$\bar{E}_{12} = \sqrt{3}\,\bar{E} \cos (\alpha - 150°) \qquad (10\text{-}12)$$

†Some synchros are designed to provide a linear output with rotor displacement.

The expression for the line voltage between terminals 2 and 3 can be similarly determined. Thus,

$$\bar{E}_{23} = \bar{E}_{2n} + \bar{E}_{n3} = \bar{E}_{2n} - \bar{E}_{3n} \tag{10-13}$$

Substituting Eqs. (10-4) and (10-8) leads to

$$\bar{E}_{23} = \left(\frac{3}{2} \cos \alpha + \frac{\sqrt{3}}{2} \sin \alpha\right) \bar{E} \tag{10-14}$$

or

$$\bar{E}_{23} = \sqrt{3}\, \bar{E} \cos (\alpha - 30°) \tag{10-15}$$

The line voltage between terminals 1 and 3 is

$$\bar{E}_{31} = \bar{E}_{3n} + \bar{E}_{n1} = \bar{E}_{3n} - \bar{E}_{1n} \tag{10-16}$$

Upon inserting Eqs. (10-6) and (10-8) there results

$$\bar{E}_{31} = \bar{E} \cos (\alpha + 120°) - \bar{E} \cos (\alpha - 120°)$$
$$= -\sqrt{3}\, \bar{E} \sin \alpha \tag{10-17}$$

or

$$\bar{E}_{31} = \sqrt{3}\, \bar{E} \cos (\alpha + 90°) \tag{10-18}$$

Equation (10-17) is put in the cosine form in order to facilitate comparison with Eqs. (10-12) and (10-15) to show that these line voltages are such that the arithmetic sum of two of them must be equal to the third.

As a result of the variations in the line voltages that can be effected by displacement of the rotor, it is possible to feed these voltages to the stator of a control transformer, a receiver, or even a differential synchro and thereby produce useful effects. Such applications are discussed in the next section.

10-3 APPLICATIONS

The use of synchros in servomechanisms represents one of the biggest areas of applications. Shown in Fig. 10-5 is the arrangement most often found. This configuration constitutes that part of the servomechanisms called the *error detector*. Its chief function is to convert a command in the form of a mechanical displacement of the cx rotor to an electrical signal appearing as a voltage at the rotor winding terminals of the control transformer. It offers the distinct advantage of furnishing information about shaft displacement by means of wires, which allows the control transformer to be located remotely from the control transmitter.

What is the mechanism by which this is brought about? Keep in mind that the rotor of the control transmitter is excited with an a-c voltage so that an alternating field is created whose axis coincides with the cx rotor axis. The time-varying nature of the air-gap flux and its linkage with the stator coils induce in them corresponding voltages. Thus, if α equals 30° measured in the counterclockwise direction, it follows that the voltages induced in the stator phases are as specified in Fig. 10-5. When the cx stator terminals are connected to the ct stator terminals in the fashion

Figure 10-5 Synchro arrangement for converting a CX displacement to a corresponding voltage signal at the rotor of the CT.

shown in Fig. 10-5, the transmitter is made to supply magnetizing currents in the CT stator windings, which in turn create an alternating flux field in its own air gap. The values of the CT stator phase currents must be such that the resulting air-gap flux induces voltages that are equal and opposite to those prevailing in the stator of the transmitter. This denotes the equilibrium condition satisfying Kirchhoff's voltage law in the circuit between the stator windings of the two synchros. Accordingly, the direction of the resultant flux produced by the CX stator phase currents is forced to take a position of exact correspondence to that of the rotor axis of the control transmitter.

If the CT rotor is assumed to be held fast at its electrical zero position as depicted in Fig. 10-5, then the rms voltage induced in the rotor can be expressed as

$$e = E_r \sin \alpha \qquad (10\text{-}19)$$

where E_r denotes the maximum rms voltage induced by the CT air-gap flux when the coupling with the rotor winding is greatest and α is the displacement angle of the rotor of the control transmitter. The sinusoidal function appears because the distributed stator winding produces a sinusoidally distributed flux density along the air gap. The polarity markings are used to indicate those sides of the stator coils that have the same instantaneous polarity. For equilibrium to be established in the stator circuits it is therefore necessary that both phase and magnitude conditions be satisfied.

It is important to note that the value of the CT rotor voltage is in essence an indication of a lack of correspondence between the rotor positions of the transmitter and the transformer. Thus, if the CT rotor were manually turned counterclockwise in Fig. 10-5 by an amount $\alpha = 30°$, then clearly the rotor emf would become zero. In this manner exact correspondence is provided with the transmitter measured from the respective electrical zero positions of each synchro.

A general expression for the instantaneous value of the voltage (e) induced in the CT rotor when the displacement of the CX is α_x and that of the rotor of the CT is

α_T is

$$e = E_r \sin \omega t \sin (\alpha_x - \alpha_T) \qquad (10\text{-}19a)$$

Figure 10-6 is the schematic diagram showing how the synchro pair of Fig. 10-5 is used in a servomechanism. If it is desirable here for the load to be moved a specified amount in response to a command applied in the form of a displacement of the CX rotor, an error voltage e should initially exist because of the inertia on the output shaft associated with the servomotor and the load. As time elapses, the servomotor will begin to respond to this error voltage and thus displace the load. As the output displacement increases, the mechanical connection between the servomotor shaft and that of the CT rotor will cause the CT rotor to advance in the direction to reduce the error voltage e. The servomotor will cease to turn when a position of exact correspondence occurs—at which time e is zero. It is worthwhile to note that with such a control system a large load (e.g., a 100-ft radar dish antenna) can be controlled with very little effort applied at the rotor of the CX synchro.

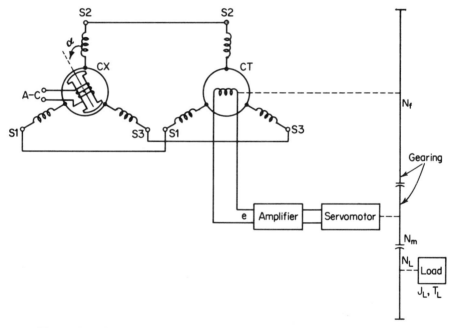

Figure 10-6 Servomechanism illustrating the use of synchros as the error detector. Dashed lines denote mechanical connections.

A second application of synchros is for *torque transmission* over an appreciable distance without the use of a rigid mechanical connection. The connection diagram is illustrated in Fig. 10-7. The arrangement requires the use of a transmitter and a receiver unit. The manner in which the system works is quite straightforward. When the rotor of the transmitter is displaced, say, by $\alpha = 30°$, the stator behaves like the secondary of a transformer and delivers current to the three-phase stator winding of the receiver; this produces an ampere-conductor distribution whose axis is fixed by

Figure 10-7 Illustrating the use of synchros for remote position indication.

α. By exciting the rotor of the receiver from the same a-c source as the rotor of the transmitter, a flux field is created which interacts with the stator ampere-conductor distribution to produce an electromagnetic torque. The action of this torque brings the rotor axis of the receiver, which is free to turn, to a position of correspondence with the rotor of the transmitter. Hence the CR rotor revolves by α degrees in the counterclockwise direction. Note that at this position the induced voltages in the stator windings of the receiver ideally bear the same phase and magnitude as those prevailing in the stator coils of the transmitter.

The use of a differential synchro to add or subtract an angle in the torque transmission system is illustrated in Fig. 10-8. In part (a) of this figure the voltage distribution in the stator phase windings of each synchro is shown corresponding to a CX rotor displacement of $\alpha_X = 30°$ in the positive (or counterclockwise) direction. The differential synchro is assumed to be kept at the electrical zero position in this instance and so does not alter the transmission characteristic from that represented by Fig. 10-7. The rotor of the receiver, of course, responds by turning counterclockwise by an amount equal to an α_X of 30°. It is helpful to note that a counterclockwise displacement of the CX rotor yields a counterclockwise displacement of the CR rotor.

Depicted in Fig. 10-8(b) is the effect of introducing a displacement of α_D in the differential synchro superimposed upon the situation represented in Fig. 10-8(a). For the convenience of illustration α_D is taken equal to $-60°$. The minus sign means that the rotor of the differential synchro is turned clockwise. Since the axis of the flux field in the differential synchro is determined by α_X, the displacement of the CD rotor by $-60°$ means that the induced emf in stator winding 2 must be zero, while in the remaining windings the rms value of the induced emf is $(\sqrt{3}/2)\,E$ having the polarities indicated. Immediately after the displacement of the CD rotor, a voltage unbalance exists between the rotor of the differential and the stator of the receiver. This unbalance causes currents to flow so that the resulting ampere-conductor distribution in the CR stator produces an electromagnetic torque that brings the CR rotor to the position that ensures voltage balance. An examination of Fig. 10-8(b) discloses that the resulting equilibrium position of the CR rotor is a position that is 90° displaced counterclockwise from the electrical zero. Note that a *clockwise* displacement of the CD rotor yields a *counterclockwise* displacement of the CR rotor. A mathematical

Figure 10-8 Illustrating the use of a control differential synchro in adding an angle to a torque transmission system.

459

description of the transmitter-differential-receiver configuration of Fig. 10-8 is given by

$$\alpha_R = a_X - \alpha_D \qquad (10\text{-}20)$$

where α_R is the net displacement of the CR synchro corresponding to an α_X displacement of the transmitter and an α_D displacement of the differential synchros. Positive values of the angles expressed in degrees are used for counterclockwise displacements and negative values for clockwise displacements.

10-4 THE ADVANTAGE OF SINUSOIDALLY DISTRIBUTED WINDINGS

Rather than consider separately the various schemes that are used for synchro windings, attention is directed in this section to show that no harmonic induced emf's can occur in a sinusoidally distributed winding even though such harmonics do exist in the field distribution created by the field winding. Synchro residual voltages originating from this source can accordingly be kept to a minimum by employing winding arrangements that approximate the sinusoidal distribution. The concentric type of winding is such a scheme and so is frequently employed in synchro design.

Depicted in Fig. 10-9 is the sinusoidally distributed armature winding† of a synchro. Let

$$N = \text{turns/pole} \qquad (10\text{-}21)$$

$$N_\alpha = N_p \sin \alpha = \text{sinusoidal distribution of turns} \qquad (10\text{-}22)$$
$$\text{(in turns/rad)}$$

These quantities are related by

$$N = \int_0^\pi N_\alpha \, d\alpha \qquad (10\text{-}23)$$

Performing the indicated integration then yields

$$N = N_p[-\cos \alpha]_0^\pi = 2N_p \qquad (10\text{-}24)$$

Hence the distribution of turns may be expressed in turns per pole as

$$N_\alpha = \frac{N}{2} \sin \alpha \qquad \text{turns/rad} \qquad (10\text{-}25)$$

In general the field distribution produced by the CT synchro field winding contains a large fundamental component plus odd harmonics. No even harmonics exist because the field patterns beneath the north and south poles of the synchros are mirror images of one another. The expression for the nth harmonic flux density is given by

$$b_n(\alpha) = B_n \sin (n\alpha - \gamma) \qquad (10\text{-}26)$$

where B_n is the amplitude of the nth harmonic field component. In the case of the transmitter synchro, B_n is a sinusoidal function of the line frequency ω applied to the rotor winding. Also, α denotes a space angle measured in the manner illustrated in Fig. 10-9. Moreover, the quantity γ is proportional to the CX rotor speed ω_r. A

---

†Armature winding is the stator for the CX synchro and the rotor for the CT synchro.

(a)

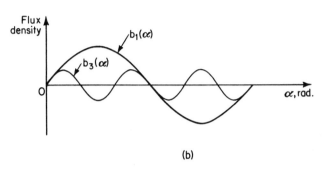

(b)

Figure 10-9 (a) Sinusoidally distributed armature winding; (b) air-gap field distribution, showing the fundamental and the third harmonic components.

comparison of Eq. (10-26) with Eq. (4-8) discloses that both are expressions for traveling waves.

Because the induced emf in the synchro armature winding is determined by the time rate of change of the flux linkage, it is worthwhile to find the expression for the flux linkage for both the fundamental field component and the harmonic field components. But first the equation for the flux is needed. For the nth harmonic this is

$$\phi_n = \int_\alpha^{\alpha+\pi} b_n(\alpha) lr \, d\alpha \tag{10-27}$$

where r denotes the rotor radius and l is the axial length. For simplicity a unit axial length is considered. Equation (10-27) may then be written as

$$\phi_n = \int_\alpha^{\alpha+\pi} B_n \sin(n\alpha - \gamma) r \, d\alpha = \frac{2B_n r}{n} \cos(n\alpha - \gamma) \tag{10-28}$$

The associated expression for the flux linkage then becomes

$$\lambda_n = \int_0^\pi N_\alpha \, d\alpha \, \phi_n(\alpha) = \frac{NB_n r}{n} \int_0^\pi \sin\alpha \cos(n\alpha - \gamma) \, d\alpha \tag{10-29}$$

Inserting the trigonometric identity

$$\sin\alpha \cos(n\alpha - \gamma) = \tfrac{1}{2}\sin[(n+1)\alpha - \gamma] - \tfrac{1}{2}\sin[(n-1)\alpha - \gamma] \tag{10-30}$$

leads to

$$\lambda_n = \frac{NB_n r}{2n} \int_0^\pi \{\sin[(n+1)\alpha - \gamma] - \sin[(n-1)\alpha - \gamma]\} \, d\alpha \tag{10-31}$$

The flux linkage of the fundamental field component is found upon substituting $n = 1$ into Eq. (10-31) and then performing the indicated integration. This leads to

$$\lambda_1 = \frac{B_1 N \pi r}{2} \sin \gamma \qquad (10\text{-}32)$$

Keep in mind that γ varies with the angular position of the CX rotor and that superimposed upon this is a sinusoidal line-frequency variation of B_1, the amplitude of the fundamental field component. It is also worthwhile to note that the fundamental flux linkage by Eq. (10-32) is not identically equal to zero.

A more useful expression for the flux linkage of the nth harmonic is obtained by integrating Eq. (10-31) and then inserting the specified limits. Thus

$$\lambda_n = \frac{N B_n r}{2n} \left(\left\{ \frac{-1}{n+1} \cos \left[(n+1)\alpha - \gamma \right] \right\}_0^\pi + \left\{ \frac{1}{n-1} \cos \left[(n-1)\alpha - \gamma \right] \right\}_0^\pi \right)$$

$$(10\text{-}33)$$

or

$$\lambda_n = \frac{N B_n r}{2n} \left[\frac{-1}{n+1} (\cos \gamma - \cos \gamma) + \frac{1}{n-1} (\cos \gamma - \cos \gamma) \right] \equiv 0 \qquad (10\text{-}34)$$

This last result is a significant one. It states that, for a sinusoidally distributed armature winding, the induced emf's of all the field harmonics are zero despite their presence in the field distribution.†

10-5 STATOR MMF OF THE CONTROL TRANSFORMER IN THE CX-CT MODE

In Sec. 10-3 it is pointed out that the magnetizing currents that flow in the transformer of the CX-CT configuration are such that the axis of the resultant air-gap flux follows the position of the CX rotor axis. This conclusion was reached on the basis of satisfying Kirchhoff's voltage law in the circuit involving the three stator phases of each synchro. Now attention is directed to a verification of this result in appropriate mathematical formulations. For the sake of generality the analysis is carried out for the nth harmonic. This procedure offers the advantage of making the general result available in the next section to determine the static errors caused by the CT space harmonics.

To focus attention solely on the stator mmf harmonics, several simplifying assumptions are made concerning the control transformer. These are: no saturation of the iron, uniform air gap between rotor and stator, and balanced phase windings. In addition, the control transmitter is assumed to be perfect, which means that it is equipped with a sinusoidally distributed winding as well as balanced stator windings and a uniform air gap. In Fig. 10-10 is a typical stator mmf wave for the actual winding distribution employed in synchros. Specifically only the mmf curve of winding 2 is illustrated here. A Fourier-series analysis applied to this stepwise function makes it clear that mmf space harmonics do exist and must be of odd order because of half-wave symmetry. Also depicted in Fig. 10-10 is the location of the rotor axis of the

---

†Another way of describing this result is to state that all the pitch and distribution factors of the harmonics are zero.

Figure 10-10 (a) Magnetomotive force distribution in phase 2 of the control transformer, showing one pole; (b) transmitter rotor position that produces the magnetizing current leading to the mmf distribution of (a).

transmitter measured from a zero reference line. Note that displacements measured in the CT air gap are denoted by Γ, while the CX displacements are represented by α.†

The expression for the mmf per pole for winding 2 in the stator of the control transformer for the nth harmonic is

$$f_{n2} = F_n \sin \omega t \cos n\Gamma \qquad (10\text{-}35)$$

where F_n is the amplitude of the mmf harmonic contained within the stepped wave shape. The $\sin \omega t$ factor represents the alternating nature of the mmf wave; the angle Γ permits finding the value of the mmf at any fixed point in space for a specified time instant. The harmonic amplitude F_n can be expressed in terms of the winding constants and the effective value of the stator phase current. In balanced polyphase machines used in conjunction with three-phase voltages this effective current is a fixed quantity. However, in the case of the control transformer this current is a variable quantity dependent upon the value of α. If I denotes the maximum rms current in winding 2 corresponding to the maximum applied voltage and if N_e denotes an equivalent number of effective turns per pole, then Eq. (10-35) can be more specifically written as

$$f_{n2} = (N_e I) \cos \alpha \sin \omega t \cos n\Gamma \qquad (10\text{-}36)$$

Keep in mind that Γ and α are independent quantities. For given values of α and ωt we may choose different values of Γ.

By introducing the fact that windings 1 and 3 are located, respectively, ahead and behind winding 2 by 120°, the expressions for the mmf of windings 1 and 3 are correspondingly

$$f_{n1} = (N_e I) \sin \omega t \cos (\alpha + 120°) \cos n(\Gamma + 120°) \qquad (10\text{-}37)$$

and

$$f_{n3} = (N_e I) \sin \omega t \cos (\alpha - 120°) \cos n(\Gamma - 120°) \qquad (10\text{-}38)$$

To find the resultant mmf in the stator of the control transformer, it is merely necessary

†This α is not the same as that used in Sec. 10-4.

to add the last three equations. Thus

$$f_{nr} = (N_e I) \sin \omega t \, [\cos \alpha \cos n\Gamma + \cos (\alpha - 120°) \cos (n\Gamma - 120°n)$$
$$+ \cos (\alpha + 120°) \cos (n\Gamma + 120°n)] \qquad (10\text{-}39)$$

Inserting the trigonometric identity

$$\cos x \cos y = \tfrac{1}{2} \cos (x - y) + \tfrac{1}{2} \cos (x + y) \qquad (10\text{-}40)$$

into Eq. (10-39) yields

$$f_{nr} = (N_e I) \sin \omega t \, \{\tfrac{1}{2} \cos (\alpha - n\Gamma) + \tfrac{1}{2} \cos (\alpha + n\Gamma)$$
$$+ \tfrac{1}{2} \cos [(\alpha - n\Gamma) - (n - 1)120°] + \tfrac{1}{2} \cos [(\alpha + n\Gamma) + (n + 1)120°]$$
$$+ \tfrac{1}{2} \cos [(\alpha - n\Gamma) + (n + 1)120°] + \tfrac{1}{2} \cos [(\alpha + n\Gamma) - (n + 1)120°]\}$$
$$(10\text{-}41)$$

After some manipulation the last equation reduces to

$$f_{nr} = \frac{N_e I}{2} \sin \omega t \, [A \cos (\alpha - n\Gamma) + B \cos (\alpha + n\Gamma)] \qquad (10\text{-}42)$$

where

$$A = 1 + 2 \cos (n - 1)120° \qquad (10\text{-}43)$$
$$B = 1 + 2 \cos (n + 1)120° \qquad (10\text{-}44)$$

Equation (10-42) can be further simplified by determining the values of A and B for various values of n. In this connection recall that n is restricted to odd values. Accordingly,

for $n = 1$, $A = 3$

                              $B = 0$

        $n = 3$, $A = 0$

                              $B = 0$

        $n = 5$, $A = 0$

                              $B = 3$

or more succinctly

$$A = 3 \qquad \text{for } n = 3k + 1 \qquad (10\text{-}45)$$

where $k = 0, 2, 4, 6, \cdots$,
and

$$B = 3 \qquad \text{for } n = 3k - 1 \qquad (10\text{-}46)$$

where $k = 2, 4, 6, \cdots$.

Therefore, the expression for the resultant mmf may be written as

$$f_{nr} = \tfrac{3}{2}(N_e I) \sin \omega t \cos (\alpha - n\Gamma) \qquad \text{for } n = 3k + 1 \qquad (10\text{-}47a)$$
$$= \tfrac{3}{2}(N_e I) \sin \omega t \cos (\alpha + n\Gamma) \qquad \text{for } n = 3k - 1 \qquad (10\text{-}47b)$$

where the k's are defined as in Eqs. (10-45) and (10-46). Examination of Eq. (10-47) discloses that a single expression can be used provided that k is allowed to take on negative as well as positive values. Accordingly, we may write

$$\boxed{f_{nr} = \tfrac{3}{2}(N_eI) \sin \omega t \cos (\alpha - n\Gamma)} \tag{10-48a}$$

where

$$n = 3k + 1 \tag{10-48b}$$

and

$$k = 0, \pm 2, \pm 4, \pm 6, \pm \cdots \tag{10-48c}$$

When a negative number is used for k, it yields a negative value for n. This sign must be carried along with the magnitude of n. When it is, Eq. (10-47b) immediately results.

The use of the negative sign offers still another advantage. If for the moment the cx rotor is assumed to be driven at a constant velocity ω_r, it follows that $\alpha = \omega_r t$. Insertion of this result into Eq. (10-48a) yields the equation of a traveling wave having a fixed amplitude of $\tfrac{3}{2}(N_eI)$. The use of $k = -2$ in Eq. (10-48a) leads to

$$f_{5r} = \tfrac{3}{2}(N_eI) \sin \omega t \cos (\omega_r t + 5\Gamma) \tag{10-49}$$

Examination of the last expression discloses that it is the equation of a wave traveling in the *negative* Γ direction. Therefore the minus sign that results from Eq. (10-48b) corresponding to $k = -2$ can be used to denote those harmonic components of the mmf wave that travel in the negative Γ direction.

How can Eq. (10-48) be used to show that the fundamental component of the resultant field in the air gap of the control transformer follows exactly the cx rotor displacements? First let us write the expression for this fundamental term. Thus,

$$f_{1r} = \tfrac{3}{2}(N_eI) \sin \omega t \cos (\alpha - \Gamma) \tag{10-50}$$

Now assume that an observer is placed at $\Gamma = 0$ degrees in the air gap of the control transformer and stays fixed there. Then with the cx rotor at its zero position (i.e., $\alpha = 0°$), Eq. (10-50) states that the observer at $\Gamma = 0$ degrees sees the amplitude of the field mmf. Of course, this amplitude will be at its maximum value when the cT stator phase current is a maximum. Next assume that the cx rotor is displaced by $\alpha = 30°$. In accordance with Eq. (10-50) the observer at $\Gamma = 0$ degrees now sees a value of the mmf which is given by

$$f_{r1} = \tfrac{3}{2}(N_eI) \sin \omega t \cos \alpha = \tfrac{3}{2}(N_eI) \sin \omega t \cos 30° \tag{10-51}$$

A comparison of Eq. (10-51) with Eq. (10-50) reveals a reduction by the factor $\cos \alpha$. Hence one can conclude that the axis of the mmf in the air gap of the control transformer (i.e., the location of its amplitude) has revolved in the positive direction by the amount $\alpha = 30°$. *Therefore the stator field of the control transformer follows exactly the position of the transmitter rotor axis.*

Perhaps a more convincing demonstration of this conclusion can be had by placing the observer at a point $\Gamma = 30°$. Then when α is $0°$, the largest mmf seen by the observer is $\sqrt{3}/2$ times the peak value. However, when α is increased to $30°$, Eq. (10-48) indicates that the observer now sees the peak mmf. Hence the axis of the air-gap flux must have moved in the positive† Γ direction by the amount α.

†If the wave would have moved in the negative Γ direction, the observer would see only one-half the peak mmf.

10-6 DISPLACEMENT ERRORS CAUSED BY CT SPACE HARMONICS

The presence of harmonics in the stator mmf of the CT synchro is responsible for the existence of a displacement error in the CX-CT configuration. The variation of this error can be shown to be a sixth-harmonic function of the CX rotor displacement angle. In this section consideration is first given to a physical demonstration of why the presence of harmonics leads to such a displacement error. A mathematical description then follows.

The physical explanation requires the use of a suitable zero reference from which the CX and CT rotor displacements are measured. For convenience this reference is taken as the previously defined electrical zero positions. At this point it is assumed that the rotors of the CX and CT synchros are in exact correspondence. This situation can be readily illustrated through a graphical interpretation of Eq. (10-48). In the interests of simplifying the representation, only the fundamental component and the fifth and seventh harmonics are assumed present in the CT stator mmf wave. At the electrical zero position, α is zero. Similarly, to place the CT rotor at electrical zero it is assumed that Γ is also zero. Equation (10-48) then yields the following results:

$n = 1$,

$$f_{1r} = \frac{3}{2}(N_e I) \sin \omega t \cos (\alpha - \Gamma) = \frac{3}{2}(N_e I) \sin \omega t \cos 0°$$

$$= \frac{3}{2}(N_e I) \sin \omega t \equiv F_1 \tag{10-52}$$

$n = -5$,

$$f_{5r} = \frac{3}{2}\left(\frac{N_e I}{5}\right) \sin \omega t \cos (0° + 5 \times 0°)$$

$$= \frac{3}{2}\left(\frac{N_e I}{5}\right) \sin \omega t \equiv F_5 \tag{10-53}$$

$n = +7$,

$$f_{7r} = \frac{3}{2}\left(\frac{N_e I}{7}\right) \sin \omega t \cos (0° - 7 \times 0°)$$

$$= \frac{3}{2}\left(\frac{N_e I}{7}\right) \sin \omega t \equiv F_7 \tag{10-54}$$

where F_1, F_5, and F_7 are defined as indicated. A glance at these expressions shows that all three components are in *space* phase as well as in *time* phase. The space-phase condition is graphically illustrated in Fig. 10-11(a). Note that the line of action of the resultant mmf F_R coincides with that of the fundamental F_1 because the harmonic terms are in phase at this point too. The coupling with the CT rotor winding is therefore zero so that no emf occurs. The control transformer is accordingly at the null† position.

---

†The null position is the position of minimum voltage in the CT rotor winding, which in this case is zero.

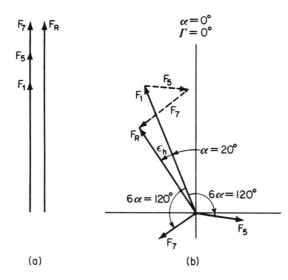

(a)   (b)

Figure 10-11 Graphical representation of the displacement error caused in a CX-CT synchro configuration by the CT mmf harmonics: (a) condition at electrical zero; (b) location of the resultant mmf when $\alpha = 20°$ and the CT rotor is displaced by $\Gamma = \alpha = 20°$.

Next consider the case where the cx rotor is displaced by 20° in the positive counterclockwise direction. Then imagine that the rotor of the ct synchro is displaced by a like amount in the same direction, i.e., $\Gamma = 20°$. If the resultant field mmf continues to be in space phase with the fundamental component, then a null voltage occurs at $\Gamma = 20°$ and so no displacement error exists. On the other hand, if the resultant mmf, which produces the net field flux, does not lie along the 20° line, then some coupling with the ct rotor winding exists and so causes a small voltage to appear between its terminals. To reduce this voltage to zero it will then be necessary to displace the ct rotor by an additional amount, ϵ_h. The cx-ct synchro configuration is then said to have a displacement error of ϵ_h at $\alpha = 20°$. By resorting once again to Eq. (10-48) we can ascertain which condition prevails. Accordingly, we get

$n = 1$,

$$f_{1r} = \frac{3}{2}(N_e I) \sin \omega t \cos (\alpha - \Gamma) = F_1 \cos (0 - 0)$$

$$= F_1 \cos 0° \tag{10-55}$$

$n = 5$,

$$f_{5r} = \frac{3}{2}\left(\frac{N_e I}{5}\right) \sin \omega t \cos (\alpha + 5\Gamma) = F_5 \cos [20 + 5(20)]$$

$$= F_5 \cos 120° \tag{10-56}$$

$n = 7$,

$$f_{7r} = \frac{3}{2}\left(\frac{N_e I}{7}\right) \sin \omega t \cos (\alpha - 7\Gamma) = F_7 \cos [20 - 7(20)]$$

$$= F_7 \cos (-120°) \tag{10-57}$$

A graphical representation of these results is shown in Fig. 10-11(b). The fifth and seventh harmonics are now no longer in phase with the fundamental component as can be seen by the difference in the argument of the cosine terms in the foregoing equations. Keep in mind that the angle Γ in these equations denotes the amount by which the rotor of the CT synchro is displaced. To determine whether or not a null results, Γ is set equal to α. Hence for the fundamental component the argument of the cosine term becomes identically zero. The argument of the cosine term can thus be considered to be an angle that is measured relative to the position of the CX rotor axis. The zero-degree value for the fundamental means that there is exact correspondence between the axis of the fundamental CT field and the CX rotor axis. In the case of the fifth harmonic the argument is $120°$, which means that the axis of the fifth-harmonic mmf lies $120°$ displaced from the location of the fundamental component. In the diagram, F_5 is depicted *behind* F_1 by $120°$ because the fifth harmonic is known to rotate in a direction opposite to that of the fundamental component. However, note that F_5 is displaced $100°$ from the electrical zero reference line, which is consistent with the value $n\Gamma = -5(20) = -100°$. For the seventh harmonic, which travels in the same direction† as the fundamental, F_7 is shown ahead of F_1 by $120°$. Note again that this is consistent with the fact that F_7 is displaced by $n(20) = 7(20) = 140°$ in the positive direction from the zero reference line. It is also worthwhile to note that the positions of F_5 and F_1 are *sixth*-harmonic functions of the CX rotor displacement angle α.

In general the sum of the fundamental mmf component and the harmonics yields a resultant mmf that is displaced from the fundamental by the angle ϵ_h. Accordingly, if the CX-CT arrangement is used as an error detector in a servomechanism and a command is applied at the transmitter in the form of $\alpha = 20°$, the servomechanism achieves a position of equilibrium (i.e., a null position) when the CT synchro is displaced through the mechanical feedback connection by an amount equal to $\alpha + \epsilon_h$. Therefore, the output follows the command but with an error of ϵ_h.

Mathematical Analysis

An equation for the displacement error ϵ_h can be found by first determining the expression for the null voltage. This can be conveniently done by finding the voltage induced in the transformer rotor winding by each harmonic component of the revolving mmf. Keeping in mind that at the electrical zero position the axis of the CT rotor winding is displaced by 90 electrical degrees from the assumed zero position of the stator mmf axis (i.e., the $\Gamma = 0$ degree point), the voltage induced in the rotor winding by the fundamental field component is

$$E_1 = K_1 \cos\left[\alpha - \left(\Gamma - \frac{\pi}{2}\right)\right] \tag{10-58}$$

where

$$K_1 = 4.44 f N_r K_{wr} \Phi_1 \tag{10-59}$$

Note that K_1 actually denotes the rms voltage induced in the rotor winding by the fundamental field component corresponding to maximum coupling. Accordingly, Eq. (10-58) describes the rms value of the voltage induced in the rotor winding for a given CX rotor displacement α and a CT rotor displacement Γ. The cosine term of

†The minus sign in the argument of the cosine term of Eq. (10-48) represents positive rotation.

this equation is simply a factor that indicates the degree of coupling that exists between the fundamental field flux ϕ_1 produced by the fundamental field mmf and the CT rotor winding. Equation (10-58) may be rewritten as

$$E_1 = -K_1 \sin(\alpha - \Gamma) \qquad (10\text{-}60)$$

This result can now be used to find the voltage induced in the rotor winding when the CT rotor is placed at the null position. It is important to keep in mind that in general the coupling of the fundamental flux component with the rotor winding will not be zero at the null point. This is graphically illustrated in Fig. 10-11, where the rotor winding is at null when its axis is in quadrature with the resultant mmf. However, note that there is some small amount of coupling of the winding with the fundamental field component even though the coupling with the resultant field is zero. Since the null position is found by displacing the CT rotor by $\Gamma = \alpha + \epsilon_h$ rather than by $\Gamma = \alpha$, it follows that the rms value of the voltage induced by the fundamental field at null is

$$E_{1\text{ null}} = -K_1 \sin(\alpha - \Gamma) = -K_1 \sin(\alpha - \alpha - \epsilon_h) = K_1 \sin \epsilon_h \qquad (10\text{-}61)$$

Because such displacement errors are often very small, the last expression may be written as

$$E_{1\text{ null}} = K_1 \epsilon_h \qquad (10\text{-}62)$$

A similar procedure can be followed to find the emf induced in the CT rotor by the fifth harmonic of the field mmf. Thus the general expression for the rms value of this voltage is

$$E_5 = K_5 \cos\left[\alpha + 5\left(\Gamma - \frac{\pi}{2}\right)\right] \qquad (10\text{-}63)$$

where

$$K_5 = 4.44 f N_r K_{wr5} \Phi_5 \qquad (10\text{-}64)$$

Here K_{wr5} is the rotor winding factor for the fifth harmonic and Φ_5 denotes the fifth-harmonic flux per pole. In Eq. (10-63) the $\pi/2$ quantity is treated in the same way as Γ because it is a space angle associated with the CT rotor. By means of an appropriate trigonometric identity Eq. (10-63) can be rewritten as

$$E_5 = K_5 \sin(\alpha + 5\Gamma) \qquad (10\text{-}65)$$

When the CT rotor is assumed to be placed at its null position, Γ takes on the value $\alpha + \epsilon_h$, which upon insertion into Eq. (10-65) yields

$$E_{5\text{ null}} = K_5 \sin(\alpha + 5\Gamma) = K_5 \sin(\alpha + 5\alpha + 5\epsilon_h) = K_5 \sin(6\alpha + 5\epsilon_h) \qquad (10\text{-}66)$$

An important observation to make at this point is that the voltage induced by the fifth-harmonic field mmf is of the *same* frequency as that of the fundamental component. Moreover, it is in *time phase* with the fundamental. Therefore the fundamental and the harmonic voltage contributions can be added directly in order to obtain the resultant voltage.

The expression for the rms value of the induced voltage for the seventh harmonic is found to be

$$E_7 = K_7 \sin(\alpha - 7\Gamma) \qquad (10\text{-}67)$$

where K_7 is defined in a manner similar to that for K_5. The corresponding equation for the null voltage associated with this harmonic is then

$$E_{7 \text{ null}} = -K_7 \sin (6\alpha + 7\epsilon_h) \tag{10-68}$$

It is interesting to note that the argument in Eqs. (10-66) and (10-68) involves a sixth-harmonic function of α. It is reasonable to expect therefore that extension of the foregoing procedure to the eleventh and thirteenth mmf harmonics leads to expressions that involve twelfth-harmonic functions of α. Specifically it can be shown that

$$E_{11 \text{ null}} = K_{11} \sin (12\alpha + 11\epsilon_h) \tag{10-69}$$

$$E_{13 \text{ null}} = K_{13} \sin (12\alpha + 13\epsilon_h) \tag{10-70}$$

Of course, as higher order harmonics are considered, their importance diminishes because the magnitudes of the K coefficients vary inversely with the order of the harmonic. Hence the harmonics that bear the greatest influence in causing displacement errors are the fifth and seventh.

The complete expression for the null voltage that occurs when the CT rotor is placed at the position $\Gamma = \alpha + \epsilon_h$ may be written as

$$E_{\text{null}} = E_{1 \text{ null}} + E_{5 \text{ null}} + E_{7 \text{ null}} + \cdots \tag{10-71}$$

An algebraic sum is permitted because these voltages are of the same frequency and in time phase. Inserting Eqs. (10-62), (10-66), (10-68), (10-69), and (10-70) yields

$$E_{\text{null}} = \epsilon_h K_1 + K_5 \sin (6\alpha + 5\epsilon_h) - K_7 \sin (6\alpha + 7\epsilon_h)$$
$$+ K_{11} \sin (12\alpha + 11\epsilon_h) - K_{13} \sin (12\alpha + 13\epsilon_h) + \cdots \tag{10-72}$$

This expression can be simplified by noting that the quantity $n\epsilon_h$ appearing in the arguments of the sine terms can be neglected with little or no error. Accordingly, we have

$$E_{\text{null}} = \epsilon_h K_1 + (K_5 - K_7) \sin 6\alpha + (K_{11} - K_{13}) \sin 12\alpha + \cdots \tag{10-73}$$

However, with the CT rotor placed at the null position the resultant emf induced is zero. Hence

$$E_{\text{null}} = 0 = \epsilon_h K_1 + (K_5 - K_7) \sin 6\alpha + (K_{11} - K_{13}) \sin 12\alpha + \cdots \tag{10-74}$$

from which it follows that the expression for the displacement error is

$$\boxed{\epsilon_h = \frac{K_5 - K_7}{K_1} \sin 6\alpha + \frac{K_{11} - K_{13}}{K_1} \sin 12\alpha + \cdots} \tag{10-75}$$

An examination of Eq. (10-75) discloses that the magnitude of the displacement error caused by the space mmf harmonics in a CX-CT synchro configuration can be controlled by properly manipulating the K coefficients. Thus, if K_5 and K_7 are made equal, the sixth harmonic displacement error is entirely eliminated. This result can be achieved in some synchros through the judicious selection of the winding factors for these harmonics. When cancellations cannot be accomplished, the winding design is chosen so that the winding factors for the harmonics are kept as small as possible.

The accuracy of a servomechanism can be no better than the synchro pair that is used as the error detector. Hence the static displacement errors must be kept as small as manufacturing tolerances will allow, consistent of course with economical production. A typical static error curve for a CX-CT synchro error detector is depicted in Fig. 10-12. This curve is obtained by displacing the rotor of the transmitter by an amount α and then displacing the CT rotor to that position which yields a minimum voltage across its rotor winding terminals. The difference between the CT rotor displacement Γ and the CX rotor displacement α constitutes the static displacement error. In well-designed synchros this error rarely exceeds $\frac{1}{2}$ degree. A study of Fig. 10-12 reveals the predominance of a six-cycle error. As already described in the preceding section, this is due to the effect of the mmf harmonics in the stator of the control transformer. Note that elimination of this source of error goes a long way towards increasing the accuracy of the CX-CT synchro pair.

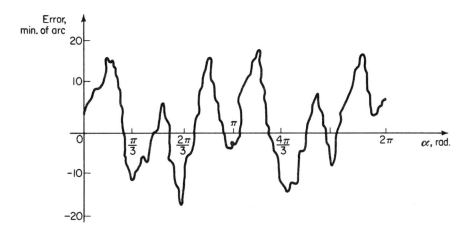

Figure 10-12 Typical displacement error curve for a CX-CT synchro pair. Note the predominance of a six-cycle and a two-cycle error variation.

Another dominant component that can readily be distinguished in the static error curve is a double frequency error of the CX rotor displacement α. The source of this second harmonic term lies in the slight unbalance that necessarily occurs in the stator phase windings of the synchros. It can be shown that, if the resistive and reactive components of the phase impedances are not identical, then a displacement error exists which varies as a second harmonic function of α. The nonidentical nature of the windings is very often attributable to manufacturing difficulties. These often involve one or more of the following factors: slight deviations in the turns used for each phase, varying tensions on the wire spools during the winding process, slight deviations in the cross-sectional area of the wires obtained from different spools or different manufacturers, slight differences in the leakage reactance of each phase winding, and finally the effect of tooth iron saturation. The relationship between the static error and the unbalanced impedances discloses that unbalances in the reactive parts of the impedances exert a greater influence on the static error than unbalance in the resistive parts.

Another source of a double frequency error in the error curve is the absence of a perfectly uniform air gap. This can happen for several reasons: ellipsing of the rotor, ellipsing of the stator, an eccentrically placed rotor, or a skewed rotor axis. Usually production methods are such that there is no difficulty in getting a perfectly round rotor. However, the same is not true for the stator structure; there is a greater tendency for the stator to come off the production line slightly ellipsed. Actually the ellipsing need not be very large to cause a significant static error. For example, a difference between the major and minor axes of the stator of only 0.0001 in. can cause a static error spread of as much as 3 to 4 minutes of arc.

Besides static displacement errors, the CX-CT synchro pair has the additional shortcoming of residual voltages at null. Whenever residual voltages do in fact exist, they cannot contain a component of fundamental frequency that is in phase with the applied line voltage. After all, the null position is achieved whenever the in-phase fundamental frequency component is zero. Hence, any voltage that exists at null must be due to other sources. Figure 10-13 depicts a typical curve showing how the residual voltage varies with α (as the transformer is placed at the null position for each value of α).

Figure 10-13 Typical variation of a residual voltage content in a CX-CT configuration at null.

Analysis shows that the residual voltage contains two components: time harmonics of the fundamental frequency and quadrature voltages. The time harmonics in turn arise from two sources, i.e., the saturation of the iron used in the magnetic circuit of the synchros as well as the nonuniform field distribution in the air gap of the control transmitter. The quadrature voltage is defined as the minimum voltage of fundamental frequency existing at null. Its presence is caused either by unbalanced stator phase windings or a nonuniform air gap or by a combination of both. In each instance, however, it can be shown that the quadrature voltage is a double frequency function of α. A glance at Fig. 10-13 shows the prevalence of the second harmonic term. Moreover, analysis also shows that resistance unbalance in the synchro phase windings exerts a greater influence than reactance unbalance does. It is for this reason that switches in the stator line connections of the CX-CT synchro pair should be avoided if at all possible.

Good synchro design focuses on keeping the residual voltages as small as is economically possible especially when these units are used as the error detectors in servo systems. Unusually large residual voltages can easily cause saturation of servoamplifiers which provide the control voltages that drive the output devices. When saturation sets in, the system gain is sharply reduced, which adversely affects the total system behavior. Moreover, although the residual voltages are incapable of producing any useful output in such systems, they can nevertheless cause currents to flow in the output devices that create unnecessary heat losses.

PROBLEMS

10-1. Explain why an umbrella-like rotor construction keeps to a minimum the magnetizing current drawn by the control transformer.

10-2. Prepare a rough sketch to show how the magnetizing current of a control transformer equipped with a salient-pole rotor varies with rotor displacement.

10-3. When the rotor of the synchro depicted in Fig. P10-3 is at the electrical zero position, the rms value of the phase voltage induced in winding 2 is measured to be 52 V. Determine the voltage induced in each winding from neutral-to-line and from line-to-line when the rotor is at a position 60° from electrical zero in a counterclockwise direction. Clearly indicate the polarity markings for each voltage.

Figure P10-3

10-4. Repeat Prob. 10-3 for the instance where the rotor occupies the following positions:
(a) 180° counterclockwise.
(b) 90° clockwise.

10-5. A control transmitter, a control differential, and a control receiver are connected in the manner shown in Fig. 10-8(a). The control transmitter is displaced 60° counterclockwise and fastened. The control receiver is displaced 30° clockwise and fastened. In which direction and by what amount does the differential synchro move? In your solution, redraw the diagram for the new position of the synchro, and clearly indicate the rms value of induced emf in each synchro winding. Assume that the maximum rms voltage from line-to-neutral in each unit is E.

10-6. The output voltage of a control transmitter is fed to the stator of a differential synchro whose rotor output is then applied to the stator of a control transformer. Assume that, initially, each synchro is at its proper electrical zero position. If the rotor of the transmitter is displaced 90° counterclockwise and the rotor of the differential synchro is turned 60° counterclockwise, determine the value of the induced voltage at the rotor terminals of the control transformer, expressed in terms of its maximum voltage E.

10-7. The output voltage of a control transmitter is fed to the stator of a differential synchro whose output is then applied to the stator of a control transformer.
(a) Draw the schematic diagram of the system with each unit at the electrical zero positions.
(b) If the rotor of the transmitter is displaced 30° counterclockwise and that of the differential synchro is turned 60° clockwise, find the value of the induced voltage at the rotor terminals of the CT in terms of its maximum voltage E.

10-8. A synchro transmitter has its stator connected to the stator of a differential synchro whose rotor, in turn, is connected to the stator of a synchro control transformer. The rotor of the transmitter is properly energized and all three synchros are adjusted to their zero positions. The transmitter rotor is then turned 15° clockwise, and the differential rotor is turned 75° counterclockwise.
(a) Find the angle and direction of the flux axis in each synchro.
(b) Determine the voltage across the rotor terminals of the control transformer in terms of the maximum possible value of this voltage E.

10-9. Repeat Prob. 10-8 for the case where the rotor of the CX is turned 45° clockwise and that of the CD is displaced 90° counterclockwise.

10-10. With the aid of Eq. (10-48a) demonstrate that the resultant mmf associated with the fifth harmonic travels in the negative Γ direction. Assume that an observer is placed at $\Gamma = 30°$.

10-11. Determine the maximum value of the sixth harmonic error in a CX-CT synchro transmission unit where the effect of the fifth space harmonic is 20% and that of the seventh harmonic is 14% of the fundamental. Assume that all other errors are negligible.

10-12. The mmf distribution in the air-gap of the control transformer in a CX-CT synchro transmission configuration is known to have a fundamental component as well as fifth and seventh harmonics. The fifth harmonic flux per pole is one-fifth that of the fundamental, and the seventh harmonic flux per pole is one-seventh that of the fundamental. Moreover, the rotor winding factor is one-fourth and one-fifth for the fifth and seventh harmonics, respectively. Assume for convenience that the displacement error due to the space harmonics is zero at the electrical zero position.
(a) For a CX rotor displacement of 30°, compute the value of displacement error that occurs when the CT is turned to the null position.
(b) Repeat part (a) for a CX displacement of 240°.
(c) Prepare a rough diagram of the conditions prevailing in part (b).

10-13. An ideal control transmitter is used with a control transformer in a typical CX-CT voltage indicating system. The air-gap flux of the CT contains a fifth and seventh harmonic component in addition to the fundamental. The magnitude of the flux per pole of the harmonics is inversely proportional to the order of the harmonic. A rotor winding is used for the CT unit, which causes a 5 per cent reduction in effective induced voltage. This winding also yields identical winding factors for the harmonics. Compute the harmonic winding factor that corresponds to a sixth harmonic displacement error which never exceeds 1°.

10-14. The predominant space harmonics in the control transformer of a CX-CT arrangement are known to be the 5th, 7th, 11th, and 13th. The winding factors for the fundamental and the 11th and 13th harmonics are each 0.96, while for the 5th and 7th they are 0.25 and 0.2, respectively. Assume that the harmonic displacement error is zero at the electrical zero of the CT.

    (a) Compute the value of the harmonic displacement error for the following values of the CX rotor displacement: 5°, 10°, 15°, 20°, 25°, and 30°.

    (b) Sketch the general nature of the curve that passes through these points.

10-15. Refer to the CX-CT synchros of Prob. 10-11, and assume that the harmonic displacement function passes through zero at the electrical zero position of the CT unit. Also, assume that the maximum rms voltage induced by the fundamental flux is E_1 V.

    (a) What is the value of the voltage induced in the rotor of the CT by the fundamental flux, when the CT is at the null position corresponding to a CX rotor displacement of 10°?

    (b) What is the value of the fifth harmonic induced voltage in part (a)?

    (c) Is the quantity computed in part (b) of the same frequency as the fundamental? Explain.

10-16. Repeat Prob. 10-15 for the conditions specified in Prob. 10-12.

DYNAMICS
OF ELECTRIC
MACHINES

Dynamics

In this chapter a study is made of the dynamic behavior of electromechanical devices. In all the preceding chapters the analysis was concerned with steady-state performance only. However, nowadays, a knowledge of the behavior of electromechanical devices in the transient state is at least equally important because this equipment is often found as part of a total system. For example, generators and motors are frequently found in the same interconnected electric power grid system. As loads change and shift, it is the dynamic characteristic of the individual machines taken together that will determine whether the system can react in a stable manner to such disturbances. Similarly, in many feedback control systems such as the servomechanism, it is not uncommon to find d-c motors, two-phase servomotors, pilot generators, amplidynes, and other electromechanical devices as important components of such systems. The overall behavior and stability of the entire control system is often greatly influenced by the dynamic characteristics of these devices. An understanding of the steady-state characteristics alone is not sufficient to provide a useful study of total performance in these situations. After all, if the system of interest fails to achieve steady-state operation because of poor dynamics, an understanding of the steady-state performance is really only academic.

The *transfer function* is the vehicle by which the dynamic characteristics of electromechanical devices are described. This matter is discussed and illustrated in Sec. 11-1. However, the development proceeds on the assumption that the reader already has a background in Laplace transforms.

11-1 THE TRANSFER FUNCTION

As the name implies, the *transfer function* is a mathematical formulation that relates the output variable of a device to the input variable. For linear devices the transfer function is dependent solely upon the parameters of the device plus any operations of time, such as differentiation and integration, which it may possess. It is independent of the input quantity—both its magnitude and variation.

The general procedure for deriving the transfer function is always the same irrespective of the type component or engineering device under study. It involves the following three steps: (1) determine the governing equation for the device expressed in terms of the output and input variables; (2) Laplace transform the governing equation, assuming all initial conditions are zero; and (3) rearrange the equation to formulate the ratio of the output to input variable.

An example is useful at this point to illustrate the procedure and to show how the dynamic behavior is thereby described. Refer to Fig. 11-1. Assume that the input quantity is the voltage applied to the field winding of a machine, V_f, and that the

Figure 11-1 Field winding circuit of an electromechanical energy converter. Example of a first-order system.

desired output quantity is the field winding current, I_f. The equation that relates I_f to V_f in terms of the appropriate parameters of the circuit is clearly

$$V_f = I_f R_f + L_f \frac{dI_f}{dt} \tag{11-1}$$

Equation (11-1) is the governing equation and specifically is a first-order differential because it involves one energy-storing element, the inductance L_f. Laplace transforming Eq. (11-1) for zero initial field winding current leads to the following transformed differential equation

$$V_f(s) = I_f(s)R_f + L_f s I_f(s) \tag{11-2}$$

where $V_f(s)$ denotes the Laplace transform of V_f, $I_f(s)$ denotes the Laplace transform of I_f and the Laplace operator s denotes the first derivative of time. Upon formulating the ratio of output to input variable in the last equation, the desired transfer function results. Thus

$$\frac{I_f(s)}{V_f(s)} = \frac{1}{R_f + sL_f} = \frac{1}{R_f} \frac{1}{1 + s\tau_f} \tag{11-3}$$

where

$$\tau_f = \frac{L_f}{R_f} \tag{11-4}$$

and is the time constant of the field winding.

Inspection of the right side of Eq. (11-3) discloses that the transfer function is solely dependent on the circuit parameters R_f and L_f and the derivative operator s. Moreover, the dynamic behavior of this circuit is completely determined by the form of the *denominator of the transfer function*. This is because the denominator of the transfer function is the *characteristic equation* of the governing differential equation when set equal to zero. Reference to Eq. (11-1) makes this readily apparent. To obtain the characteristic equation by the classical method of solving differential equations, it is necessary first to set the forcing function V_f equal to zero and then to put the remaining terms in operator form. Thus

$$0 = R_f I_f + L_f s I_f \qquad (11\text{-}5)$$

or

$$R_f + s L_f = 0 \qquad (11\text{-}6)$$

A comparison of Eq. (11-6) with the denominator of Eq. (11-3) reveals the two expressions to be identical.

The complete time description of the output variable can always be found from the transfer function. This includes steady-state as well as transient behavior. In the case of the R-L circuit, e.g., the expression for the transformed solution of the field current can be written from Eq. (11-3) as

$$I_f(s) = \frac{1}{R_f + s L_f} V_f(s) \qquad (11\text{-}7)$$

If V_f is now identified for convenience as a step voltage of magnitude V_f applied to the field winding, the corresponding Laplace transform is

$$V_f(s) = \frac{V_f}{s} \qquad (11\text{-}8)$$

Inserting this expression into Eq. (11-7) yields

$$I_f(s) = \frac{V_f}{L_f s (s + R_f/L_f)} \qquad (11\text{-}9)$$

A partial fraction expansion allows Eq. (11-9) to be written as

$$I_f(s) = \frac{V_f}{L_f s (s + R_f/L_f)} = \frac{K_0}{s} + \frac{K_1}{s + R_f/L_f} \qquad (11\text{-}10)$$

Evaluation of K_0 and K_1 then leads to

$$I_f(s) = \frac{V_f}{R_f} \left(\frac{1}{s} - \frac{1}{s + R_f/L_f} \right) \qquad (11\text{-}11)$$

By employing a table of Laplace-transform pairs the corresponding time solution is found to be

$$i_f(t) = \frac{V_f}{R_f} (1 - \epsilon^{-t/\tau_f}) \qquad (11\text{-}12)$$

where $\tau_f = L_f/R_f$.

A comparison of Eq. (11-12) with Eq. (11-11) shows that the transient term $-\epsilon^{-t/\tau_f}$ is directly related to that term of the transformed version of the solution that results from the transfer function [the second term of Eq. (11-11)].

Sec. 11-1   The Transfer Function

481

As a matter of convenience the transfer function is frequently represented graphically by the block diagram notation depicted in Fig. 11-2(a). Sometimes the *signal flow diagram* shown in Fig. 11-2(b) is used. The output quantity of the block is always the Laplace transformed version of the output variable and the input quantity is the Laplace transformed version of the input variable.

(a) (b)

Figure 11-2 Graphical representation of the transfer function of the circuit of Fig. 11-1: (a) block diagram form; (b) signal flow diagram form.

Figure 11-3 Block diagram for a feedback system.

Systems often are represented by a block diagram showing the interconnected transfer functions of the individual components of which they are comprised. Large complex systems contain many direct transfer functions of the type depicted in Fig. 11-2, but they may also contain feedback transfer functions. Whenever a feedback transfer function appears around a direct transfer function,† the total transfer function of the combination is readily found by applying the feedback formula which is now derived. Refer to the block diagram appearing in Fig. 11-3. The quantity $G(s)$ denotes the direct transmission function between the reference input signal $R(s)$ and the output signal $C(s)$. The quantity $H(s)$ represents the feedback function. Note that the output quantity $C(s)$ is operated on by this function and then *subtracted* from the reference input $R(s)$ to generate the actual input signal $E(s)$ to the direct transfer function $G(s)$. The quantity $E(s)$ is often called the *actuating signal* of the feedback system. The reference input signal and the feedback signal must be of opposite signs. This is a desirable condition for suitable operation of the feedback system and is commonly termed *negative* feedback. Whenever a feedback path such as $H(s)$ is wrapped around a direct path such as $G(s)$, the significant transfer function is the one that relates the output $C(s)$ to the input $R(s)$. The ratio of these two quantities can be readily expressed in terms of $H(s)$ and $G(s)$. A glance at Fig. 11-3 makes it apparent that the output quantity can be expressed as

$$E(s)G(s) = C(s) \tag{11-13}$$

But

$$E(s) = R(s) - H(s)C(s) \tag{11-14}$$

so that upon its insertion into Eq. (11-13) there results

$$[R(s) - H(s)C(s)]G(s) = C(s) \tag{11-15}$$

This last equation now involves just the input and output variables. Accordingly, after terms are collected and the ratio of output to input is formulated, the desired *closed-loop transfer function T(s)* results. Thus

†A direct transfer function of a system component is one that appears in the direct path linking the input and output variables.

$$T(s) = \frac{G(s)}{1 + H(s)G(s)} \qquad (11\text{-}16)$$

The plus sign appears in the denominator whenever negative feedback is used. The quantity $H(s)G(s)$ is called the *loop transfer function* since it is obtained by taking the product of the transfer functions encountered in traversing a closed loop.

In dealing with systems that involve several energy-storing elements identified in the form of Eq. (11-3), it is sometimes useful to represent the known transfer function in terms of an equivalent transfer function having a direct transfer function that is pure integration (as denoted by the Laplace operator $1/s$). The advantage of such a block diagram representation is that the respective output quantities of all the pure integrators in the direct transmission path are then in fact the *state variables*† of the total system in terms of which the complete system behavior can be described. In the simple case of the first-order system that leads to Eq. (11-3), it is possible through appropriate manipulation to rewrite the direct transfer function in the form of Eq. (11-16). Keeping in mind that it is desirable to identify $G(s)$ as $1/s$ in the equivalent formulation of Eq. (11-16), it is helpful to divide numerator and denominator of Eq. (11-3) by $s\tau$. This leads to

$$\frac{1}{R_f}\frac{1}{1+s\tau_f} = \frac{1}{R_f}\frac{1/s\tau}{(1/s\tau)+1} = \frac{1}{R_f\tau}\left[\frac{1/s}{1+(1/s)(1/\tau)}\right] \qquad (11\text{-}17)$$

A block diagram representation of the bracketed term of Eq. (11-17) appears in Fig. 11-4(b). Note that it is identical with the representation depicted in Fig. 11-4(a), but the former does identify a pure integration block in the direct transmission path.

The transformed equation associated with the block diagram of Fig. 11-4(b) by inspection can be written as

$$V_f(s)\frac{1}{R_f\tau_f} - \frac{1}{\tau_f}I_f(s) = sI_f(s) \qquad (11\text{-}18)$$

The corresponding time equation is obtained by replacing s with d/dt and by dropping

---

†When dealing with systems containing energy-storing elements, it is convenient to use as the dependent variables those variables by which the system's energy state can be described at any specified time. For example, in an R-L circuit the particular state of the circuit can always be conveniently identified in terms of the current irrespective of the past history of the circuit. This is because a knowledge of the current immediately yields the energy state by the relationship $\frac{1}{2}Li^2$. A similar situation prevails for the capacitor. Here, however, it is convenient to choose the instantaneous value of the voltage across the capacitor v_c as the state variable. Once v_c is known, the state of the network containing C is known through the relationship $\frac{1}{2}Cv_c^2$. The state function depends solely upon the final state of the element.

The principal advantage of working with state variables is that it makes it possible to represent the state of a system containing n independent energy-storing elements in terms of n *state equations*, which are always first-order differential equations. Thus, if a system contains three independent energy-storing elements, the total system behavior can be described by three state equations. In each instance the state equation is a first-order differential equation involving one of the state variables and possibly an external forcing function. In contrast, if the total system behavior were to be represented by a single governing equation, it would be a third-order differential equation. The outstanding advantage of the state approach to system analysis is that it lends itself naturally to easy simulation on the analog and digital computers.

(a)

(b)

Figure 11-4   Alternative block diagram representations of transfer functions: (a) direct representation; (b) equivalent representation, bringing into evidence pure integration in the direct transmission path.

the "of s" notation.[†] Accordingly, we get

$$\frac{dI_f}{dt} = \frac{V_f}{R_f \tau_f} - \frac{1}{\tau_f} I_f \tag{11-19}$$

Equation (11-19) is called the *state equation* for this system since it involves a relationship between the state variable I_f and its first derivative. A little reflection of course discloses that this equation is identical with the original governing differential equation [see Eq. (11-1)].

11-2 DYNAMIC BEHAVIOR OF THE D-C MOTOR WITH CONNECTED MECHANICAL LOAD

As already pointed out in the preceding section, the dynamic behavior of electro-mechanical devices is determined by first finding the appropriate transfer function and then applying the techniques of the Laplace transform theory. In the process suitable block diagram representations are developed and the proper state equations identified in order to enhance a computer study of the machine dynamics for those readers who are so inclined. A study of the dynamics of the d-c motor is made for two cases—one where the armature winding leakage inductance[‡] is negligibly small and the other where it is large enough to be considered.

Figure 11-5 gives the schematic diagram of the d-c motor with its connected load. The combined inertia of the load and the rotor of the motor is denoted by J.[§] The equivalent viscous friction of the motor and the load is denoted by F. The opposing load torque is called T_L. The quantity L_a is the armature leakage inductance which is

---

[†]Although it is customary to use lower-case letters for time functions and capital letters for the Laplace-transformed functions, that practice is not followed in this chapter because of the desire to preserve the identity of the variables with the notation used in Part I of this book. As a substitute this distinction is made by the absence or presence of the "of s" notation.

[‡]This refers to the part of the flux that can be considered produced by the armature winding mmf and not linked with the field winding.

[§]A convenient set of units is assumed throughout this discussion for J, F, and K. For example, J would be expressed in lb-ft-sec^2, F in lb-ft-sec, and K in lb-ft/rad.

Figure 11-5 Schematic diagram of the d-c motor with mechanically connected load. Field current is maintained constant.

initially assumed equal to zero; R_a is the armature winding resistance; and ω denotes the speed of the rotor expressed in radians per second.

Assume we wish to find the manner in which the motor speed responds to changes in the applied armature winding voltage for constant field current. In short, we want to find the transfer function that relates ω to V_t. To do this, it is necessary to establish a governing equation that relates ω to V_t. At the motor shaft the electromagnetic torque of the motor must be equal to the sum of the opposing torques. Thus

$$T = J\frac{d\omega}{dt} + F\omega + T_L \tag{11-20}$$

or, expressed more conveniently,

$$T - T_L = J\frac{d\omega}{dt} + F\omega \tag{1-121}$$

It is helpful to note at this point that Eq. (11-21) is a first-order differential equation; the sole energy-storing element is denoted by the total inertia J. Furthermore, by Eq. (6-4) the electromagnetic torque is expressed as

$$T = K_T\Phi I_a \tag{11-22}$$

where Φ denotes the air-gap flux. Since flux is held fixed in the system of Fig. 11-5, Eq. (11-22) can be more succinctly written as

$$T = K_t I_a \tag{11-23}$$

where K_t is a torque constant expressed in units of torque per ampere. Also, the armature current is related to the terminal voltage by

$$I_a R_a = V_t - E_a \tag{11-24}$$

which follows from Eq. (6-5) for R_s equal to zero. Moreover, the armature induced emf is directly proportional to the speed as shown by Eq. (6-3). Thus

$$E_a = K_E\Phi n = K_\omega\omega \tag{11-25}$$

where K_ω is a speed constant expressed in units of volts per radian per second. Inserting Eq. (11-25) into Eq. (11-24) yields

$$I_a R_a = V_t - K_\omega\omega \tag{11-26}$$

It is interesting to note that the right-hand side of the last equation involves both the input variable V_t and the output variable ω of the transfer function being sought.

Accordingly, Eq. (11-26) lends itself to the partial block diagram representation of Fig. 11-6, expressed as a negative feedback arrangement (see solid portion of the diagram). Note further that, if the quantity $I_a R_a$ is multiplied by $1/R_a$, the armature current I_a results. Subsequent multiplication by K_t then yields the developed electromagnetic torque. The last two statements are represented in Fig. 11-6 by the dashed-line blocks.

Figure 11-6 The block diagram representation of Eq. (11-26) is drawn with solid lines.

In order to complete the block diagram, it is necessary to relate torque to speed. Since this relationship is already available in Eq. (11-21), the desired result is found by Laplace transformation. Thus

$$T(s) - T_L(s) = Js\omega(s) + F\omega(s) \tag{11-27}$$

or

$$\frac{\omega(s)}{T(s) - T_L(s)} = \frac{1}{sJ + F} = \frac{1}{F}\frac{1}{1 + s\tau_m} \tag{11-28}$$

where

$$\tau_m = \text{mechanical time constant} = \frac{J}{F} \tag{11-29}$$

Upon adding the block diagram representation of Eq. (11-28) to the partial block diagram of Fig. 11-6, the complete block diagram shown in Fig. 11-7 is obtained. Finally, by applying the feedback formula of Eq. (11-6) to the configuration of Fig. 11-7, the transfer function relating speed to terminal voltage results. Accordingly,

$$\frac{\omega(s)}{V_t(s)} = \frac{(K_t/R_a F)[1/(1 + s\tau_m)]}{1 + (K_t K_\omega/R_a F)[1/(1 + s\tau_m)]} \tag{11-30}$$

Simplifying and collecting terms yields

$$\boxed{\frac{\omega(s)}{V_t(s)} = \frac{K_t}{R_a F + K_t K_\omega}\frac{1}{1 + s\tau_m'}} \tag{11-31}$$

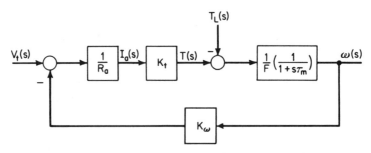

Figure 11-7 Complete block diagram of the system of Fig. 11-5.

where

$$\tau_m' = \frac{JR_a}{R_aF + K_tK_\omega} \tag{11-32}$$

It is possible of course to derive the desired transfer function without first developing the block diagram. A single equation relating the input and output variables can be obtained by substituting the expression for I_a from Eq. (11-26) into Eq. (11-23) and then inserting the latter expression into Eq. (11-21). This leads to

$$\frac{K_t}{R_a}V_t = J\frac{d\omega}{dt} + F\omega + \frac{K_tK_\omega}{R_a}\omega + T_L \tag{11-33}$$

If we now let T_L be zero for convenience and then Laplace transform Eq. (11-33), we get

$$\frac{K_t}{R_a}V_t(s) = Js\omega(s) + F\omega(s) + \frac{K_tK_\omega}{R_a}\omega(s) \tag{11-34}$$

Hence

$$\frac{\omega(s)}{V_t(s)} = \frac{K_t/R_a}{sJ + F + (K_tK_\omega/R_a)} = \frac{K_t}{FR_a + K_tK_\omega}\frac{1}{1 + s\tau_m'} \tag{11-35}$$

A comparison of Eq. (11-35) with Eq. (11-31) shows the two to be identical.

What is the dynamic response of the motor speed to a step change in the applied armature voltage V_t for zero load torque? Assuming the step change in applied voltage is V_t, it follows that $V_t(s) = V_t/s$. From the transfer function of Eq. (11-31), the Laplace transformed solution for the speed becomes

$$\omega(s) = \frac{K_tV_t}{R_aF + K_tK_\omega}\frac{1}{s}\frac{1}{1 + s\tau_m'} = \frac{K_tV_t}{JR_a}\frac{1}{s}\frac{1}{s + 1/\tau_m'} \tag{11-36}$$

or

$$\omega(s) = \frac{K_tV_t}{JR_a}\frac{1}{s}\frac{1}{s + 1/\tau_m'} = \frac{K_0}{s} + \frac{K_1}{s + 1/\tau_m'} \tag{11-37}$$

Evaluation of the coefficients of the partial fraction expansion leads to

$$\omega(s) = \frac{K_tV_t}{FR_a + K_tK_\omega}\left(\frac{1}{s} - \frac{1}{s + 1/\tau_m'}\right) \tag{11-38}$$

The corresponding time solution is found to be

$$\omega(t) = \frac{K_tV_t}{FR_a + K_tK_\omega}(1 - \epsilon^{-t/\tau_m'}) \tag{11-39}$$

Examination of the last expression reveals that the speed rises as depicted in Fig. 11-8 in accordance with a time constant τ_m' to a steady-state speed given by the coefficient on the right side of Eq. (11-39).

How does one obtain the dynamic response of the motor speed to a step change in load torque for a constant applied terminal voltage? The answer is readily obtained once the transfer function between speed and load torque is known. A glance at Fig. 11-7 provides this information. Note that between $T_L(s)$ and $\omega(s)$ there is a direct transmission function of $(1/F)[1/(1 + s\tau_m)]$. The corresponding feedback function in this instance is merely K_tK_ω/R_a. Hence the use of Eq. (11-16) yields as the appropriate

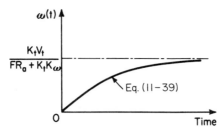

Eq. (11-39)

Figure 11-8 Response of motor speed to a step change in V_t for the system of Fig. 11-5.

transfer function

$$\frac{\omega(s)}{-T_L(s)} = \frac{(1/F)[1/(1 + s\tau_m)]}{1 + (K_t K_\omega/R_a F)[1/(1 + s\tau_m)]} = \frac{R_a}{R_a F + K_t K_\omega}\left(\frac{1}{1 + s\tau_m'}\right) \quad (11\text{-}40)$$

where τ_m' is defined by Eq. (11-32). Then by following a procedure similar to that employed for a change in V_t the complete expression for $\omega(t)$ is obtained. The details are left as an exercise for the reader.

The block diagram of Fig. 11-7 can be redrawn by employing the manipulation illustrated in Eq. (11-17) to bring into evidence pure integration in the direct transmission path of the system block diagram. This leads to the configuration shown in Fig. 11-9. Since the output of a pure integrator is a state variable, it follows that the

Figure 11-9 Modification of the block diagram of Fig. 11-7, to bring into evidence pure integration in the direct transmission path.

motor speed ω is the state variable. The corresponding state equation is found by writing the equation that applies at the summing junction of the integrator. Thus for Fig. 11-9 we get (neglecting T_L for convenience)

$$\frac{T(s)}{F\tau_m} - \frac{1}{\tau_m}\omega(s) = s\omega(s) \quad (11\text{-}41)$$

The associated time-domain equation is

$$\frac{d\omega(t)}{dt} = \frac{T}{F\tau_m} - \frac{1}{\tau_m}\omega(t) \quad (11\text{-}42)$$

DYNAMICS CHAP. 11

Inserting the appropriate expression for T leads to

$$\frac{d\omega(t)}{dt} = \frac{V_t - K_\omega\omega(t)}{R_a} K_t \frac{1}{F\tau_m} - \frac{1}{\tau_m}\omega$$

$$= \frac{K_t}{R_a F\tau_m}V_t - \left(\frac{1}{\tau_m} + \frac{K_\omega K_t}{R_a F\tau_m}\right)\omega(t) \qquad (11\text{-}43)$$

The last expression is the state equation. Note that it involves the state variable $\omega(t)$ and its first derivative. Of course the solution of this equation has already been shown to be Eq. (11-39).

Motor Dynamics with Armature Leakage Inductance Not Negligible

Since the general solution procedure has already been outlined above, attention here is focused only on those parts of the analysis where the presence of L_a bears an influence. Actually this occurs in just one place, Kirchhoff's voltage law for the armature circuit. Thus, instead of the difference between the terminal voltage and the armature induced emf being equal to the $I_a R_a$ drop alone, it is now equal to the armature impedance drop. Expressed mathematically,

$$I_a R_a + L_a \frac{dI_a}{dt} = V_t - E_a \qquad (11\text{-}44)$$

Laplace transformed, this becomes

$$I_a(s)R_a(1 + s\tau_a) = V_t(s) - K_\omega\omega(s) \qquad (11\text{-}45)$$

where

$$\tau_a \equiv \frac{L_a}{R_a} \qquad (11\text{-}46)$$

or

$$\frac{I_a(s)}{V_t(s) - K_\omega\omega(s)} = \frac{1}{R_a(1 + s\tau_a)} \qquad (11\text{-}47)$$

Equation (11-47) discloses that, to obtain the armature current from the difference between the terminal voltage and the counter armature emf, this difference must be operated upon, not by $1/R_a$, but rather by $1/[R_a(1 + s\tau_a)]$. Hence the block diagram of Fig. 11-7 becomes that illustrated in Fig. 11-10 when $L_a \neq 0$.

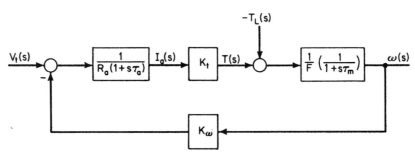

Figure 11-10 Block diagram of the system of Fig. 11-5 when $L_a \neq 0$.

Application of the feedback formula of Eq. (11-16) to Fig. 11-10 discloses that the motor speed is now related to a change in terminal voltage by

$$\frac{\omega(s)}{V_t(s)} = \frac{\dfrac{K_t}{R_a F(1 + s\tau_a)(1 + s\tau_m)}}{1 + \dfrac{K_t K_\omega}{R_a F(1 + s\tau_a)(1 + s\tau_m)}} \tag{11-48}$$

which simplifies to

$$\frac{\omega(s)}{V_t(s)} = \frac{K_t}{\tau_a \tau_m R_a F} \frac{1}{s^2 + \dfrac{\tau_a + \tau_m}{\tau_a \tau_m}s + \dfrac{1}{\tau_a \tau_m} + \dfrac{K_t K_\omega}{\tau_a \tau_m R_a F}} \tag{11-49}$$

A glance at Eq. (11-49) indicates that now the characteristic equation is second order. This is not surprising in view of the presence of an additional energy-storing element in the form of the armature leakage inductance. The nature of the dynamic response to a step change in applied voltage depends upon the character of the roots of the denominator quadratic in Eq. (11-49). If they are complex conjugate roots with negative real parts, the response is a damped oscillation. If the roots are both negative real, the solution consists of a fixed term plus two exponentially decaying terms.

Appearing in Fig. 11-11 is the block diagram of Fig. 11-10 redrawn to bring into evidence the two pure integration terms associated with the second-order system of Fig. 11-5 when $L_a \neq 0$. The two state variables are the output quantities of the two integrators, armature current I_a and motor speed ω. The two state equations, which are linear first-order differential equations, are obtained by writing the expression for the input signal to each integrator, multiplying by $1/s$, and then equating to the specific state variable. Thus at the first integrator in Fig. 11-11 the transformed version of the state equation is

$$sI_a(s) = \frac{1}{R_a \tau_a}[V_t(s) - K_\omega \omega(s)] - \frac{1}{\tau_a}I_a(s) \tag{11-50}$$

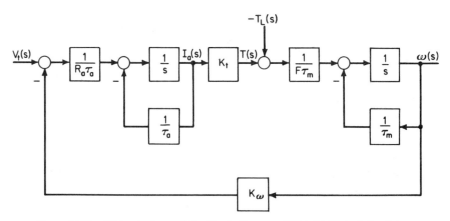

Figure 11-11 Alternate form of the block diagram of Fig. 11-10, to bring into evidence pure integrators in the direct transmission path. The output quantities of the integrators are the state variables of the system.

At the second integrator the result is

$$s\omega(s) = \frac{1}{F\tau_m}[K_tI_a(s) - T_L(s)] - \frac{1}{\tau_m}\omega(s) \tag{11-51}$$

The corresponding time-domain form of the state equations is clearly

$$\frac{dI_a}{dt} = -\frac{1}{\tau_a}I_a - \frac{K_\omega}{R_aL_a}\omega + \frac{1}{R_a\tau_a}V_t \tag{11-52}$$

$$\frac{d\omega}{dt} = \frac{K_t}{F\tau_m}I_a - \frac{1}{\tau_m}\omega - \frac{1}{F\tau_m}T_L \tag{11-53}$$

In both equations, I_a and ω are functions of time. In matrix notation the time-domain formulation of the state equations may be written as follows:

$$\begin{bmatrix} \dfrac{dI_a}{dt} \\[2mm] \dfrac{d\omega}{dt} \end{bmatrix} = \begin{bmatrix} -\dfrac{1}{\tau_a} & -\dfrac{K_\omega}{R_aL_a} \\[2mm] \dfrac{K_t}{F\tau_m} & -\dfrac{1}{\tau_m} \end{bmatrix} \begin{bmatrix} I_a \\[2mm] \omega \end{bmatrix} + \begin{bmatrix} \dfrac{1}{R_a\tau_a} & 0 \\[2mm] 0 & -\dfrac{1}{F\tau_m} \end{bmatrix} \begin{bmatrix} V_t \\[2mm] T_L \end{bmatrix} \tag{11-54}$$

Equation (11-54) is called the *normal* form of the state equations describing the complete behavior of the system depicted in Fig. 11-5.

11-3 D-C GENERATOR DYNAMICS

How does the armature induced emf of a d-c generator respond to a change in field current? How does the armature winding current of the d-c generator respond to changes in field current? These are questions that are treated in this section. As with motor dynamics, the answers lie in establishing the appropriate transfer function relating the variables of interest. Two cases of generator dynamic behavior are treated: one where no external load is applied to the generator (i.e., switch S in Fig. 11-12 is open), and the other where external load is applied (S closed).

Figure 11-12 Direct-current generator with connected load.

No External Generator Load

Consider the generator of Fig. 11-12 driven at constant speed by the prime mover. Let it be desirable to find the transfer function relating the armature induced voltage to the field winding voltage. The differential equation for the field winding circuit is given by

$$V_f = I_f R_f + L_f \frac{dI_f}{dt} \tag{11-55}$$

In general I_f is a function of time whenever V_f or R_f is changing. The Laplace transformation of Eq. (11-55) for zero initial conditions is

$$V_f(s) = I_f(s)R_f + L_f s I_f(s) \tag{11-56}$$

where $I_f(s)$ denotes the Laplace transform of the time function I_f and $V_f(s)$ is the Laplace transform of V_f. Upon formulating the ratio of $I_f(s)$ to $V_f(s)$, the transfer function between these quantities results. Thus

$$\frac{I_f(s)}{V_f(s)} = \frac{1}{R_f(1 + s\tau_f)} \tag{11-57}$$

where

$$\tau_f \equiv \frac{L_f}{R_f} = \text{field winding time constant} \tag{11-58}$$

The relationship between the field current and the armature induced emf is described by Eq. (6-3), which is repeated here for convenience. Thus

$$E_a = K_E \Phi n = K_f I_f \tag{11-59}$$

The right side of Eq. (11-59) is a valid equivalence provided that no saturation occurs. For consistency the units of K_f are volts per field ampere. Expressed in Laplace notation, Eq. (11-59) becomes merely

$$E_a(s) = K_f I_f(s) \tag{11-60}$$

so that the transfer "function" relating field current to the corresponding induced emf is just a constant. Accordingly,

$$\frac{E_a(s)}{I_f(s)} = K_f \tag{11-61}$$

The complete transfer function is then obtained by combining Eqs. (11-61) and (11-57). Thus

$$\boxed{\frac{E_a(s)}{V_f(s)} = \left[\frac{I_f(s)}{V_f(s)}\right]\left[\frac{E_a(s)}{I_f(s)}\right] = \frac{K_f}{R_f(1 + s\tau_f)}} \tag{11-62}$$

The block diagram representation of the last equation appears in Fig. 11-13.

Figure 11-13 Block diagram of the system of Fig. 11-12, with load switch open.

Depicted in Fig. 11-14 is the form of the block diagram that brings into evidence the pure integrator in the direct transmission path. The output of this integrator is the state variable, namely, field current. The corresponding state equation in transformed notation is found by taking the input to this integrator and setting it equal to $sI_f(s)$. Thus

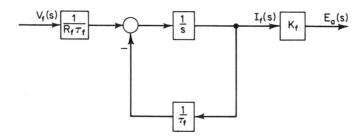

Figure 11-14 Block diagram of Fig. 11-13 modified to show the pure integrator. I_f is the state variable.

$$sI_f(s) = \frac{1}{R_f\tau_f}V_f(s) - \frac{1}{\tau_f}I_f(s) \qquad (11\text{-}63)$$

The corresponding time-domain form is then clearly

$$\frac{dI_f}{dt} = \frac{1}{R_f\tau_f}V_f - \frac{1}{\tau_f}I_f \qquad (11\text{-}64)$$

Because state equations are only associated with energy-storing elements, E_a does not appear in Eq. (11-64). However, the transition from the state variable I_f to the desired output variable E_a is readily achieved through multiplication by a constant, K_f.

Generator Dynamics Including the Effect of the Load

This situation is represented in Fig. 11-12 by closing switch S. The generator load is assumed to contain inductance as well as resistance. Also, R_a and L_a are respectively the armature winding resistance and leakage inductance.

The transfer function that relates $E_a(s)$ to $V_f(s)$ is identical with Eq. (11-62). Always keep in mind that transfer functions are dependent upon the system parameters —in this instance K_f, R_f, and L_f—and these have not been altered. If it is now desirable to describe how the armature current is influenced by changes in field voltage, then an additional expression is needed that relates E_a to I_a. Clearly this is readily available through the application of Kirchhoff's voltage law to the armature circuit. In differential equation form we have

$$E_a = I_aR_a + L_a\frac{dI_a}{dt} + I_aR + L\frac{dI_a}{dt} \qquad (11\text{-}65)$$

where R and L are the resistance and inductance of the load respectively. Laplace transforming and collecting terms yield

$$E_a(s) = I_a(s)(R_a + R)(1 + s\tau_A) \qquad (11\text{-}66)$$

where

$$\tau_A = \text{armature circuit time constant} = \frac{L_a + L}{R_a + R} \qquad (11\text{-}67)$$

The transfer function then becomes

$$\frac{I_a(s)}{E_a(s)} = \frac{1}{(R_a + R)(1 + s\tau_A)} \qquad (11\text{-}68)$$

It is worthwhile to note here that the existence of s in the last equation is occasioned by the presence of a second energy-storing element in the system of Fig. 11-12 with the load connected.

The total transfer function relating the armature current to the field voltage is determined by formulating the product of Eqs. (11-62) and (11-68). Accordingly,

$$\frac{I_a(s)}{V_f(s)} = \frac{K_f}{R_f(R_a + R)} \frac{1}{1 + s\tau_f} \frac{1}{1 + s\tau_A} \qquad (11\text{-}69)$$

Again note that the denominator of the complete transfer function is now second order in s, which is consistent with the fact that there are now two independent energy-storing elements in the system. The block diagram of the system is depicted in Fig. 11-15.

The modified form of the block diagram of Fig. 11-15 is illustrated in Fig. 11-16. Two pure integrators appear because the system is second order. Hence there are two state variables, which a glance at Fig. 11-16 discloses to be the field current and the

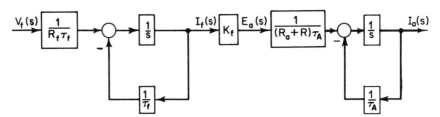

Figure 11-15 Block diagram of the system of Fig. 11-12 when the load switch is closed.

Figure 11-16 Modified version of the block diagram of Fig. 11-15 to bring into evidence the state variables, I_f and I_a.

armature current. The two state equations are found by employing the procedure previously described. In the s notation these are

$$sI_f(s) = -\frac{1}{\tau_f} I_f(s) + \frac{1}{R_f \tau_f} V_f(s) \qquad (11\text{-}70)$$

$$sI_a(s) = -\frac{1}{\tau_A} I_a(s) + \frac{K_f}{(R_a + R)\tau_A} I_f(s) \qquad (11\text{-}71)$$

The corresponding time-domain versions are

$$\frac{dI_f}{dt} = -\frac{1}{\tau_f} I_f + \frac{1}{R_f \tau_f} V_f \qquad (11\text{-}72)$$

$$\frac{dI_a}{dt} = \frac{K_f}{(R_a + R)\tau_A} I_f - \frac{1}{\tau_A} I_a \qquad (11\text{-}73)$$

In matrix notation these state equations are written as

$$\begin{bmatrix} \dfrac{dI_f}{dt} \\[2ex] \dfrac{dI_a}{dt} \end{bmatrix} = \begin{bmatrix} -\dfrac{1}{\tau_f} & 0 \\[2ex] +\dfrac{K_f}{(R_a + R)\tau_A} & -\dfrac{1}{\tau_A} \end{bmatrix} \begin{bmatrix} I_f \\[2ex] I_a \end{bmatrix} + \begin{bmatrix} \dfrac{1}{R_f\tau_f} \\[2ex] 0 \end{bmatrix} V_f \tag{11-74}$$

Example 11-1

The d-c generator depicted in Fig. 11-12 has a field winding resistance of 40 Ω and a field inductance of 8 H. The generated emf per field ampere is 100, and the magnetization curve is linear. Moreover, the armature winding resistance and leakage inductance are respectively 0.1 Ω and 0.2 H. The load resistance is 5 Ω and the associated load inductance is 2.35 H.

Determine the time solution for the armature current when a field voltage of 102 V is applied to the field winding. Assume that the prime mover is running at rated speed and that the load switch is closed.

Solution The transfer function that relates the armature current to the field voltage is given by Eq. (11-69). Thus

$$\frac{I_a(s)}{V_f(s)} = \frac{K_f}{R_f(R_a + R)} \frac{1}{1 + s\tau_f} \frac{1}{1 + s\tau_A}$$

$$= \frac{100}{40(0.1 + 5)} \frac{1}{1 + (s/5)} \frac{1}{1 + (s/2)} \tag{11-75}$$

where

$$\tau_f = \frac{L_f}{R_f} = \frac{8}{40} = \frac{1}{5} \text{ s} \tag{11-76}$$

and

$$\tau_A = \frac{L_a + L}{R_a + R} = \frac{0.2 + 2.35}{0.1 + 5} = \frac{1}{2} \text{ s} \tag{11-77}$$

The Laplace transform of the field voltage for a step change of 102 V is described by

$$V_f(s) = \frac{V_f}{s} = \frac{102}{s} \tag{11-78}$$

Hence the transformed form of the solution for the armature current can be written as

$$I_a(s) = \frac{102}{s}\left(\frac{1000}{204} \frac{1}{s + 5} \frac{1}{s + 2}\right) = \frac{500}{s(s + 5)(s + 2)} \tag{11-79}$$

Evaluation of the last equation is facilitated by a partial fraction expansion. Accordingly,

$$I_a(s) = \frac{500}{s(s + 5)(s + 2)} = \frac{K_0}{s} + \frac{K_1}{s + 5} + \frac{K_2}{s + 2} \tag{11-80}$$

These coefficients are found to be

$$K_0 = \left[\frac{500}{(s + 5)(s + 2)}\right]_{s=0} = 50 \tag{11-81}$$

$$K_1 = \left[\frac{500}{s(s + 2)}\right]_{s=-5} = +33.3 \tag{11-82}$$

$$K_2 = \left[\frac{500}{s(s + 5)}\right]_{s=-2} = -83.3 \tag{11-83}$$

Insertion into Eq. (11-80) allows the transformed solution to be written as

$$I_a(s) = \frac{50}{s} + \frac{33.3}{s+5} - \frac{83.3}{s+2} \qquad (11\text{-}84)$$

The corresponding time solution as obtained from a suitable table of Laplace-transform pairs becomes

$$I_a(t) = 50 + 33.3\epsilon^{-5t} - 83.3\epsilon^{-2t} \qquad (11\text{-}85)$$

which is the desired solution. This last expression completely describes the process of armature current buildup when a field voltage of 102 V is suddenly applied to the field circuit. Note that the steady-state (or final-state) value of the armature current is 50 A—a result that can be verified through use of the procedure described in Chapter 6 for finding steady-state performance.

11-4 DYNAMICS OF A GENERATOR-MOTOR SYSTEM

A method of speed adjustment that is employed in industry whenever control over a wide speed range is required is shown in Fig. 11-17. Speed adjustment above base speed† is obtained by field current (I_{fm}) variation with rated voltage applied to the armature terminals. Below base speed the speed is adjusted by changing the armature voltage below its rated value. A convenient way to control the armature voltage is to control the field current of a supplying generator as depicted in Fig. 11-17. Large amounts of power at the load can be controlled by easy adjustment of the generator field current. This method of speed control is referred to as the *Ward-Leonard system*.

Figure 11-17 System diagram for studying dynamic behavior of the Ward-Leonard generator and motor system of speed control.

A little thought should make it apparent that adjustments of the field current in the system of Fig. 11-17 do not translate immediately into appropriate speed changes. The change in speed must occur *smoothly* from one commanded level to another. The change cannot occur abruptly because of the presence of energy-storing elements. Examination of this system reveals the presence of three energy-storing elements, each of which influences system dynamics. These are generator field inductance, generator

†Base speed corresponds to operation with rated armature voltage and rated field current at rated motor output.

and motor armature winding leakage inductances, and the inertia of the load and the motor.

A description of the system's dynamic behavior lies in identifying the transfer function that properly relates the motor speed ω to the generator field voltage V_f. As a first step in deriving this transfer function, we write the differential equation for the generator field winding circuit. Thus

$$V_f = I_f R_f + L_f \frac{dI_f}{dt} \tag{11-86}$$

Laplace transforming yields

$$V_f(s) = I_f(s)R_f(1 + s\tau_f) \tag{11-87}$$

where $\tau_f = L_f/R_f$. Formulating the ratio of field current to field voltage leads to

$$\frac{I_f(s)}{V_f(s)} = \frac{1}{R_f(1 + s\tau_f)} \tag{11-88}$$

The generated armature voltage is related to the field current by

$$\frac{E_g(s)}{I_f(s)} = K_f \quad \text{V/A} \tag{11-89}$$

In turn the generated emf is related to armature current by Kirchhoff's voltage law. Specifically,

$$E_g = I_a R_g + L_g \frac{dI_a}{dt} + I_a R_m + L_m \frac{dI_a}{dt} + E_m \tag{11-90}$$

where R_g and R_m are the armature winding resistances of the generator and motor, L_g and L_m denote the leakage inductances of the armature windings, and E_m denotes the counter emf of the motor. More explicitly E_m can be written as

$$E_m = K_\omega \omega \tag{11-91}$$

where K_ω is expressed in volts per radian per second and the motor speed ω is expressed in radians per second. Inserting the last expression into Eq. (11-90) and Laplace transforming gives

$$I_a(s) = \frac{E_g(s) - K_\omega \omega(s)}{R(1 + s\tau_A)} \tag{11-92}$$

where

$$R = R_g + R_m \tag{11-93}$$

$$\tau_A = \frac{L_g + L_m}{R_g + R_m} \tag{11-94}$$

It is helpful at this point to rewrite Eq. (11-92) as

$$\frac{I_a(s)}{E_g(s) - K_\omega \omega(s)} = \frac{1}{R(1 + s\tau_A)} \tag{11-95}$$

This is in the form of a transfer function that lends itself to block diagram representation. In particular, note that the input to the transfer function block consists then of the difference between the generator source voltage $E_g(s)$ and a quantity that is a function of the output variable $K_\omega\omega(s)$. Hence the latter quantity can be considered as coming from a feedback path originating from the output.

Continuing the development in a logical stepwise fashion, we next note that the armature current combines with the motor field flux to produce an electromagnetic torque that can be expressed as

$$T = K_t I_a \qquad (11\text{-}96)$$

where K_t bears units of torque per armature ampere. The transfer function notation is simply

$$\frac{T(s)}{I_a(s)} = K_t \qquad (11\text{-}96\text{a})$$

Finally, to relate the developed torque to the motor speed, we note that the former must be equal to the sum of all opposing torques. If we now let J denote the total inertia of the load and the motor and F denote the total viscous friction, we can write

$$T = J\frac{d\omega}{dt} + F\omega \qquad (11\text{-}97)$$

The load torque is omitted in this analysis because it is assumed to be constant. Our interest here, remember, is to find the manner in which the speed changes in response to changes in the generator field current (or field voltage). Upon Laplace transforming Eq. (11-97) and formulating the ratio of speed to torque, we have

$$\frac{\omega(s)}{T(s)} = \frac{1}{F(1 + s\tau_m)} \qquad (11\text{-}98)$$

where

$$\tau_m = \text{mechanical time constant} = \frac{J}{F} \qquad (11\text{-}99)$$

A successive block diagram representation of Eqs. (11-88), (11-89), (11-95), (11-96), and (11-98) taken in order leads to the complete system block diagram depicted in Fig. 11-18(a). By applying the feedback formula of Eq. (11-16) to the feedback loop the complete expression of speed to generated voltage is obtained. Thus

$$\frac{\omega(s)}{E_g(s)} = \frac{\dfrac{K_t}{RF(1 + s\tau_A)(1 + s\tau_m)}}{1 + \dfrac{K_t K_\omega}{RF(1 + s\tau_A)(1 + s\tau_m)}} = \frac{K_t}{RF(1 + s\tau_A)(1 + s\tau_m) + K_\omega K_t}$$

$$= \frac{K_t}{RF + K_\omega K_t}\left[\frac{1}{s^2\dfrac{\tau_A\tau_m RF}{RF + K_\omega K_t} + s\dfrac{(\tau_A + \tau_m)RF}{RF + K_\omega K_t} + 1}\right] \qquad (11\text{-}100)$$

This result can also be found analytically without the use of the block diagram by inserting Eq. (11-92) into Eq. (11-96) and then substituting for $T(s)$ in Eq. (11-98). Accordingly,

$$\frac{\omega(s)}{\dfrac{E_g(s) - K_\omega\omega(s)}{R(1 + s\tau_A)}K_t} = \frac{1}{F(1 + s\tau_m)} \qquad (11\text{-}101)$$

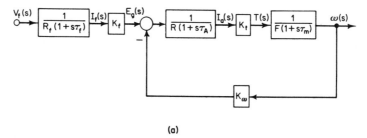

(a)

Figure 11-18(a) Block diagram of the generator motor system of Fig. 11-17.

or

$$RF(1 + s\tau_A)(1 + s\tau_m)\omega(s) = K_t E_g(s) - K_\omega K_t \omega(s) \qquad (11\text{-}102)$$

from which

$$\frac{\omega(s)}{E_g(s)} = \frac{K_t}{RF(1 + s\tau_A)(1 + s\tau_m) + K_\omega K_t} \qquad (11\text{-}103)$$

Comparison with the intermediate step of Eq. (11-100) discloses the two expressions to be identical.

The complete transfer function is determined of course by multiplying Eq. (11-100) by Eqs. (11-88) and (11-89). Hence

$$\frac{\omega(s)}{V_f(s)} = \frac{K_f K_t}{R_f(RF + K_\omega K_t)} \frac{1}{1 + s\tau_f} \left[\frac{1}{s^2 \dfrac{\tau_A \tau_m RF}{RF + K_\omega K_t} + s \dfrac{(\tau_A + \tau_m)RF}{RF + K_\omega K_t} + 1} \right]$$

(11-104)

The characteristic equation of the governing differential equation of this system is obtained by setting the denominator of Eq. (11-104) equal to zero. This yields a third-order equation which indicates the presence of three time lags in the system, a result which is entirely consistent with the presence of three energy-storing elements.

Appearing in Fig. 11-18(b) is the modified version of the block diagram of Fig. 11-18(a) in order to show explicitly the pure integrators, the outputs of which identify

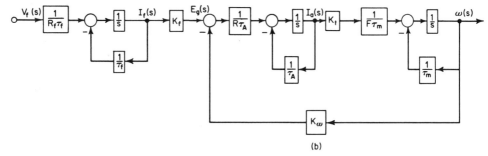

(b)

Figure 11-18(b) Alternate form of the block diagram of Fig. 11-17, to show explicitly the pure integrators of the system. There are three; hence the number of state variables is also three.

the state variables. Accordingly, the state variables of this system are field current, armature current, and motor speed. The state equations are found as described in the preceding sections. This leads to the following transformed forms:

$$sI_f(s) = -\frac{1}{\tau_f}I_f(s) + \frac{1}{R_f\tau_f}V_f(s) \tag{11-105}$$

$$sI_a(s) = \frac{K_f}{R\tau_A}I_f(s) - \frac{1}{\tau_A}I_a(s) - \frac{K_\omega}{R\tau_A}\omega(s) \tag{11-106}$$

$$s\omega(s) = \frac{K_t}{F\tau_m}I_a(s) - \frac{1}{\tau_m}\omega(s) \tag{11-107}$$

Expressed in the time domain the equations are

$$\frac{dI_f}{dt} = -\frac{1}{\tau_f}I_f + \frac{1}{R_f\tau_f}V_f \tag{11-108}$$

$$\frac{dI_a}{dt} = \frac{K_f}{R\tau_A}I_f - \frac{1}{\tau_A}I_a - \frac{K_\omega}{R\tau_A}\omega \tag{11-109}$$

$$\frac{d\omega}{dt} = \frac{K_t}{F\tau_m}I_a - \frac{1}{\tau_m}\omega \tag{11-110}$$

Finally, the matrix formulation becomes

$$\begin{bmatrix} \dfrac{dI_f}{dt} \\[2mm] \dfrac{dI_a}{dt} \\[2mm] \dfrac{d\omega}{dt} \end{bmatrix} = \begin{bmatrix} -\dfrac{1}{\tau_f} & 0 & 0 \\[2mm] \dfrac{K_f}{R\tau_A} & -\dfrac{1}{\tau_A} & -\dfrac{1}{\tau_A} \\[2mm] 0 & \dfrac{K_t}{F\tau_m} & -\dfrac{1}{\tau_m} \end{bmatrix} \begin{bmatrix} I_f \\[2mm] I_a \\[2mm] \omega \end{bmatrix} + \begin{bmatrix} \dfrac{1}{R_f\tau_f} \\[2mm] 0 \\[2mm] 0 \end{bmatrix} V_f \tag{11-111}$$

There are three state equations associated with the three independent energy-storing elements.

11-5 D-C MOTOR TRANSFER FUNCTION

Two forms of the d-c motor transfer function are of interest. One is the armature-controlled mode with fixed field current; the other is the field-controlled mode with constant armature current flowing.

Armature-Controlled D-C Motor

Figure 11-19 depicts the armature-controlled d-c motor. The motor speed is made to respond to variations in the applied motor armature voltage V_t. For simplicity, armature winding leakage inductance is assumed to be negligible. Our concern here is to develop a transfer function that relates motor speed to applied armature voltage. Let J denote the total inertia of the load and the rotor of the motor. Let F denote the total viscous friction of the motor and load. A good starting point in the development is to write the torque equation as it applies at the motor shaft. Accordingly,

I_{fm}

Figure 11-19 Armature controlled d-c motor, field current maintained constant.

$$T = K_t I_a$$

$$= J\frac{d\omega}{dt} + F\omega \tag{11-112}$$

where T is the motor electromagnetic torque and ω is the motor speed. The expression for the armature current is given by

$$I_a = \frac{V_t - E_a}{R_a}$$

$$= \frac{V_t - K_\omega \omega}{R_a} \tag{11-113}$$

where E_a is the motor counter-emf, which is also expressed as a speed voltage constant K_ω in volts per radian per second. In the interest of leading to the appropriate block diagram representation, it is worthwhile to write Eq. (11-113) in Laplace notation. Keeping in mind that both I_a and ω are the variables, we have

$$I_a(s) = [V_t(s) - K_\omega \omega(s)]\frac{1}{R_a} \tag{11-114}$$

The right-hand side has been written as indicated in order to emphasize that the reference input quantity $V_t(s)$ and a quantity derived from the output variable, $K_\omega \omega(s)$, are being compared and that the difference is then multiplied by $1/R_a$ to yield the armature current variable. Hence the symbol for a summing point followed by a block marked $1/R_a$ can be made to represent Eq. (11-114). It is interesting to note that the counter-emf of a motor can always be treated as a negative feedback quantity.

Once $I_a(s)$ is available, the transfer function relating it to the electromagnetic torque is

$$\frac{T(s)}{I_a(s)} = K_t \tag{11-115}$$

Finally, the transfer function that describes the variation of the motor speed with torque is obtained from Eq. (11-112) after it is Laplace transformed. Thus

$$T(s) = Js\omega(s) + F\omega(s) \tag{11-116}$$

or

$$\frac{\omega(s)}{T(s)} = \frac{1}{F(1 + s\tau_m)} \tag{11-117}$$

where

$$\tau_m = \text{mechanical time constant} = \frac{J}{F} \qquad (11\text{-}118)$$

The successive block-diagram notation of Eqs. (11-114), (11-115), and (11-117) leads to the complete block diagram depicted in Fig. 11-20.

Figure 11-20 Block diagram representation of Fig. 11-19. The output variable is motor velocity.

The desired transfer function between the motor speed and the motor applied voltage is found by applying Eq.(11-16) to the block diagram of Fig.11-20. Accordingly,

$$\frac{\omega(s)}{V_t(s)} = \frac{(K_t/R_aF)[1/(1 + s\tau_m)]}{1 + (K_\omega K_t/R_aF)[1/(1 + s\tau_m)]}$$

or

$$\frac{\omega(s)}{V_t(s)} = \frac{K_t}{K_t K_\omega + R_aF} \frac{1}{1 + s\tau_m'} \qquad (11\text{-}119)$$

where

$$\tau_m' = \tau_m \frac{R_aF}{R_aF + K_tK_\omega} = \frac{JR_a}{R_aF + K_tK_\omega} \qquad (11\text{-}120)$$

Equation (11-120) discloses that the feedback path that appears around the direct transmission path of Fig. 11-20 is responsible for reducing the motor time constant over the nominal value of J/F. The practical aspect of this observation is that the motor responds dynamically more quickly than one would expect by looking at τ_m alone.

If it is desirable to derive the transfer function of Eq. (11-119) without resorting to the block diagram representation followed by the application of the feedback formula, this may be readily accomplished by inserting Eq. (11-113) into Eq. (11-112) and then rearranging terms. Thus

$$\frac{V_t - K_\omega \omega}{R_a} K_t = J \frac{d\omega}{dt} + F\omega \qquad (11\text{-}121)$$

or

$$V_t = \frac{R_aF}{K_t}\left[\frac{J}{F}\frac{d\omega}{dt} + \frac{R_aF + K_\omega K_t}{R_aF}\omega\right]$$

$$= \frac{R_aF + K_\omega K_t}{K_t}\left[\tau_m \frac{R_aF}{R_aF + K_\omega K_t}\frac{d\omega}{dt} + \omega\right] \qquad (11\text{-}122)$$

Laplace transforming and formulating the ratio of output to input yields a result that is identical with Eq. (11-119).

Sometimes the d-c motor of Fig. 11-19 is used as the output actuator in feedback control systems where the controlled variable may be position rather than velocity. Since velocity is the derivative of the position variable c, the transfer function relating the two is obtained as follows:

$$\frac{dc}{dt} = \omega \tag{11-123}$$

Laplace transforming for zero initial conditions gives

$$c(s) = \frac{1}{s}\omega(s) \tag{11-124}$$

The complete block diagram now becomes that shown in Fig. 11-21. Moreover, the complete transfer function between the motor displacement variable and the applied armature voltage becomes

$$\boxed{\frac{c(s)}{V_t(s)} = \frac{K_t}{K_\omega K_t + R_a F} \frac{1}{s(1 + s\tau'_m)}} \tag{11-125}$$

where τ'_m is specified by Eq. (11-120).

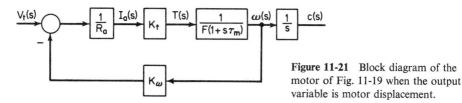

Figure 11-21 Block diagram of the motor of Fig. 11-19 when the output variable is motor displacement.

Appearing in Fig. 11-22 is the form of the block diagram of Fig. 11-21 that allows convenient identification of the state variables, which in this case are clearly the motor velocity and the motor displacement. The first state equation in transformed form is found by taking the input to the first integrator and setting it equal to $s\omega(s)$. Hence

$$s\omega(s) = \frac{K_t V_t(s)}{R_a F \tau_m} - \frac{K_\omega K_t}{R_a F \tau_m}\omega(s) - \frac{1}{\tau_m}\omega(s)$$

$$= \frac{K_t}{R_a F \tau_m}V_t(s) - \left(\frac{K_\omega K_t}{R_a F \tau_m} + \frac{1}{\tau_m}\right)\omega(s) \tag{11-126}$$

The second state equation is given by Eq. (11-124). The corresponding time-domain expressions are

Figure 11-22 Alternate form of Fig. 12-21, drawn to facilitate writing the appropriate state equations.

$$\frac{d\omega}{dt} = \frac{K_t}{R_a F \tau_m} V_t - \left(\frac{K_\omega K_t}{R_a F \tau_m} + \frac{1}{\tau_m}\right)\omega \qquad (11\text{-}127)$$

$$\frac{dc}{dt} = \omega \qquad (11\text{-}128)$$

The matrix notation in normal form is then

$$\begin{bmatrix} \dfrac{d\omega}{dt} \\[2.5ex] \dfrac{dc}{dt} \end{bmatrix} = \begin{bmatrix} -\left(\dfrac{K_\omega K_t}{R_a F \tau_m} + \dfrac{1}{\tau_m}\right) & 0 \\[2.5ex] 1 & 0 \end{bmatrix} \begin{bmatrix} \omega \\[2.5ex] c \end{bmatrix} + \begin{bmatrix} \dfrac{K_t}{R_a F \tau_m} \\[2.5ex] 0 \end{bmatrix} V_t \qquad (11\text{-}129)$$

Field-Controlled D-C Motor

The circuit configuration for this mode of operation is shown in Fig. 11-23. The armature is supplied with a constant current from a suitable current source. In light of this fact the variation of electromagnetic torque is dependent upon the variation

Figure 11-23   Field-controlled d-c motor.

in field current. This is apparent upon recalling that $T = K_T \Phi I_a$ and that the air-gap flux Φ is directly related to the field current for the assumed condition of no saturation. Therefore, the torque in this instance may be written simply as

$$T = K_{tf} I_f \qquad (11\text{-}130)$$

where K_{tf} is a motor constant expressed in units of torque per field ampere. The torque equation at the motor shaft then becomes

$$K_{tf} I_f = J\frac{d\omega}{dt} + F\omega \qquad (11\text{-}131)$$

where J and F have the same meanings as above. Laplace transforming and formulating the ratio of output to input yields the transfer function that relates motor velocity to field current. Thus

$$\frac{\omega(s)}{I_f(s)} = \frac{K_{tf}}{F(1 + s\tau_m)} \qquad (11\text{-}132)$$

In turn the field current is related to the applied field voltage by the usual transfer function. Accordingly,

$$\frac{I_f(s)}{V_f(s)} = \frac{1}{R_f(1 + s\tau_f)} \qquad (11\text{-}133)$$

where $\tau_f = L_f R_f$. By combining the last two expressions, the transfer function between speed and field voltage results:

$$\frac{\omega(s)}{V_f(s)} = \frac{I_f(s)}{V_f(s)} \frac{\omega(s)}{I_f(s)} = \frac{K_{tf}}{R_f F} \frac{1}{1 + s\tau_f} \frac{1}{1 + s\tau_m} \qquad (11\text{-}134)$$

Finally, to obtain the relationship between motor displacement and applied field voltage, it is merely necessary to multiply the last equation by Eq. (11-124) expressed in the form $[c(s)/\omega(s)] = 1/s$. This yields

$$\frac{c(s)}{V_f(s)} = \frac{K_{tf}}{R_f F} \frac{1}{s(1 + s\tau_f)(1 + s\tau_m)} \qquad (11\text{-}135)$$

A glance at the denominator of Eq. (11-135) discloses that now the characteristic equation is third order. Figure 11-24 depicts the block diagram of Eq. (11-135).

Figure 11-24   Block diagram of the system of Fig. 11-23.

Consistent with the third-order nature of the characteristic equation, there are three state equations which describe the complete dynamic behavior of this system. These equations are more easily identified by referring to the alternate block diagram illustrated in Fig. 11-25. Clearly, the three state variables are field current, motor

Figure 11-25   Alternate form of Fig. 11-24.

speed, and motor displacement, which are respectively the output quantities of each of the integrators in Fig. 11-25. The transformed versions of the three state equations as associated with the three integrators are then

$$sI_f(s) = \frac{1}{R_f \tau_f} V_f(s) - \frac{1}{\tau_f} I_f(s) \qquad (11\text{-}136)$$

$$s\omega(s) = \frac{K_{tf}}{F\tau_m} I_f(s) - \frac{1}{\tau_m} \omega(s) \qquad (11\text{-}137)$$

$$sc(s) = \omega(s) \qquad (11\text{-}138)$$

The corresponding first-order differential equations then become

$$\frac{dI_f}{dt} = \frac{1}{R_f \tau_f} V_f - \frac{1}{\tau_f} I_f \qquad (11\text{-}139)$$

$$\frac{d\omega}{dt} = \frac{K_{tf}}{F\tau_m} I_f - \frac{1}{\tau_m} \omega \qquad (11\text{-}140)$$

$$\frac{dc}{dt} = \omega \qquad (11\text{-}141)$$

Finally the matrix formulation is as follows:

$$\begin{bmatrix} \dfrac{dI_f}{dt} \\[2ex] \dfrac{d\omega}{dt} \\[2ex] \dfrac{dc}{dt} \end{bmatrix} = \begin{bmatrix} -\dfrac{1}{\tau_f} & 0 & 0 \\[2ex] \dfrac{K_{tf}}{F\tau_m} & -\dfrac{1}{\tau_m} & 0 \\[2ex] 0 & 1 & 0 \end{bmatrix} \begin{bmatrix} I_f \\[2ex] \omega \\[2ex] c \end{bmatrix} + \begin{bmatrix} \dfrac{1}{R_f\tau_f} \\[2ex] 0 \\[2ex] 0 \end{bmatrix} V_f \qquad (11\text{-}142)$$

11-6 TWO-PHASE SERVOMOTOR TRANSFER FUNCTIONS

The steady-state characteristics of the two-phase servomotor are discussed in Chapter 7. Attention is directed here to the dynamic behavior of these motors as described by their transfer functions.

Although the servomotor transfer functions could be derived by following the procedure outlined for the d-c motor in the preceding section, in the interest of versatility an alternate approach is employed using the external torque-speed characteristic. Typical torque-speed characteristics such as those depicted in Fig. 7-13 are assumed to be available. Furthermore, as indicated in Fig. 11-26, the servomotor is assumed to be operated at a fixed reference voltage. The input variable in this arrangement is the control winding voltage V_c. What quantity then shall we consider to be the output variable? Shall it be velocity ω or displacement c? Examination of Fig. 11-26 certainly leaves no doubt that the application of a given control winding voltage manifests itself as an output velocity. However, this alone is not necessarily the determining factor in establishing what the output variable shall be. Keep in mind that servomotors are used in feedback control systems so that a more important consideration is the function which the servomotor will fulfill. If the output of the control system is to be velocity, then it might be appropriate to derive a transfer function that relates velocity to the control winding voltage. On the other hand, if the motor output is to be used to drive a *position-sensitive* transducer such as a potentiometer or a resolver or a synchro, then clearly the motor displacement should be considered the output variable.

The primary concern at the moment is the general form that the servomotor transfer function assumes. Consequently, the torque-speed curves shown in Fig. 7-13 are assumed to be linear for simplicity. A study of these curves reveals that an increased

Figure 11-26 Two-phase servomotor with fixed reference winding voltage.

DYNAMICS CHAP. 11

motor torque comes about either by an increase in control winding voltage or by a decrease in speed for a fixed V_c. Expressed mathematically the developed motor torque is

$$T = K_M V_c - F_M \frac{dc}{dt} \qquad (11\text{-}143)$$

where K_M = motor torque constant in units of torque/volt

F_M = motor equivalent viscous-friction constant in units of torque/rad/sec

It is helpful to note that F_M is nothing more than the slope of the torque-speed curve at constant V_c. Also, once the torque-speed curves are known, the parameter K_M is merely the change in torque per unit of change in control voltage at constant speed. Equating this torque to the sum of the opposing load torques yields the governing differential equation that relates motor displacement to control voltage. Thus

$$K_M V_c - F_M \frac{dc}{dt} = J \frac{d^2c}{dt^2} + F_L \frac{dc}{dt} \qquad (11\text{-}144)$$

where J = motor inertia + load inertia referred to motor shaft

F_L = viscous friction of load referred to motor shaft

Rearranging and Laplace transforming leads to

$$K_M V_c(s) = (Js^2 + sF)c(s) \qquad (11\text{-}145)$$

where

$$F = F_M + F_L \qquad (11\text{-}146)$$

Hence the desired transfer function becomes

$$\boxed{\frac{c(s)}{V_c(s)} = \frac{K_m}{s(1 + s\tau_m)}} \qquad (11\text{-}147)$$

where

$$K_m = \frac{K_M}{F} \qquad (\text{V-s})^{-1} \qquad (11\text{-}148)$$

$$\tau_m = \frac{J}{F} \qquad \text{s} \qquad (11\text{-}149)$$

Frequently the motor time constant τ_m plays an important role in determining the character of the dynamic response of the control system of which it is a part. This is because the magnitude of τ_m is established in terms of the inertial properties of physical elements such as the size of the servomotor rotor and the reflected load inertia. These elements are such that they cannot be reduced to negligence primarily because the size of J goes hand in hand with the required level of output power. Accordingly, in many cases the motor inertia is the determining factor that establishes the extent to which the system will be capable of following variations of the command signal with time.

Effect of the Control Winding Time Constant

The result of Eq. (11-147) is approximate because it neglects the time lag caused by the inductance of the control winding. A more exact description of the servomotor transfer function can be obtained by recognizing that the control winding voltage really produces a control winding field current I_c, which in turn produces the flux that enables the motor to develop torque. The torque equation as it applies at the motor shaft then becomes

$$K_{ct}I_c - F_M \frac{dc}{dt} = J \frac{d^2c}{dt^2} + F_L \frac{dc}{dt} \qquad (11\text{-}150)$$

where K_{ct} denotes a torque constant expressed in units of torque per field ampere. The corresponding transfer function is given by

$$\frac{c(s)}{I_c(s)} = \frac{K_{ct}}{F} \frac{1}{s(1 + s\tau_m)} \qquad (11\text{-}151)$$

Next compare Fig. 11-27 with Fig. 7-13. At a fixed speed the input impedance at the control winding terminals remains constant. If operation is say at V_{c2}, then a sudden increase in control voltage to V_{c3} at the same speed increases the torque along an ordinate line. If the field winding time constant is not negligible then, as V_{c2} increases to V_{c3}, so too does the control current increase from I_{c2} (in Fig. 11-27) to I_{c3}, but with a time delay.

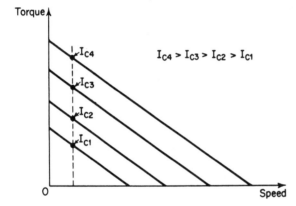

Figure 11-27 Linearized torque-speed curves showing variation of control winding current with control winding voltage at constant speed. Each straight line corresponds to a fixed control voltage: the higher the line, the larger the control voltage.

The field current is therefore related to the control voltage by the usual transfer function for an R-L circuit. Thus

$$\frac{I_c(s)}{V_c(s)} = \frac{1}{R_c(1 + s\tau_c)} \qquad (11\text{-}152)$$

where $\tau_c = L_c/R_c$. Upon combining Eqs. (11-151) and (11-152), the desired transfer function results:

$$\frac{c(s)}{V_c(s)} = \frac{K_{ct}}{FR_c} \frac{1}{s(1 + s\tau_m)(1 + s\tau_c)}$$

or

$$\boxed{\frac{c(s)}{V_c(s)} = \frac{K_m}{s(1 + s\tau_m)(1 + s\tau_c)}} \qquad (11\text{-}153)$$

It is interesting to note in this development that

$$\frac{K_{ct}}{R_c} = K_M \qquad (11\text{-}154)$$

A comparison of Eq. (11-153) with Eq. (11-147) indicates that the difference does indeed lie in the time constant of the control field winding.

Figure 11-28 depicts the block diagram representation of Eq. (11-153). Note again that, whenever the motor output variable is taken to be displacement and the motor field winding time constant is included in the analysis, the result is a third-order characteristic equation. Of course this means that there are three state variables and three state equations. Reference to Fig. 11-29 reveals these state equations in transformed form to be as follows:

$$sI_c(s) = \frac{1}{R_c\tau_c}V_c(s) - \frac{1}{\tau_c}I_c(s) \tag{11-155}$$

$$s\omega(s) = \frac{K_{ct}}{F\tau_m}I_c(s) - \frac{1}{\tau_m}\omega(s) \tag{11-156}$$

$$sc(s) = \omega(s) \tag{11-157}$$

Figure 11-28   Block diagram of the system of Fig. 11-26.

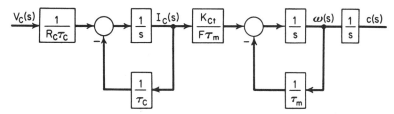

Figure 11-29   Alternate form of Fig. 11-28, modified to show explicitly the pure integrators, the outputs of which identify the state variables of the system.

The corresponding first-order differential equations can be written as

$$\frac{dI_c}{dt} = \frac{1}{R_c\tau_c}V_c - \frac{1}{\tau_c}I_c \tag{11-158}$$

$$\frac{d\omega}{dt} = \frac{K_{ct}}{F\tau_m}I_c - \frac{1}{\tau_m}\omega \tag{11-159}$$

$$\frac{dc}{dt} = \omega \tag{11-160}$$

The matrix formulation is then:

$$\begin{bmatrix} \dfrac{dI_c}{dt} \\[2mm] \dfrac{d\omega}{dt} \\[2mm] \dfrac{dc}{dt} \end{bmatrix} = \begin{bmatrix} -\dfrac{1}{\tau_c} & 0 & 0 \\[2mm] \dfrac{K_{ct}}{F\tau_m} & -\dfrac{1}{\tau_m} & 0 \\[2mm] 0 & 1 & 0 \end{bmatrix} \begin{bmatrix} I_c \\[2mm] \omega \\[2mm] c \end{bmatrix} + \begin{bmatrix} \dfrac{1}{R_c\tau_c} \\[2mm] 0 \\[2mm] 0 \end{bmatrix} V_c \tag{11-161}$$

The amplidyne generator is a power-amplifying device capable of generating high output levels, usually in the range of 1 to about 50 kW. In control system applications it often serves as a power stage that drives a d-c motor. The control field winding of the amplidyne (of which there are several sections) is usually driven from the output stage of an electronic amplifier.

The amplidyne is basically a d-c generator. It is driven at constant speed by a suitable motor which serves as a source of energy for the unit. The magnitude of its output voltage is controlled by the amount of field current flowing through the control winding. Since the armature winding of a d-c motor is frequently placed across the output terminals of the amplidyne, it is important to account for the motor loading effect whenever the transfer function is derived.

The principle of operation of the amplidyne is straightforward. A small current flowing in the control winding is made to create a flux Φ_d directed as illustrated in Fig. 11-30. Because the armature winding is being driven at full speed, a voltage E_q

Figure 11-30 Amplidyne generator, showing windings and connected d-c motor and load.

is induced and it is collected at the brushes marked q-q. By short-circuiting these brushes and designing an armature winding of low resistance, very large armature currents can be made to flow. For example, if E_q is 1 V and the armature winding resistance is 0.01 Ω, an armature current of 100 A results. As a consequence of the huge armature reaction associated with such a large current, a very strong flux field Φ_q is produced and directed in quadrature to the original field Φ_d. By placing a second set of brushes, d-d, normal to the Φ_q field, a high-level voltage source is made available that can be applied to the d-c motor, which in turn provides the desired output at the mechanical load. In practice it is customary to short-circuit the brushes q-q through a series field which enables the required values of Φ_q to be established with smaller quadrature-axis armature current. Another modification involves the inclusion of a compensating winding placed in the direct-axis circuit. Its purpose is to prevent the armature reaction flux produced by the armature current I_d from diminishing the original field Φ_d. The compensating winding is designed to furnish 100% compensation, a condition that is essential in view of the small values that Φ_d assumes during normal operation.

The transfer function of the amplidyne as an isolated unit—without its connected load—is found by considering the direct-axis induced voltage E_d as the output quantity and the control voltage V_c as the input quantity. The derivation that follows is based on the assumptions that there is zero coupling between the flux in the direct axis and the flux in the quadrature axis and that the magnetization curve is linear. The differential equation for the field winding is as usual

$$V_c = I_c R_c + L_c \frac{dI_c}{dt} \tag{11-162}$$

where R_c = control winding resistance, Ω
L_c = control winding inductance, H

Expressed in transformed form, this equation becomes

$$\frac{I_c(s)}{V_c(s)} = \frac{1}{R_c(1 + s\tau_c)} \tag{11-163}$$

where $\tau_c = L_c/R_c$ and denotes the control winding time constant.

The presence of the control winding current produces the direct-axis flux Φ_d which in turn causes an emf to appear across the q-q brushes and which specifically is expressed as

$$E_q = K_1 \Phi_d \tag{11-164}$$

where K_1 includes winding factors as well as the speed of rotation of the armature. A more convenient form of Eq. (11-164) results by recalling that Φ_d is proportional to I_c so that we can write

$$E_q = K_q I_c \tag{11-165}$$

where K_q is a proportionality factor for the number of volts generated per control-winding ampere. The Laplace transformed version of this expression is simply

$$\frac{E_q(s)}{I_c(s)} = K_q \tag{11-166}$$

The governing differential equation for the quadrature-axis circuit is readily seen from Fig. 11-30 to be

$$E_q = I_q R_q + L_q \frac{dI_q}{dt} \tag{11-167}$$

where R_q = total resistance in the quadrature axis circuit, Ω
L_q = total inductance in the quadrature axis circuit, H
I_q = current flowing in the quadrature axis, A

The transformed version of the last equation expressed as a ratio then becomes

$$\frac{I_q(s)}{E_q(s)} = \frac{1}{R_q(1 + s\tau_q)} \tag{11-168}$$

where

$$\tau_q = \frac{L_q}{R_q} \tag{11-169}$$

Finally, with the amplidyne output circuit open it follows that the output voltage appearing at the direct-axis brushes is simply

$$E_d = K_d I_q \tag{11-170}$$

where K_d denotes the number of volts induced in the direct axis per quadrature-axis ampere. The Laplace transformed expression is

$$\frac{E_d(s)}{I_q(s)} = K_d \tag{11-171}$$

The transfer function that applies for the unloaded amplidyne generator readily follows by multiplying Eqs. (11-163), (11-166), (11-168), and (11-171). Thus

$$\boxed{\frac{E_d(s)}{V_c(s)} = \frac{K_q K_d}{R_c R_q} \frac{1}{(1 + s\tau_c)(1 + s\tau_q)}} \tag{11-172}$$

The characteristic equation is second order in this mode of operation. Clearly the usefulness of this result is limited to those applications where the load current of the amplidyne, I_d, is negligible. This is certainly not the case in the situation depicted in Fig. 11-30 when the load switch is closed.

Amplidyne Transfer Function with Connected Load

The transfer function for this arrangement must be found by considering the d-c motor as part of the direct-axis circuitry of the amplidyne and by this means accounting for the loading effect. The resulting transfer function then provides the correct description of the dynamics of the amplidyne-motor combination. For this derivation the additional assumption is made that the output circuit is compensated perfectly, which ensures that the load current I_d has no effect on the control flux Φ_d.

A few more equations are needed to complete the analysis. The first is obtained by applying Kirchhoff's voltage law to the direct-axis circuit. Thus

$$E_d = K_q I_c = I_d R_d + K_\omega \omega \tag{11-173}$$

where R_d is the total resistance found in the direct-axis path, ω is the motor speed, and K_ω denotes the number of volts per radian per second generated by the d-c motor. For simplicity the inductance in the direct axis is assumed negligibly small. Expressed in the Laplace transform notation this equation becomes

$$\frac{E_d(s) - K_\omega \omega(s)}{R_d} = I_d(s) \tag{11-174}$$

Keep in mind that $E_d(s)$ is the Laplace transform of E_d, $I_d(s)$ is the Laplace transform of I_d, and $\omega(s)$ is the Laplace transform of ω. Also, Eq. (11-174) is written as indicated on the left-hand side in order to emphasize the need for a summing point in the block diagram representation. Recall that $\omega(s)$ is treated as a feedback quantity in the formulation of Eq. (11-174).

A second required equation relates the motor-developed torque to the direct-axis current through the torque constant K_t. In Laplace transform notation we have

$$\frac{T(s)}{I_d(s)} = K_t \tag{11-175}$$

Moreover, the relationship between torque and speed follows from writing the torque equation as it applies to the motor shaft. Thus

$$T = J\frac{d\omega}{dt} + F\omega \qquad (11\text{-}176)$$

Here again J denotes the total inertia reflected to the motor shaft and F denotes the total viscous friction reflected to the motor shaft. The transfer function for Eq. (11-176) thus becomes

$$\frac{\omega(s)}{T(s)} = \frac{1}{F(1 + s\tau_m)} \qquad (11\text{-}177)$$

where $\tau_m = J/F$. Finally, in those applications where displacement rather than velocity is of primary interest, use is made of

$$\frac{c(s)}{\omega(s)} = \frac{1}{s} \qquad (11\text{-}178)$$

The total system block diagram can now be drawn by employing a successive block representation of Eqs. (11-163), (11-166), (11-168), (11-170), (11-174), (11-175), (11-177), and (11-178). The result is depicted in Fig. 11-31. The one feedback path exists because of the presence of the d-c motor and its associated counter-emf during running conditions. By employing Eq. (11-16) the single transfer function that relates motor speed to the amplidyne induced emf in the direct axis is obtained. Accordingly,

$$\frac{\omega(s)}{E_d(s)} = \frac{K_t/[R_d F(1 + s\tau_m)]}{1 + [K_\omega K_t/R_d F(1 + s\tau_m)]}$$

$$= \frac{K_t}{F R_d[1 + (K_\omega K_t/F R_d)]} \frac{1}{1 + s\tau'_m} \qquad (11\text{-}179)$$

where

$$\tau'_m = \tau_m \frac{1}{1 + (K_\omega K_t/F R_d)} \qquad (11\text{-}180)$$

Figure 11-31 Block diagram representation of the amplidyne motor system of Fig. 11-30.

The complete amplidyne-motor-system transfer function can now be written by formulating the product of Eqs. (11-163), (11-166), (11-168), (11-170), (11-179), and (11-178). The result is

$$\boxed{\frac{c(s)}{V_c(s)} = \frac{K_q K_d}{R_c R_q} \frac{K_t}{F R_d[1 + (K_\omega K_t/F R_d)]} \frac{1}{s(1 + s\tau_c)(1 + s\tau_q)(1 + s\tau'_m)}} \qquad (11\text{-}181)$$

A comparison of this expression with Eq. (11-172) reveals the precise manner in which the presence of the motor modifies the transfer function. It introduces a change in direct transmission gain along with pure integration and a simple time lag as represented by τ'_m. If the loading effect that the motor has on the amplidyne generator had been neglected, the result would have been a larger mechanical time constant and a different direct gain.

The magnitudes of the amplidyne time constants τ_c and τ_q vary with the size of the unit, but often they lie in a range of 0.05 s for smaller units to 0.5 s for the larger ones. The value of τ'_m of course depends upon the size of the motor and the nature of the connected mechanical load, but it easily can be of the order of several seconds. The use of such a combination in control systems leads to stability problems because of the time lags associated with the composite transfer function.

A glance at Eq. (11-181) discloses the characteristic equation of the amplidyne motor system to be of fourth order. Accordingly, there are also four state variables and four state equations. Figure 11-32 reveals the state variables to be the control

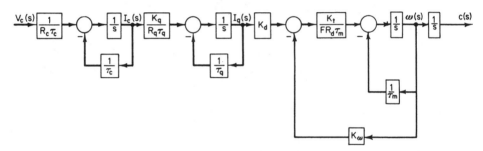

Figure 11-32 Alternate form of the block diagram of the amplidyne motor combination. The output of the pure integrators discloses the state variables.

winding current I_c, quadrature axis current I_q, motor speed ω, and motor displacement c. The state equations in Laplace notation are found by setting the input signal to each of the four integrators equal to s times the output of the respective integrator. This leads to

$$sI_c(s) = \frac{1}{R_c\tau_c}V_c(s) - \frac{1}{\tau_c}I_c(s) \tag{11-182}$$

$$sI_q(s) = \frac{K_q}{R_q\tau_q}I_c(s) - \frac{1}{\tau_q}I_q(s) \tag{11-183}$$

$$s\omega(s) = \frac{K_dK_t}{FR_d\tau_m}I_q(s) - \left(\frac{K_\omega K_t}{FR_d\tau_m} + \frac{1}{\tau_m}\right)\omega(s) \tag{11-184}$$

$$sc(s) = \omega(s) \tag{11-185}$$

The corresponding time-domain equations expressed in the matrix formulation appear below:

$$\begin{bmatrix} \dfrac{dI_c}{dt} \\[2ex] \dfrac{dI_q}{dt} \\[2ex] \dfrac{d\omega}{dt} \\[2ex] \dfrac{dc}{dt} \end{bmatrix} = \begin{bmatrix} -\dfrac{1}{\tau_c} & 0 & 0 & 0 \\[2ex] \dfrac{K_q}{R_q\tau_q} & -\dfrac{1}{\tau_q} & 0 & 0 \\[2ex] 0 & \dfrac{K_d K_t}{FR_d\tau_m} & -\left(\dfrac{K_\omega K_t}{FR_d\tau_m}+\dfrac{1}{\tau_m}\right) & 0 \\[2ex] 0 & 0 & 1 & 0 \end{bmatrix} \begin{bmatrix} I_c \\[2ex] I_q \\[2ex] \omega \\[2ex] c \end{bmatrix} + \begin{bmatrix} \dfrac{1}{R_c\tau_c} \\[2ex] 0 \\[2ex] 0 \\[2ex] 0 \end{bmatrix} V_c \quad (11\text{-}186)$$

11-8 A FEEDBACK VOLTAGE CONTROL SYSTEM

To illustrate how knowledge of the steady-state and dynamic performance of a d-c generator can be applied in a general way, attention is directed to the elementary voltage control system depicted in Fig. 11-33. The purpose of this control system is to provide a terminal voltage across the load that will remain essentially invariant in spite of changes in load current. Here the load current and the armature current are assumed to be the same because the current drawn by the 10-kΩ potentiometer is negligible for all practical purposes. Also, assume that the generator external characteristics are properly represented by Fig. 11-34. The curves are linearized for simplicity. These characteristics state, e.g., that, if the generator were operating isolated and at a constant field current of I_{f3}, a 20-V drop would occur in terminal voltage for a load demand corresponding to 20 A. This is consistent with an armature circuit resistance R_a of 1 Ω for the generator. Of course the purpose of the arrangement of Fig. 11-33 is to prevent this drop from occurring and to maintain stability in doing so.

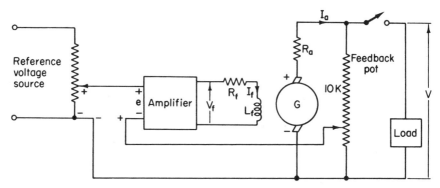

Figure 11-33 Elementary feedback voltage system.

Before proceeding with a description of how this automatic regulating action is achieved, let us first calculate the reference voltage needed to make available a specified no-load voltage. Assume the remaining system parameters are as follows:

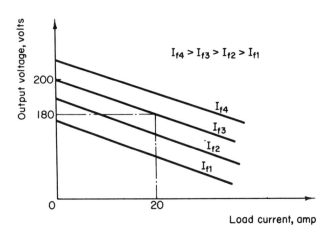

Figure 11-34 Generator output characteristics.

K_a = amplifier gain constant = 2000 V/V
R_f = generator field winding resistance = 200 Ω
K_f = generated emf constant = 300 V/field ampere
h = feedback factor = 0.1 numeric

The generated emf constant is valid for a specified constant prime mover speed. The feedback factor refers to the slider-arm setting of the feedback potentiometer. Assuming the desired output at no-load to be 200 V, an amplifier input signal e_o is needed which satisfies the equation

$$\frac{e_o K_a}{R_f} K_f = 200 \text{ V} \tag{11-187}$$

or

$$\frac{e_o 2000}{200} 300 = 200$$

Hence

$$e_o = \tfrac{1}{15} = 0.065 \text{ V}$$

It should be apparent that Eq. (11-187) merely identifies how the amplifier input signal at no-load is operated upon to yield the desired no-load terminal voltage in steady state. Thus $e_o K_a$ denotes the amplifier output voltage that is applied to the field circuit to produce a field current. The steady-state value of this field current is found by dividing the field voltage by R_f, which yields $e_o K_a / R_f$. Finally, by multiplying this quantity by K_f the desired no-load voltage is obtained. The required reference voltage is found from the equation that indicates how the amplifier input voltage is obtained. A glance at Fig. 11-33 discloses that it is the *difference* between the reference voltage and the portion of the output voltage that is fed back. Expressing this mathematically we have generally

$$E_R - hV = e \tag{11-188}$$

Therefore for the case at hand

$$E_R = e_o + hV = 0.065 + (0.1)200 = 20.065 \text{ V} \tag{11-189}$$

This expression states that, if the slider arm of the potentiometer at the reference source is set at 20.065 V, the generated output voltage at no-load will be 200 V. Obviously, as a matter of convenience, the reference potentiometer can be calibrated to convey this information.

Let us now describe how the closed-loop system of Fig. 11-33 automatically furnishes a self-correcting action as load current to the extent of 20 A is permitted to flow. From Eq. (6-5) we know that the terminal voltage is related to the generated emf by

$$V = E_a - I_a R_a \tag{11-190}$$

Hence, for $I_a = 20$ and $R_a = 1$, the terminal voltage attempts to fall to a value of 180 V. However, as it moves in this direction the feedback quantity hV becomes correspondingly smaller, which increases the amplifier input signal as indicated by Eq. (11-188). The increased amplifier input means a greater field current, which in turn generates an increased E_a to offset the drop in terminal voltage. Complete neutralization does not occur in this system because some small drop in terminal voltage must be allowed in order for the corrective action to exist.

The block diagram of this system may be developed by following the procedure employed in the preceding sections. The transfer function of each part of the system is found and these are then combined to yield the total system transfer function and block diagram. The Laplace transformed version of the equation that provides the actuating signal for the amplifier is obtained from Eq. (11-188). Thus

$$E_R(s) - hV(s) = e(s) \tag{11-191}$$

Since $V(s)$ is the output variable, a summing point is used to identify this equation in the block diagram with $hV(s)$ coming from a feedback path. The transfer function for the amplifier is clearly just the associated gain factor. Hence

$$\frac{V_f(s)}{e(s)} = K_a \tag{11-192}$$

where $V_f(s)$ denotes the amplifier output voltage which is applied to the field circuit. Next, because of the R-L nature of the field circuit, our previous experience† tells us that field current and field voltage are related by

$$\frac{I_f(s)}{V_f(s)} = \frac{1}{R_f(1 + s\tau_f)} \tag{11-193}$$

where $\tau_f = L_f/R_f$. Proceeding on through the system, we next note that the field current of the last equation yields a generated emf described by the transfer function

$$\frac{E_a(s)}{I_f(s)} = K_f \tag{11-194}$$

The final step in the development is the Laplace transformed version of Eq. (11-190), which provides the relationship between the $E_a(s)$ of the last equation and the output quantity $V(s)$. Thus

$$E_a(s) - I_a(s)R_a = V(s) \tag{11-195}$$

A successive block interpretation of Eqs. (11-191) to (11-195) leads to the complete

†See derivation of Eq. (11-57).

block diagram depicted in Fig. 11-35. Note that the increase in armature current is treated as a negative input disturbance that acts to reduce the output voltage.

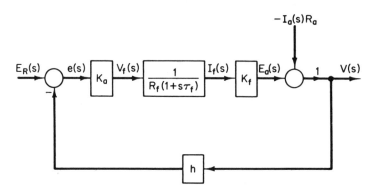

Figure 11-35 Block diagram for the system of Fig. 11-34.

It is important to understand that from the system block diagram it is now possible to write conveniently expressions which describe the manner in which the output voltage responds to changes in either the reference signal or load disturbances or in both. Moreover, these expressions can be made to reveal information about dynamic behavior as well as steady-state performance. To illustrate, let us first write the transfer function relating output voltage to reference input. Application of Eq. (11-16) to the block diagram yields

$$\frac{V(s)}{E_R(s)} = \frac{[K_a K_f / R_f (1 + s\tau_f)]}{1 + [hK_a K_f / R_f (1 + s\tau_f)]} = \frac{K}{1 + hK} \frac{1}{1 + s\tau_f'} \qquad (11\text{-}196)$$

where

$$K = \frac{K_a K_f}{R_f} = \text{direct transmission gain} \qquad (11\text{-}197)$$

$$\tau_f' = \frac{\tau_f}{1 + hK} \qquad (11\text{-}198)$$

In a similar fashion the transfer function between the output voltage and the armature voltage drop can be shown to be given by

$$\frac{V(s)}{-R_a I_a(s)} = \frac{1}{1 + [hK_a K_f / R_f (1 + s\tau_f)]} = \frac{1}{1 + hK} \frac{1 + s\tau_f}{1 + s\tau_f'} \qquad (11\text{-}199)$$

The complete expression for the output voltage corresponding to E_R and I_a is found by superposing the results obtained from Eqs. (11-196) and (11-199). Accordingly, we have

$$\boxed{V(s) = \frac{K}{1 + hK} \frac{1}{1 + s\tau_f'} E_R(s) - \frac{R_a I_a(s)}{1 + hK} \frac{1 + s\tau_f}{1 + s\tau_f'}} \qquad (11\text{-}200)$$

To illustrate the use of the last equation, let it be desirable to find the value of the output voltage for a reference input of 20.065 V and zero armature current. The Laplace transform of the reference input is given by

$$E_R(s) = \frac{20.065}{s} \qquad (11\text{-}201)$$

Hence

$$V(s) = \frac{1}{1 + hK} \frac{1}{1 + s\tau'_f} \frac{20.065}{s} \qquad (11\text{-}202)$$

Keep in mind that Eq. (11-202) furnishes total information about the response, dynamic as well as steady-state. Since for the moment our interest is focused on the steady-state value, applying the final value theorem to Eq. (11-202) yields the desired result. Thus

$$V = \lim_{s \to 0} sV(s) = \frac{K}{1 + hK} 20.065 = \frac{3000}{301} 20.065 = 200 \text{ V} \qquad (11\text{-}203)$$

This, of course, is exactly the value used previously in connection with $E_R = 20.065$ V. Refer to Eq. (11-189).

If, in addition to E_R, it is further assumed that a current of 20 A is delivered to the load, then application of the final value theorem to Eq. (11-200) discloses that the new steady-state value is

$$V = \frac{K}{1 + hK} 20.065 - \frac{20}{1 + hK} = 200 - \frac{20}{301} = 200 - 0.0655 = 199.9345 \text{ V}$$
$$(11\text{-}204)$$

Clearly, only a very slight drop in output voltage occurs in the closed-loop system. Note that the placement of the generator in a closed-loop system caused the normal 20-V drop in terminal voltage to be reduced to 0.0655, a reduction by the factor $1/(1 + hK)$ where hK is the loop gain factor. What is more, all of this has been achieved automatically.

If it should be desirable to describe the dynamic behavior of the system in response to $E_R(s)$ and $I_a(s)$, the usual procedure must be applied. First, the appropriate expressions for $E_R(s)$ and $I_a(s)$ are introduced into Eq. (11-200). This is then followed by a partial fraction expansion and subsequent evaluation of the associated coefficients. Evaluation of the inverse Laplace transforms by means of a table of Laplace transform pairs furnishes the complete time solution for the output voltage. In this instance the output voltage will be seen to rise in accordance with the usual buildup that takes place in any system containing a single energy-storing element. See Eq. (11-12) for example.

11-9 DYNAMICS OF THE SYNCHRONOUS GENERATOR–PRIME MOVER SYSTEM

In Sec. 5-10 a differential equation is derived that describes the dynamical behavior of a prime mover coupled to a synchronous generator in response to a change in the share of the bus load. The result is a second-order linearized differential equation as displayed by Eq. (5-99) and repeated here for convenience.

$$T_{\text{sh}} = k_j \ddot{\Delta} + k_d \dot{\Delta} + k_s \Delta = T \qquad (11\text{-}205)$$

The quantity $T_{\text{sh}} = T$ denotes the shaft torque delivered to the synchronous generator by the prime mover; the parameters k_J, k_d, k_s are defined in Sec. 5-10; and Δ represents the generator power angle expressed in electrical degrees.

The transfer function of this system is readily determined by applying to Eq. (11-205) the procedure outlined in Sec. 11-1. This leads to

$$\frac{\Delta(s)}{T(s)} = \frac{1/k_J}{s^2 + (k_d/k_J)s + k_s/k_J} \tag{11-206}$$

The denominator expression is the characteristic polynomial. When set equal to zero, it becomes the characteristic equation. Compare with Eq. (5-100). Information about the dynamical behavior or about the total solution to an input force $T(s)$, originating from the prime mover, can be obtained by following the procedure outlined in Example 5-6.

Attention in this section, however, is directed chiefly to the matter of obtaining a solution to the original nonlinear version of Eq. (11-205), which is given by Eq. (5-97) and repeated here for convenience. Thus

$$T_{\text{sh}} = k_J\ddot{\Delta} + k_d\dot{\Delta} + k_s \sin \Delta = T \tag{11-207}$$

The presence of the transcendental term makes this equation more difficult to solve in a direct analytical fashion. But the solution of nonlinear differential equations by the digital computer is a fairly routine procedure. We now undertake to outline such a process for Eq. (11-207) primarily because the previous linearized results are valid solely for small values of the power angle (usually $< 15°$). However, when the prime mover–synchronous generator system operates under stress, such as during fault conditions, these angles can assume much larger values in which case the linearized differential equation can easily lead to erroneous results.

Although FORTRAN can be used to describe an algorithm to solve Eq. (11-207), we employ instead an IBM program known as CSMP, which is an acronym for *continuous system modeling program*. CSMP was originally developed for the System 360/370 machines in order to facilitate the solution by the digital computer of problems that are formulated in terms of transfer functions, block diagrams, signal flow graphs, and differential equations. The ease with which programs can be written using CSMP is attributable to the availability of numerous subroutines that are callable by descriptive symbolic names such as those listed in Table 11-1. Item 1 in this table shows that to perform integration in the time domain subject to an initial condition, it is necessary merely to use the command Y = INTGRL(IC,X) and to supply within the parentheses the value of the initial condition and the identity of the integrand. The symbolic name INTGRL calls the subroutine that performs the integration process. Item 3 provides a process for solving a first-order differential equation subject to an initial condition. Item 5 makes available the opportunity to obtain the solution to a second-order linear differential equation with a single command to the computer when equipped with CSMP. As a matter of fact the solution of Eq. (11-206) can easily be programmed with the use of this command. See Example 11-2. Listed in Table 11-2 are two of the forcing functions commonly used in examining the dynamic behavior of systems.

The CSMP language has several noteworthy features. It allows liberal use of

TABLE 11-1 A FEW OF THE COMMON SUBROUTINES AVAILABLE IN CSMP

	General form	Simulated function
1	Y = INTGRL (IC, X) Y(0) = IC INTEGRATOR	$y = \int_0^t x\, dt + IC$ Equivalent Laplace Transform: $\dfrac{1}{s}$
2	Y = DERIV (IC, X) x (t = 0) = IC DERIVATIVE	$y = \dfrac{dx}{dt}$ Equivalent Laplace Transform: s
3	Y = REALPL (IC, P, X) Y(0) = IC First Order LAG (Real Pole)	$P\dot{y} + y = x$ Equivalent Laplace Transform: $\dfrac{1}{1 + sP} = \dfrac{y}{x}$
4	Y = LEDLAG (P_1, P_2, X) LEAD-LAG	$P_2\dot{y} + y = P_1\dot{x} + x$ $\dfrac{y}{x} = \dfrac{1 + sP_1}{1 + sP_2}$
5	Y = CMPXPL(IC_1, IC_2, P_1, P_2, X) Y(0) = IC; $\dot{y}$(0) = IC_2	$\ddot{y} + 2P_1 \cdot P_2\dot{y} + P_2^2 y = x$ $\dfrac{y}{x} = \dfrac{1}{s^2 + 2P_1P_2s + P_2^2}$

symbolic names to represent variables and functions subject to the restriction that they start with a letter and be limited to six characters. It also permits use of the FORTRAN operators and functions. Execution control statements that direct the simulation run can also be accomplished in a simple way. A single timer statement allows the management of the independent variable such as setting the step size in the integration process or the print interval for print plotting. An especially attractive feature of the program is the use of the output control statement, PRTPLT, which allows any number of variables to be print plotted in a convenient and magnitude-scaled format.

TABLE 11-2 TWO COMMON SIGNAL SOURCES AVAILABLE IN CSMP

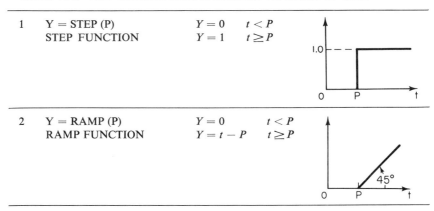

1	Y = STEP (P) STEP FUNCTION	$Y = 0 \quad t < P$ $Y = 1 \quad t \geq P$
2	Y = RAMP (P) RAMP FUNCTION	$Y = 0 \qquad t < P$ $Y = t - P \quad t \geq P$

Example 11-2

Using CSMP write a program to obtain the solution to the linear differential equation that describes the dynamic response of the prime mover–synchronous generator system of Example 5-6.

Solution The equation to be simulated is Eq. (11-206). Since the denominator is a second-order characteristic polynomial, item 5 in Table 11-1 is used. But because this subroutine does not provide a numerator gain adjustment, the equation to be simulated is rearranged to read as follows:

$$\frac{A}{W} \times \frac{W}{T} = \frac{1}{k_j}\left[\frac{1}{s^2 + (k_d/k_j)s + k_s/k_j}\right] \tag{11-208}$$

where A is selected to represent the power angle in the simulation and T denotes the input shaft torque. The function

$$\frac{W}{T} = \frac{1}{s^2 + (k_d/k_j)s + k_s/k_j} \tag{11-209}$$

is now ready at least in form to be simulated by the subroutine of item 5 in Table 11-1. Inserting the various values listed for these parameters, we get

$$\frac{k_d}{k_j} = \frac{11.2}{1.6} = 7 \qquad \frac{k_s}{k_j} = \frac{440}{1.6} = 275$$

and accordingly Eq. (11-209) can be written as

$$\frac{W}{T} = \frac{1}{s^2 + 7s + 275} = \frac{1}{s^2 + 2(0.21)16.6s + (16.6)^2} \tag{11-210}$$

The second form of the last expression now exactly corresponds to the form called for by item 5. Observe that P_2 of item 5 is the square root of the constant term of the characteristic polynomial and parameter P_1 is determined as the coefficient of s in the characteristic polynomial divided by twice P_2.

In accordance with the manipulation displayed in Eq. (11-208), two structure statements are needed to build the computer model of the prime mover–synchronous generator system. The first is represented by Eq. (11-210) and the second is simply

$$\frac{A}{W} = \frac{1}{k_j} = \frac{1}{1.6} = 0.625 \tag{11-211}$$

The complete CSMP listing looks as follows:

```
LABEL DYNAMICAL BEHAVIOR OF PM/SG SYSTEM: LINEAR SOLUTION
T= 7964*STEP(0.)
W = CMPXPL(10.,0.,.21,16.6,T)
A = 0.625*W
TIMER DELT= 0.00001,FINTIM= 1.4, OUTDEL= 0.03
PRTPLT A
END
STOP
ENDJOB
```

A few comments about this program describe additional features of CSMP. The LABEL card followed by a space is used whenever a print plotted output is desired; the label also appears as a heading at the top of each page. The second statement, T = 7964, is the value in N-m of the new commanded input specified in Example 5-6. The third statement is the first structure statement of the computer model and it results directly from item 5. Observe that the first entry inside the parentheses is 10, which is the initial value of the power angle. The third entry on this line is the damping ratio of the characteristic equation, 0.21; the fourth entry of this line is the corresponding natural frequency of 16.6. The fifth entry within the parentheses is the new prime mover torque that drives the system. Because T is specified, the iteration represented by the W statement can immediately proceed on activating the program. The fourth statement in the listing provides the actual value of the power angle in degrees. The TIMER card furnishes the computer with important information about how the data are to be managed. The first entry on this card following the space key is the increment size to be used by the computer in solving the differential equation. If this quantity is not specified, a default value of FINTIM $\times$ 10^{-7} s is used. The second entry elicits a termination time for the computations. It is usually taken to be slightly larger than the expected settling time of the system response. FINTIM must be supplied or the program will not run. The third entry, OUTDEL, instructs the computer as to which computed points are to be print plotted. In this case those points separated by 0.03 s will be print plotted. The customary practice is to choose OUTDEL so that the recorded transient response is uniformly spread on a standard $8\frac{1}{2}$ by 11 in. sheet. In many printers this requires 50 points, so here OUTDEL is taken as 1.5/50 or 0.03. The PRTPLT statement is followed by a space and then by a listing of all the variables that appear in the program that are to be print plotted. Here only the power angle A is so identified.

The actual computer readout obtained by this program is shown in Fig. 11-36. Note the general agreement with the results found by analysis in Example 5-6.

Example 11-3

Obtain the solution for the dynamical response of the prime mover–synchronous generator system for the conditions described in Example 5-6 when the exact differential equation is used.

Solution The solution is found by means of the digital computer through use of the simplifying features of CSMP. A little thought should make it apparent that item 5 in Table 11-1 is no longer viable. Rather we now use a technique of solving Eq. (11-207), which is standard procedure in analog computer simulations. The strategy merely requires that the nth-order differential equation be integrated successively n times. In this case two integrations are required and these are performed by the subroutine of item 1 in Table 11-1. However, before proceeding with this task, it is first necessary to modify the synchronizing torque coefficient k_s. In Eq. (11-205) this parameter carries units of N-m/electrical degrees, which makes it consistent with its appearance as a multiplier of the power angle in electrical degrees. But in the nonlinear version of the system equation it is $\sin \Delta$ that appears. Hence a modification is needed which here is

TIME	A	I		I
0.0	1.0000E 01	+		
3.0000E-02	1.0917E 01	---+		
6.0000E-02	1.3231E 01	--------------+		
9.0000E-02	1.6171E 01	-------------------------+		
1.2000E-01	1.8976E 01	--+		
1.5000E-01	2.1059E 01	---+		
1.8000E-01	2.2100E 01	---+		
2.1000E-01	2.2069E 01	--+		
2.4000E-01	2.1176E 01	---+		
2.7000E-01	1.9780E 01	--+		
3.0000E-01	1.8280E 01	--+		
3.3000E-01	1.7026E 01	--------------------------------------+		
3.6000E-01	1.6244E 01	----------------------------------+		
3.9000E-01	1.6016E 01	---------------------------------+		
4.2000E-01	1.6286E 01	----------------------------------+		
4.5000E-01	1.6903E 01	-------------------------------------+		
4.8000E-01	1.7664E 01	---+		
5.1000E-01	1.8376E 01	---+		
5.4000E-01	1.8893E 01	---+		
5.7000E-01	1.9139E 01	--+		
6.0000E-01	1.9111E 01	--+		
6.3000E-01	1.8867E 01	---+		
6.6000E-01	1.8502E 01	---+		
6.9000E-01	1.8118E 01	---+		
7.2000E-01	1.7802E 01	---+		
7.5000E-01	1.7612E 01	--+		
7.8000E-01	1.7564E 01	--+		
8.1000E-01	1.7643E 01	--+		
8.4000E-01	1.7806E 01	---+		
8.7000E-01	1.8003E 01	---+		
9.0000E-01	1.8184E 01	--+		
9.3000E-01	1.8312E 01	---+		
9.6000E-01	1.8369E 01	---+		
9.9000E-01	1.8357E 01	---+		
1.0200E 00	1.8291E 01	---+		
1.0500E 00	1.8195E 01	--+		
1.0800E 00	1.8097E 01	---+		
1.1100E 00	1.8018E 01	---+		
1.1400E 00	1.7972E 01	---+		
1.1700E 00	1.7963E 01	---+		
1.2000E 00	1.7985E 01	---+		
1.2300E 00	1.8028E 01	---+		
1.2600E 00	1.8079E 01	---+		
1.2900E 00	1.8125E 01	---+		
1.3200E 00	1.8156E 01	---+		
1.3500E 00	1.8170E 01	---+		
1.3800E 00	1.8165E 01	---+		

Figure 11-36 Computer solution for dynamic response for Example 11-2. Observe that the new steady-state value of the power angle is 18.17° and the peak value is 22.1°.

$$k'_s = \frac{k_s \times 10}{\sin 10°} = \frac{4400}{\sin 10°} = 25,338.6 \text{ N-m}$$

Inserting this value into Eq. (11-207) along with the other parameters and the new value of the prime mover shaft torque, the nonlinear differential equation becomes

$$T = 7964 = 1.6\frac{d^2\Delta}{dt^2} + 11.2\frac{d\Delta}{dt} + 25{,}388.6\sin\Delta \qquad (11\text{-}212)$$

Rearranging and solving for the highest derivative in preparation for successive integration, we get

$$\frac{d^2\Delta}{dt^2} = 4977.5 - 7\frac{d\Delta}{dt} - 15{,}836.6\sin\Delta \qquad (11\text{-}213)$$

To prepare this equation for simulation by CSMP we must assign symbolic names to the variables. Hence let A represent the power angle Δ in the computer; let the second derivative be represented by D2A; and finally let the first derivative be represented by DA. Then the computer version of the physical system can be written as

D2A = 4977.5 −7∗DA − 15,836.6∗sin(A)

The expression *sin(A)* is a FORTRAN function and as such is available to CSMP. The program proceeds as follows:

```
LABEL DYNAMICAL RESPONSE OF PM/SG SYSTEM: NONLINEAR SOLUTION
T = 4977.5*STEP(0.)
S = SIN(A)
D2A = T − 7*DA − 15,836.6*S
DA = INTGRL(0., D2A)
 A = INTGRL(10., DA)
TIMER DELT=0.00001, FINTIM=1.4, OUTDEL=0.036
PRTPLT A
END
STOP
ENDJOB
```

The computer model consists of three structure statements: D2A, DA, and A. The initial condition on the power angle is specified in the statement for A as the quantity 10. Accordingly, once the program is started, all the quantities on the right side of the D2A statement are known and the first iteration can be completed by the computer. Observe that the newly computed value of D2A serves as the integrand for the subsequent evaluation of DA. In turn DA then furnishes an integrand value for the first iterative value of the power angle A. The PRTPLT statement calls for a plot of the power angle. The transient response obtained from the computer simulation is displayed in Fig. 11-37. Observe that the nonlinear solution yields a maximum power angle of 26° versus 22.1° for the linear case.

```
                    MINIMUM                A    VERSUS TIME              MAXIMUM
                    1.7453E-01                                          4.5412E-01
  TIME         A          I                                                 I
0.0         1.7453E-01    +
3.600CE-02  3.6377E-01    ------------------------------------------------+
7.2CCOE-02  4.1203E-01    --------------------------------------------------+
1.08CCE-01  2.3927E-01    -----------+
1.440CE-01  2.9239E-01    ------------------+
1.8CCCE-01  3.9797E-01    ----------------------------------------------+
2.16CCCE-01 3.0121E-01    --------------------+
2.52CCE-01  2.6942E-01    ----------------+
2.88CCE-01  3.6122E-01    -------------------------------------------+
3.24CCE-01  3.3765E-01    --------------------------------------+
3.6CCCE-01  2.7848E-01    -----------------+
3.96CCE-01  3.2773E-01    ----------------------------------+
4.32CCE-01  3.4781E-01    --------------------------------------+
4.68CCE-01  2.9904E-01    -------------------+
5.04CCE-01  3.0886E-01    ----------------------+
5.4CCCE-01  3.4158E-01    ------------------------------+
5.76CCE-01  3.1685E-01    ------------------------+
6.12CCE-01  3.0431E-01    ---------------------+
6.48CCE-01  3.3008E-01    --------------------------+
6.84CCE-01  3.2632E-01    -------------------------+
7.2CCOE-01  3.0826E-01    ----------------------+
7.56CCE-01  3.2062E-01    -------------------------+
7.92CCE-01  3.2818E-01    --------------------------+
8.28CCE-01  3.1463E-01    -----------------------+
8.64CCE-01  3.1583E-01    -----------------------+
9.CCCCE-01  3.2573E-01    --------------------------+
9.36CCE-01  3.1961E-01    ------------------------+
9.72CCE-01  3.1512E-01    -----------------------+
1.008CE 00  3.2222E-01    -------------------------+
1.044CE 00  3.2199E-01    -------------------------+
1.08CCE 00  3.1660E-01    -----------------------+
1.116CE 00  3.1959E-01    ------------------------+
1.152CE 00  3.2221E-01    -------------------------+
1.188CE 00  3.1851E-01    ------------------------+
1.224CE 0C  3.1845E-01    ------------------------+
1.26CCE 00  3.2134E-01    ------------------------+
1.296CE 00  3.1985E-01    ------------------------+
1.332CE 00  3.1841E-01    ------------------------+
```

Figure 11-37 Computer readout for the power angle variation in the solution of the nonlinear equation of Eq. (11-213). The peak power angle is now $(0.45412)(180/\pi) = 26°$.

PROBLEMS

11-1. Obtain the transfer function relating output voltage to input voltage for each circuit shown in Fig. P11-1. Put the result in the form that involves factors such as $(1 + s\tau)$.

(a)   (b)

Figure P11-1

11-2. The values of the parameters in the circuit of Fig. P11-1 are as follows: $L = 0.5$ H, $R_1 = 2\,\Omega$, and $R_2 = 3\,\Omega$. Find the complete expression for the output voltage for a unit step input voltage.

11-3. Find the transfer function relating the output voltage to the input voltage for each circuit depicted in Fig. P11-3. Manipulate the result into the standard form.

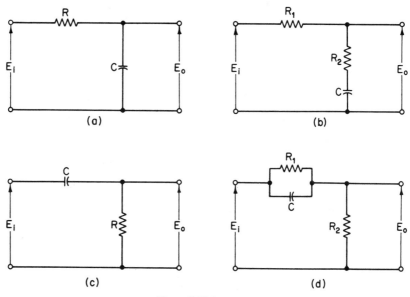

Figure P11-3

11-4. Refer to Fig. P11-4.

(a) Derive the transfer function of this circuit relating E_o to E_i.

(b) Obtain the transfer function that relates the voltage across capacitor C_2 to the input voltage E_i.

(c) Represent the transfer function between E_o and E_i by a block diagram that brings into evidence a pure integration function in the direct transmission path.

(d) Identify the state variable for this network and indicate its location in the block diagram.

Figure P11-4

11-5. The block diagram relating output to input of an electric circuit is shown in Fig. P11-5.

(a) Find the transfer function between input E_i and output E_o.

(b) Can you suggest a circuit for which this transfer function applies? The C's denote capacitance and R resistance.

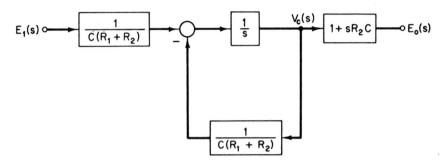

E_1(s) → $\dfrac{1}{C(R_1 + R_2)}$ → ⊖ → $\dfrac{1}{s}$ → $V_c(s)$ → $1 + sR_2C$ → E_o(s)

$\dfrac{1}{C(R_1 + R_2)}$

Figure P11-5

11-6. Refer to Fig. P11-3(d).
  (a) Derive the transfer function $E_o(s)/E_i(s)$.
  (b) Find the transfer function relating the voltage across the capacitor to the input voltage.
  (c) Represent the transfer function between input and output in terms of a pure integration in the direct transmission path.
  (d) Identify the state variable in the block diagram of part (c).

11-7. Determine the transfer function that relates the voltage E_2 to the input voltage E_i in the system depicted in Fig. P11-7. The system parameters are as follows:

Generator: $L_f = 50\ \text{H}, R_f = 50\ \Omega, R_a = 1\ \Omega, L_a = 1\ \text{H}$
emf constant: $K_g = 100\ \text{V/field ampere}$
Low-pass filter: $L = 1\ \text{H}$
$\qquad\qquad\qquad R = 1\ \Omega$

Figure P11-7

11-8. The motor of Fig. 11-5 is operated at constant field current. The motor parameters are: $R_a = 0.5\,\Omega$, $L_a = 0$, $K_\omega = 2\,V/rad/s$, $K_t = 5\,lb\text{-}ft/A$, and $F = 1\,lb\text{-}ft/rad/s$. The equivalent inertia is 42 lb-ft-s².

(a) What is the numerical expression for the transfer function that relates motor speed in radians per second to the voltage applied to the armature circuit?

(b) If the applied armature voltage is a step of 200 V magnitude, find the steady-state speed of the motor.

(c) Obtain the complete expression for the motor speed from the instant the voltage of part (b) is applied to the time that virtual steady-state is reached.

(d) Show the block diagram of this system, using the numerical values.

(e) Write the state equation for the system.

11-9. The application of 100 V to the armature of the machine of Prob. 11-8 at no-load generates an output velocity of 47.6 rad/s.

(a) If a load torque of 40 lb-ft is applied to the motor shaft, find the new operating speed.

(b) How long does it take for the new speed to be reached? Use three time constants as a satisfactory indication of settling time.

11-10. Repeat Prob. 11-8, taking into account the motor armature leakage inductance. Assume that $L_a = 0.05\,H$.

11-11. The parameters of the system of Fig. P11-11 are as follows:

$$K_t = 0.0234\ lb\text{-}ft/A = \text{torque constant}$$
$$K_n = 0.2\ V/rev/s = \text{speed constant}$$
$$K_A = 0.15\ V/V = \text{amplifier gain}$$
$$R_a = 0.2\ \Omega$$
$$L_a = 0.0$$
$$J\ \ = 2 \times 10^{-6}\ slug\text{-}ft^2$$

Figure P11-11

The viscous friction coefficient of the load is negligible.

(a) Develop the block diagram for this system, relating output velocity to amplifier input.

(b) Find the expression for the closed-loop transfer function, using the numerical values.

(c) Write the differential equation for this system.

(d) Is there an equivalent viscous friction (i.e., damping coefficient) in this system? Explain.

11-12. Find the complete solution for the motor velocity in the system of Fig. P11-11 corresponding to a unit step input to the amplifier.

11-13. Determine the differential equation of the system shown in Fig. P11-13 in terms of the letter symbols denoting the system parameters. Consider that the armature leakage inductance of the generator and motor are both negligible. Also, take the generator voltage constant as K_g, in volts per field ampere, the motor speed constant as K_ω in volts per radian per second, and the motor torque constant as K_t in lb-ft per armature ampere.

Figure P11-13

11-14. Express the dynamical behavior of the system of Fig. P11-13 in terms of the state equations expressed in normal matrix form.

11-15. A speed control system is depicted in Fig. P11-15. The generator field winding has resistance R_f and inductance L_f. Let R denote the armature circuit resistance of both the motor and generator. The armature winding leakage inductances are negligible and zero-load torque is applied. Also, the output voltage of the generator is described by

Figure P11-15

DYNAMICS   CHAP. 11

$$K_g I_f = I_a R + K_\omega \omega \qquad \text{where } K_g = \text{V/field amp}$$

and the output speed and developed motor torque are related by

$$K_t I_a = J \frac{d\omega}{dt} + F\omega \qquad \text{where } K_t = \text{lb-ft/arm. amp}$$

Moreover, β is a tachometer constant expressed in volts per radian per second.
(a) Develop the block diagram showing the transfer function of each block.
(b) What is the characteristic equation of this system?
(c) Write the state equations in the standard matrix form.

11-16. A system is represented by the block diagram shown in Fig. P11-16.
(a) Find the expression for the closed loop transfer function.
(b) What is the characteristic equation?
(c) What influence does the parameter K_e exert on the system dynamic response?

Figure P11-16

11-17. A system is represented by the block diagram depicted in Fig. P11-17.
(a) Determine the closed loop transfer function.
(b) What is the expression for the characteristic equation?
(c) Write the defining differential equation for this system.
(d) What role does K_0 play in determining the dynamic response?

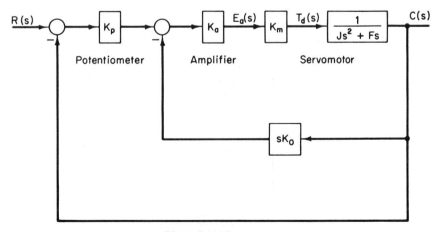

Figure P11-17

11-18. A system is represented by the block diagram depicted in Fig. P11-18. The amplifier section is designed to provide an output signal that contains a proportional as well as an integral component of the input signal.

(a) Find the closed loop transfer function.

(b) What is the characteristic equation?

(c) How does the result of part (b) compare with part (b) of Probs. 11-16 and 11-17? Explain.

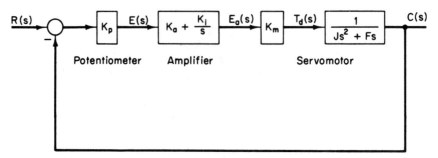

Figure P11-18

11-19. In the system of Fig. P11-19 assume that the reference and feedback tachometer generators are identical. In addition, the system has the following parameters:

$$K_a = \text{amplifier gain, V/V}$$
$$R_f = \text{generator field winding resistance, } \Omega$$
$$L_f = \text{generator field winding inductance, H}$$
$$K_f = \text{generator emf factor, V/field amp}$$
$$R \ = \text{armature circuit resistance of motor and generator}$$
$$K_t = \text{motor torque constant, lb-ft/arm. amp}$$
$$K_\omega = \text{motor speed constant, V/rad/s}$$
$$\beta \ \ = \text{d-c tachometer speed constant, V/rad/s}$$

Figure P11-19

J = system inertia in consistent units

F = system viscous friction coefficient in consistent units

(a) Draw the complete block diagram of this system, indicating clearly the transfer function of each block.

(b) Identify the state variables.

(c) Write the state equations in matrix form.

(d) Determine the characteristic equation of this system.

(e) Are oscillations possible in the dynamic state? Explain.

11-20. A 10-hp armature-controlled d-c motor drives a load whose viscous friction coefficient is 2 lb-ft/rad/s and whose angular inertia is 17.5 lb-ft-s². The field winding is separately excited and maintained fixed. The corresponding motor parameters are then as follows:

$$R_a = 0.3 \, \Omega$$
$$K_\omega = 1.0 \, \text{V/rad/s}$$
$$K_t = 1.5 \, \text{lb-ft/arm. amp}$$

The armature winding inductance is negligible.

(a) Compute the steady-state speed corresponding to a step applied armature voltage of 210 V.

(b) How long does the motor take to reach within 95% of the steady-state speed of part (a)?

(c) What is the value of the total effective viscous damping coefficient of this motor-load configuration?

11-21. The motor of Prob. 11-20 is now used as a field-controlled d-c machine in the manner illustrated in Fig. 11-23. The armature is assumed to be energized from a constant current source. The field winding parameters are: $R_f = 50 \, \Omega$ and $L_f = 20 \, \text{H}$. Moreover, the motor torque constant now expressed in lb-ft per field amperes (K_{tf}) is 60.

(a) Draw the block diagram.

(b) Calculate the value of steady-state speed for a step applied field voltage of 100 V.

(c) Approximately how long does it take the motor to reach within 95% of the final speed of part (b)?

(d) Write the state equations in matrix form, using numerical values.

11-22. A two-phase servomotor has a standstill torque at full control winding voltage of 4 oz-in. Also, at full control voltage, the no-load speed is known to be 100 rad/s. The motor torque constant is 0.1 oz-in. per V, or 4 oz-in. per control ampere. The control winding inductance is 6 H. Assume linear torque-speed curves and a rotor inertia of 0.5 oz-in.²

(a) Determine the complete numerical expressions relating motor displacement to control winding voltage.

(b) If the result of part (a) is assumed driven from an amplifier having a voltage gain of 200, write the total expression for the direct transmission function.

(c) Place a unity feedback loop around the direct transmission function of part (b), and find the closed loop transfer function.

(d) What is the characteristic equation of the system of part (c)?

(e) What is the characteristic equation of the system of part (b)?

11-23. An amplidyne generator is used as a power stage that drives the d-c motor of Prob. 11-20. The amplidyne has a control winding resistance of 400 Ω, a quadrature-axis circuit resistance of 0.1 Ω, and a direct-axis circuit resistance of 0.5 Ω. Moreover, the voltage induced in the cross-axis for each milliampere of control current is 0.1 V, and the voltage induced in the direct axis for each ampere in the quadrature axis is 5 V. The inductance of the control winding is 50 H and that of the quadrature axis is 2 mH.

(a) Determine the transfer function that relates motor output speed to amplidyne control winding voltage.

(b) For a control voltage of 5 V compute the steady-state speed of the motor.

(c) Identify the predominant factor in establishing the dynamic response of this amplidyne-motor system.

11-24. A unity feedback loop is placed around the amplidyne-motor configuration of Prob. 11-23. The feedback element is a position-sensitive device that has a transducer constant of 1 volt per radian of motor displacement.

(a) Find the closed-loop transfer function.

(b) Identify the characteristic equation.

(c) How many transient modes are there?

(d) Compute the time constant of each transient mode.

11-25. Refer to the system of Fig. 11-33, and assume that the parameters are those listed on p. 516.

(a) Calculate the no-load terminal voltage corresponding to a reference input voltage of 30.1 V.

(b) What is the value of the closed loop *gain*?

(c) Determine the complete solution for the output voltage corresponding to a step input of 30.1 V. Assume no-load.

(d) What is the time constant of this closed loop system?

11-26. In the system configuration of Fig. 11-33 find the complete solution for the output voltage for a step reference input of 30.1 V and the simultaneous load current demand of 25 A.

11-27. A speed control system is designed to have the configuration shown in Fig. P11-27. The system parameters have the following identification:

Figure P11-27

K_a = amplifier gain, V/V

R_f = generator field winding resistance, Ω

L_f = generator field winding inductance, H

K_f = emf constant of generator, V/field amp

R = combined armature circuit resistance of generator and motor, Ω

K_t = motor torque constant, units of torque/arm. amp

J = equivalent inertia of motor and load

F = viscous friction coefficient of the load, assume negligible

K_ω = motor speed-voltage constant, V/rad/s

K_0 = d-c tachometer speed-voltage constant, V/rad/s

Develop the complete block diagram for this system, and clearly show the transfer function for each block.

11-28. Refer to the system of Prob. 11-27, and find the expression for the no-load steady-state speed in terms of the system parameters for a given value of reference voltage.

11-29. Refer to the system of Prob. 11-27, and determine the expression for the final change in speed in terms of the system parameters corresponding to an applied load torque of T_L units of torque.

11-30. Determine the state equations in matrix form for the system of Fig. P11-27. Assume that the system is operating in response to an applied reference voltage E_R and an applied load torque of T_L.

IV

INTRODUCTION TO ELECTRIC POWER SYSTEMS

12

Transmission Line Parameters

The transmission lines of an electric power system serve as the channels by which energy, generated at a remote location, is supplied at a distribution level to regions where there exist industrial and commercial needs as well as the needs of the population at large. Even in situations where the power generation takes place within the region where electric energy is consumed, transmission lines nevertheless exist in order to provide interconnection with other power systems which are members of a common power pool. The transmitting voltages used on transmission lines are frequently quite high. The usual range for lines classified as *high voltage* (HV) is from 115 kV to 230 kV; and the most often used values in this range are 115 kV, 138 kV, and 230 kV. For lines that must span larger distances, transmitting voltages selected from the category of *extra high voltages* (EHV) are used. The common quantities in this group are 345 kV, 500 kV, and 765 kV. Although the equipment costs of an electric power system rise with the use of higher transmitting voltages, nonetheless the higher voltages are indispensable if operating costs are to be optimized and line capacity is to be preserved. Keep in mind that I^2R losses are incurred in the conductors that constitute the transmission line. For a given power transmission these losses can be appreciably reduced by diminishing the current flowing in the line conductors, a result that is the immediate consequence of higher transmitting voltages. The smaller current also offers the advantage of reduced voltage drop in the line between the sending and receiving ends.

The generation of these high and extra-high transmission line voltage levels is efficiently and conveniently accomplished through use of the transformer. In fact, this is one of the principal reasons for the preference of a-c transmission over d-c transmission, although in situations where very large distances are involved (> 600 km) d-c transmission is the preferred means.

The present state of the art in the design of synchronous generators puts the maximum voltage rating at 24 kV. Many generators however carry voltage ratings of 14 kV and 18 kV. Clearly, then, a three-phase transformer or banks of three single-phase transformers are needed to get to the selected transmission level. Invariably, the connection of the synchronous generator to the transmission-line bus is achieved through a Δ-Y arrangement of the transformers (see Fig. 12-1). At the receiving end of the transmission line, transformer banks are again employed to reduce the transmission line voltage to a more moderate level and this can take place at a number of places called substations. The operating voltage range here can run anywhere from 34.5 kV to as high as 138 kV. Large consumers of electric energy are often supplied directly from such substations but the chief function of these substations is to branch the energy to the distribution substations, which are represented in Fig. 12-1 by buses 3 and 4. When electric energy arrives at the distribution substations from the bulk-power substations, transformer banks are again used to provide a further reduction in voltage to the distribution levels. Here, the voltages are often in the range from 4.16 kV to 13.8 kV as they appear at buses 5 and 6. Many industrial and commercial loads are supplied from these lines. Small consumers obtain their energy at the output of the secondaries of the last bank of transformers depicted in Fig. 12-1. In the case of a three-phase four-wire distribution system, the popular voltages encountered are 208/120 V and 480/277 V. For the single-phase three-wire system the voltages are 240/120 V.

The longest section by far of the connections that join the output of the synchronous generator to the consumers at the distribution points is the transmission line. The manner in which the transmission line is treated analytically is dependent upon its length. If it is 50 km or less only the resistance and inductance of the line conductors are considered and these are treated as a series impedance linking bus 1 to the step-down transformers at bus 2 in Fig. 12-1. When the transmission line length falls between 50 and 160 km, it is customary to take into account the capacitive shunting effect that exists between the lines. But this is done with lumped parameters. Finally, if the line exceeds 160 km, a more sophisticated representation of the transmission section is necessary in order to account for the distributed nature of the parameters all along the length of the line. The lumped model is no longer accurate enough to be acceptable. The first part of this chapter focuses on the lumped-parameter representation of the transmission line by describing how the inductance and capacitance properties of the line per unit length can be determined. In the latter part long-line theory is applied in order to arrive at a more realistic representation of the effect of the long transmission line on total electric power system performance.

12-1 LINE INDUCTANCE

As a review of Sec. 1-8 reminds us, inductance is a circuit parameter that is related to the medium as well as to the geometry associated with current-carrying conductors. We also know that this characteristic of the circuit manifests itself only when the

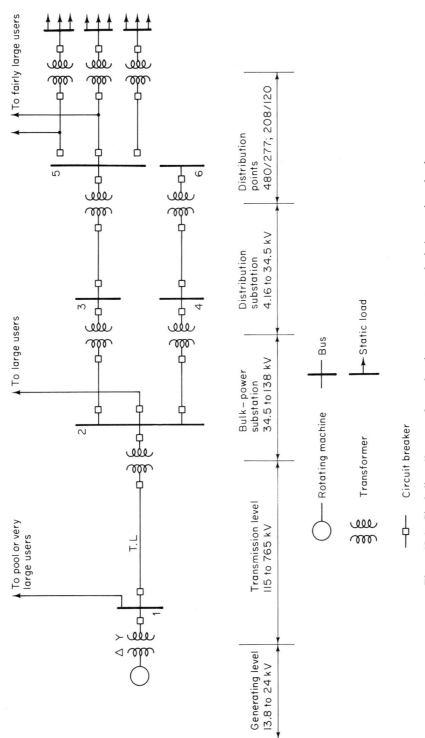

Figure 12-1 Single-line diagram of an electric power system depicting various levels distinguished by operating voltage values.

541

current is undergoing change. Accordingly, when a transmission line is designed to transmit power at d-c voltages, line inductance is not an adverse factor during steady-state operations. But in view of the fact that the vast majority of electric energy is transmitted with alternating currents (ac), the calculation methods and the effects of line inductance are matters of primary importance in the study of transmission lines and the overall behavior of electric power systems. Our concern in this section is to describe the procedures used to obtain the inductance per unit length of transmission lines placed in various geometrical configurations.

Internal Inductance of Line Conductor

The self inductance of a single line is computed in two parts. One is the inductance L_i associated with the internal geometry of the conductor. The other is the inductance L_o associated with the geometry that is external to the current-carrying conductor. In typical situations the former is much smaller than the latter and, frequently, it does not exceed more than 5% of the total self-inductance.

The inductance of the inside portion of the solid conductor carrying the instantaneous current I is best found through application of the relationship

$$L_i = \frac{\psi_r}{I} \qquad (12\text{-}1)$$

where ψ_r denotes the total flux linkage of the conductor extended out to its radius r. As one moves along the radius of the conductor, the encountered flux linkages do not vary in a simple linear fashion. Hence the expression for the total flux linkage of the conductor with its internal flux is really dependent upon integration of an appropriate expression for the differential flux linkage at a distance x from the center. The situation is depicted in Fig. 12-2. Here, an annular ring of thickness dx is drawn at radius x. Our interest is with a cylinder of unit length (1 m for SI units) having the cross-sectional area represented by the annular ring. There is a differential flux $d\phi_x$ that penetrates the longitudinal cross-sectional area of this cylinder which can be expressed as

$$d\phi_x = B_x A_c = \mu_r \mu_0 H_x \,(dx \times 1\text{ m}) \qquad (12\text{-}2)$$

Here μ_r is the relative permeability of the medium, which for transmission line conductors is usually aluminum, μ_0 is the permeability of free space, and H_x is the value

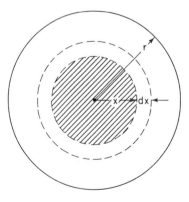

Figure 12-2 Cross section of a solid round transmission line conductor for calculation of the partial flux linkage of radius x. The conductor has a radius r.

of the internal field intensity at the radius x. The flux $d\phi_x$ is produced by that portion of the current I that is associated with the shaded area shown in Fig. 12-2. The expression for the field intensity H_x is readily determined by applying Ampere's circuital law along the inner circumference of the annular ring. Thus,

$$\int_0^{2\pi x} H_x \, dl = \text{enclosed current} = \frac{x^2}{r^2} I \qquad (12\text{-}3)$$

or

$$H_x = \frac{I}{2\pi r^2} x \qquad \text{A-t/m} \qquad (12\text{-}4)$$

Inserting this expression into Eq. (12-2) then yields

$$d\phi_x = \frac{\mu_r \mu_0 I}{2\pi r^2} x \, dx \qquad (12\text{-}5)$$

The partial differential flux linkage associated with this differential flux is given by

$$d\psi_x = N_x \, (d\phi_x) \qquad (12\text{-}6)$$

where N_x denotes the number of turns through which the enclosed current of Eq. (12-3) flows. Because we are dealing here with a single conductor (which for convenience is assumed to have a return conductor remotely located), N_x is described by that fraction of one turn represented by the shaded area of Fig. 12-2. Accordingly,

$$N_x = \frac{\pi x^2}{\pi r^2}(1) = \frac{x^2}{r^2} \qquad (12\text{-}7)$$

Introducing Eqs. (12-5) and (12-7) into Eq. (12-6), the desired expression for the partial differential flux linkage existing at radius x becomes

$$d\psi_x = \frac{\mu_r \mu_0 I}{2r^4} x^3 \, dx \qquad (12\text{-}8)$$

The total flux linkage of the inside of the conductor can now be found as

$$\psi_r = \int_0^r d\psi_x = \frac{\mu_r \mu_0 I}{2r^4} \int_0^r x^3 \, dx = \frac{\mu_r \mu_0}{8\pi} I \qquad (12\text{-}9)$$

Upon dividing both sides of the last expression by I, the result for the internal inductance of the conductor becomes

$$\boxed{L_i = \frac{\psi_r}{I} = \frac{\mu_r \mu_0}{8\pi}} \qquad \text{H/m} \qquad (12\text{-}10)$$

Note the striking verification that inductance is dependent on the medium ($\mu_r \mu_0$) and the geometrical configuration. It is also important to keep in mind that this result is applicable to a unit length of 1 m.

External Inductance of Line Conductor

The flux that is related to the self-inductance of the conductor outside the wire is more directly obtained because a typical flux path such as the one shown in Fig. 12-3 encloses the conductor current completely. For the sake of argument let it be desired

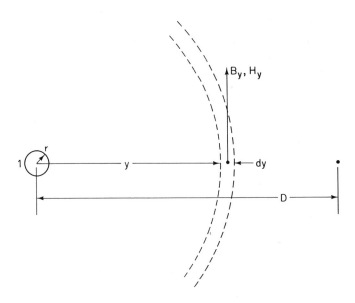

Figure 12-3 Geometrical configuration for finding the inductance for the flux linkage outside the conductor.

that the outside inductance, L_o, be found per unit length of the conductor out to a distance D measured radially from the wire center. For the moment it is assumed that the return conductor is not to be found in the region up to distance D. The differential flux linkage associated with the flux density, B_y, at a distance y from the conductor for an axial length of wire of 1 m is

$$d\psi_y = N(d\phi_y) = N(B_y A_y) \qquad (12\text{-}11)$$

But for a single conductor $N = 1$ and for a 1-m axial length $A_y = 1 \times dy$. Hence the last equation can be rewritten as

$$d\psi_y = B_y \, dy = \mu_r \mu_0 H_y \, dy \qquad (12\text{-}12)$$

The circuital law applied to the closed flux line at the radius y yields

$$\int_0^{2\pi y} H_y \, dl = \text{enclosed current} = I \qquad (12\text{-}13)$$

from which it follows that

$$H_y = \frac{1}{2\pi y} I \qquad (12\text{-}14)$$

Inserting this result into Eq. (12-12), the required expression for the differential flux linkage is seen to be

$$d\psi_y = \frac{\mu_r \mu_0 I}{2\pi} \frac{dy}{y} \qquad (12\text{-}15)$$

To obtain the total flux linked with the conductor on the outside from its surface to the distance D, it is necessary to integrate Eq. (12-15) from r to $D - r$. Thus,

$$\psi_o = \frac{\mu_r \mu_0}{2\pi} I \int_r^{D-r} \frac{dy}{y} = \frac{\mu_r \mu_0 I}{2\pi} \ln \frac{D-r}{r} \qquad (12\text{-}16)$$

where ψ_o denotes the flux linking the conductor outside the wire. For the often encountered situation where $D \gg r$, the last equation can be rewritten as

$$\psi_o = \frac{\mu_r \mu_0}{2\pi} I \ln \frac{D}{r} \tag{12-17}$$

Since inductance is flux linkage per unit enclosed current, it follows that the expression for the outside inductance up to distance D is simply

$$L_o = \frac{\psi_o}{I} = \frac{\mu_r \mu_0}{2\pi} \ln \frac{D}{r} \qquad \text{H/m} \tag{12-18}$$

Again take note of the dependence of the inductance upon the medium and the parameters associated with the geometry of the configuration. It is useful to note too that allowing D to take on values approaching infinity does not lead to exorbitant inductance. This is because the return conductor will eventually be enclosed at some distance thereby yielding a net enclosed current which is zero. In turn this means that a flux line enclosing both conductors cannot exist.

The total self-inductance of the single conductor is simply the sum of the internal and external values. Thus, summing and collecting the terms of Eqs. (12-10) and (12-17), the total inductance per meter length is determined to be

$$L = L_i + L_o = \frac{\mu_r \mu_0}{2\pi} \left(\frac{1}{4} + \ln \frac{D}{r} \right) \qquad \text{H/m} \tag{12-19}$$

By use of the identity $\frac{1}{4} = \ln \epsilon^{1/4}$, Eq. (12-19) can be rewritten as

$$L = L_i + L_o = \frac{\mu_r \mu_0}{2} \left(\ln \epsilon^{1/4} + \ln \frac{D}{r} \right) = \frac{\mu_r \mu_0}{2\pi} \ln \frac{D}{\epsilon^{-1/4} r} \tag{12-20}$$

The quantity $\epsilon^{-1/4}$ has the value 0.7788. If the actual solid conductor of radius r is now replaced by a hollow cylindrical conductor of radius r', where

$$r' = 0.7788r = \epsilon^{-1/4} r \tag{12-21}$$

the inductance of this equivalent geometrical configuration can be more simply expressed as

$$L = \frac{\mu_r \mu_0}{2\pi} \ln \frac{D}{r'} \qquad \text{H/m} \tag{12-22}$$

The use of Eq. (12-21) permits the internal inductance to be treated in terms of an equivalent radius. The conductor radius in Eq. (12-22) is sufficiently reduced so that the natural logarithm of the slightly greater ratio D/r' exactly accounts for the effect of the internal inductance of the conductor.

Introducing the result displayed in Eq. (1-4) for the permeability of free space, the expression for the total conductor inductance in air becomes

$$L = \frac{\mu_r (4\pi \times 10^{-7})}{2\pi} \ln \frac{D}{r'} = 2 \times 10^{-7} \ln \frac{D}{r'} \qquad \text{H/m} \tag{12-23}$$

where the relative permeability of air is taken to be unity. If the axial reference length of the conductor is taken to be a kilometer rather than a meter, the conductor inductance equation becomes

$$L = 2 \times 10^{-4} \ln \frac{D}{r'} \qquad \text{H/km} \qquad (12\text{-}24)$$

Total Line Inductance (Including Return Wire)

Assume next that an identical return conductor is placed at the distance D in Fig. 12-3. The complete expression for the self inductance of the first conductor is given by the last equation in terms of henrys per kilometer. The self inductance of the return conductor is also described by Eq. (12-24) since the procedures and the geometrical considerations are the same. Hence the total inductance of the line is twice that of the single conductor. Accordingly,

$$L_T = 2L = 4 \times 10^{-4} \ln \frac{D}{r'} \qquad \text{H/km (mH/m)} \qquad (12\text{-}25)$$

where D denotes the separation distance between the centers of the two conductors, r' is the equivalent radius that accounts for the inside inductance of the wires, and L_t denotes the total inductance. When the line transmits electric energy at a frequency of 60 Hz, the associated reactance is readily determined to be

$$X = \omega L_T = 2\pi(60)4 \times 10^{-4} \ln \frac{D}{r'} = 0.1508 \ln \frac{D}{r'} \qquad \Omega/\text{km} \qquad (12\text{-}26)$$

This result can also be expressed in ohms per mile as

$$X = 0.2426 \ln \frac{D}{r'} \qquad \Omega/\text{mi} \qquad (12\text{-}27)$$

Manufacturers' tables that describe various characteristics of transmission line conductors continue the use of English units. Accordingly, familiarity with these alternative expressions can be helpful.

Example 12-1

A transmission line consists of two solid aluminum conductors having a radius of 2.5 cm and separated on centers by a distance of 3.6 m.

(a) Find the total line inductance in millihenrys (mH) per meter (m).

(b) Repeat part (a) for the case where the separation is 7.2 m.

(c) Determine the reactance of the transmission line in part (a) expressed in ohms per kilometer and ohms per mile. Assume that the frequency of the conductor current is 60 Hz.

Solution (a) The total line reactance is given by Eq. (12-25). Accordingly,

$$L_T = 4 \times 10^{-4} \ln \frac{3.6}{(0.7788)(0.025)} = 4 \times 10^{-4} \ln 184.9$$

$$= 4 \times 10^{-4}(5.22) = 0.0209 \text{ mH/m}$$

(b) Applying the same formula yields

$$L_T = 4 \times 10^{-4} \ln \frac{7.2}{0.01947} = 4 \times 10^{-4} \ln 369.8 = 4 \times 10^{-4}(5.912)$$

$$= 0.02365 \text{ mH/m}$$

A comparison with the result of part (a) shows that doubling the separation between the conductors causes only about a 13% rise in inductance. The presence of the logarithmic function, of course, is responsible for this reduced sensitivity to the separation dimension.

(c) Application of Eq. (12-26) gives

$$X = \omega L_T = (0.1508)(5.22) = 0.787 \ \Omega/\text{km}$$

while Eq. (12-27) puts the reactance at the value

$$X = \omega L_T = (0.2426)(5.22) = 1.266 \ \Omega/\text{mi}$$

Effect of a Second Conductor per Line

Next let us assume that conductor 1 in Fig. 12-3 is joined by a second conductor 2 displaced by a distance d_{12} from conductor 1 in the manner illustrated in Fig. 12-4. These two conductors are further assumed to make up one side of the single phase line. The return conductor configuration is not shown. Our aim now is to write the expression for the flux linkage of conductor 1 relative to a point sufficiently far removed from conductors 1 and 2 so that the inequality $D_p \gg d_{12}$ is valid. However, D_p is not to be so large that it encloses the return conductors. When the current I_1 is considered to act alone, the flux linkge associated with the self inductance of conductor 1 is given with the aid of Eq. (12-22) by

$$\psi_{11} = \frac{\mu I_1}{2\pi} \ln \frac{D_p}{r'} \qquad (12\text{-}28)$$

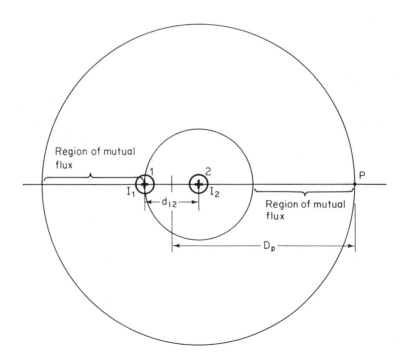

Figure 12-4 Illustrating the region where flux produced by I_2 links with conductor 1.

where $\mu = \mu_r \mu_0$ and ψ_{11} denotes the total self-flux linking conductor 1 out to distance D_p. When the current I_2 is assumed to act alone, it produces a flux field which links with conductor 1 over the region from d_{12} to D_p. Figure 12-4 illustrates this statement as the region between the two circles. Application of Eq. (12-22) in a general way permits the mutual flux produced by I_2 and linking conductor 1 to be expressed as

$$\psi_{12} = \frac{\mu I_2}{2} \ln \frac{D_p}{d_{12}} \tag{12-29}$$

Hence the total flux linking conductor 1 in the configuration of Fig. 12-4 with both currents flowing and the return current neglected is

$$\psi_1 = \underbrace{\frac{\mu I_1}{2\pi} \ln \frac{D_p}{r'}}_{\substack{\text{self flux} \\ \text{component}}} + \underbrace{\frac{\mu I_2}{2\pi} \ln \frac{D_p}{d_{12}}}_{\substack{\text{mutual flux} \\ \text{component}}} \tag{12-30}$$

On the assumption that the two conductors are identical and carry equal current, Eq. (12-30) can be written as

$$\psi_1 = \frac{\mu I_1}{2\pi} \ln \frac{D_p^2}{r' d_{12}} \tag{12-31}$$

so that the inductance of conductor 1 becomes

$$L_I = \frac{\psi_1}{I_1} = \frac{\mu}{2\pi} \ln \frac{D_p^2}{r' d_{12}} \tag{12-32}$$

Now that the effect of current in an adjacent conductor is determined, attention is next focused on the more practical situation where the return conductors are considered.

12-2 SINGLE-PHASE LINE WITH MULTICONDUCTORS: BUNDLING

Refer to the configuration shown in Fig. 12-5. Line A is assumed to be composed of two identical conductors 1 and 2. Line B is made up of three conductors a, b, and c. For convenience, further assume that $I_a = I_b = I_c$ and $I_1 = I_2$. Because at any time one line serves as the sending line while the other serves as the return line,

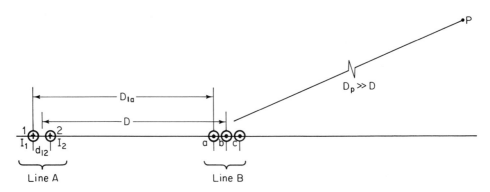

Figure 12-5 Bundled single-phase line and a point P far removed from the conductor array.

Kirchhoff's current law requires that

$$I_a + I_b + I_c = -(I_1 + I_2) \tag{12-33}$$

The first item to be investigated with this new configuration is the effect that *bundling* of the line has on the per length inductance of any single conductor.

Bundling is the term that is used to describe the composition of a line with two or more conductors. Bundling of the transmission line conductors offers several notable advantages. One is the reduction in skin effect. The use of three or four or even five conductors to replace a single conductor of comparable cross-sectional area means a significant reduction in line resistance at the transmitting frequency. This matter is examined further in Sec. 12-3. Another advantage of bundling is that it reduces the likelihood that corona will develop. *Corona* describes the breakdown or ionization of the air around the conductor. This usually occurs at electric field intensities of approximately 3000 kV/m. When a single conductor is used in the extra-high transmitting voltage range, field intensities of this magnitude can exist at the conductor surface. By replacing the single conductor by several conductors, the surface electric field intensity diminishes because of the increased surface area. The third advantage of bundling results as a direct consequence of the analysis that follows which shows that the line reactance diminishes. In addition to reducing the line reactance drop, there is also an increase in the transmitting capacity of the line.

When currents are assumed to be flowing in each of the conductors of the array depicted in Fig. 12-5, the expression for the total flux linkage of conductor 1 is conveniently determined by applying the procedure that leads to the result displayed in Eq. (12-30). Point P is assumed to be remotely situated relative to the conductor array, i.e., $D_p \gg D$, where D_p denotes the distance from the array to point P, and D is defined as shown in Fig. 12-5. Accordingly, we get

$$\psi_1 = \frac{\mu I_1}{2\pi} \ln \frac{D_p}{r_1'} + \frac{\mu I_2}{2\pi} \ln \frac{D_p}{d_{12}} + \frac{\mu I_a}{2\pi} \ln \frac{D_p}{D_{1a}} + \frac{\mu I_b}{2\pi} \ln \frac{D_p}{D_{1b}} + \frac{\mu I_c}{2\pi} \ln \frac{D_p}{D_{1c}} \tag{12-34}$$

The first term on the right side of this last expression is recognized as the total self-flux associated with conductor 1 out to point P. The second term is the mutual flux contributed by current I_2 in conductor 2; and, of course, this mutual flux is identified in the region from d_{12} to D_p. Similarly, the third term describes the mutual flux contributed by the flow of I_a in conductor a; and again this mutual flux is defined in the region extending from D_{1a} to the distance D_p that separates point P from the array. It is helpful next to rewrite Eq. (12-34) as follows:

$$\psi_1 = \frac{\mu}{2\pi}(I_1 + I_2 + I_a + I_b + I_c) \ln D_p + I_1 \ln \frac{1}{r_1'} + I_2 \ln \frac{1}{d_{12}}$$
$$+ I_a \ln \frac{1}{D_{1a}} + I_b \ln \frac{1}{D_{1b}} + I_c \ln \frac{1}{D_{1c}} \tag{12-35}$$

By KCL the sum of all conductor currents is zero. Hence

$$\psi_1 = \frac{\mu}{2\pi}\left(I_1 \ln \frac{1}{r_1'} + I_2 \ln \frac{1}{d_{12}} + I_a \ln \frac{1}{D_{1a}} + I_b \ln \frac{1}{D_{1b}} + I_c \ln \frac{1}{D_{1c}}\right) \tag{12-36}$$

Now let

$$I_1 = I_2 = \frac{I}{2} \quad \text{and} \quad I_a = I_b = I_c = -\frac{I}{3} \tag{12-37}$$

where I denotes the total current of line A or line B. The expression for the total flux linkage of conductor 1 thus becomes

$$\psi_1 = \frac{\mu}{2\pi}\left(\frac{I}{2}\ln\frac{1}{r_1'd_{12}} - \frac{I}{3}\ln\frac{1}{D_{1a}D_{1b}D_{1c}}\right)$$

Combining terms then yields

$$\psi_1 = \frac{\mu I}{2\pi}\ln\frac{\sqrt[3]{D_{1a}D_{1b}D_{1c}}}{\sqrt{r_1'd_{12}}} \tag{12-38}$$

At this point, it is useful to write a generalized version of the inductance per meter length.

Generalized Version

Let the two conductors of line A in Fig. 12-5 be replaced by n identical conductors and let line B be replaced by m identical conductors. We continue to let I denote the total current of either line. In the case of line A, however, we have the relationship $I = nI_n$ where I_n is the current per conductor. Similarly, for line B we can write $I = mI_m$. In this general configuration of a bundled single-phase transmission line, the expression for the total flux linkage of conductor 1 with all the other conductors in the array is plainly

$$\psi_1 = \frac{\mu I}{2\pi}\ln\frac{\sqrt[m]{D_{1a}D_{1b}\cdots D_{1m}}}{\sqrt[n]{r_1'd_{12}\cdots d_{1n}}} \qquad \text{H/m} \tag{12-39}$$

To find the total inductance of conductor 1 it is necessary to formulate the ratio ψ_1/I_1. Because $I_1 = I_n = I/n$, the last equation may be rewritten to yield the desired result. Thus,

$$L_1 = \frac{\psi_1}{I_1} = \frac{\mu}{2\pi}n\ln\frac{\sqrt[m]{D_{1a}D_{1b}\cdots D_{1m}}}{\sqrt[n]{r_1'd_{12}\cdots d_{1n}}} \tag{12-40}$$

By repeating the foregoing procedure for conductor 2, it is easy to show the result to be

$$L_2 = \frac{\psi_2}{I_2} = \frac{\mu}{2\pi}n\ln\frac{\sqrt[m]{D_{2a}D_{2b}\cdots D_{2m}}}{\sqrt[n]{r_2'd_{21}d_{23}\cdots d_{2n}}} \tag{12-41}$$

In general the argument to the logarithmic terms in the last two equations will differ somewhat. Hence, it is customary in performing these calculations, to evaluate an *average* value of the conductor inductance associated with each line bundle. Thus for the bundle of n conductors of line A we have

$$L_{av} = \frac{L_1 + L_2 + \cdots + L_n}{n} \tag{12-42}$$

Of course, the parameter of principal interest in these situations is the total inductance per meter length of the line itself. Because the conductors in the bundle of line A are all connected in parallel, it follows that the resultant line inductance is merely

$$L_A = \frac{L_{av}}{n} = \frac{L_1 + L_2 + \cdots + L_n}{n^2} \tag{12-43}$$

Substituting expressions such as Eqs. (12-40) and (12-41) into Eq. (12-43) reveals the total line inductance to take the form

$$L_A = \frac{\mu}{2\pi}\frac{1}{n}\left(\ln\frac{\sqrt[m]{D_{1a}D_{1b}\cdots D_{1m}}}{\sqrt[n]{r_1'd_{12}\cdots d_{1n}}} + \ln\frac{\sqrt[m]{D_{2a}D_{2b}\cdots D_{2m}}}{\sqrt[n]{r_2'd_{21}d_{23}d_{2n}}}\right.$$

$$\left. + \cdots + \ln\frac{\sqrt[m]{D_{na}D_{nb}\cdots D_{nm}}}{\sqrt[n]{r_n'd_{n1}\cdots d_{n-1}}}\right) \tag{12-44}$$

Combining the logarithmic terms leads to

$$L_A = \frac{\mu}{2\pi}\ln\frac{\sqrt[mn]{(D_{1a}\cdots D_{1m})(D_{2a}\cdots D_{2m})\cdots(D_{na}\cdots D_{nm})}}{\sqrt[mn]{(r_1'd_{12}\cdots d_{1n})(r_2'd_{21}\cdots d_{2n})\cdots(r_n'd_{n1}\cdots d_{n-1})}} \tag{12-45}$$

This end result for the bundled line has a striking similarity to the result found for the single-phase line composed of a total of two conductors. A comparison with Eq. (12-22) bears this out. Whereas in Eq. (12-22) the quantity D is the distance between the two conductors of the single-phase line, in the bundled case of Eq. (12-45) the separation distance between the bundles is represented by the mnth root of the product of mn terms. Each distance term is taken in sequence from the first conductor of the bundle n to each conductor of the bundle m with the process continued to the last conductor in bundle n to each conductor of bundle m. The net effect is to unveil a distance quantity which is the *geometric mean distance*, D_g, that separates the two bundles. Thus,

$$D_g \equiv \sqrt[mn]{(D_{1a}D_{1b}\cdots D_{1m})(D_{2a}D_{2b}\cdots D_{2m})\cdots(D_{na}D_{nb}\cdots D_{nm})}$$

$$\tag{12-46}$$

Analysis leads to a similar conclusion regarding the denominator of the argument of the logarithm in Eq. (12-45). Observe that the distances involved here are all within the n bundle and the operation exhibits the n^2 root of the product of n^2 distance terms. This leads to a quantity which is properly interpreted as a *geometric mean radius*, R_g, of the bundle. Thus,

$$R_{gn} \equiv \sqrt[n^2]{(r_1'd_{12}\cdots d_{1n})(r_2'd_{21}d_{23}\cdots d_{2n})\cdots(r_n'd_{n1}\cdots d_{n-1})}$$

$$\tag{12-47}$$

The second subscript is introduced to identify bundle n. Upon introducing the last two expressions into Eq. (12-45), the inductance of line A can be written simply as

$$L_A = \frac{\mu}{2\pi}\ln\frac{D_g}{R_{gn}} \qquad \text{H/m} \tag{12-48}$$

For the case of aluminum conductors in air this becomes

$$L_A = 2\times 10^{-7}\ln\frac{D_g}{R_{gn}} \qquad \text{H/m} \tag{12-49a}$$

Or, when expressed in millihenrys per kilometer the result is

$$L_A = 0.2\ln\frac{D_g}{R_{gn}} \qquad \text{mH/km} \tag{12-49b}$$

Sec. 12-2   Single-Phase Line with Multiconductors: Bundling

The net inductance of line B, which is assumed to consist of m conductors, cannot be calculated from Eq. (12-49) because its geometric mean radius is different. The geometric mean distance, however, is the same. Therefore, to find L_B we must first obtain R_{gm} from

$$R_{gm} = \sqrt[m^2]{(r'_a d_{ab} \cdots d_{am})(r'_b d_{ba} d_{bc} \cdots d_{bm}) \cdots (r'_m d_{ma} \cdots d_{m(m-1)})} \quad (12\text{-}50)$$

where again the second subscript denotes the m bundle. The inductance of line B is then found from

$$L_B = 2 \times 10^{-7} \ln \frac{D_g}{R_{gm}} \quad \text{H/m} \quad (12\text{-}51)$$

The inductance of the entire line is

$$L_T = L_A + L_B = 2 \times 10^{-7} \ln \frac{D_g^2}{R_{gm} R_{gn}} \quad (12\text{-}52)$$

Finally, in the more typical practical situation where the bundle configuration of each line is the same, the total line inductance is given simply as

$$L_T = 4 \times 10^{-7} \ln \frac{D_g}{R_g} \quad \text{H/m} \quad (12\text{-}53)$$

The 60-Hz reactance of this inductance expressed in more practical units follows.

$$\boxed{X = 0.1508 \ln \frac{D_g}{R_g}} \quad \Omega/\text{km} \quad (12\text{-}54)$$

or

$$\boxed{X = 0.2426 \ln \frac{D_g}{R_g}} \quad \Omega/\text{mi} \quad (12\text{-}55)$$

Example 12-2

A single-phase transmission line uses the bundled arrangement shown in Fig. 12-6. Line A has its conductors situated at the corners of an equilateral triangle. All conductors are identical and each has a radius of 2 cm.

(a) Find the inductance of the individual lines as well as the total inductance. Express the results in millihenrys per kilometer.

(b) For operation of the line at a frequency of 60 Hz, determine the total line reactance in ohms per kilometer and ohms per mile.

Solution (a) The geometric mean distance that separates the two bundles is found by executing the instruction mathematically displayed in Eq. (12-46). For the spacing dimensions indicated in Fig. 12-6, we get

$$D_g = \sqrt[9]{(6.1)^2 (6.2)^2 (6.3)6(6.05)(6.15)(6.25)} = 6.15 \text{ m}$$

The geometric mean radius of the equilateral arrangement of line A is found to be

$$R_{gA} = \sqrt[9]{(0.015576)^3 (0.1)^6} = 0.0538 \text{ m}$$

Here the first term beneath the radical is obtained from

$$r' = 0.7788r = (0.7788)(0.02) = 0.015576 \text{ m}$$

(a)

(b)

Figure 12-6 Diagram for Example 12-2: (a) actual conductor configuration of a bundled, single-phase transmission line; (b) equivalent representation in terms of a separation which is the geometric mean distance and the geometric mean radius of each bundle.

Observe that the geometric mean radius is quite a bit larger than r', owing to the separation between conductors in the bundle. The use of this increased quantity in an equation such as Eq. (12-49) is what accounts for the beneficial reduction of line reactance associated with the bundling of the conductors in place of a single, larger conductor.

Because line B uses a different conductor configuration, its geometric mean radius is different than that of line A. The result here is

$$R_{gB} = \sqrt[9]{(0.015576)^3(0.1)^4(0.2)^2} = 0.0628 \text{ m}$$

It is interesting and instructive to note at this point that the actual configuration of Fig. 12-6(a) can now be replaced by the two equivalent hollow conductors each with its own geometric mean radius and separated by the geometric mean distance as illustrated in Fig. 12-6(b).

Line A inductance can now be computed by inserting the foregoing results into Eq. (12-49). Thus,

$$L_A = 2 \times 10^{-7} \ln \frac{6.15}{0.0538} = 9.48 \times 10^{-7} \text{ H/m} = 0.948 \text{ mH/km}$$

Similarly, for line B

$$L_B = 2 \times 10^{-7} \ln \frac{6.15}{0.0628} = 9.17 \times 10^{-7} \text{ H/m} = 0.917 \text{ mH/km}$$

Therefore the total line inductance is

$$L_T = L_A + L_B = 1.865 \text{ mH/km}$$

(b) At 60 Hz the total line reactance becomes

$$X = \omega L = 377 \times 1.865 \times 10^{-3} = 0.703 \text{ }\Omega/\text{km}$$

On a per-unit mile basis this result is

$$X = 1.609(0.703) = 1.131 \ \Omega/\text{mi}$$

12-3 LINE INDUCTANCE OF THREE-PHASE TRANSMISSION SYSTEMS

Attention is next directed from single-phase lines to three-phase lines. The foregoing results are indispensable in analyzing the three-phase case because these results are directly applicable subject only to the constraints peculiar to the three-phase situation.

Equilateral Spacing

We begin by considering a nonbundled three-phase transmission line designed with equilateral spacing. In other words, the lines are placed at the corners of an equilateral triangle in the manner depicted in Fig. 12-7 and the neutral wire is omitted for now. To find the expression for the line inductance of phase A, we start by writing the flux linkage of phase A conductor caused by current in each phase and expressed relative to a distance point P. Thus,

$$\psi_a = \frac{\mu}{2\pi}\left(I_a \ln\frac{D_p}{r'_a} + I_b \ln\frac{D_p}{D_{ab}} + I_c \ln\frac{D_p}{D_{ac}}\right) \qquad (12\text{-}56)$$

$$\underset{C}{\bigcirc}$$

Figure 12-7 Nonbundled three-phase transmission line with equilateral spacing.

But for $D_{ab} = D_{ac} = D$ this expression becomes

$$\psi_a = \frac{\mu}{2\pi}\left[(I_a + I_b + I_c)\ln D_p + I_a \ln\frac{1}{r'_a} + I_b \ln\frac{1}{D} + I_c \ln\frac{1}{D}\right] \qquad (12\text{-}57)$$

Upon introducing the boundary condition for the three-phase line, namely, $I_a + I_b + I_c = 0$ there results

$$\begin{aligned}
\psi_a &= \frac{\mu}{2\pi}\left[I_a \ln\frac{1}{r'_a} + (I_b + I_c)\ln\frac{1}{D}\right] \\
&= \frac{\mu}{2\pi}\left(I_a \ln\frac{1}{r'_a} - I_a \ln\frac{1}{D}\right) \\
&= \frac{\mu I_a}{2\pi}\ln\frac{D}{r'_a} \qquad \text{H/m}
\end{aligned} \qquad (12\text{-}58)$$

Consequently, the line inductance of phase A is

$$L_a = \frac{\psi_a}{I_a} = \frac{\mu}{2\pi}\ln\frac{D}{r'_a} = 2 \times 10^{-7}\ln\frac{D}{r'_a} \qquad \text{H/m} \qquad (12\text{-}59)$$

A repetition of the foregoing procedure for each of the remaining phases unveils identical expressions except for the subscripts. Thus,

$$L_b = 2 \times 10^{-7} \ln \frac{D}{r_b'} \quad \text{and} \quad L_c = 2 \times 10^{-7} \ln \frac{D}{r_c'} \tag{12-60}$$

Because the results that describe the phase line inductances are identical, it is clear that the mutual coupling between any one phase and the other two is the same. This is not the case, for example, when parallel spacing of the phase conductors is employed.

Parallel Spacing in a Plane (Unsymmetrical Spacing)

The construction cost of building a three-phase transmission line with equilateral spacing is greater than for the use of parallel spacing such as that shown in Fig. 12-8.

Figure 12-8 Nonbundled three-phase transmission line with parallel spacing.

The analysis of this case proceeds by placing emphasis on the determination of expressions for the flux linkage of each phase per meter length rather than the line inductances for reasons soon to be evident. Again, resorting to a point remotely located relative to the conductor array, the flux linkage of phase A is given by

$$\psi_a = \frac{\mu}{2\pi} \left(I_a \ln \frac{D_p}{r_a'} + I_b \ln \frac{D_p}{D} + I_c \ln \frac{D_p}{2D} \right) \tag{12-61}$$

or

$$\psi_a = \frac{\mu}{2\pi} \left[(I_a + I_b + I_c) \ln D_p + I_a \ln \frac{1}{r_a'} + I_b \ln \frac{1}{D} + I_c \ln \frac{1}{2D} \right]$$

$$= \frac{\mu}{2\pi} \left[I_a \ln \frac{1}{r_a'} + I_b \ln \frac{1}{D} + I_c \ln \frac{1}{2D} \right] \tag{12-62}$$

It is instructive to note here that the expression on the right side cannot be manipulated to isolate I_a as was done for equilateral spacing where the mutual coupling effects were the same because of the symmetrical spacing. Nonetheless some simplification can be effected by inserting $I_c = -(I_a + I_b)$ and collecting the terms. The result is

$$\psi_a = \frac{\mu}{2\pi} \left(I_a \ln \frac{2D}{r_a'} + I_b \ln 2 \right) \tag{12-63}$$

The flux linkage for phase B can be written initially as

$$\psi_b = \frac{\mu}{2\pi} \left(I_a \ln \frac{D_p}{D} + I_b \ln \frac{D_p}{r_b'} + I_c \ln \frac{D_p}{D} \right) \tag{12-64}$$

or

$$\psi_b = \frac{\mu}{2\pi} \left[(I_a + I_c) \ln \frac{1}{D} + I_b \ln \frac{1}{r_b'} \right]$$

Note the presence of symmetry as regards phase B. As a result ψ_b may be written more compactly as

$$\psi_b = \frac{\mu}{2\pi} \left(I_b \ln \frac{D}{r_b'} \right) \quad \text{H/m} \tag{12-65}$$

A repetition of this procedure for phase C leads to a result identical to that of phase A except that subscript a is replaced by c. Thus,

$$\psi_c = \frac{\mu}{2\pi}\left(I_b \ln 2 + I_c \ln \frac{2D}{r'_c}\right) \qquad \text{H/m} \qquad (12\text{-}66)$$

Here too dissymmetry exists. In fact by putting Eqs. (12-63), (12-65), and (12-66) in matrix form the overall dissymmetry associated with parallel spacing is impressively unveiled. Thus,

$$\begin{bmatrix} \psi_a \\ \psi_b \\ \psi_c \end{bmatrix} = \frac{\mu}{2\pi} \begin{bmatrix} \ln\dfrac{2D}{r'} & \ln 2 & 0 \\ 0 & \ln\dfrac{2D}{r'} & 0 \\ 0 & \ln 2 & \ln\dfrac{2D}{r'} \end{bmatrix} \begin{bmatrix} I_a \\ I_b \\ I_c \end{bmatrix} \qquad (12\text{-}67)$$

where $r' = r'_a = r'_b = r'_c$. Observe that the inductance matrix is not diagonal as is the case for equilateral spacing. A line that is permitted to operate with parallel spacing will incur unbalanced voltage drops in the line even for balanced line currents. To avoid this shortcoming, such lines are strung in a *transposed* fashion.

Transposed, Parallel-Spaced Transmission Line

A three-phase transmission line is said to be transposed when it is crossed to occupy the normal position of each of the other two phases for a distance equal to one-third the total line length. The situation is illustrated in Fig. 12-9. The flux linkage of phase A can now be determined over a unit length by taking into account the fact that over the first third of this length phase A conductor occupies its normal position in which case the flux linkage is described by Eq. (12-63). Over the next third of the length conductor A occupies the position normally filled by phase B, in which case the flux linkage is described by Eq. (12-65). Of course, when writing this equation to apply to phase A, it is necessary to replace I_b by I_a. Finally, over the last third of the length, conductor A takes up the normal position of phase C corresponding to which the flux linkage is described by Eq. (12-66). It is important to note, however, that in this section of the line I_c takes on the role of I_b and I_a plays the role of I_c as Fig. 12-9

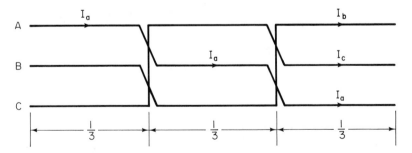

Figure 12-9 Illustrating the transposition of the phases that is used in parallel spacing in order to realize a diagonal inductance matrix.

makes obvious. Accordingly, these changes must be made when applying Eq. (21-66) in the transposed case for evaluating the flux linkage of phase A. With these points in mind the expression for the flux linkage of phase A for a transposed line can readily be written as

$$\psi_a = \frac{1}{3}\frac{\mu}{2\pi}\left(I_a \ln \frac{2D}{r'} + I_b \ln 2\right) + \frac{1}{3}\frac{\mu}{2\pi}\left(I_a \ln \frac{D}{r'}\right) + \frac{1}{3}\frac{\mu}{2\pi}\left(I_c \ln 2 + I_a \ln \frac{2D}{r'}\right)$$

(12-68)

Collecting terms leads to

$$\psi_a = \frac{\mu}{2\pi}\frac{1}{3}\left[2I_a \ln \frac{2D}{r'} + I_a \ln \frac{D}{r'} + (I_2 + I_3) \ln 2\right]$$

(12-69)

Introducing $I_b + I_c = -I_a$ and simplifying yields finally

$$\psi_a = \frac{\mu I_a}{2\pi} \ln \frac{\sqrt[3]{2}\,D}{r'} \qquad \text{H/m}$$

(12-70)

A quick examination of this expression reveals the striking result that the flux linkage of phase A is now entirely and solely expressible in terms of the current of phase A unlike the case of the untransposed line. In fact, this feature of the transposed line is similar to the result obtained for the phase symmetrical equilateral spacing. Accordingly, Eq. (12-70) leads us to the important and useful conclusion that a transposed parallel-spaced transmission line can be treated as an equivalent line with equilateral spacing.

A simple expression for the inductance of line A readily follows from Eq. (12-70) as

$$L_a = \frac{\psi_a}{I_a} = \frac{\mu}{2\pi} \ln \frac{\sqrt[3]{2}\,D}{r'}$$

(12-71)

By repeating the foregoing procedure for each remaining phase the same result is found. Thus for phases B and C we get

$$L_b = \frac{\psi_b}{I_b} = \frac{\mu}{2\pi} \ln \frac{\sqrt[3]{2}\,D}{r'}$$

(12-72)

and

$$L_c = \frac{\psi_c}{I_c} = \frac{\mu}{2\pi} \ln \frac{\sqrt[3]{2}\,D}{r'}$$

(12-73)

The corresponding matrix representation of the flux linkages involved in the last three equations for the transposed line takes on the form

$$
\begin{bmatrix} \psi_a \\ \psi_b \\ \psi_c \end{bmatrix} = \frac{\mu}{2\pi}
\begin{bmatrix} \ln \dfrac{\sqrt[3]{2}\,D}{r'} & 0 & 0 \\ 0 & \ln \dfrac{\sqrt[3]{2}\,D}{r'} & 0 \\ 0 & 0 & \ln \dfrac{\sqrt[3]{2}\,D}{r'} \end{bmatrix}
\begin{bmatrix} I_a \\ I_b \\ I_c \end{bmatrix}
$$

(12-74)

The symmetry of the transposed line is exhibited in the diagonal character of the inductance matrix. A diagonal matrix is especially important because it permits analysis of electric power systems to proceed on a per phase basis. The phases are uncoupled from each other.

Effect of the Neutral Conductor

Three-phase transmission lines are equipped with overhead neutrals for several reasons. The most obvious reason is that it provides a path for current flow when the system operates under unbalanced conditions. It also serves the important role of providing the transmission with a shield to protect against lightning. The neutral wire is always connected to the support towers and is thereby grounded. The size of the neutral conductor is generally substantially smaller than the phase conductors. The neutral conductor assumes no role when the electric power system operates under balanced conditions. However, in situations of extreme unbalance, as for example when a line-to-ground fault occurs, the neutral assumes a significant role and it is important that we know how to deal with it properly. This matter is developed further in the following chapter. Our interest here is to prepare expressions that are useful at that time.

For the sake of current considerations we assume a three-phase line configured in the equilateral mode and equipped with an overhead neutral wire in the manner illustrated in Fig. 12-10. Our goal is to obtain an expression for the flux linkage of phase A when current is simultaneously flowing in each of the other system conductors, including the neutral wire. As always we begin by writing the expressions for self and mutual coupling with respect to a remote point P located a distance D_p from the transmission line. Thus,

$$\psi_a = \frac{\mu}{2\pi}\left(I_a \ln \frac{D_p}{r_a'} + I_b \ln \frac{D_p}{D_{ab}} + I_c \ln \frac{D_p}{D_{ac}} + I_n \ln \frac{D_p}{D_{an}}\right) \qquad (12\text{-}75)$$

Expanding and collecting terms leads to

$$\psi_a = \frac{\mu}{2\pi}\left[(I_a + I_b + I_c + I_n) \ln D_p + I_a \ln \frac{1}{r_a'} + I_b \ln \frac{1}{D_{ab}}\right.$$
$$\left. + I_c \ln \frac{1}{D_{ac}} + I_a \ln \frac{1}{D_{an}}\right] \qquad (12\text{-}76)$$

n
O

A O   O B

O
C

Figure 12-10 Three-phase, equilaterally spaced, transmission line equipped with an overload neutral.

Introducing the boundary conditions, $I_a + I_b + I_c + I_n = 0$, $r'_a = r'$, and $D = D_{ab} = D_{ac}$, the last expression can be rewritten as

$$\psi_a = \frac{\mu}{2\pi}\left(I_a \ln\frac{1}{r'} + I_b \ln\frac{1}{D} + I_c \ln\frac{1}{D} + I_n \ln\frac{1}{D_{an}}\right) \qquad (12\text{-}77)$$

Upon eliminating I_n and rearranging terms, we can express the flux linkage of phase A for unbalanced conditions as

$$\psi_a = \frac{\mu}{2\pi}\left(I_a \ln\frac{D_{an}}{r'} + I_b \ln\frac{D_{an}}{D} + I_c \ln\frac{D_{an}}{D}\right) \qquad (12\text{-}78)$$

Observe that under unbalanced conditions a decoupled formulation of the flux linkage of phase A does not result from the present analysis despite equilateral spacing.

When treating the transmission line under fault conditions, the expression for the flux linkage of the neutral wire is also of particular interest. By proceeding as we did for phase A, the flux linkage of the neutral conductor can be expressed as

$$\psi_n = \frac{\mu}{2\pi}\left(I_a \ln\frac{D_p}{D_{na}} + I_b \ln\frac{D_p}{D_{nb}} + I_c \ln\frac{D_p}{D_{nc}} + I_n \ln\frac{D_p}{r'_n}\right) \qquad (12\text{-}79)$$

which upon rearrangement and the use of the boundary conditions becomes

$$\psi_n = \frac{\mu}{2\pi}\left(I_a \ln\frac{1}{D_{na}} + I_b \ln\frac{1}{D_{nb}} + I_c \ln\frac{1}{D_{nc}} + I_n \ln\frac{1}{r'_n}\right) \qquad (12\text{-}80)$$

Use is made of Eqs. (12-77) and (12-80) in Chapter 13 in order to determine the line drops incurred in a phase line as well as the neutral wire under fault conditions.

Example 12-3

A bundled and transposed three-phase transmission line has the conductor configuration depicted in Fig. 12-11. The identical conductors have a radius of 0.74 cm. The spacing between phase conductors is 30 cm.

Figure 12-11 Bundled, transposed, three-phase transmission line of Example 12-3.

(a) Find the inductance per phase expressed in mH/km and mH/mi.
(b) Determine the line reactance in ohms per mile for each phase at 60 Hz.

Solution Because this parallel-spaced line is transposed, the effect of mutual coupling is the same for each phase. Hence the line inductance which is determined for phase A also applies to phases B and C.

(a) Expressed in meters, the equivalent radius of the equivalent hollow conductor is

$$r' = (0.7788)(0.0074) = 0.00576 \text{ m}$$

The geometric mean radius of each phase is described by

$$R_g = \sqrt[4]{(r')^2(0.3)^2} = 0.0416 \text{ m}$$

A convenient way to find the geometric mean distance of this transmission line is to find the geometric distance between the respective phases and then apply the GMD formula using the new separation distance. Thus, the geometric mean distance between the conductors of phases A and B is

$$D_{gAB} = \sqrt[4]{6^2(6.3)(5.7)} = 5.996 \text{ m}$$

Similarly,

$$D_{gBC} = \sqrt[4]{6^2(6.3)(5.7)} = 5.996 \text{ m}$$

and

$$D_{gCA} = \sqrt[4]{12^2(12.3)(11.7)} = 11.998 \text{ m}$$

It is not uncommon for these distances to be the same as the actual separation of the phase conductors measured to their centers. Hence in situations of this kind the geometrical mean distance is simply taken as

$$D_g = \sqrt[3]{D_{ab}D_{bc}D_{ca}}$$

where the double-subscripted quantities are merely the distances between the centers of the phase conductors. In this specific case the result is simply

$$D_g = \sqrt[3]{(6)(6)12} = 7.56 \text{ m}$$

The inductance per phase can now be found upon invoking Eq. (12-49b) directly. Thus

$$L = 0.2 \ln \frac{D_g}{R_g} = 0.2 \ln \frac{7.56}{0.0416} = 1.041 \text{ mH/km}$$

or

$$L = (1.609)(1.041) = 1.674 \text{ mH/mi}$$

(b) The line reactance for each phase then becomes

$$X = \omega L 377(1.674)10^3 = 0.6313 \text{ } \Omega/\text{mi per phase}$$

12-4 THE TRANSMISSION LINE CAPACITANCE PARAMETER: SINGLE-PHASE

When delivering energy, the transmission-line conductors are in a state where there is a flow of charge in conductors which are associated with a large potential difference from line to neutral. Our study of electrostatics reminds us that the ratio of charge, in coulombs to potential difference, in volts, defines a quantity that is called *capacitance*. Thus

$$C = \frac{Q}{V} \quad \text{C/V or F} \quad (12\text{-}81)$$

where Q denotes the charge and V the potential difference between the two conducting materials. We also know that the value of this parameter is dependent upon the medium in which the conductors are placed as well as the geometrical configuration of the conducting surfaces. Recall that this result is not unlike that encountered for inductance. Because there is charge residing along the entire length of the transmission line along with potential difference, it follows that the transmission line represents a capacitor of rather large dimensions. Since capacitance is a shunt element, it acts to drain current all along the length of the line. Placing the conductors farther apart and higher above the ground, as the transmitting voltage of a transmission line is increased, serves to diminish this current drain of the line. Of course, the greater the length of the line, the more serious the current drain problem can become. Usually, however, for short length lines (less than 50 km) the capacitive effect of the line can be entirely neglected without noticeable consequence. For lines of medium length (50 to 160 km) line capacitance is taken into consideration but it is done so as a lumped parameter. In fact it is the purpose of the treatment here to develop suitable formulas to calculate this capacitive effect. For very long transmission lines it is necessary to use an analytical procedure which recognizes that the line behaves in a manner consistent with the presence of distributed parameters rather than a lumped parameter. This is the concern of Sec. 12-6. Finally, it is notable to remark that the drain problem associated with line capacitance becomes enormous when underground high-voltage cables are used to transmit power. The close proximity of the conductors means high capacitance with the attendant decrease in capacitive reactance. In some cases the current loss can be so large that the useful line length is very small indeed.

The Underlying Theory

Because charge is distributed along the entire length of the conductor in a transmission line, it is more appropriate to work with charge per meter length, q, rather than with total charge, Q. Figure 12-12 depicts a section of a transmission line and also shows the electric field lines that emanate from the charge source. The amount of electric flux density, $\mathfrak{D}$, that emanates from a one meter length of the conductor at a distance D away from the charge is readily obtained from Gauss's law. This law states that the product of the electric flux and the area it penetrates is equal to the enclosed charge. Hence for a 1-m length of line we can write

$$q = \mathfrak{D}(2\pi D)(1) \tag{12-82}$$

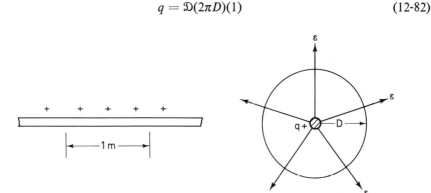

Figure 12-12 Illustrating the distribution of charge along a transmission line along with its associated radial electric field intensity.

so that the expression for the electric flux density becomes

$$\mathcal{D} = \frac{q}{2\pi D} \quad \text{C/m}^2 \tag{12-83}$$

which clearly carries units of coulombs per square meter. The electric field intensity, $\mathcal{E}$, that is associated with this electric flux density, is obtained by introducing the influence of the medium through the use of the permittivity factor $\epsilon = \epsilon_r \epsilon_0$. Here ϵ_0 is the permittivity of free space and ϵ_r denotes the relative value of the permittivity of a given medium to free space. For air ϵ_r is essentially unity. Accordingly, we may write as the expression for electric field intensity

$$\mathcal{E} = \frac{\mathcal{D}}{\epsilon} = \frac{q}{2\pi\epsilon D} \quad \text{V/m or N/C} \tag{12-84}$$

This result is valid for cylindrical geometries such as we have here. The units of ϵ are commonly expressed in volts per meter, but an alternative description is in newtons per coulomb.

The electric field intensity is instrumental in understanding potential difference. In this regard refer to Fig. 12-13 and assume for the moment that only line charge q, is present. The electric field intensity produced by q, has the value $\mathcal{E}_1 = q_1/2\pi\epsilon D_1$ at point P. If a unit positive charge is assumed to be brought from infinity to point P *against* the action of the electric field of q_1, then the work done is described by an integration of the electric field intensity over the distance from infinity to D_1. Expressed algebraically we can write

$$\text{work per unit charge} = -\int_{D_\infty}^{D_1} \mathcal{E} \, dD \tag{12-85}$$

The negative sign appears because work is being done against the action of the electric field rather than by the field. Also, D denotes direction along a radius centered at q_1. When $\mathcal{E}$ is treated in units of volts per meter, then clearly the result of the integration yields a quantity that bears units of volts. However, when $\mathcal{E}$ is used in units of newtons per coulomb, integration over distance adds the distance dimension to the result so

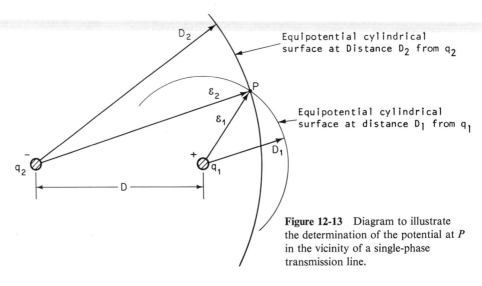

Figure 12-13 Diagram to illustrate the determination of the potential at P in the vicinity of a single-phase transmission line.

that it now bears units of newton-meters per coulomb. But this is precisely work per unit charge and is alternatively called potential and expressed also in volts. Surely the name is apt for the unit charge does possess a potential energy equal to the work done against the electric field to bring it from infinity to distance D_1. An expression for the potential at P due to q_1 is readily determined by inserting Eq. (12-84) into Eq. (12-85) and performing the integration. Thus,

$$V_{p1} = -\int_{D_\infty}^{D_1} \frac{q_1}{2\pi\epsilon} \frac{dD}{D} = \frac{q_1}{2\pi\epsilon} \ln \frac{D_\infty}{D_1} \quad \text{V} \tag{12-86}$$

Because our interest is with situations where the sum of all charges in the system is zero, i.e., $\sum_{i=1}^{n} q_i = 0$, we next introduce into the geometry of Fig. 12-12 a second line charge q_2 which is equal and opposite to q_1. By superposition q_2 also produces an electric field intensity at point P. In this case, however, the field $\mathcal{E}_2$ acts to release energy to the unit positive charge at P. By employing the procedure displayed in Eq. (12-86) the contribution of the potential at point P in Fig. 12-13 caused by the presence of q_2 is accordingly

$$V_{p2} = \frac{q_2}{2\pi\epsilon} \ln \frac{D_\infty}{D_2} \tag{12-87}$$

The net potential at P caused by the action of both q_1 and q_2 is found by summation of the contribution of each charge. Thus

$$V_p = V_{p1} + V_{p2} = \frac{1}{2\pi\epsilon}\left(q_1 \ln \frac{D_\infty}{D_1} + q_2 \ln \frac{D_\infty}{D_2}\right) \tag{12-88}$$

Expanding and collecting terms leads to

$$V_p = \frac{1}{2\pi\epsilon}\left[(q_1 + q_2) \ln D_\infty + q_1 \ln \frac{1}{D_1} + q_2 \ln \frac{1}{D_2}\right] \tag{12-89}$$

Introducing the boundary condition for conservation of charge, namely,

$$\sum_{i=1}^{2} q_i = q_1 + q_2 = 0 \tag{12-90}$$

it follows that Eq. (12-89) becomes more simply

$$\boxed{V_p = \frac{q_1}{2\pi\epsilon} \ln \frac{1}{D_1} + \frac{q_2}{2\pi\epsilon} \ln \frac{1}{D_2}} \quad \text{V} \tag{12-91}$$

The individual terms appearing on the right side are called *logarithmic potentials*. They serve a very useful purpose because the general application of the procedure displayed in Eq. (12-91) permits the potential at any point in space, under the influence of any number of conductors carrying line charges, to be written in a simple and direct manner. This is illustrated presently, but first let us continue with the analysis of the two conductor case.

Capacitance of the Single-Phase Line

To identify the capacitance of the single-phase line, it is necessary to obtain the ratio of the charge on the line to the potential difference between the two lines. To find the potential difference between the lines employing the logarithmic potentials, it is useful to keep in mind that Eq. (12-91) is a general expression for the two-wire case for the potential at a point relative to infinity. In other words, it describes the net work that must be done to bring a unit charge from infinity to point P. But P can be anywhere, even on the surface of a conductor. Accordingly, the potential of conductor 1 in Fig. 12-13, can be found by placing P on the surface of the conductor, which is assumed to have a radius r_1. Thus

$$V_1 = \frac{q_1}{2\pi\epsilon} \ln \frac{1}{r_1} + \frac{q_2}{2\pi\epsilon} \ln \frac{1}{D} \qquad (12\text{-}92)$$

But

$$q_2 = -q_1$$

Hence

$$V_1 = \frac{q_1}{2\pi\epsilon} \ln \frac{D}{r_1} \qquad \text{V} \qquad (12\text{-}93)$$

By placing P on conductor 2, the potential of this conductor can similarly be expressed as

$$V_2 = \frac{q_1}{2\pi\epsilon} \ln \frac{1}{D} + \frac{q_2}{2\pi\epsilon} \ln \frac{1}{r_2}$$

$$= \frac{-q_2}{2\pi\epsilon} \ln \frac{1}{D} + \frac{q_2}{2\pi\epsilon} \ln \frac{1}{r_2}$$

$$= \frac{q_2}{2\pi\epsilon} \ln \frac{D}{r_2} \qquad \text{V} \qquad (12\text{-}94)$$

It then follows that the potential difference between the lines is simply the difference of these two potentials. Hence

$$V_{12} = V_1 - V_2 = \frac{1}{2\pi\epsilon} \left(q_1 \ln \frac{D}{r} - q_2 \ln \frac{D}{r} \right) \qquad \text{V} \qquad (12\text{-}95)$$

Inserting $q_2 = -q_1$, leads to

$$V_{12} = \frac{q_1}{2\pi\epsilon} \ln \frac{D^2}{r_1 r_2} \qquad \text{V}$$

$$= \frac{q_1}{\pi\epsilon} \frac{1}{2} \ln \frac{D^2}{r_1 r_2} = \frac{q_1}{\pi\epsilon} \ln \frac{D}{\sqrt{r_1 r_2}} \qquad \text{V} \qquad (12\text{-}96)$$

Then by formulating the ratio of line charge q_1 to the associated potential difference V_{12} the desired expression for the capacitance of a single phase line is determined. Therefore

$$\boxed{C_{12} = \frac{\pi\epsilon}{\ln \dfrac{D}{\sqrt{r_1 r_2}}}} \qquad \text{F/m} \qquad (12\text{-}97)$$

Since the derivation is based on the use of line charges per meter length, the capacitance parameter is also expressed per meter length. Also notable is the fact that Eq. (12-97) involves use of the actual radius of the conductor. This is not to be confused with the use of equivalent conductor radius that was encountered in connection with the line inductance parameter.

In the case where the two line conductors are identical, $r_1 = r_2 = r$. Equation (12-97) can then be rewritten as

$$C_{12} = \frac{\pi\epsilon}{\ln (D/r)} \qquad \text{V} \qquad (12\text{-}98)$$

Upon introducing $\epsilon = 8.854 \times 10^{-12}$ F/m, the line capacitance expressed in nanofarads per kilometer (nF/km) becomes

$$C_{12} = \frac{27.82}{\ln (D/r)} \qquad \text{nF/km} \qquad (12\text{-}99)$$

Or, expressed in nanofarads per mile, it is

$$C_{12} = \frac{44.76}{\ln (D/r)} \qquad \text{nF/mi} \qquad (12\text{-}100)$$

Moreover, when the line transmits 60-Hz power, then the capacitive reactance is represented by the formula

$$X_c = \frac{1}{\omega C} = \frac{10^9}{377(44.76)} \ln \frac{D}{r} = 59.26 \times 10^3 \ln \frac{D}{r} \qquad \Omega\cdot\text{mi} \qquad (12\text{-}101)$$

Observe that the units of this last equation are ohm-miles. For a 1-mi line the capacitive reactance is exactly given by Eq. (12-101). However, for a 100-mi line this quantity is reduced to one one-hundredth of its value. Accordingly, the current drain between wires will be a good deal greater.

12-5 CAPACITANCE PARAMETER OF THREE-PHASE LINES

Attention is focused in this section on the application of the method of logarithmic potentials to determine the capacitance parameter of three-phase transmission line systems. The procedure is applied to balanced three-phase lines with equilateral spacing and unsymmetrical spacing as well as to bundled lines. Finally, an expression is obtained for conditions of unbalance in preparation for the material treated in the next chapter.

Equilateral Spacing and Balanced Operation

The configuration for this case is illustrated in Fig. 12-14. The constraint of balanced operation means that there is no charge distribution along the neutral wire.

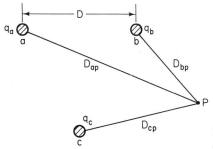

Figure 12-14   Three-phase line with equilateral spacing.

Employing logarithmic potentials, the potential of the point P in the space around the three-phase line is simply

$$V_p = \frac{1}{2\pi\epsilon}\left(q_a \ln\frac{1}{D_{ap}} + q_b \ln\frac{1}{D_{bp}} + q_c \ln\frac{1}{D_{cp}}\right) \qquad (12\text{-}102)$$

where the distances are measured between the two points identified by the subscripts. To find the potential of line a, Eq. (12-102) is modified so that it is consistent with the placement of P on the surface of conductor a. Thus,

$$V_a = \frac{1}{2\pi\epsilon}\left(q_a \ln\frac{1}{r} + q_b \ln\frac{1}{D} + q_c \ln\frac{1}{D}\right) \qquad (12\text{-}103)$$

Observe that D_{ap} becomes equal to the radius of conductor a and simultaneously D_{bp} and D_{cp} take on the value D, which is the actual separation between the line conductors. Furthermore, with no charge assumed to be residing on the neutral wire, it follows that $V_n = 0$. Accordingly, the potential difference between line a and the neutral n is

$$V_{an} = V_a - V_n = V_a \qquad (12\text{-}104)$$

Equation (12-103) can be rewritten as

$$V_a = \frac{1}{2\pi\epsilon}\left[q_a \ln\frac{1}{r} + (q_b + q_c) \ln\frac{1}{D}\right] \qquad (12\text{-}105)$$

and upon introducing $(q_b + q_c) = -q_a$ this expression simplifies to

$$V_a = V_{an} = \frac{q_a}{2\pi\epsilon} \ln\frac{D}{r} \qquad \text{V} \qquad (12\text{-}106)$$

The capacitance from line to neutral immediately results upon formulating the ratio of charge to potential difference. Thus

$$\boxed{C_{an} = \frac{q_a}{V_{an}} = \frac{2\pi\epsilon}{\ln(D/r)}} \qquad \text{F/m} \qquad (12\text{-}107)$$

Expressed in terms of nanofarads/km, when air is the surrounding medium, yields

$$C_{an} = \frac{55.63}{\ln(D/r)} \qquad \text{nF/km} \qquad (12\text{-}108)$$

If the reference length is taken to be the mile, the last expression is modified to

$$C_{an} = \frac{89.53}{\ln(D/r)} \qquad \text{nF/mi} \qquad (12\text{-}109)$$

Moreover, the corresponding reactance associated with this capacitance at a frequency of 60 Hz becomes

$$X_c = \frac{1}{\omega C_{an}} = 29.63 \times 10^3 \ln\frac{D}{r} \qquad \Omega \cdot \text{mi} \qquad (12\text{-}110)$$

Three-Phase Line with Unsymmetrical Spacing

The configuration is shown in Fig. 12-15. The potential of point P for charges q_a, q_b, and q_c residing on the three line conductors is by the use of logarithmic potentials represented by

$$V_p = \frac{1}{2\pi\epsilon}\left(q_a \ln\frac{1}{D_{ap}} + q_b \ln\frac{1}{D_{bp}} + q_c \ln\frac{1}{D_{cp}}\right) \qquad \text{V} \qquad (12\text{-}111)$$

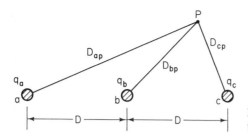

Figure 12-15 Three-phase line with unsymmetrical spacing.

Placing P on conductor a reveals the potential of this surface to be

$$V_a = \frac{1}{2\pi\epsilon}\left(q_a \ln \frac{1}{r} + q_b \ln \frac{1}{D} + q_c \ln \frac{1}{2D}\right) \qquad (12\text{-}112)$$

which for $q_c = -(q_a + q_b)$ yields

$$V_a = \frac{1}{2\pi\epsilon}\left(q_a \ln \frac{2D}{r} + q_b \ln 2\right) \qquad V \qquad (12\text{-}113)$$

It is instructive at this point to compare this result with Eq. (12-63), which describes the flux linkage of conductor a for the unsymmetrical three-phase line. The correspondence is striking and the comparison also stresses that the resulting capacitance matrix for this case is not diagonal. Repeating the foregoing procedure for lines b and c yields the following results:

$$V_b = \frac{q_b}{2\pi\epsilon} \ln \frac{D}{r} \qquad V \qquad (12\text{-}114)$$

and

$$V_c = \frac{1}{2\pi\epsilon}\left(q_b \ln 2 + q_c \ln \frac{2D}{r}\right) \qquad V \qquad (12\text{-}115)$$

Thus the matrix representation of the last three expressions takes the form

$$\begin{bmatrix} V_a \\ V_b \\ V_c \end{bmatrix} = \begin{bmatrix} \ln\dfrac{2D}{r} & \ln 2 & 0 \\ 0 & \ln\dfrac{D}{r} & 0 \\ 0 & \ln 2 & \ln\dfrac{2D}{r} \end{bmatrix} \begin{bmatrix} q_a \\ q_b \\ q_c \end{bmatrix} \qquad (12\text{-}116)$$

Effect of Line Transposition

By transposing the line conductors in the manner illustrated in Fig. 12-9, the same beneficial effect that appeared in the line inductance matrix also appears for capacitance, namely, a diagonal matrix. This means that each line voltage expression is uncoupled from the charge quantities on the other lines which in turn allows a simple expression for the capacitance to be written. By following the procedure used in the case of line inductance, it is easy to show that the resulting equation for the line voltage at conductor a is

$$V_a = \frac{q_a}{2\pi\epsilon} \ln \frac{D_g}{r} \qquad (12\text{-}117)$$

where D_g is the geometric mean distance of the conductor arrangement. For the three-phase unbundled case

$$D_g = \sqrt[3]{D \cdot D \cdot 2D} = \sqrt[3]{2}\,D \qquad (12\text{-}118)$$

so that the final expression for the capacitance parameter of a transposed, unsymmetrical, three-phase line is

$$\boxed{C_{an} = \frac{V_a}{q_a} = \frac{2\pi\epsilon}{\ln(\sqrt[3]{2}\,D/r)}} \quad \text{F/m} \qquad (12\text{-}119)$$

The equation for conductors placed in air and expressed in units of nanofarads per mile is

$$C_{an} = \frac{89.53}{\ln(\sqrt[3]{2}\,D/r)} \quad \text{nF/mi} \qquad (12\text{-}120)$$

corresponding to which is a capacitive reactance of

$$X_c = 29.63 \times 10^3 \ln \frac{\sqrt[3]{2}\,D}{r} \quad \Omega \cdot \text{mi} \qquad (12\text{-}121)$$

Observe that these expressions are the same as those that apply for the three-phase line with equilateral spacing except for the argument of the logarithm.

Effect of Bundling

How is the expression for the line capacitance affected when the phase conductors of a three-phase line are arranged in bundles? For convenience we examine the situation by assuming that each line is composed of two identical conductors and, furthermore, that each conductor carries one-half the total charge of that line. Figure 12-16 depicts the arrangement. Again by writing the logarithmic potential associated with each line charge the total potential of P can be described as

$$V_p = \frac{1}{2\pi\epsilon}\left(\frac{q_a}{2}\ln\frac{1}{D_{ap}} + \frac{q_a}{2}\ln\frac{1}{D_{a'p}} + \frac{q_b}{2}\ln\frac{1}{D_{bp}} + \frac{q_b}{2}\ln\frac{1}{D_{b'p}}\right.$$
$$\left. + \frac{q_c}{2}\ln\frac{1}{D_{cp}} + \frac{q_c}{2}\ln\frac{1}{D_{c'p}}\right) \qquad (12\text{-}122)$$

Assuming $D \gg d$ and placing point P on the surface of conductor a, the potential of this point can then be expressed as

$$V_a = \frac{1}{2\pi\epsilon}\left(\frac{q_a}{2}\ln\frac{1}{r} + \frac{q_a}{2}\ln\frac{1}{d} + q_b\ln\frac{1}{D} + q_c\ln\frac{1}{2D}\right) \qquad (12\text{-}123)$$

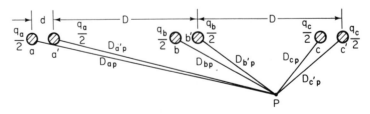

Figure 12-16 Bundled three-phase transmission line.

This can be rewritten as

$$V_a = \frac{1}{2\pi\epsilon}\left[\frac{q_a}{2}\left(\ln\frac{1}{rd}\right) + q_b \ln\frac{1}{D} - (q_a + q_b)\ln\frac{1}{2D}\right]$$

$$= \frac{1}{2\pi\epsilon}\left(q_a \ln\frac{1}{\sqrt{rd}} + q_b \ln 2 - q_a \ln\frac{1}{2D}\right)$$

$$= \frac{1}{2\pi\epsilon}\left(q_a \ln\frac{2D}{\sqrt{rd}} + q_b \ln 2\right) \qquad (12\text{-}124)$$

Keep in mind that this result is for a bundled but untransposed line. A comparison with Eq. (12-113), which applies to the unbundled, untransposed line, reveals that the sole effect of bundling is to replace the radius of the single conductor line r by an equivalent radius, R_g, which is simply the geometric mean radius of the bundle.

Consequently, it follows that if a three-phase line is bundled as well as transposed, then the capacitance parameter from line to neutral is described by a formula which is a generalized version of Eq. (12-119), namely,

$$\boxed{C_{an} = \frac{2\pi\epsilon}{\ln(D_g/R_g)}} \qquad \text{F/m} \qquad (12\text{-}125)$$

where D_g is the mean geometric distance of the line arrangement and R_g is the mean geometric radius of the individual bundles. Implied in the derivation of Eq. (12-125) is a bundle configuration which is identical for each line and a neutral location which is remote to the conductors and free of charge.

Example 12-4

Refer to the bundled, transposed, three-phase line of Example 12-3 (see Fig. 12-11) and determine the following: (a) the line-to-neutral capacitance expressed in nF/km and nF/mi; (b) the capacitive reactance of the line expressed in ohms-miles per phase: (c) the capacitive reactance expressed in ohms per phase for a transmission line length of 30 mi and 150 mi. Assume operation at 60 Hz.

Solution (a) The capacitance in this case is given by Eq. (12-125). The geometric mean radius R_g is found to be

$$R_g = \sqrt{rd} = \sqrt{(0.0074)(0.3)} = 0.0471 \text{ m}$$

When making this calculation take care to use the actual conductor radius and not r' which is used for inductance calculations. The geometric mean distance, however, is the same as that found in Example 12-3. Thus

$$D_g = \sqrt[3]{6\cdot6\cdot12} = 7.56 \text{ m}$$

Hence the line-to-neutral capacitance may be expressed as

$$C_{an} = \frac{55.63}{\ln(7.56/0.0471)} = 10.95 \text{ nF/km} - 17.63 \text{ nF/mi}$$

(b) The capacitive reactance follows from

$$X_c = 29.63 \times 10^3 \ln\frac{D_g}{R_g} = 29.63 \times 10^3 \ln\frac{7.56}{0.0471} - 150{,}500 \ \Omega\cdot\text{mi}$$

(c) For a 30-mi-long line the capacitive reactance in ohms per phase is

$$(X_c)_{30} = \frac{150{,}500 \ \Omega \cdot \text{mi}}{30 \ \text{mi}} = 5016.7 \ \Omega/\text{phase}$$

When the line length is 100 mi, this quantity reduces to

$$(X_c)_{100} = 1505 \ \Omega/\text{phase}$$

Consequently, even more current drain occurs before it reaches its destination point.

Effect of Charge on the Neutral Conductor

When the three-phase line operates under conditions of severe unbalance as happens for unsymmetrical line faults, there can be a substantial current flow in the neutral wire. In turn the corresponding presence of charge along the neutral wire must now be considered in evaluating the line-to-neutral voltage drop. The situation to be analyzed is illustrated in Fig. 12-17 where there are now four line charges present, subject to the constraint, however, that

$$q_a + q_b + q_c + q_n = 0 \tag{12-126}$$

Moreover, for convenience, both equilateral spacing and single conductors per line are assumed. Then by the method of logarithmic potentials the potential at point P can be written as

$$V_p = \frac{1}{2\pi\epsilon}\left(q_a \ln \frac{1}{D_{ap}} + q_b \ln \frac{1}{D_{bp}} + q_c \ln \frac{1}{D_{cp}} + q_n \ln \frac{1}{D_{np}}\right)$$

Eliminating q_n by means of the constraint condition, we get

$$V_p = \frac{1}{2\pi\epsilon}\left(q_a \ln \frac{D_{np}}{D_{ap}} + q_b \ln \frac{D_{np}}{D_{bp}} + q_c \ln \frac{D_{np}}{D_{cp}}\right)$$

Figure 12-17 Equilaterally spaced, three-phase line with charge residing on the neutral.

When point P is placed on the surface of conductor a, the distance quantities in this last expression assume the following values:

$$D_{ap} = r_a = r$$
$$D_{bp} = D_{cp} = D$$
$$D_{np} = D_{na}$$

Consequently, the potential of line a becomes

$$V_a = \frac{1}{2\pi\epsilon}\left(q_a \ln \frac{D_{na}}{r_a} + q_b \ln \frac{D_{na}}{D} + q_c \ln \frac{D_{na}}{D}\right) \qquad (12\text{-}127)$$

In a similar way the potential of the neutral line can be found by placing point P onto the surface of the neutral wire. Accordingly,

$$V_n = \frac{1}{2\pi\epsilon}\left(q_a \ln \frac{1}{D_{na}} + q_b \ln \frac{1}{D_{nb}} + q_c \ln \frac{1}{D_{nc}} + q_n \ln \frac{1}{r_n}\right) \qquad (12\text{-}128)$$

where r_n denotes the radius of the neutral wire. The insertion of $q_n = -(q_a + q_b + q_c)$ then leads to

$$V_n = \frac{1}{2\pi\epsilon}\left(q_a \ln \frac{r_n}{D_{na}} + q_b \ln \frac{r_n}{D_{nb}} + q_c \ln \frac{r_n}{D_{nc}}\right) \qquad (12\text{-}129)$$

The potential difference between line a and the neutral wire is obtained by subtracting V_n from V_a. Thus, the result is

$$V_{an} = V_a - V_n = \frac{1}{2\pi\epsilon}\left(q_a \ln \frac{D_{na}^2}{r_a r_n} + q_b \ln \frac{D_{na}D_{nb}}{D_{ab}r_n} + q_c \ln \frac{D_{na}D_{nc}}{D_{ac}r_n}\right) \qquad (12\text{-}130)$$

Clearly this expression does not lend itself to manipulation that leads to a simple expression for the line-to-neutral capacitance. Simplification must await the use of symmetrical components for the treatment of unbalanced conditions. This matter is addressed further in the next chapter.

12-6 LINE RESISTANCE

Transmission lines are made of conducting material of finite dimensions and as such exhibit resistance to the flow of current. The influence of resistance in determining voltage drops along the line for operation at 60 Hz is very small in comparison to the effect of the line inductance. However, line resistance certainly exerts a key role with regard to the transmission line losses.

Transmission lines are generally constructed of stranded aluminum conductors which are partially spiraled around center steel wires for strength. Figure 12-18 shows the cross-sectional view of six aluminum conductors wrapped around a steel wire. These lines are referred to as aluminum conductors, steel reinforced and are repre-

Figure 12-18   Cross-sectional view of an ACSR line with six aluminum conductors and one steel wire for support.

sented by the acronym ACSR. There are also other categories, such as all-aluminum conductors (AAC), all-aluminum alloy conductors (AAAC), and aluminum conductor alloy reinforced (ACAR). A stranded configuration of the line conductor is employed in an effort reduce the increase in resistance caused by skin effect.† Keep in mind that transmission lines are generally designed to carry fairly large currents despite the use of high transmitting voltages. In order to maintain current densities at tolerable levels, it becomes necessary for the total cross-sectional area of the conductor to be quite large. If a single, solid conductor were to be used, the skin effect could cause the resistance to increase by appreciable amounts. As a result, operation of the transmission line becomes more costly, less efficient and may even lead to disruption of service. The latter can happen when transmission line conductors operate at such elevated temperatures that they exceed their elastic limit and sag. Often, the sagging line begins to interfere with trees and other objects, and, before long, a fault occurs. Although stranding of the line conductor does not eliminate these shortcomings entirely, it certainly goes a long way toward diminishing their effects. Stranding proves to be effective in reducing skin effect because there is generally not enough cross-sectional area to a single strand to cause significant differences in the flux linkages between an internal point and one that is closer to the surface of the strand.

The size and characteristics of transmission line conductors are available in tables prepared by various organizations, such as the Aluminum Association, Westinghouse, and others. A part of the published features of several ASCR wires is listed in Table 12-1. The classification is in accordance with the cross-sectional area of the aluminum expressed in circular mils. The circular mil is a unit that is used to describe the cross-sectional area of round conductors. By definition the circular mil is the area of a circle having a diameter of 0.001 in. or 1 mil. To find the area of any round conductor in circular mils, it is necessary merely to find the diameter in inches, express this in mils by multiplying by 1000 and then to square this number. The procedure is readily illustrated by referring to the first entry in the table. It is necessary first to determine the diameter d_e of an equivalent solid conductor that has the same area as the 26 strands. Thus,

$$\frac{\pi}{4}d_e^2 = \frac{\pi}{4}(0.1074)^2 \cdot 26$$

TABLE 12-1   DATA ON SELECTED COMMERCIAL ASCR BARE CONDUCTORS

Entry no.	Aluminum area (cmil)	Strands (Al/St)	Strand diameter (in.)	Approx. current rating (A)	Resistance (Ω/mi)		
					dc, 20°C	60 Hz, 20°C	60 Hz, 50°C
1	300,000	26/7	0.1074	490	0.300	0.307	0.337
2	477,000	30/7	0.1261	670	0.1874	0.1919	0.2107
3	566,500	26/7	0.1463	730	0.1621	0.1663	0.1826
4	1,590,000	54/19	0.1716	1,380	0.0570	0.0612	0.0667

†For an explanation see page 000.

where the figure 0.1074 is the diameter of each strand. Hence

$$d_e = (0.1074)\sqrt{26} = 0.54764 \text{ in.} = 547.64 \text{ mils}$$

Therefore the area of this conductor expressed in circular mils is

$$A = (547.64)^2 = 299{,}910$$

which compares favorably with the listed value for entry 1.

An indication of the extent of the skin effect in the stranded conductor of entry 1 of the table can be obtained by comparing the d-c resistance with the corresponding 60 Hz resistance at the same temperature. The table reveals this ratio to be $0.307/0.300 = 1.023$. Therefore the skin effect is a little in excess of 2%. The skin effect, of course, increases as the size of the strand gets larger. In the case of entry 4 this ratio becomes $0.0612/0.057 = 1.074$, so that now the increase is in excess of 7%.

When the transmission line delivers power on a continuous basis, the losses in the line resistance cause the conductor temperature to rise, which in turn means a further increase in resistance. For entry 4 of the table, the increase in resistance associated with a temperature rise of $30°C$ is described by the ratio $0.0667/0.612 = 1.09$. Thus the temperature rise of the aluminum conductor is responsible for a 9% increase in resistance, which is slightly in excess of the change caused by the skin effect. However, it should be noted that for the smaller conductors temperature rise exerts a more important influence than does skin effect.

12-7 TRANSMISSION LINE EQUIVALENT CIRCUITS

The determination of the transmission line parameters, as outlined in the preceding sections of this chapter, is tied to a convenient unit length of line. Thus formulas for the calculation of line inductance and capacitance are expressed in suitable units expressed per meter length or per kilometer or even per mile length. Once the total length of the transmission line is known in any specific case, these parameters can be modified to apply to the entire line length. In this way it is possible to represent the transmission line by an equivalent circuit with appropriate circuit parameters. The particular manner in which the circuit elements are arranged, however, is determined by the length of the line. Experience with these circuit representations reveals that a simple series impedance with negligible capacitive shunting from line to ground is entirely acceptable for lines that are 50 km or less in length. These are called *short lines*. When the transmission line length lies between 50 and 160 km or so, the capacitive effect becomes sufficiently important to be included in the equivalent circuit usually in the form of two lumped parameters. One is placed at the sending end terminals and the other at the receiving end. Lines in this group are often referred to as *medium lines*. Finally, when the transmission line exceeds lengths of 160 km, it becomes necessary to treat that line as one that is characterized by a continuous distribution of elemental circuit parameters. Such lines are called *long lines*. They are described by long-line theory, which involves the solution of a pair of differential equations. In this section attention is focused on describing the equivalent circuit for each group and then illustrating the associated computations.

The Short-Line Equivalent Circuit

For an overhead transmission line with a total length that is less than 50 km, the shunt capacitance effects are usually negligible while line inductance definitely is not. In such situations the circuit representation of the transmission line is simply that of an impedance $\bar{Z}$ that joins the sending end of the line to the receiving end in the manner illustrated in Fig. 12-19. Here $\bar{Z}$ denotes the total impedance of the line treated in terms of the total line length l and the unit line impedance $\bar{z}$ expressed in units of ohms per kilometer or ohms per mile. Thus

$$\bar{Z} = \bar{z}l \quad \Omega \tag{12-131}$$

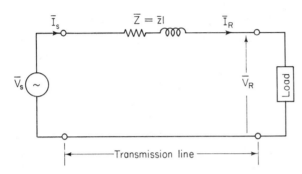

Figure 12-19 Small-line equivalent circuit.

where

$$\bar{z} = (r + j\omega L) \quad \Omega/\text{km or } \Omega/\text{mi} \tag{12-132}$$

or

$$\bar{Z} = rl + j\omega Ll - R \mid jX_L \tag{12-133}$$

where R is the total short-line resistance and X_L is the total inductive reactance of the line.

In dealing with transmission line equivalent circuits, it is customary to assume that rated megavolt-amperes (MVA) is delivered to the load at the receiving end. Application of circuit analysis to the appropriate equivalent circuit then allows us to determine the expressions for the corresponding sending-end current, voltage, and MVA. In the case of the short line the results are quite apparent and by inspection of Fig. 12-19 can be written to be

$$\bar{I}_s = \bar{I}_R \tag{12-134}$$

$$\bar{V}_s = \bar{V}_R + \bar{I}_R\bar{Z} \tag{12-135}$$

$$\bar{S} = 3\bar{V}_s\bar{I}_s^* \tag{12-136}$$

where $\bar{V}_R$ denotes the line-to-neutral voltage at the receiving end and is usually taken as the reference phasor, $\bar{S}$ denotes the sending-end megavolt-amperes, and $\bar{I}_s^*$ is the complex conjugate of $\bar{I}_s$. Of course, $\bar{Z}$ represents the total line impedance in ohms. The appearance of 3 in the last equation serves to remind us that emphasis here is on three-phase transmission.

The Medium-Line Equivalent Circuit

When the length of an overhead transmission line falls in the medium range, it becomes necessary to account for the current flow associated with the shunt capacitance that exists all along the line length. Experience shows that in these cases the transmission line is accurately represented by the use of a π-type equivalent circuit employing lumped parameters as depicted in Fig. 12-20. The total line impedance $\bar{Z}$ is computed exactly as indicated in Eqs. (12-131) and (12-133). The quantity $\bar{Y}$ is the reciprocal of the capacitive reactance of the line computed as outlined in Example 12-4(c) for the entire line length. Thus,

$$\bar{Y} = \frac{1}{-j(X_c)_l} = j\frac{1}{(X_c)_l} = \frac{1}{(X_c)_l}\,\underline{/90^\circ} \tag{12-137}$$

Figure 12-20 π-type equivalent circuit for medium-length transmission lines.

where the second subscript on X_c is included to remind us that the per unit length quantity X_c must be adjusted to account for the entire line length. It is customary in the analysis of transmission lines to neglect completely the conductance between lines and from line to neutral. Hence the admittance $\bar{Y}$ is actually treated as a susceptance. Observe too that the total line admittance is treated in two parts. One-half the admittance is placed at the sending end and the other half is at the receiving end.

Based on the assumption that $\bar{V}_R$ and $\bar{I}_R$ are known, the corresponding sending-end quantities are readily obtained by application of Kirchhoff's voltage and current laws to the circuit of Fig. 12-20. By KCL the current that flows through the line impedance $\bar{Z}$ is

$$\bar{I}_Z = \bar{I}_R + \bar{V}_r\left(\frac{\bar{Y}}{2}\right) \tag{12-138}$$

Then by KVL the sending-end voltage is found to be

$$\bar{V}_s = \bar{V}_R + \bar{I}_Z\bar{Z} \tag{12-139}$$

where $\bar{I}_Z$ is defined as indicated in Fig. 12-20. In turn this last item permits the sending-end current to be determined from

$$\bar{I}_s = \bar{I}_Z + \bar{V}_s\left(\frac{\bar{Y}}{2}\right) \tag{12-140}$$

Finally, the real power and reactive power delivered by the source readily follows upon substituting Eqs. (12-139) and (12-140) into Eq.(12-136). It is also instructive to note in these last few equations that insertion of $\bar{Y} = 0$ yields the results of the short line.

Example 12-5

A three-phase, 60-Hz transmission line is constructed of ACSR conductors and described by entry 3 of Table 12-1. For the specific spacing used in the line construction the per mile line impedance is given by

$$\bar{z} = r + j337L = 0.1826 + j0.784 \ \Omega/\text{mi per phase}$$

and the unit capacitive reactance is known to be

$$\bar{X}_c = 185.5 \times 10^3 \underline{/90°} \ \Omega \cdot \text{mi per phase}$$

(a) Find the total line impedance and admittance for a line length of 100 mi.

(b) The load at the receiving end draws 200 MVA at a line-to-line voltage of 230 kV and unity power factor. Determine the sending end voltage, current, and real and reactive power.

(c) Compute the voltage regulation.

Solution (a) The total line impedance is

$$\bar{Z} = (0.1826 + j0.784)100 = 18.26 + j78.4 = 80.5\underline{/76.80°} \ \Omega/\text{phase}$$

The line admittance for 100 mi is

$$\bar{Y} = \frac{1}{(X_c)_l} \underline{/90°} = \frac{1}{\dfrac{185.5 \times 10^3}{100}} \underline{/90°} = 0.5391 \times 10^{-3} \underline{/90°} \ \Omega/\text{phase}$$

(b) At the receiving end we have

$$\bar{I}_R = \frac{200 \times 10^6}{\sqrt{3}(230 \times 10^3)} = 502 \text{ A}; \qquad \bar{I}_R = 502\underline{/0°}$$

Also,

$$\bar{V}_R = \frac{230,000}{\sqrt{3}} \underline{/0°} = 132,800\underline{/0°} = 132.8\underline{/0°} \text{ kV}$$

From Eq. (12-138)

$$\bar{I}_Z = \bar{I}_R + \bar{V}_R\left(\frac{\bar{Y}}{2}\right) = 502\underline{/0°} + (132,800\underline{/0°})(0.27 \times 10^{-3}\underline{/90°})$$

$$= 502 + j35.86 = 503.3\underline{/4.09°} \text{ A}$$

The sending-end voltage then follows from Eq. (12-139). Thus,

$$\bar{V}_s = 132.8\underline{/0°} + (0.5033\underline{/4.09°})(80.5\underline{/76.89°})$$

$$= 132.8 + 40.514\underline{/80.98°}$$

$$= 139.152 + j40.01 = 144.79\underline{/16.04°} \text{ kV}$$

This yields a magnitude for the line to line voltage of 250.784 kV. The corresponding sending-end current is obtained from Eq. (12-140).

$$\bar{I}_s = \bar{I}_Z + \bar{V}_s\left(\frac{\bar{Y}}{2}\right) = 502 + j35.86 + (144.79\underline{/16.04°})(0.27\underline{/90°})$$

$$= 502 + j35.86 + 39.09\underline{/106.04°}$$

$$= 502 - 10.8 + j(35.86 + 37.6)$$

$$= 491.2 + j73.46 = 496.7\underline{/8.5°} \text{ A}$$

Accordingly,

$$\bar{S} = 3(144.79)(0.4967)\underline{/16.04° - 8.5°} = 213.88 + j28.31 \text{ MVA}$$

where $\theta_s = \underline{/\bar{V}_s} - \underline{/\bar{I}_s} = 16.04° - 8.5° = 7.54°$.

This result for $\bar{S}$ also makes it apparent that the line copper loss is approximately given by the sending-end MW less the receiving-end MW which here is 13.88 MW. This figure compares quite favorably with the results obtained from an I^2R determination of the line loss.

(c) The voltage regulation is the change in voltage from the sending end to the receiving end using $\bar{V}_R$ as the base quantity. Thus,

$$\text{percent regulation} = \frac{|V_s| - |V_R|}{|V_R|} \times 100 = \frac{144.79 - 132.8}{132.8} \times 100 = 9\%$$

The Long-Line Equivalent Circuit

As the length of the transmission line begins to exceed 160 km, the use of the series line impedance $\bar{Z}$ and the shunt line admittance $\bar{Y}$ causes noticeable error. Consequently, there arises a need to introduce corrective factors in these circuit parameters to make the equivalent circuit more representative of actual line conditions. But first a more exact analysis of the transmission line is necessary; one which takes into account the distributed nature of the inductance and capacitance parameters of the line. The situation is illustrated in Fig. 12-21 at a point along the transmission line located at a distance x from the receiving end. Measurements are initiated from the receiving end because we want to establish specified conditions at the load.

The current and voltage of the transmission line at distance x are assumed to be $\bar{I}$ and $\bar{V}$, respectively. Although these are sinusoidally, time-varying quantities, we

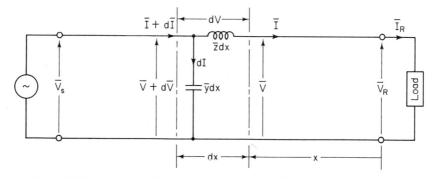

Figure 12-21 Representation of a transmission line illustrating the distributed character of inductance and capacitance at a distance x from the receiving end.

work with rms values only and thereby focus our attention solely on the effect of a change in distance. Thus, for an incremental change in distance by dx, there is a corresponding differential change in line voltage caused by the flow of current $\bar{I}$ in the differential line impedance $\bar{z}\,dx$. Expressed mathematically,

$$d\bar{V} = \bar{I}(\bar{z}\,dx) \tag{12-141}$$

Similarly, the presence of two charge-carrying conductors in the space associated with this differential distance means that there is a differential current flow through the differential capacitance existing between the two conductors. Hence, across the differential section dx, there occurs a differential current change which is described by

$$d\bar{I} = \bar{V}(\bar{y}\,dx) \tag{12-142}$$

where $\bar{y}$ is the admittance expressed per unit length. This equation is written on the valid assumption that second-order effects are negligible.

Our immediate goal is to obtain general expressions for the voltage and current at any point along the line expressed in terms of the receiving-end quantities $\bar{I}_R$ and $\bar{V}_R$. Accordingly, let us differentiate Eq. (12-141) with respect to x to get

$$\frac{d^2\bar{V}}{dx^2} = \bar{z}\left(\frac{d\bar{I}}{dx}\right) \tag{12-143}$$

and then use Eq. (12-142) to eliminate $\bar{I}$. Thus,

$$\frac{d^2\bar{V}}{dx^2} = \bar{z}\frac{d\bar{I}}{dx} = \bar{z}\bar{y}\bar{V}$$

or

$$\frac{d^2\bar{V}}{dx^2} - \bar{z}\bar{y}\bar{V} = 0 \tag{12-144}$$

where $\bar{V}$ represents the value of the transmission line voltage at x and in fact is a function of x all along the line. The same procedure can be used to eliminate $\bar{V}$ from Eqs. (12-141) and (12-142) in which case the resulting differential equation becomes

$$\frac{d^2\bar{I}}{dx^2} - \bar{z}\bar{y}\bar{I} = 0 \tag{12-145}$$

These last two equations are the transmission line equations which, upon solution, permit accounting for the effects of distributed capacitance and inductance.

Equations (12-144) and (12-145) are second-order differential equations whose solutions must be exponential functions of time and whose characteristic equation has eigenvalues given by

$$\gamma = \pm\sqrt{\bar{z}\bar{y}} \tag{12-146}$$

Because our interest in this section is with equivalent circuits, we omit the details of the solution process for these equations and simply cite the results. Hence it can be shown that

$$\bar{V} = \bar{V}_R \cosh \gamma x + \bar{I}_R \bar{Z}_c \sinh \gamma x \tag{12-147}$$

$$\bar{I} = \bar{I}_R \cosh \gamma x + \frac{\bar{V}_R}{\bar{Z}_c} \sinh \gamma x \tag{12-148}$$

where

$$\bar{Z}_c = \sqrt{\frac{\bar{z}}{\bar{y}}} = \text{characteristic impedance of transmission line} \qquad (12\text{-}149)$$

The solution for the sending-end voltage and the sending-end current follows forthwith from the general solutions by merely inserting $x = l$. Thus

$$\bar{V}_s = \bar{V}_R \cosh \gamma l + \bar{I}_R \bar{Z}_c \sinh \gamma l \qquad (12\text{-}150)$$

$$\bar{I} = \bar{I}_R \cosh \gamma l + \frac{\bar{V}_R}{\bar{Z}_c} \sinh \gamma l \qquad (12\text{-}151)$$

It is now possible to represent this last set of equations, which involves the current and voltage quantities at both ends of the transmission line, by a π equivalent circuit. It is helpful in this connection first to return to Eqs. (12-138) to (12-140). By eliminating $\bar{I}_Z$ from the expressions for $\bar{V}_s$ and $\bar{I}_s$, general equations for these quantities result. These are

$$\bar{V}_s = \bar{V}_R \left(1 + \frac{\bar{Z}\bar{Y}}{2}\right) + \bar{I}_R \bar{Z} \qquad (12\text{-}152)$$

$$\bar{I}_s = \bar{I}_R \left(1 + \frac{\bar{Z}\bar{Y}}{2}\right) + \bar{V}_R \bar{Y} \left(1 + \frac{\bar{Y}\bar{Z}}{4}\right) \qquad (12\text{-}153)$$

A π-type equivalent circuit representation of the long transmission line that is valid for specified *terminal* conditions can be derived by setting the coefficients of $\bar{V}_R$ and $\bar{I}_R$ in the last two equations equal to the coefficients that appear for these same quantities in Eqs. (12-150) and (12-151). In general, this requires that the quantities $\bar{Z}$ and $\bar{Y}$ in the last pair of equations be modified for application to a long line which is characterized by distributed parameters. The need for an adjustment is readily demonstrated by comparing the coefficients of $\bar{I}_R$ in Eqs. (12-152) and (12-150). Clearly, one condition for equivalency is that $\bar{Z} = \bar{Z}_c \sinh \gamma l$. Therefore, it follows that the series impedance to be used in the π equivalent circuit of the long line is not the same as the value used to represent short and medium lines. To stress this matter, it is helpful to rewrite Eqs. (12-152) and (12-153) using $\bar{Z}'$ and $\bar{Y}'$ in place of $\bar{Z}$ and $\bar{Y}$. Accordingly, a π-type equivalent circuit for a long transmission line can be described by the equation set

$$\bar{V}_s = \bar{V}_R \left(1 + \frac{\bar{Z}'\bar{Y}'}{2}\right) + \bar{I}_R \bar{Z}' \qquad (12\text{-}154)$$

$$\bar{I}_s = \bar{I}_R \left(1 + \frac{\bar{Z}'\bar{Y}'}{2}\right) + \bar{V}_R \bar{Y} \left(1 + \frac{\bar{Z}'\bar{Y}'}{4}\right) \qquad (12\text{-}155)$$

It is of course desirable to identify the quantities $\bar{Z}'$ and $\bar{Y}'$ for terminal equivalence in terms of the original values Z and Y, which are readily available for a given transmission line. Because only $\bar{Z}'$ and $\bar{Y}'$ are unknown in the equivalent configuration of the long line, expressions for these quantities can be determined by equating the coefficients of $\bar{I}_R$ and $\bar{V}_R$ in Eq. (12-154) to those appearing in Eq. (12-150). Thus we have

$$\bar{Z}' = \bar{Z}_c \sinh \gamma l \qquad (12\text{-}156)$$

and

$$1 + \frac{\bar{Y}'\bar{Z}'}{2} = \cosh \gamma l \qquad (12\text{-}157)$$

By proper algebraic manipulation of Eq. (12-156) it is possible to express $\bar{Z}'$ in terms of Z. Accordingly,

$$\bar{Z}' = \sqrt{\frac{\bar{z}}{\bar{y}}} \sinh \gamma l = \frac{\bar{z}}{\sqrt{\bar{z}\bar{y}}} \sinh \gamma l = \frac{\bar{z}l}{\gamma l} \sinh \gamma l = \bar{Z} \frac{\sinh \gamma l}{\gamma l} \qquad (12\text{-}158)$$

In a similar way it can be shown from Eq. (12-157) that the shunt elements in the π equivalent circuit take on the identity

$$\frac{\bar{Y}'}{2} = \frac{1}{\bar{Z}'}(\cosh \gamma l - 1) = \frac{1}{\bar{Z}_c} \tanh \frac{\gamma l}{2} = \frac{\bar{Y}}{2}\left[\frac{\tanh (\gamma l/2)}{\gamma l/2}\right] \qquad (12\text{-}159)$$

where use is made of the equality $1/\bar{Z}_c = \bar{Y}/\gamma l$.

The results displayed in the last two equations are significant because these expressions state that the long line with distributed parameters can be represented, insofar as terminal conditions are concerned, by the same balanced π equivalent circuit that was used for the medium-length line. It is necessary merely to apply corrective factors to $\bar{Z}$ and $\bar{Y}$ to account for the distributed effects of inductance and capacitance. These corrective factors are respectively $\sinh \gamma l/\gamma l$ and $[\tanh (\gamma l/2)]/(\gamma l/2)$. Once the length of line is known, these quantities can be determined. However, as a glance at Eq. (12-146) makes obvious, γ is often a complex quantity. Consequently, the evaluation of these adjustments is not as direct as we might like. In such cases the expansion formulas for the hyperbolic sine and cosine functions are used. Thus, if we let

$$\gamma = \alpha + j\beta \qquad (12\text{-}160)$$

then

$$\sinh \gamma l = \sinh (\alpha l + j\beta l) = \sinh \alpha l \cos \beta l + j \cosh \alpha l \sin \beta l \qquad (12\text{-}161)$$

$$\tanh \frac{\gamma l}{2} = \frac{\cosh \gamma l - 1}{\sinh \gamma l} \qquad (12\text{-}162)$$

and

$$\cosh \gamma l = \cosh (\alpha l + j\beta l) = \cosh \alpha l \cos \beta l + j \sinh \alpha l \sin \beta l \qquad (12\text{-}163)$$

Keep in mind that the arguments of these trigonometric functions are frequently expressed in radians. The inconvenience of these computations can be avoided by the use of charts such as those developed by L. F. Woodruff.[†] These are included in Appendix F. The corrective factors are readily obtained from these charts once the complex quantity $\bar{Z}\bar{Y}$ is determined.

The π-type equivalent circuit configuration as it applies to the long transmission line is displayed in Fig. 12-22. The sole difference between this arrangement and that of the medium-length line is in the use of $\bar{Z}'$ for Z and $\bar{Y}'$ for $\bar{Y}$.

[†]L. F. Woodruff, Complex Hyperbolic Function Charts, *Elec. Eng.*, Vol. 54, pp. 550–554, May 1935.

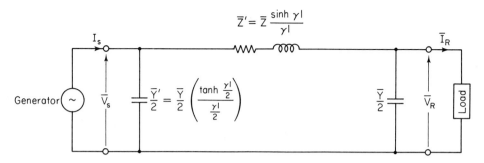

$$\overline{Z}' = \overline{Z}\,\frac{\sinh \gamma l}{\gamma l}$$

$$\frac{\overline{Y}'}{2} = \frac{\overline{Y}}{2}\left(\frac{\tanh \frac{\gamma l}{2}}{\frac{\gamma l}{2}}\right)$$

$$\frac{\overline{Y}}{2}$$

Figure 12-22 Long line π equivalent circuit showing adjusted impedance $\overline{Z}'$ and admittance $\overline{Y}'$ to account for the distributed nature of the parameters. The equivalence is valid for terminal conditions.

Example 12-6

Assume that the transmission line of Example 12-5 is extended to cover a distance of 200 mi. Determine (a) the sending-end voltage; (b) the sending-end current; (c) the sending-end real power and reactive power; and (d) the voltage regulation. Assume conditions at the load to be the same as for Example 12-5.

Solution   One method of solution is to work with Eqs. (12-150) and (12-151) directly. This requires that the cosh γl and sinh γl be evaluated by Eqs. (12-161) and (12-163). The alternative procedure is to use the charts to obtain the corrective factors that yield $\overline{Z}'$ and $\overline{Y}'$ followed by a routine circuit analysis of the equivalent circuit. Here we choose the latter method.

   (a) From the data of Example 12-5 we have

$$\bar{z} = r + j\omega L = 0.1826 + j0.784 \;\Omega/\text{mi per phase}$$

$$\bar{y} = \frac{1}{X_c\underline{/-90°}} = \frac{1}{185.5 \times 10^3}\underline{/90°} = 5.391 \times 10^{-6}\underline{/90°} \;\text{mho/mi per phase}$$

Then for a 200-mi line length the total values are

$$\overline{Z} = \bar{z}l = (0.1826 + j0.784)(200) = 36.52 + j156.8 = 160.99\underline{/76.89°}$$

$$\overline{Y} = \bar{y}l = (5.391 \times 10^{-6}\underline{/90°})(200) = 1.078 \times 10^{-3}\underline{/90°}$$

To enter the charts in Appendix F, it is necessary to have both the magnitude and phase of $\overline{Z}\overline{Y}$. Hence

$$\overline{Z}\overline{Y} = (160.99\underline{/76.89°})(1.078 \times 10^{-3}\underline{/90°}) = 0.1736\underline{/166.89°}$$

Using this value with Chart II, there results a value for the corrective factor of

$$\frac{\sinh \gamma l}{\gamma l} = 0.972\underline{/0.37°}$$

so that the adjusted value of the line impedance becomes

$$Z' = Z\frac{\sinh \gamma l}{\gamma l} = (160.99\underline{/76.89°})(0.972\underline{/0.37°}) = 156.48\underline{/76.52°}$$

From Chart III we find

$$\frac{\tanh (\gamma l/2)}{\gamma l/2} = 1.0144\underline{/-0.19°}$$

which yields

$$\frac{\bar{Y}'}{2} = \frac{\bar{Y}}{2}\left[\frac{\tanh(\gamma l/2)}{\gamma l/2}\right] = (0.539 \times 10^{-3}\underline{/90°})(1.0144\underline{/-0.19°})$$

$$= 0.5476 \times 10^{-3}\underline{/89.81°}$$

Figure 12-23 shows the circuit configuration which represents the long line for terminal equivalency.

Figure 12-23 Equivalent circuit showing all parameter values for the long-line system of Example 12-6.

By KCL we have

$$\bar{I}_{Z'} = \bar{I}_R + \bar{V}_R\frac{\bar{Y}'}{2} = 502\underline{/0°} + (132,800\underline{/0°})(0.5476 \times 10^{-3}\underline{/89.81°})$$

$$= 502.24 + j72.72 = 507.5\underline{/8.24°} \text{ A}$$

Also,

$$\bar{V}_s = \bar{V}_R + \bar{I}_{Z'}\cdot\bar{Z}' = 132,800\underline{/0°} + 507.5\underline{/8.24°}\,(156.48\underline{/76.52°})$$

$$= 140,052 + j79,078 = 160,835\underline{/29.45°} \text{ V}$$

(b) Again by KCL the sending-end current is

$$\bar{I}_s = \bar{I}_{Z'} + \bar{V}_s - \frac{\bar{Y}'}{2} = 502.24 + j72.72 + (160.835)(0.5476)\underline{/29.45° + 89.81°}$$

$$= 459.19 + j149.56 = 482.93\underline{/18.04°} \text{ A}$$

(c) The expression for complex power then follows as

$$\bar{S} = 3\bar{V}_s\bar{I}_s^* = 3(160.835)(0.48293)\underline{/29.45° - 18.04°}$$

$$= 228.41 + j46.1 \text{ MVA}$$

Thus the sending-end power is 228.41 MW and the reactive power is 46.1 MVA reactive. Comparison of the real power with that delivered to the load reveals a line loss of 28.41 MW.

(d) The voltage regulation of the transmission line is

$$\text{voltage regulation} = \frac{160.835 - 132.8}{132.8} \times 100 = 21.1\%$$

Observe that doubling the line length more than doubles the voltage regulation.

TRANSMISSION LINE PARAMETERS CHAP. 12

PROBLEMS

12-1. A 60-Hz, single-phase transmission line consists of two conductors each having a radius of 10 cm and spaced 4 m apart. Find the inductive reactance per mile.

12-2. A 60-Hz, single-phase, two-conductor transmission line has a total inductive reactance per phase of 1.32 Ω/mi. If each conductor has a radius of 3.67 cm, what is the spacing between centers of the two conductors?

12-3. An alternative derivation of Eq. (12-10) is based on an energy viewpoint. Recall that the magnetic field energy density is generally expressed as $w_f = \int H \, dB$. If H denotes magnetic field intensity in A/m and B is flux density in teslas, then w_f is in J/m³.

    (a) Find the expression for w_f for the conductor of Fig. 12-2 in terms of the total conductor current and distance x.

    (b) Obtain the expression for the total differential field energy associated with a conductor length of one meter.

    (c) Derive the expression for the total conductor inductance by determining the total field energy in a one meter volume of the conductor and invoking Eq. (1-33).

12-4. A 60-Hz, single-phase line has the configuration illustrated in Fig. P12-4. Each conductor has a radius of 1.5 cm.

Figure P12-4

    (a) Find the geometric mean radius of each line.

    (b) Find the geometric mean distance.

    (c) Determine the total inductive reactance in ohms/km and ohms/mi.

12-5. A 60-Hz, single-phase transmission line is arranged as depicted in Fig. P12-5. Line A conductors have a radius of 2 cm; line B conductors have a radius of 3.5 cm.

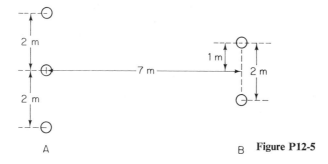

Figure P12-5

    (a) Find the geometric mean radius for each line.

    (b) Compute the geometric mean distance.

    (c) Determine the total inductance of line A in mH/km and mH/mi.

    (d) Find the total line inductance per mile.

12-6. The bundled conductors of a line are arranged as shown in Fig. P12-6. Each conductor has a radius r. Find the expression for the mean geometric radius in terms of r.

Figure P12-6   Figure P12-7

12-7. Repeat Prob. 12-6 for the conductor arrangement shown in Fig. P12-7.

12-8. Each line in the arrangement of Fig. P12-8 consists of three conductors placed at the corners of an equilateral triangle with 0.5 cm separation between centers.

Figure P12-8

(a) Compute the inductance of this line in mH/km and mH/mi.
(b) Find the inductive reactance in ohms per mile for a frequency of 60 Hz.

12-9. A three-phase line is designed for equilateral spacing with a separation on each side of 4 m. Each conductor has a radius of 0.5 cm. Determine the inductance per phase in mH/mi.

12-10. Each line of a three-phase transmission system consists of three conductors of 1 cm radius placed at the corners of an equilateral triangle with 30 cm separation between conductors. Moreover, each line is also placed at the corners of an equilateral triangle with a spacing of 8 m between the centers of each bundle. Obtain the value of the inductance per phase expressed in mH/km and mH/mi.

12-11. The lines of the transmission system described in Prob. 12-10 are constructed with parallel spacing in the horizontal plane using a separation of 10 m between the centers of each triangular bundle. The lines are transposed. Determine the inductance per phase expressed in mH/km and mH/mi.

12-12. A three-phase, transposed transmission line has the configuration shown in Fig. P12-12. Find the expression for the inductance for each phase expressed in H/m. Assume $d \ll D$. Also use r_a on the radius of each strand of line A, r_b for line B, and r_c for line C.

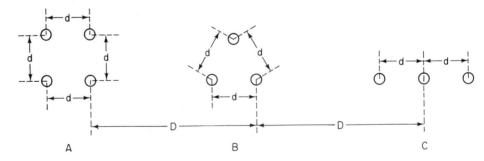

Figure P12-12

12-13. Determine the inductive reactance in Ω/mi of the three-phase, 60-Hz, transposed transmission line shown in Fig. P12-13. All conductors are identical and have a radius of 2 cm. The spacing between conductors in each bundle is 25 cm.

A   B   C

Figure P12-13

12-14. A three-phase power line and a telephone line are supported on the same towers in the manner illustrated in Fig. P12-14. Neither the power line nor the telephone line is transposed. The telephone line is located directly below phase 2. Derive the expression for the mutual inductance between the power line and the telephone line.

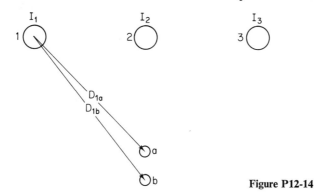

Figure P12-14

12-15. A 60-Hz, single-phase line uses conductors of 1 cm radius. If the spacing between the lines is 5 m, determine **(a)** the capacitance of the line in nF/km as well as nF/mi, and **(b)** the capacitive reactance in $\Omega \cdot$ mi.

12-16. Each line of a 60-Hz, single-phase system consists of two identical conductors of one cm radius and placed 10 cm apart. The separation of the bundle centers is 5 m. Compute the capacitive reactance of this line for a length of 50 mi.

12-17. The length of the transmission line of Prob. 12-15 is 50 mi. It is desired that the total capacitive reactance be at least 10,000 Ω. Determine the seperation distance needed between lines to achieve this result.

12-18. A 60-Hz, three-phase, transposed line has its line conductors arranged as shown in Fig. P12-18. Each conductor has a diameter of 1 cm. The line-to-line voltage is 161 kV.

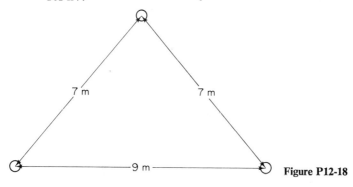

Figure P12-18

(a) Find the line-to-neutral capacitance.

(b) Determine the capacitive reactance of this line for a 1-mi length.

(c) Compute the capacitive reactance to neutral for a 100-mi length.

(d) What is the current drain associated with the total line-to-neutral capacitive reactance of the line?

12-19. Repeat Prob. 12-18 for the case where each line consists of two parallel conductors each with a radius of 0.35 cm and a separation between conductors in the bundle of 10 cm. The distance between the lines remains the same.

12-20. A three-phase line has the equilateral hexagonal spacing shown in Fig. P12-20. The conductors carry the line charges indicated. Also, conductors carrying like charges are connected in parallel. Each conductor comes with a radius r.

 Figure P12-20

(a) Write the expression for the potential of conductor 1 in terms of the specified charges and distance D.

(b) Repeat part (a) for conductor 2.

(c) Find the expression for the potential difference between conductors 1 and 2.

(d) Determine the expression for the capacitance per phase.

12-21. Refer to the transmission line of Fig. P12-18. Determine the maximum line length needed to ensure that the total current drain of the transmission line does not exceed 50 A.

12-22. A three-phase transmission line is arranged with parallel spacing in a horizontal plane. Each conductor has a radius of 2 cm. Line B is displaced 4 m from line A; line C is displaced 4 m from line B. At a particular instant in its cyclic variation the charge on line A is known to be $q_A = 40$ nC/m (nanocoulombs/meter) and the corresponding values on lines B and C are -20 nC/m. Neglect the effect of the ground.

(a) Employing logarithmic potentials, find the potential of line B expressed in terms of q_A, r and D.

(b) Repeat part (a) for line C.

(c) Compute the potential difference between lines B and C in volts.

(d) Propose a change in the construction of the transmission line that will yield a zero result in part (c).

12-23. Refer to entry 2 in Table 12-1 and answer the following questions.

(a) Find the diameter expressed in mils of an equivalent solid conductor.

(b) What is the percent increase in resistance caused by the skin effect of 60-Hz current for this stranded conductor (of entry 2)?

(c) What is the percent increase in resistance associated with a 30°C increase in operating temperature?

12-24. A stranded conductor has a total cross-sectional area of 477,000 circular mils. Determine the diameter in cm of each strand if the total number of strands is to be 30.

12-25. A three-phase transmission line uses ACSR stranded conductors having a cross-sectional area of 300,000 circular mils for each phase. Compute the total power dissipated in the conductors for a 50-mi line length when the line is delivering 500 A. Assume 60-Hz operation and a stabilized operating temperature of 50°C.

12-26. The conductor size and spacing used for a 60-Hz, three-phase, transposed transmission line of 100 km length yields a series line impedance of $0.211 + j0.852 \ \Omega/\text{mi}$ and a shunt admittance of $2 \times 10^5 \ \Omega \cdot \text{mi}$. Determine the parameters to be used in the π-type equivalent circuit of this line. Draw the equivalent circuit with each parameter clearly labeled.

12-27. The load at the receiving end of the transmission line which is described in Prob. 12-26 draws 150 mW at a lagging power factor of 0.8. The rated line-to-line voltage at the receiving end is 161 kV. Determine the following:
(a) Sending-end voltage.
(b) Sending-end current.
(c) Sending-end real and reactive power.
(d) Total power dissipated in the transmission line conductors.

12-28. A 60-Hz, three-phase, transposed transmission line has the following total parameters:

$$\bar{Z} = 13.114 + j52.952$$
$$\bar{Y} = j0.311 \times 10^{-3}$$

The sending-end line-to-line voltage is 161 kV. The total receiving-end load is represented by the load impedance

$$\bar{Z}_L = 143\underline{/36.8°}$$

(a) Determine the receiving-end line-to-line voltage.
(b) Find the voltage regulation.
(c) Compute the sending-end current.
(d) Calculate the real and reactive power at the sending-end.
(e) How much power is delivered at the receiving end?
(f) What is the transmission efficiency?

12-29. Assume that the transmission line described in Example 12-5 is extended to cover a length of 400 km. Find the parameters of the π-type equivalent circuit that correctly represents this long transmission line for terminal equivalence.

12-30. Consider that the load at the receiving end of the transmission line of Prob. 12-29 receives 150 MW at unity pf and rated line-to-line voltage of 161 kV. Determine:
(a) Sending-end voltage and current.
(b) Sending-end real and reactive power.
(c) Power loss in the transmission line.

12-31. A 60-Hz, three-phase transmission line is 175 mi long. It has a total series line impedance of $35 + j140 \ \Omega$ and a shunt admittance of $930 \times 10^{-6} \ \text{S}$. It delivers 40 MW at 220 kV and 90% pf lagging. Find the voltage of the sending end using long-line theory.

Parameters of Electric Power Systems

A Description in Terms of Symmetrical Components

An electric power system that is subjected to a fault is under severe stress and as such operates in an unbalanced state. Analysis in such circumstances is greatly facilitated by replacing the unbalanced conditions by an equivalent system that is composed of balanced sets of voltages and currents. This goal is achieved through the use of symmetrical components. By the applications of an appropriate matrix transformation the unbalanced conditions arising on a faulted power system are replaced by a symmetrical, balanced set of positive- and negative-sequence voltages and a zero-sequence single-phase voltage set. Although the symmetrical components offer the considerable advantage of employing the simpler circuit analysis associated with symmetrical, balanced systems, it is first necessary to examine the behavior of the major power system components to these various sequences. The responses are not always the same to each sequence set. For example, the transmission line presents an impedance to the zero sequence set that is appreciably larger than that presented to the positive- and negative-sequence sets. Similar situations prevail for synchronous generators and motors and three-phase transformer connections. It is the purpose of this chapter to describe the method of symmetrical components as well as to obtain the parameters of various power system components to these symmetrical balanced sequences.

13-1 PER-UNIT CALCULATIONS FOR ELECTRIC POWER SYSTEMS

The electric power system frequently involves the interconnection of many electrical devices and components of different ratings and specifications. A very simple representation of the situation is illustrated in Fig. 13-1. The principal components that comprise this power system are the two synchronous generators, the two synchronous motors, the four banks of three-phase transformers, the three transmission lines spanning different lengths, and an assortment of circuit breakers and fuses. Of course any typical power system also supplies energy to many static loads and hosts of induction motors of all sizes. However, these are not included here both for simplicity and because their influence is of minor consequence when investigating the behavior of the power system under fault conditions.

Figure 13-1 Four-bus electric power system. Ratings data are for the individual components indicated.

The information that is listed alongside each major component of this power system is the nameplate rating of that individual item taken by itself. Since the volt-ampere and voltage ratings of these units are different, it is necessary to exercise care in the way the reactance parameters are used. This is especially important when an equivalent circuit of the entire power system, as expressed in reference to a single base voltage and volt-ampere rating, is desired. Such a formulation is required if the effect of each major component on the behavior of the power system is to be correctly described by means of a *single* line circuit operating at a suitably selected voltage. Although the reference voltage can be chosen arbitrarily, it is frequently taken to be a transmission line voltage or the rated voltage of a major generator. For example, in the system of Fig. 13-1, if the rated voltage of generator 1 is selected as the reference voltage, then a single-line circuit representation of the entire system can be obtained provided that all the components of the power system are referred back to the generator line side. For consistency the individual reactance parameters must be properly modified when reflected into the generator line circuit. It is the purpose of this section to describe this procedure.

The reactance parameters that are cited for the generators in Fig. 13-1 refer to the transient synchronous reactance of these machines. This is the reactance that is of importance when representing the synchronous generator after the first few cycles following a power system fault. If the effect of fault conditions are to be investigated during these first few cycles, then use is made of the subtransient synchronous reactances both for the generators as well as for the synchronous motors. The reactance parameter for transformers is the equivalent leakage reactance expressed in percent. Of course, if this quantity is divided by 100, the per-unit value of this leakage reactance is obtained. Thus, a transformer with a 5% leakage reactance has a per-unit value of 0.05. The advantages associated with the use of per-unit quantities are explained at some length in Sec. 2-8. One attractive feature of the per-unit notation, especially where transformers are involved, is that the *same* value is used whether one views the transformer from the high-voltage side or the low-voltage side. In either case a p-u $X = 0.05$ means that 5% of the rated voltage of that side appears as a reactance voltage drop across the leakage reactance at rated current. Because transformers are used in many places throughout an electric power system, the use of per-unit quantities is an indispensable tool in fault analysis.

Electric power systems deal with enormous amounts of power at high voltages. The real and reactive powers are frequently measured in megawatts (MW) and megavolt-amperes reactive (Mvars), while the line voltage is measured in kilovolts (kV). Because of the importance of the per-unit notation and its dependence upon base quantities, it is helpful at this point to obtain an equation for the base impedance in terms of a base voltage expressed in kV and a base current which is replaced by its equivalent form involving the rated megavolt-amperes of the device. Hence, proceeding on the assumption that the component of interest (e.g., a generator, transformer, motor, etc.) is designed for three-phase operation, the base impedance is defined as

$$Z_B = \frac{V_p}{I_p} \qquad (13\text{-}1)$$

where V_p denotes the rated phase voltage and I_p is the rated phase current. By assuming that the three-phase component is represented by an equivalent Y configuration, it follows that

$$V_p = \frac{V_L}{\sqrt{3}} \qquad \text{and} \qquad I_p = \frac{VA}{\sqrt{3}\,V_L} \qquad (13\text{-}2)$$

where VA denotes volt-amperes. Inserting Eqs. (13-2) into Eq. (13-1) yields

$$\boxed{Z_B = \frac{V_L^2}{VA} = \frac{(kV)_L^2}{MVA}} \qquad \Omega \qquad (13\text{-}3)$$

where MVA denotes megavolt-amperes. It is important to remember to use the line-to-line voltage when applying Eq. (13-3).

Example 13-1

Refer to the system of Fig. 13-1.

(a) Find the base impedance of the generator GI and determine the ohmic value of the transient reactance.

(b) Obtain the base impedance of transformers T4 at bus 4 referred to the low side and calculate the ohmic value of the leakage reactance on the low side as well as the high side.

Solution (a) Employing Eq. (13-3) directly gives

$$Z_{BG} = \frac{(13.8)^2}{50} = 3.81 \ \Omega$$

Therefore

$$X_G^1 = (\text{p-u } X')Z_{BG} = \left(\frac{20}{100}\right)(3.81) = 0.762 \ \Omega$$

(b) The low-voltage rating of transformers T4 is 6.9 kV. Hence

$$Z_{BL} = \frac{(6.9)^2}{20} = 2.381 \ \Omega$$

Hence

$$X_{eL} = (\text{p-u } X_e)Z_{BL} = (0.08)(2.381) = 0.19 \ \Omega$$

On the high side this becomes

$$X_{eH} = a^2 X_{eL} = \left(\frac{69}{6.9}\right)^2 (0.19) = 19 \ \Omega$$

where a is the ratio of transformation from the high side to the low side.

Expressing P-U Quantities to a Common Reference Base

The results of Example 13-1 should serve to make it apparent that different components of the same power system have per-unit values for their respective reactances which are determined from different base values. If it is desired to represent the power system with a single-line, consistent equivalent circuit, then a reconciliation of these quantities is necessary. This can be achieved by expressing the ohmic version of the individual components in terms of a common reference base impedance. This reference quantity may be that of any of the power system components or an arbitrarily selected one.

We now derive the equation that permits a per-unit quantity based on one base impedance to be converted to a per-unit value based on the reference base impedance. Let the subscript o denote quantities that are based on the original base impedance and the subscript r represent those same quantities referred to the reference base impedance. Then for the original component we have as the ohmic value of the reactance parameter

$$X_o = \left(\frac{\%X}{100}\right)Z_{BO} \qquad \Omega \qquad (13-4)$$

where Z_{BO} is the base impedance of the component based on its rated phase voltage and phase current. Upon invoking Eq. (13-3), the last equation can be rewritten as

$$X_o = \frac{(\%X_o)}{100} \frac{(kV)_o^2}{(MVA)_o} \quad \Omega \tag{13-5}$$

Next, if we call the chosen reference voltage $(kV)_r$ and the chosen reference megavolt-amperes $(MVA)_r$, then the expression for the reference base impedance becomes

$$Z_{Br} = \frac{(kV)_r^2}{(MVA)_r} \quad \Omega \tag{13-6}$$

The ohmic value of the reactance X_o can now readily be expressed as a per-unit quantity using the reference base impedance by simply dividing Eq. (13-5) by Eq. (13-6). Therefore,

$$\boxed{\text{p-u } X_{or} = \frac{X_o(\text{ohms})}{Z_{Br}(\text{ohms})} = \frac{(\%X_o)}{(100)} \frac{(kV)_o^2}{(kV)_r^2} \frac{(MVA)_r}{(MVA)_o}} \tag{13-7}$$

where X_{or} denotes the original reactance parameter expressed in per unit and based on a reference base impedance. In situations where step-up and step-down transformers are present, as is invariably the case in electric power systems, care must be taken that the specified reference voltage be properly transferred to the appropriate voltage level consistent with the use of the transformer turns ratio. In other words, the voltage ratio appearing in this equation must always be formulated on the *same* side of the transformers. For example, if the reference voltage in the system of Fig. 13-1 were selected to be 69 kV, then the application of Eq. (13-7) to G2 must employ a reference voltage of 13.8 kV. This is the value obtained when 69 kV is transformed to the low side by the action of transformer bank T2.

Example 13-2

Determine the per-unit values of the reactances of generators G1 and the transformers at bus 4. Use as the reference base: 100 MVA at 69 kV.

Solution The reference voltage selection of 69 kV is the transmission line voltage. Generator GI, however, is on the low-voltage side of transformer bank T1, so the proper value of the reference voltage to use for the conversion of the transient reactance of G is 13.8 kV. A direct application of Eq. (13-7) yields

$$\text{p-u } X'_{sr} = \left(\frac{20}{100}\right)\left(\frac{13.8}{13.8}\right)^2 \frac{100}{50} = 0.4$$

Observe that this p-u value is larger than the original value by a factor of 2. To recover the original ohmic value of this reactance, it is necessary to work with the reference base impedance evaluated on the low-voltage side. Thus,

$$Z_{BL} = \frac{13.8^2}{100} = 1.9 \ \Omega$$

Accordingly,

$$X'_s = (0.4)Z_{BL} = 0.4(1.9) = 0.762 \ \Omega$$

which agrees with the value found in Example 13-1.

For the transformers at bank T4 the equivalent leakage reactance based on the selected reference is

$$\text{p-u } X_e = \left(\frac{8}{100}\right)\left(\frac{69}{69}\right)^2\left(\frac{100}{20}\right) = 0.4$$

Example 13-3

The purpose of this example is to illustrate how a single-line, equivalent circuit is obtained for a two-bus electric power system employing ohmic values of the parameters as well as ideal transformers. It is then demonstrated that the same result can be achieved more easily through the use of per-unit values and Eq. (13-7).

A two-bus electric power system is depicted in Fig. 13-2(a) along with the individual ratings of the system components. The length of the transmission line, whose reactance is 0.85 Ω/mi per phase, is 40 mi. The load at bus 2 consists of a synchronous motor operating at unity power factor and drawing 30 MVA at 13.8 kV and a resistive load that draws 20 MVA.

(a) Determine an equivalent circuit representation of this power system using ohmic values for all the parameters that are involved.

(b) Convert the circuit of part (a) to one where the parameters are all expressed on a per-unit basis referenced to a base of 13.8 kV and 50 MVA.

(c) Obtain the result in part (b) directly through use of Eq. (13-7).

(d) If the voltage at bus 2 is the rated value of 13.8 kV when 50 MVA is drawn from the line, determine the generator excitation voltage.

Solution (a) The ohmic values of each of the components are obtained as follows:

Generator: $\qquad Z_{BG} = \dfrac{(13.8)^2}{50} = 3.81\ \Omega$

$$X'_g = (0.35)Z_{BG} = 1.33\ \Omega$$

Transformers: $\qquad Z_{BH} = \dfrac{(138)^2}{50} = 380.9\ \Omega$

$$X_{1H} = (0.08)380.9 = 30.5\ \Omega$$

$$X_{1L} = \dfrac{30.5}{(138)^2/13.8} = 0.305\ \Omega$$

Here the first subscript on X denotes leakage and the second refers to the high- and low-voltage sides.

Transmission line: $\qquad X_{TL} = (0.85)(40) = 34\ \Omega$

Synchronous motor: $\qquad Z_{BM} = \dfrac{(13.8)^2}{30} = 6.35\ \Omega$

$$\therefore\quad X'_s = (0.1)6.35 = 0.635\ \Omega$$

Resistive load at bus 2: At 20 MVA the current drawn by R_L is given by $I_L = 20{,}000/\sqrt{3}\,(13.8) = 836.74$ A. Hence the value of the resistive element that represents this load on a per-phase basis is

$$R_L = \dfrac{13{,}800}{\sqrt{3}\,(836.74)} = 9.52\ \Omega$$

Of course, this value can also be found by noting that for resistive loads,

$$\dfrac{(kV)^2}{R_L} = \text{MVA}$$

Hence here

$$R_L = \dfrac{(kV)^2}{\text{MVA}} = \dfrac{(13.8)^2}{20} = 9.52\ \Omega$$

Figure 13-2 Development of the impedance diagram of a two-bus transmission line: (a) line diagram; (b) equivalent circuit using ohmic values of the parameters and ideal transformers; (c) all parameters reduced to the common reference base of the generator; (d) circuit representation of the transmission line expressed entirely in per unit.

The circuit for this power system can now be represented as indicated in Fig. 13-2(b). Several points are worth noting. The transformers are represented on a per-phase basis by ideal transformers in series with their equivalent leakage reactances. However, for the transformer at bus 1 the low-side leakage reactance value is used whereas for the transformers at bus 2 the high-side value is used. The synchronous motor is represented by its transient synchronous reactance placed in series with its excitation emf $\bar{E}_{fm}$.

A single-line diagram is readily established by referring the quantities that appear in the high-voltage section of Fig. 13-2(b) to the low-voltage level of 13.8 kV. This means dividing the parameters $j34$ and $j30.5$ by 100, which is the turns ratio of the transformers squared. The final result is depicted in Fig. 13-2(c).

(b) An equivalent circuit expressed entirely in per-unit values can now be determined by dividing each element value that appears in Fig. 13-2(c) by the reference base impedance which in this case is the same as Z_{BG}. The result is shown in Fig. 13-2(d). Note that the per-unit value of the load resistor by this procedure comes out to be

$$\text{p-u } R_L = \frac{9.52}{3.81} = 2.5$$

which is consistent with the fact that R_L draws 20 MVA/50 MVA = 0.4 of the per-unit load current. Since the p-u value of the voltage at the load is unity, it follows that p-u $R_L = 1/0.4 = 2.5$.

(c) The circuit of Fig. 13-2(d) can be found directly without first drawing Figs. 13-2(b) and (c). Because the reference base quantities are those of the generator, the circuit is begun at the generator using its specified p-u value of X_{gi}^{1}. Next the bus 1 transformer leakage reactance is referred to the generator base by means of Eq. (13-7). Thus,

$$\text{p-u } X_{tr} = (0.08)\left(\frac{13.8}{13.8}\right)^{2}\left(\frac{50}{50}\right) = 0.08$$

Because the transmission line reactance was specified in ohms, the convenient way to put it into per-unit form is simply to divide by the reference base impedance. But first the ohmic value must be referred to the reference base voltage. Thus, for a ratio of transmission line voltage (138 kV) to reference base voltage (13.8 kV) of 10, the per-unit value of transmission line reactance is

$$\text{p-u } X_{TLr} = \frac{34/100}{3.81} = \frac{34}{381} = 0.089$$

The per-unit value of the load resistor is found in the same manner as in part (b). Since the voltage across this resistor is the same as the reference base value, there is no need for a referral adjustment in this case. Finally, the per-unit value of the transient reactance of the synchronous motor is found to be

$$\text{p-u } X_{sr}' = (0.1)\left(\frac{13.8}{13.8}\right)^{2}\left(\frac{50}{30}\right) = 0.167$$

Upon series connecting the foregoing parameters, the circuit of Fig. 13-2(d) is obtained directly.

(d) The generator excitation voltage required to yield a voltage of 13,800 V at the load when rated 50 MVA is delivered can be found either from the circuit of Fig. 13-2(c) or Fig. 13-2(d). When Fig. 13-2(c) is used, it is necessary to work with the rated current at the load in amperes. Thus,

$$I_g = \frac{50 \text{ MVA}}{(\sqrt{3})13.8 \text{ kV}} - 2091.85 \text{ A}$$

Applying KVL to the circuit of Fig. 13-2(c) using per phase values yields

$$\bar{E}_g = \frac{13,800}{3}\underline{/0°} + j(1.33 + 0.305 + 0.34 + 0.305)2091.85$$

$$= 7967.43 + j(2.28)(2091.85) = 9286\underline{/30.9°}$$

or, expressed line-to-line, this quantity becomes $\sqrt{3}\,(9286) = 16,084$ V.

When Fig. 13-2(d) is used, the per-unit value of the excitation voltage can be written simply as

$$\text{p-u } \bar{E}_g = 1.0\underline{/0°} + j(0.35 + 0.08 + 0.089 + 0.08)(\text{p-u } I_g)$$

$$= 1.0 + j(0.599)1 = 1.1657\underline{/30.9°}$$

Therefore the line-to-line voltage is $1.1657(13,800) = 16,086$ V, which compares favorably with the preceding result.

13-2 SYMMETRICAL COMPONENTS FOR UNBALANCED THREE-PHASE SYSTEMS

When a power system is placed in a state of stress as a result of a serious fault occurring somewhere in the system, the system is forced to operate in a severe condition of unbalance. The determination of the ensuing currents and voltages that exist in all the components of the system including the transmission lines, transformers, generators, motors, and static loads is considerably simplified by treating the unbalanced situation in terms of equivalent balanced systems. The procedure was devised by C. L. Fortescue and published as long ago as 1918.† In essence the method permits any system of unbalanced three-phase quantities to be replaced by two sets of balanced three-phase quantities of opposite phase sequence and a single-phase set of equal magnitude for each phase of the unbalanced set. Such a replacement means that answers can be more readily obtained by merely applying balanced three-phase circuit theory to the unbalanced electric power network. It is the purpose of this section to describe the theory that underlies this equivalency and then to obtain the equations that provide the conversions of the unbalanced quantities into corresponding balanced sets.

We begin by recalling to mind that a balanced three-phase set of voltages, is one in which each phase has the same magnitude but is displaced from the other phasors by 120 electrical degrees. The set ($\bar{V}_a$, $\bar{V}_b$, $\bar{V}_c$) that is depicted in Fig. 13-3(a) clearly does not satisfy these conditions and accordingly is labeled as an unbalanced set. Recall, too, that the positive direction of rotation of these phasor quantities is counterclockwise and that the phase sequence represented here is a, b, c. There are two kinds of unbalanced voltage sets that arise: those whose resultant adds to zero and those whose resultant does not. The unbalanced voltage system of Fig. 13-3(a) belongs to the latter category as is obvious from a glance at Fig. 13-3(b). The first step in replacement of an unbalanced set of voltages or currents by symmetrical components is the removal of this nonzero resultant, which is identified as $3\bar{V}_0$. Here $\bar{V}_0$ is treated as a *single-phase* voltage. By removing it from each of the phases of the original set in

†C. L. Fortescue, Method of Symmetrical Coordinates Applied to the Solution of Polyphase Networks, *Trans. AIEE*, Vol. 37, pp. 1027–1140, 1918.

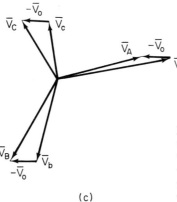

Figure 13-3 Unbalanced voltage sets:
(a) original unbalanced three-phase
voltage set; (b) identifying the resultant
voltage as $3\bar{V}_0$; (c) identifying a new
unbalanced voltage set $(\bar{V}_A, \bar{V}_B, \bar{V}_C)$
whose resultant voltage is zero.

the manner illustrated in Fig. 13-3(c), there results a new unbalanced voltage set $(\bar{V}_A,$
$\bar{V}_B, \bar{V}_C)$ which is distinguished from the original set $(\bar{V}_a, \bar{V}_b, \bar{V}_c)$ by having a resultant
that *is* zero. The quantity $\bar{V}_0$ is called the *zero-sequence* component of the original
unbalanced set. The voltage set $(\bar{V}_A, \bar{V}_B, \bar{V}_C)$ can now properly be called an *unsymmetrical balanced set* in contrast to the original *unsymmetrical unbalanced set* $(\bar{V}_a,$
$\bar{V}_b, \bar{V}_c)$. Finally, from Fig. 13-2(b) it follows that the value of $\bar{V}_0$ is readily determined
as one-third the sum of the original unsymmetrical unbalanced voltage set. Thus

$$\bar{V}_0 = \tfrac{1}{3}(\bar{V}_a + \bar{V}_b + \bar{V}_c) \tag{13-8}$$

Attention is next focused on replacing the unsymmetrical balanced set by two
symmetrical balanced voltage sets. In this connection then consider the two balanced
three-phase sets depicted in Fig. 13-4(a). One set is identified as $\bar{V}_{a1}, \bar{V}_{b1}, \bar{V}_{c1}$ and the
other as $\bar{V}_{a2}, \bar{V}_{b2}, \bar{V}_{c2}$. Initially, the respective phase voltages are displaced from one
another, but each set for the moment is assumed to have the same phase sequence,
i.e., *a, b, c*. It is instructive to note that the phasor addition of these two balanced
three-phase sets yields a resultant set which is also balanced because each phase voltage of the one set is affected in exactly the same way by the symmetrical additions
of each phase voltage of the second set. This situation is illustrated in Fig 13-4(b).
Clearly, such a combination can not be useful in replacing an unsymmetrical balanced voltage set such as the one in Fig. 13-3(c). To be useful, the two balanced three-

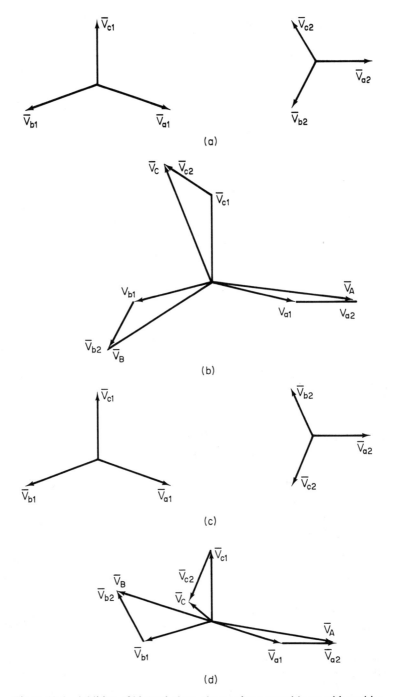

Figure 13-4 Addition of blanced, three-phase voltage sets: (a) sets with positive phase sequence; (b) addition of sets in (a) yielding a resultant which is symmetrical and balanced; (c) a positive and a negative sequence of balanced voltage sets; (d) addition of sets in (c) yielding a resultant which is unsymmetrical and balanced.

PARAMETERS OF ELECTRIC POWER SYSTEMS   CHAP. 13

phase voltage sets must yield, in general, an unsymmetrical balanced set upon addition. If we examine why the addition illustrated in Fig. 13-4(b) fails to meet this condition, we conclude that it is because the second set is introduced in a symmetrical manner to the first set thereby preserving symmetry. Further thought discloses that the way to achieve a sum that meets the condition of an unsymmetrical balanced set is to select the second balanced three-phase voltage with a reversed phase sequence, i.e., *acb* rather than *abc*. The situation is illustrated in Fig. 13-4(c). A subsequent addition of the positive-sequence set $(\bar{V}_{a1}, \bar{V}_{b1}, \bar{V}_{c1})$ yields the unsymmetrical balanced set that is depicted in Fig. 13-4(d). Thus, the general statement can be made that any unsymmetrical balanced three-phase voltages (or currents) can be represented by a *positive-sequence* balanced three-phase set and a *negative*-sequence balanced three-phase set. If in addition a zero-sequence component is included, then it is possible to represent any unsymmetrical, unbalanced set of voltages and currents.

It follows from the results appearing in Fig. 13-4(d) that the unsymmetrical, balanced three-phase voltage set may be expressed in terms of the positive- and negative-sequence sets as

$$\bar{V}_A = \bar{V}_{a1} + \bar{V}_{a2}$$
$$\bar{V}_B = \bar{V}_{b1} + \bar{V}_{b2} \qquad (13\text{-}9)$$
$$\bar{V}_C = \bar{V}_{c1} + \bar{V}_{c2}$$

Because both the positive-sequence and negative-sequence sets are symmetrical, balanced, and three-phase, it is possible to express the voltages of the *b* and *c* phases in terms of an operator *a* whose sole function is to provide a net angular displacement of 120 electrical degrees. Thus by defining the operator *a* as

$$a = 1\underline{/120°} \qquad (13\text{-}10)$$

and

$$a^2 = 1\underline{/240°} = 1\underline{/-120°} \qquad (13\text{-}11)$$

it is possible to identify the following equalities:

$$\bar{V}_{b1} = a^2 \bar{V}_{a1} \qquad \text{and} \qquad \bar{V}_{c1} = a\bar{V}_{a1} \qquad (13\text{-}12)$$

and

$$\bar{V}_{b2} = a\bar{V}_{a2} \qquad \text{and} \qquad \bar{V}_{c2} = a^2 \bar{V}_{a2} \qquad (13\text{-}13)$$

Accordingly, Eqs. (13-9) may be expressed solely in terms of $\bar{V}_{a1}$, $\bar{V}_{a2}$, and *a*. Thus

$$\bar{V}_A = \bar{V}_{a1} + \bar{V}_{a2}$$
$$\bar{V}_B = a^2 \bar{V}_{a1} + a\bar{V}_{a2} \qquad (13\text{-}14)$$
$$\bar{V}_C = a\bar{V}_{a1} + a^2 \bar{V}_{a2}$$

By restoring the zero-sequence component $\bar{V}_0$ to each of the foregoing expressions, it becomes possible to represent the original unsymmetrical and unbalanced voltage set $[\bar{V}_a, \bar{V}_b, \bar{V}_c$ of Fig. 13-3(a)] by

$$\bar{V}_a = \bar{V}_A + \bar{V}_0 = \bar{V}_{a1} + \bar{V}_{a2} + \bar{V}_0$$
$$\bar{V}_b = \bar{V}_B + \bar{V}_0 = a^2 \bar{V}_{a1} + a\bar{V}_{a2} + \bar{V}_0 \qquad (13\text{-}15)$$
$$\bar{V}_c = \bar{V}_C + \bar{V}_0 = a\bar{V}_{a1} + a^2 \bar{V}_{a2} + \bar{V}_0$$

At this point it is no longer necessary to carry the subscript a in the identity of the symmetrical components; hence the last equation may be written more succinctly as

$$\boxed{\begin{aligned} \bar{V}_a &= \bar{V}_1 + \bar{V}_2 + \bar{V}_0 \\ \bar{V}_b &= a^2\bar{V}_1 + a\bar{V}_2 + \bar{V}_0 \\ \bar{V}_c &= a\bar{V}_1 + a^2\bar{V}_2 + \bar{V}_0 \end{aligned}} \qquad (13\text{-}16)$$

Equations (13-16) represent the essence of the symmetrical components method as it applies to unsymmetrical and unbalanced three-phase systems. They state that any set of unsymmetrical, unbalanced, three-phase voltages (or currents) can be replaced by a symmetrically balanced, positive-sequence, three-phase set; a symmetrically balanced, negative-sequence, three-phase set; and a zero-sequence component. The latter is a single-phase component that is identical in each phase.

The procedure for finding the magnitude and phase of the zero-sequence component $\bar{V}_0$ in terms of the given unsymmetrical, unbalanced set $(\bar{V}_a, \bar{V}_b, \bar{V}_c)$ is displayed in Eq. (13-8). To obtain expressions to allow the evaluation of positive-sequence voltage $\bar{V}_1$ and the negative-sequence voltage $\bar{V}_2$, it is necessary merely to manipulate Eqs. (13-16) with the intention of isolating $\bar{V}_1$ and $\bar{V}_2$ successively. The isolation of $\bar{V}_1$ is readily achieved by multiplying the second equation of the set represented by Eqs. (13-16) by a, the third equation by a^2 and then summing. Thus

$$\begin{aligned} \bar{V}_a &= \bar{V}_1 + \bar{V}_2 + \bar{V}_0 \\ a\bar{V}_b &= a^3\bar{V}_1 + a^2\bar{V}_2 + a\bar{V}_0 \\ a^2\bar{V}_c &= a^3\bar{V}_1 + a^4\bar{V}_2 + a^2\bar{V}_0 \\ \hline \bar{V}_a + a\bar{V}_b + a^2\bar{V}_c &= 3\bar{V}_1 + 0 + 0 \end{aligned} \qquad (13\text{-}17)$$

where use is made of the identities

$$a^3 = 1\underline{/0°} \qquad a^4 = a \qquad 1 + a^2 + a = 0 \qquad (13\text{-}18)$$

Therefore the equation for $\bar{V}_1$ becomes

$$\boxed{\bar{V}_1 = \tfrac{1}{3}(\bar{V}_a + a\bar{V}_b + a^2\bar{V}_c)} \qquad (13\text{-}19)$$

This last equation states that $\bar{V}_1$ is determined by first obtaining the sum of $\bar{V}_a$, $\bar{V}_b$ advanced in phase by 120 electrical degrees, and $\bar{V}_c$ advanced in phase by 240 electrical degrees; and then taking one-third of the resulting sum. These details are illustrated in Example 13-4.

By following a similar procedure for $\bar{V}_2$, an appropriate expression for its evaluation in terms of the original unbalanced set is established. The result is

$$\boxed{\bar{V}_2 = \tfrac{1}{3}(\bar{V}_a + a^2\bar{V}_b + a\bar{V}_c)} \qquad (13\text{-}20)$$

A comparison with $\bar{V}_1$ indicates that the difference lies in the way $\bar{V}_b$ and $\bar{V}_c$ are rotated by the displacement operator a in generating the sum appearing in the parentheses.

Example 13-4

Determine the positive-, negative-, and zero-sequence components that can be used to replace the unsymmetrical unbalanced voltage set described by

$$\bar{V}_a = 140\underline{/0^\circ} \qquad \bar{V}_b = 100\underline{/-90^\circ} \qquad \bar{V}_c = 80\underline{/90^\circ}.$$

Solution Application of Eq. (13-8) yields the zero-sequence component. Thus,

$$\bar{V}_0 = \tfrac{1}{3}(\bar{V}_a + \bar{V}_b + \bar{V}_c) = \tfrac{1}{3}(100 - j100 + j80) = \tfrac{1}{3}(140 - j20)$$
$$= \tfrac{1}{3}(141.42)\underline{/-8.13^\circ} = 47.14\underline{/-8.13^\circ}$$

Similarly, the application of Eq. (13-19) gives the magnitude and phase of the positive-sequence component of $\bar{V}_a$ as

$$\bar{V}_1 = \tfrac{1}{3}(\bar{V}_a + a^2\bar{V}_b + a\bar{V}_c) = \tfrac{1}{3}(140 + 100\underline{/-90^\circ + 120^\circ} + 80\underline{/90^\circ - 120^\circ})$$
$$= \tfrac{1}{3}(295.9 + j10) = 98.7\underline{/1.94^\circ}$$

Keep in mind that there is a positive-sequence component of $\bar{V}_b$ which is given by $a^2\bar{V}_1 = 98.7\underline{/1.94^\circ} - 120^\circ = 98.7\underline{/-118.06^\circ}$. Correspondingly, there is also a positive-sequence component in $\bar{V}_c$ of $a\bar{V}_1 = 98.7\underline{/1.94^\circ} + 120^\circ = 98.7\underline{/121.94^\circ}$. Although the positive-sequence components in each phase have different angular arguments, it is really necessary to deal solely with the component $\bar{V}_1$ which is associated with phase a. This is because, as a symmetrical, balanced system, the effect of the positive sequence in phases b and c is identical except for a known displacement angle of ± 120 electrical degrees. Herein lies the advantage of the use of symmetrical components. The calculations can be performed on a per-phase basis even in the presence of severely unbalanced voltages.

The negative-sequence component results from Eq. (13-20) and is found to be

$$\bar{V}_2 = \tfrac{1}{3}(\bar{V}_a + a\bar{V}_b + a^2\bar{V}_c) = \tfrac{1}{3}(140 + 100\underline{/-90^\circ + 240^\circ} + 80\underline{/90^\circ + 120^\circ})$$
$$= \tfrac{1}{3}(-15.9 + j10) = \tfrac{1}{3}(18.8)\underline{/147.83^\circ} = 6.26\underline{/147.83^\circ}$$

Observe that the negative-sequence component is small compared to the positive-sequence component. This is not an atypical occurrence in such circumstances.

As a check on the foregoing evaluations, we now determine $\bar{V}_a$ in terms of the sequence components. Thus,

$$\bar{V}_a = \bar{V}_1 + \bar{V}_2 + \bar{V}_0 = 98.7\underline{/1.94^\circ} + 6.26\underline{/147.83^\circ} + 47.14\underline{/-8.13^\circ}$$
$$= 98.64 + j3.34 - 5.3 + j3.33 + 46.67 - j6.67$$
$$= 139.94 \approx 140\underline{/0^\circ} \qquad\qquad \text{Q.E.D.}$$

Matrix Representation

The equation set displayed by Eq. (13-16) lends itself to a matrix representation. We can rewrite these expressions as

$$\begin{bmatrix} \bar{V}_a \\ \bar{V}_b \\ \bar{V}_c \end{bmatrix} = \begin{bmatrix} 1 & 1 & 1 \\ a^2 & a & 1 \\ a & a^2 & 1 \end{bmatrix} \begin{bmatrix} \bar{V}_1 \\ \bar{V}_2 \\ \bar{V}_0 \end{bmatrix} \qquad (13\text{-}21)$$

Proceeding on the assumptions that $\bar{V}_a, \bar{V}_b, \bar{V}_c$ are the phase voltages, we can define the phase voltage vector as

$$\bar{\mathbf{V}}_p = \begin{bmatrix} \bar{V}_a \\ \bar{V}_b \\ \bar{V}_c \end{bmatrix} \qquad (13\text{-}22)$$

Similarly, we define the symmetrical sequence voltage vector $\bar{V}_s$ as

$$\bar{\mathbf{V}}_s = \begin{bmatrix} \bar{V}_1 \\ \bar{V}_2 \\ \bar{V}_0 \end{bmatrix} \qquad (13\text{-}23)$$

Finally, to exploit the compactness in formulation offered by the use of matrices, we define

$$\mathbf{T} \equiv \begin{bmatrix} 1 & 1 & 1 \\ a^2 & a & 1 \\ a & a^2 & 1 \end{bmatrix} \qquad (13\text{-}24)$$

Accordingly, the expressions represented by Eq. (13-21) can now be written in the simple and succinct form

$$\boxed{\bar{\mathbf{V}}_p = \mathbf{T}\bar{\mathbf{V}}_s} \qquad (13\text{-}25)$$

where it then becomes apparent that $\mathbf{T}$ serves as a matrix transformation which fulfills the useful purpose of permitting the original unbalanced voltage set to be represented by balanced positive-, negative-, and zero-sequence sets.

13-3 COMPLEX POWER

Electric power systems are capable of managing enormous amounts of real power and reactive power. Real power is often measured in units of megawatts and is customarily computed on a per-phase basis by taking the product of the voltage and the in-phase component of the associated phase current. Reactive power is frequently measured in units of megavolt-amperes reactive and is determined by taking the product of the phase voltage and the quadrature component of the phase current. It is possible to express the phasor sum of these two quantities on a per-phase basis in terms of a simple, single product of the phase voltage and a suitably identified version of the phase current. The derivation starts by examining the situation that is depicted in Fig. 13-5.

The circuit and the circuit quantities are expressed on a per-phase basis. The phasor diagram shows the current lagging the voltage by α degrees. Both phasor quantities, however, have their instantaneous angular displacements measured with respect to a reference time line. On this basis the expressions for the phasor voltage and current become

$$\bar{V} = V\underline{/\phi} = V\epsilon^{j\phi} \qquad (13\text{-}26)$$

$$\bar{I} = I\underline{/\theta} = I\epsilon^{j\theta} \qquad (13\text{-}27)$$

Because our goal is to uncover a simple expression for complex power, we might

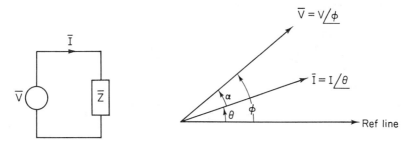

Figure 13-5 Phasor diagram.

explore the obvious suggestion that presents itself of formulating the product of the two quantities. This yields

$$\bar{V}\bar{I} = VI\underline{/\phi + \theta} = VI\epsilon^{j(\phi+\theta)} \tag{13-28}$$

Our experience with single-phase analysis makes it quickly apparent that Eq. (13-28) is not a useful formulation. Although the product of the magnitudes of the voltage and current is meaningful, the presence of the sum of the phasor angles is without meaning regarding real power and reactive power. Our experience with circuits of this kind tells us that it is the *difference* of the angle between voltage and current that is significant. Fortunately, the difference angle can conveniently be made to enter the formulation by replacing the actual current (or voltage) by its complex conjugate. The use of a complex conjugate means the use of the negative of the phase angle. Thus, for the current phasor we have

$$\bar{I}* = I\epsilon^{-j\theta} \tag{13-29}$$

where * is used to denote the complex conjugate of a quantity. If we now formulate the product of $\bar{V}$ and $\bar{I}*$, there results

$$\bar{S} \equiv \bar{V}\bar{I}* = VI\epsilon^{j(\phi-\theta)} = VI\epsilon^{j\alpha}$$

$$= VI \cos \alpha + jVI \sin \alpha$$

$$= P + jQ \tag{13-30}$$

The real part of Eq. (13-30) now properly represents the real power present in the circuit of Fig. 13-5 and the *j* part denotes correctly the reactive volt-amperes. Observe, too, that a lagging reactive volt-ampere is represented by a positive value for Q, which is also consistent with custom.

A useful application of the result displayed in Eq. (13-30) is the representation of the complex power that exists in an electric power system under fault conditions. Thus, if $\bar{V}_a, \bar{V}_b, \bar{V}_c$ are the unbalanced phase voltages and $\bar{I}_a, \bar{I}_b, \bar{I}_c$ are the corresponding phase currents, then real and reactive power in the system may be represented by

$$\bar{S} = \bar{V}_a\bar{I}_a^* + \bar{V}_b\bar{I}_b^* + \bar{V}_c\bar{I}_c^* \tag{13-31}$$

In matrix form the last equation becomes

$$\bar{S} = [\bar{V}_a \quad \bar{V}_b \quad \bar{V}_c] \begin{bmatrix} \bar{I}_a \\ \bar{I}_b \\ \bar{I}_c \end{bmatrix}^* = \bar{V}_p^T\bar{I}_p^* \tag{13-32}$$

where $\bar{\mathbf{V}}_p^T$ denotes the transposed phase voltage vector [i.e., the column representation of Eq. (13-22) is written as a row] and $\bar{\mathbf{I}}_p^*$ denotes the complex conjugate of the phase current vector.

Because of the simplicity associated with calculations that involve symmetrical, balanced voltages and currents, it is worthwhile to express the complex power in faulted power systems in terms of these symmetrical components. This can be accomplished by proceeding as outlined below. In accordance with the general result that appears in Eq. (13-25), the complex conjugate of the phase current vector expressed in terms of its corresponding symmetrical components $(\bar{I}_1, \bar{I}_2, \bar{I}_0)$ is

$$\bar{\mathbf{I}}_p^* = \bar{\mathbf{T}}^* \bar{\mathbf{I}}_s^* \tag{13-33}$$

where

$$\bar{\mathbf{I}}_s = \begin{bmatrix} \bar{I}_1 \\ \bar{I}_2 \\ \bar{I}_0 \end{bmatrix} \tag{13-34}$$

Inserting Eqs. (13-25) and (13-33) into Eq. (13-32), the expression for complex power becomes†

$$\bar{\mathbf{S}} = [\mathbf{T}\bar{\mathbf{V}}_s]^T [\mathbf{T}^* \bar{\mathbf{I}}_s^*] = \mathbf{V}_s^T \mathbf{T}^T \mathbf{T}^* \mathbf{I}_s^* \tag{13-35}$$

Further development of this equation depends upon evaluation of $\mathbf{T}^T \mathbf{T}^*$.

The transpose of the transformation matrix $\mathbf{T}$ is readily obtained upon letting the columns of $\mathbf{T}$ serve as the rows of $\mathbf{T}^T$. Thus

$$\mathbf{T}^T = \begin{bmatrix} 1 & a^2 & a \\ 1 & a & a^2 \\ 1 & 1 & 1 \end{bmatrix} \tag{13-36}$$

Moreover, the conjugate of $\mathbf{T}$ is found by simply replacing a by a^2 and a^2 by a wherever they occur in the expression for $\mathbf{T}$. Accordingly,

$$\mathbf{T}^* = \begin{bmatrix} 1 & 1 & 1 \\ a & a^2 & 1 \\ a^2 & a & 1 \end{bmatrix} \tag{13-37}$$

Finally, by applying the rule for the multiplication of two matrices, the result is found to be

$$\mathbf{T}^T \mathbf{T}^* = \begin{bmatrix} 1 & a^2 & a \\ 1 & a & a^2 \\ 1 & 1 & 1 \end{bmatrix} \begin{bmatrix} 1 & 1 & 1 \\ a & a^2 & 1 \\ a^2 & a & 1 \end{bmatrix} = \begin{bmatrix} 3 & 0 & 0 \\ 0 & 3 & 0 \\ 0 & 0 & 3 \end{bmatrix} = 3 \begin{bmatrix} 1 & 0 & 0 \\ 0 & 1 & 0 \\ 0 & 0 & 1 \end{bmatrix} = 3\mathbf{U} \tag{13-38}$$

where $\mathbf{U}$ denotes the identity matrix with all ones on the principal diagonal and zeros elsewhere.

Returning to Eq. (13-35) with Eq. (13-38) inserted, we now get

$$\bar{\mathbf{S}} = 3\bar{\mathbf{V}}_s^T \mathbf{U} \bar{\mathbf{I}}_s^* = 3[\bar{V}_1 \quad \bar{V}_2 \quad \bar{V}_0] \begin{bmatrix} 1 & 0 & 0 \\ 0 & 1 & 0 \\ 0 & 0 & 1 \end{bmatrix} \begin{bmatrix} \bar{I}_1^* \\ \bar{I}_2^* \\ \bar{I}_0^* \end{bmatrix} \tag{13-39}$$

†From matrix algebra we have the property $(\mathbf{AB})^T = \mathbf{B}^T \mathbf{A}^T$.

Carrying out the operations indicated, the result for the expression of the complex power expressed solely in terms of the symmetrical components of the voltage and the conjugate of the current is

$$\bar{S} = 3\bar{V}_1\bar{I}_1^* + 3\bar{V}_2\bar{I}_2^* + 3\bar{V}_0\bar{I}_0^* \qquad (13\text{-}40)$$

This result is strikingly simple and even intuitively expected. It states that the system complex power is the sum of the complex power of each sequence. The factor 3 is consistent with the use of phase quantities for three-phase systems.

13-4 POSITIVE-, NEGATIVE-, AND ZERO-SEQUENCE PARAMETERS OF TRANSMISSION LINES

When a fault condition occurs in an electric power system, the system is put into a state of severe unbalance. However, through the use of symmetrical components, the unbalanced state can be analyzed in terms of a symmetrical balanced positive sequence, a symmetrical balanced negative sequence, and a zero sequence of voltage. It is important now to examine how the various components of the electric power system appear to these different sets of voltages. In this section we investigate this matter from the viewpoint of the transmission line inductance and capacitance.

Line Inductance

The preparatory work for this investigation is done in Sec. 12-3 where expressions are derived for the flux linkage of one line of a three-phase system equipped with a neutral that is conducting current as well as for the flux linkage of the neutral wire under the same conditions. If the instantaneous values of the three line currents i_a, i_b, i_c and the neutral current i_n are used, the resulting expression for the instantaneous flux linkage ψ_a can be written as

$$\psi_a = \frac{\mu}{2\pi}\left(i_a \ln\frac{1}{r'} + i_b \ln\frac{1}{D_{ab}} + i_c \ln\frac{1}{D_{ac}} + i_n \ln\frac{1}{D_{an}}\right) \qquad (13\text{-}41)$$

where r' is the effective radius of line conductor a, D_{ab} is the spacing between conductors a and b, D_{ac} is the spacing between conductors a and c, and D_{an} is the spacing between conductors a and n measured from conductor center to center. Equation (13-41) follows directly from Eq. (12-77). Similarly, by means of Eq. (12-8), the expression for the instantaneous total flux linking the neutral wire when current flows in all conductors is

$$\psi_n = \frac{\mu}{2\pi}\left(i_a \ln\frac{1}{D_{na}} + i_b \ln\frac{1}{D_{nb}} + i_c \ln\frac{1}{D_{nc}} + i_n \ln\frac{1}{r'_n}\right) \qquad (13\text{-}42)$$

Here r'_n is the effective radius of the neutral wire. Our interest in writing these equations is to find an appropriate expression for the voltage drop that occurs in the line conductors as well as in the neutral wire. First, however, it is helpful to remove i_n from Eqs. (13-41) and (13-42) by applying the boundary condition

$$i_a + i_b + i_c + i_n = 0 \qquad (13\text{-}43)$$

Accordingly, these equations can then be expressed as

$$\psi_a = \frac{\mu}{2\pi}\left(i_a \ln \frac{D_{an}}{r'} + i_b \ln \frac{D_{an}}{D_{ab}} + i_c \ln \frac{D_{an}}{D_{ac}}\right) \tag{13-44}$$

and

$$\psi_n = \frac{\mu}{2\pi}\left(i_a \ln \frac{r'_n}{D_{an}} + i_b \ln \frac{r'_n}{D_{nb}} + i_c \ln \frac{r'_n}{D_{nc}}\right) \tag{13-45}$$

The voltage drop in the inductance of line a is found by applying Faraday's law. Thus, for sinusoidal currents, the result is

$$\frac{d\psi_a}{dt} = j\frac{\omega\mu}{2\pi}\left(\bar{I}_a \ln \frac{D_{an}}{r'} + \bar{I}_b \ln \frac{D_{an}}{D_{ab}} + \bar{I}_c \ln \frac{D_{an}}{D_{ac}}\right) \tag{13-46}$$

The corresponding voltage drop in the neutral wire expressed also in terms of the phasor currents of lines a, b, c becomes

$$\frac{d\psi_n}{dt} = j\frac{\omega\mu}{2\pi}\left(\bar{I}_a \ln \frac{r'_n}{D_{an}} + \bar{I}_b \ln \frac{r'_n}{D_{nb}} + \bar{I}_c \ln \frac{r'_n}{D_{nc}}\right) \tag{13-47}$$

All line currents plus the neutral current are assumed to be flowing from the source to the load as displayed by Eq. (13-43), which of course is simply a statement of Kirchhoff's current law. Figure 13-6 depicts the situation on a per-phase basis. Observe that for these assumed directions of the line current i_a and neutral current i_n, the voltage drops in these lines have the polarities indicated. Hence, if we are interested in describing the change in voltage that occurs between terminals a and n, it is necessary to subtract Eq. (13-47) from Eq. (13-46). Moreover, because the expressions for the flux linkages apply for the per-unit meter length, it follows that this change is a change per unit length of line. Accordingly, we have

$$\frac{\Delta \bar{V}_{an}}{l} = \frac{d\psi_a}{dt} - \frac{d\psi_n}{dt} = j\frac{\omega\mu}{2\pi}\left(\bar{I}_a \ln \frac{D_{an}^2}{r'r'_n} + \bar{I}_b \ln \frac{D_{an}D_{nb}}{D_{ab}r'_n} + \bar{I}_c \ln \frac{D_{an}D_{nc}}{D_{ac}r'_n}\right) \tag{13-48}$$

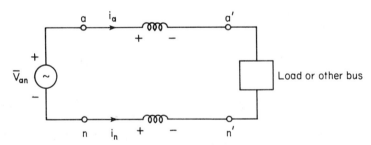

Figure 13-6 Per-phase representation of transmission line with all currents flowing toward the receiving end. Polarity markings indicate voltage drops in line conductors for assumed current directions.

where l is the length of the transmission line. We now assume that for a transposed line the average distance from each line to neutral is the same. Hence we set $D_{an} = D_{bn} = D_{cn} = D_n$. If we further assume that the distance between conductors is the same and specifically equal to D, the last equation simplifies to

$$\frac{\Delta \bar{V}_{an}}{l} = j\frac{\omega\mu}{2\pi}\left(\bar{I}_a \ln \frac{D_n^2}{r'r'_n} + \bar{I}_b \ln \frac{D_n^2}{Dr'_n} + \bar{I}_c \ln \frac{D_n^2}{Dr'_n}\right) \tag{13-49}$$

A similar procedure applied to lines b and c yields

$$\frac{\bar{V}_{bn}}{l} = j\frac{\omega\mu}{2\pi}\left(\bar{I}_a \ln \frac{D_n^2}{Dr'_n} + \bar{I}_b \ln \frac{D_n^2}{r'r'_n} + \bar{I}_c \ln \frac{D_n^2}{Dr'_n}\right) \qquad (13\text{-}50)$$

and

$$\frac{\Delta\bar{V}_{cn}}{l} = j\frac{\omega\mu}{2\pi}\left(\bar{I}_a \ln \frac{D_n^2}{Dr'_n} + \bar{I}_b \ln \frac{D_n^2}{Dr'_n} + \bar{I}_c \ln \frac{D_n^2}{r'r'_n}\right) \qquad (13\text{-}51)$$

By introducing

$$z_s \equiv j\frac{\omega\mu}{2\pi} \ln \frac{D_n^2}{r'r'_n} \qquad (13\text{-}52)$$

and

$$z_m \equiv j\frac{\omega\mu}{2\pi} \ln \frac{D_n^2}{Dr'_n} \qquad (13\text{-}53)$$

Eqs. (13-49) to (13-51) can be expressed in matrix form as

$$\frac{\Delta\mathbf{V}_p}{l} = \frac{1}{l}\begin{bmatrix} \Delta\bar{V}_a \\ \Delta\bar{V}_b \\ \Delta\bar{V}_c \end{bmatrix} = \begin{bmatrix} z_s & z_m & z_m \\ z_m & z_s & z_m \\ z_m & z_m & z_s \end{bmatrix}\begin{bmatrix} \bar{I}_a \\ \bar{I}_b \\ \bar{I}_c \end{bmatrix} \equiv \bar{\mathbf{Z}}_p\bar{\mathbf{I}}_p \qquad (13\text{-}54)$$

where $\bar{\mathbf{Z}}_p$ denotes the transmission line impedance matrix, which is defined by

$$\bar{\mathbf{Z}}_p = \begin{bmatrix} z_s & z_m & z_m \\ z_m & z_s & z_m \\ z_m & z_m & z_s \end{bmatrix} \quad \Omega/\text{m} \qquad (13\text{-}55)$$

A description of the transmission line impedance matrix in terms of symmetrical components can now be obtained by introducing Eq. (13-25) for voltage (and an analogous expression for the current) into Eq. (13-54). Thus,

$$\mathbf{T}\frac{\Delta\bar{\mathbf{V}}_s}{l} = \bar{\mathbf{Z}}_p\mathbf{T}\bar{\mathbf{I}}_s \qquad (13\text{-}56)$$

multiplying both sides by the inverse of $\mathbf{T}$ gives

$$\frac{\Delta\bar{\mathbf{V}}_s}{l} = \mathbf{T}^{-1}\bar{\mathbf{Z}}_p\mathbf{T}\bar{\mathbf{I}}_s \qquad (13\text{-}57)$$

or

$$\frac{\Delta\bar{\mathbf{V}}_s}{l} = \bar{\mathbf{Z}}_s\bar{\mathbf{I}}_s \qquad (13\text{-}58)$$

where

$$\bar{\mathbf{Z}}_s = \mathbf{T}^{-1}\bar{\mathbf{Z}}_p\mathbf{T} \qquad (13\text{-}59)$$

This last expression provides the instructions that convert the transmission line unit inductive reactance matrix into a form that is consistent with the use of positive-, negative-, and zero-sequence voltages ($\bar{\mathbf{V}}_s$) and currents ($\bar{\mathbf{I}}_s$).

In order to evaluate the symmetrical components impedance matrix for the transmission line, $\bar{\mathbf{Z}}_s$, it is necessary first to determine the inverse of $\mathbf{T}$ as it is defined by Eq. (13-24). The application of matrix algebra leads to the following result:

$$\mathbf{T}^{-1} = \frac{1}{3} \begin{bmatrix} 1 & a & a^2 \\ 1 & a^2 & a \\ 1 & 1 & 1 \end{bmatrix} \qquad (13\text{-}60)$$

where a denotes the rotational operator defined by Eq. (13-10). It is now possible to write Eq. (13-59) as

$$\bar{\mathbf{Z}}_s = \frac{1}{3} \begin{bmatrix} 1 & a & a^2 \\ 1 & a^2 & a \\ 1 & 1 & 1 \end{bmatrix} \begin{bmatrix} z_s & z_m & z_m \\ z_m & z_s & z_m \\ z_m & z_m & z_s \end{bmatrix} \begin{bmatrix} 1 & 1 & 1 \\ a^2 & a & 1 \\ a & a^2 & 1 \end{bmatrix} \qquad (13\text{-}61)$$

or

$$\bar{\mathbf{Z}}_s = \begin{bmatrix} z_s - z_m & 0 & 0 \\ 0 & z_s - z_m & 0 \\ 0 & 0 & z_s + 2z_m \end{bmatrix} = \begin{bmatrix} Z_1 & 0 & 0 \\ 0 & Z_2 & 0 \\ 0 & 0 & Z_0 \end{bmatrix} \qquad (13\text{-}62)$$

where

$$\bar{Z}_1 = z_s - z_m = \text{positive-sequence impedance} \qquad (13\text{-}63)$$

$$\bar{Z}_2 = z_s - z_m = \text{negative-sequence impedance} \qquad (13\text{-}64)$$

$$\bar{Z}_0 = z_s + 2z_m = \text{zero-sequence impedance} \qquad (13\text{-}65)$$

For the values of z_s and z_m indicated by Eqs. (13-52) and (13-53), the expressions for the symmetrical sequence impedances become

$$\boxed{\bar{Z}_1 = \bar{Z}_2 = z_s - z_m = j\frac{\omega\mu}{2\pi}\ln\frac{D}{r'}} \qquad \Omega/\text{m} \qquad (13\text{-}66)$$

and

$$\boxed{\bar{Z}_0 = z_s + 2z_m = j\frac{\omega\mu}{2\pi}\ln\frac{D_n^6}{D^2 r'(r'_n)^3}} \qquad \Omega/\text{m} \qquad (13\text{-}67)$$

Several important conclusions may be drawn from the last two expressions. One, the impedance "seen" by the positive- and negative-sequence waves as they move along the transmission line is the same. This conclusion, of course, is entirely consistent with a physical argument. The reversal of the phase sequence of voltages applied to a static component (the transmission line) can not be expected to experience a different impedance because these waves encounter the *same* geometry of the lines irrespective of the sequence. Second, the impedance presented by the transmission line to the zero-sequence voltage component is substantially larger than that confronting the positive- and negative-sequence voltages. Finally, it is useful to note that the operations called for in Eq. (13-59) lead to an impedance matrix for symmetrical components which is *diagonal*. That is, $\bar{\mathbf{Z}}_s$ has non zero elements appearing only on the principal diagonal. This is important because it means that the positive-, negative-, and zero-sequence impedance drops may be treated separately, i.e., uncoupled from one another.

It is useful to keep in mind that the foregoing results are quite general. For example, they are applicable to a bundled, unsymmetrical line provided that D is used as the geometric mean distance and r' is replaced by a corresponding geometric mean radius.

Example 13-5

A three-phase transmission line with equilateral spacing has a neutral wire located at the centroid of the equilateral configuration. Each line including the neutral consists of simple conductors having a radius of 0.74 cm. The spacing between line conductors is 6 m. Determine the values of the positive-, negative-, and zero-sequence impedances of this transmission line expressed in units of ohms per kilometer and ohms per mile. The line frequency is 60 Hz.

Solution From geometrical considerations it can easily be shown that the neutral wire is located at a distance from each line conductor of $D_n = D/\sqrt{3} = 6/\sqrt{3} = 3.464$ m. Furthermore, the effective radius of each conductor is $r' = r'_n = (0.7788)(0.0074) = 0.00576$ m. By Eq. (13-66) the values of the positive and negative sequences are found to be equal to

$$\bar{Z}_1 = \bar{Z}_2 = j\frac{377 \times (1/411)0.10}{2\pi} \ln\frac{6}{0.00576} = j377(2 \times 10^{-7})1041.1$$

$$= j0.524 \; \Omega/\text{km} = j0.843 \; \Omega/\text{mi}$$

Similarly, by Eq. (13-67) the zero-sequence impedance is determined to be

$$\bar{Z}_0 = j(377)(2 \times 10^{-7}) \ln\frac{(3.464)^6}{(0.00576)^4 6^2} = j1.847 \; \Omega/\text{km} = j2.972 \; \Omega/\text{mi}$$

A comparison reveals that in this case the zero-sequence impedance of the transmission line is more than three times larger than the positive- or negative-sequence impedance.

Line Capacitance

The analysis that leads to expressions for the positive-, negative-, and zero-sequence capacitive reactances (or admittances) of a transmission line begins with the result displayed by Eq. (12-130). There it is shown that the line-to-neutral voltage per unit length of line expressed in terms of the respective charges residing on the line conductors is given by

$$v_{an} = v_a - v_n = \frac{1}{2\pi\epsilon}\left(q_a \ln\frac{D_{na}^2}{r_a r_n} + q_b \ln\frac{D_{na}D_{nb}}{D_{ab}r_n} + q_c \ln\frac{D_{na}D_{nc}}{D_{ac}r_n}\right) \qquad (13\text{-}68)$$

The line-to-neutral voltages from lines b and c can similarly be expressed as

$$v_{bn} = v_b - v_n = \frac{1}{2\pi\epsilon}\left(q_a \ln\frac{D_{nb}D_{na}}{D_{ba}r_n} + q_b \ln\frac{D_{nb}^2}{r_b r_n} + q_c \ln\frac{D_{nb}D_{nc}}{D_{bc}r_n}\right) \qquad (13\text{-}69)$$

$$v_{cn} = v_c - v_n = \frac{1}{2\pi\epsilon}\left(q_a \ln\frac{D_{nc}D_{na}}{D_{ca}r} + q_b \ln\frac{D_{nc}D_{nb}}{D_{cb}r_n} + q_c \ln\frac{D_{nc}^2}{r_c r_n}\right) \qquad (13\text{-}70)$$

In the interest of working with more manageable expressions, the foregoing equations are modified to apply to the equilaterally spaced transmission line with the neutral wire located at the centroid. Also, each conductor is assumed to have the same radius. The results then are more simply expressed as

$$v_{an} = \frac{1}{2\pi\epsilon}\left(q_a \ln \frac{D^2}{3r^2} + q_b \ln \frac{D}{3r} + q_c \ln \frac{D}{3r}\right) \tag{13-71}$$

$$v_{bn} = \frac{1}{2\pi\epsilon}\left(q_a \ln \frac{D}{3r} + q_b \ln \frac{D^2}{3r^2} + q_c \ln \frac{D}{3r}\right) \tag{13-72}$$

$$v_{cn} = \frac{1}{2\pi\epsilon}\left(q_a \ln \frac{D}{3r} + q_b \ln \frac{D}{3r} + q_c \ln \frac{D^2}{3r^2}\right) \tag{13-73}$$

where D is the distance between conductors and r is the radius of each conductor. Employing matrix formulation, these last three equations can further be written as

$$\mathbf{v}_p = \begin{bmatrix} v_{an} \\ v_{bn} \\ v_{cn} \end{bmatrix} = \frac{1}{2\pi\epsilon} \begin{bmatrix} \ln \frac{D^2}{3r^2} & \ln \frac{D}{3r} & \ln \frac{D}{3r} \\ \ln \frac{D}{3r} & \ln \frac{D^2}{3r^2} & \ln \frac{D}{3r} \\ \ln \frac{D}{3r} & \ln \frac{D}{3r} & \ln \frac{D^2}{3r^2} \end{bmatrix} \begin{bmatrix} q_a \\ q_b \\ q_c \end{bmatrix} = \mathbf{Sq} \tag{13-74}$$

where the matrix $\mathbf{S}$ is defined by the quantity in brackets. Thus,

$$\mathbf{S} = \frac{1}{2\pi\epsilon} \begin{bmatrix} \ln \frac{D^2}{3r^2} & \ln \frac{D}{3r} & \ln \frac{D}{3r} \\ \ln \frac{D}{3r} & \ln \frac{D^2}{3r^2} & \ln \frac{D}{3r} \\ \ln \frac{D}{3r} & \ln \frac{D}{3r} & \ln \frac{D^2}{3r^2} \end{bmatrix} \equiv \begin{bmatrix} \mathcal{S}_\sigma & \mathcal{S}_m & \mathcal{S}_m \\ \mathcal{S}_m & \mathcal{S}_\sigma & \mathcal{S}_m \\ \mathcal{S}_m & \mathcal{S}_m & \mathcal{S}_\sigma \end{bmatrix} \tag{13-75}$$

where

$$\mathcal{S}_\sigma \equiv \frac{1}{2\pi\epsilon} \ln \frac{D^2}{3r^2} \tag{13-76}$$

$$\mathcal{S}_m \equiv \frac{1}{2\pi\epsilon} \ln \frac{D}{3r} \tag{13-77}$$

The matrix of Eq. (13-75) may now be expressed in terms of symmetrical components by applying the usual transformation procedure. Hence

$$\mathbf{S}_s = \mathbf{T}^{-1}\mathbf{S}\mathbf{T} \tag{13-78}$$

where $\mathbf{S}_s$ denotes the symmetrical-components version of the matrix $\mathbf{S}$ and $\mathbf{T}$ is defined by Eq. (12-10). Upon performing the operations indicated in this last expression, the result is once again a diagonal matrix, which here is expressed as

$$\mathbf{S}_s = \mathbf{T}^{-1}\mathbf{S}\mathbf{T} = \begin{bmatrix} \mathcal{S}_1 & 0 & 0 \\ 0 & \mathcal{S}_2 & 0 \\ 0 & 0 & \mathcal{S}_0 \end{bmatrix} \tag{13-79}$$

where

$$\mathcal{S}_1 = \mathcal{S}_\sigma - \mathcal{S}_m = \frac{1}{2\pi\epsilon} \ln \frac{D}{r} = \mathcal{S}_2 \tag{13-80}$$

and

$$\mathcal{S}_0 = \mathcal{S}_\sigma - 2\mathcal{S}_m = \frac{1}{2\pi\epsilon} \ln \frac{D^4}{27r^4} \tag{13-81}$$

Our chief concern, however, is still with capacitance and its associated capacitive reactance or admittance. Fundamentally, we know capacitance involves charge per unit voltage. With the decoupling that is characteristic of the formulation associated with symmetrical components, we are now in a position to determine the capacitance parameter in decoupled form. By introducing $\mathbf{v}_p = \mathbf{T}\mathbf{v}_s$ and $\mathbf{q} = \mathbf{T}\mathbf{q}_s$ into Eq. (13-74), we get an expression in terms of the symmetrical components of voltage ($\mathbf{v}_s$) and charge ($\mathbf{q}_s$). Thus

$$\mathbf{T}\mathbf{v}_s = \mathbf{s}\mathbf{T}\mathbf{q}_s \qquad (13\text{-}82)$$

or

$$\mathbf{v}_s = \mathbf{T}^{-1}\mathbf{s}\mathbf{T}\mathbf{q}_s = \mathbf{s}_s\mathbf{q}_s \qquad (13\text{-}83)$$

This last expression is analogous to Eq. (13-58) for inductance except that here we are still dealing with instantaneous values of voltage and charge. It is helpful next to isolate charge by multiplying through by $\mathbf{s}_s^{-1}$. This yields

$$\mathbf{q}_s = \mathbf{s}_s^{-1}\mathbf{v}_s \qquad (13\text{-}84)$$

But from the result displayed in Eq. (13-79) the inverse matrix is simply given as

$$\mathbf{s}_s^{-1} = \begin{bmatrix} \dfrac{1}{\mathcal{S}_1} & 0 & 0 \\[2mm] 0 & \dfrac{1}{\mathcal{S}_2} & 0 \\[2mm] 0 & 0 & \dfrac{1}{\mathcal{S}_0} \end{bmatrix} \qquad (13\text{-}85)$$

Inserting the last equation into Eq. (13-84) permits writing

$$\begin{bmatrix} q_1 \\[2mm] q_2 \\[2mm] q_0 \end{bmatrix} = \begin{bmatrix} \dfrac{1}{\mathcal{S}_1} & 0 & 0 \\[2mm] 0 & \dfrac{1}{\mathcal{S}_2} & 0 \\[2mm] 0 & 0 & \dfrac{1}{\mathcal{S}_0} \end{bmatrix} \begin{bmatrix} v_1 \\[2mm] v_2 \\[2mm] v_0 \end{bmatrix} \qquad (13\text{-}86)$$

where the subscripts denote the usual positive-, negative-, and zero-sequence components. Because of the decoupling that exists in this formulation, the line capacitance for each sequence is readily identified. The equations represented by the last expression are

$$q_1 = \left(\frac{1}{\mathcal{S}_1}\right)v_1 \qquad q_2 = \left(\frac{1}{\mathcal{S}_2}\right)v_2 \qquad q_0 = \left(\frac{1}{\mathcal{S}_0}\right)v_0 \qquad (13\text{-}87)$$

Therefore, the positive-sequence capacitance is

$$C_1 = \frac{q_1}{v_1} = \frac{1}{\mathcal{S}_1} = \frac{2\pi\epsilon}{\ln(D/r)} \quad \text{F/m} \qquad (13\text{-}88)$$

For the negative sequence the result is

$$C_2 = \frac{q_2}{v_2} = \frac{1}{\mathcal{S}_2} = \frac{2\pi\epsilon}{\ln(D/r)} \quad \text{F/m} \qquad (13\text{-}89)$$

which in this case is the same as C_1. The zero-sequence capacitance is

$$C_0 = \frac{q_0}{v_0} = \frac{1}{S_0} = \frac{2\pi\epsilon}{\ln\left(D^4/27r^4\right)} \qquad \text{F/m} \qquad (13\text{-}90)$$

For sinusoidal steady-state analysis the admittances associated with these capacitive parameters expressed in S/m become

$$\boxed{\bar{Y}_1 = \bar{Y}_2 = j\frac{2\pi\epsilon\omega}{\ln\left(D/r\right)} = \begin{array}{l}\text{positive- and negative-sequence}\\ \text{admittance per meter length}\end{array}} \qquad (13\text{-}91)$$

and

$$\boxed{\bar{Y}_0 = j\frac{2\pi\epsilon\omega}{\ln\left(D^4/27r^4\right)} = \begin{array}{l}\text{zero-sequence admittance per}\\ \text{meter length}\end{array}} \qquad (13\text{-}92)$$

These formulas are useful in determining the π-equivalent circuit to be used under fault conditions to analyze the behavior of the electric power system when the line is of medium or long length.

Example 13-6

Find the parameters of the π-equivalent circuits for the transmission line of Example 13-5 for a length of 100 mi.

Solution The π-equivalent circuit of a medium length line is determined. The positive- and negative-sequence impedances associated with the line inductance follows from the result obtained in Example 13-5, where a value of 0.843 Ω/mi was determined. Hence the total value of this impedance for 100 mi is

$$\bar{Z}_1 = \bar{Z}_2 = j(0.843)(100) = j84.3 \ \Omega$$

The result for the zero-sequence impedance is

$$\bar{Z}_0 = j(2.972)100 = j297.2 \ \Omega$$

The admittance quantity to be used with the positive (and negative)-sequence equivalent circuit follows from Eq. (13-91). In order to obtain the total value, Eq. (13-91) is multiplied by 1000 to get km, then by 1.609 to get miles, and finally by 100 for the line length. Thus

$$\bar{Y}_1 = \bar{Y}_2 = j\frac{2\pi(8.854 \times 10^{-12})(377)}{\ln\left(6/0.0074\right)}(1.609 \times 10^5) = j5.038 \times 10^{-4} \text{ S}$$

Keep in mind that one-half this value is placed at the sending-end terminals and the other half is placed at the receiving end as illustrated in Fig. 13-7(a). The total value of the zero-sequence admittance is found from Eq. (13-92) after adjusting for the line length. Thus,

$$\bar{Y}_0 = j\frac{2\pi(8.854 \times 10^{-12})377}{\ln\left[6^4/27(0.0074)^4\right]}(1.609 \times 10^5) = j(1.4362) \times 10^{-4} \text{ S}$$

The zero-sequence π-equivalent circuit now involves $\bar{Y}_0/2$ at both ends of the transmission line along with a series impedance of $\bar{Z}_0 = j297.2 \ \Omega$. The equivalent circuit for this case is depicted in Fig. 13-7(b).

$$j\,84.3\,\Omega$$

$$\frac{\bar{Y}_1}{2}=\frac{\bar{Y}_2}{2}$$
$$=j\,2.519\times10^{-4}$$

$$\frac{\bar{Y}_1}{2}=\frac{\bar{Y}_2}{2}$$
$$=j\,2.519\times10^{-4}\ S$$

(a)

$$j\,297.2\,\Omega$$

$$\frac{\bar{Y}_0}{2}$$
$$=j\,0.7181\times10^{-4}$$

$$\frac{\bar{Y}_0}{2}$$
$$=j\,0.7181\times10^{-4}\ S$$

(b)

Figure 13-7 π-equivalent circuit for the medium-length transmission line of Example 13-6: (a) the positive- and negative-sequence circuit; (b) the zero-sequence circuit.

13-5 SEQUENCE IMPEDANCES OF TRANSFORMERS

The general procedure for identifying the positive-, negative-, and zero-sequence impedances of transformers, as they are found in electric power systems, can be determined by analyzing the three-phase Y-Δ transformer arrangement depicted in Fig. 13-8. Three single-phase transformers are used in this connection and the primary neutral n is assumed connected to reference ground through its impedance $\bar{Z}_n$. The quantity $\bar{Z}_e$ denotes the equivalent impedance of each transformer and it represents the winding resistance as well as the leakage reactance of both primary and secondary windings. Moreover, the magnetizing current for each transformer is assumed to be negligibly small. This means that the magnetizing impedance is considered to be infinite. For the current directions indicated in Fig. 13-8 the expression for the differ-

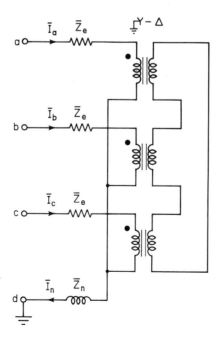

Figure 13-8 Circuit configuration used to identify positive-, negative-, and zero-sequence impedances of three-phase transformer connections.

ence between each line-to-neutral voltage and its associated induced emf is

$$\Delta \bar{V}_a = \bar{I}_a \bar{Z}_e + \bar{I}_n \bar{Z}_n$$
$$\Delta \bar{V}_b = \bar{I}_b \bar{Z}_e + \bar{I}_n \bar{Z}_n \qquad (13\text{-}93)$$
$$\Delta \bar{V}_c = \bar{I}_c \bar{Z}_e + \bar{I}_n \bar{Z}_n$$

Upon replacing $\bar{I}_n$, these expressions can be rewritten as

$$\Delta \bar{V}_a = (\bar{Z}_e + \bar{Z}_n)\bar{I}_n + \bar{Z}_n \bar{I}_b + \bar{Z}_n \bar{I}_c$$
$$\Delta \bar{V}_b = \bar{Z}_n \bar{I}_a + (\bar{Z}_e + \bar{Z}_n)\bar{I}_b + \bar{Z}_n \bar{I}_c \qquad (13\text{-}94)$$
$$\Delta \bar{V}_c = \bar{Z}_n \bar{I}_a + \bar{Z}_n \bar{I}_b + (\bar{Z}_e + \bar{Z}_n)\bar{I}_c$$

Using matrix notation, we get

$$\begin{bmatrix} \Delta \bar{V}_a \\ \Delta \bar{V}_b \\ \Delta \bar{V}_c \end{bmatrix} = \begin{bmatrix} \bar{Z}_e + \bar{Z}_n & \bar{Z}_n & \bar{Z}_n \\ \bar{Z}_n & \bar{Z}_e + \bar{Z}_n & \bar{Z}_n \\ \bar{Z}_n & \bar{Z}_n & \bar{Z}_e + \bar{Z}_n \end{bmatrix} \begin{bmatrix} \bar{I}_a \\ \bar{I}_b \\ \bar{I}_c \end{bmatrix} \qquad (13\text{-}95)$$

By operating upon the impedance matrix appearing in this last equation, in accordance with the general instructions displayed in Eq. (13-59), it is possible to identify the desired results. Thus, we begin with

$$\bar{Z}_s = \mathbf{T}^{-1} \bar{\mathbf{Z}} \bar{\mathbf{T}} \qquad (13\text{-}96)$$

The left side of Eq. (13-96) is a diagonal matrix with the positive-, negative-, and zero-sequence elements appearing on the principal diagonal. The quantity $\mathbf{T}$ is the symmetrical component transformation that serves to convert the impedance matrix $\bar{\mathbf{Z}}$ into its symmetrical component parts. Thus,

$$\bar{\mathbf{Z}}_s = \begin{bmatrix} 1 & a & a^2 \\ 1 & a^2 & a \\ 1 & 1 & 1 \end{bmatrix} \begin{bmatrix} \bar{Z}_e + \bar{Z}_n & \bar{Z}_n & \bar{Z}_n \\ \bar{Z}_n & \bar{Z}_e + \bar{Z}_n & \bar{Z}_n \\ \bar{Z}_n & \bar{Z}_n & \bar{Z}_e + \bar{Z}_n \end{bmatrix} \begin{bmatrix} 1 & 1 & 1 \\ a^2 & a & 1 \\ a & a^2 & 1 \end{bmatrix}$$

$$= \begin{bmatrix} \bar{Z}_e & 0 & 0 \\ 0 & \bar{Z}_e & 0 \\ 0 & 0 & \bar{Z}_e + 3\bar{Z}_n \end{bmatrix} \qquad (13\text{-}97)$$

or

$$\begin{bmatrix} \bar{Z}_1 & 0 & 0 \\ 0 & \bar{Z}_2 & 0 \\ 0 & 0 & \bar{Z}_0 \end{bmatrix} = \begin{bmatrix} \bar{Z}_e & 0 & 0 \\ 0 & \bar{Z}_e & 0 \\ 0 & 0 & \bar{Z}_e + 3\bar{Z}_n \end{bmatrix} \qquad (13\text{-}98)$$

Therefore,

$$\boxed{\begin{aligned} \bar{Z}_1 &= \bar{Z}_e = \text{positive-sequence impedance} \\ \bar{Z}_2 &= \bar{Z}_e = \text{negative-sequence impedance} \\ \bar{Z}_0 &= \bar{Z}_e + 3\bar{Z}_n = \text{zero-sequence impedance} \end{aligned}} \qquad (13\text{-}99)$$

Equation (13-99) displays the identity of the sequence impedances of transformers. It states that the positive- and negative-sequence impedances are the same.

Where static devices are involved, this result is entirely expected. The transformer as an impedance looks no different to the negative-sequence voltages than it does to the positive-sequence voltages. Of course, specifically, this sequence impedance is equal to the equivalent leakage impedance of the transformer. When viewed from the high-voltage side, the impedance is best represented by $\bar{Z}_{e\mathrm{H}}$ where the second subscript denotes high side. Similarly, when the transformer is viewed from the low side, this impedance is called $\bar{Z}_{e\mathrm{L}}$ and is smaller than $\bar{Z}_{e\mathrm{H}}$ by the ratio of transformation squared. If these impedances are expressed on a per-unit basis, then the values are identical.

The zero-sequence impedance includes the equivalent leakage impedance but its value can be augmented by three times whatever impedance is placed between the transformer neutral point and reference ground. When $\bar{Z}_n$ is made nearly zero by using a solid connection to ground, the zero-sequence impedance for the case depicted in Fig. 13-8 becomes equal to the positive- and negative-sequence impedance. However, it is important to understand at this point that the situation just described for the zero-sequence impedance is valid only for the special circumstances of this configuration, namely, a primary Y connection with a neutral and a Δ secondary connection. If a fault were to occur on the primary (or Y) side, then the effect of the zero sequence, in the analysis of the unbalanced conditions of the line, would be correctly represented by the use of the zero-sequence impedance as described by Eq. (13-99). On the other hand, if a fault were to occur on the Δ side of the transformer connections, then the zero-sequence impedance no longer plays a role. In other words the zero-sequence impedance viewed from the Δ side is in essence infinity. Keep in mind that the zero-sequence current is a single-phase component and as such *its existence is dependent upon a closed path that must be completed through reference ground*. Because the Δ side cannot have an intentional physical connection to the reference ground, no zero-sequence current can flow. Hence the zero-sequence impedance is infinite from this side. Consequently, the manner of drawing an equivalent circuit for the zero-sequence impedance is now very much dependent upon whether or not a particular three-phase transformer connection permits a zero-sequence current to flow to reference ground as viewed from either side of the transformer connections. We explore this matter further below but first let it be said that this situation does not apply to the positive- and negative-sequence sets. Because the positive and negative sets are balanced sequences, the effects they produce on either side of the transformer connections are symmetrical.

An equivalent circuit for the zero-sequence impedance that is consistent with the conclusions described in the foregoing is shown in Fig. 13-9. Here P denotes the primary side (or Y in this case) and S denotes the secondary (or Δ) side. Observe that this circuit does permit a zero-sequence current to flow to ground reference upon application of a zero-sequence, single-phase voltage at terminal P. Applying this zero-sequence voltage to the secondary terminal, S, however, produces no result. If the primary neutral is assumed to be removed, then the sole means of zero-sequence current to reach the ground reference is also eliminated. Hence no zero-sequence current can flow, which means that the zero-sequence impedance is infinite. This situation is depicted in Fig. 13-8(b) by showing an open circuit between P and ground reference.

A further exploration of zero-sequence current flow in three-phase transformer

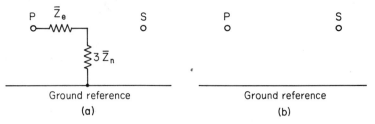

Figure 13-9 Equivalent circuit for zero-sequence impedance of the Y-Δ connection of Fig. 13-8: (a) Y primary is equipped with a primary neutral; (b) no primary neutral.

connections can be made by examining the situation illustrated in Fig. 13-10. Let it be required that an equivalent circuit be drawn for this section of an electric power system for the zero-sequence impedance. For simplicity, assume that all transformers are identical. The analysis begins with a question. Can this arrangement permit a zero-sequence current to flow? In other words, if a zero-sequence voltage is assumed applied simultaneously to each line of the Y primaries P, does a closed path to ground reference exist to allow zero-sequence current to flow? Keep in mind that in order that a current flow in the primaries it is necessary for a low impedance closed path to exist in the corresponding secondary circuits. Examination of the circuit in Fig. 13-10 shows the presence of a neutral connection on the S secondaries. However,

Figure 13-10 Zero-sequence equivalent circuit of a section of an electric power system: (a) Y-Y equipped primary and secondary neutrals and Y-Δ without a primary neutral; (b) zero-sequence equivalent circuit; (c) zero-sequence equivalent circuit when transformers at P' are equipped with a neutral connection to ground through $\bar{Z}_n$.

there is no neutral on the P′ primaries of the second bank of transformers. This means that the impedance from line to neutral at P′ is the magnetizing impedance which we have assumed to be infinite. Hence the impedance seen by the secondaries at S is also infinite, and this in turn is reflected into the primaries at P as an infinite impedance. Accordingly, no zero-sequence current can be caused by a zero-sequence voltage assumed applied at the primary terminals at P. This situation is represented by the equivalent circuit appearing in Fig. 13-10(b). The ground connection at the secondaries at S in Fig. 13-10(a) permits joining P to S through the transformer equivalent imped-ance as shown in Fig. 13-10(b). Also, S is connected to P′ through the zero-sequence impedance of the transmission line. Point S′ remains isolated because a zero-sequence voltage applied to the Δ side fails to produce any zero-sequence current. Because this equivalent circuit is unconnected at any point to ground reference, it is not possible for zero-sequence current to flow in the arrangement of Fig. 13-10(a) as it stands.

How is the equivalent circuit of Fig. 13-10(b) changed when the transformers at P′ are supplied with a neutral connection $\bar{Z}_n$ to ground reference? The presence of this neutral connection to ground now permits a closed path to exist between each secondary winding at S and the corresponding primary windings at P′. With the primaries at P energized by a zero-sequence voltage, a current flows in the Y primaries at P′. Corresponding in-phase currents then flow in the secondaries of the Δ-con-nected transformers. The effect in the primaries at P′ is to give the appearance of a short circuit except for the leakage impedance and the neutral impedance Z_n. The net effect is to modify the equivalent circuit of Fig. 13-10(b) to that shown in Fig. 13-10(c) by placing P′ to ground reference through the combined impedance $(\bar{Z}_e + 3\bar{Z}_n)$.

13-6 SEQUENCE IMPEDANCES OF SYCHRONOUS GENERATORS AND MOTORS

The synchronous generator and motor under balanced conditions of operations is primarily the subject matter of Chapter 5. In this section the focus is on a description of these machines under the unbalanced conditions that prevail when the synchronous machine is in a fault mode. For the unbalanced case the vector matrix equation that describes the line-to-neutral voltages through Kirchhoff's voltage law is

$$\bar{\mathbf{E}}_p = \bar{\mathbf{V}}_p + \bar{\mathbf{Z}}_u \bar{\mathbf{I}}_p \tag{13-100}$$

Here $\bar{\mathbf{E}}_p$ denotes the induced emf vector expressed per phase of the generator, $\mathbf{V}_p$ is the phase terminal voltage vector, and $\mathbf{I}_p$ is the phase current vector. The impedance $\bar{\mathbf{Z}}_u$ is a 3 × 3 impedance matrix which in general contains nine nonzero elements repre-sentative of the unbalanced state of affairs in the generator. Of course, when the generator operates under balanced conditions, this impedance matrix is a simple diagonal matrix with just the synchronous reactance appearing in each position on the principal diagonal. We will not attempt to particularize the element values of $\bar{\mathbf{Z}}_u$ for the unbalanced case as it appears in Eq. (13-100) because it is preferable to deal with this matter after Eq. (13-100) is converted to symmetrical components through use of the symmetrical component transformation $\mathbf{T}$. Upon rearranging, the last equation becomes

$$\bar{\mathbf{V}}_p = \bar{\mathbf{E}}_p - \bar{\mathbf{Z}}_u\bar{\mathbf{I}}_p \qquad (13\text{-}101)$$

It is useful to keep in mind that the generated emf vector is always expressed as

$$\bar{\mathbf{E}}_p = \begin{bmatrix} \bar{E}_a \\ \bar{E}_b \\ \bar{E}_c \end{bmatrix} = \begin{bmatrix} \bar{E}_a \\ a^2\bar{E}_a \\ a\bar{E}_a \end{bmatrix} \qquad (13\text{-}102)$$

This is a balanced voltage vector which is a *designed* characteristic of the generator. It prevails at all times whether or not a fault exists.

To express Eq. (13-101) in terms of symmetrical components, we next introduce

$$\bar{\mathbf{V}}_p = \mathbf{T}\bar{\mathbf{V}}_s \qquad \text{and} \qquad \bar{\mathbf{I}}_p = \mathbf{T}\bar{\mathbf{I}}_s \qquad (13\text{-}103)$$

to get

$$\mathbf{T}\bar{\mathbf{V}}_s = \bar{\mathbf{E}}_p - \bar{\mathbf{Z}}_u\mathbf{T}\bar{\mathbf{I}}_s \qquad (13\text{-}104)$$

or

$$\bar{\mathbf{V}}_s = \mathbf{T}^{-1}\bar{\mathbf{E}}_p - \mathbf{T}^{-1}\bar{\mathbf{Z}}_u\mathbf{T}\bar{\mathbf{I}}_s \qquad (13\text{-}105)$$

The first term on the right side of this last equation evaluates to be

$$\bar{\mathbf{T}}^{1}\bar{\mathbf{E}}_p = \bar{\mathbf{E}}_s = \frac{1}{3}\begin{bmatrix} 1 & a & a^2 \\ 1 & a^2 & a \\ 1 & 1 & 1 \end{bmatrix}\begin{bmatrix} \bar{E}_a \\ a^2\bar{E}_a \\ a\bar{E}_a \end{bmatrix} = \frac{1}{3}\begin{bmatrix} (1 + a^3 + a^3)\bar{E}_a \\ (1 + a + a^2)\bar{E}_a \\ (1 + a^2 + a)\bar{E}_a \end{bmatrix}$$

$$= \frac{1}{3}\begin{bmatrix} 3\bar{E}_a \\ 0 \\ 0 \end{bmatrix} = \begin{bmatrix} \bar{E}_a \\ 0 \\ 0 \end{bmatrix} \qquad (13\text{-}106)$$

This is an entirely expected result. It states that the generator has only a positive-sequence voltage set with a per-phase value of $\bar{E}_a$. Both the negative-sequence and zero-sequence set are zero. The generator by design only produces induced voltages that are positive sequence. It cannot generate either of the other two sets.

The second term on the right side of Eq. (13-105) contains the now-familiar form of the symmetrical component impedance matrix, which has previously been demonstrated to be diagonal. The general form therefore is

$$\bar{\mathbf{Z}}_s = \mathbf{T}^{-1}\bar{\mathbf{Z}}_u\mathbf{T} = \begin{bmatrix} \bar{Z}_1 & 0 & 0 \\ 0 & \bar{Z}_2 & 0 \\ 0 & 0 & \bar{Z}_0 \end{bmatrix} \qquad (13\text{-}107)$$

where $\bar{Z}_1$, $\bar{Z}_2$, and $\bar{Z}_0$ represent the positive-, negative-, and zero-sequence components of $\bar{Z}_u$, respectively. We now invoke our understanding of the synchronous machine, as obtained from Chapter 5, to draw general conclusions concerning the composition of these sequence impedances.

The Positive-Sequence Generator Impedance

The positive-sequence impedance of a device is in essence a parameter (or operator) that allows a positive-sequence current to be derived from the action of an imposed positive-sequence set of voltages. However, this situation is already analyzed in detail in Sec. 5-9. When a *sustained* positive-sequence voltage appears at the generator terminals, the impedance that limits the current is the synchronous reactance jx_s. If interest is centered on a description of the positive-sequence current for the first few cycles following a fault, then the impedance associated with the positive sequence is the subtransient synchronous reactance. As described in Sec. 5-9, this reactance takes into account the effect of both the field winding of the generator as well as the damper winding both of which act to increase the current produced by a positive-sequence voltage set. An equivalent circuit of this version of the positive-sequence impedance is displayed in Fig. 5-30. In those instances where the purpose of the fault analysis is to determine the size of fault currents that exist between the expirations of the first few cycles and the condition of the sustained short circuit, the more appropriate value to assign to the positive-sequence impedance is the transient synchronous reactance. This quantity differs from the subtransient value because the effect of the damper winding is no longer present. Table 13-1 summarizes the foregoing remarks and includes other useful information.

The Negative-Sequence Generator Impedance

The negative-sequence impedance is an operator which, when applied to the negative-sequence voltage set, provides information about the corresponding negative-sequence current produced. When an unsymmetrical fault occurs somewhere along an electric power system, the result is a set of unbalanced voltages. Through symmetrical components this unbalanced set at the point of the fault may be considered replaced by balanced sets of positive and negative sequences and a zero-sequence set. It is appropriate and useful to inquire at this point about the reaction of the generator (which is operating by design in the positive-sequence mode) to the imposition of a negative-sequence voltage set appearing at its armature terminals. With a negative-sequence voltage set so applied, there is produced in the generator a negative-sequence flux field which with respect to the generator stator coils revolves at line frequency. But with respect to the rotor of the positive-sequence generator this field revolves at double the line frequency since the field and the rotor travel in opposite directions. This means that for 60 Hz the negative-sequence flux field covers a pole pitch in 1/120 s. Accordingly, the flux field will find itself alternately (and in quick succession) in the direct axis and the quadrature axis. When the flux is in the direct axis, the sequence impedance is indistinguishable from that of the positive sequence because the effects are the same. In other words, reaction is elicited from both the field winding and the damper winding. However, when the negative-sequence flux field acts in the quadrature axis, the generator field winding has no influence because the coupling of the field winding with the flux is zero. Consequently, only the damper winding of the two rotor windings produces an effect in the quadrature axis. Hence, there is no distinction between the transient and subtransient reactances in the quadrature axis as there is in the direct axis. The approximate circuit configuration for the quadrature axis subtransient reactance is illustrated in Fig. 13-11. When the effect of the damper winding ceases, i.e., when x_D in Fig. 13-11 disappears, then the transient reactance in the quadrature axis becomes the synchronous reactance in that axis. In short, $x_q' = x_q$.

Because of the rapid alternation of the negative-sequence flux field between

TABLE 13-1 POSITIVE-SEQUENCE IMPEDANCES OF SYNCHRONOUS GENERATORS AND MOTORS[a]

Impedance	Approximate equivalent circuit	Portion of transient period during which circuit is valid	Typical range of values in p-u	
			Salient-pole generator	Turbogenerator
$Z_1 = jx_d'' =$ subtransient reactance		First few cycles	0.20–0.30	0.09–0.14
$Z_1 = jx_d' =$ transient reactance		Beyond first few cycles up to several hundred cycles	0.35–0.40	0.15–0.25
$Z_1 = jx_d =$ synchronous reactance		Beyond several hundred cycles; during sustained short circuit	1.0–1.25 (unsaturated)	1.0–1.2

[a]x_l denotes leakage reactance per phase of the armature winding. Range is 0.05 to 0.10 p-u.
x_f denotes leakage reactance of the field winding.
x_D denotes leakage reactance of the damper winding.
x_{ad} denotes armature reaction reactance of the armature winding.

Figure 13-11 Equivalent circuit for the subtransient reactance in the quadrature axis.

the direct axis and the quadrature axis, it is reasonable to use a value of the negative-sequence impedance which is the average of the values found in the direct-axis configuration and the quadrature axis. It is important to understand, however, that that value which is appropriate is the equivalent circuit for the positive-sequence impedance for the damper leakage reactance $\bar{X}$ is not proper for use in the circuit of Fig. 13-11. The reason lies primarily in the greatly increased skin effect associated with the double frequency currents that are induced in the damper winding by the negative-sequence flux field. The end result of this averaging and adjustment for skin effect is to yield values for the negative-sequence impedance of the generator which are essentially comparable to the subtransient reactance of the positive sequence. That is,

$$\boxed{Z_2 \approx jx_d''}$$

(13-108)

Manufacturers of synchronous machines usually provide specific values for this quantity. In the absence of such information, it would not be overly unsettling to employ the relationship of the last expression.

The Zero-Sequence Generator Impedance

The zero-sequence impedance represents the response of the generator to a single-phase zero-sequence voltage applied to the line terminals in the manner illustrated in Fig. 13-12. On the assumption that the generator neutral is connected to ground reference through $\bar{Z}_n$, there exist closed paths through each phase coil to ground. Hence zero-sequence currents can flow. However, the question that arises is: What impedance does the zero-sequence voltage source "see"? In searching for the answer, it is useful first to keep in mind that the ensuing zero-sequence currents $\bar{I}_0$ flowing in each phase are of the same magnitude and in time phase with each other. Moreover, these currents flow in phase coils which are space displaced from each other by 120 electrical degrees. The result is that the net air-gap flux that is produced by these currents is zero. Therefore no magnetizing impedance can exist from any line to neutral. The sole factor that limits this zero-sequence current aside from the external impedance $\bar{Z}_n$ is the leakage impedance of each phase coil. Of course, the leakage flux can not cancel because to do so the phase coils would have to occupy the very same space. Accordingly, disregarding secondary effects, the zero-sequence impedance of the generator can be described approximately by

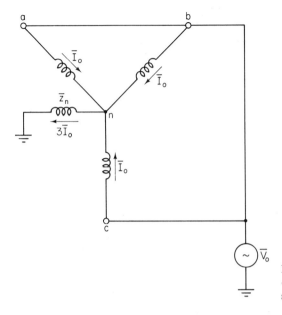

Figure 13-12 Illustrating the action of the zero sequence voltage on the synchronous generator.

$$\bar{Z}_0 = \bar{Z}_e + 3\bar{Z}_n \qquad (13\text{-}109)$$

where $\bar{Z}_e$ is leakage impedance per phase. For all practical purposes then $\bar{Z}_e \approx jx_l$.

Now that the elements of $\bar{Z}_e$ in Eq. (13-107) are determined, we can return to Eq. (13-105) in order to complete the formulation of the voltage equations of the generator for unbalanced operations in terms of its symmetrical components. Thus,

$$\bar{\mathbf{V}}_s = \bar{\mathbf{E}}_s - \bar{\mathbf{Z}}_s \bar{\mathbf{I}}_s \qquad (13\text{-}110)$$

or

$$\begin{bmatrix} \bar{V}_1 \\ \bar{V}_2 \\ \bar{V}_0 \end{bmatrix} = \begin{bmatrix} \bar{E}_a \\ 0 \\ 0 \end{bmatrix} - \begin{bmatrix} \bar{Z}_1 & 0 & 0 \\ 0 & \bar{Z}_2 & 0 \\ 0 & 0 & \bar{Z}_0 \end{bmatrix} \begin{bmatrix} \bar{I}_1 \\ \bar{I}_2 \\ \bar{I}_0 \end{bmatrix} \qquad (13\text{-}111)$$

where $\bar{Z}_1$ is selected from Table 13-1 in accordance with the portion of the transient time period which is of interest, $\bar{Z}_2$ is given by Eq. (13-108), and $\bar{Z}_0$ by Eq. (13-109).

Equations (13-110) may be expressed individually as

$$\bar{V}_1 = \bar{E}_a - \bar{Z}_1 \bar{I}_1 \qquad (13\text{-}112)$$

$$\bar{V}_2 = \qquad - \bar{Z}_2 \bar{I}_2 \qquad (13\text{-}113)$$

$$\bar{V}_0 = \qquad - \bar{Z}_0 \bar{I}_0 \qquad (13\text{-}114)$$

The circuit representation of Eq. (13-112) appears in Fig. 13-13(a) and it is the positive-sequence network of the generator. The circuit interpretation of Eq. (13-113) is shown in Fig. 13-13(b) and it represents the negative-sequence network of the generator. Finally, Fig. 13-13(c) depicts the zero-sequence network that derives directly from Eq. (13-114).

Figure 13-13 Sequence networks for the synchronous generator: (a) positive-sequence network; (b) negative-sequence network; (c) zero-sequence network.

PROBLEMS

13-1. In the electric power system of Fig. 13-1, determine the ohmic value of the transient reactance of generator G2. Also, find the equivalent leakage reactance of transformer bank T2 expressed in ohms on the high-voltage side.

13-2. The transmission line TL1 in the power system of Fig. 13-1 has a line reactance per phase of 24 Ω. Find the value of this impedance as it appears in the circuit of generator G2. Express the result in ohms and in per unit.

13-3. Two three-phase generators and their associated transformer banks are joined by a transmission line operating at 69 kV. The voltage drop across the line inductance is 3.98 kV when rated 500 A flows.
(a) What is the reactance of the line expressed in percent?
(b) What is the corresponding MVA rating associated with this percent value?

13-4. Develop the single-line equivalent circuit of the electric power system shown in Fig. P13-4 using per unit values throughout. Use a reference base of 200 MVA at 138 kV.

G1: 100 MVA   T1: 150 MVA   T2: 200 MVA   G2: 200 MVA
     20 kV   20/138 kV   138/13.8 kV   13.2 kV
   $x_s'' = 15\%$ $x_e = 10\%$ $x_e = 8\%$ $x_s'' = 12\%$

Figure P13-4

13-5. In the electric power system of Fig. P13-5 system components G1, G2, T1, and T2 have the same ratings as for the system of Fig. P13-4. Draw the single-line equivalent circuit of this system using per-unit values. Use a reference base of 200 MVA at 138 kV.

Figure P13-5

13-6. In the system of Fig. P13-4 assume that generator G1 is operating as a synchronous motor drawing rated MVA from the line at 20 kV and a lagging power factor of 0.9.
(a) Determine the excitation voltage of generator G2 per phase.
(b) Find the terminal line voltage of G2.
(c) What is the value of the line voltage at bus 2?
Make all computations using a reference base of 200 MVA at 138 kV.

13-7. A fault occurs on a 60-Hz, three-phase transmission line and causes the following unbalanced line currents to exist:

$$\bar{I}_a = -j583.2 \text{ A}$$
$$\bar{I}_b = j162 \text{ A}$$
$$\bar{I}_c = j162 \text{ A}$$

Find the positive-, negative-, and zero-sequence current components that can be used to replace this unbalanced current set.

13-8. The occurrence of a fault on a three-phase transmission line gives rise to the following line-to-neutral voltages at a particular location along the power system:

$$\bar{V}_a = 100\underline{/-41.5°} \text{ kV}$$
$$\bar{V}_b = 100\underline{/-138.5°} \text{ kV}$$
$$\bar{V}_c = 133\underline{/90°} \text{ kV}$$

Determine the positive-, negative-, and zero-sequence components of this unbalanced set of voltages.

13-9. At the point at which a fault occurs on a three-phase transmission line the respective sequence currents are known to be as follows:

$$\bar{I}_1 = -j1538.4 \text{ A}$$
$$\bar{I}_2 = j1538.4 \text{ A}$$
$$\bar{I}_0 = 0$$

where the subscripts denote positive, negative, and zero sequence.

(a) Find the corresponding values of the unbalanced line currents that flow.

(b) On the basis of the results of part (a), what conclusion can you draw regarding the type of fault that has occurred?

13-10. The line-to-neutral voltages at the point at which a fault occurs on a three-phase transmission line are described by

$$\bar{V}_a = 200\underline{/0°} \text{ kV}$$
$$\bar{V}_b = 100\underline{/-180°} \text{ kV}$$
$$\bar{V}_c = 100\underline{/-180°} \text{ kV}$$

(a) Determine the corresponding sequence components of these unbalanced voltages.

(b) In view of the value found for the zero component in part (a), what conclusion can you come to concerning the role of the ground in the type of fault that leads to this unbalanced voltage set?

13-11. The condition of unbalance produced by a fault on a three-phase transmission line leads to the following sequence components:

$$\bar{I}_1 = -j813.8 \text{ A}$$
$$\bar{I}_2 = j180.6 \text{ A}$$
$$\bar{I}_0 = j191.4 \text{ A}$$

Determine the unbalanced set of line currents represented by these symmetrical sequence components.

13-12. Following the occurrence of a fault at a specific location along a three-phase electric power system, an unbalanced set of line currents is known to flow which is described by

$$\bar{I}_a = -j617.4 \text{ A}$$
$$\bar{I}_b = 1788\underline{/153.4°} \text{ A}$$
$$\bar{I}_c = 1788\underline{/26.6°} \text{ A}$$

Find the symmetrical sequence components that can be used to replace this set.

13-13. A fault occurs on a three-phase transmission line. It causes unbalanced line currents to exist at a synchronous generator which is part of the electric power system. These currents are described by

$$\bar{I}_a = 1154\underline{/-180°} \text{ A}$$
$$\bar{I}_b = 1154\underline{/-180°} \text{ A}$$
$$\bar{I}_c = 2308\underline{/0°} \text{ A}$$

Compute the symmetrical sequence components that can be used to represent this unbalanced set.

13-14. A three-phase voltage is applied to an unbalanced three-phase impedance load. It produces the following unbalanced line currents:

$$\bar{I}_a = 40.17\underline{/-75°}\text{ A}$$
$$\bar{I}_b = 10.77\underline{/164.9°}\text{ A}$$
$$\bar{I}_c = 36\underline{/90°}\text{ A}$$

Determine the positive-, negative-, and zero-sequence components of this unbalanced set.

13-15. The application of a balanced three-phase voltage set to an unbalanced load impedance produces the following set of unbalanced line-to-neutral voltages:

$$\bar{V}_{an} = 179.6\underline{/-11.6°}\text{ V}$$
$$\bar{V}_{bn} = 48.16\underline{/228.3°}\text{ V}$$
$$\bar{V}_{cn} = 161\underline{/63.4°}\text{ V}$$

Find the positive-, negative-, and zero-sequence components.

13-16. The data cited in Probs. 13-14 and 13-15 are associated with the same unbalanced three-phase load impedance.
(a) Compute the real and reactive power delivered to the load by employing Eq. (13-31).
(b) Repeat part (a) but now work with the sequence components in conjunction with Eq. (13-40).

13-17. A three-phase transmission line is in a faulted state. The unbalanced set of line currents and the line-to-neutral voltages at the terminals of a synchronous generator are described by

$$\bar{I}_a = 3105\underline{/-90°}\text{ A} \qquad \bar{V}_a = 8.71\underline{/-41.5°}\text{ kV}$$
$$\bar{I}_b = 3105\underline{/90°}\text{ A} \qquad \bar{V}_b = 8.71\underline{/-138.5°}\text{ kV}$$
$$\bar{I}_c = 0 \qquad\qquad\quad \bar{V}_c = 0$$

(a) Obtain the complex power by using Eq. (13-31).
(b) Determine the symmetrical sequence components of these unbalanced sets.
(c) Find the complex power employing the results of part (b).

13-18. A 60-Hz, three-phase, transposed transmission line employs the unsymmetrical parallel spacing depicted in Fig. P13-18. Each conductor including the neutral has a radius of 2 cm.
(a) Determine the values of the positive- and negative-sequence inductive reactance of this transmission line expressed in Ω/mi.
(b) Find the value of the zero-sequence inductive reactance in Ω/mi.

Figure P13-18

13-19. Derive the expressions for the positive-, negative-, and zero-sequence shunt admittances for a three-phase, transposed transmission line having the arrangement shown in Fig. P13-18. Obtain the results in terms of the conductor radius r, the separation between conductors D, and the average distance of each transposed line from the neutral.

13-20. Assume that the transmission line of Fig. P13-18 has a length of 120 mi. Find all the parameters of the π-equivalent circuit of this line. Show the equivalent circuit with all parameter values clearly indicated.

13-21. A 60-Hz, three-phase, transposed transmission line uses a parallel spacing arrangement similar to that shown in Fig. P13-18 except that each line including the neutral consists of two conductors placed 20 cm apart. The distance between bundle centers remains 4 m but each conductor now has a radius of 1.5 cm.
   (a) Find the positive- and negative-sequence inductive reactance of this line in Ω/mi.
   (b) Determine the value of the zero-sequence inductive reactance in Ω/mi.
   (c) Find the positive- and negative-sequence capacitive reactance of this line in Ω/mi.
   (d) Repeat part (c) for the zero sequence.
   (e) Draw the π-equivalent circuit for a line length of 120 mi.

13-22. Refer to the arrangement shown in Fig. P13-22. The following data apply:

$$X_{0g} = \text{zero-sequence inductive reactance of } \Delta\text{-connected}$$
$$\text{generator per phase}$$

$$X_{0\text{TL}} = \text{zero-sequence inductive reactance of transmission}$$
$$\text{line per phase}$$

$$X_{0t} = \text{zero-sequence inductive reactance of the}$$
$$\text{transformers per phase}$$

   (a) Draw the zero-sequence equivalent circuit for this system.
   (b) Does the application of a zero-sequence voltage at the generator terminals produce a zero-sequence current in the generator? Explain.
   (c) Does the voltage of part (b) produce a zero-sequence current along the transmission line? Explain.
   (d) Which impedances of those listed are active in the answer to part (c)?

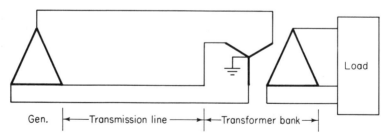

Gen. |←——Transmission line——→|←—Transformer bank—→|

Figure P13-22

13-23. Repeat Prob. 13-22 for the case where the generator phase coils are Y-connected without a primary neutral.

13-24. Repeat Prob. 13-22 for the case where the generator phase coils are Y-connected and equipped with a solid connection to reference ground.

13-25. An electric power system has the configuration illustrated in Fig. P13-25.

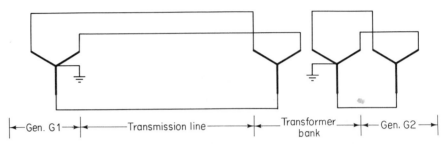

Figure P13-25

(a) Draw the equivalent circuit that applies to the flow of zero-sequence current.

(b) If a zero-sequence voltage appears across the line terminals of generator G1, can a zero-sequence current exist in the transmission lines? Explain.

(c) Does the application of a zero-sequence voltage across the line terminals of generator G2 produce a zero-sequence current in the transmission lines? Explain.

(d) Does the answer to part (b) change when the neutral of the Y connection of the transformer bank is solidly grounded? Why?

(e) Does the answer to part (c) change when the neutral of the Y connection of the transformer bank is grounded? Explain.

(f) Draw the equivalent circuit for the zero-sequence when the neutrals of all Y connections are solidly grounded.

14

Equivalent Circuits
For Fault Connections

There are many reasons why faults occur in electric power systems. Some of these reasons have their origins in natural disturbances such as violent wind and snow storms that can uproot trees and bring them crashing upon bare transmission lines. Earthquakes and earth tremors cause similar problems by bringing down poles that support the distribution lines that supply electric power to rural and urban areas. Other causes are related to equipment failures such as the breakdown of insulation in the large electric power synchronous generators, power transformers, and protective switchgear distributed all throughout the electric power system. Still other troublesome sources relate to small animals and birds as well as to people, who through airplane crashes and other vehicular collisions with transmission line towers and poles can cause severe difficulties. The most common cause of power interruption is the single line-to-ground fault on the transmission line. This is followed in succession by the double line-to-ground fault, the line-to-line fault and finally by the three-phase symmetrical short circuit. Although the latter is the least frequent to occur, it is the first of the fault conditions to be studied in this chapter because it is the easiest.

A fault represents a highly disruptive condition for the power system. Hence, if appropriate action is not taken immediately upon its occurrence, extreme damage can take place in the generators, transformers, and other connected loads of the system. Besides the enormous expense involved in repairing and/or replacing such equipment, it may also be necessary to be without electric service for a period long enough to cause discomfort. Therefore, all electric power systems must be equipped with strategically located circuit breakers and other switchgear in order to disconnect sensitive equipment before irreparable damage takes place. It is the purpose of fault analsyis, such as the type described in this chapter, to provide the information that leads to the correct selection of the switchgear to provide this protection to the electric power system.

The analytical procedures developed in this chapter are based on a few simplifying assumptions which are useful to keep in mind. They are listed in an order that carries no significance regarding importance. When a fault occurs, it is assumed to be a solid fault which means that the impedance at the fault is taken to be zero. Prefault load currents are generally assumed to be negligible compared to the fault currents unless specifically stated otherwise. All transformer magnetizing currents and core losses are neglected. All sequence impedances are treated as purely reactive quantities which means that the associated resistance components are treated as negligible. All line shunt capacitances are neglected unless otherwise directed. Finally, wherever applicable, circuit time constants are assumed to be sufficiently small so that the use of sinusoidal steady-state analysis is appropriate to meet the goals of the analysis.

14-1 SEQUENCE NETWORKS FOR FAULT CALCULATIONS

The presence of a fault at any particular point in an electric power system means generally that the system is put into an unbalanced state of operation. Our study of symmetrical components tells us that this unbalanced condition can be replaced by a symmetrically balanced positive-sequence set, a symmetrically balanced negative-sequence set, and a single-phase zero-sequence set. Thus, when the fault occurs, it is just as if these three sequence quantities are injected into the power system replacing the unbalanced set of voltages and/or currents. To make a correct calculation of the postfault currents and voltages throughout the power system, it is then necessary to determine the response of the power system to each sequence set. But to achieve this objective, a proper circuit representation of the power system as it appears to *each* sequence set must be drawn. Fortunately, with the background of Chapter 13, the drawing of these circuit diagrams becomes a routine matter because the positive-, negative-, and zero-sequence characterizations of each of the major components of the power system are now known.

We choose to illustrate the procedure for drawing these sequence networks by dealing with the relatively simple two-bus electric power system illustrated in Fig. 14-1. All ratings and parameter values listed in this figure are individual ratings and

Figure 14-1 Two-bus electric power system.

are not yet referred to a common reference base. Of course, as outlined in Sec. 13-1, before a single line equivalent circuit can be drawn, it is necessary to choose a reference base and then, through use of Eq. (13-7), express all system parameters to this reference base. For convenience, we choose a reference base for this power system which is the same as that of generator 1, namely:

reference base quantities: 50 MVA; 20 kV

A glance at the parameter data for the two generators in this system shows a listing of the subtransient synchronous reactances. This means that the ensuing sequence networks are to be used to get information about the fault currents that flow during the first few cycles following the occurrence of a fault.

Since generator G1 and transformer bank T1 have the same MVA and voltage ratings as the reference base quantities, no adjustments are needed in the percent parameter values of these components. However, adjustments are needed for generator G2 and transformer bank T2. It is useful to note here that no adjustment is needed for voltage in the application of Eq. (13-7) for G2 and T2 because the turns ratio of the transformers are so specified that 20 kV at G1 corresponds to 69 kV on the transmission line. In turn at G2 this voltage becomes 12 kV. However, an adjustment is needed to reconcile the MVA ratings. Accordingly, the subtransient and leakage reactances of generator G2 become

$$(x_d'')_{G2} = (0.2)\frac{50}{70} = 0.144 \text{ p-u} \qquad (14\text{-}1)$$

$$(x_l)_{G2} = (0.06)\frac{50}{70} = 0.0432 \text{ p-u} \qquad (14\text{-}2)$$

The modified value of the equivalent leakage reactance for the transformers at T2 is

$$(X_e)_{T2} = (0.1)\frac{50}{70} = 0.072 \text{ p-u} \qquad (14\text{-}3)$$

Because the transmission line parameters are expressed in ohms, the corresponding per-unit values can readily be found once the reference base impedance is evaluated. Since the reference base voltage at the transmission line level is 69 kV, it follows that the corresponding value of this base impedance is

$$Z_{BH} = \frac{(kV)_r^2}{(MVA)_r} = \frac{69^2}{50} = 95.22 \ \Omega \qquad (14\text{-}4)$$

where the second subscript H denotes the high-voltage side. The passive nature of the transmission line means that the positive- and negative-sequence impedances are the same. Hence the per-unit value which applies to both is

$$(x_{TL})_1 = (x_{TL})_2 = \frac{10}{95.22} = 0.105 \text{ p-u} \qquad (14\text{-}5)$$

where the subscripts 1 and 2 denote positive and negative sequence, respectively. For the zero-sequence impedance of the transmission line the per-unit value is

$$(x_{TL})_0 = \frac{35}{95.22} = 0.3675 \text{ p-u} \qquad (14\text{-}6)$$

Of course the transmission line ohmic impedances could be referred to the 20 kV side of the power system, in which case the values become

$$(x_{TL1})_{\text{low side}} = (x_{TL2})_{\text{low side}} = \frac{10}{(69/20)^2} = 0.84\ \Omega \qquad (14\text{-}7)$$

and

$$(x_{TL0})_{\text{low side}} = \frac{35}{(69/20)^2} = 2.94\ \Omega \qquad (14\text{-}8)$$

Subsequent division of these quantities by $Z_{BL} = 8\ \Omega$, which is the base impedance referred to the 20-kV side, yields identical per-unit values.

All the parameters in the power system of Fig. 14-1 are now expressed in per unit and all are specifically referred to the reference base quantity of 50 MVA and 20 kV. We can now proceed with the development of each sequence network.

The Positive-Sequence Network

The development of the equivalent circuit of the power system of Fig. 14-1 as seen by the positive-sequence voltage set can begin with generator G1. It is shown in Sec. 13-6 that the positive-sequence representation of a generator always consists of its generated emf per phase placed in series with the positive-sequence parameter of the generator as displayed in Fig. 13-12(a). With interest focused on the subtransient period we also know from Sec. 13-6 that the generator positive-sequence parameter is given by the subtransient synchronous reactance x_d'', which in this case has the value $j0.14$ p-u. Thus, the generator G1 is represented by a voltage source E_{ag1} in series with the impedance $j0.14$ as indicated in Fig. 14-2(a) between terminals marked A and N_1. Here A corresponds to a point on the generator side of transformer bank T1 in Fig. 14-1. Also, N_1 denotes the neutral of the positive-sequence equivalent circuit of the electric power system. The potential between the neutral of the generator and the ground is zero for balanced voltage systems such as the positive-sequence set.

As we move along the power system in Fig. 14-1 from point A to bus 1, the impedance that is encountered between these points is simply the positive-sequence impedance of the transformer. Equation (13-99) tells us that this positive-sequence impedance is just the leakage impedance of the transformer, which in turn can be taken as essentially equal to the transformer leakage reactance. Thus, there appears between terminals A and bus 1 in Fig. 14-2(a) the per-unit value of this leakage reactance, which has the value $j0.08$. No shunt magnetizing impedance appears for the transformer because this is assumed to be infinite.

The impedance that appears between bus 1 and bus 2 over the entire length of the transmission line is the line reactance. This quantity has already been computed and found to be $j0.105$ p-u [see Eq. (14-5)]. Next, between bus 2 and point B, there appears the leakage impedance of transformer T2. This value is found in Eq. (14-3) to be $j0.072$ p-u. Finally, generator 2 is shown in Fig. 14-2(a) to be represented by the same circuit configuration used to represent G1, namely, a voltage source in series with the generator positive-sequence impedance, which was found in Eq. (14-1) to have the p-u value $j0.144$. This completes the development of the sequence network for the positive-sequence voltage set.

The way in which this circuit is used in making fault calculations depends upon where the fault occurs and also upon the type of fault. We address the latter point

(a)

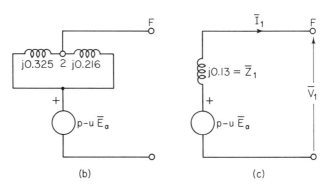

(b)   (c)

Figure 14-2 (a) Positive-sequence network of the electric power system of Fig. 14-1; (b) and (c) Thévenin equivalent circuit for a fault that occurs at bus 2 and is indicated by F.

in succeeding sections. For now let us prepare the specific circuits to be used in the various fault calculations by making appropriate circuit reductions consistent with the particular location of the fault. In this connection assume that a fault occurs at bus 2. We consider the fault to involve a *short circuit* to ground reference without at this point being concerned with the specific type of fault. Our interest here is to prepare circuits which can then be suitably arranged for application to the various fault types. The two generators in the system are assumed to be operating at no load and at rated voltage of $1/\underline{0^\circ}$ per unit. Accordingly, just prior to the occurrence of the fault at bus 2, the voltage is clearly p-u$\bar{E}_{ag1}$ = p-u$\bar{E}_{ag2}$ = $1/\underline{0^\circ}$. In fact, this quantity conveniently can be made to represent the open-circuit voltage source to be used in a Thévenin equivalent circuit of the positive-sequence network looking into the circuit at the fault point F. The Thévenin equivalent impedance is then found by taking the parallel combination of the impedance to the left of the fault with that appearing on the right as illustrated in Fig. 14-2(b). Observe that the total impedance to the left is $j0.325$ and that to the right is $j0.216$. The final form of the equivalent circuit of the electric power system for evaluating the effect of the positive-sequence set at the fault point is displayed in Fig. 14-2(c). The equivalent positive-sequence impedance of the given electric power system for a fault at bus 2 in $\bar{Z}_1 = j0.13$ p-u.

The Negative-Sequence Equivalent Circuit

In accordance with the results found in Sec. 13-6 and displayed in Fig. 13-12(b) the negative-sequence circuit of the generator, as it appears between its line to neutral terminals, consists solely of its negative-sequence impedance. There is no generated negative-sequence voltage because by design the generator can only produce positive-sequence emfs. Moreover, the value of this negative-sequence impedance for subtransient power system studies is generally equal to the subtransient synchronous reactance as indicated by Eq. (13-108). Hence in Fig. 14-3(a), which shows the negative-sequence network of the power system, generator G1 is now represented merely as the subtransient reactance appearing between terminals A and the neutral N_2. Here the subscript 2 is used to remind us that we are now dealing with the negative-sequence network representation of the power system.

In moving along the power system from A to bus 1, the negative-sequence impedance of transformer bank T1 is encountered. Equation (13-99) tells us that this quantity is equal to the leakage impedance, which here is taken to be simply the total leakage reactance of the transformer. Its per-unit value is $j0.08$. Next, between bus 1 and bus 2, the negative-sequence impedance of the transmission line appears. This has the same value as the positive-sequence impedance, namely, $j0.105$. Transformer bank T2 is represented between terminals 2 and B by its negative-sequence impedance $j0.072$, which is the same as for the positive sequence. Finally, as was the case with generator 1, generator 2 also is represented by its negative-sequence impedance.

The circuit of Fig. 14-3(a) represents the electric power system of Fig. 14-1 as it appears to a set of negative-sequence voltages. Moreover, when a fault is assumed to appear at bus 2, this circuit reduces to the simple representation depicted in Fig. 14-3(b). Here $\bar{Z}_2 = j0.13$ denotes the effective negative-sequence impedance looking into the power system at the fault point. The absence of a source voltage in this circuit should not be disconcerting because, when a particular fault type is identified, the boundary conditions associated with the given fault contain the instructions about how this circuit is to be connected to the other sequence networks.

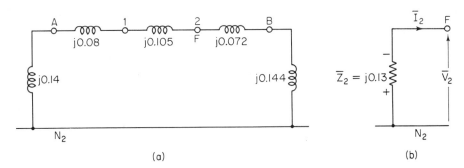

Figure 14-3 (a) Negative-sequence equivalent circuit of the power system of Fig. 14-1; (b) equivalent negative-sequence impedance of the power system for a fault at bus 2.

The Zero-Sequence Network of the Power System

The zero-sequence impedance of the generator G1 is represented in the zero-sequence network of the power system of Fig. 14-4(a) by the per-unit value of the generator leakage reactance. This is in accordance with the results displayed in Eqs. (13-114) and (13-109) and Fig. 13-2(c).

Figure 14-4 (a) Zero-sequence equivalent circuit of the power system of Fig. 14-1; (b) equivalent zero-sequence impedance of the power system for a fault at bus 2.

The zero-sequence network of the power system of Fig. 14-4(a) shows a gap (or open-circuit) between terminal A and bus 1. The gap is included in recognition of the fact that, if a fault were to occur at point A in this power system, zero-sequence current *can* flow through the neutral of generator 1 to ground reference. However, it cannot find a path to ground reference on the right side of A because of the Δ arrangement of the transformer bank T1. A review of Sec. 13-5 at this point is helpful.

The impedance of the transmission line between bus 1 and bus 2 is the zero-sequence impedance of the line which our experience with Chapter 13 reveals to be appreciably different than the positive-sequence value. This conclusion is apparent upon comparing Eqs. (13-66) and (13-67). In this case the per-unit value of the zero-sequence reactance of the transmission line is $j0.3675$, which is three and a half times greater than the positive- and negative-sequence quantities.

When a fault takes place on the Y side of transformer bank T2, a zero-sequence current can flow because of the presence of the neutral connection to ground on the Y side and the Δ connection. Keep in mind that the closed Δ arrangement represents a short circuit (except for leakage reactance) to the Y-side transformer coils for a zero-sequence voltage when the neutral is grounded. Hence the transformer zero-sequence impedance, which in this case is simply the leakage reactance ($\bar{Z}_n = 0$), is shown connected in Fig. 13-4(a) to the ground reference N_0. A gap is again included at terminal B of generator G2 for the same reason that was cited for generator G1.

The complete circuit of Fig. 14-4(a) is a representation of the manner in which the power system of Fig. 14-1 is affected by the zero-sequence component at various places along the electric power system. For example, if a fault occurs at the terminals of generator G2 at point B in the system, then a zero-sequence effect only occurs in generator 2 and nowhere else in the system. The Δ connection of the transformers on the B side ensures this isolation. However, if a fault occurs at bus 2, which is the case of our specific interest in this chapter, then zero-sequence quantities can exist all along the transmission line as well as in the coils of both transformer banks. The generators, however, are isolated from the effects of the zero-sequence for this fault location. The equivalent zero-sequence impedance of the electric power system of Fig. 14-1 for a fault located at bus 2 is represented in Fig. 14-4(b). It has a per-unit value of $j0.0621$.

14-2 THE BALANCED THREE-PHASE FAULT

The entire two-bus electric power system depicted in Fig. 14-1 is represented in terms of positive-, negative-, and zero-sequence components by the equivalent circuits displayed respectively in Figs. 14-2(c), 14-3(b), and 14-4(b) for a fault that is assumed to occur at bus 2. The manner in which these circuits are connected to each other is determined by the instructions that derive from the boundary conditions associated with each fault type. In this section our attention is focused on the balanced three-phase fault.

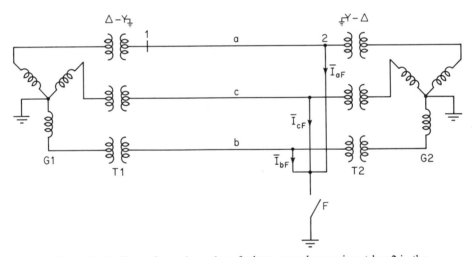

Figure 14-5 Illustrating a three-phase fault to ground occurring at bus 2 in the electric power system of Fig. 14-1.

The occurrence of a balanced three-phase fault at bus 2 means that each of the three transmission lines is brought solidly into contact with the ground and kept there in the manner illustrated in Fig. 14-5. The boundary conditions associated with this symmetrical three-phase short circuit to ground at bus 2 are represented by the following statements:

$$\bar{V}_a = \bar{V}_b = \bar{V}_c = 0 \tag{14-9}$$

$$\bar{I}_a + \bar{I}_b + \bar{I}_c = 0 \tag{14-10}$$

$$\bar{I}_b = a^2 \bar{I}_a \qquad \bar{I}_c = a \bar{I}_a \qquad (14\text{-}11)$$

The first mathematical statement simply recognizes the fact that when all three lines are brought to ground their line-to-neutral values become zero. Also, Eqs. (14-10) and (14-11) state that a symmetrical short circuit on the lines results in a balanced current set. It then follows that both the negative-sequence and the zero-sequence components are zero. Thus, from Eqs. (13-8) and (13-20), which are modified to apply to currents, we have

$$\bar{I}_0 = \tfrac{1}{3}(\bar{I}_a + \bar{I}_b + \bar{I}_c) = 0 \qquad \text{at F} \qquad (14\text{-}12)$$

and

$$\bar{I}_2 = \tfrac{1}{3}(\bar{I}_a + a^2 \bar{I}_b + a \bar{I}_c) = \tfrac{1}{3}(\bar{I}_a + a \bar{I}_a + a^2 \bar{I}_a) = 0 \qquad \text{at F} \qquad (14\text{-}13)$$

Similarly, the positive-sequence current at the fault point is found to be

$$\bar{I}_1 = \tfrac{1}{3}(\bar{I}_a + a \bar{I}_b + a^2 \bar{I}_c) = \tfrac{1}{3}(\bar{I}_a + a^3 \bar{I}_a + a^3 \bar{I}_a) = \bar{I}_a \qquad (14\text{-}14)$$

which is entirely expected because of the assumed balanced nature of the short circuit. It also follows from the direct application of Eqs. (13-8), (13-9), and (13-20) that

$$\bar{V}_1 = 0 \qquad \bar{V}_2 = 0 \qquad \bar{V}_0 = 0 \qquad (14\text{-}15)$$

Hence each sequence voltage at the point of the fault in this case is zero.

This last statement provides the instructions for the use of the equivalent circuits of the preceding section to be consistent with conditions associated with the three-phase short circuit. Thus, the positive-, negative-, and zero-sequence equivalent circuits must now be drawn as illustrated in Fig. 14-6. Because by Eq. (14-15) $\bar{V}_1$ is zero, it is necessary, in applying the circuit of Fig. 14-2(c) to the three-phase fault, to connect the output terminal F of this circuit to neutral. This ensures that V_1 is zero. The negative- and zero-sequence circuits must also be drawn in this way. Note too, in Fig. 14-6, that the negative- and zero-sequence networks are shown isolated (i.e., unconnected) from the positive-sequence circuit. This is justified by Eqs. (14-12) and (14-13) which state that the negative- and zero-sequence currents are zero. Accordingly, these networks are not to be joined to a circuit containing an active source. Therefore, the balanced three-phase fault at bus 2 is simply represented by

Figure 14-6 Sequence networks as they apply to the power system of Fig. 14-1 at bus 2 for a balanced three-phase fault. Observe that the negative- and zero-sequence networks are not connected to an active source and can produce no currents.

the positive-sequence equivalent circuit. All calculations that allow the determination of currents and voltages throughout the complete power system originate from this equivalent circuit. The details of this procedure are described below.

Since the positive-sequence equivalent circuit is the only active one, the positive-sequence current has a value given by

$$\bar{I}_1 = \frac{1\underline{/0^\circ}}{j0.13} = -j7.692 \text{ p-u} \tag{14-16}$$

Accordingly the line currents in per unit that flow through the fault at bus 2 are

$$\bar{I}_{aF} = \bar{I}_1 = -j7.692 = 7.692\underline{/-90^\circ}$$
$$\bar{I}_{bF} = a^2\bar{I}_a = 7.692\underline{/150^\circ} \tag{14-17}$$
$$\bar{I}_{cF} = a\bar{I}_a = 7.692\underline{/30^\circ}$$

To express these currents in amperes, it is necessary to multiply the p-u values by the reference base current evaluated at 69 kV which in this case is

$$\bar{I}_{B_{TL}} = \frac{50000}{3(69)} = 418.4 \text{ A} \tag{14-18}$$

Hence

$$\bar{I}_{aF} = 7.692 \times 418.4 \underline{/-90^\circ} = 3218.31\underline{/-90^\circ} \text{ A}$$
$$\bar{I}_{bF} = 3218.3\underline{/150^\circ} \text{ A} \tag{14-19}$$
$$\bar{I}_{cF} = 3218.3\underline{/30^\circ} \text{ A}$$

Although knowledge of these currents is of interest, it is surely more important to know the corresponding currents distributed throughout the electric power system when such fault currents flow. Thus once the current of the fault is determined with the aid of the equivalent circuit of Fig. 14-6, the next step is to determine that portion of this current that comes from the power system to the *left* of the fault point and that portion that originates from the *right* of the fault. In this connection it is helpful to use a more detailed equivalent circuit than merely the positive-sequence impedance $\bar{Z}_1$. Here the circuit of Fig. 14-2(a) is more suitable because it contains terminal points between which specific elements of the power system can be readily identified. This circuit is repeated in Fig. 14-7 for convenience. By applying the current division principle, we see that the positive-sequence fault current divides so that 40% emanates from the left side of the fault and the remaining 60% comes from the right side. Because the entire length of the transmission line is located on the left side of the fault, it follows that the transmission line currents can now conveniently be found as

$$(\bar{I}_{aF})_{TL} = (0.4)7.692\underline{/-90^\circ} = 3.077\underline{/-90^\circ} \text{ p-u}$$
$$(\bar{I}_{bF})_{TL} = 3.077\underline{/150^\circ} \text{ p-u} \tag{14-20}$$
$$(\bar{I}_{cF})_{TL} = 3.077\underline{/30^\circ} \text{ p-u}$$

Expressed in amperes these currents are

$$\bar{I}_1 = 7.692\underline{/-90°}$$

Figure 14-7 Detailed positive-sequence equivalent network showing the current division factors in parentheses.

$$(\bar{I}_{aF})_{TL} = 3.077(418.3)\underline{/-90°} = 1287\underline{/-90°}\ \text{A}$$
$$(\bar{I}_{bF})_{TL} = 1287\underline{/150°}\ \text{A} \qquad\qquad (14\text{-}21)$$
$$(\bar{I}_{cF})_{TL} = 1287\underline{/30°}\ \text{A}$$

Keep in mind that these are transmission line currents associated with a fault at bus 2 for subtransient conditions, i.e., occurring during the first few cycles immediately following the balanced three-phase short circuit.

What are the currents that flow through the phase coils on the Y and Δ sides of the transformer bank at bus 1? A glance at the detailed schematic diagram of the two-bus power system displayed in Fig. 14-8 makes it immediately obvious that on the high-voltage side (i.e., the Y side) the coil currents are identical to the transmission line currents appearing in Eqs. (14-21). The corresponding currents that flow in the associated Δ connection can readily be found by equating the volt-amperes of matching phase coils. Thus the phase current that flows in coil a of the Δ is

$$\bar{I}_{a\Delta} = \frac{69/\sqrt{3}}{20}(1287.4)\underline{/-90°} = 2564.3\underline{/-90°}\ \text{A} \qquad (14\text{-}22)$$

Similar results are obtained for coils b and c. Thus,

$$\bar{I}_{b\Delta} = 2564.3\underline{/150°} \quad\text{and}\quad \bar{I}_{c\Delta} = 2564.3\underline{/30°}\ \text{A} \qquad (14\text{-}23)$$

The associated line currents on the Δ side, which are also the line currents of generator G1, can be found by application of KCL at each corner of the Δ. Hence from the notation employed on the Δ side of G1 in Fig. 14-8 we have for line a generator current

$$(\bar{I}_a)_{G1} = \bar{I}_{a\Delta} - \bar{I}_{c\Delta} = 2564.3\underline{/-90°} - 2564.3\underline{/30°}$$
$$= 2564.3\left(-j - \frac{\sqrt{3}}{2} - j\frac{1}{2}\right) = 4441.3\underline{/-120°}\ \text{A} \qquad (14\text{-}24a)$$

Figure 14-8 Power system schematic diagram showing the magnitude of all currents for a balanced three-phase fault at bus 2 for subtransient conditions.

For the remaining two lines the expressions are

$$(\bar{I}_b)_{G1} = \bar{I}_{b\Delta} - \bar{I}_{a\Delta} = 4441.3\underline{/120°} \text{ A} \tag{14-24b}$$

$$(\bar{I}_c)_{G1} = \bar{I}_{c\Delta} - \bar{I}_{b\Delta} = 4441.3\underline{/0°} \quad \text{A} \tag{14-24c}$$

It is interesting to note here that these generator currents are in excess of 300% of the rated generator current, which means that about nine times the allowable heating is being developed. Clearly the need for interruption of these exorbitant currents is obvious if severe damage to equipment is to be avoided.

A Digression

There is an alternative procedure for finding the generator line currents where Y-Δ transformer connections are used in an electric power system. But first it is necessary to be aware of the 30° phase displacement that exists between the coil voltages when three single-phase transformers are arranged in Δ on one side and in Y on the other side. This situation is illustrated by the arrangement of Fig. 14-9. Assume that the low-voltage side is denoted by the lowercase letters and the high-voltage side by the uppercase letters. Figure 14-9(a) thus shows the low side connected in Δ and the high side arranged in Y. The phasor diagram of Fig. 14-9(b) represents the situation in the transformer coils as they apply to the positive-sequence $a\,b\,c$. Observe that $\bar{V}_{ab}$ and $\bar{V}_{AN}$ are in phase because they are produced by the same flux in transformer a-A. Also, $\bar{V}_{AN}$ is given a longer length than $\bar{V}_{ab}$ because it represents the high voltage. The line voltage on the high side, $\bar{V}_{AB}$, is the phasor sum of $\bar{V}_{AN}$ and $\bar{V}_{NB}$. In particular $\bar{V}_{AB}$ has a magnitude that is $\sqrt{3}$ times greater than $\bar{V}_{AN}$ and *leads* $\bar{V}_{AN}$ by 30°. Since $\bar{V}_{AN}$ and $\bar{V}_{ab}$ are in time phase, it follows that the line voltage between A and B on the Y side *leads* the line voltage on the Δ side by 30°.

When the foregoing analysis is repeated for negative-sequence voltages, the phasor diagram depicted in Fig. 14-9(c) results. In this case the line voltage on the Y side *lags* the line voltage on the Δ side. As a result of this 30° characteristic displacement angle that exists between the line voltages of the Y-Δ connection, the ASA† has adopted the following standard: the phases of three-phase transformer connections shall be so labeled that the positive-sequence quantity on the high side is to *lead* the corresponding quantity on the low side by 30°; similarly, the negative-sequence quantity on the high side is to *lag* the corresponding negative-sequence quantity on the low side by 30°.

The alternative procedure for finding the line currents in G1 can now be demonstrated. The currents in the high side coils of the transformers in bank T1 are the same as those in the transmission line, which are represented in per unit by Eq. (14-20). Of course the per-unit value of the transformer currents do not change by being referenced to the high- or low-voltage side. It will be recalled that this is a distinct advantage of the use of per-unit quantities. However, it is necessary to modify the angular component of this per-unit quantity. Because we are moving from the high side to the low side, it is necessary to introduce a 30° phase lag. Accordingly, the per-unit value of the generator positive-sequence current becomes that of the trans-

†ASA is the American Stardard Association.

(a)

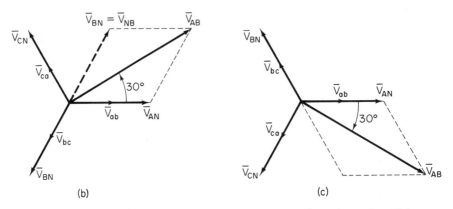

(b)   (c)

Figure 14-9 Illustrating the 30° displacement between line voltages in a Y-Δ connection of transformers: (a) phase coil connections with the high voltage side in Y; (b) the phasor diagram for the positive sequence set; (c) the phasor diagram for the negative-sequence set.

mission line current minus an additional 30°. Thus,

$$(\bar{I}_1)_{G1} = 3.077\underline{/-90°} - 30° = 3.077\underline{/-120°} \text{ p-u} \qquad (14\text{-}25)$$

Expressed in amperes, this becomes

$$(\bar{I}_1)_{G1} = 3.077(1443.4)\underline{/-120°} = 4441.3\underline{/-120°} \text{ A} \qquad (14\text{-}26)$$

Consequently, the generator line currents are

$$(\bar{I}_a)_{G1} = (\bar{I}_1)_{G1} = 4441.3\underline{/-120°}$$

$$(\bar{I}_b)_{G1} = a^2(\bar{I}_1)_{G1} = 4441.3\underline{/120°} \qquad (14\text{-}27)$$

$$(\bar{I}_c)_{G1} = a(\bar{I}_1)_{G1} = 4441.3\underline{/0°}$$

These results are in complete agreement with those displayed in Eqs. (14-24). The quantity (1443.4) used in Eq. (14-26) is the base current evaluated for 50 MVA and 20 kV.

At transformer bank T2 the value of the per-unit current is

$$(\bar{I}_1)_R = 0.6(7.692)\underline{/-90°} = 4.6152\underline{/-90°} \text{ p-u} \qquad (14\text{-}28)$$

where the subscript R denotes right side of the fault. But, as the circuitry of Fig. 14-7 makes obvious, the same p-u positive-sequence current flows through generator G2 as flows through the transformers in T2. The sole adjustment needed therefore is the 30° lag associated with the Y-Δ arrangement in going from the high-voltage side to the low-voltage side. Thus,

$$(\bar{I}_1)_{G2} = 4.6152\underline{/-90° - 30°} = 4.6152\underline{/-120°} \text{ p-u} \qquad (14\text{-}29)$$

To convert this quantity to amperes, it is necessary to use a base current that is consistent with the 12-kV rating of generator G2. This yields a value of 2405.6 A. Hence, the line currents for G2 are

$$(\bar{I}_a)_{G2} = (\bar{I}_1)_{G2} = 4.6152(2405.6)\underline{/-120° - 30°}$$
$$= 11,102.6\underline{/-120°} \text{ A}$$
$$(\bar{I}_b)_{G2} = a^2(\bar{I}_1)_{G2} = 11,102.6\underline{/120°} \text{ A} \qquad (14\text{-}30)$$
$$(\bar{I}_c)_{G2} = a(\bar{I}_1)_{G2} = 11,102.6\underline{/0°} \text{ A}$$

In view of the fact that the rated current of G2 is 3367.9 A, these currents represent considerable overload.

The values of the line voltages as they exist along the power system, under fault conditions, can also be found by employing the circuitry of Fig. 14-7. As an illustration let it be desired that the line voltages of generator 1 be determined when a three-phase balanced fault occurs at bus 2. The solution in this instance is obtained by first finding the potential difference between point A and ground in Fig. 14-7. Proceeding in terms of per-unit values, this potential difference is readily found to be

$$(\bar{V}_1)_A = 1 - j(0.14)(-j3.077) = 1 - 0.4308 = 0.5692 \text{ p-u} \qquad (14\text{-}31)$$

However, because the generator is on the Δ side of the T1 transformers, it is now necessary to add −30° adjustment in the result displayed in the last equation in order that it properly apply to the generator. Accordingly,

$$(\bar{V}_1)_{G1} = 0.5692\underline{/-30°} \text{ p-u} \qquad (14\text{-}32)$$

The corresponding phase voltages of the generator thus become

$$(\bar{V}_a)_{G1} = 0.5692\underline{/-30°} \text{ p-u}$$
$$(\bar{V}_b)_{G1} = a^2(V_1)_{G1} = 0.5692\underline{/-150°} \text{ p-u} \qquad (14\text{-}33)$$
$$(\bar{V}_c)_{G1} = a(V_1)_{G1} = 0.5692\underline{/90°} \text{ p-u}$$

The conversion to voltages expressed in volts is obtained by introducing the rated voltage per phase of generator 1, which is $20/\sqrt{3}$ or 11.547 kV. The phase voltages are therefore

$$(\bar{V}_a)_{G1} = 6573\underline{/-30°} \text{ V}$$

$$(\bar{V}_b)_{G1} = 6573\underline{/-150°} \text{ V} \qquad (14\text{-}34)$$

$$(\bar{V}_c)_{G1} = 6573\underline{/90°} \text{ V}$$

These values indicate that a rather severe voltage drop accompanies this fault.

14-3 THE SINGLE LINE-TO-GROUND FAULT

The single line-to-ground fault is the most commonly encountered electric power system fault. It is represented schematically for our two-bus system in Fig. 14-10 based on the assumption that the fault occurs at bus 2. The boundary conditions for this fault condition are readily identified at the fault point F as

$$\bar{I}_{aF} \neq 0 \qquad \bar{I}_{bF} = \bar{I}_{cF} = 0 \qquad \bar{V}_{aF} = 0 \qquad (14\text{-}35)$$

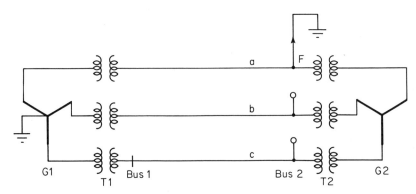

Figure 14-10 Schematic diagram illustrating a single line to ground fault occurring at bus 2.

The instructions for the interconnection of the positive-, negative-, and zero-sequence networks as they apply for a fault at bus 2 are contained in these boundary conditions and may be deduced by examining the expressions for each of the sequence currents. Accordingly, we have

$$\bar{I}_{1F} = \tfrac{1}{3}(\bar{I}_{aF} + a\bar{I}_{bF} + a^2\bar{I}_{cF}) = \tfrac{1}{3}(\bar{I}_{aF} + 0 + 0) = \tfrac{1}{3}\bar{I}_{aF}$$

$$\bar{I}_{2F} = \tfrac{1}{3}(\bar{I}_{aF} + a^2\bar{I}_{bF} + a\bar{I}_{cF}) = \tfrac{1}{3}(\bar{I}_{aF} + 0 + 0) = \tfrac{1}{3}\bar{I}_{aF} \qquad (14\text{-}36)$$

$$\bar{I}_{0F} = \tfrac{1}{3}(\bar{I}_{aF} + \bar{I}_{bF} + \bar{I}_{cF}) = \tfrac{1}{3}(\bar{I}_{aF} + 0 + 0) = \tfrac{1}{3}\bar{I}_{aF}$$

where the F subscript serves as a reminder that these are the sequence currents as they flow in the fault. Because each sequence current in the single line-to-ground fault is the *same*, it follows that each sequence equivalent circuit must be *series* connected in the manner illustrated in Fig. 14-11(a). The principal contribution made by the arrangement of this circuit is to permit ready evaluation of the fault current $\bar{I}_{aF}$. However, as already demonstrated for the symmetrical three-phase fault, a primary objective of fault analysis is to obtain information about the magnitude of the fault currents that exist in the various components of the electric power system. To achieve this goal we

Figure 14-11 Equivalent circuit of the two-bus power system of Fig. 14-1 for a line to ground fault at bus 2: (a) the simplified equivalent circuit used to find $\bar{I}_{aF}$; (b) the detailed equivalent circuit used to find the distribution of currents and voltages throughout the power system. The quantities in parentheses denote the division of current in the parallel branches.

must have recourse to the more detailed circuitry for each sequence equivalent circuit but still arranged to be consistent with the particular boundary conditions. This version of the equivalent circuit for the entire power system is depicted in Fig. 14-11(b). Finally, it is instructive to note that this series connection of the sequence equivalent circuits is returned to ground potential in response to the boundary condition on the line voltage $\bar{V}_{aF}$ expressed in the form

$$\bar{V}_{aF} = \bar{V}_{1F} + \bar{V}_{2F} + \bar{V}_{0F} = 0 \tag{14-37}$$

The fault current that flows into the ground reference from line a is simply determined from the circuit of Fig. 14-11(a). Since each sequence current takes on the same value, we have

$$\bar{I}_{1F} = \bar{I}_{2F} = \bar{I}_{0F} = \frac{1/0°}{\bar{Z}_1 + \bar{Z}_2 + \bar{Z}_0} = \frac{1/0°}{j0.13 + j0.13 + j0.0621} = 3.105/-90° \text{ p-u} \tag{14-38}$$

so the fault current takes on the per-unit value

$$\bar{I}_{aF} = \bar{I}_{1F} + \bar{I}_{2F} + \bar{I}_{0F} = 3\bar{I}_{0F} = 3(3.105)/-90° = 9.315/-90° \text{ p-u} \tag{14-39}$$

In order to find the value of this current in amperes, we must use a base current that is consistent with the reference MVA evaluated at the transmission line voltage of 69 kV. Its value in this case is 418.4 A, which in turn yields a fault current of

$$\bar{I}_{aF} = 9.315(418.4)/-90° = 3897.2/-90° \text{ A} \tag{14-40}$$

This current flows from line a at bus 2 into the ground reference and is so indicated in the system diagram displayed in Fig. 14-12.

The transmission line currents. Since the fault takes place at bus 2 the transmission line fault currents are those currents that flow from the left side of the fault point. Referring to Fig. 14-11(b) and employing the current division factor shown in parentheses, the per-unit value of the positive-, negative-, and zero-sequence components that constitute these transmission line currents are readily found to be

$$\bar{I}_{1L} = 0.4\bar{I}_{1F} = (0.4)(3.105/-90°) = 1.242/-90° \text{ p-u}$$
$$\bar{I}_{2L} = 0.4\bar{I}_{2F} = 0.4(3.105/-90°) = 1.242/-90° \text{ p-u} \tag{14-41}$$
$$\bar{I}_{0L} = 0.139\bar{I}_{0F} = 0.139(3.105/-90°) = 0.432/-90° \text{ p-u}$$

where the second subscript, L, is used to denote that these are the respective sequence currents emanating from the left (L) side of the fault. The transmission line currents then become

$$\begin{bmatrix} \bar{I}_a \\ \bar{I}_b \\ \bar{I}_c \end{bmatrix}_{TL} = \begin{bmatrix} 1 & 1 & 1 \\ a^2 & a & 1 \\ a & a^2 & 1 \end{bmatrix}\begin{bmatrix} \bar{I}_1 \\ \bar{I}_2 \\ \bar{I}_0 \end{bmatrix}_L = \begin{bmatrix} 1 & 1 & 1 \\ a^2 & a & 1 \\ a & a^2 & 1 \end{bmatrix}\begin{bmatrix} 1.242/-90° \\ 1.242/-90° \\ 0.432/-90° \end{bmatrix} = \begin{bmatrix} 2.916/-90° \\ -0.18 /-90° \\ -0.81 /-90° \end{bmatrix} \text{ p-u}$$

$$= \begin{bmatrix} 1220.1/-90° \\ -338.9/-90° \\ -338.9/-90° \end{bmatrix} \text{ A} \tag{14-42}$$

Figure 14-12 Schematic diagram of the two-bus system of Fig. 14-1 showing the magnitudes of all the currents throughout the system for a line a to ground fault at bus 2.

where the results are expressed both in per unit and in amperes. To get amperes, the p-u values are multiplied by 418.4 A, which is the reference base current at the transmission line voltage of 69 kV.

Transformer bank T1 currents. A glance at Fig. 14-12 makes it obvious that the transmission line currents are also the phase currents of the transformer coils in the Y connection on the high-voltage side. Keep in mind that when the fault at bus 2 is assumed to occur, the power system at prefault is in a quiescent state of no load. Hence the currents that exist in the system postfault are entirely the result of the fault. Accordingly, when a fault current of $I_{aF} = 3897.2$ A $= 9.315$ p-u is computed to flow to the ground reference then clearly there must exist a return path for this current. The only permissible paths for this return must be through the neutrals on the Y side of transformer banks T1 and T2. No return paths through the generator neutrals are possible because the Δ-connected transformers to which these generators are joined prohibit the flow of the zero-sequence current. And, as already indicated in Eq. (14-39), this fault current at F is given by three times the zero-sequence current. In fact this relationship holds generally at each neutral where the zero-sequence current is permitted to exist. Thus for the situation at transformer bank T1 we can write from Eq. (14-42) the following equations:

$$(\bar{I}_a)_{TL} = \bar{I}_{1L} + \bar{I}_{2L} + \bar{I}_{0L} = 2\bar{I}_{1L} + \bar{I}_{0L}$$
$$(\bar{I}_b)_{TL} = a^2\bar{I}_{1L} + a\bar{I}_{2L} + \bar{I}_{0L} = (a^2 + a)\bar{I}_{1L} + \bar{I}_{0L} = -I_{1L} + I_{0L} \quad (14\text{-}43)$$
$$(\bar{I}_c)_{TL} = a\bar{I}_{1L} + a^2\bar{I}_{2L} + \bar{I}_{0L} = (a + a^2)\bar{I}_{1L} + \bar{I}_{0L} = -\bar{I}_{1L} + \bar{I}_{0L}$$

Taking the sum then yields

$$3\bar{I}_{0L} = (\bar{I}_a)_{TL} + (\bar{I}_b)_{TL} + (\bar{I}_c)_{TL} \quad (14\text{-}44)$$

Therefore the current that flows from ground reference into the neutral at T1 can be determined by evaluating

$$3\bar{I}_{0L} = 3(0.432)\underline{/-90°} = 1.296\underline{/-90°} = 542.3\underline{/-90°} \text{ A} \quad (14\text{-}45)$$

or else simply summing the three line currents whose values are displayed in Eq. (14-42). Thus,

$$3\bar{I}_{0L} = (\bar{I}_a)_{TL} + (\bar{I}_b)_{TL} + (\bar{I}_c)_{TL} = 1220.1\underline{/-90°} - 338.9\underline{/-90°} - 338.9\underline{/-90°}$$
$$= 542.3\underline{/-90°} \text{ A} \quad (14\text{-}46)$$

It is instructive to note too that the current in coil a of transformer bank T1 is actually composed of the arithmetic sum of the currents in phases b and c (the negative sign in these currents means that the actual current flow is reversed) and the total zero-sequence current that flows from ground reference.

The corresponding currents that flow in the Δ side of T1 are readily determined by equating the volt-amperes of the secondary phase coil to the primary phase coil. This procedure leads to the current magnitudes that are indicated in Fig. 14-12.

Generator G1 line currents. Of course once the Δ-side currents are evaluated by the volt-ampere rule, the corresponding line currents on the Δ side of T1 are obtained by applying Kirchhoff's current law. These results are also shown in Fig. 14-12.

The generator line currents can also be found directly from knowledge of the Y side line currents without having to determine first the individual phase currents of the Δ configuration. We must be careful, however, to modify the Y high-side line quantities by ∓30°. Accordingly, we can write for the positive- and negative-sequence components of the line currents of generator 1

$$(\bar{I}_1)_{G1} = 1.242\underline{/-90°} - 30° = 1.242\underline{/-120°}\ \text{p-u} \qquad (14\text{-}47)$$

$$(\bar{I}_2)_{G1} = 1.242\underline{/-90°} + 30° = 1.242\underline{/-60°}\ \text{p-u} \qquad (14\text{-}48)$$

Observe that the adjustment is positive for the negative sequence. The location of these currents in the detailed equivalent circuit are indicated in Fig. 14-11(b). Note, too, in this diagram that the zero-sequence current in the generator line cannot exist. Hence the current matrix for G1 can be expressed as

$$
\begin{bmatrix} \bar{I}_a \\ \bar{I}_b \\ \bar{I}_c \end{bmatrix}_{G1}
=
\begin{bmatrix} 1 & 1 & 1 \\ a^2 & a & 1 \\ a & a^2 & 1 \end{bmatrix}
\begin{bmatrix} \bar{I}_1 \\ \bar{I}_2 \\ 0 \end{bmatrix}_{G1}
=
\begin{bmatrix} 2.151\underline{/-90°} \\ 2.151\underline{/90°} \\ 0 \end{bmatrix}\text{p-u}
=
\begin{bmatrix} 3105\underline{/-90°} \\ 3105\underline{/90°} \\ 0 \end{bmatrix}\text{A} \qquad (14\text{-}49)
$$

Here the conversion from per unit to amperes is achieved through use of the base current at 20 kV, which is 1443.4 A.

Transformer bank T2 currents.   The sequence currents that flow through the fault from the right side can be written forthwith by referring to Fig. 14-11(b) and using the indicated current division factors there listed.

$$\bar{I}_{1R} = (0.6)(3.105)\underline{/-90°} = 1.863\underline{/-90°}\ \text{p-u}$$

$$\bar{I}_{2R} = (0.6)3.105\underline{/-90°} = 1.863\underline{/-90°}\ \text{p-u} \qquad (14\text{-}50)$$

$$\bar{I}_{0R} = (0.861)3.105\underline{/-90°} = 2.673\underline{/-90°}\ \text{p-u}$$

The transformer T2 high-side line currents are, correspondingly,

$$(\bar{I}_a)_{T2} = \bar{I}_{1R} + \bar{I}_{2R} + \bar{I}_{0R} = 6.399\underline{/-90°}\ \text{p-u}$$

$$(\bar{I}_b)_{T2} = a^2\bar{I}_{1R} + a\bar{I}_{2R} + \bar{I}_{0R} = -\bar{I}_{1R} + \bar{I}_{0R} = 0.81\underline{/-90°}\ \text{p-u} \qquad (14\text{-}51)$$

$$(\bar{I}_c)_{T2} = a\bar{I}_{1R} + a^2\bar{I}_{2R} + \bar{I}_{0R} = -\bar{I}_{1R} + \bar{I}_{0R} = 0.81\underline{/-90°}\ \text{p-u}$$

The matrix formulation is

$$
\begin{bmatrix} \bar{I}_a \\ \bar{I}_b \\ \bar{I}_c \end{bmatrix}_{T2}
=
\begin{bmatrix} 1 & 1 & 1 \\ a^2 & a & 1 \\ a & a^2 & 1 \end{bmatrix}
\begin{bmatrix} \bar{I}_1 \\ \bar{I}_2 \\ \bar{I}_0 \end{bmatrix}_{R}
=
\begin{bmatrix} 6.399\underline{/-90°} \\ 0.81\underline{/-90°} \\ 0.81\underline{/-90°} \end{bmatrix}\text{p-u}
=
\begin{bmatrix} 2677.6\underline{/-90°} \\ 338.9\underline{/-90°} \\ 338.9\underline{/-90°} \end{bmatrix}\text{A} \qquad (14\text{-}52)
$$

On the low-voltage Δ side the current in phase coil a is found from equality of the volt-amperes in the transformer between the low and high voltage side. Thus,

$$(\bar{I}_{a\Delta})_{T2} = \frac{69/\sqrt{3}}{12}(2677.6)\underline{/-90°} = 8889\underline{/-90°}\ \text{A} \qquad (14\text{-}53)$$

Similarly, the currents in coils b and c are found to be $1125.1\underline{/-90°}$ A are so indicated in Fig. 14-12.

The ground current that enters the Y-connected transformers can be found directly by invoking an expression analogous to that appearing in Eq. (14-45). Accordingly, we get

$$3(\bar{I}_0)_{T2} = 3\bar{I}_{0R} = 3(2.673\underline{/-90°}) = 8.019\underline{/-90°} \text{ p-u} = 3355.8\underline{/-90°} \text{ A} \quad (14\text{-}54)$$

It is useful to note in the current distribution diagram of Fig. 14-12 that the sum of the two currents that enter the neutrals of the Y of transformer banks T1 and T2 is equal to the fault current that enters the ground at the fault, i.e.,

$$\bar{I}_{aF} = 3(\bar{I}_{0L}) + 3(\bar{I}_{0R}) = 1.296\underline{/-90°} + 8.019\underline{/-90°} = 9.315\underline{/-90°} \text{ p-u} \quad (14\text{-}55)$$

The line currents at generator G2. Once again by referring to the power system equivalent circuit of Fig. 14-11(b) we find that the appropriate positive- and negative-sequence components of these line currents are the same as $\bar{I}_{1R}$ and $\bar{I}_{2R}$ of Eq. (14-50) but modified to account for the $\mp30°$ displacement that is caused by the Y-Δ arrangement of the transformers. Accordingly, these sequence components become

$$(\bar{I}_1)_{G2} = 1.863\underline{/-90°} - 30° = 1.863\underline{/-120°} \text{ p-u}$$
$$(\bar{I}_2)_{G2} = 1.863\underline{/-90°} + 30° = 1.863\underline{/-60°} \text{ p-u} \quad (14\text{-}56)$$

The zero-sequence component is of course zero. Hence the matrix expression for the line currents at G2 is

$$\begin{bmatrix} \bar{I}_a \\ \bar{I}_b \\ \bar{I}_c \end{bmatrix}_{G2} = \begin{bmatrix} 1 & 1 & 1 \\ a^2 & a & 1 \\ a & a^2 & 1 \end{bmatrix} \begin{bmatrix} \bar{I}_1 \\ \bar{I}_2 \\ 0 \end{bmatrix}_{G2} = \begin{bmatrix} 3.2268\underline{/-90°} \\ 3.2268\underline{/90°} \\ 0 \end{bmatrix} \text{ p-u} = \begin{bmatrix} 7763.9\underline{/-90°} \\ 7763.9\underline{/90°} \\ 0 \end{bmatrix} \text{ A} \quad (14\text{-}57)$$

The base current used to get amperes in this last equation is that referred to a 12-kV base, or 2405.8 A.

The line voltages at generator G1. The power system equivalent circuit of Fig. 14-11(b) may also be used to calculate the line voltages at any point along the electric power system. We illustrate the procedure by computing the line voltages at the terminals of G1 when a single line-to-ground fault occurs at bus 2. Our experience with symmetrical components tells us that the unbalanced line voltages can readily be found once we know the sequence components as they pertain to G1. These sequence components are identified in Fig. 14-11(b) as $(\bar{V}_1)_{G1}$ and $(\bar{V}_2)_{G2}$. The zero-sequence component is zero because of the transformers.

The expression that leads to the evaluation of the positive-sequence voltage component is seen from Fig. 14-11(b) to be

$$(\bar{V}_1)_{G1Y} = 1\underline{/0°} - j(0.14)\bar{I}_{1L} = 1 - j(0.14)(1.242\underline{/-90°}) = 0.826\underline{/0°} \text{ p-u} \quad (14\text{-}58)$$

where I_{1L} is the left-side positive-sequence transmission line current and is evaluated in Eq. (14-41). The subscript Y is added to remind us that this expression is valid on the high-voltage side. Similarly, the negative-sequence voltage on the left side of the fault is given by

$$(\bar{V}_2)_{G1Y} = -j(0.14)\bar{I}_{2L} = -j(0.14)(1.242)\underline{/-90°} = -0.1739 \text{ p-u} \quad (14\text{-}59)$$

Although the last two equations correctly represent the magnitudes of these sequence voltages, the phase angles are in need of adjustment when referred to the generator because of the Y-Δ connections. Hence, before Eqs. (14-48) and (14-59) can be used for calculations on the line side of the Δ connection at transformer bank T1, the phase angle of the positive-sequence voltage must be diminished by 30° and that of the negative-sequence must be augmented by the same amount. Thus, we have

$$(\bar{V}_1)_{G1} = 0.826\underline{/-30°} \quad \text{and} \quad (\bar{V}_2)_{G1} = -0.1739\underline{/30°} \text{ p-u} \quad (14\text{-}60)$$

The line-to-neutral phase voltages at the generator are then determined in matrix form to be

$$
\begin{bmatrix} \bar{V}_a \\ \bar{V}_b \\ \bar{V}_c \end{bmatrix}_{G1}
=
\begin{bmatrix} 1 & 1 & 1 \\ a^2 & a & 1 \\ a & a^2 & 1 \end{bmatrix}
\begin{bmatrix} \bar{V}_1 \\ \bar{V}_2 \\ \bar{V}_0 \end{bmatrix}_{G1}
=
\begin{bmatrix} 1 & 1 & 1 \\ a^2 & a & 1 \\ a & a^2 & 1 \end{bmatrix}
\begin{bmatrix} 0.826\underline{/-30°} \\ -0.1739\underline{/30°} \\ 0 \end{bmatrix}
$$

$$
=
\begin{bmatrix} 0.7543\underline{/-41.5°} \\ 0.7543\underline{/-138.5°} \\ 1.0\underline{/90°} \end{bmatrix} \text{p-u}
=
\begin{bmatrix} 8710\underline{/-41.5°} \\ 8710\underline{/-138.5°} \\ 11,547\underline{/90°} \end{bmatrix} \text{V} \quad (14\text{-}61)
$$

Observe that line c voltage is not disturbed at all by a line-to-ground fault on line a. A glance at Fig. 14-12 makes clear the reason for this. Note that the phase current in line c is zero which means that there are no impedance drops in this phase. Therefore the phase voltage remains at its rated value of $20,000/\sqrt{3} = 11,547$ V.

14-4 THE DOUBLE LINE-TO-GROUND FAULT

The schematic representation of this fault type is depicted in Fig. 14-13. Again, in order to use forthwith the sequence equivalent circuits derived in Sec. 14-1 for this two-bus power system, the double line-to-ground fault is assumed to occur at bus 2. The corresponding boundary conditions are then clearly represented by

$$\bar{V}_{bF} = \bar{V}_{cF} = 0 \quad \text{and} \quad \bar{I}_{aF} = 0 \quad (14\text{-}62)$$

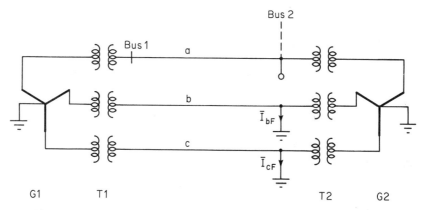

Figure 14-13 Schematic diagram illustrating a double line to ground fault occurring at bus 2.

These boundary conditions contain the instructions for the interconnection of the sequence equivalent circuits. To uncover these instructions we first make use of the boundary condition on voltage because there are two bits of information involved here. We use this information to obtain expressions for the sequence voltages as they exist at the fault point. Invoking Eqs. (13-19), (13-20), and (13-18) in matrix form we can write

$$
\begin{bmatrix} \bar{V}_1 \\ \bar{V}_2 \\ \bar{V}_0 \end{bmatrix}_F = \frac{1}{3} \begin{bmatrix} 1 & a & a^2 \\ 1 & a^2 & a \\ 1 & 1 & 1 \end{bmatrix} \begin{bmatrix} \bar{V}_a \\ \bar{V}_b \\ \bar{V}_c \end{bmatrix}_F = \frac{1}{3} \begin{bmatrix} 1 & a & a^2 \\ 1 & a^2 & a \\ 1 & 1 & 1 \end{bmatrix} \begin{bmatrix} \bar{V}_{aF} \\ 0 \\ 0 \end{bmatrix}
\tag{14-63}
$$

It then follows that

$$
\bar{V}_{1F} = \bar{V}_{2F} = \bar{V}_{0F} = \tfrac{1}{3}\bar{V}_{aF}
\tag{14-64}
$$

Hence, for a double line-to-ground fault, the sequence equivalent circuits for a fault at bus 2 [which are repeated for convenience in Fig. 14-14(a)] must be arranged in *parallel* in the manner illustrated in Fig. 14-14(b) in order to meet the requirements of Eq. (14-64). Furthermore, observe that the currents are defined as positive as they flow to the upper node in conformity with the boundary condition on current which states that

$$
\bar{I}_{1F} + \bar{I}_{2F} + \bar{I}_{0F} = \bar{I}_{aF} = 0
\tag{14-65}
$$

(a)

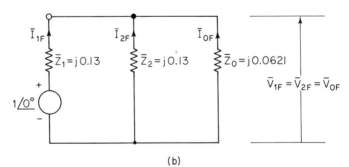

(b)

Figure 14-14 Equivalent circuit of the power system of Fig. 14-13 for a double line to ground fault at bus 2: (a) sequence networks for a fault at bus 2; (b) the specific arrangement of the sequence networks for consistency with the boundary conditions of a double line to ground fault.

From the equivalent circuit of Fig. 14-14(b) it is now possible to perform calculations that yield information about the actual currents and voltages that exist anywhere in the electric power system. However, to facilitate this process, it is necessary to have recourse to the more detailed version of the circuit of Fig. 14-14(b). This detailed version appears in Fig. 14-15.

Figure 14-15 Detailed equivalent circuit of the two-bus power system for use in computing all line voltages and currents throughout the system for a double line to ground fault at bus 2.

Unlike the situation that prevailed in the previous two cases, the flow of current into the ground at the point of the fault does not appear in Fig. 14-14(b). The reason is that the only line current that is represented by this circuit is $\bar{I}_{aF}$, which we know to be zero. Of course fault currents from lines b and c do exist and they do flow into the ground. But they do not have an explicit direct presence in Fig. 14-14(b). Their presence appears here only *indirectly* through the three sequence currents and the appropriate rotational operators a and a^2.

Our aim in the remainder of this section is to show how to find the remaining line currents and line voltages at the location of the fault. The determination of the currents and voltages throughout the rest of the power system is left as a series of exercises for the reader in solving a number of associated problems listed at the end of this chapter.

The computation procedure begins by applying conventional circuit analysis to determine the three sequence currents. The positive-sequence current is easily found from

$$\bar{I}_{1F} = \frac{\bar{E}_a}{\bar{Z}_1 + \dfrac{\bar{Z}_2\bar{Z}_0}{\bar{Z}_2 + \bar{Z}_0}} = \frac{1/0°}{j0.13 + \dfrac{(j0.13)(j0.0621)}{j0.1921}} = \frac{1}{j0.172} = -j5.812 \text{ p-u}$$

(14-66)

The negative-sequence current component is obtained by employing the current division principle. Thus,

$$\bar{I}_{2F} = -\frac{\bar{Z}_0}{\bar{Z}_1 + \bar{Z}_0}\bar{I}_{1F}$$

(14-67)

Here the negative sign is introduced because the actual current flow for the negative sequence is in a direction opposite to the assumed positive direction. Accordingly, its value becomes

$$\bar{I}_{2F} = -\frac{0.0621}{0.1921}(-j5.813) = j1.879 \text{ p-u}$$

(14-68)

Similarly, the zero-sequence component at the fault is found to be

$$\bar{I}_{0F} = -\frac{0.13}{0.1921}\bar{I}_{1F} = j3.934 \text{ p-u}$$

(14-69)

Line currents at fault F. In terms of its symmetrical components the fault current in line a is

$$\bar{I}_{aF} = \bar{I}_{1F} + \bar{I}_{2F} + \bar{I}_{0F} = -j5.813 + j1.879 + j3.934 = 0$$

This result obviously is entirely in keeping with the boundary condition as expressed in Eqs. (14-62) and (14-65). It is evaluated here merely as a check on the determination of the sequence currents. The fault current that flows into ground from line b is given by the expression

$$\bar{I}_{bF} = a^2\bar{I}_{1F} + a\bar{I}_{2F} + \bar{I}_{0F} = \quad 5.813/\underline{150°} + 1.879/\underline{210°} + 3.934/\underline{90°}$$
$$= -6.6615 + j5.901 = 8.9/\underline{138.5°} \text{ p-u} \quad (14\text{-}70)$$

Similarly, for line c there results

$$\bar{I}_{cF} = a\bar{I}_{1F} + a^2\bar{I}_{2F} + \bar{I}_{0F}\,5 = 5.813/\underline{0°} + 1.879/\underline{-30°} + j3.934$$
$$= 6.6615 + j5.901 = 8.9/\underline{41.5°} \text{ p-u} \quad (14\text{-}71)$$

A comparison of the last two equations reveals that the net current that flows into ground from lines a and b is purely reactive and has a total value of $j11.802$ p-u. Of course this current must return to the power system and our experience in the preceding section tells us that for this power system the return takes place through the neutral connections made to reference ground at the Y side of transformer banks T1 and T2.

Placed in the shorthand notation of matrices, the foregoing results can be summarized to read as follows:

$$\begin{bmatrix} \bar{I}_a \\ \bar{I}_b \\ \bar{I}_c \end{bmatrix}_F = \begin{bmatrix} 1 & 1 & 1 \\ a^2 & a & 1 \\ a & a^2 & 1 \end{bmatrix}\begin{bmatrix} \bar{I}_1 \\ \bar{I}_2 \\ \bar{I}_0 \end{bmatrix}_F = \begin{bmatrix} 1 & 1 & 1 \\ a^2 & a & 1 \\ a & a^2 & 1 \end{bmatrix}\begin{bmatrix} -j5.813 \\ j1.879 \\ j3.934 \end{bmatrix}$$

$$= \begin{bmatrix} 0 \\ 8.9/\underline{138.5°} \\ 8.9/\underline{41.5°} \end{bmatrix} \text{ p-u} \quad (14\text{-}72)$$

Line voltages at fault F.   Since by Eq. (14-64) each sequence voltage has the same value, the individual value can be conveniently computed as

$$\bar{V}_{2F} = -\bar{Z}_2 \bar{I}_2 = -(j0.13)(j1.879) = 0.2443 \text{ p-u} \qquad (14\text{-}73)$$

Then again by Eq. (14-64) it follows that line a voltage at the point of the fault is

$$\bar{V}_{aF} = 3\bar{V}_{2F} = 3(0.2443) = 0.7329 \text{ p-u} \qquad (14\text{-}74)$$

The matrix representation for the line voltages thus becomes

$$\begin{bmatrix} \bar{V}_a \\ \bar{V}_b \\ \bar{V}_c \end{bmatrix}_F = \begin{bmatrix} 1 & 1 & 1 \\ a^2 & a & 1 \\ a & a^2 & 1 \end{bmatrix} \begin{bmatrix} \bar{V}_1 \\ \bar{V}_2 \\ \bar{V}_0 \end{bmatrix}_F = \begin{bmatrix} 1 & 1 & 1 \\ a^2 & a & 1 \\ a & a^2 & 1 \end{bmatrix} \begin{bmatrix} 0.2443 \\ 0.2443 \\ 0.2443 \end{bmatrix} = \begin{bmatrix} 0.7329 \\ 0 \\ 0 \end{bmatrix} \text{ p-u.} \qquad (14\text{-}75)$$

Note the 27% reduction that occurs in line a voltage. Although there is no current flowing out of line a to reference ground at the fault point, a current does flow throughout line conductor a during the fault.

14-5 THE LINE-TO-LINE FAULT

The boundary conditions associated with a line-to-line fault are readily deduced from the schematic diagram of Fig. 14-16, which shows the fault occurring at bus 2. Expressed mathematically these boundary conditions are

$$\bar{I}_{aF} = 0 \qquad \bar{I}_{bF} = -\bar{I}_{cF} \qquad \bar{V}_{bF} = \bar{V}_{cF} \qquad (14\text{-}76)$$

Figure 14-16 Schematic diagram illustrating a line to line fault occurring at bus 2.

How the sequence networks must be treated in a manner consistent with these conditions is deduced by examining the consequences of these boundary conditions on the sequence currents and voltages as they exist at the fault point. We focus first on the currents. Thus for the positive-sequence current at the fault we have

$$\bar{I}_{1F} = \tfrac{1}{3}(\bar{I}_{aF} + a\bar{I}_{bF} + a^2\bar{I}_{cF}) = \tfrac{1}{3}(0 + a\bar{I}_{bF} + a^2\bar{I}_{bF}) = \tfrac{1}{3}(a - a^2)\bar{I}_{bF} \qquad (14\text{-}77)$$

For the negative sequence the result is

$$\bar{I}_{2F} = \tfrac{1}{3}(\bar{I}_{aF} + a^2\bar{I}_{bF} + a\bar{I}_{cF}) = \tfrac{1}{3}(0 + a^2\bar{I}_{bF} + a\bar{I}_{bF}) = -\tfrac{1}{3}(a - a^2)\bar{I}_{bF} \qquad (14\text{-}78)$$

For the zero sequence we get

$$\bar{I}_{0F} = \tfrac{1}{3}(\bar{I}_{aF} + \bar{I}_{bF} + \bar{I}_{cF}) = \tfrac{1}{3}(0 + \bar{I}_{bF} - \bar{I}_{bF}) = 0 \qquad (14\text{-}79)$$

Equations (14-77) and (14-78) provide the important information that $\bar{I}_{1F} = -\bar{I}_{2F}$. In turn, this tells us the arrangement of the positive-sequence network and the negative-sequence network must be such that the current $\bar{I}_{1F}$ flows in a direction opposite to the defined positive direction of flow of the negative-sequence current. This means of course a *series* connection of the positive- and negative-sequence networks. Moreover, Eq. (14-79) states that zero-sequence current is zero and so there is no connection between it and the other two sequence networks. In short, the zero-sequence network is isolated.

An examination of the corresponding sequence voltages at the fault furnishes further insight into the configuration of the equivalent circuit for this fault condition. Proceeding as for the currents, we get

$$\bar{V}_{1F} = \tfrac{1}{3}(\bar{V}_{aF} + a\bar{V}_{bF} + a^2\bar{V}_{cF}) = \tfrac{1}{3}[\bar{V}_{aF} + (a + a^2)\bar{V}_{bF}] = \tfrac{1}{3}(\bar{V}_{aF} - \bar{V}_{bF}) \quad (14\text{-}80)$$

$$\bar{V}_{2F} = \tfrac{1}{3}(\bar{V}_{aF} + a^2\bar{V}_{bF} + a\bar{V}_{cF}) = \tfrac{1}{3}[\bar{V}_{aF} + (a + a^2)\bar{V}_{bF}] = \tfrac{1}{3}(\bar{V}_{aF} - \bar{V}_{bF}) \quad (14\text{-}81)$$

$$\bar{V}_{0F} = \tfrac{1}{3}(\bar{V}_{aF} + \bar{V}_{bF} + \bar{V}_{cF}) = \tfrac{1}{3}(\bar{V}_{aF} + 2\bar{V}_{bF}) \qquad (14\text{-}82)$$

Since the first two expressions are the same, they indicate a *parallel* connection of the positive- and negative-sequence networks. Moreover, this arrangement is also completely in agreement with the fact that the positive- and negative-sequence currents are equal in magnitude. Accordingly, by connecting the positive and negative sequences in the manner displayed in Fig. 14-17, we have a circuit that is in complete accord with Eqs. (14-77) and (14-78) as well as with Eqs. (14-80) and (14-81). Hence this is the equivalent circuit of the two-bus system for a fault at bus 2. Because the zero-sequence current is zero, so too is the zero-sequence voltage. Consequently, it follows from Eq. (14-82) that the voltage of line a at the fault is related to the voltages on lines b and c by the expression

$$\bar{V}_{aF} = -2\bar{V}_{bF} = -2\bar{V}_{cF} \qquad (14\text{-}83)$$

It is certainly not surprising that the zero sequence plays no role in this type fault because there is, after all, no connection to reference ground.

The detailed version of the equivalent circuit of Fig. 14-17 for use in computing voltages and currents throughout the power system is depicted in Fig. 14-18. Again note the isolation of the zero-sequence network from the rest of the circuit.

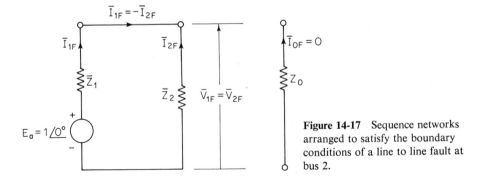

Figure 14-17 Sequence networks arranged to satisfy the boundary conditions of a line to line fault at bus 2.

Line currents and voltages at the fault point. The value of the positive-sequence current at the fault is now found to be

$$\bar{I}_{1F} = -\bar{I}_{2F} = \frac{1\underline{/0°}}{\bar{Z}_1 + \bar{Z}_2} = \frac{1\underline{/0°}}{j0.13 + j0.13} = -j3.846 \text{ p-u} \qquad (14\text{-}84)$$

and

$$\bar{I}_{2F} = j3.846 \text{ p-u}$$

The current from line b in the fault connection (see Fig. 14-16) is then

$$\bar{I}_{bF} = a^2\bar{I}_{1F} + a\bar{I}_{2F} = 3.846\underline{/150°} + 3.846\underline{/210°}$$

$$= -3.33 + j1.923 - 3.33 - j1.923 = -6.66 \text{ p-u} \qquad (14\text{-}85)$$

Similarly, the current from line c in the fault connection is

$$\bar{I}_{cF} = -\bar{I}_{bF} = 6.66 \text{ p-u} \qquad (14\text{-}86)$$

In this case the matrix formulation takes the form

$$\begin{bmatrix} \bar{I}_a \\ \bar{I}_b \\ \bar{I}_c \end{bmatrix}_F = \begin{bmatrix} 1 & 1 & 1 \\ a^2 & a & 1 \\ a & a^2 & 1 \end{bmatrix}\begin{bmatrix} \bar{I}_1 \\ \bar{I}_2 \\ \bar{I}_0 \end{bmatrix}_F = \begin{bmatrix} 1 & 1 & 1 \\ a^2 & a & 1 \\ a & a^2 & 1 \end{bmatrix}\begin{bmatrix} -j3.846 \\ j3.846 \\ 0 \end{bmatrix} = \begin{bmatrix} 0 \\ -6.66 \\ 6.66 \end{bmatrix} \text{p-u} \qquad (14\text{-}87)$$

To find the line voltages at the fault point, it is necessary first to determine the sequence voltages. From Fig. 14-17 it is clear that

$$\bar{V}_{1F} = \bar{V}_{2F} = -\bar{Z}_2\bar{I}_{2F} = -j(0.13)(j3.846) = 0.5 \text{ p-u} \qquad (14\text{-}88)$$

Hence the line voltages at the fault thus become

$$\begin{bmatrix} \bar{V}_a \\ \bar{V}_b \\ \bar{V}_c \end{bmatrix}_F = \begin{bmatrix} 1 & 1 & 1 \\ a^2 & a & 1 \\ a & a^2 & 1 \end{bmatrix}\begin{bmatrix} \bar{V}_1 \\ \bar{V}_2 \\ \bar{V}_0 \end{bmatrix} = \begin{bmatrix} 1 & 1 & 1 \\ a^2 & a & 1 \\ a & a^2 & 1 \end{bmatrix}\begin{bmatrix} 0.5 \\ 0.5 \\ 0 \end{bmatrix} = \begin{bmatrix} 1.0 \\ -0.5 \\ -0.5 \end{bmatrix} \text{p-u} \qquad (14\text{-}89)$$

Observe that $\bar{V}_{aF}$ has a fault value which is identical to the prefault value. This, too, is not unexpected in view of the fact that line a current is zero.

Of the four fault cases studied in this chapter, the line-to-line fault is the only one that does not involve the reference ground. Consequently, because of its pedagogical value, we treat this case with the same detail that applied to the line-to-ground fault. We begin with the computation of the transmission line currents.

Transmission line currents. In this connection refer to the circuitry of Fig. 14-18 where the transmission line is represented by the impedance $j0.105$ appearing between nodes 1 and 2. By current division the part of the positive-sequence fault current that flows through this impedance is

$$\bar{I}_{1L} = (0.4)\bar{I}_{1F} = (0.4)(-j3.846) = -j1.5384 \text{ p-u} \qquad (14\text{-}90)$$

where the subscript L again denotes the left side of the fault point. Similarly, for the negative sequence we have

$$\bar{I}_{2L} = (0.4)\bar{I}_{2F} = j1.5384 \text{ p-u} \qquad (14\text{-}91)$$

Figure 14-18 Detailed equivalent circuit of the two-bus power system for use in calculating currents and voltages throughout the system.

The matrix for the transmission line currents then becomes

$$\begin{bmatrix} \bar{I}_a \\ \bar{I}_b \\ \bar{I}_c \end{bmatrix}_{TL} = \begin{bmatrix} 1 & 1 & 1 \\ a^2 & a & 1 \\ a & a^2 & 1 \end{bmatrix} \begin{bmatrix} \bar{I}_1 \\ \bar{I}_2 \\ \bar{I}_0 \end{bmatrix}_L = \begin{bmatrix} 1 & 1 & 1 \\ a^2 & a & 1 \\ a & a^2 & 1 \end{bmatrix} \begin{bmatrix} -j1.5384 \\ j1.5384 \\ 0 \end{bmatrix} = \begin{bmatrix} 0 \\ -2.664 \\ 2.664 \end{bmatrix} \text{p-u} \quad (14\text{-}92)$$

These currents are shown on the system diagram of Fig. 14-19.

Y-side line currents in transformer bank T2. Here the current division factor of the parallel circuit associated with the positive-sequence impedance $\bar{Z}_1$ of Fig. 14-18 is 0.6. Accordingly, the positive-sequence current coming from the right side (R) of the fault point is

$$\bar{I}_{1R} = 0.6\bar{I}_{1F} = 0.6(-j3.846) = -j2.3076 \text{ p-u} \quad (14\text{-}93)$$

For the negative sequence it is

$$\bar{I}_{2R} = 0.6\bar{I}_{2F} = 0.6(j3.846) = j2.3076 \text{ p-u} \quad (14\text{-}94)$$

The matrix for the corresponding line currents is then

$$\begin{bmatrix} \bar{I}_a \\ \bar{I}_b \\ \bar{I}_c \end{bmatrix}_{T2} = \begin{bmatrix} 1 & 1 & 1 \\ a^2 & a & 1 \\ a & a^2 & 1 \end{bmatrix} \begin{bmatrix} \bar{I}_1 \\ \bar{I}_2 \\ \bar{I}_0 \end{bmatrix}_R = \begin{bmatrix} 1 & 1 & 1 \\ a^2 & a & 1 \\ a & a^2 & 1 \end{bmatrix} \begin{bmatrix} -j2.3076 \\ j2.3076 \\ 0 \end{bmatrix} = \begin{bmatrix} 0 \\ -3.997 \\ 3.997 \end{bmatrix} \text{p-u} \quad (14\text{-}95)$$

These currents are also indicated in the system diagram of Fig. 14-19.

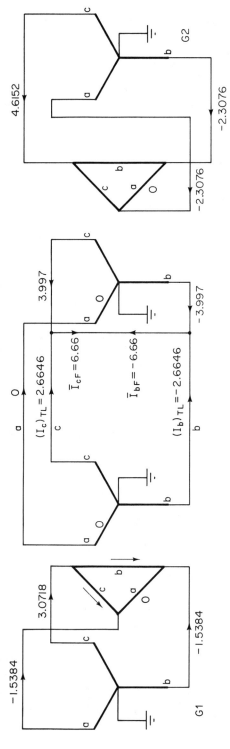

Figure 14-19 Two-bus electric power system diagram showing all line currents expressed in per unit.

Line currents at generator G1. The positive- and negative-sequence currents on the high-voltage side of the Y-connected transformers of T1 are the same as in Eqs. (14-90) and (14-91). Because the transformer bank T1 is in a Δ-Y arrangement, these positive- and negative- sequence components must be adjusted for the $\mp 30°$ displacement before they can be used on the low side of the transformer bank for the evaluation of the generator line currents. Upon inserting these adjustments, the positive- and negative-sequence currents in per unit, as they apply to the generator, thus become

$$(\bar{I}_1)_{G1} = 1.5384\underline{/-90° - 30°} = 1.5384\underline{/-120°} \text{ p-u} \qquad (14\text{-}96)$$

$$(\bar{I}_2)_{G1} = 1.5384\underline{/90° + 30°} = 1.5384\underline{/120°} \text{ p-u} \qquad (14\text{-}97)$$

This leads to the following matrix for the line currents of generator G1:

$$
\begin{bmatrix} \bar{I}_a \\ \bar{I}_b \\ \bar{I}_c \end{bmatrix}_{G1} =
\begin{bmatrix} 1 & 1 & 1 \\ a^2 & a & 1 \\ a & a^2 & 1 \end{bmatrix}
\begin{bmatrix} \bar{I}_1 \\ \bar{I}_2 \\ \bar{I}_0 \end{bmatrix}_{G1} =
\begin{bmatrix} 1 & 1 & 1 \\ a^2 & a & 1 \\ a & a^2 & 1 \end{bmatrix}
\begin{bmatrix} 1.5384\underline{/-120°} \\ 1.5384\underline{/120°} \\ 0 \end{bmatrix} =
\begin{bmatrix} -1.5384 \\ -1.5384 \\ 3.0768 \end{bmatrix} \text{ p-u}
$$
$$(14\text{-}98)$$

Refer to the system diagram of Fig. 14-19 for a schematic distribution of these currents.

Line currents at generator G2. Generator G2 is located on the right side of the fault. Accordingly, the positive- and negative-sequence currents for G2 can be obtained by applying the phase angle adjustment to the expressions for $\bar{I}_{1R}$ and $\bar{I}_{2R}$ that appear in Eqs. (14-93) and (14-94). Thus,

$$(\bar{I}_1)_{G2} = 2.3076\underline{/-90° - 30°} = 2.3076\underline{/-120°} \text{ p-u} \qquad (14\text{-}99)$$

$$(\bar{I}_2)_{G2} = 2.3076\underline{/90° + 30°} = 2.3076\underline{/120°} \text{ p-u} \qquad (14\text{-}100)$$

Therefore the values of the line currents at G2 during the fault condition are

$$
\begin{bmatrix} \bar{I}_a \\ \bar{I}_b \\ \bar{I}_c \end{bmatrix}_{G2} =
\begin{bmatrix} 1 & 1 & 1 \\ a^2 & a & 1 \\ a & a^2 & 1 \end{bmatrix}
\begin{bmatrix} 2.3076\underline{/-120°} \\ 2.3076\underline{/120°} \\ 0 \end{bmatrix} =
\begin{bmatrix} -2.3076 \\ -2.3076 \\ 4.6152 \end{bmatrix} \text{ p-u} \qquad (14\text{-}101)
$$

An examination of the current distributions for the line-to-line fault as they are displayed in Fig. 14-19 is instructive. Observe that the current in the fault connection flows from line c to line b with a value of 6.66 p-u. At the junction point at line b this current splits two ways. One part, 2.646 p-u, flows to phase coil b on the Y side of transformer bank T1 and then continues to complete a loop by flowing through phase coil c followed by a flow along the transmission line back to the fault point. The second part, 3.997 p-u, flows through phase coils b and c on the Y side of transformer bank T2 and then also completes its loop by returning to the fault point. Of course the existence of these sinusoidal, time-varying currents in the Y-sides of the transformers at banks T1 and T2 produces corresponding currents in the Δ sides of these transformers which in turn cause line currents to exist in the phase coils of the generators. At generator G1 observe that the current in line c is twice that in lines a or b. In fact, the current in line c flows through coils b and c on the

Δ side of T1 splitting equally and then leaves the Δ connection without ever passing through phase coil *a*. Each split quantity moves along lines *a* and *b* to the phase *a* and phase *b* coils of the generator. Then at the neutral of the generator the split currents combine to yield the doubled current in line *c*. Of course a similar situation takes place at the other end of the electric power system between the Δ side of transformer bank T2 and generator G2.

14-6 PROCEDURE FOR COMPUTERIZED FAULT CALCULATIONS

When fault analysis is to be made on a power system of low dimensionality (i.e., four or five buses or so), the preceding procedure based on conventional circuit analysis is entirely manageable. However, when the power system consists of many buses, other analytical procedures are used in order to exploit the advantages of the digital computer. It is the purpose of this section to outline briefly the method employed by the computer to provide the value of all fault currents and bus voltages throughout the power system.

In the treatment that follows we continue to replace the unbalanced power system that results upon the occurrence of a fault by the appropriate sequence networks. Then each sequence circuit, which is associated with an *n*-bus system, is formulated as an *n*-port network in terms of self and transfer impedances. Expressed mathematically we can write for each sequence network the following general equations:

$$
\begin{aligned}
\bar{V}_1 &= \bar{z}_{11}\bar{I}_1 + \bar{z}_{12}\bar{I}_2 + \cdots + \bar{z}_{1n}\bar{I}_n \\
\bar{V}_2 &= \bar{z}_{21}\bar{I}_1 + \bar{z}_{22}\bar{I}_2 + \cdots + \bar{z}_{2n}\bar{I}_n \\
&\;\vdots \\
\bar{V}_n &= \bar{z}_{n1}\bar{I}_1 + \bar{z}_{n2}\bar{I}_2 + \cdots + \bar{z}_{nn}\bar{I}_n
\end{aligned} \tag{14-102}
$$

where all subscripts refer to bus numbers ranging from 1 to *n*. Here $\bar{z}_{11}$ denotes the self-impedance and corresponds to the ratio of the voltage from bus 1 to reference to the current that flows into bus 1 when the current in all other buses is zero. Similar interpretations apply to each of the other impedance entries in Eq. (14-102). In matrix notation, the last equation becomes

$$
\begin{bmatrix} \bar{V}_1 \\ \bar{V}_2 \\ \vdots \\ \bar{V}_n \end{bmatrix} = \begin{bmatrix} \bar{z}_{11} & \bar{z}_{12} & \cdots & \bar{z}_{1n} \\ \bar{z}_{21} & \bar{z}_{22} & \cdots & \bar{z}_{2n} \\ \vdots & & & \\ \bar{z}_{n1} & \bar{z}_{n2} & \cdots & \bar{z}_{nn} \end{bmatrix} \begin{bmatrix} \bar{I}_1 \\ \bar{I}_2 \\ \vdots \\ \bar{I}_n \end{bmatrix} \tag{14-103}
$$

The left side is the bus voltage vector and has the dimension $n \times 1$. The current matrix is also a vector of dimension $n \times 1$. The impedance matrix has a dimension of $n \times n$ and is often called the *bus* impedance matrix for obvious reasons. Equation (14-103) can be written more compactly as

$$
\bar{V}_{\text{bus}} = \bar{Z}_{\text{bus}}\bar{I}_{\text{bus}} \tag{14-104}
$$

where the general entry for the bus impedance matrix can be expressed as

$$\bar{z}_{ij} = \frac{\bar{V}_i}{\bar{I}_j} \qquad (14\text{-}105)$$

Here $\bar{V}_i$ is the voltage appearing at the ith bus when a current I_j flows at the jth bus with all other bus currents assumed equal to zero. Observe that when $i \neq j$ the transfer impedance results whereas when $i = j$ the result is the self-impedance at the bus.

A bus impedance matrix can be identified for each sequence. It is instructive at this point to illustrate the application of the foregoing formulation for the single line-to-ground fault that is analyzed in Sec. 14-3. We focus attention solely on the zero-sequence network portion of the equivalent circuit that is displayed in Figs. 14-11(b) and 14-4(a). The element values of the bus impedance matrix are now derived based on the results obtained by the conventional circuit analysis that is used in Sec. 14-3. Recall that this is a two-bus system, i.e., $n = 2$. Hence the bus impedance matrix is 2×2. One entry in the zero-sequence bus matrix is $\bar{z}_{22}$, which relates the voltage at bus 2 to the current that flows at bus 2. Our previous analysis found this current to be $-j3.105$ p-u as shown in Eq. (14-38). The net voltage that is associated with the zero-sequence network is accordingly equal to the product of this current and the net impedance of the zero-sequence network as seen from bus 2. This impedance is the parallel combination of $(j0.072)$ and $(j0.08 + j0.3675)$.

$$\bar{z}_{22_0} = \frac{\bar{V}_{2_0}}{\bar{I}_{2_0}} = \frac{\left[\dfrac{j(0.072)(j0.4475)}{j0.5195}\right]\bar{I}_{2_0}}{\bar{I}_{2_0}} = j0.062 \qquad (14\text{-}106)$$

where the second subscript is used to denote zero-sequence quantities.

It follows by analogy that the $\bar{z}_{11}$ entry of the bus impedance matrix is the parallel combination of $(j0.08)$ with $(j0.3675 + j0.072)$ or

$$\bar{z}_{11_0} = \frac{\bar{V}_{1_0}}{\bar{I}_{1_0}} = \frac{j(0.08)(j0.4395)}{j0.5195} = j0.0677 \qquad (14\text{-}107)$$

Note here that actual values for $\bar{V}_{1_0}$ and $\bar{I}_{1_0}$ are not used because these quantities were not determined in Sec. 14-3. The foregoing element value is useful when the fault occurs at bus 1.

To find element $\bar{z}_{12}$ we formulate the ratio of the voltage at bus 1, $\bar{V}_{1_0}$, associated with a current flow at bus 2 of $\bar{I}_{2_0}$ based on the assumption that at bus 1, $\bar{I}_{1_0}$ is zero. The current that flows at bus 1 for a fault that occurs at bus 2 is obtained by using the current division factor shown in parentheses in Fig. 14-11(b). Accordingly, the voltage of bus 1 relative to its neutral is expressed as

$$\bar{V}_{1_0} = -j(0.08)[(0.139)\bar{I}_{2_0}] = -j(0.08)(0.139)\bar{I}_{2_0} = -j0.0111\bar{I}_{2_0}$$
$$= -\bar{z}_{12}\bar{I}_{2_0} \qquad (14\text{-}108)$$

The quantity in brackets represents the branch current where bus 1 is located. It follows then that

$$\bar{z}_{12_0} = j0.0111 \text{ p-u} \qquad (14\text{-}109)$$

A similar procedure is employed to find $\bar{z}_{21_0}$ except that now a fault is assumed to occur at bus 1 with zero current flowing out of bus 2. If we call the zero-sequence current that flows out of bus 1, $\bar{I}_{1_0}$, then the corresponding voltage of bus 2 can be written in terms of $\bar{I}_{1_0}$ by again using current division. Thus,

$$\bar{V}_{2_0} = -j(0.072)\left(\frac{0.08}{0.5195}\bar{I}_{1_0}\right) = -j(0.072)(0.154)\bar{I}_{1_0} = -j0.0111\bar{I}_{1_0}$$

$$= -\bar{z}_{21}\bar{I}_{1_0} \qquad (14\text{-}110)$$

Here the bracket quantity denotes the branch current where bus 2 is located. Hence,

$$\bar{z}_{21_0} = j0.0111 \text{ p-u} \qquad (14\text{-}111)$$

Consequently, the zero-sequence bus impedance matrix for this two-bus system can be represented as

$$\bar{\mathbf{Z}}_{\text{bus}_0} = \begin{bmatrix} \bar{z}_{11} & \bar{z}_{12} \\ \bar{z}_{12} & \bar{z}_{22} \end{bmatrix} = j\begin{bmatrix} 0.0677 & 0.0111 \\ 0.0111 & 0.062 \end{bmatrix} \qquad (14\text{-}112)$$

Whenever knowledge of a bus matrix such as this one is available, the bus voltages can easily be determined from use of the relationship displayed in Eq. (13-109). Thus, for the zero-sequence network we have

$$\bar{\mathbf{V}}_{\text{bus}_0} = -\bar{\mathbf{Z}}_{\text{bus}_0}\bar{\mathbf{I}}_{\text{bus}_0} \qquad (14\text{-}113)$$

The negative sign is included to be consistent with the notation of Fig. 14-4(b). Alternatively,

$$\begin{bmatrix} \bar{V}_1 \\ \bar{V}_2 \end{bmatrix}_0 = -\begin{bmatrix} \bar{z}_{11} & \bar{z}_{12} \\ \bar{z}_{21} & \bar{z}_{22} \end{bmatrix}\begin{bmatrix} \bar{I}_1 \\ \bar{I}_2 \end{bmatrix}_0 \qquad (14\text{-}114)$$

If we now assume a fault to occur at bus 2, then Eq. (14-114) is particularized to represent this situation by rewriting it as

$$\begin{bmatrix} \bar{V}_1 \\ \bar{V}_2 \end{bmatrix}_0 = -\begin{bmatrix} z_{11} & z_{12} \\ z_{21} & z_{22} \end{bmatrix}_0\begin{bmatrix} 0 \\ \bar{I}_2 \end{bmatrix}_0 = -j\begin{bmatrix} 0.0677 & 0.0111 \\ 0.0111 & 0.062 \end{bmatrix}_0\begin{bmatrix} 0 \\ \bar{I}_2 \end{bmatrix}_0 \qquad (14\text{-}115)$$

In other words, a current is assumed to flow *into* bus 2 alone; there is no current flowing *into* any of the unfaulted buses. Accordingly, the zero-sequence voltages at buses 1 and 2 can be conveniently represented as

$$\bar{V}_{1_0} = -j0.0111\bar{I}_{2_0} = -\bar{z}_{12_0}\bar{I}_{2_0} \qquad (14\text{-}116)$$

$$\bar{V}_{2_0} = -j0.062\bar{I}_{2_0} = -\bar{z}_{22_0}\bar{I}_{2_0} \qquad (14\text{-}117)$$

It is important to understand the distinction between Eqs. (14-116) and (14-108) even though they represent the same quantity. Equation (14-108) is written based on the premise that the *branch* current that involves bus 1 is known and therefore can be used to find the bus 1 voltage. On the other hand, Eq. (14-116) is written based on the assumption that the bus impedance matrix can be determined *independently of the use of conventional circuit theory*, i.e., there is no need to determine branch currents before the bus voltages are found. Rather, it is necessary merely to find the zero-sequence current at the faulted bus and then to multiply by the appropriate transfer impedances of the bus matrix in order to identify properly the various bus voltages. In fact, a computer solution of the currents and voltages appearing throughout a faulted power system is based on the independently determined bus impedance matrices for the zero, positive, and negative sequences.

The foregoing procedure, which was confined to the zero-sequence network, is equally applicable to the negative-sequence and positive-sequence networks. In other words, it is possible to identify an appropriate negative-sequence bus impedance matrix $\bar{Z}_{bus_2}$ and a positive-sequence bus impedance matrix $\bar{Z}_{bus_1}$. Frequently, these two matrices are identical. Moreover, the Kirchhoff voltage law for the negative sequence quantities is expressed by

$$\bar{V}_{bus_2} = -\bar{Z}_{bus_2}\bar{I}_{bus_2} \qquad (14\text{-}118)$$

where the subscript 2 denotes negative sequence and where again the minus sign is introduced in recognition of the way in which the positive directions for the negative-sequence voltage and current were defined [see Fig. 14-3(b)]. In the case of positive-sequence quantities the expression for the bus vector voltage is given by

$$\bar{V}_{bus_1} = \bar{E} - \bar{Z}_{bus_1}\bar{I}_{bus_1} \qquad (14\text{-}119)$$

This equation is written on the assumption that all prefault currents are negligible. Consequently, all the generated voltages in the power system have the same magnitude and phase which thus permits them to be placed in parallel and represented by a single voltage source.

Algorithm for the independent determination of the bus impedance matrix. We now demonstrate how the zero-sequence bus impedance matrix for the two-bus case displayed in Eq. (14-112) can be found by the use of algorithms that allow the power system equivalent circuit to be built for each sequence by starting with a single bus and branch. We choose not to derive these algorithms but rather simply to cite them in the interest of brevity. The building process is characterized by a step-by-step procedure which admirably befits it for computer evaluation by programming.

In the description that follows the expression for the bus impedance matrix must be modified to account for the addition of branches and buses as the equivalent circuit for a given sequence is constructed. The process is continued until all buses and branches are included. As each step is introduced, a new modified bus impedance matrix results. To distinguish one modification from the next, use is made of the notation $\bar{Z}_{mod_1}$, $\bar{Z}_{mod_2}$, etc. Here $\bar{Z}_{mod_2}$ denotes the second modification. It is evaluated in terms of the elements of $\bar{Z}_{mod_1}$ and the effects of introducing additional branches and/or buses.

For the case at hand we start the process of building the bus impedance matrix by placing the branch of $j0.08$ between bus 1 and the neutral reference as called for by the zero-sequence network of Fig. 14-4(a) for the two-bus system. This first step in the building process is illustrated in Fig. 14-20(a). Of course for a single element the bus impedance is a scalar quantity represented by

$$\bar{Z}_{start} = j0.08 = \bar{Z}_{pre_1} \qquad (14\text{-}120)$$

Addition of a new bus with branch to reference. Next we add the second bus, 2, as well as branch, $j0.072$, connected to reference in the manner depicted in Fig. 14-20(b). The algorithm that applies for this type of modification is given by

Figure 14-20 Illustrating the steps in the determination of the zero-sequence bus impedance matrix for the two-bus system of Fig. 14-1.

$$\bar{\mathbf{Z}}_{\text{mod}_1} = \begin{bmatrix} \bar{\mathbf{Z}}_{\text{pre}_1} & 0 \\ \hline 0 & \bar{\mathbf{Z}}_b \end{bmatrix} \tag{14-121}$$

where $\bar{\mathbf{Z}}_b$ denotes the impedance of the branch that is being added from the new bus to reference, and $\mathbf{Z}_{\text{pre}_1}$ refers to the impedance matrix of the *preceding* step. Thus the updated version of the bus impedance matrix now becomes

$$\bar{\mathbf{Z}}_{\text{mod}_1} = \begin{bmatrix} j0.08 & 0 \\ 0 & j0.072 \end{bmatrix} \tag{14-122}$$

Addition of a branch between two existing buses. For our simple two-bus case there is just one more modification needed to complete the zero-sequence equivalent circuit. It is to add a branch, $j0.3675$, between the two existing buses 1 and 2. The algorithm for a modification of this type is expressed by

$$\bar{\mathbf{Z}}_{\text{mod}_2} = \bar{\mathbf{Z}}_{\text{mod}_1} + \frac{\mathbf{AB}}{\bar{\mathbf{Z}}_b + \bar{z}_{ii} + \bar{z}_{jj} + 2\bar{z}_{ij}} \tag{14-123}$$

where

$$\mathbf{A} = \begin{bmatrix} \bar{z}_{ii} & -\bar{z}_{ij} \\ \cdot & \cdot \\ \cdot & \cdot \\ \cdot & \cdot \\ \bar{z}_{ki} & -\bar{z}_{kj} \end{bmatrix} \tag{14-124}$$

and

$$\mathbf{B} = \bar{z}_{j1} - \bar{z}_{i1}, \ldots, \bar{z}_{jk} - \bar{z}_{ik} \tag{14-125}$$

In Eq. (14-123) the subscripts i and j denote the two existing buses to which the new branch is connected. Moreover, the quantities $\bar{z}_{ii}$, $\bar{z}_{jj}$, and $\bar{z}_{ij}$ are the corresponding elements of $\bar{\mathbf{Z}}_{\text{mod}_1}$.

Before evaluating the elements of $\bar{\mathbf{Z}}_{\text{mod}_2}$, we identify the following data:

$$\left.\begin{aligned} i &= 1 \qquad j = 2 \\ \bar{z}_{ii} &= \bar{z}_{11} = j0.08 \\ \bar{z}_{ij} &= \bar{z}_{12} = 0 = \bar{z}_{21} \\ \bar{z}_{jj} &= \bar{z}_{22} = j0.072 \\ \bar{\mathbf{Z}}_b &= j0.3675 \end{aligned}\right\} \text{from } \bar{\mathbf{Z}}_{\text{mod}_1} \tag{14-126}$$

Furthermore,

$$\mathbf{A} = \begin{bmatrix} \bar{z}_{11} - \bar{z}_{12} \\ \bar{z}_{21} - \bar{z}_{22} \end{bmatrix} = \begin{bmatrix} j0.08 \\ -j0.072 \end{bmatrix} \qquad (14\text{-}127)$$

and

$$\mathbf{B} = [\bar{z}_{21} - \bar{z}_{11}, \bar{z}_{22} - \bar{z}_{12}] = [-j0.08 \quad j0.072] \qquad (14\text{-}128)$$

Also,

$$\bar{Z}_b + \bar{z}_{11} + \bar{z}_{22} - 2\bar{z}_{12} = j(0.3675 + 0.08 + 0.072) = j0.5195 \qquad (14\text{-}129)$$

Inserting the results displayed in Eqs. (14-122), (14-127), (14-128), and (14-129) into Eq. (14-123) yields

$$\bar{Z}_{mod_2} = \bar{Z}_{bus_0} = j\begin{bmatrix} 0.08 & 0 \\ 0 & 0.072 \end{bmatrix} + j\frac{1}{0.5195}\begin{bmatrix} 0.08 \\ -0.072 \end{bmatrix}[-0.08 \quad 0.072]$$

$$= j\begin{bmatrix} 0.08 & 0 \\ 0 & 0.072 \end{bmatrix} + j\begin{bmatrix} -0.0123 & 0.0111 \\ 0.0111 & -0.01 \end{bmatrix}$$

$$= j\begin{bmatrix} 0.0677 & 0.0111 \\ 0.0111 & 0.062 \end{bmatrix} \qquad (14\text{-}130)$$

A comparison of Eq. (14-130) with Eq. (14-112) shows the results to be identical. Therefore, this is the zero-sequence bus impedance matrix which can then be used in conjunction with Eq. (14-113) to evaluate all the zero-sequence bus voltages. The only additional information needed to perform these calculations is the zero-sequence current at the faulted bus which is routinely determined. The advantage of having all the bus voltages available is that the branch current for the given sequence is readily evaluated from the simple expression

$$\bar{I}_{q_0} = \frac{\bar{V}_{i_0} - \bar{V}_{j_0}}{\bar{Z}_{q_0}} \qquad (14\text{-}131)$$

where i and j denote two buses between which is connected the impedance $\bar{Z}_{q_0}$.

Again keep in mind that the foregoing procedures apply with equal validity to the determination of the bus impedance matrices for the positive- and negative-sequence equivalent circuits. Once these sequence components are known at any point in the system, the corresponding actual currents and voltages are routinely determined.

Algorithm for addition of a new branch from a new bus to an existing bus. The simplicity of the two-bus case just concluded did not require invoking two additional algorithms to cover cases that invariably occur in constructing the bus impedance matrix of complex power systems. One is the computational procedure associated with the addition of a new branch between a new bus and an already existing bus. In such cases the algorithum is described by

$$\bar{Z}_{mod} = \begin{bmatrix} & & & \bar{z}_{1J} \\ & \bar{Z}_{pre} & & \bar{z}_{2J} \\ & & & \vdots \\ & & & \bar{z}_{kJ} \\ \hline \bar{z}_{J1}\bar{z}_{J2} & \cdots & \bar{z}_{Jk} & \bar{z}_{JJ} + \bar{z}_B \end{bmatrix} \qquad (14\text{-}132)$$

where j denotes the new bus and $\mathbf{Z}_{\text{pre}}$ is the bus impedance matrix as it appears in the preceding step of the construction process.

Algorithm for addition of new branch from existing bus to reference. In this situation the appropriate algorithm is described by

$$\bar{\mathbf{Z}}_{\text{mod}} = \bar{\mathbf{Z}}_{\text{pre}} - \frac{1}{\bar{z}_{jj} + \bar{Z}_b} \begin{bmatrix} \bar{z}_{1j} \\ \cdot \\ \cdot \\ \cdot \\ \bar{z}_{nj} \end{bmatrix} [\bar{z}_{j1} \quad \cdots \quad \bar{z}_{jn}] \tag{14-133}$$

where j is the existing bus to which the branch is joined.

Application of bus impedance matrix to evaluation of single line-to-ground fault. As a means of illustrating the use of the computer in making fault calculations, attention is here directed to the case of the single line-to-ground fault. Of course in any complete program for fault calculations all four fault cases are programmed and suitably identified for callout.

One of the first tasks of the program is to provide for the topological determination by the computer of the positive-, negative- and zero-sequence bus impedance matrices $\mathbf{Z}_{\text{bus}_1}$, $\mathbf{Z}_{\text{bus}_2}$, and $\mathbf{Z}_{\text{bus}_0}$ by implementing the procedures already outlined. Then proceeding on the assumption that a solid single line-to-ground fault occurs at bus k, the computer is further programmed to evaluate the sequence currents at the point of the fault in the fashion described by Eq. (13-38) but now generalized as

$$\bar{I}_{k_1} = \bar{I}_{k_2} = \bar{I}_{k_0} = \frac{\bar{E}}{\bar{z}_{kk_1} + \bar{z}_{kk_2} + \bar{z}_{kk_0}} \tag{14-134}$$

Observe that only the self-impedance element of each sequence matrix at the kth bus is employed in this computation and that each sequence current is the same. Next the program is extended to include the evaluation of all bus voltages for each sequence. Thus, the positive-sequence set is found by invoking Eq. (14-119), which leads to

$$\bar{\mathbf{V}}_{\text{bus}_1} = \bar{\mathbf{E}} - \begin{bmatrix} \bar{z}_{11_1} & \cdots & \bar{z}_{1k_1} & \cdots & \bar{z}_{1n_1} \\ \cdot & & \cdot & & \cdot \\ \cdot & & \cdot & & \cdot \\ \cdot & & \cdot & & \cdot \\ \bar{z}_{k1_1} & \cdots & \bar{z}_{kk_1} & \cdots & \bar{z}_{kn_1} \\ \cdot & & \cdot & & \cdot \\ \cdot & & \cdot & & \cdot \\ \cdot & & \cdot & & \cdot \\ \bar{z}_{n1_1} & \cdots & \bar{z}_{nk_1} & \cdots & \bar{z}_{nn_1} \end{bmatrix} \begin{bmatrix} 0 \\ \cdot \\ \cdot \\ \cdot \\ \bar{I}_{k_1} \\ \cdot \\ \cdot \\ \cdot \\ 0 \end{bmatrix} \tag{14-135}$$

Note that for the current vector all elements are zero except for the bus at which the fault occurs. Upon performing the matrix multiplication and inserting Eq. (14-134) for I_{k_1}, the last expression becomes

$$\bar{\mathbf{V}}_{i_1} = \mathbf{E} - \frac{\bar{z}_{ik_1}\bar{\mathbf{E}}}{\bar{z}_{kk_1} + \bar{z}_{kk_2} + \bar{z}_{kk_0}} = \frac{\bar{z}_{kk_1} + \bar{z}_{kk_2} + \bar{z}_{kk_0} - \bar{z}_{ik_1}}{\bar{z}_{kk_1} + \bar{z}_{kk_2} + \bar{z}_{kk_0}} \mathbf{E} \tag{14-136}$$

Here i is an index that goes from 0 to n.

In a similar manner the set of negative-sequence bus voltages is obtained from Eq. (14-118). Thus,

$$\bar{V}_{bus_2} = - \begin{bmatrix} \bar{z}_{11_2} & \cdots & \bar{z}_{1k_2} & \cdots & \bar{z}_{1n_2} \\ \cdot & & \cdot & & \cdot \\ \cdot & & \cdot & & \cdot \\ \cdot & & \cdot & & \cdot \\ \bar{z}_{k1_2} & \cdots & \bar{z}_{kk_2} & \cdots & \bar{z}_{kk_2} \\ \cdot & & \cdot & & \cdot \\ \cdot & & \cdot & & \cdot \\ \cdot & & \cdot & & \cdot \\ \bar{z}_{n1_2} & \cdots & \bar{z}_{nk_2} & \cdots & \bar{z}_{nn_2} \end{bmatrix} \begin{bmatrix} 0 \\ \cdot \\ \cdot \\ \cdot \\ \bar{I}_{k_2} \\ \cdot \\ \cdot \\ \cdot \\ 0 \end{bmatrix} \qquad (14\text{-}137)$$

Or, after expansion and collecting of terms, there results

$$\bar{V}_{i_2} = -\bar{z}_{ik_2}\bar{I}_{k_2} = \frac{-\bar{z}_{ik_2}\bar{E}}{\bar{z}_{kk_1} + \bar{z}_{kk_2} + \bar{z}_{kk_0}} \qquad (14\text{-}138)$$

Finally from Eq. (14-113) the set of zero-sequence bus voltages can be expressed as

$$\bar{V}_{i_0} = -\bar{z}_{ik_0}\bar{I}_{k_0} = \frac{-\bar{z}_{ik_0}\bar{E}}{\bar{z}_{kk_1} + \bar{z}_{kk_2} + \bar{z}_{kk_0}} \qquad (14\text{-}139)$$

The branch currents for each sequence can next be determined by invoking Eq. (14-131). This equation is specifically written for the zero sequence, but it is equally applicable for the positive and negative sequences upon modifying the second subscript.

In those parts of the electric power system that contain Y-Δ transformer connections the computer program must be furnished with a routine that permits the ±30° phase adjustments to be made in the line currents. For further details concerning this matter, consult the bibliography in Appendix G.

14-7 PROTECTION FOR ELECTRIC POWER SYSTEM COMPONENTS

Experience with the preceding material in this chapter makes it apparent that currents many times the rated values of system components can exist for the various fault types. Unless such currents are interrupted soon after their appearance, severe damage to expensive equipment can occur as well as prolonged interruptions of service, which can be costly to all involved. Consequently, a primary objective of the installation of protective devices on a power system is to prevent damage to equipment, but it is equally important that these devices be so designed and arranged that they isolate only the faulted section of the system in order to minimize the extent of the outage caused by the fault.

Power system protection is achieved through the use of devices which are capable of detecting abnormal conditions on the electric power system. These abnormalities can take the form of excessive current flow (frequently referred to as an overload or overcurrent), overvoltage, undervoltage, and underfrequency, or a combination of these conditions. Detection is customarily fulfilled by the action of appropriately selected relays which are preset to interrupt a given part of the system whenever the threshold value of the system variable is exceeded. Of course the power system variables of chief interest are current, voltage and frequency. Two additional, derived variables are power and impedance. Relays are available which are designed to respond to each of these quantities. However, these relays are not constructed to respond directly to the elevated voltages and currents that reside throughout the power system. Rather, the relays are supplied from the output terminals of instrument transformers which, besides simplifying the design of the relays, also provides the necessary isolation of the instrumentation from the power system itself.

Instrument transformers. There are two types of instrument transformers: the current transformer and the potential transformer. It is the purpose of the current transformer to provide information about the current flow at any location in the power system in conjunction with a suitable overcurrent relay situated at that location. In construction the current transformer is not unlike the units studied in Chapter 2. However, there are differences in details. For example, the primary winding frequently consists of a single turn which is series connected with the equipment it is selected to protect. Refer to Fig. 14-12(a). The secondary coil consists of many turns; the actual number depends on the desired turns ratio. The secondary coil is always rated for 5 A. Hence, if the primary current to be detected is 1000 A, a turns ratio of 200:1 is needed. As a general rule instrument transformers are larger in size, heavier, and more costly than comparable conventional transformers of the same volt-ampere rating. Special care is taken in the design of these units to reduce the effect of the winding resistance drop, the leakage reactance drops, and the magnetizing current. Accordingly, the windings are designed for lower current density which means more copper, lower leakage reactance achieved through additional subdivision of the coils, and lower magnetizing current, which is accomplished by the use of higher-permeability iron of greater cross-sectional area. All this adds to cost and size.

The potential transformer furnishes information about the magnitude of voltage at any point in the power system where it is installed. The secondary winding has a standard secondary voltage rating of 120 V and it is used in the manner illustrated in Fig. 14-21(b). The number of turns of the primary winding depends upon the size of primary voltage to be measured. Standard turns ratios up to 4500 are commercially available; hence primary voltages up to 500 kV can be conveniently measured.

L1

CT

5–A ammeter
or relay coil

(a)

L1

120–V voltmeter
or relay coil

L2   PT

(b)

Figure 14-21 Instrument transformers: (a) the current transformer; (b) the potential transformer.

Relays

A relay is a switching device which becomes activated when a preset value of the response variable is exceeded. Relays that respond to excessive currents are called *overcurrent* or *overload* relays. There are also voltage relays that react to an overvoltage or to an undervoltage condition depending upon its particular design. Also to be found in the role of protecting an electric power system are such special purpose relays as the *differential* relay, the *directional* relay and the *distance* relay.

In order to be successful in its role as a "silent sentinel" protecting the power system, a relay must exhibit two salient features. One, it must be reliable. This means that it must be capable of performing its assigned function even after years of idleness. Two, it must be selective so that it responds to truly serious abnormalities and not to ephemeral, harmless ones.

Overcurrent protection. An example of protection of the stator winding of a synchronous generator through the use of current transformers, an overcurrent relay, and a circuit breaker is illustrated in Fig. 14-22. Observe that a CT is used for each line. The overcurrent relay is preset to protect the generator against the harmful effects of an overload current, which is usually determined after fault analyses are made. When the preset current magnitude is exceeded and a preadjusted time delay takes place, one or several OC relay coils are energized. This in turn closes one or more of the contactors A, B, C, which then applies excitation to the trip coil of the main circuit breaker CB. Once energized, the line contactors of the circuit breaker are opened, thereby isolating the generator from the remainder of the power system.

Figure 14-22 Illustrating overcurrent protection of a synchronous generator through use of CTs, overcurrent relay, and a circuit breaker.

Short-circuit protection. How can the synchronous generator be protected from difficulty that arises internally rather than externally? For example, how can the synchronous generator be isolated should a short circuit occur in phase *c* coil? Protection from this type of fault is readily available by the use of a differential relay. The arrangement is illustrated in Fig. 14-23 but the contactors of the circuit breaker are not shown for simplicity. The output from two CTs is used to drive currents through the relay coil in opposite directions. Since the current coming from the current transformers is the *same* phase current, the net effect in the relay coil R is cancellation. Hence contactor R remains in the normally opened position. When a short circuit occurs in the phase coil, e.g., as a result of insulation breakdown, the output currents of the CTs will be unbalanced, which in turn causes relay coil R to be energized, thus closing contactor R and subsequently tripping the line circuit breaker.

Figure 14-23 Illustrating the application of a differential relay to provide short-circuit protection to a synchronous generator. Details for one phase only are shown.

This technique of employing the differential relay in the manner just described can also be used to detect short circuits in the primary and secondary coils of transformers. One CT is placed on the primary side and the other on the secondary side. An adjustment is made for the turns ratio of the power transformer by using a different tap position on one of the CTs in order that equal but opposite currents be fed to the differential relay under normal circumstances. This procedure can also be extended for the protection of buses and similar equipment.

Coordinated protection of a power system: use of directional and distance relays. The weakest link in an electric power system from the viewpoint of susceptibility to faults is the transmission line. Because these lines must traverse terrains of all kinds and be subjected to abuses from nature, people, and animals, the majority of power system faults originate with the transmission lines. Accordingly, special consideration is directed toward their protection in the interest of limiting to the greatest extent possible the interruption of service along the system. This matter can be conveniently illustrated by considering the simple electric power system depicted in Fig. 14-24. It is a radial two-bus system with a long transmission line segment that

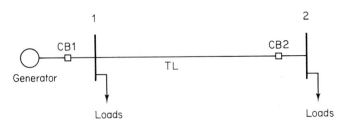

1   2

CB1   CB2

Generator   TL

Loads   Loads

Figure 14-24   Line diagram of a two-bus power system.

connects bus 1 to bus 2. Overload protection can be realized by employing overcurrent relays with circuit breakers CB1 and CB2. It should be obvious that if a fault occurs at bus 2, it would be unwise to have both circuit breakers respond by tripping. Such action would needlessly remove electric power service to the loads connected at bus 1. On the other hand, it is prudent to alert CB1 to the presence of trouble at bus 2 so that it may take action in the event of the failure of CB2 to clear the fault. Because CB2 is closer to the fault, matters are arranged to allow it to trip almost instantaneously to clear the fault. If it fails to do so, then CB1 can assume the task through the action of a *distance relay*.

The distance relay is designed to respond to the ratio of voltage to current, which is impedance. For this reason these relays are also called impedance or *ratio relays*. Because both voltage and current are needed, such relays require the use of a potential transformer as well as a current transformer. The distance relay is designed on the premise that in general the closer the CB is located to a fault point, the lower is the impedance between it and the fault location. Referral to the fault equivalent circuit diagrams of the preceding sections makes this statement self-evident. In application the distance relay is preset to a value $\bar{Z}_p$. Whenever a fault impedance $|\bar{Z}_F|$ is less than $|\bar{Z}_p|$, the circuit breaker trips. Thus, when a fault occurs at bus 2 in the system of Fig. 14-24, the distance relay associated with CB2 measures an impedance $|\bar{Z}_F|_{CB2}$ while simultaneously the distance relay of CB1 measures an impedance of $|\bar{Z}_F|_{CB1}$. Because $|\bar{Z}_F|_{B2} < |\bar{Z}_F|_{B1} < |\bar{Z}_p|$, the control circuitry of the relays is arranged to allow CB2 the first opportunity to clear the fault. If CB2 fails to function properly, then, after a predetermined time delay, CB1 is permitted to accomplish the task. The use of distance relays provides the power system with backup protection. In short, there is a primary as well as a secondary level of protection.

The coordination of power system protection, however, is not complete without the additional use of *directional relays*. The situation depicted in Fig. 14-25(a) illustrates the problem. A fault that occurs to the left of bus 2 at F easily activates the overcurrent relays that control circuit breakers CB2, CB2′, and CB2″. It is certainly undesirable and unnecessary for CB2′ and CB2″ to be tripped since this means removing service to loads L2 and L3. Because of the proximity of the fault to each of these circuit breakers, the distance relays can not provide the required discrimination. However, by combining the action of a directional relay with that of the overcurrent relay, the desired selectivity can be achieved.

(a)

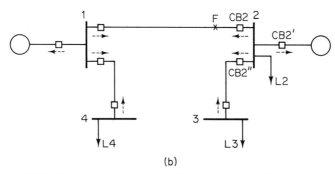

(b)

Figure 14-25 Four-bus power system: (a) system protection includes overcurrent relays and distance relays; (b) system protection extended to include directional relays as well. The dashed-line arrows indicate the zones of protection selectively provided by the directional relays.

The design of the directional relay is based on the phase of the fault current relative to the line voltage at the location of the protective device. The current is considered to be positive when the fault occurs at such a location that a lagging current ensues in the forward direction, i.e., toward its zone of protection. The choice of lagging current is consistent with the inductive reactive nature of a faulted power system. The action of such a current succeeds in energizing the directional relay. However, when a fault occurs on a side that is away from its zone of protection, the associated current relative to the line voltage at the relay coil takes on a reversed phase displacement. This current fails to energize the relay coil. In this way the relay is made to operate with a directional bias.

In Fig. 14-25(b), the forward directions of the zones of protection associated with the directional relays are indicated by the dashed-line arrows. With the introduction of this additional feature in the protection of this power system, it is plain that a fault that occurs at F acts to trip only circuit breaker CB2 at bus 2. Figure 14-26 depicts a more detailed diagram that illustrates the coordinated action of the overcurrent (OC) relay with the directional (D) relay. Observe that selectivity now involves the logic AND operation of Boolean algebra.

Figure 14-26 Circuit diagram illustrating coordinated control of an overcurrent relay with a directional relay. OC, overcurrent relay; D, directional relay; CB, circuit breaker.

PROBLEMS

14-1. In the system of Fig. 14-1, determine the per-unit values of all parameters for a reference base of MVA = 100 and kV = 69.

14-2. Repeat Prob. 14-1 for reference base quantities of MVA = 100 and kV = 100.

14-3. The single-line diagram of a two-bus power system is shown in Fig. P14-3. The transmission line has a length of 50 mi. The positive-sequence line reactance in Ω/mi is known to be 0.75. The corresponding zero-sequence value is 2.4. The other system components are rated as follows:

Generator: 100 MVA,   24 kV,   15% for x_d'', 5% for x_l
Motor: 50 MVA,   13.2 kV, 20% for x_d'', 6% for x_l
Transformer bank T1: 150 MVA, 161/24 kV,   10%
Transformer bank T2: 100 MVA, 168/14.4 kV, 8%

(a) Find all the parameter values of this power system expressed in per unit using a reference base of 100 MVA, 161 kV on the high-voltage side.

(b) Draw the positive-sequence network for this system.

Figure P14-3

(c) Draw the negative-sequence network.

(d) Draw the zero-sequence network. Assume that transformer bank T1 is connected in Δ-Y with grounded neutral on the high-voltage side and that T2 is connected Y-Y both with a neutral connection to ground.

14-4. In the four-bus system depicted in Fig. P14-4, all component parameters are expressed in ohms using reference base quantities of 100 MVA, 110 kV. Observe that no generator ground connections are used. Assume that all generated emfs are equal and in-phase, and that all resistances and all shunt admittances (including passive loads) are negligible.

(a) Express all the ohmic parameter values in per unit.

(b) Develop the positive-sequence equivalent circuit for this system showing all parameter values in per unit.

(c) Determine the equivalent positive-sequence impedance for a fault at bus 2.

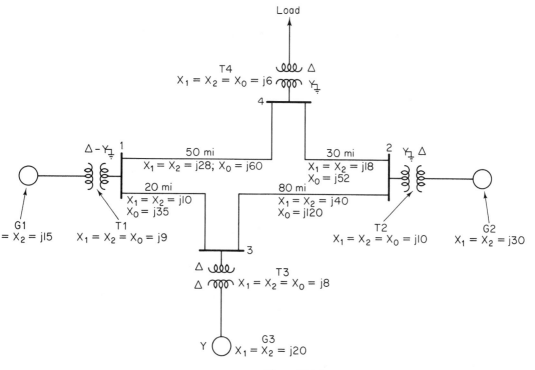

Figure P14-4

14-5. Refer to the power system depicted in Fig. P14-4.

(a) Develop the negative-sequence equivalent circuit.

(b) Find the equivalent negative-sequence impedance for a fault at bus 2.

(c) Repeat part (b) when a fault occurs at bus 1.

14-6. Refer to the electric power system shown in Fig. P14-4 and determine the following:

(a) The zero-sequence equivalent circuit.

(b) The zero-sequence equivalent impedance for a fault at bus 2.

(c) The zero-sequence equivalent impedance for a fault at bus 1.

14-7. Assume that the motor in the system of Fig. P14-3 is operating as a generator having the same magnitude and phase as generator G1. A symmetrical three-phase short circuit occurs at bus 2. Find the values of the three line currents at the fault in per unit as well as amperes. Neglect all prefault currents and assume the generated emfs to be $1/0°$ p-u.

14-8. A symmetrical three-phase short circuit occurs at bus 2 in the power system of Fig. P14-4.

(a) Calculate the currents that flow from each line to the fault point.

(b) Obtain the value of the voltage appearing at bus 4 corresponding to the fault condition.

14-9. The single-line diagram of a two-bus three-phase system is depicted in Fig. P14-9. The transmission line has a series reactance of 100 Ω. The remaining equipment is rated as follows:

Generator: 20 MVA, 15 kV, $x_d'' = 30\%$

Motor: 10 MVA, 12 kV, $x_d'' = 25\%$

Transformer bank T1: 20 MVA, 13/130 kV, $X_e = 10\%$

Transformer bank T2: 10 MVA, 7.5 kV/75 kV, $X_e = 15\%$

Passive load: 10 MVA, 12 kV, pf $= 1$

Draw the positive-sequence equivalent circuit with all quantities including the passive elements expressed in per unit. Assume that line 2 is operated at 12 kV and that the system reference base is 100 MVA at 10 kV.

Figure P14-9

14-10. A symmetrical three-phase short-circuit occurs on the high-voltage side of transformer bank T1 located at bus 2. Assume that the parameter values expressed in terms of a reference base of 100 MVA at 10 kV are as follows:

Generator: $x_d'' = 0.3$ p-u

Transformer T1: $X_e = 0.1$ p-u

Transformer T2: $X_e = 0.1$ p-u

Motor: $x_d'' = 0.4$ p-u

Transmission line: $X = 0.1$ p-u

Consider further that the passive load is disconnected and that the motor is drawing 10 MVA with a lagging power factor of 0.8 at a bus voltage of 12 kV immediately before the fault occurs.

(a) Find the fault current in amperes.

(b) Compute the current that flows through circuit breaker A in amperes.

(c) Compute the current through circuit breaker B.

14-11. A single line-to-ground fault occurs at bus 2 in the system of Fig. P14-3 when the motor is arranged to operate as a generator in a fashion identical to G1 and having a prefault emf of $1/0°$.

(a) Find the line-to-ground fault current.

(b) Determine the transmission line currents during the fault condition.

(c) Compute generator G1 line currents and determine the degree of overload that occurs in this power system component expressed in percent.

(d) Repeat part (c) for the other generator.

(*Note:* This problem is best solved in conjunction with Prob. 14-3.)

14-12. (*Note:* This problem is best assigned after Probs. 14-4 to 14-6.) A single line-to-ground fault occurs at bus 2 in the power system of Fig. P14-4.

(a) Find the magnitude of the fault current.

(b) Determine the transmission line currents along the 30-mi span between bus 2 and bus 4.

(c) Repeat part (b) for the transmission line currents between bus 3 and bus 2.

(d) How much of the total fault current enters the power system at the neutral of transformer bank T2?

14-13. The equivalent circuit of the power system of Fig. 14-1 for a double line-to-ground fault is shown in Figs. 14-14 and 14-15. The line currents at the fault are computed and displayed in Eqs. (14-70) and (14-71). In the interest of further developing the analysis of this case, determine the following items:

(a) The transmission line currents in p-u and amperes.

(b) The transformer currents on the high-voltage side of bank T2.

(c) The total current that enters the power system through the ground-to-neutral connection at the Y of T2.

(d) The total current that enters the power system through the ground-to-neutral connection at the Y configuration of T1.

(e) The relationship between the results found in parts (c) and (d) and the total fault current.

(f) The line currents associated with generator G1.

14-14. In the case of a solid line-to-line fault at bus 2 in the system of Fig. 14-1, the per-unit values of the line currents of generator G1 were found in Eqs. (14-96) and (14-97) by employing the phase angle adjustment in the positive- and negative-sequence currents associated with Δ-Y configurations. Verify these results by first finding the phase currents in the Δ side at T1 and then applying Kirchhoff's current law at the corners of the connection to obtain the line currents.

14-15. Using the results contained in Eqs. (14-98) and (14-101), calculate the currents in amperes that flow in the lines to G1 and G2 in the system of Fig. 14-1 for a line-to-line fault.

14-16. The system represented in Fig. P14-16 has all parameter values expressed in terms of a reference base system of 100 MVA at 115 kV.

(a) For a phase *a*-to-ground fault at bus 2, determine the fault current in amperes. Assume a generated emf of unity and negligible effects from prefault currents.

(b) Find the currents in the three-phase line at bus 1.

Figure P14-16

14-17. A three-bus electric power system has the configuration illustrated in Fig. P14-17. The indicated ratings and parameters refer to the individual components.
   (a) Convert all parameters to per-unit values evaluated for a reference base of 100 MVA at 69 kV.
   (b) Determine the positive-, negative-, and zero-sequence networks for this power system.

Figure P14-17

14-18. A double line-to-ground fault occurs in the system of Fig. P14-17 at F.
   (a) Find the equivalent impedances of the positive-, negative-, and zero-sequence sets as they exist at F.
   (b) Calculate each sequence current component as it exists at the fault point.
   (c) Determine the currents that flow through the solid fault connection expressed in amperes for each line.
   (d) How much current flows into the ground reference at the fault point?

14-19. All the parameters shown in Fig. P14-19 are referred to a reference base of 20 MVA at 69 kV. A line-to-ground fault occurs at bus 3. Calculate the ground currents in amperes and show where they flow back into the power system and by what amounts. Assume that $\bar{E}_a = 1\underline{/90°}$.

14-20. In the system of Fig. P14-19 a line-to-ground fault occurs on line a of generator G1.
   (a) Find the fault current in amperes.
   (b) In view of the result found in part (a), what is the net effect of such a fault in this situation?
   (c) Repeat part (a) for the case where the neutral connection of the phase coils of G1 is grounded.
   (d) For the condition of part (c), determine the phase a-to-neutral voltage that exists at bus 3.

14-21. Assume that the positive-, negative-, and zero-sequence matrices of an n-bus power system are known. Derive the general expressions for the line-to-neutral voltages

Figure P14-19

that exist at the *i*th bus for each sequence when a double line-to-neutral fault occurs at bus *k*.

14-22. Repeat Prob. 14-21 for the case of a line-to-line fault at bus *k*.

14-23. Repeat Prob. 14-21 for the case of a symmetrical three-phase short circuit at bus *k*.

APPENDICES

A. UNITS AND CONVERSION FACTORS

TABLE A-1 UNITS

Quantity	Symbol	Name of unit	Dimension
Fundamental			
Length	l, L	meter	L
Mass	m, M	kilogram	M
Time	t	second	T
Current	i, I	ampere	I
Mechanical			
Force	F	newton	MLT^{-2}
Torque	T	newton-meter	ML^2T^{-2}
Angular displacement	θ	radian	—
Velocity	v	meter/second	LT^{-1}
Angular velocity	ω	radian/second	T^{-1}
Acceleration	a	meter/second^2	LT^{-2}
Angular acceleration	α	radian/second^2	T^{-2}
Spring constant (translation)	K	newton/meter	MT^{-2}
Spring constant (rotational)	K	newton-meter	ML^2T^{-2}
Damping coefficient (translational)	D, F	newton-second/meter	MT^{-1}
Damping coefficient (rotational)	D	newton-meter-second	ML^2T^{-1}
Moment of inertia	J	kilogram-meter^2	ML^2
Energy	W	joule (watt-second)	ML^2T^{-2}
Power	P	watt	ML^2T^{-3}
Electrical			
Charge	q, Q	coulomb	TI
Electric potential	v, V, E	volt	$ML^2T^{-3}I^{-1}$
Electric field intensity	$\mathcal{E}$	volt/meter (or newton/coulomb)	$MLT^{-3}I^{-1}$
Electric flux density	D	coulomb/meter^2	$L^{-2}TI$
Electric flux	ψ, Q	coulomb	TI
Resistance	R	ohm	$ML^2T^{-3}I^{-2}$
Resistivity	ρ	ohm-meter	$ML^3T^{-3}I^{-2}$
Capacitance	C	farad	$M^{-1}L^{-2}T^4I^2$
Permittivity	ϵ	farad/meter	$M^{-1}L^{-3}T^4I^2$
Magnetic			
Magnetomotive force	$\mathcal{F}$	ampere(-turn)	I
Magnetic field intensity	H	ampere(-turn)/meter	$L^{-1}I$
Magnetic flux	ϕ	weber	$ML^2T^{-2}I^{-1}$
Magnetic flux density	B	tesla	$MT^{-2}I^{-1}$
Magnetic flux linkages	λ	weber-turn	$ML^2T^{-2}I^{-1}$
Inductance	L	henry	$ML^2T^{-2}I^{-2}$
Permeability	μ	henry/meter	$MLT^{-2}I^{-2}$
Reluctance	$\mathcal{R}$	ampere/weber	$M^{-1}L^{-2}T^2I^2$

TABLE A-2 CONVERSION FACTORS

Quantity	Multiply number of	by	to obtain
Length	meters	100	centimeters
	meters	39.37	inches
	meters	3.281	feet
	inches	0.0254	meters
	inches	2.54	centimeters
	feet	0.3048	meters
Force	newtons	0.2248	pounds
	newtons	10^5	dynes
	pounds	4.45	newtons
	pounds	4.45×10^5	dynes
	dynes	10^{-5}	newtons
	dynes	2.248×10^{-6}	pounds
Torque	newton-meters	0.7376	pound-feet
	newton-meters	10^7	dyne-centimeters
	pound-feet	1.356	newton-meters
	dyne-centimeters	10^{-7}	newton-meters
Energy	joules (watt-seconds)	0.7376	foot-pounds
	joules	2.778×10^{-7}	kilowatt-hours
	joules	10^7	ergs
	joules	9.480×10^{-4}	British thermal units
	foot-pounds	1.356	joules
	joules	1.6×10^{-19}	electron-volts
Power	watts	0.7376	foot-pounds/second
	watts	1.341×10^{-3}	horsepower
	horsepower	745.7	watts
	horsepower	0.7457	kilowatts
	foot-pounds/second	1.356	watts

B. PITCH AND DISTRIBUTION FACTORS FOR MACHINE WINDINGS

In the interest of making more effective use of the copper and iron as well as to discriminate severely against the presence of space harmonics in the air-gap field distribution, the windings of a-c machines are intentionally distributed and pitched so that all the phase turns do not fully link the total air-gap flux. In a three-phase machine, the distribution of the winding is such that each phase is made to occupy a region of 60 electrical degrees beneath each pole. Moreover, each phase coil is made to span less than 180 electrical degrees; i.e., the span is less than a pole pitch. Such a coil is referred to as a *fractional-pitch coil*. Although the use of the fractional-pitch coil means some reduction in the value of emf induced by the fundamental component of the field, the attenuation of the space harmonics is so much greater that it easily justifies the cost.

The Pitch Factor

This factor serves as a measure of the effectiveness of a selected coil pitch in reducing the influence of field harmonics on the emf induced in the armature winding of the machine. To understand how this comes about, attention is directed first to the reduction caused in the fundamental value of the induced emf compared to the case of a full-pitch coil. Refer to Fig. B-1, and assume that the coil consists of the two coil sides

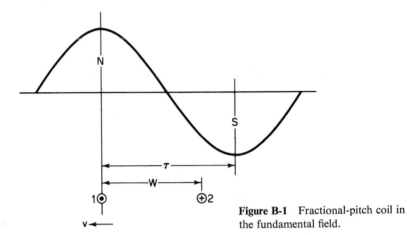

Figure B-1 Fractional-pitch coil in the fundamental field.

1 and 2 which are moving toward the left relative to the stationary north- and south-pole fundamental field component. The voltage induced in coil side 1 is therefore out of the paper, and that induced in 2 is into the paper. The instantaneous induced emf in these coil sides can conveniently be represented graphically by means of Fig. B-2.

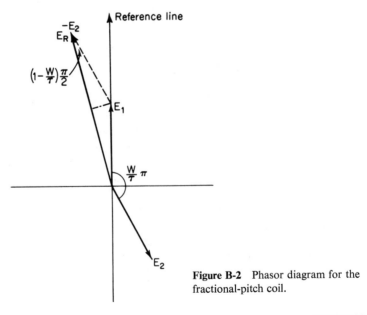

Figure B-2 Phasor diagram for the fractional-pitch coil.

The voltages E_1 and E_2 denote the maximum values of emf associated with each coil side. Accordingly, the instantaneous values are found by obtaining the projection of these peak voltages on the vertical reference line. Of course, each coil side has the same peak voltage. Placing E_1 coincident with the reference line implies that coil side 1 is instantaneously under the influence of the maximum flux density. Figure B-2 is the phasor diagram representation of the situation depicted in Fig. B-1. The projection of E_2 onto the reference line is negative, which is consistent with the fact that coil side 2 is under the influence of a pole of reversed polarity from that of coil side 1. Because the end connection that joins coil sides 1 and 2 provides addition of the coil voltages, the resultant coil voltage is found in Fig. B-2 by adding $-E_2$ to E_1. It is worthwhile to keep in mind here that the maximum value of this resultant coil voltage remains fixed. The instantaneous value, which varies with the position of the coil relative to the flux field, is readily found as the vertical projection of the E_R phasor. It is important to note that, because of the use of a fractional-pitch coil, the resultant voltage can never be equal to the algebraic sum of the coil side voltages. As a matter of fact, the pitch factor is simply defined as the *ratio of the resultant coil voltage to the absolute sum of the voltages induced per coil side.*

A convenient analytical expression for the pitch factor for the fundamental induced voltage can be derived easily from the geometry of Fig. B-2. Let W denote the coil width in suitable units, and let τ denote the pole pitch in the same units. The angle between E_1 and E_2 is then $W\pi/\tau$. Drawing the perpendicular bisector from the tip of E_1 onto E_R reveals two right triangles. The resultant coil voltage can then be expressed in terms of the coil side voltage by

$$E_R = 2E_1 \cos\left(1 - \frac{W}{\tau}\right)\frac{\pi}{2} \tag{B-1}$$

The expression for the pitch factor for the fundamental is therefore

$$K_{p1} = \frac{E_R}{2E_1} = \cos\left(1 - \frac{W}{\tau}\right)\frac{\pi}{2} = \sin\frac{\pi}{2}\sin\frac{W}{\tau}\frac{\pi}{2}$$

$$= \sin\frac{W}{\tau}\frac{\pi}{2} \tag{B-2}$$

For a machine that has nine slots per pole, the use of a coil width of seven slots leads to a value of $K_{p1} = 0.940$. Thus there is just a 6% reduction in fundamental induced emf. However, the corresponding reduction in the harmonics is much greater. This is demonstrated next.

If n denotes the order of a space field harmonic, a little thought discloses that the phasor diagram differs by the amount of rotation associated with one coil side voltage relative to the other. Thus in Fig. B-2 the angle that $-E_2$ makes with the vertical is increased by n times. Accordingly, the expression for the nth harmonic pitch factor becomes

$$K_{pn} = \frac{E_{Rn}}{2E_{cn}} = \cos n\left(1 - \frac{W}{\tau}\right)\frac{\pi}{2} = \sin n\frac{\pi}{2}\sin n\frac{W}{\tau}\frac{\pi}{2} \tag{B-3}$$

For the same machine with nine slots per pole, the value of fifth harmonic pitch factor becomes

$$K_{p5} = -0.174 \tag{B-4}$$

Note that the emf induced by the fifth harmonic field component is reduced by approximately 83% compared to 6% for the fundamental. The minus sign in Eq. (B-4) indicates a phase reversal for the seventh harmonic emf, brought about by a negative flux linkage caused by the chording of the winding.

The Distribution, or Breadth, Factor

In order to focus solely upon the effect of distributing the winding of an a-c machine over sixty electrical degrees, it is assumed that only full-pitch coils are used. Once again the derivation is carried out first for the fundamental field wave. In this connection refer to Fig. B-3, which shows the four coils that are assumed to constitute

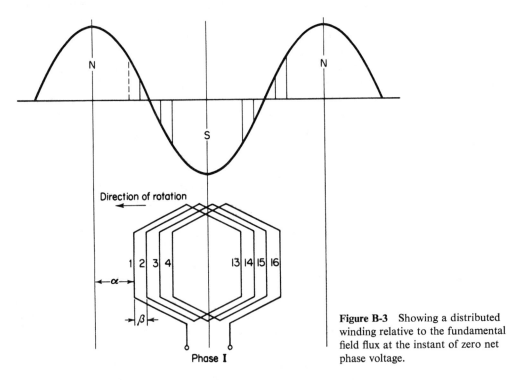

Figure B-3 Showing a distributed winding relative to the fundamental field flux at the instant of zero net phase voltage.

all of phase I of a single-layer, two-pole, three-phase machine. It is helpful to note that the instant is shown which corresponds to a net phase voltage of zero. This means that in the phasor diagram of Fig. B-4 the phasor denoting the phase voltage (AE) must have a zero projection on the reference line. In Fig. B-4 the quantity AB denotes the peak value of the voltage induced in coil 1-14. Note that this coil is displaced from the position of maximum voltage by the angle α. At the same time, because of the distribution pattern, the voltage of coil 2-14 (denoted by BC) is away from its peak value by the angle $(\alpha + \beta)$, where β denotes the angle between slots. It is interesting to observe that the projection of BC onto the reference line is smaller

Reference line

$AB = BC = CD = DE$ = peak coil voltage = E_c
$AE = E_R$ = peak phase voltage

Figure B-4 Derivation of the distribution factor. Illustration is for four slots per pole per phase.

than that of *AB*. A glance at Fig. B-3 indicates consistency with the physical relationship these coils bear to the field distribution. Coil 3-15 is represented by *CD* in Fig. B-4. Its projection on the time line is negative, thus indicating that coil sides 3 and 15 are under the influence of poles of reversed respective polarity than those for coils 1-13 and 2-14. Coil 4-16 is represented by *DE* in Fig. B-4. Each successive coil voltage is displaced by an additional amount (equal to the slot angle β) from the preceding coil. The effect is that a resultant phase voltage is obtained that is *less* than the absolute sum of the individual coil voltage. In fact, the distribution factor is defined as the *ratio of the resultant voltage to the absolute sum of the coil voltages comprising the resultant.*

An examination of Fig. B-4 leads to a convenient analytical expression for the distribution factor. By drawing the perpendicular bisectors of each of the coil voltages represented by *AB*, *BC*, etc., the center of a circle passing through points *A*, *B*, *C*, *D*, and *E* can be found. The quantities *AB*, *BC*, *CD*, and *DE* are then also equal chords of this circle. Accordingly, it follows that

$$\sphericalangle AOE = Q\beta \tag{B-5}$$

where Q denotes the number of slots per pole per phase. Moreover,

$$\sphericalangle AOB = \beta \tag{B-6}$$

$$\sphericalangle AOG = \frac{\beta}{2} \tag{B-7}$$

and

$$\sphericalangle AOF = \frac{Q\beta}{2} \tag{B-8}$$

The expression for the peak value of the resultant phase voltage becomes

$$E_R = AE = 2(AF) = 2(OA)\sin\frac{Q\beta}{2} \tag{B-9}$$

By definition, the distribution factor is then

$$K_d \equiv \frac{E_R}{QE_c} = \frac{2(OA)\sin\frac{Q\beta}{2}}{Q2(AG)} = \frac{2(OA)\sin\frac{Q\beta}{2}}{Q2(OA)\sin\beta/2}$$

$$= \frac{\sin\frac{Q\beta}{2}}{Q\sin\beta/2} \tag{B-10}$$

For the machine shown in Fig. B-4 the values of the quantities in Eq. (B-10) are $Q = 4$, $\beta = 15°$. Hence the value of the distribution factor for the fundamental is 0.958, thus representing less than a 5% reduction. The value of the distribution factor for the fifth harmonic for the same machine is found to be 0.205, thus representing an 80% reduction.

The net winding factor for a winding that is pitched, as well as distributed, is equal to the product of the pitch and distribution factors. Expressed mathematically,

$$K_{wn} = K_{pn}K_{dn} \tag{B-11}$$

where the second subscript n denotes the nth harmonic. When K_{pn} and K_{dn} are low for a given harmonic, the net winding factor assumes such small values as to render that particular harmonic inconsequential.

C. DERIVATION OF THE PEAK MMF PER POLE OF A POLYPHASE WINDING

To begin simply, consider a two-pole machine with three phases and a total of six slots. Each phase therefore consists of a single coil, which for a full-pitch winding means that the coil spans from slot 1 to slot 4, as depicted in Fig. C-1. For a uniform air-gap machine, any current that flows through this coil produces a rectangular mmf distribution. The total coil mmf is distributed equally between slots 1 and 4 (say for flux entering the air-gap) and slots 4 and 1 (where the flux is assumed leaving the air-gap for the same instant). Attention for the moment is directed to the mmf distribution of one phase. The effect of the remaining phases is considered presently.

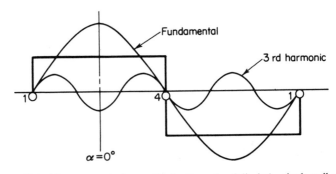

Figure C-1   Magnetomotiveforce distribution of a full-pitch, single-coil phase winding in a uniform air-gap machine.

If the phase coil is assumed to be carrying a sinusoidally varying current of rms value I, then the height of the rectangular distribution alternates between a positive maximum and a negative maximum value, in accordance with the following expression:

$$\mathfrak{F} = \frac{\sqrt{2}}{2} z_s I \sin \omega t \tag{C-1}$$

where z_s denotes the conductors per slot. Note that when the argument is $\pi/2$, the coil mmf has a positive maximum value. In the interest of obtaining a result that is applicable generally, i.e., to the harmonics as well as to the fundamental contained in the rectangular distribution of Fig. C-1, a Fourier analysis is applied to this wave. The equation for the mmf distribution for phase I, expressed as a function of time and space displacement from the indicated reference position $\alpha = 0°$, then becomes

$$\mathfrak{F}_I(\alpha) = \sum_{n=1,3}^{\infty} a_n \cos n\alpha \tag{C-2}$$

where

$$a_n = \frac{1}{\pi} \int_0^{2\pi} \mathfrak{F} \cos n\alpha \, d\alpha \tag{C-3}$$

and $\mathfrak{F}$ is given by Eq. (C-1).

Inserting Eq. (C-1) into Eq. (C-3) and noting that the given wave possesses quarter-wave symmetry, we may write

$$a_n = \frac{4}{\pi} \int_0^{\pi/2} \frac{\sqrt{2}}{2} I z_s \sin \omega t \cos n\alpha \, d\alpha = 0.9 z_s I \sin \omega t \left(\frac{1}{n} \sin n\frac{\pi}{2}\right) \tag{C-4}$$

Substituting this result into Eq. (C-2) then yields

$$\mathfrak{F}_I(\alpha) = 0.9 z_s I \sin \omega t \, (\cos \alpha - \tfrac{1}{3} \cos 3\alpha + \tfrac{1}{5} \cos 5\alpha - \tfrac{1}{7} \cos 7\alpha + \cdots) \tag{C-5}$$

Now, if instead of one slot per pole per phase the machine is equipped with Q slots per pole per phase, the foregoing equation can be made to apply by introducing two modifications: replacing z_s by $z_s Q$ and including the distribution factor of each harmonic. Accordingly, we get

$$\mathfrak{F}_I(\alpha) = 0.9 z_s Q I \sin \omega t \left(K_{d1} \cos \alpha - \frac{K_{d3}}{3} \cos 3\alpha + \frac{1}{5} K_{d5} \cos 5\alpha - \cdots \right) \tag{C-6}$$

Furthermore, if each coil is assumed to be fractional pitch, then each term of Eq. (C-6) must be further modified by the pitch factor for each harmonic. Thus the complete expression that describes the mmf distribution of one phase of a fractional-pitch distributed winding is

$$\mathfrak{F}_I(\alpha) = 0.9 z_s Q I \sin \omega t \left(K_{w1} \cos \alpha - \frac{K_{w3}}{3} \cos 3\alpha + \frac{K_{w5}}{5} \cos 5\alpha - \cdots \right) \tag{C-7}$$

where $K_{wn} = K_{dn} K_{pn}$ is the winding factor obtained as the product of the individual distribution and pitch factors.

Since our interest is chiefly in the resultant mmf produced by three-balanced windings cooperating to produce a revolving mmf in the manner described in Sec. 4-1, we need first the expressions for the mmf of phases II and III. These are readily obtained from Eq. (C-7) by introducing the required 120° displacement in time and space. Hence

$$\mathfrak{F}_{II}(\alpha) = 0.9z_sQI \sin(\omega t - 120°) \left[K_{w1} \cos(\alpha - 120°) - \frac{K_{w3}}{3} \cos(3\alpha - 360°) \right.$$

$$\left. + \frac{K_{w5}}{5} \cos(5\alpha + 600°) - \cdots \right] \qquad \text{(C-8)}$$

$$\mathfrak{F}_{III}(\alpha) = 0.9z_sQI \sin(\omega t - 240°) \left[K_{w1} \cos(\alpha - 240°) - \frac{K_{w3}}{3} \cos 3(\alpha - 240°) \right.$$

$$\left. + \frac{K_{w5}}{5} \cos 5(\alpha - 240°) - \cdots \right] \qquad \text{(C-9)}$$

It is helpful to note that the mmf of each individual phase has an alternating character. This is true of the harmonic as well as the fundamental components. However, the resultant fundamental or harmonic waves are found to revolve with time. For any particular harmonic this is readily corroborated by summing the contributions for each phase. Thus for the fundamental the result is found to be

$$\mathfrak{F}_{r1}(\alpha) = \tfrac{3}{2}(0.9)z_sQIK_{w1} \sin(\omega t - \alpha) \qquad \text{(C-10)}$$

As already demonstrated in Sec. 4-1, Eq. (C-10) is the equation of a traveling sine wave moving in the positive α direction. It follows therefore that the *amplitude* of this resultant wave is the coefficient of the trigonometric term. Hence

$$\mathfrak{F}_A = \tfrac{3}{2}(0.9)z_sQK_{w1}I \qquad \text{(C-11)}$$

The factor 3 appears because we are dealing with three phases. For a q-phase machine, Eq. (C-11) becomes

$$\mathfrak{F}_A = \frac{q}{2}(0.9)z_sQK_{w1}I \qquad \text{(C-12)}$$

It is often desirable to express this quantity in terms of the total number of turns per phase (N). This is possible to do upon recognizing that

$$N\left(\frac{\text{turns}}{\text{phase}}\right) = \frac{z_s}{2}\left(\frac{\text{turns}}{\text{slot}}\right)Q\left(\frac{\text{slots}}{\text{pole} \times \text{phase}}\right)p(\text{poles}) \qquad \text{(C-13)}$$

Accordingly,

$$\frac{z_sQ}{2} = \frac{N}{p} \qquad \text{(C-14)}$$

Insertion of the last expression into Eq. (C-12) yields the final and desired form of the mmf amplitude of a polyphase winding, namely,

$$\mathfrak{F}_A = 0.9q\frac{NK_{w1}}{p}I \qquad \text{(C-15)}$$

A similar procedure can be followed to find the expression of the peak mmf of the harmonics in the resultant revolving field. Adding the third harmonic components of each of the three phases yields zero—a result that is obvious from an inspection of Eqs. (C-7), (C-8), and (C-9). This conclusion also applies to all multiples of the third harmonic. Accordingly, there is no resultant third harmonic component in the traveling mmf wave of a balanced three-phase winding.

The addition of the fifth harmonic contributions in each of the three phases leads to the following result:

$$\mathcal{F}_{rs}(\alpha) = \frac{3}{2}(0.9)z_s QI\frac{K_{w5}}{5}\sin(\omega t + \alpha) \qquad \text{(C-16)}$$

Note the form of the argument is such as to indicate that the fifth harmonic component of the resultant wave moves in the negative α direction. The amplitude of the fifth harmonic wave is then clearly

$$\mathcal{F}_{A5} = 3(0.9)\frac{Qz_s}{2}\frac{K_{w5}}{5}I = 3(0.9)\frac{N}{p}\frac{K_{w5}}{5}I \qquad \text{(C-17)}$$

For a q-phase machine this becomes,

$$\mathcal{F}_{A5} = 0.9q\frac{NK_{w5}}{5p}I \qquad \text{(C-18)}$$

A little thought concerning the derivation of the last expression readily reveals that the general expression for the amplitude of the resultant traveling wave of the nth harmonic of a q-phase winding is

$$\mathcal{F}_{An} = 0.9q\frac{NK_{wn}}{np}I \qquad \text{(C-19)}$$

Upon inserting $n = 1$, Eq. (C-15) is obtained.

D. REDUCTION FACTORS FOR POLYPHASE WINDINGS

A *reduction factor* is a numerical quantity which, when applied to a rotor quantity such as voltage, current, impedance, etc., refers that quantity to the stator. This means that the effect produced by the referred quantity in the stator is identical to the effect caused by the original quantity in the rotor. The reduction factors are different for the various electrical quantities, and they also differ depending upon whether the rotor is of the wound or squirrel-cage type.

The Wound Rotor

There are three reduction factors of interest: one to handle voltage, one to handle current, and the third to deal with resistance, reactance, and impedance. As described in Chapter 2, a reduction factor is essentially a transformation ratio that converts the actual rotor winding into an equivalent rotor winding having the same number of turns as the stator winding. If we let E_1 denote the standstill emf induced per phase in the stator winding by the air-gap flux, and E_2 denote the standstill emf induced per phase in the rotor winding by the same air-gap flux, then by Eqs. (4-9) and (4-10) the ratio of these two quantities is

$$\frac{E_1}{E_2} = \frac{N_1 K_{w1}}{N_2 K_{w2}} = a_v \qquad \text{(D-1)}$$

Equation (D-1) is the reduction factor for referring *any* voltage associated with the rotor winding back to the stator. The convention used in this book to denote the referred quantity is the *prime* notation. Accordingly, to refer the quantity E_2 to the stator, we write

$$E_2' = a_v E_2 \tag{D-2}$$

A comparison with Eq. (D-1) shows E_2' is identical to E_1, and they should be since both quantities are related to a winding having the same effective turns cut by the same gap flux.

The reduction factor for current can be derived by recalling how the rotor current is made to flow. Ideally, at no-load and for an induction motor with negligible rotational losses, the rotor current is zero. The slip correspondingly is also zero. As load is alppied to the rotor shaft, the rotor structure slows down, thus causing an emf to be induced in the rotor winding, which in turn produces enough rotor current to supply the required energy to the load. As rotor current flows through the q_2-phases of the rotor, it presents a total rotor mmf of magnitude $q_2 N_2 K_{w2} I_2$ that acts in a direction to oppose the original stator mmf, $q_1 N_1 K_{w1} I_m$. The latter quantity is the effective stator mmf that produces the air-gap flux, the value of which is essentially determined by the applied stator voltage. To offset this opposing action of the rotor mmf and thus preserve the value of air-gap flux demanded by the applied stator voltage, there must occur an increase in stator mmf equal and opposite to the total rotor mmf. Calling the increased stator current I_2', at equilibrium we can then write

$$q_1 I_2' N_1 K_{w1} = q_2 I_2 N_2 K_{w2} \tag{D-3}$$

Or, the required increase in stator current becomes

$$I_2' = \frac{q_2 N_2 K_{w2}}{q_1 N_1 K_{w1}} I_2 = a_i I_2 \tag{D-4}$$

where

$$a_i \equiv \frac{q_2 N_2 K_{w2}}{q_1 N_1 K_{w1}} = \text{current reduction factor} \tag{D-5}$$

The application of a_i to any rotor phase current gives the equivalent effect of that current in the stator winding. It is interesting to note that when the rotor and stator phases are the same, the reduction factor for current is the inverse of the reduction factor for voltage.

The manner of transferring the effect of a given rotor resistance to the stator can be found from the relationship that preserves either copper loss or per-unit resistance drop. For example, for a rotor winding resistance r_2 and rotor current I_2 the per-unit resistance drop is $I_2 r_2 / E_2$. When placed in the stator, the effect of this rotor resistance (now called r_2') must be the same as it produces in the rotor itself, in spite of the fact that in the stator the r_2' carries the current I_2' and must be associated with the stator voltage E_1 (or E_2'). Equivalence is preserved by satisfying the following equation:

$$\frac{I_2' r_2'}{E_2'} = \frac{I_2 r_2}{E_2} \tag{D-6}$$

It then follows that the rotor resistance as it must appear in the stator is

$$r_2' = \frac{I_2}{I_2'} \frac{E_2'}{E_2} r_2 \tag{D-7}$$

Inserting Eqs. (D-1), (D-2), and (D-4) yields

$$r_2' = \frac{q_1}{q_2}\left(\frac{N_1 K_{w1}}{N_2 K_{w2}}\right)^2 r_2 = a_z r_2 \qquad \text{(D-8)}$$

where

$$a_z = \frac{q_1}{q_2}\left(\frac{N_1 K_{w1}}{N_2 K_{w2}}\right)^2 = \frac{\text{reduction factor for resistance,}}{\text{reactance and impedance}} \qquad \text{(D-9)}$$

It is a simple matter to show that Eq. (D-9) applies with equal validity to reactance and impedance quantities. This is left as an exercise for the reader.

The Squirrel-Cage Rotor

It is helpful to keep in mind that the squirrel-cage rotor consists of solid copper bars that are often driven through the slots and connected at both ends by means of copper end rings. As a result of this construction each bar can be considered to be a separate and individual phase of a q-phase system, reckoned in terms of the traditional definition of what constitutes a phase. Since each bar (or conductor) is a phase, it follows that the number of turns per phase must be $\frac{1}{2}$ (it takes two conductors to make one turn), and the winding factor must be identically unity.

The voltage reduction factor for the squirrel-cage rotor is readily determined by inserting $N_2 = \frac{1}{2}$ and $K_{w2} = 1$ into Eq. (D-1), which is a general result. Thus

$$a_v = 2N_1 K_{w1} \qquad \text{(D-10)}$$

When the number of rotor bars is not divisible by the pole pairs through an integer, the number of rotor phases is equal to the number of bars. Hence, under these circumstances for the squirrel-cage rotor, Eq. (D-3) becomes

$$q_1 I_2' N_1 K_{w1} = S_2(\tfrac{1}{2})(1)I_2 \qquad \text{(D-11)}$$

or

$$I_2' = \left(\frac{S_2}{2q_1 N_1 K_{w1}}\right)I_2 \qquad \text{(D-12)}$$

where S_2 denotes the number of rotor slots, and the quantity in parentheses is the current reduction factor.

The reduction factor for resistance is found again by employing Eq. (D-6). For the squirrel-cage rotor this leads to

$$r_2' = \frac{E_2'}{E_2}\frac{I_2}{I_2'}r_2 \qquad \text{(D-13)}$$

Introducing the appropriate current and voltage reduction factors into the last equation yields

$$r_2' = \left[\frac{4q_1(N_1 K_{w1})^2}{S_2}\right]r_2 \qquad \text{(D-14)}$$

where the quantity in brackets identifies the desired result. As before, this reduction factor is applicable to reactance and impedance as well as resistance.

E. SLIP TEST FOR SALIENT-POLE SYNCHRONOUS MACHINES

Information about the direct-axis synchronous reactance and the quadrature-axis synchronous reactance of synchronous machines can be obtained from a relatively simple no-load test known as the *slip test*. The reason for the name will be apparent presently. Figure E-1 depicts the wiring arrangement for the test, including the instruments that permit calculations of the reactances.

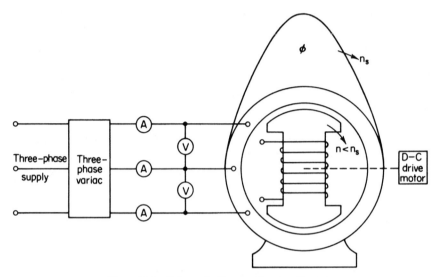

Figure E-1   Schematic diagram for the slip test.

The three-phase variac is needed in order to apply an adjustable voltage of rated frequency to the three-phase stator winding, which is set appreciably below the rated machine voltage. Application of this voltage creates a revolving set of magnetic poles in the air-gap of the synchronous generator, traveling at synchronous speed n_s. At the same time, a separate source is used to drive the salient-pole rotor structure at a speed slightly different from synchronous speed and in the *same* direction as the revolving stator mmf. A quick and easy way to ascertain that the direction is proper is to put a voltmeter across the open-field winding terminals. If a small reading is registered the direction is proper. (Why?) The use of a d-c motor is convenient as the separate drive source because of the ease with which the speed may be adjusted.

By driving the rotor structure at a speed either slightly below or slightly above synchronous speed, the rotor appears to be slipping either behind or ahead of the rotating magnetic poles of the stator. Consequently, there will be a period of time when the stator magnetic poles are in a position of alignment with the rotor iron pole structure, thus presenting a minimum reluctance to the stator mmf and resulting in a minimum magnetizing current, as indicated by the line ammeters, *A*, in Fig. E-1. At a slightly later instant the direct-axis of the rotor pole structure will be in a position of quadrature with respect to the axes of the stator magnetic poles. Accordingly, the air-gap is very large, and so a large magnetizing current must flow to establish essentially the same air-gap flux demanded by the applied voltage. The increase in current is quite appreciable and is easily registered by the line ammeters. Keep in mind that this variation of the current from a minimum to a maximum level occurs at a frequency that is dependent directly upon $n_s - n$. Obviously, one technique required in this test is to keep this frequency small enough so that the true maximum and minimum values can be recorded. However, at the same time, this frequency must not be made so small that the rotor structure is pulled into synchronism with the stator magnetic poles through the action of the reluctance power. As a matter of fact, the reason for performing this test at reduced voltage is precisely to minimize the influence of the reluctance torque to pull the rotor into synchronism. Unfortunately, a lower limit must be placed on this applied voltage; otherwise, the computed value of synchronous reactance will be nearer to the unsaturated value, which we already know can be misleading.

Because the internal impedance of the variac is not likely to be negligible, the wide swing in the minimum and maximum values of the line ammeters will cause some fluctuations in the voltmeter reading. These too should be recorded. Also note that the maximum voltage reading is associated with the minimum line current, and the minimum voltage reading goes with the maximum current.

The direct-axis synchronous reactance is found by using V_{max} and I_{min} since these are the readings that are registered when the rotor-pole axis and the magnetic poles of the stator are in alignment. Assuming a Y-connected stator winding, the expression for the direct-axis synchronous reactance thus becomes

$$x_d = \frac{V_{max}}{\sqrt{3}\,I_{min}} \tag{E-1}$$

Average values of the three ammeters and two voltmeters should be used.

By a similar argument, it should be clear that the quadrature-axis synchronous reactance can be computed from

$$x_q = \frac{V_{min}}{\sqrt{3}\,I_{max}} \tag{E-2}$$

The reliability of the results obtained from Eqs. (E-1) and (E-2) can be improved by employing oscillographic methods to record and measure the variations in line currents and line voltages. This permits a larger slip frequency to be used, which in turn allows a higher applied voltage to be used without the danger of pull-in. Furthermore, the maximum and minimum values can be measured much more accurately because the inertia effects of the indicating ammeters and voltmeters are eliminated entirely.

F. COMPLEX HYPERBOLIC FUNCTION CHARTS

$$\theta = \sqrt{ZY}$$

Angle of ZY

Size of $\overline{ZY}$

Angle of cosh θ

Size of cosh θ

Chart I Hyperbolic cosine of θ.

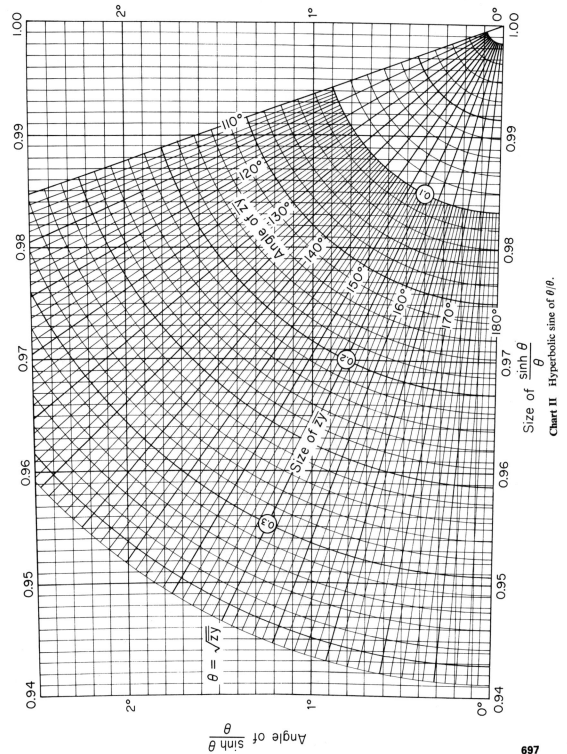

Chart II   Hyperbolic sine of θ/θ.

697

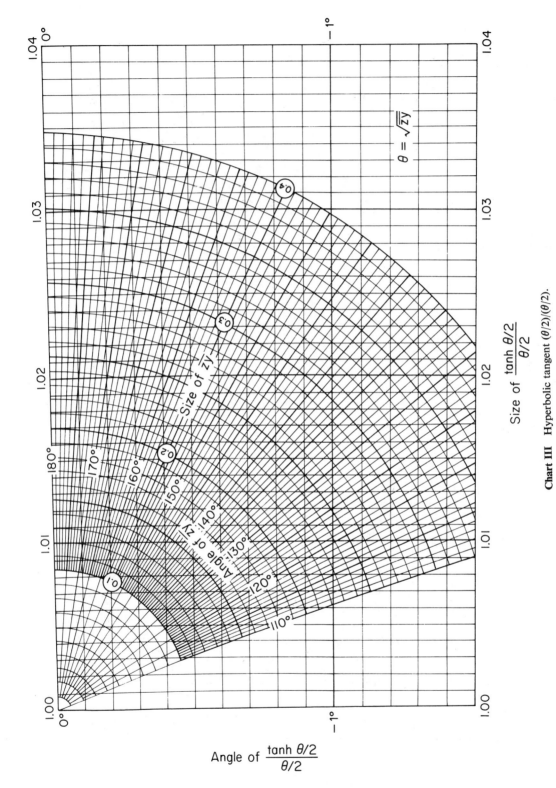

Angle of $\dfrac{\tanh \theta/2}{\theta/2}$

Size of $\dfrac{\tanh \theta/2}{\theta/2}$

$\theta = \sqrt{zy}$

Size of zy

Angle of zy

Chart III Hyperbolic tangent $(\theta/2)/(\theta/2)$.

G. A SHORT BIBLIOGRAPHY

The following is a short selected list of books that the reader can find useful as a supplementary source of information in the study of this textbook.

Electric Machines

FITZGERALD, A. E., C. KINGSLEY, JR., and S. D. UMANS, *Electric Machinery*, 4th ed. (New York: McGraw-Hill Book Company, 1983).

LIWSCHITZ-GARIK, M., and C. C. WHIPPLE, *Electric Machinery*, vols. I and II (Princeton, N.J.: D. Van Nostrand Co., Inc., 1946).

Electric Power Systems

ELGERD, O. I., *Electric Energy Systems Theory*, 2nd ed. (New York: McGraw-Hill Book Company, 1982).

GROSS, C. A., *Power System Analysis* (New York: John Wiley & Sons, Inc., 1979).

STEVENSON, W. D., JR., *Elements of Power Systems Analysis*, 4th ed. (New York: McGraw-Hill Book Company, 1982).

Index

Index